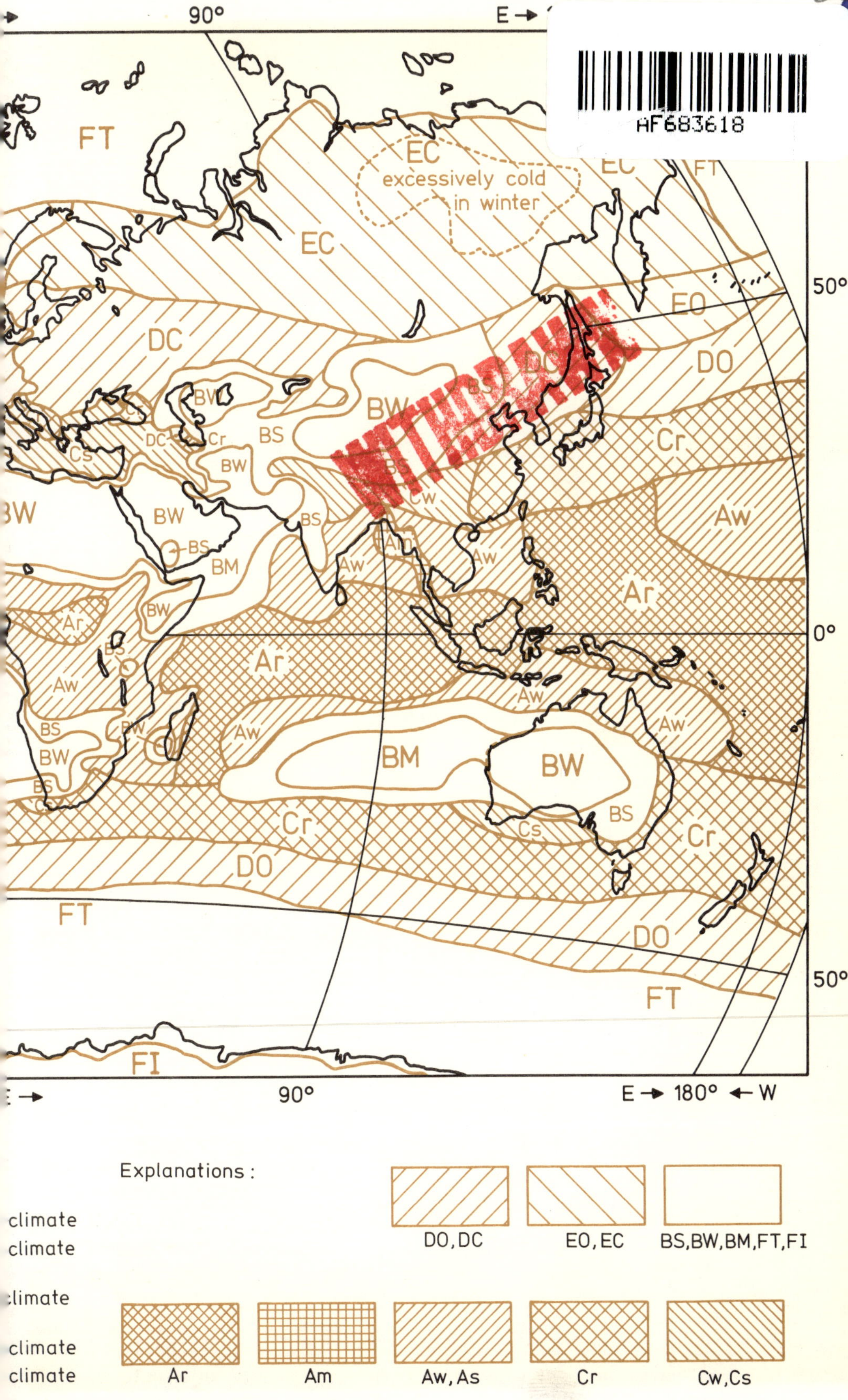
90°
E →
FT
EC
EC
excessively cold
in winter
EC
FT
FO
DC
DO
DO
BW
BS
Cr
Cs
DC
BW
BS
BW
BW
BS
BS
BM
Am
Aw
Aw
Aw
Ar
Ar
Ar
BW
Aw
Aw
Aw
BS
BW
BS
BM
BW
BS
Cs
Cr
Cr
DO
DO
FT
FT
FI
50°
0°
50°
90°
E → 180° ← W
Explanations :
climate
climate
climate
climate
climate
DO, DC
EO, EC
BS, BW, BM, FT, FI
Ar
Am
Aw, As
Cr
Cw, Cs

Willy Rudloff
World-Climates

World-Climates

with tables of climatic data and practical suggestions

by Willy Rudloff, Buxtehude

50 figures, 1474 climatic tables, 116 hygrothermal diagrams,
30 other tables and reviews,
1 bicoloured map on the front-papers

BOOKS OF THE JOURNAL
NATURWISSENSCHAFTLICHE RUNDSCHAU
Editor Prof. Dr. Wolfgang Höll, München

WVG

Wissenschaftliche Verlagsgesellschaft mbH Stuttgart 1981

CIP-Kurztitelaufnahme der Deutschen Bibliothek

Rudloff, Willy
World-Climates : with tables of climatic data
and practical suggestions / by Willy Rudloff.
– Stuttgart : Wissenschaftliche Verlagsgesellschaft,
1981.
(Bücher der Zeitschrift Naturwissenschaftliche
Rundschau)
ISBN 3-8047-0509-X
ISSN 0077-6157

Satz: Waldburg Hädicke, Composersatz, 7000 Stuttgart
Druck: Schwetzinger Verlagsdruckerei GmbH, 6830 Schwetzingen

To my wife

Preface

This book was written at the suggestion of the editor, Mr. H. Rotta. Both editor and author are confident of a need on the part both of the general public and tourism, which now covers the globe, for a work on climatology suitable for practical use.

It is nearly half a century since the second edition of Köppen's *Outlines of Climatology* was published in 1931. At that time, his climate classification system was the most commonly used all over the world.

In the meantime, meteorological observation has greatly intensified, and the resulting data are available in numerous annual and other publications for the use of specialists.

However, there is still a lack of a world-wide collection of basic data providing general information about climatic conditions throughout the world. These are presented in the following volume, together with qualitative conclusions by the author concerning heat stress and clothing appropriate to climate.

The introductory chapters provide information on the definition of weather and climate, the principal elements of weather, the global atmospheric circulation, bioclimatic relations, and a classification of climate which the author considers a revised and consistent development of Köppen's.

Several of the arguments contained in the book have been published in German in articles in various publications of the German Weather Service (Deutscher Wetterdienst), to whom I am greatly indebted for permission for re-publication and for permission to complete some climatic tables by data which could be extracted from the archive of oversea's climatology of the Seewetteramt in Hamburg.

Last but not least I pay tribute to Dr. Colin Boone who carefully polished up my English version of this book.

I favour the ways and means he found to express precisely what I had in mind – even in thoughtful sentences of relatively few words. I hope that it will prove to the user's advantage.

Altea (Alicante)/Spain, December 1980 *Willy Rudloff*

Contents

Illustrations:

All world maps are turned about 90° in order to
1. prevent some maps to become badly defined by using a smaller scale;
2. to facilitate an easy comparison between two maps.

1. Introduction

1.1. Weather, general weather and climate as definitions

Ideal holidays demand ideal weather!

To most people, "ideal weather" means fine sunny weather, rendered temperate by an agreeable breeze, and of course settled, too. Thus the initial sentence ought rather to read: "Ideal holidays demand ideal general weather".

"General weather" will be taken to refer to the mean weather at any particular place during a certain period which may last for some hours, days or weeks, or even several months or years.

"Weather" will be understood as a confluence of certain phenomena in the atmosphere and in the upper layer of the land or sea at a particular place at a given moment.

This definition of "weather" becomes clear if we define the phenomena of the atmosphere, which brings us to meteorology, the science of the weather.

There are different ways of treating the abundant events in the earth's atmosphere that fall under the heading of "weather". The meteorologist tries to penetrate the secrets of the weather by methods derived from mathematics and physics, and to apply his knowledge in weather forecasting. His "differential" approach contrasts with the "integral" one used by the climatologist, who considers the occurrences in the atmosphere in larger areas and over longer periods, in a way that can be called more geographical.

By applying both methods, it is hoped, it is possible to make statements about human influences on atmospheric processes with greater confidence. Problems like air pollution, the increasing scarcity of drinking-water, and steppe formation are urgently demanding solution, and towards this the science of meteorology can make important contributions.

In one of his celebrated conversations with *Eckermann*, in 1806, *Goethe* said of meteorology: "The objects observed by meteorologists are indeed similar to the living ones which we can daily observe in the living and creating process. They proceed from the assumption of a synthesis, but one so manifold in its aspects that the human observer cannot encompass it."

Since then, a century and a half have passed; scientists are still endeavouring strenuously to penetrate the secrets of the atmosphere by physical methods, and even venture to forecast the weather, in spite of the gaps in their knowledge of its laws. Doubtless it will never be possible completely to avoid erroneous forecasts, although the increase in our knowledge of the complicated atmospheric processes is bringing about a reduction in them. At any rate, in spite of all the problems involved, nearly every country in the world has a meteorological service; the minimal proportion of their budgets spent on this is rewarded many times over by the profit drawn from weather hazard warnings, and meteorological consulting and forecasting services. This is a circumstance of which the general public remains unaware, though it is supplied several times a day by press, radio and television with the most important product of the meteorological services, the weather forecast. It is also aware of the fact that the forecasts are not infallible. It can always happen that forecasts do not coincide with the actual weather.

The worst critics of the forecasters are often the meteorologists themselves; as a rule, however, it turns out that a check based on the current state of knowledge in this field confirms that the forecast weather was more likely than that which in fact ensued.

Perhaps meteorologists should add to their forecasts the degree of probability or uncertainty; some of them, on television, do in fact indicate it either directly or indirectly – with success, in my opinion. I certainly take the view that the audience welcome such indications.

Given our present state of knowledge, forecasting the weather over longer periods of time is definitely a ticklish venture; and I believe it will always be a hazardous enterprise to give a weather forecast in the form of a calendar. What can be of positive assistance to all is an all-round knowledge of the processes involved in the weather; this can enable people to make their own forecasts, or modifications of the forecasts, if the development of the weather situation is uncertain, if, say, a weather centre changes its velocity, or if the weather displays remarkable differences over short distances.

That may, for the present, be enough on forecasts, important as they are, especially for tourists in regions where variable weather prevails. After all, on holiday so many, if not all, of our activities are dependent on the weather. The best advice is that contained in the old saying: "There is no such thing as bad weather, providing you wear the right clothes". If one is supplied with a suitable repertoire of clothes, only really appalling weather can be a worry; and such weather is fortunately rare, on the whole, throughout the world.

Let us now consider the weather from the clothing angle. This will help us to arrive at a further useful objectivisation of the notion of "weather". It is true that different individuals respond differently to air temperature, which defines the heat conditions of given weather situations in nature; but it may be assumed that most people have some idea of the mean air temperature of a given season. A general or particular weather situation involving heat conditions corresponding to the mean temperature of the place concerned may be termed "normally tempered"; here heat conditions are "normal". Higher degrees of heat would be called "warm", lower ones "cold". These three grades will be abbreviated below as follows: w = warm; n = normal; c = cold. The double letters ww or cc can be used for conditions of great heat or cold.

Besides heat, which is of paramount importance for all living creatures, including man, water is a further factor of equal importance, whether descending in the form of precipitation, or remaining in the air in gaseous form. As to precipitation, we shall differentiate between the following conditions, using the abbreviations indicated: a = dry; i = some precipitation, scattered showers; o = unsettled weather, recurrent showers; u = rainy weather, precipitation of long duration. If the precipitation falls as snow, we add the letter "e" to the corresponding vowel, thus: ie, oe, ue. This codification is proposed here for those who like to keep a diary of their own weather observations.

In addition to these remarks about air temperature and precipitation, some remarks on cloudiness are appropriate. It is sufficient to distinguish three grades: 1. cloudless, or with very few clouds (sunny by day, clear at night); 2. cloudy; 3. with heavy cloud, overcast. The corresponding abbreviations are: s = sunny/clear; l = cloudy, v = overcast. It is tempting to add indications of further weather elements, but these should be avoided except in special cases, to avoid straining the code. Squalls, gales or storm are certainly important and should be mentioned, as should fog, black ice, snow cover, thunderstorms (with squalls, hail, or soft hail). If necessary, these will be referred to under the collective name "special weather phenomena".

Those mentioned so far are only a selection from the abundance of phenomena which are registered in the observations and weather reports of the meteorological services. Let us add here only some of the very interesting optical phenomena, such as rainbow, corona or aureole, a halo round the sun or moon. The aurora borealis or australis and St. Elmo's fire, being electric phenomena, do not belong here; the thunderstorm, on the other hand, does, its lightning being accompanied by squalls, cloudbursts, or hailstorms, to say nothing of the thunder itself. Distant lightning indicates a thunderstorm afar off. In some tropical regions thunderstorms are a regular phenomenon in the rainy season.

At this point, it will be convenient to present the first world map (see Figure 1: "World-wide thunderstorm activity"). It demonstrates in simplified form the mean number of days with thunderstorms, as given in 1956 by the World Meteorological Organization. A day is counted as a day with thunderstorms if at least one thunderstorm is observed at the station. Thus the number of days with thunderstorms does not correspond

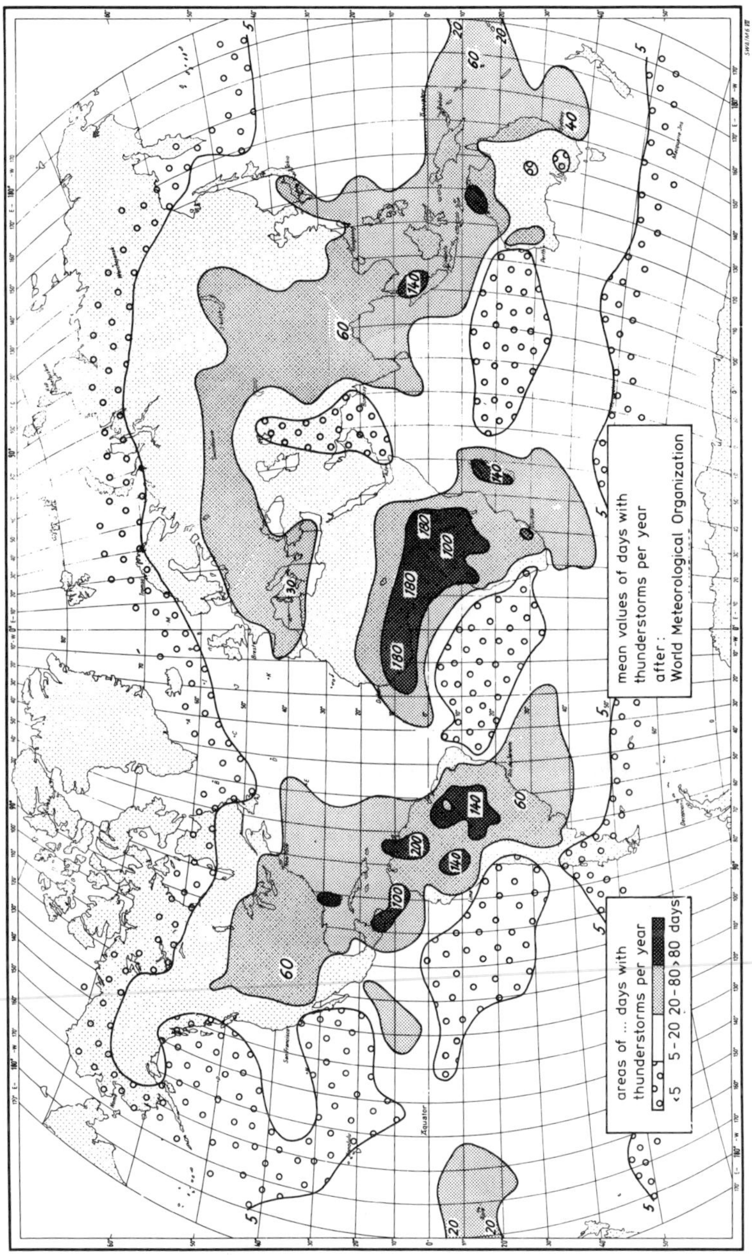

Fig. 1. World-wide thunderstorm activity

to the number of individual thunderstorms passing the station. It is not possible to show small-scale differences because of the small scale of the map and the lack of sufficient observations. (In Germany, such differences range from 15 to 35 days.)

The world record for days with thunderstorms is 322 days, observed at the station of Buitenzorg in Java; this is greatly in excess of the figures for the surrounding area, and indicates extremely favourable conditions for the formation of thunderstorms at this place. There are very few thunderstorms in polar regions. Extreme figures of 100 days or more are to be found in the continental parts of the tropics. South America and Africa have larger areas with such high figures than Asia. During the rainy season in the monsoon regions thunderstorm activity is strikingly lower than at the beginning and end of this period; this is because the particular vertical structure of the air masses in the monsoon do not favour the development of thunderstorms. For thunderstorms to develop, it is necessary not only for the air to be warm and humid but also for the rising air to be warmer than its environment up to great heights. The summits of thunderstorm clouds in these regions can reach up 6–12 km, in extreme cases over high tropical mountains even 15–18 km.

To return to the "special weather phenomena" which we were considering above, the addition of sandstorms and snowstorms should be made to strong winds and gales. In polar regions these cause drifting snow, developing into a snowstorm, when snow cover is present, and in deserts they raise dust and sand, to form sandstorms. Further "special weather phenomena" include black ice, and a special form of liquidifying of the water vapour in the atmosphere which we call fog, mist or haze. This is formed by small droplets floating in the air and reducing the visibility below 1 km in the case of fog. If the visibility is between 1 and 2 km, we call it "mist" or "haze", according to whether it is produced by condensed water particles or by solid foreign matter such as dust or smoke.

In the climatic descriptions in Chapter 4, reference will be made to the following special weather phenomena mentioned above: gales, squalls, thunderstorms, fog, and snow cover, where significant.

After having thus defined the phenomena falling within our definition of "weather", let us take one step further on the path to generalisation, with the definition:

"Climate is the mean and normal course of weather at any given place."

Wladimir Köppen, famous German meteorologist of the German Naval Observatory at Hamburg, where he worked from 1875–1919, and internationally recognised patriarch of climatology, once said: "The weather changes, while the climate stays."

This statement is not strictly correct, inasmuch as the climate can change over the course of several decades. The notion of "climate" as a definition is in a progressive scale along with that of "weather" and "general weather". These are qualitatively equal, i.e. they share certain main elements in characteristics that describe particular phenomena. The main elements of "weather" are at the same time those of "climate". The three notions differ from each other in a gradual way, in that "weather" is related to a fixed moment in time, "general weather" to a fixed period, and "climate" to a 30-year span.

The characteristics of our present definitions are less extensive than those used in meteorological science and in the meteorological services. There "weather" is defined as containing the main elements temperature, wind, precipitation, relative air humidity, form and quantity of clouds, atmospheric pressure and others. The figures for its main elements are referred to statistical computations of mean values, standard deviations, frequencies and a number of further figures. As a rule, a collection of such data is sufficient to provide information about weather and climate, where observations made over a number of years are available, in meteorological yearbooks published by the national meteorological services. There is international agreement on the use of a 30-year span for the computation of climatological normals. It is important to take figures for the same period where comparisons are to be made. The standard period for which most of the normals were computed is 1901–1930.

In 1960, the Commission of Climatology of the World Meteorological Organization decided that climatological normals should also be computed for the new standard period 1931–1960. The figures were published in 1962 and 1971 in the publication WMO Nr. 117.TP52, quoted below as *CLINO*.

Differences between the mean values of one period and any other – which are frequently very small – are called "climatic changes".

The meteorological data published in the tables in the yearbooks show just which meteorological phenomena are included in the scientific definition of "weather". "Weather" is a notion that is defined by a group of characteristics. For our purposes, it can be said that these characteristics delimit an area of the environment. Put briefly: *weather is the meteorological environment of a given place at a fixed moment; general weather is the meteorological environment of a given place during a certain period; climate is the meteorological environment of a given place.* It has to be borne in mind that this is valid only for a particular period of 30 years, or about half the duration of a human life.

Clearly, different people, in taking into account various characteristics of "weather", do so according to their various individual interests and feelings. This may lead to misunderstandings, when people use the same words with differing definitions. To the navigator and aviator, wind and visibility are the most important weather elements. Aviation in particular is very sensitive to certain bad weather conditions. Before landing at an airport, a plane is informed about wind, visibility, the amount and height of clouds at the airport. Farmers, of course, are not interested in such details, but in the main weather elements temperature, wind, precipitation and sunshine, which affect their harvest. It should be added that these occupations are the last to give grounds for misunderstanding in referring to the weather; generally they are the best and most objective observers of the weather, because of their continual intimate contact with it. Such contact is often not given in the case of many people, for whom contact with the weather is restricted to their journeys to and from work, weekends and holidays. For these, it is the general weather that is decisive in regard to the activities they plan.

Who has not given vent to his feelings on a holiday spoilt by rain by cursing the weather? This simple example makes it clear that the weather acts directly upon man's mental and physical condition. When the rainy days are succeeded by serene and sunny ones, we quickly cheer up again.

In many regions of the world, the weather remains largely unaltered over a period of many weeks, and there the general weather of a few years is identical with the climate. This is the case with some tropical regions. The more one approaches the middle latitudes, the more changeable the weather becomes. Here, only the observations of several decades suffice to define the climate.

1.2 The principal elements of the weather

1.2.1. Temperature

The ultimate cause of the fact just referred to is to be found in the conditions of solar radiation. These are of paramount importance to all processes in the atmosphere, and all vital functions on earth, which are necessarily interconnected. On an annual average, the incoming short-wave radiative solar energy received in the tropical zone between latitude 30° N and S – which represents half the surface of the earth – exceeds the outgoing long-wave energy. At the two polar caps, however – representing the other half of the earth's surface – the incoming energy is less than the outgoing. Let us call radiative temperature the temperature occurring in the atmosphere if only solar radiation is taken into account. If we compare this with the real temperature – calculated by averaging the observed temperatures – and take the difference, it turns out that this difference functions almost exactly as the balance of radiation along the lines of latitude. This relationship shows that the difference between real and radiative temperature is ultimately caused by the radiative conditions, which stir the atmosphere into global circulation.

First, let us treat the diurnal variation of solar radiation to be found on the equator at the equinox outside the earth's atmosphere. This variation can be represented by a simple harmonic curve with a maximum at 12 o'clock and zero values at 6 o'clock and 18 o'clock. During the night there is no solar radiation (see Figure 2: "Diurnal variation of solar radiation".) At night there is only the outgoing infrared radiation, with its cooling effect. If there were no atmosphere, the diurnal variation of the temperature on the upper ground would be represented by an asymmetrical curve. Both the extreme values would occur some hours later, the maximum early in the afternoon, the minimum at early dawn. This can be explained by the exchange of heat in the ground. The soil temperatures observed at Tiflis in summer (June–August) show a variation of 14 °C (Figure 3: "Diurnal variation of soil temperature"). By contrast, the temperature of the upper layer of the Atlantic Ocean on the equator is almost constant, as observed by the expeditions of the Meteor. There is a diurnal variation of less than 1 °C (Figure 4: "Diurnal variation of sea surface temperature on the equator"). On land, the temperature differences between day and night become larger when the ground is dry. The form of the surface also influences temperature – wheter the site is horizontal or on the slope, a trough, a peak or summit. Sites convex in form have small diurnal variations, while concave ones have larger ones. The same is true of the temperature of the air near the ground. The largest diurnal variations in air temperature occur in deserts; near the equator they are smothered by the influence of clouds and the high content of water and water vapour of the air.

The annual variation in air temperature – generally defined as the difference between the mean air temperatures of the warmest and the coldest month – also has its maximum value over deserts. They are 15 ° – 20 °C in the sub-tropics. Minimum values are to be found near the equator, whilst continental polar regions have extreme maximum values – Canada about 45 °C, Siberia up to 60 °C!

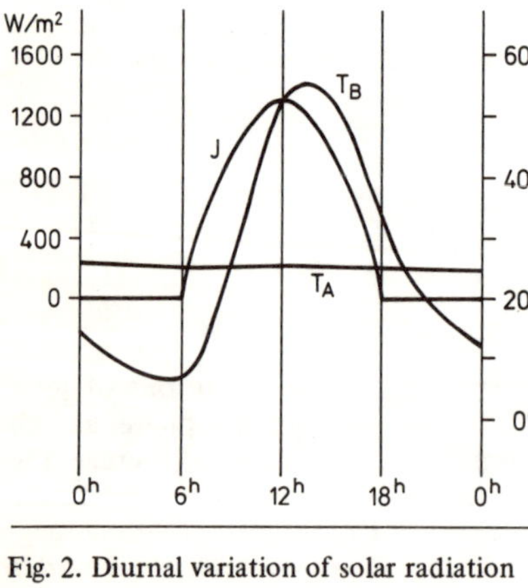

Fig. 2. Diurnal variation of solar radiation

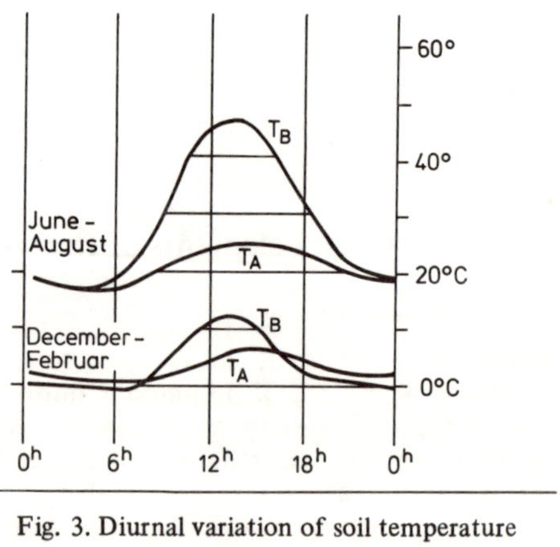

Fig. 3. Diurnal variation of soil temperature

This annual variation is a characteristic of the annual course of the air temperature, which is further characterised by the time of the occurrence of the extreme values. Over the continents they occur earlier than over the oceans. The retardation of these extremes compared with those of solar radiation is smaller over land than over the sea. Over land the temperature remains above its mean value longer than over the sea. The extent of the annual variation in temperature is influenced not only by the geographical latitude and the distance from the ocean, but also by the height above sea level and other local circumstances. Generally it diminishes with increasing height. It is also reduced near great continental lakes.

The principal forms of the annual variation of air temperature, according to *Köppen* and others, are:

1. normal type: so called because it predominates over large parts of the continents of the temperate zones, and reaches far into the polar and equatorial zones. The maxima occur in July in the northern hemisphere, and in January in the southern, with the minima conversely in January and July respectively.

2. Cape Verde type: this is characterised by a retardation of the maximum in the northern hemisphere until September. The minimum may be retarded until March. It occurs above all in the Cape Verde Islands, Senegal and California. As *Köppen* observes, the sucking-in of cool oceanic air into the heated interior keeps down the warmth of summer, so that the warmth does not reach its maximum until the on-shore wind weakens.

3. Indian type: this is characterised by a premature maximum. In the northern hemisphere it occurs in April, May or June, and in the southern hemisphere in October, November or December. This is because the rise in temperature is interrupted immediately after the monsoon breaks, where upon cool oceanic air invades the land.

4. Sudan type: this has a minimum at the solstice. It occurs in the northern hemisphere in the west of Togo, the east of Ethiopia, and in the West Ghats and south Malabar Coast of India.

5. equatorial type: this is notable less for the form of the annual variation than for its amount, which is less than 3 °C. There are sometimes double variations, due to radiation or to a rainy season or the predominance of a certain air current. Near the ground, air temperature generally decreases with height, about 0.5 °C per 100 m elevation. If we compare this decrease in the vertical plane with that in the horizontal from equator to pole – which is about 0.75 °C per degree of latitude, i.e. 111 km – we find that an elevation of 150 m is needed to balance a movement of one degree of latitude away from the equator. The air temperature changes vertically 740 times as fast as horizontally.

In the free atmosphere the temperature of the air is generally about 2 °C warmer than on hill and mountain summits of the same height, where the long-wave irradiation off the ground is more intensive than in the free atmosphere and causes a greater heat loss. On the windward side of a mountain, the air temperature does not decrease as rapidly as on the leeward side. Tropical mountains demonstrate this effect particularly. The decrease

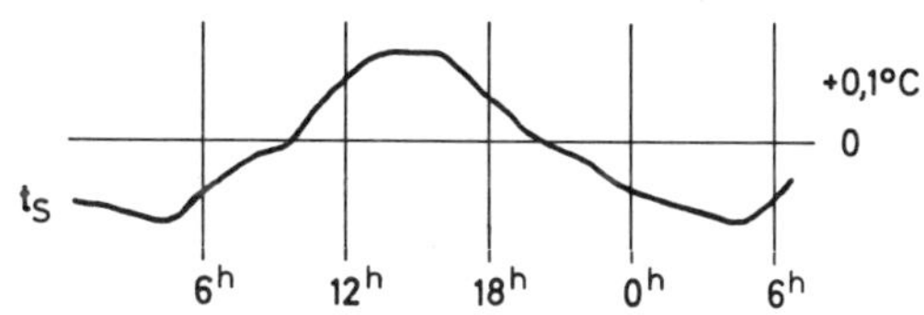

Fig. 4. Diurnal variation of the sea surface temperature at the equator

on the windward, wet side is remarkably less than on the dry leeward side. In Sri Lanka a decrease of 0.55 °C per 100 m was observed on the windward side, and one of 0.80 °C on the leeward.

The general decrease in temperature in the atmosphere with height is caused by gravitation. The air in the lower layers has to support the air above it, which presses upon it and thus heats it. The decrease in temperature extends up to the tropopause, which is at a height of 9 km in polar regions and 16 km in equatorial regions.

The part of the atmosphere between the ground and the tropopause is called the troposphere (see Figure 5); it is the layer in which the weather processes happen, especially those of condensation and evaporation. The layer above the tropopause is called the stratosphere. In it the temperature is nearly constant. It extends up to about 20–25 km; thereafter the temperature increases considerably in what is called the ozone layer, up to 50–60 km, where the ionosphere begins.

An air particle that moves up and down without loss or gain of heat in the atmosphere undergoes an decrease or increase in temperature, by reason of adiabatic expansion or pressure, of 1 °C per 100 m of change in height in the lower layers. Where the movement is not adiabatic, i.e. if, for example, there is a supply of heat, then conditions will change. When humid saturated air rises, water vapour is transformed into liquid water and heat which had arisen as vaporisation heat on the evaporation of the water is set free by condensation. This serves to warm the air, so that the vertical decrease in temperature is less in rising humid saturated air than in dry air.

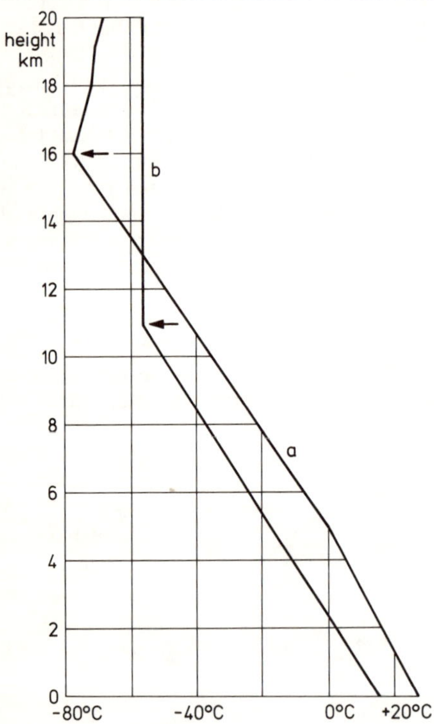

Fig. 5. Structure of the atmosphere

1.2.2. Atmospheric pressure and wind

Atmospheric pressure also decreases with height. From sea level, the vertical pressure difference is 1 millibar per 8 m height difference. This is called height of unit decrease in pressure. The change in pressure according to height is shown in the following table. The values are valid at a temperature of 0 °C, and increase with a rise in temperature, or decrease with a fall in temperature, by about 0.4% per 1 °C.

Atmospheric pressure in mbar	1000	950	900	850	800	750	700	650	600	550	500
Height of unit decrease in pressure	8.0	8.4	8.9	9.4	10.0	10.6	11.4	12.3	13.4	14.6	16.0

(After *Chromow* 1942, Einführung in die synoptische Wetteranalyse, p. 38.)

The air temperature determines decisively the height of unit decrease of pressure. The International Organization of Civil Aviation has defined a Standard Atmosphere in the interests of aviation safety; this is valid for the temperate zones of the earth, and is used in particular to calibrate altimeters in aircraft. In 1959 the Indian meteorologist Pisharoty defined a tropical standard atmosphere.

The values of both these standard atmospheres are represented in the following table:

In temperate zones			In the tropics	
temperature °C	pressure mbar	altitude km	temperature °C	pressure mbar
+ 15.0	1013.25	0	+ 27.0	1013.25
+ 8.5	898.76	1	+ 21.6	903.30
+ 2.0	795.01	2	+ 16.2	803.58
- 4.5	701.21	3	+ 10.8	713.29
- 11.0	616.60	4	+ 5.4	631.70
- 17.5	540.48	5	0.0	558.12
- 24.0	472.18	6	- 7.0	491.93
- 30.5	411.05	7	- 14.0	431.72
- 36.9	356.52	8	- 21.0	377.72
- 43.4	308.01	9	- 28.0	329.22
- 49.9	265.00	10	- 35.0	285.82
- 56.4	227.00	11	- 42.0	247.09
- 56.5	193.99	12	- 49.0	212.65
- 56.5	165.80	13	- 56.0	182.15
- 56.5	141.70	14	- 63.0	155.23
- 56.5	121.12	15	- 70.0	131.57
- 56.5	103.53	16	- 77.0	110.88
- 56.5	88.50	17	- 75.0	93.24
- 56.5	75.65	18	- 72.0	78.54
- 56.5	64.67	19	- 71.0	66.27
- 56.5	55.29	20	- 69.0	56.01

The unequal distribution of temperature on earth caused by the differing radiational conditions in the different latitudes, and also longitudes, cause first of all differences in density and atmospheric pressure in the horizontal. These differences give rise to air currents, i.e. winds, tending to compensate for the differences.

The horizontal distribution of atmospheric pressure near the equator is irregular and flat. Here the semi-diurnal oscillation of atmospheric pressure predominates. Maxima are at 10 o'clock in the morning and evening, minima at 4 o'clock in the morning and afternoon. The amplitude of this oscillation is 2.5 times as great as that of the diurnal oscillation. The latter has its maximum over the interior of the continents at about 3 o'clock in the morning and its minimum at 3 o'clock in the afternoon. It is caused by the solar heating of the atmosphere during the daytime and the cooling by irradiation during the night. Over the open sea, this oscillation hardly exists, because the sea-surface temperature is almost constant. It is not easy to explain the semi-diurnal oscillation of atmospheric pressure by the diurnal variation in solar radiation.

There is a permanent input of water vapour into the atmosphere from the surface of the sea, especially in the regions of the trade winds, which cover 31% of the area of the oceans. The trade winds have great constancy, owing to the very constant pressure gradient between the equatorial trough and the subtropical high-pressure belt. The mean wind velocity is between 4 and 8 metres per second. Within the precincts of the equatorial trough itself, the winds are weak and changeable; there are also often calms.

In the northern hemisphere, the trade winds blow from the north-east, and in the southern from the south-east, both in the direction of the equator. Because of the convergence of the air currents and the associated rising of the air, the atmosphere here becomes rich in water vapour, especially the layers between 5 and 10 km in altitude. The molecules of water vapour absorb a remarkable amount of solar radiation, reaching a maximum of heating shortly after mid-day. The nocturnal irradiation from the upper cloud surfaces causes an increase in shower activity over the tropical oceans in the precincts of the intertropical convergence zone, were the trade winds of the two hemispheres meet. The heating of the tropical atmosphere by condensation seems to attain its maximum in the second part of the night. In this way, the tropical atmosphere is heated twice a day. This causes a circulation of the atmosphere corresponding to the semi-diurnal oscillation of pressure at the bottom of it.

In his first paper on mountain wind and valley breeze, the famous Austrian meteorologist *Hann* made the following statement:
"In all the mountains of the earth the phenomenon of diurnal wind oscillation can be observed. During the day-time, from about 9 or 10 o'clock until sunset, a wind blows up the valleys. During the night, an air current moves downwards out of the valleys, lasting until the early morning." Generally, the valley breeze is more marked than the mountain wind (see Fig. 6: "Day's wind systems"). The same relation exists between the sea breeze or on-shore wind and the off-shore wind. On-shore winds, like valley breezes, can sometimes reach gale force. By contrast, it can happen that a powerful monsoon is completely suppressed by an on-shore wind.

"Air which is more or less in motion we call wind, and very violently moving air a gale or hurricane"; this definition is to be found in the "Lehrbuch der Meteorologie" (Textbook of Meteorology) by *Hann-Süring* (Leipzig 1940, p. 481). Movement of air is characterised by the distance covered in relation to the time required to cover it. The ratio distance divided by time is called velocity; this is a measure of movement of which the unit nowadays is m/s (metres per second). This value alone does not suffice to characterise the movement of the air in total. It is also necessary to know the direction from which the movement is coming. It is important to note here that in oceanography and navigation a current is denominated according to the direction in which it flows. We may use the following formulation as an aid to memory: in the case of WIND, we ask where the wind is coming from, since then we can estimate some of the qualities of the approaching air; with CURRENTS at sea, we ask where it will take our ship if we let it drift.

The denomination of the winds according to the quarters was introduced by *Charlemagne* (742–814). Eight principal directions are used: north (N), north-east (NE), east (E), south-east (SE), south (S), south-west (SW), west (W), and north-west (NW). Occa-

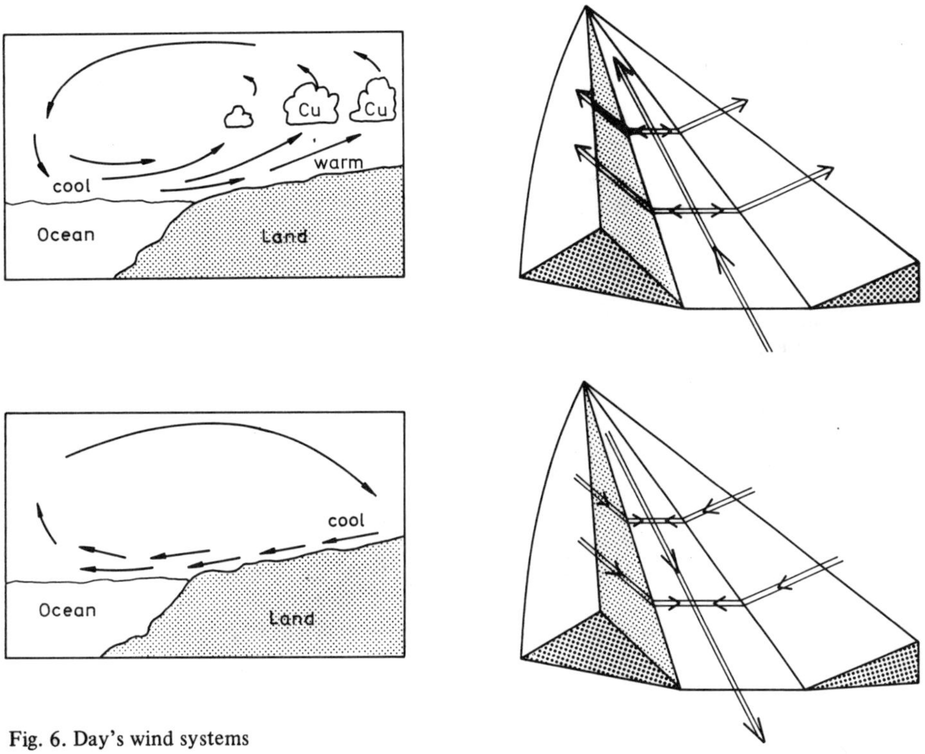

Fig. 6. Day's wind systems

sionally a sub-division is made, using north-north-east (NNE) and so on. Most wind observations of early times are estimations using a numerical scale first introduced by the British admiral *Beaufort* in 1805, and still in use, the Beaufort scale of wind force. It has been modified several times. By comparing estimations and actual measurements of wind movement, it was possible to relate the effects of wind on objects on land or sea more precisely to the velocities measured. This is still important, as it is not always possible to use an anemometer. Estimations of wind should always be made in such a way that they represent the general air current of the region in question. If this is done precisely, wind estimations may be of more value than actual wind measurements made in a place which is not representative of a wider area.

Wind force is estimated on a numerical scale ranging from 0 = calm to 12 = hurricane. In 1949, the attempt was made to extend the scale to 17, but was discontinued, since it proved impossible to make meaningful estimates for such high degrees. Thus today the highest measurement on the Beaufort scale remains 12, which includes all winds above 32.6 m/s.

Relation between the Beaufort degrees of wind force and the effects of the wind inland

Beaufort degree	Name	effects
0	Calm	smoke rises vertically.
1	light air	Direction of wind shown by smoke drift, but not by wind vanes.
2	light breeze	Wind felt on face; leaves rustle; vane moved by wind.
3	gentle breeze	Leaves and small twigs in constant motion; wind extends light flag.
4	moderate breeze	Raises dust and loose paper; small branches are moved.
5	fresh breeze	Small trees in leaf begin to sway; crested wavelets form on inland waters.
6	strong breeze	Large branches in motion; whistling heard in telegraph wires; umbrellas used with difficulty.
7	moderate gale	Whole trees in motion; inconvenience felt when walking against wind.
8	fresh gale	Breaks twigs off trees; generally impedes progress.
9	strong gale	Slight structural damage occurs (chimney pots and slates removed).
10	whole gale	Seldom experienced inland; trees uprooted; considerable structural damage occurs.
11	storm	very rarely experienced; accompanied by widespread damage.
12	hurricane	

When the impact of the atmosphere in motion increases to such a degree that our word "gale" is no longer able to describe the phenomena attendant upon it, we use the word of Caribbean origin "hurricane". In the night of 16th–17th February 1962, a westerly wind, with hurricane-like squalls up to 40 m/s (144 km/h) whipped the waters of the North Sea up the river Elbe, causing a catastrophic tidal wave in Hamburg. Velocities as high as this and higher may be expected on the coast at times during stormy weather conditions. In the open sea in the middle latitudes, it happens that the wind reaches such velocities, not only for short periods of minutes, but for several hours, with squalls surpassing at the maximum 60 m/s (216 km/h). Likewise, over the warm seas in the precincts of tropical revolving storms, such velocities are not unusual. The almost circular or elliptical area of a hurricane – as the natives of the Caribbean Sea named mature revolving storms – can have a diameter of more than 350 km (see Fig. 7: "The structure of a hurricane"). A wind velocity of at least 33 m/s (118 km/h) is defined as a hurricane. Maximum squalls of 50 m/s (180 km/h) are not rare. They can even reach more than 80 m/s (288 km/h), as estimated from damage found in areas hit by a hurricane.

It is necessary to differentiate strictly between the wind velocity as such and the velocity which a hurricane tracks. At first this is often only 15–25 km/h, increasing to 50 km per h, sometimes 80–100 km/h. There is one phenomenon that surpasses the tropical

Beaufort degrees, lower and upper limits of speed and mean wind pressure

Beaufort degree	velocity m/s	km/h	mean wind pressure mbar
0	0.0 – 0.2	0 – 1	0.00 – 0.00
1	0.3 – 1.5	1 – 5	0.00 – 0.01
2	1.6 – 3.3	6 – 11	0.01 – 0.07
3	3.4 – 5.4	12 – 19	0.07 – 0.18
4	5.5 – 7.9	20 – 28	0.18 – 0.39
5	8.0 – 10.7	29 – 38	0.39 – 0.71
6	10.8 – 13.8	39 – 49	0.71 – 1.17
7	13.9 – 17.1	50 – 61	1.17 – 1.80
8	17.2 – 20.7	62 – 74	1.80 – 2.63
9	20.8 – 24.4	75 – 88	2.63 – 3.66
10	24.5 – 28.4	89 – 102	3.66 – 4.95
11	28.5 – 32.6	103 – 117	4.95 – 6.52
12	32.7 – 36.9	118 – 133	6.52 – 8.35
13	37.0 – 41.4	134 – 149	8.55 – 10.51
14	41.5 – 46.1	150 – 166	10.35 – 13.03
15	46.2 – 50.9	167 – 183	13.03 – 15.88
16	51.0 – 56.0	184 – 201	15.88 – 19.21
17	56.1 and more	202 and more	19.21 and more

The pressure due to the wind on any object exposed to it arises from the impact of the air on the windward side and suction on the leeward side; the mean pressure depends on the shape and size of the object.

hurricane in violence: the tornado or grand waterspout. These are called tornadoes, in the great plains of the middle west of the United States of America, where they appear most frequently, especially in the states of Kansas, Iowa, Texas, Oklahoma and Arkansas. On average, there are well over 150 a year. They are frequent between April and July. Tornadoes mostly have a diameter of less than 1500 m and hang down from the clouds. They track with a mean velocity of between 10 and 40 km/h, and last from several minutes to a few hours. In them, wind velocities occur such as are never observed in other phenomena. Speeds of more than 200 km/h have been measured, but damage permits the assumption that 600 km/h can be reached.

Catastrophic as they are, the material effects of tornadoes are smaller in the United States than those of hurricanes; the latter cause on average damage amounting to about 50,000,000 US dollars a year, while that caused by tornadoes amounts to about $ 14,000,000. On the other hand, the loss of human life from tornadoes is higher, at 230 a year compared to 140 from hurricanes, since the latter can be better forecast.

In spite of considerable investment in scientific research into tornadoes and hurricanes, a number of questions remain unsolved. Researchers in the USA showed by film in 1962 that there is a slow downward movement of air in the inner part of the tube of a grant spout, whilst in the immediate vicinity of the fringe of the tube there is an intensive upward air movement coupled with spinning round the axis of the tube with the velocities mentioned above.

What inside a tornado can be seen Mr. *Will Keller*, a farmer near Greensburg, Kans., is the man to whom reference is made in the following story (see Justice, *Alonzo A.* 1930, Monthly Weather Review p 205–206: Seeing the inside of a tornado):

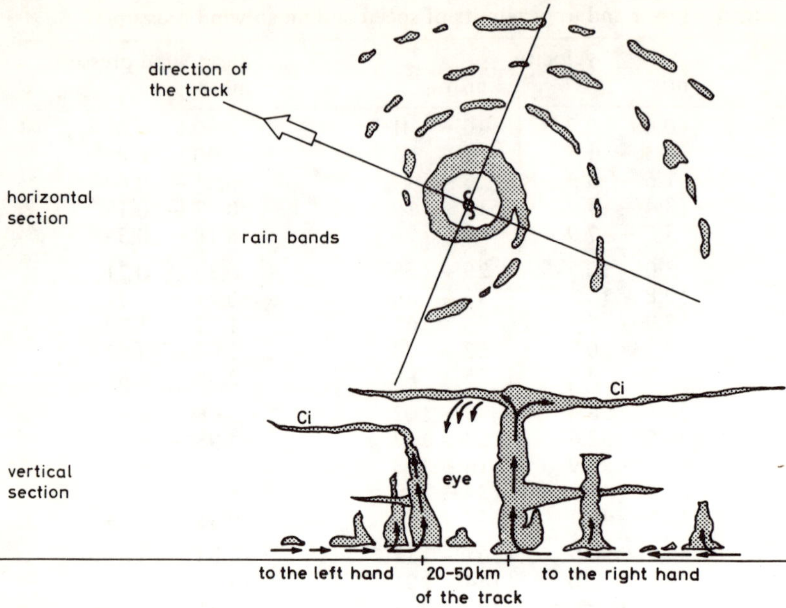

Fig. 7. Structure of a hurricane

„It was on the afternoon of June 22, 1928, between 3 and 4 o'clock. I was out in my field with my family looking over the ruins of our wheat crop which just had been completely destroyed by a hailstorm. I noticed an umbrella-shaped cloud in the west and southwest and from its appearance suspected that there was a tornado in it. The air had that peculiar oppressiveness which nearly always precedes the coming of a tornado.

But my attention being on other matters, I did not watch the approach of the cloud. However, its nearness soon caused me to take another look at it. I saw at once that my suspicions were correct, for hanging from the greenish-black base of the cloud was not just one tornado but three.

One of the tornadoes was already perilously near and apparantly headed directly for our place. I lost no time therefore in hurrying with my family to our cyclone cellar.

The family had entered the cellar and I was in the doorway when I decided that I would take a last look at the approaching tornado. I have seen a number of these things and have never become panic-striken when near them. So I did not lose my head now, though the approaching tornado was indeed an impressive sight.

The surrounding country is level and there was no thing to obstruct the view. There was little or no rain falling from the cloud. Two of the tornadoes were some distance away and looked to me like great ropes dangling from the clouds, but the near one was shaped more like a funnel with ragged clouds surrounding it. It appeared to be much larger and more energetic than the others and it occupied the central position of the cloud, the great cumulus dome being directly over it.

As I paused to look I saw that the lower end which had been sweeping the ground was beginning to rise. I knew which that meant, so I kept my position. I knew that I was comparatively safe and I knew that if the tornado again dipped I could drop down and close the door before any harm could be done.

Steadily the tornado came on, the end gradually rising above the ground. I could have stood there only a few seconds but so impressed was I with what was going on that it seemed a long time. At last the shaggy end of the funnel hung directly overhead. Everything was still as death. There was a strong gassy odor and it seemed that I could not breathe. There was a screaming, hissing sound coming directly from the end of the funnel. I looked up and to my astonishment I saw right up into the heart of the tornado. There was a circular opening in the center of the funnel, about 50 or 100 feet in diameter, and extending straight upward for a distance of at least one half mile, as best I could judge from the circumstances. The walls of the opening were of rotating clouds and the whole was made brillantly visible by constant flashes of lightning which zigzagged from side to side. Had it not been for the lightning I could not have seen the opening, not any distance up into it anyway.

Around the lower rim of the great vortex small tornadoes were constantly forming and breaking away. These looked like tails as they writhed their way around the end of the funnel. It was these that made the hissing noise.

I noticed that the direction of rotation of the great whirl was anticlock-wise, but the small twisters rotated both ways – some one way and some another.

The opening was entirely hollow except for something which I could not exactly make out, but suppose that it was a detached wind cloud. This thing was in the center and was moving up and down. The tornado was not travelling at a great speed. I had planty of time to get a good view of the whole thing, inside and out. It came from the direction of Greensburg, which town is 3 miles west and 1 mile north of the place. Its course was not in a straight line, but it zigzagged across the country, in a general northeasteerly direction.

After it passed my place it again dipped and struck and demolished the home and barn of a farmer by the name of Evans. The Evans family, like ourselves, had been out looking over their hailed-out wheat and saw the tornado coming. Not having time to reach their cellar they took refuge under a small bluff that faced to the leeward of the approaching tornado. They lay down flat on the ground and caught hold of some plume bushes which fortunately grew within their reach. As it was, they felt themselves lifted from the ground. Mr. Evans said, that he could see the wreckage of his house, among it being the cook stove, going round and round over his head. The eldest child, a girl of 17, being the most exposed, had her clothing completely torn off. But none of the family were hurt.

I am not the first one to lay claims to having seen the inside of a tornado. I remember that in 1915 a tornado passed near Mullinville and a hired man on a farm over which the tornado passed had taken refuge in the barn. As the tornado passed over the barn, the door was blown open and the man saw up into it, and this one like the one I saw, was hollow and lite up by lightning. As the hired man was not well known, no one paid much attention to what he said." –

Tropical revolving storms and tornadoes display great differences in diameter, between about one and several hundred km. Common to them all, however, is the slow downward air movement in the inner part with the most violent upward movement very close to this. The descending air in the hurricane is to be found in its so-called eye, which has a diameter of about 25 km, and in which are found weak winds or a calm, often scattered clouds or even sunshine.

The following story is the eye-witness report of a meteorologist describing the eye of a supertyphoon (see *Simpson, R.H.* 1952, Bull. Am. Met. Soc. p 286–298, Fig 11 "Exploring Eye of Typhoon 'MARGE' ", 1951):

„Here was one of Nature's most spectacular displays. "Marge's" eye was a vast coliseum of clouds, 40 miles in diameter, whose walls rose like galleries in a great opera house to a height of approximately 35,000 feet where the upper rim of the clouds was smoothly rounded off against a background of deep blue sky. The sea surface was obscured by a stratocumulus undercast except for two circular openings on the east and west sides of the eye respectively. Clouds in the undercast layer were grouped in bands which spiralled cyclonically about each of these openings, or clear spots, both of which were approximately five miles wide. The horizontal alignment of clouds suggested the possibility that two seperate small eddy circulations were present within the eye envelope.

In the geometric center of the eye the stratocumulus overcast bulged upward in a domelike fashion to a height of 8,000 feet. Light turbulence in the top of the dome was comparable to that in ordinary ocean cumulus.

The walls of the eye on the west side were steep, either vertical overhanging, and had a soft stratiform appearance. On the east side, however, clouds were more a cumuliform type with a hard cauliflowery appearance. In this sector the walls of the eye rose with a gradual concave slope to the upper rimm. The overall appearance indicated that the axis of symmetry (in the vertical plane) was tilted to the east or northeast."

There is still no hypothesis which accounts satisfactorily for these two phenomena (typhoon or hurricane and tornado) as a whole.

1.2.3. Atmospheric humidity

After radiation and temperature, and pressure and wind, the next element to be treated is humidity. Humidity and heat are of vital importance to life on earth. In the tropics, where heat is present in abundance, humidity is the decisive element (except for the high mountains, notably the Andes in South America and the Himalayas in Asia).

Atmospheric humidity depends directly on air temperature, which determines the degree and limit of saturation. The degree of atmospheric humidity can be represented by vapour pressure, saturation deficit, dew-point, absolute humidity, relative humidity, wet-bulb temperature, or equivalent temperature. Under normal pressure conditions there is a fixed relation between these seven phenomena, as expressed in a hygrothermal diagram; hence, as a general principle, it is sufficient to use one of them in order to represent the state of atmospheric humidity, always together with air temperature. Which of these measurements priority should be given to depends on the problem to be treated. In many cases, relative humidity is used, especially when living plants or organic matter are being considered. Wet-bulb or equivalent temperature seem best fitted to gauge the oppressiveness or sultriness of the air. In other cases, vapour pressure, dew-point, saturation deficit or absolute humidity are given preference. Saturation pressure E (given in mbar) can be expressed in the following empirical formula, derived by *Magnus,* as a function of air temperature t: $E = 6.67 \cdot \exp(17.1544\, t/(234.7+t))$. The definitions of the above-mentioned measures of humidity and saturation pressure are:

1. Vapour pressure e is the partial pressure of the water vapour carried by the humid air. It is measured in millibar (mbar).
2. Saturation pressure E is the water vapour pressure of saturated air.
3. Saturation deficit E-e is the difference between the saturation pressure E and the water vapour pressure e.
4. Relative humidity r is the percentage ratio of the actual water vapour pressure e and the saturation pressure E. $r = 100\, e/E$.
5. Absolute humidity a is the weight in grams of the water vapour in a cubic metre of humid air. (This is sometimes erroneously taken by technicians for the mixing ratio, which is the weight in grams of the water vapour in a kilogram of dry air. The weight in grams of the water vapour in a kilogram of humid air is called specific humidity.)
6. The dew-point t_d is the temperature at which humid air is fully saturated.

7. Equivalent temperature t_e is the temperature to which the air rises if all the water vapour in it is condensed, and the heat released goes to heat the air, at constant atmospheric pressure. A formula of approximation is: $t_e = t + 1.5$ e, e being expressed in mbar.
8. Wet-bulb temperature t_w is the temperature to which the air sinks if its water vapour content is increased to saturation point, and the necessary heat for vaporisation is supplied by the air itself, at constant pressure. (This temperature is nearly equal to that which a wet-bulb thermometer acquires if ventilated.) The term "heat content" used in engineering is equivalent to "enthalpy" or "heat function" in physics; this is the heat to be supplied at constant pressure to a system in order to obtain its observed temperature. If this term is used, one can make the following definitions:

7a. Equivalent temperature t_e is the temperature of dry air with a heat content equal to that of the humid air under consideration ($t_e \times 1.005$ is the heat content of the air expressed in kJ/kg).

8a. The wet-bulb temperature t_w is the temperature of saturated air with a heat content equal to that of the humid air under consideration. (It is measured according to definition 8, which almost corresponds to this definition.)

These measures of humidity can best be demonstrated by means of a diagram. The hygrothermal diagram (Fig. 8: "Hygrothermal diagram") shows the relation between temperature (linear abscissa in °C) and vapour pressure (linear ordinate in mbar, but also in mm Hg, because some tables containing this unit are still used) of the air under consideration. Saturation pressure depends exponentially on temperature; this is demonstrated by the saturation curve indicated by a relative humidity of 100%. The other lines of relative humidity are constructed by a corresponding sub-division of the values in the ordinate.

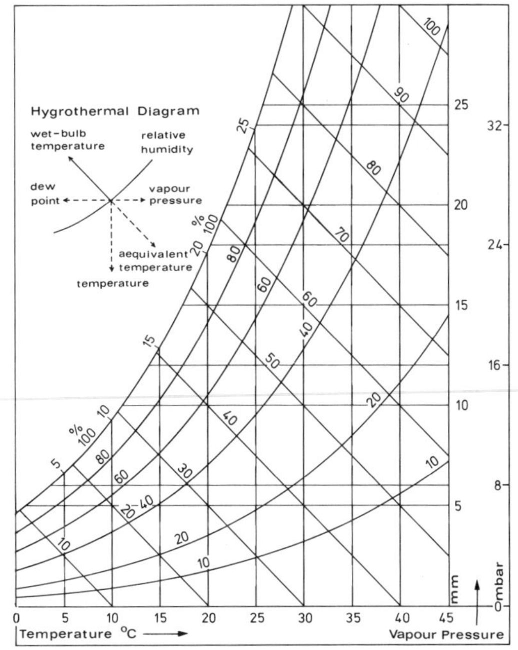

Fig. 8. The hygrothermal diagram

Besides this band of exponential curves there is a band of straight lines inclined to the right below an angle of 45°. This results from the selection of the units of temperature and vapour pressure: 5 mm Hg vapour pressure corresponds to 10 °C air temperature. Thus the straight lines approximately represent, under normal pressure conditions, lines of equivalent temperature; their values correspond to the temperature values given at the intersections with the axis of temperature (at vapour pressure = zero). The lines are numbered correspondingly between the curves of relative humidity of 40% and 50%. The temperature values of the intersections of these straight lines with the curve of saturation pressure give the corresponding wet-bulb temperature. The dew-point is only a function of the vapour pressure. The horizontal lines therefore represent at the same time lines of equal dew-point, the value of which is given by the temperature at the intersections of these horizontal lines with the curve of saturation pressure.

The wet-bulb temperature is determined by the equivalent temperature and the dew-point by the vapour pressure. Independent quantities in the diagram are: 1. temperature t; 2. vapour pressure e; 3. equivalent temperature t_e; 4. relative humidity r.

A given state of atmospheric humidity is sufficiently characterised by two of these quantities. The diagram can serve to estimate the two remaining quantities. The relations are illustrated in more detail in a sketch in the upper left portion of the diagram, and in the following examples:

1. Given	$t = 12.0\ °C$,	$e = 10.0$ mm,	
then	$r = 95\%$,	$t_e = 32\ °C$,	$t_w = 11.5\ °C$.
2. Given	$t = 26.0\ °C$,	$r = 40\%$,	
then	$e = 10.0$ mm,	$t_e = 46\ °C$,	$t_w = 17.0\ °C$.
3. Given	$t = 18.0\ °C$,	$t_w = 17.7\ °C$,	
then	$e = 15.0$ mm,	$r = 97\%$,	$t_e = 48\ °C$.
4. Given	$t_w = 20.3\ °C$,	$e = 10.0$ mm,	
then	$t = 36.0\ °C$,	$r = 22\%$,	$t_e = 56\ °C$.
5. Given	$t_e = 56\ °C$,	$r = 59\%$,	
then	$t = 26.0\ °C$,	$t_w = 20.3\ °C$,	$e = 15.0$ mm.
6. Given	$e = 20.0$ mm,	$r = 50\%$,	
then	$t = 34.0\ °C$,	$t_e = 74\ °C$,	$t_w = 25.3\ °C$.

The atmospheric water vapour is introduced into the lowest, very thin layer of air by evaporation from a water surface. Then it is vertically distributed by the wind, or more precisely by the turbulent motion of the air. The principal sources of water vapour are, of course, the oceans; they cover 71% of the earth's surface, that is, 361 Mm^2 (square megametres). The continents occupy only 29% or 150 Mm^2. The evaporation is proportional to the difference between the vapour pressure at the surface of the water and the atmospheric vapour pressure, and the speed of ventilation, and inversely proportional to the atmospheric pressure. The actual evaporation depends directly on the liquid water available for evaporation. Direct measurements made to date have made it possible to estimate the course of the absolute values, but the values are rendered imprecise by the influence of the experiments themselves. Only recently did it become possible to estimate the evaporation of open water surfaces by the balance of energy, measuring the incoming and outgoing energy of the water.

One important quantity is the areal evaporation. In relation to a large area, the following equation is valid: $N = A + V + S$, where N is the precipitation falling on the area, A is the runoff, V the evaporation, and S the water stored in the ground, i.e. the change in its water content. This equation tells us that precipitation is equal to the sum of runoff, evaporation and storage. If the storage is small, the evaporation can be estimated. This has been done for the great basins of the Congo, the Ganges and the Amazon. The values are:

	N mm	A mm	V mm	temperature in °C
Congo	1400	370	1030	22.5
Ganges	1700	553	1147	23.0
Amazon	1800	521	1279	25.5

In the tropics, the evaporation is equivalent to several times the runoff. The water evaporates from all water surfaces of the earth, from humid land, from plants, and also from snow and ice. The water molecules re-unite in liquid or solid form in the free atmosphere by forming clouds, or dew or hoar-frost on solid surfaces. Rising air cools, and its water content condensates as soon as the dew-point is reached.

1.2.4. Hydrometeors

The liquid and solid water in the atmosphere takes on many forms. Before water, having left the earth's surface in gaseous form, returns to it as precipitation, it lingers for some time in the atmosphere, as a visible cloud. The transition from gaseous to liquid form is called condensation. This happens in the atmosphere when humid air is forced to rise; its pressure decreases, it expands and cools (by about 1 °C for every 100 m increase in altitude). The colder the air, the less gaseous water it can contain. At a certain temperature at a given altitude, the air is saturated with water. This temperature is called dew-point temperature, or briefly dew-point. In the atmosphere there are always a number of nuclei of condensation. They have a diameter of 0.2–2 my (1 my = 1 micrometre, i.e. the millionth part of a metre) (see Fig. 9: "Comparison between the dimensions, concentrations and final velocities of some particles participating in the process of condensation

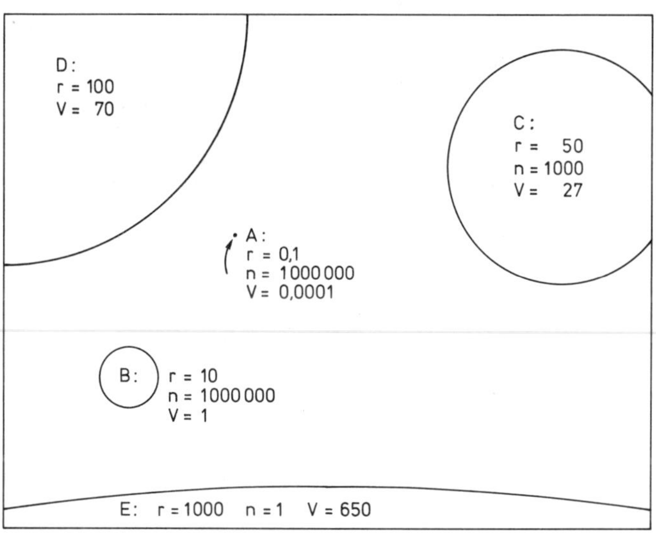

Fig. 9. Comparison between the dimensions, concentrations and final velocities of some particles participating in the process of condensation and precipitation

and precipitation"). In one litre of air, there are a million nuclei. Liquid water is condensed on them, forming drops of water with a diameter of 20 my, so small that they float in the air. By coagulation and collision, larger cloud droplets form, and finally raindrops. The limit between these drops is said arbitrarily to lie at a diameter of 200 my. Large cloud droplets are about 100 my in diameter, 1000 to a litre or 1 to a cubic centimetre. The larger the drops grow, the greater is the final velocity with which they fall; the greater the lifting velocity, the larger the size of cloud drops the air can bear. When the velocity of fall increases by the drops' growing or the slackening of the lift, it starts to rain.

A typical raindrop has a diameter of 2000 my (= 2 mm) and a final velocity of 6.5 m/s. Drops falling at more than 10 m/s are unable to cohere and are scattered. The diameter of such final drops may be 7 mm. Super-cooled water drops can also be found well below freezing point, down to about −35 °C. Pure water only freezes when it encounters minute ice crystals, or freezing nuclei as they are known; then freezing occurs within a few seconds.

In addition to these, there are sublimation nuclei. These permit a direct transition from the gaseous state to the solid state of water, by-passing the liquid state. These are subsumed with the last-mentioned as ice nuclei. They should have a diameter of at least 0.1 my. They seem normally to be rare, since super-cooled clouds can often be encountered. The study of this area of cloud physics is far from completed, and many problems have to be solved before we know exactly how precipitation comes into being in detail.

When large raindrops freeze in strongly supercooled air at great altitudes, they are transformed into soft hail. In this process, crystal skeletons are generated, on which ice and snow crystals form, soon filling the whole space above freezing point. This process happens in supercooled clouds. The vapour pressure of supercooled water is somewhat greater than that of ice at the same temperature, so that the molecules of water vapour contained in the air adhere to the ice crystals, which begin to grow. In the end, all the liquid water is transformed into ice. It is estimated that there must be about one ice nucleus to ten litres of cloud in order that precipitation, going through the ice phase, and subsequently liquifying at a temperature above freezing point, results in large raindrops. It can happen that in thunderclouds pieces of ice form of such magnitude that they do not have time to melt while falling, and thus reach the ground as hail or soft hail. The largest drops of a shower are the first to reach the ground, as can be observed during a thunderstorm in summer; the smallest drops are the last to arrive. The duration of the rain as observed on the ground is normally longer than the time taken by the raindrops to generate. The advent of the largest drops first during rain enables the approach of rain to be recognised by the noise which the falling drops produce on reaching the plants. In the tropics, where there are showers in the afternoon and evening accompanied by thunder in the rainy season, this is an everyday phenomenon. Where 15 litres of rain per square metre fall in 5 minutes, we talk of a cloudburst. Rain of such intensity also is not unusual out of the tropics, but not very frequent. On June 2nd, 1908, 13 litres to the square metre fell within 2 minutes at Torfhaus in the Harz Mountains, Germany, and on May 25th, 1920, 126 litres in 8 minutes at Bannwaldsee in Bavaria. The world record is held by Opids Camp in California, where the incredible quantity of 26 litres per square metre within half a minute was recorded. In one cubic metre of air there can be about 50 grammes of liquid water. This is a hundred times the amount normally found inside a cloud.

The various forms of precipitation are called hydrometeors. Dew and hoar-frost are products of the condensation of atmospheric water vapour on or near the surface of the earth or on objects on it. Rain, hail, sleet and snow are such products falling from a height. The conditions in which they form can be very different from those near the ground.

The bigger the drops of rain, the higher they were generated. Fine rain or drizzle has very small droplets of about half a millimetre in diameter, which raindrops often exceed.

The large drops of rain-showers are often melted hailstones. Apart from hail, solid precipitation takes the form of grains of ice, sleet, and snow. Soft hail consists of opaque white grains 2–5 mm in diameter. Small hail is balls of snow embedded in a sheet of ice. Hailstones are pieces of ice 5–50 mm in diameter. One centimetre of freshly fallen snow corresponds to a height of 1 mm of water or, in other words, to one litre per square metre.

In the tropics there is a pronounced diurnal variation of precipitation, with as a rule a maximum in the afternoon over continents and islands. The extreme of this type is found in the mountains of the equatorial zone. In the large valleys and extended plains, thunderstorms, often descending from the mountains, do not break until night, as Alexander von Humboldt noted over a century ago. Rain in the afternoon is very often accompanied by heavy thunderstorms, so that more than 50% of the diurnal precipitation often falls within a few hours. In San Jose de Costa Rica, in the rainy season from May to October, 106 mm has been observed between 0 and 12 hours, and 1634 between 12 and 24 hours, of which 783 fell between 15 and 18 hours. In Quetta (Pakistan) 80% of the rain has been observed to fall between 14 and 20 hours. In many coastal regions of the tropics, night rain predominates, as in Cameroun, Borneo, parts of New Guinea and elsewhere.

Especially on windward coasts, there is a tendency to intensification of rain at night during the monsoon season. In Djakarta, night rain predominates in January and February, at the climax of the monsoon. At the beginning in October, and going through November into December, the normal tropical afternoon maximum still predominates.

Precipitation is generated when air rises. A rising air movement is to be found

1. over mountains: orographic rain is caused by the passive rising of the air masses over the sloping ground;
2. at the boundary of air masses of differing temperature, which causes cyclonic or frontal rain;
3. over areas with intensive heating of the land, when the heat of condensation occurs at a great height, causing brief heavy showers.

In the tropics, orographic rain often occurs as trade wind rain. The trade winds blow with the greatest constancy, and are strongest, in winter. They cause rainfall on the windward side of mountains in their path. This adds to the normal monsoon or summer rain. Many examples can be found in Malaysia, Borneo, Celebes, the Philippines, or Taiwan. Frontal rains predominate outside the tropics, and in the boundary regions of the tropics and middle latitudes, where they occur in winter. They are often accompanied by thunderstorms. Such winter rains are limited to the western sides of the continents.

In the precincts of the equator, the air is involved in a rising motion more or less throughout the year, which is intensified in those latitudes where the sun is almost at the zenith at midday. At places between the tropics, the sun is at the zenith twice a year. Only for places up to 10–12 degrees of latitude are there two rainy seasons a year, if there is sufficient time between the zenith positions of the sun; in this case, the maximum rainfall also occurs at these times. The dryness is generally greatest when the sun is at the tropic in the other hemisphere. Outside the equatorial zone, the tendency for the air to rise is greatest when the sun is at the zenith. Then there is abundant rain and a simple tropical rainy season predominates. The simple dry season lasts longer than the dryness in the equatorial zone.

1.2.5. Clouds

In the atmosphere, there are various particular cloud levels. The low level is between 0 and 2 km in altitude, the middle level 2–5 km (in the tropics 2–8 km), and the high level from 5 to 9–12 km (in the tropics 8–18 km). At the high level there are different forms of ice clouds: cirrus (ci), the high white feather-shaped cloud; cirrocumulus (cc), the very thin fine fleecy cloud; cirrostratus (cs), the fine whitish veil-like cloud. At the middle level we find altocumulus (ac) and altostratus (as). ac clouds have a luminous border and shady dark interiors, by which they can be distinguished from cc. as clouds are uniform grey gloomy veil-like clouds, weakly transparent to the light of the sun and moon. The low level clouds are stratus (st), a sheet cloud; cumulus (cu), a woolpack cloud; stratocumulus (sc), a mixture of sheet and woolpack cloud, whitish, but often grey, subdivided into lumps or with the aspect of bales pushed together; nimbostratus (ns), a rain-cloud, a thick grey, often heavy sheet cloud precipitating continually either rain or snow; and cumulonimbus (cb), the shower cloud or thunderstorm cloud, developing from cumulus congestus (cu cong), which is the large woolpack cloud. In the tropics one can find cb block massifs with their tops at 18 km (see Fig. 10: "Cloud forms").

F.H. Ludlam, the British expert on cloud physics, drew up a classification of clouds according to their movement. Briefly, he differentiates between:

1. orographic clouds;
2. clouds due to turbulent exchange;
3. upslide clouds; and
4. clouds due to vertical convection.

Orographic clouds: it is known that in certain weather situations, not only in mountains, but also in flat country and over islands in the ocean, clouds form over or adjoining obstacles that hamper the general air current. The wind often shows an intense increase with height. Usually the clouds are thin, and often have the form of undulations or are lentiform, for which reason they are also called lenticularis. Here we may mention the cumulus clouds found over islands, although the cause oft their formation lies in the rising of heated air over the land. This is a typical phenomenon in the trade wind regions. The barrage clouds (not mentioned by *Ludlam*) can also be classified here, as the cause of their formation lies in the current. They too are typical of the trade wind regions.

Clouds due to turbulent exchange: fog can form when the air near the ground is sufficiently cooled, for example by irradiation during the night, and condensation occurs. If a wind is also blowing, its irregular stirring motion can transport the cooling upwards, so that the fog rises and forms a low-lying sheet cloud. The cooling of the surface of the fog or the stratus in turn produces a cooling in the interior of the cloud. Here an irregular mixing arises, producing a steep temperature gradient. The cloud surface appears undulating and bumpy. If the cloud is thin, clear intervals appear, and it changes character, from stratus to stratocumulus. The chequered shape is characteristic of thin layers of cloud whose surface loses heat by irradiation when the sky above is clear. Examples are altocumulus and cirrocumulus. Low sheet clouds can extend for several thousand km when strong air currents go from the tropics to the poles.

Upslide clouds: in depressions, there occurs a widespread regular rising of air along the fronts and other bad-weather systems. In the areas of precipitation, the nimbostratus often extends for several kilometres; outside them, the clouds have several levels, up to 10 km and more in altitude. High cirrus forms the fringes of such cloud systems. The rising of the air takes place at some cm/s, and can persist for several, or even many, hours.

Clouds due to vertical convection: these normally develop owing to the heating of the air from the ground, either by solar radiation or by the transport of cold air into warmer regions. Above the cloud base, the heat of condensation generally increases the lift of the air, which rises as in great bubbles. Its surface is like a spherical calotte, at the outsides of which the rising air is intensively mixed with the air of the environment. This causes the cloud particles, which were first cooled, then mixed, then sank, to evaporate. The clouds continue to grow upwards until the bubbles lose their lift by mixing.

In calm weather, the cumuli are well distributed and small, with vertical and horizontal dimensions of 1–2 km. When the weather is disturbed they cover the greater part of the sky. Some clouds have their tops at over 10 km. Anvil clouds develop, and with them heavy showers, hailstorms and thunderstorms. The upward velocity at the cloud base is about 1 m/s, but can reach 5 m/s. In thunderclouds it exceeds 5 m/s, and can attain 20 m/s or more. In less than 20 minutes cloud particles can be lifted from the base to the

Fig. 10. Forms of clouds

top. The cloud can exist as a unit for several hours, while individual towers within it grow and vanish again.

According to their level, their structure and their quantity, clouds have a varying influence on the conditions of radiation and the illumination of the ground.

One quantitative measure of cloud is the amount which is estimated in tenths, in eighths, according to international agreement, or as a percentage, of the total sky. Monthly and annual mean values of cloudiness are calculated on the basis of three observations per day, made early in the morning, at midday, and in the evening. The diurnal variation of cloudiness depends to a great extent on the type of cloud. Fog and low stratus tend to form during the night and in the morning. Cumulus and related forms, being products of rising air, occur most frequently in the middle of the day or in the afternoon. This results in the tendency to a maximum of cloudiness in the morning, with another in the afternoon, at least over continents and islands. The annual course of cloudiness is mostly parallel to that of precipitation, which varies according to the climatic region; there are also, however, regions where cloudiness is at its maximum during the dry season.

The subtropical zones are marked by a scarcity of cloud, especially in summer, which is often completely cloudless. Here the greatest cloudiness occurs in winter. But where, as at the east of the continents, the subtropical latitudes are marked by the monsoon and rain, the annual course of cloudiness is equivalent to that of the tropics. There the greatest cloudiness occurs when the sun is near the zenith, whilst the sky is always cloudless when the sun is at its lowest at midday. Here the contrast between the dry and rainy seasons is greater than anywhere else. Over the oceans, cloudiness is generally greatest. On a world map of cloudiness, there would be three belts of marked cloudiness:

1. an equatorial zone, between latitude 10° north and 10° south;
2. in the middle latitudes of the northern hemisphere, between 50° and 60° north; and
3. in the middle latitudes of the southern hemisphere, between 45° and 60° south.

There are two belts of little cloud, between 20° and 30° both north and south.

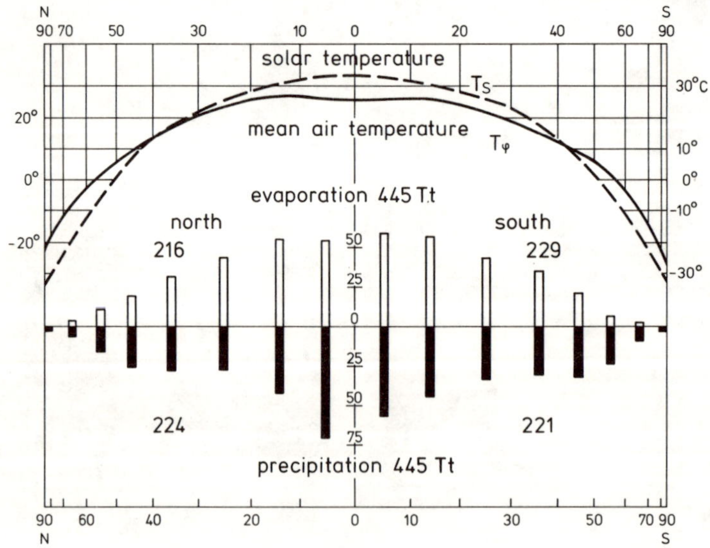

Fig. 11. Distribution of heat and water in the atmosphere

In the equatorial zone, very heavy cloudiness coincides with the rainy season on the western sides of the continents and islands. By contrast, the interiors of the continents show large areas with an almost cloudless sky all the year round. At Refresco in the Atacama desert in Chile, at latitude 25° south and longitude 70° west, cloudiness amounts to 16% in June and sinks to 3% in the almost cloudless month of January. However, on the western side of this desert, in Iquique on the coast, where practically no rain falls, cloudiness varies between 35% in March and 85% in August. Similar cloud conditions are encountered in Western Australia. In the northern hemisphere, an immense dry zone extends with little interruption from the Sahara over Egypt and Arabia to the western border of India, and far or into the deserts of China. In Aswan (Egypt), cloudiness is at a minimum of 0% in January and a maximum of 11% in March. This is probably the most cloudless region on earth. Conditions are less extreme in the deserts of Arizona and New Mexico, with minima of 8–14% in June and maxima of 24–34% in January.

Besides the almost constant ingredients of the atmosphere, i.e. nitrogen (N_2, 76% of the volume), oxygen (O_2, 20%), argon (Ar_2, 0.9%), and carbon dioxide (CO_2, 0.03%), there are varying large amounts of water (H_2O) in it, in gaseous, liquid and solid form. The changes between these states, and in the water content of the atmosphere, are of decisive importance in determining the phenomena which, taken together, form what we call the weather.

1.3. Global atmospheric circulation

1.3.1. The circulation of water

The unequal input of solar energy at different places on earth causes precipitation and evaporation to be unevenly distributed over the earth's surface. A quantitative estimate is given in Fig. 11 ("Distribution of heat and water in the atmosphere"). The temperature

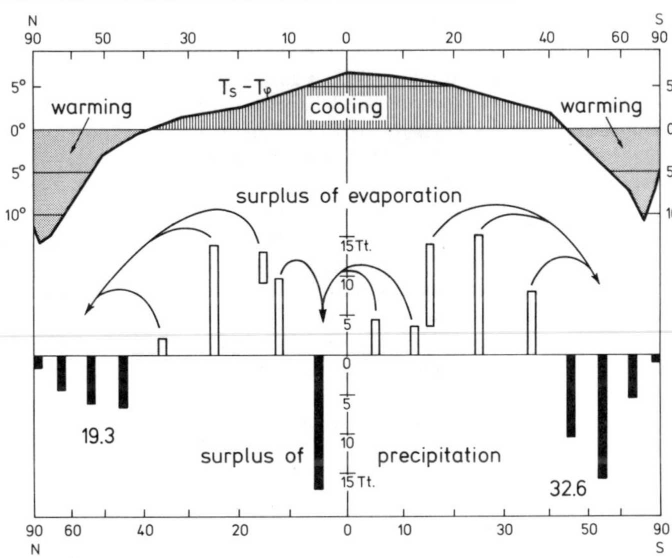

Fig. 12. Compensation of heat and water in the atmosphere

t_s given in the figure corresponds to the radiative temperature defined at the beginning of 1.2.1. above, and represents the mean annual air temperature of a line of latitude which would occur if there were no wind in the atmosphere. The atmospheric circulation resulting from the forces brought into play by the difference in density tends to compensate for the contrasts, as demonstrated in Fig. 12 ("Compensation of heat and water in the atmosphere"). Thus, vast quantities of water are transported to the polar caps, causing a surplus of precipitation over the other zones, amounting to 19.3 Tt in the northern hemisphere and 32.6 Tt in the southern, totalling 51.9 Tt. The trunk of the many-branched tree known as atmospheric circulation is the circulation initiated by the trade winds. It uses half the water budget of the atmosphere and produces a surplus of rain in the region of the inter-tropical convergence zone of 17.6 Tt. The mean position of this zone is between the equator and 10° latitude north, marked by the equatorial trough in the field of atmospheric pressure. However, in the southern summer it is located south of the equator, over South America from 40–80° longitude west and over the Indian Ocean and the western Pacific from 20° longitude east to 160° west. At that time, maximum values of vapour pressure are encountered here, that is, there is a maximum content of water vapour in the atmosphere. In the northern summer there are maximum values over the northern hemisphere.

The world map Fig. 13 ("Worldwide distribution of precipitation") shows a belt of high precipitation extending near the equator and somewhat to the south of it. It is clearly to be seen that the areas of maximum precipitation avoid the open sea. The scale of the map does not allow the pinpointing of the place with the highest precipitation record, a station on Mount Waialeale in Kauai, one of the western Hawaiian islands, where an annual mean precipitation of about 11 m is observed.

The distribution of land over the earth's surface is very unequal, and only amounts to 29% of the surface, the larger remainder of 71% being covered by water. If the annual precipitation falling on earth, with a mass of approximately 406 Tt, were re-distributed equally over land and sea, 110 Tt would fall on land and 296 Tt on the sea. As an initial rough approximation, it can be said that the mean of many years' precipitation is compensated for by evaporation over the earth as a whole. (Detailed estimates show that this is not valid for certain parts of the earth's surface.) Over the oceans, 336 Tt of water evaporate annually, and 296 Tt are precipitated. On land, 70 Tt evaporate and 110 Tt are precipitated. 40 Tt of water has to be carried in gaseous form from sea to land to produce the quantity of precipitated water observed here. Only 70 Tt, however, evaporates; the remainder of the liquid water has to run off. The mean annual runoff of a region is positive if the precipitation exceeds the evaporation. Consequently water is transported to the sea. However, where evaporation exceeds precipitation, there must be a dispatch of water vapour to higher atmospheric levels. On the map of Fig. 14 ("Global map of runoff"), arrows indicate roughly the principal directions of water vapour transport. Clearly to be seen are the extended areas of evaporation surplus over the sea in the region of the subtropical belts of high atmospheric pressure and the trade winds, and also the extended areas of precipitation surplus, with a maximum near the equator over the continents.

Water is found on earth not only as gaseous or liquid water, but also in solid form. Altogether 12,000,000 sq km, i.e. 10% of the continental surface of the earth, is covered by ice. 91% of the ice masses are in the Antarctic, and 8% in Greenland. Compared with these masses, the area of the remaining glaciers is small.

1.3.2. The oceanic wind systems

Between 30° latitude north and south lies half the surface of the earth. From a meteorological point of view this area can be called tropical. It is constantly on the equatorial side of the belts of high pressure, the centres of which are at all seasons poleward of 30° latitude. The tropics are determined astronomically by the Tropic of Cancer at 23.5°

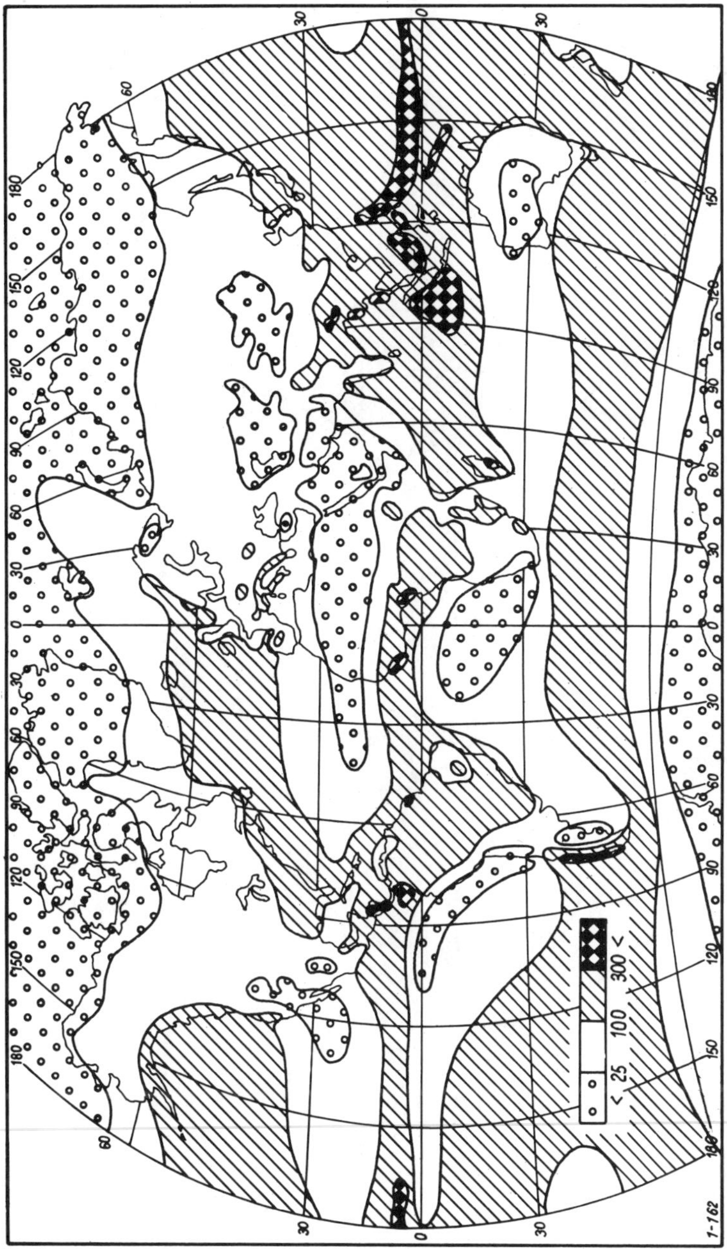

Fig. 13. World-wide distribution of precipitation

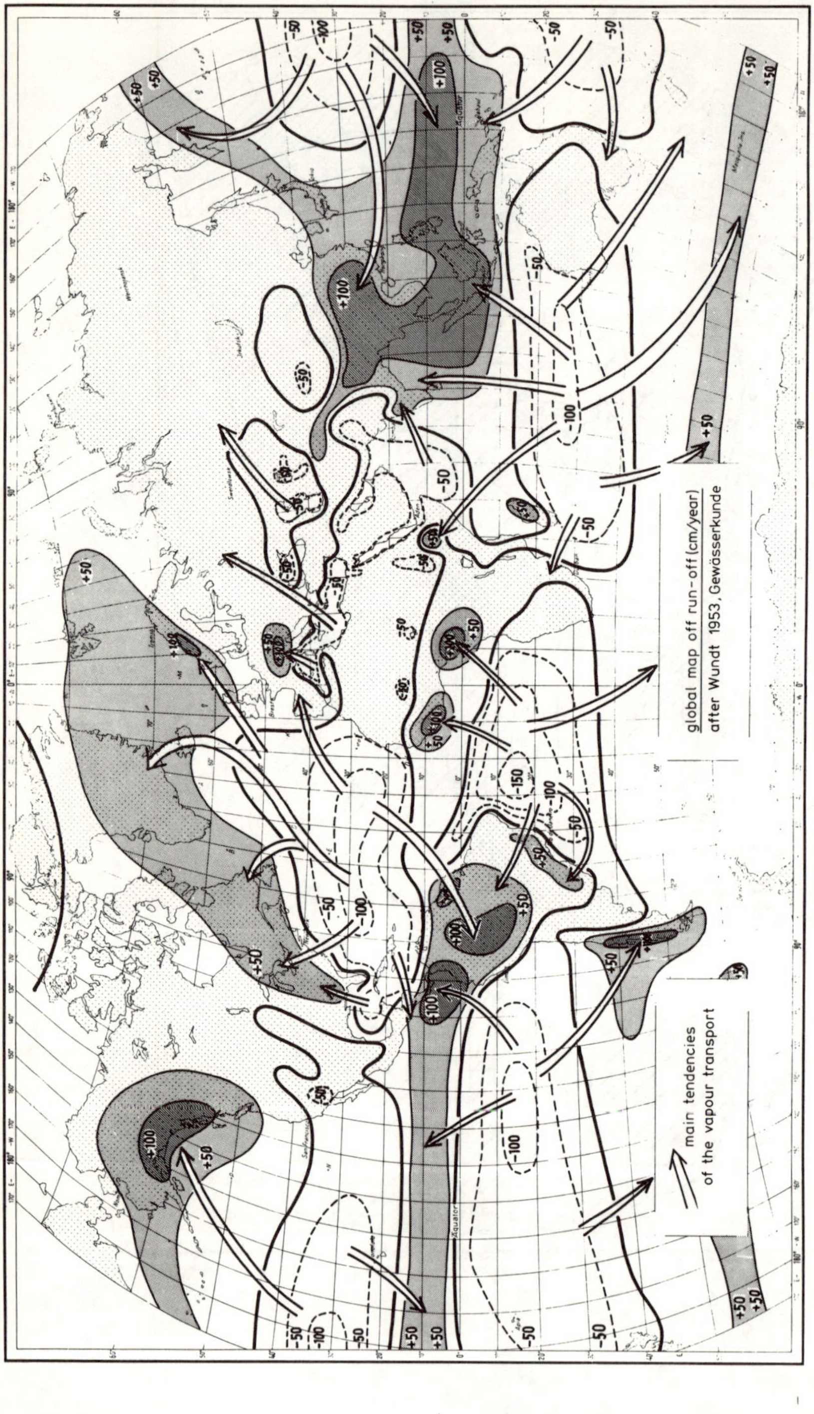

Fig. 14. Global map of run-off

north and the Tropic of Capricorn at 23.5° south. All places in the tropics have the sun at the zenith at midday at least once a year. The map in Fig. 15 ("Oceanic wind regions") – after *Köppen* – demonstrates that the wind system of the trade winds is the largest on earth. It covers 31% of the ocean surface, and taken together with the other tropical winds, doldrums and monsoons about 46%. The zones of transition in the subtropics, characterised by weak winds, cover 20% of the ocean surface, and the zones with irregular westerly winds in middle latitudes 24%. The southern hemisphere has real belts of wind extending round the globe; the west wind covers 16% and the southeasterly trade winds 20%. That is twice the sea surface covered by the corresponding wind of the northern hemisphere. The belts move to and fro and change their intensity from one season to the other less than the wind systems of the northern hemisphere.

World-wide wind belts cannot form in the northern hemisphere in particular because of the influence of the Asian continental block, including the vast tableland of Tibet and other plateaus. The alternation of the seasons causes a remarkable change in the winds: in summer there are rain-bringing south-westerly winds, called the south-west monsoon, and in winter dry north-easterly winds which dominate the southern part of the Asian continent and the adjacent seas.

With regard to air movements as a whole, it emerges that the zonal wind components – in the tropics the winds blow from east to west, and in the middle latitudes from west to east – are more important than the meridional ones. Over the oceans, the meridional wind components are directed towards the equator in the tropics, and towards the poles in the middle latitudes. The constancy of the wind varies greatly between the different regions; in the tropics, the winds blow with greater constancy than outside the tropics, where wind direction and velocity are highly variable.

1.3.3. Anticyclones and depressions

Migrating structures of the atmospheric pressure field, such as cyclones, with their bad weather fronts, and anticyclones, bringing temporary weather improvement, cause the weather as a whole to be capricious and diversified. In the tropics, too, a certain kind of weather variability can be found; but here, on the whole, a degree of monotony is predominant. In many regions, the difference in weather between one day and the next is less remarkable than the difference between the general weather of the various seasons.

Certain regions of the warm seas are affected by a weather phenomenon which is the most dangerous to mankind, the tropical revolving storm. In one of these storms alone, as much energy is transformed as in a million atom bombs of the Hiroshima type. This comparison may serve to show how powerless man is in the face of this grandiose natural event. In April 1960 a revolving storm was located by satellite for the first time, east of Australia. Since then, satellite surveillance of the tropical atmosphere has been steadily improved, and there is hope that, by means of more intensive scientific investigation, the complex of problems presented by the formation of tropical revolving storms can be solved more successfully. Several weeks after the sun has crossed the equator, the first tropical revolving storms (referred to below for brevity's sake as tropical storms) appear in the summer hemisphere, near Asia in the Bay of Bengal and the Arabian Sea, near North America in the western Caribbean and westward of Central America in the northern Pacific Ocean, near Australia in the Arafura Sea, and near Africa off Madagascar. In summer the equatorial trough and the limit of trade winds linked to it are to be found furthest from the equator in these maritime regions. There is a more or less broad zone of transition known as the "intertropical convergence zone".

In the course of the summer, the area of formation extends far to the east over the open ocean. At the same time, the number of storms increases, reaching a maximum before the sun moves to the other hemisphere, to fall in winter almost to zero. The area of formation then moves in the northern hemisphere to the west, to its starting-point. In the

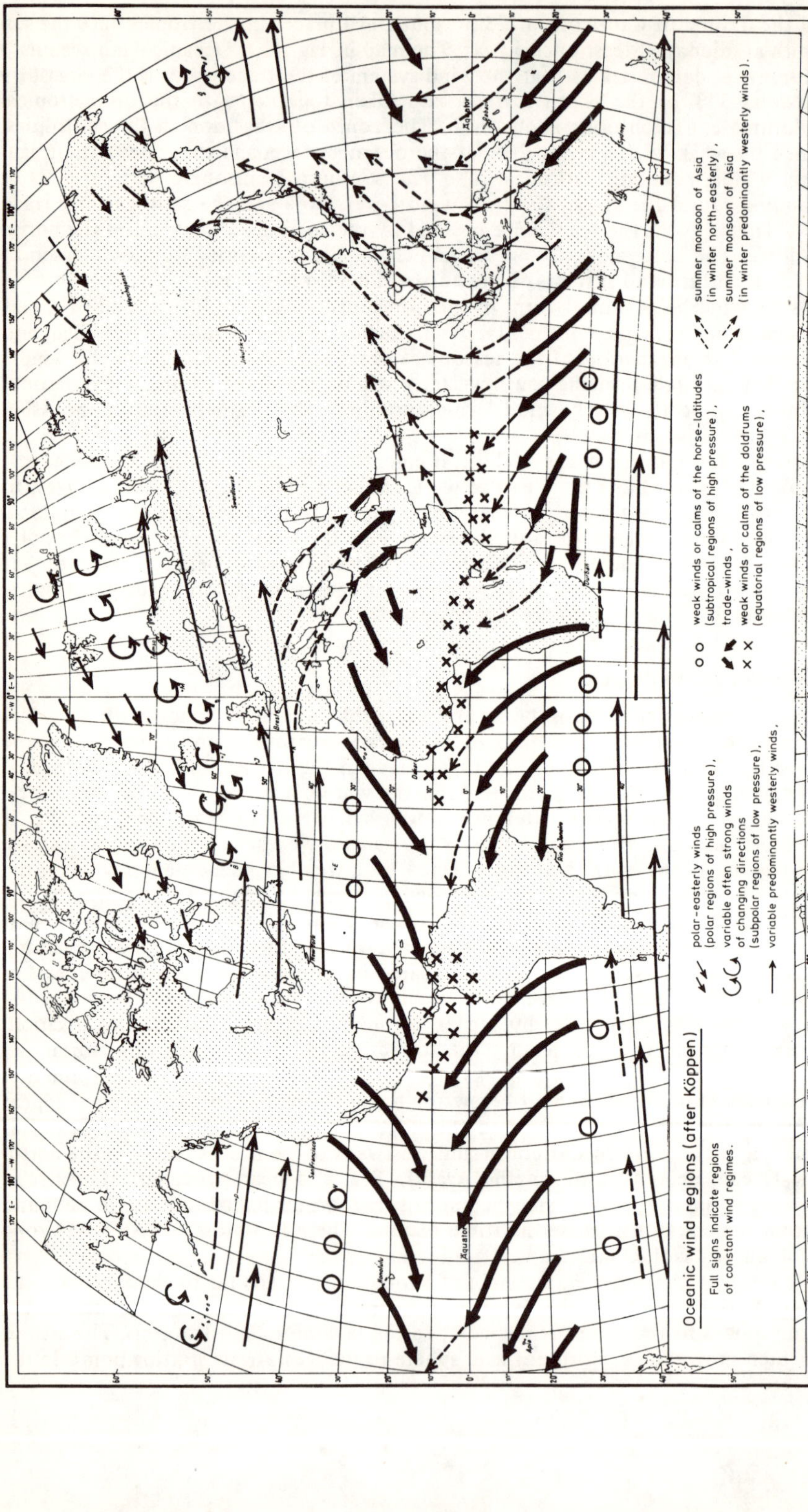

Fig. 15. Oceanic wind regions

Pacific, there are occassionally tropical storms in winter too. The map in Fig. 16 ("Tropical storms: distribution and frequency") demonstrates the distribution and frequency of tropical storms. The next figure (Fig. 17: "The number of tropical storms in the western Caribbean and the Arabian Sea) demonstrates clearly the notable course of the formation of storms in these seas, with double maxima, at the end of spring and in autumn.

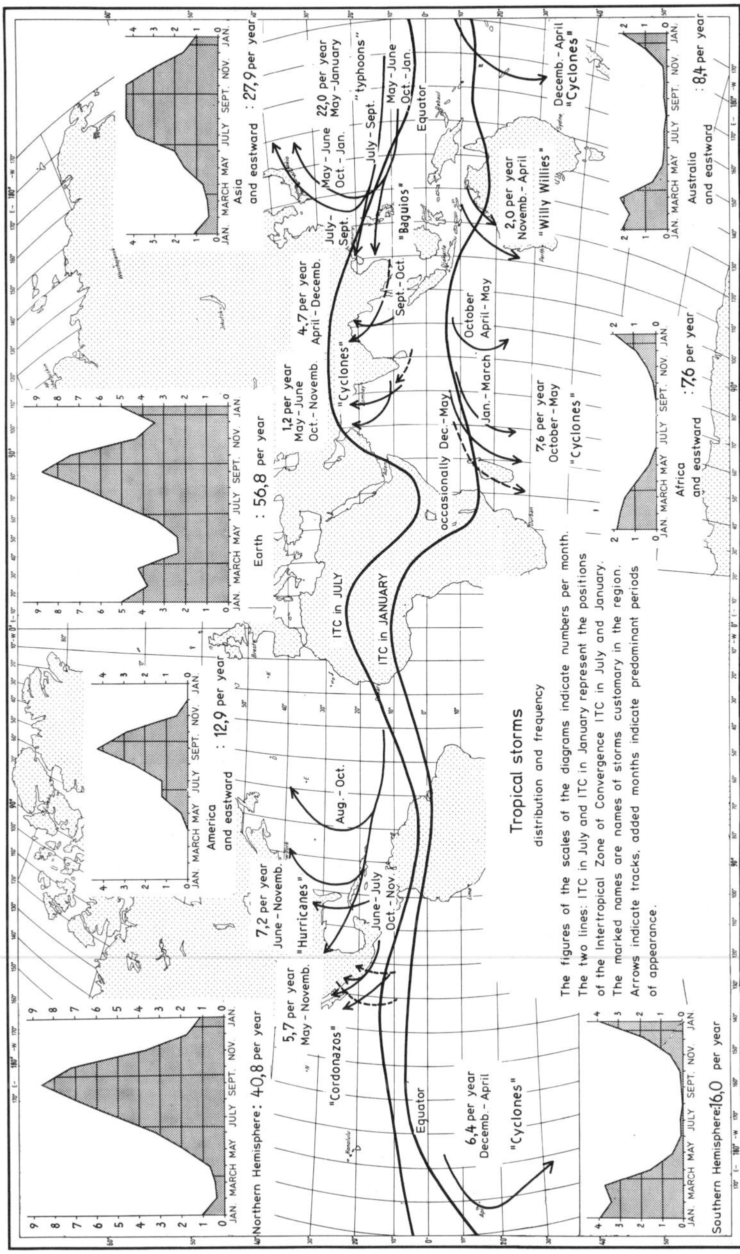

Fig. 16. Tropical storms – distribution and frequency

Tropical storms are often called cyclones. "In this restricted form the term cyclone reverts to its original meaning. Cyclonic storms are confined to very definite regions and occur on the western sides of the great oceans. The principal regions in which these storms occur and the names by which they are know are:

a) the Gulf of Mexico, the West Indies and the coast of Florida, where they are known as hurricanes;
b) the Arabian Sea and the Bay of Bengal (cyclones);
c) the China Sea and the coasts of Japan (typhoons);
d) the Indian Ocean to the east of Mauritius and Madagascar (cyclones); and
e) the Pacific Ocean to the east of Australia and Samoa (hurricanes).

In the Philippines they are known as baguios. The storms usually originate over the ocean, and their paths generally keep to the ocean; if the path crosses the land the storms soon die out and quickly lose their destructiveness" (after "The Meteorological Glossary", Meteorological Office, London, 2nd ed., 1930, p. 52).

Weather charts represent weather conditions by a map in which the weather is described by an internationally accepted set of symbols, for a selection of places at a given moment. The small scale of the map prevents the representation of the weather in detail and in every region. Nevertheless, it is often possible to draw conclusions from the distribution of atmospheric pressure and fronts as to the future development of the weather. Everybody who tries to understand the weather as a physical process needs such daily weather charts. They are indispensable for the forecaster, in the form of large working maps including many hundreds of observations.

A simple weather map is one giving isobars and fronts. Isobars are lines connecting places with the same atmospheric pressure. The numbers along the isobars are the pres-

Average number of tropical revolving storms in ten years

A Western Pacific Ocean	F Southern Pacific Ocean west of 80° E
B Northern Atlantic Ocean	G Southern Indian Ocean 80°–110° E
C Eastern Pacific Ocean north of 10° N east of 130° W	H Western Australian Waters
	I Eastern Australian Waters
D Bay of Bengal	N Northern Hemisphere
E Arabian Sea	S Southern Hemisphere
	T Total Earth

	Jan	Feb	Mar	Apr	May	Jun	Jul	Aug	Sep	Oct	Nov	Dec	10 Years
A	4	2	1	4	7	10	32	42	46	32	17	12	211
B	0	0	0	0	1	5	6	18	28	18	4	0	80
C	0	0	0	0	1	8	7	10	20	11	1	0	58
D	1	0	1	3	4	2	6	4	3	8	8	4	47
E	.5	0	0	.5	2	2	0	0	1	2	3	1	12
F	16	16	11	6	2	0	0	0	0	.5	2	6.5	60
G	2	2	4	3	1	0	0	0	0	0	2	2	16
H	5	5	6	2	.5	0	0	0	0	0	.5	2	21
I	9	8	9	3	.5	0	0	0	0	.5	1	2	23
N	5.5	2	4	7.5	15	30	51	74	98	71	33	17	408
S	32	31	30	14	4	0	0	0	0	1	5.5	12.5	130
T	37.5	33	34	21.5	19	30	51	74	98	72	38.5	29.5	538

sure values represented by the lines. To facilitate reading, the numbers should as a rule be on the side of the line towards higher pressure. Regions with higher pressure than the surrounding area are centres of high pressure areas and are marked "H". Regions with lower pressure than the surrounding area are centres of low pressure areas, and are marked "L".

"An air-mass is a considerable quantity of tropospheric air of homogenous structure, either covering the area of a stationary anticyclone, or floating for an extended period of time in a basic current of the general circulation, thus interacting with other air-masses dynamically and thermodynamically, and causing frontal precipitation and the development of cyclonic disturbances." This definition of an air-mass is to be found in *Chromow* (1942, *Einführung in die synoptische Wetteranalyse*, Springer, p. 195). It is characteristic of an air-mass that its principal qualities are arranged homogeneously and continuously. It is important whether an air-mass functions as a cold or a warm one. A boundary on the weather-map separating two air-masses is called a mass boundary. According to the air-flow, three kinds of mass boundaries can be distinguished. If there is no current, we speak of a stationary mass boundary. If warm air advances against cold air, the boundary is called a warm mass boundary, and vice versa. If a mass boundary is at the same time a line converging with the ground current – such a line is called a convergence line – the line is called a front. Particularly in the tropics, there are also convergence lines inside the air-masses. The causes of the formation and dissolution of such lines are still obscure, and forecasts are difficult in such cases. Convergence near the ground is always co-ordinated with a rising movement, and this is often bound to the formation of clouds with subse-

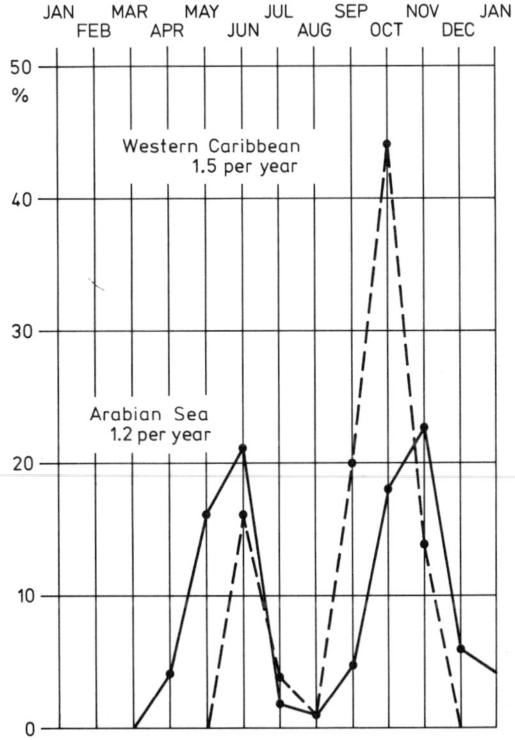

Fig. 17. The number of tropical storms in the Western Caribbean and the Arabian Sea

quent precipitation, so that we can call cold fronts weather-generating air-mass boundaries. When warm air advances towards cold air, the front is called a warm front, and vice versa. When a cold front has overtaken a warm front, the front is called an occlusion. A warm front is represented by a line containing semicircles directed into the cold air-mass. A cold front is represented by a line showing indentations directed into the warm-air-mass. An occlusion can be recognised on the weather map as a line with alternating indentations and semicircles (see Fig. 18: "Front symbols"). If the changes in temperature bound to the front happen only aloft, the indentations or semicircles stay open; otherwise they are spaced out. The indentations and semicircles always point in the direction in which the front is moving (see Fig. 19: "Isobars and Fronts in a Low").

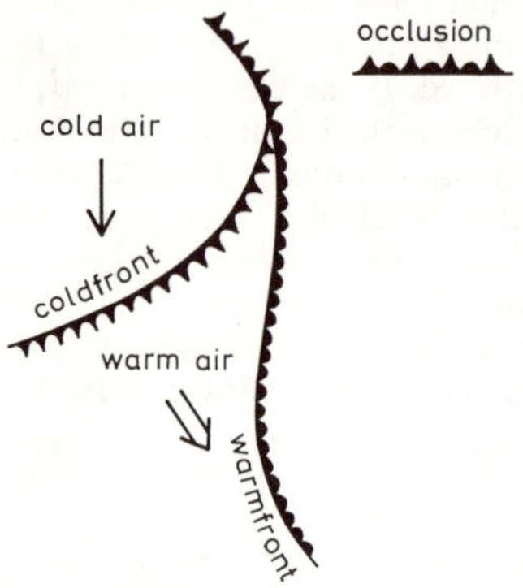

Fig. 18. Front symbols

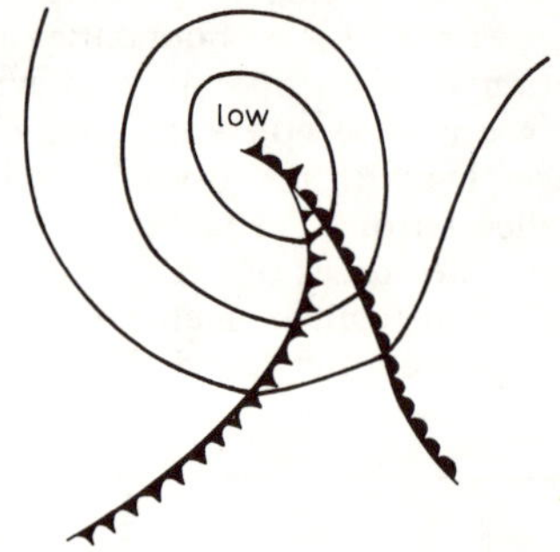

Fig. 19. Isobars and fronts in a low

The field of atmospheric pressure is intimately co-ordinated with the weather processes. The distance between the isobars is a measure of the wind at 1000 m of altitude. The closer together the isobars, the stronger the wind, the further apart the weaker it is. The direction of the isobars is approximately equal to the direction of the wind at 1000 m of altitude. Sudden shifts of wind occur as a rule only where there is a break in the isobars. This happens along fronts; the more pronounced the breaks, the stronger the change in temperature at the front. Near the front the air often rises generally. If it is sufficiently humid, it starts raining or snowing. Extended regions of precipitation are practically bound to fronts. However, precipitation in the form of showers can also happen inside air-masses as soon as, with rising air and sufficient humidity, the vertical decrease in temperature surpasses a certain limit. In such cases the structure of the atmosphere is said to be unstable, and the vertical lapse of temperature is greater than normal. This can be caused by cooling at high altitudes, or heating near the ground, or both. It is familiar that, especially in summer, when heating is intense and the air is humid, showers or even thunderstorms can develop easily. Thunderstorms often result from a very unstable stratification.

The formation of precipitation depends on an uprising movement of the air. Such movement is to be expected in low pressure areas, called cyclones or depressions. Air flowing on the surface of the earth is restrained by contact with the ground. Friction prevents it from flowing in the direction of the isobars as does the air at 1 km altitude. The air is forced by friction to flow into the area of low pressure. This happens according to Buys Ballot's Law (Professor *Buys Ballot*, of Utrecht in the Netherlands, enunciated the law in 1857), which states that if you stand in the northern hemisphere with your back to the wind, the pressure is lower on your left hand than your right, while in the southern hemisphere the reverse is true.

The convergence of air at low levels in low pressure areas forces it to lift (see Fig. 20: "Isobars with cyclonic curvature"). Although the velocity of this movement is very small compared with others (being only of a few cm/s), and very difficult to measure, it is of importance for the development of the weather. Insufficient knowledge of the processes involved is the reason for the imperfections of scientific weather forecasting. In low pressure areas, the isobars are arched in the direction of the high pressure; we then say the pressure field has a cyclonic curvature. Cyclonic curvature of the isobars, and friction, are the reasons for the fact that cyclones normally bring variable weather tending to be rainy and windy. This happens especially along fronts, and near the centres of cyclones. A further observation is that fronts quickly lose their effectivity when the pressure field in their vicinity alters and the isobars lose their cyclonic curvature.

If the isobars are arched in the direction of low pressure – as is best to be seen near the centre of high pressure areas – the friction near the ground forces the air to spread (see Fig. 21: "Isobars with anticyclonic curvature"). If the winds are such that more air flows out of a given area than into it, there is said to be divergence. The deficiency of air must be made good by a downward current from the upper layers. Such a downward current is termed subsidence. Its chief effects are absence of precipitation and frequent development of inversions with dry air above them (see Fig. 21).

When the atmospheric pressure is high over extended areas, there is generally warm and dry air, slowly subsiding from heights above 10 km approx. At the same time there are weak winds revolving around a centre situated amidst the warm air aloft. An additional indicator of anticyclones – as high pressure areas are known – is scarcity of cloud in summer and in lower latitudes. Frequently, the sun shines all day long from a blue sky dotted perhaps for a short time now and then by fine-weather cumulus. In autumn and winter in the higher latitudes, however, a decisive change takes place in the structure of the anticyclone near the ground. The heat lost by nocturnal irradiation can no longer be replaced by day as the days grow shorter. A cold ground layer of air forms, separated from the warm air higher up by a temperature inversion several hundred metres thick. When the air near the ground cools, it often reaches a relative humidity of 100%, i.e. it becomes saturated with water vapour. Dew forms or, at temperatures below freezing-point, hoarfrost, and very often fog, which can persist for a long time and increase in thickness.

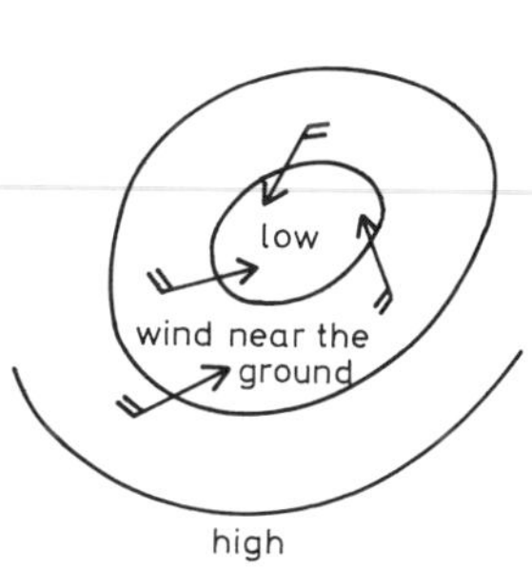

Fig. 20. Isobars with cyclonic curvature

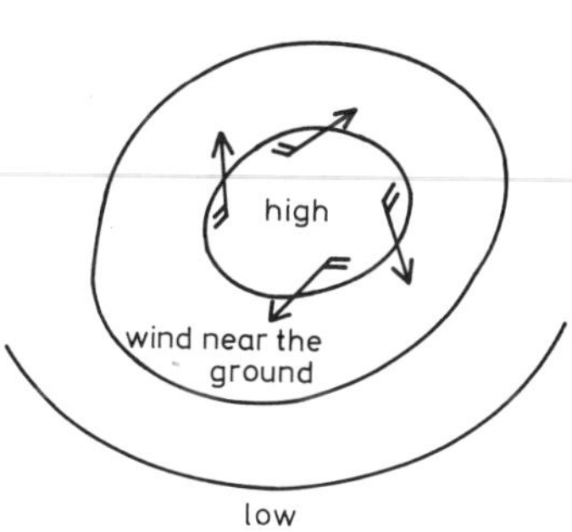

Fig. 21. Isobars with anticyclonic curvature

Taking the anticyclone as a whole, this is only a variant of its form near the ground. The rest of the above-mentioned characteristics do not change. Many people find repeated delight in the glorious sunshine experienced when flying above the cloud cover of an anticyclone in winter, with the summits of low mountains projecting through it. Many tourists spend their winter holidays in the mountains in order to escape from the gloomy wet weather of the lowlands. The inclemency of this weather is further increased by continuous industrialisation, whose air pollution is becoming a real danger to mankind. Fatal values can be reached when the air is filled with toxic waste gases that are not able to spread over a wider area both upwards and sideways. Normally there is sufficient wind near the ground to provide rapid distribution of all matter that enters the air in a natural way, or is even introduced into it by man as waste matter. If, however, pollutants are prevented by an intercepting layer from spreading upwards and diluting, the humid ground air layer is filled with toxic substances.

By fusing the words "smoke" and "fog", we arrive at the combination "smog", a concise definition of a modern phenomenon. Smog appears in areas of industrial concentration under certain weather conditions. Particularly in autumn, anticyclonic influence follows the invasion of humid polar air. Under these conditions, even a little radiative cooling near the ground causes an inversion to form, and within a short time a typical smog situation arises. Such situations can often last for several days, and as long as a week.

The processes in anticyclones are contrary to those in cyclones. Generally, cloud fields aloft tend to dissolve in anticyclones, where the wind is weak and there is as a rule no precipitation. Near the ground, where the earth's surface prevents the air from sinking, conditions can be made more complicated by the formation of an inversion (see Fig. 22: "Temperature-height diagram"). As was mentioned already, in autumn and winter there is complete cloud cover or fog beneath the inversion. Little precipitation can fall from this thin cover, which does not extend far upwards. The mountains often project into a cloudless atmosphere beyond it. In the warm season the cloudless atmosphere extends to the ground, and the kind of serene sunny days which most people prefer for their holidays predominate. In many places in the middle and high latitudes, such days make an agreeable change from the usual variable weather.

There is more than a grain of truth in the statement that the wind makes the weather; this indicates that movement is required for changes in the weather to come about. But the primary cause of all atmospheric movement is solar radiation, which causes differences in temperature and density, and thus finally a strong westerly air-flow in the middle latitudes of the revolving globe. This westerly flow would be quite regular if the earth's

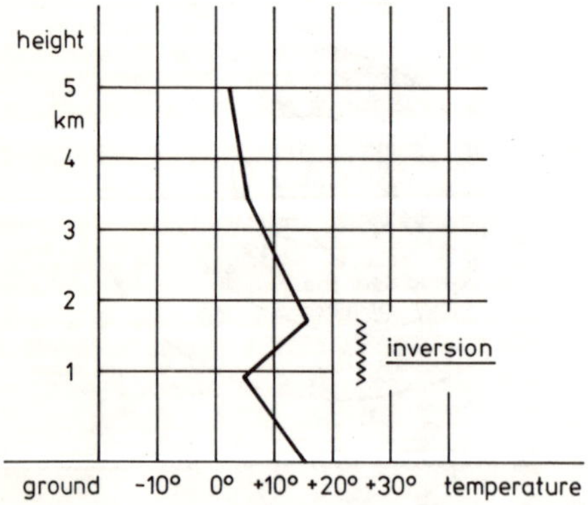

Fig. 22. Temperature-height diagram

surface were homogeneously even. The marked differences in it caused by the distribution of land and sea, however, produce great differences and irregularities in the flow, and do not allow its future behaviour to be forecast far in advance.

1.3.4. Centres of action in the atmosphere

The first pile of a path through the swamp of the unknown can be driven if we try to improve our knowledge scientifically. The irregular distribution of the continents produces "atmospheric centres of action" which may serve as such a first pile. The mean global distribution of atmospheric pressure, averaged over many years, shows striking regions of maximum and minimum pressure, which have been called "centres of action". These centres, clearly to be seen in the world map of annual atmospheric pressure for 1958, are reviewed in the following remarks, which refer to *L. Kaufeld*'s study, *The Global Field of Atmospheric Pressure at Sea Level*, published in 1972 as No. 129 of the reports of the German Weather Service.

Daily pressure figures for 1462 grid points are available for the period of the International Geophysical Year, from 1st July 1957 to 31st December 1958. Until recently, this was the best and most comprehensive body of observations, from which a number of already known facts about the behaviour of the global pressure fields could be confirmed, and some new knowledge derived. The crucial difference between the two pressure fields of the hemispheres is demonstrated by Fig. 23 ("Mean latitudinal values of atmospheric pressure"). Here the distribution of latitudinal mean values of pressure from the North Pole to 70° south are shown. *Kaufeld* distinguishes between the following latitudinal belts of atmospheric pressure:

Latitudinal belts	Denotation	Portion of earth's surface in %
1. 90.0°–67.5° N	Northern polar high-pressure zone	4
2. 67.5°–47.5° N	Northern subpolar trough	9
3. 47.5°–22.5° N	Northern subtropical high-pressure zone	18
4. 22.5° N – 17.5° S	Equatorial trough	34
5. 17.5°–47.5° S	Southern subtropical high-pressure zone	22
6. 47.5°–72.5° S	Southern subpolar trough	11
7. 72.5°–90.0° S	Southern polar high-pressure zone	2

Relatively high pressure distinguishes the northern polar region: 1018 mbar. The northern subpolar trough lies at 58° N, with a mean value of 1012 mbar. This value is still above the annual global mean of the year 1958 of 1011.15 mbar, while the equatorial trough, lying at about 5° N, has a value of 1010.1 mbar. The northern subtropical high-pressure belt between these troughs has its central position at 37° N and a mean pressure of 1016.2 mbar. It is weaker than that of the southern hemisphere, with 1018.4 mbar, and a central position at 33° S. From here, pressure decreases sharply to the southern subpolar trough at 67° S, with a mean value of 987.6 mbar. Then the pressure increases southward as far as the pole, where the mean value of 994 mbar is the maximum valid for sea level; it is a fictive value, since the actual atmosphere begins here only at a height of 2800 m.

Apart from the marked asymmetry of the pressure distribution in the two hemispheres, they also differ as to the seasonal variations. The latter are more pronounced in the northern hemisphere than the southern, where they are small beyond the tropics, as can be demonstrated by comparing the figures for the northern winter and the northern summer.

In both hemispheres, however, the distribution of the quantities "standard deviation" and "variation", shown in the lower part of Fig. 23, is remarkably similar. *Kaufeld* defined the variation as a measure of the variation in time of the daily latitudinal values of pressure. The greater the distance from the equator, the larger the variations, with maxima at the poles. The standard deviation measures how the daily grid-point values of a line of latitude are diversely distributed along it. Like variation, deviation is small along the equator; the maximum values are to be found in the middle latitudes, at about 55° N and 60° S. The mean positions of the subpolar troughs are only a few degrees nearer to the poles.

It is only on considering the important facts just outlined that one can really understand the distribution of the annual mean pressure values (see Fig. 24: "Deviation in mbar of the annual atmospheric pressure in 1958 from 1011.15 mbar (= global annual mean value)). The high and low pressure areas in this figure should not be compared with the same features of daily weather maps. They are more abstract entities, only made by summarising the daily pressure values of a long period – in this case the whole of 1958. They represent in their totality all the cyclones and anticyclones of the days in question. These actual pressure features are more or less transient, frequently passing over quickly and changing in intensity. The abstractions are different in appearance, although they indicate the areas where the migrating pressure formations of the daily weather maps underwent their greatest development in the mean.

Where the individual centres of action – defined as areas of maximum and minimum pressure in the field of annual pressure values – are presented below, it must be borne in

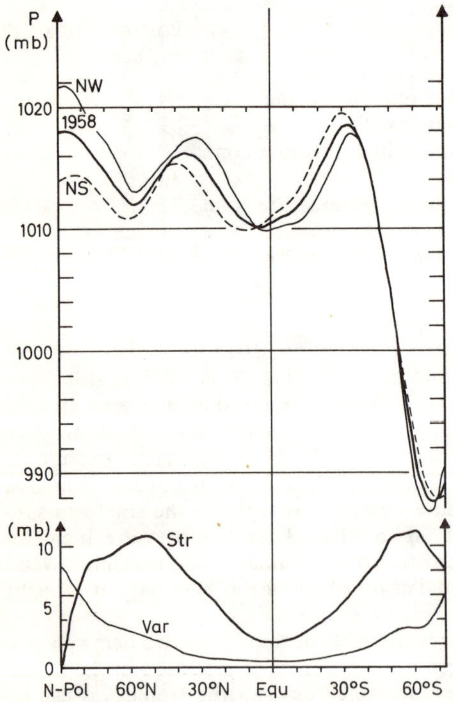

Fig. 23. Mean latitudinal values of atmospheric pressure

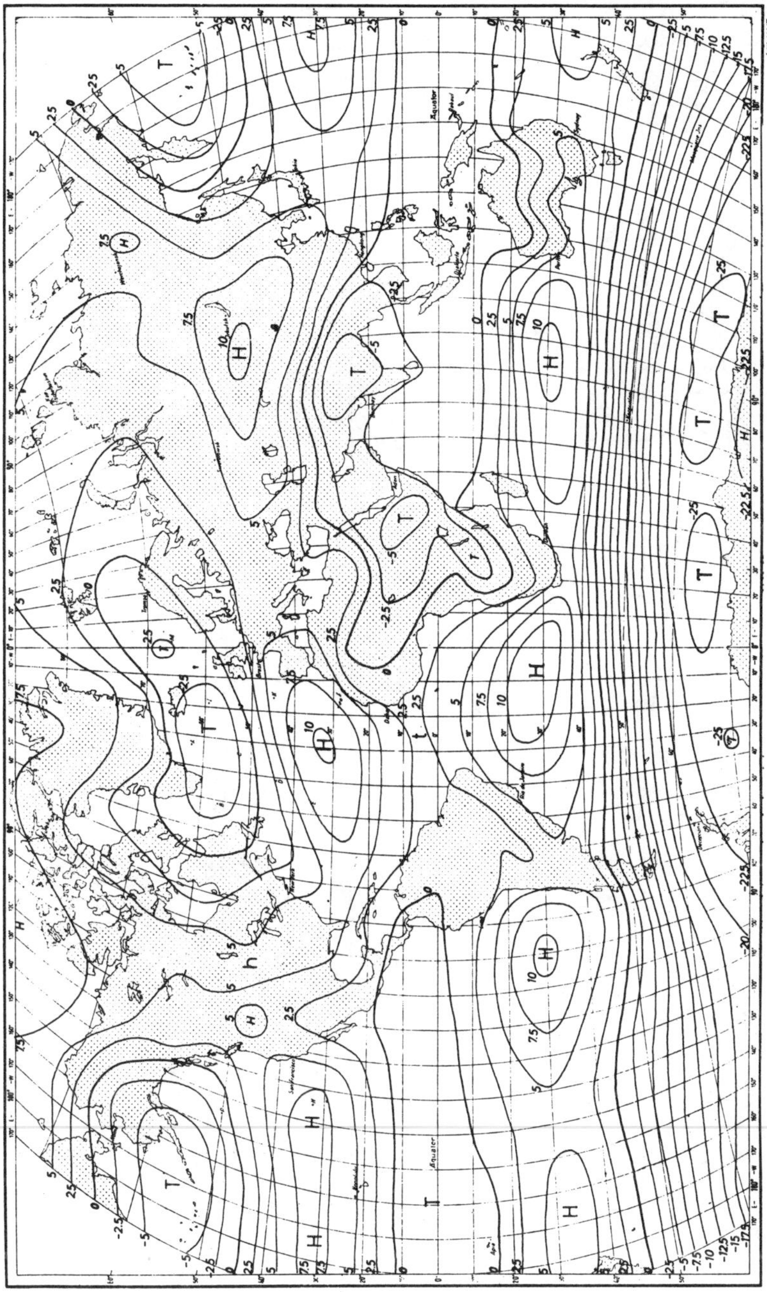

Fig. 24. Deviation in mbar of the annual atmospheric pressure in 1958 from 1011.15 mbar (= global annual mean value)

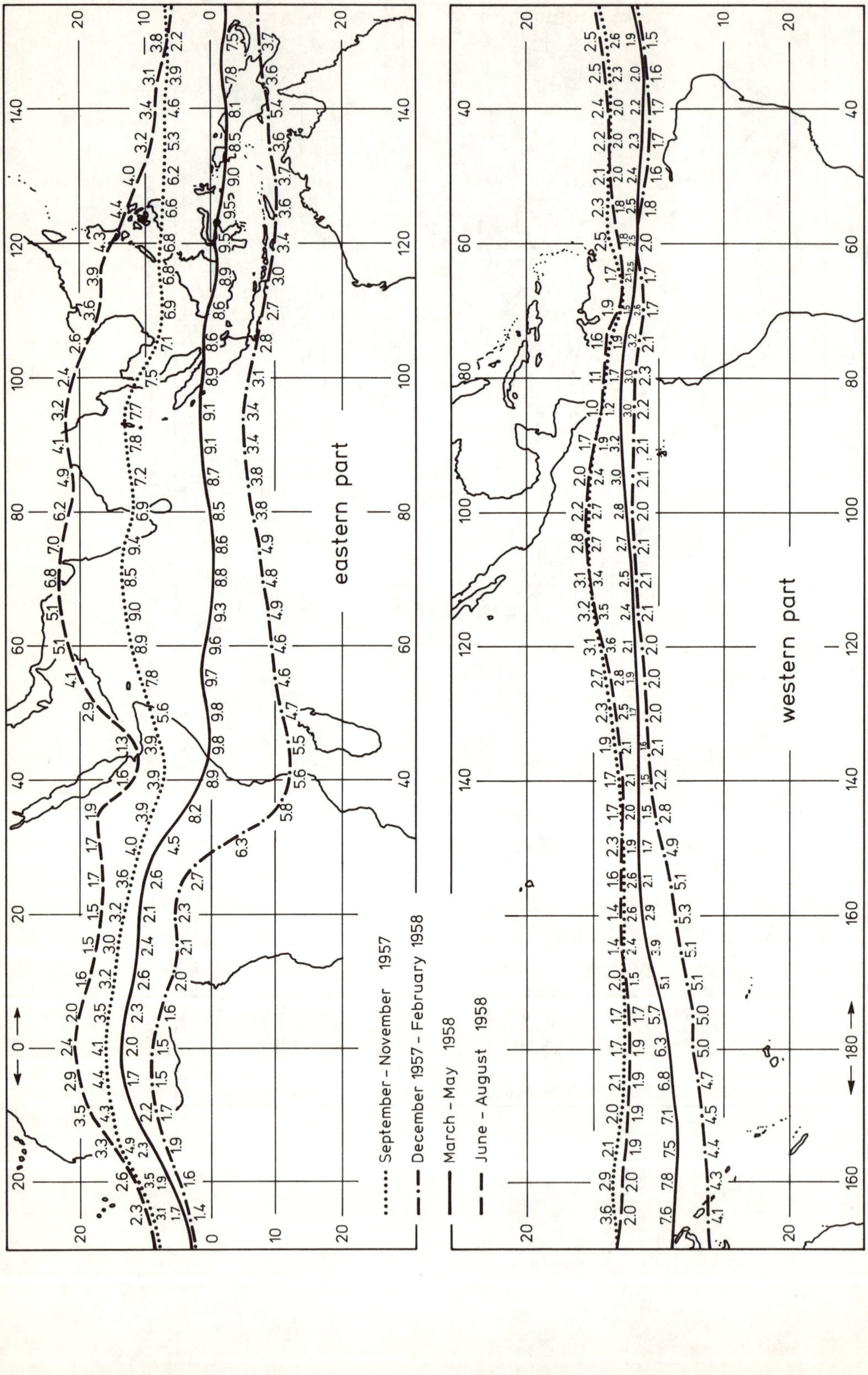

Fig. 25. Mean position of the intertropical convergence zone

mind that they are not stationary formations. Apparently, however, the equatorial trough is an exception to this rule. The tropical pressure field is quite homogeneous compared with that of the extra-tropical area, which shows marked variability. The reason for this is that in the tropics very small pressure variations are linked with considerable forces which, given an appropriate constellation of the pressure field, can cause major changes in the weather.

According to *Kaufeld*'s division into zones, we find the northern polar high pressure zone near the North Pole. Let us now present the centres of action in accordance with these zones, and the map in Fig. 24. Near the pole, there is the high of the Arctic Sea. The northern subpolar trough contains the well-known Aleutian and Icelandic lows. There are four highs in the northern subtropical pressure zones, two oceanic ones lying at about 30° N – that of the northern Pacific and northern Atlantic – and two continental ones, the Mongolian high over Asia (existing only in winter) and the weaker Appalachian high in North America. The counterpart to the Mongolian high is an Indian low existing only in northern summer; at the same time there are two weaker lows in the northern hemisphere – the Mexican low and the West Sahara low – and a weak Amazon low in the southern. All these lows are to be considered as a degeneration of the equatorial trough. This also applies to an Amazon low in southern summer, a Zambesi Highland low and a North Australian low. In northern winter there is in addition an Upper Nile low.

The equatorial trough shifts, like the intertropical convergence zone between 30° and 120° longitude east, which sometimes moves poleward of 20° S in southern summer and poleward of 30° N in northern summer (see Fig. 25: "Mean position of the intertropical convergence zone"). At 150° longitude west it shifts only between 10° N and 15° S.

In contrast to the northern high-pressure zone, the southern one appears nearly every day on the weather maps, where anticyclones can be seen migrating rapidly from west to east. In the annual pressure field, there are subtropical highs over the south-eastern Pacific, the southern Atlantic and the south-western Pacific, the latter, however, only in southern summer, whilst there is a small high over Australia in winter, a smaller counterpart of the Mongolian high. In the southern subpolar trough there is a low north of Queen Maud Land, a low north of Queen Mary Land, one in the Ross Sea and one in the Bellinghausen Sea. There is an Antarctic high near the south pole, towards the Indian Ocean.

The seasonal behaviour of the various centres of action can be seen in detail from the monthly world charts of sea level pressure for January and July 1958. First let us consider the situation of January in Fig. 26 ("Monthly atmospheric pressure in January 1958"). The Arctic Sea high here is only a wedge of the powerful Mongolian high. The Aleutian and Icelandic lows are also strongly developed, whilst the subtropical highs over the northern Atlantic and Pacific are weak. A further high lies south of the Rocky Mountains in North America.

The equatorial trough lies in the eastern Pacific and the Atlantic somewhat north of the equator. Over South America, the Amazon low is connected with a Chaco low. Over Africa, there are the Upper Nile low and the Zambesi Highland low, from which the equatorial trough extends eastwards to the Australian low and farther along 10° latitude south until it approaches the equator north of the Society Islands, crossing it at 120° W. In the southern subtropics there are the highs of the southwestern and southeastern Pacific and of the southern Atlantic, the latter being weak. In the southern subpolar trough, all the lows are present: north of Queen Maud Land, north of Queen Mary Land, the Ross Sea low, and the Bellinghausen Sea low.

Now let us consider the situation in July in Fig. 27 ("Monthly atmospheric pressure in July 1958"). The Arctic Sea high has its centre over Greenland and dominates the polar region. The Aleutian low is only fragmentary, and the Icelandic low is absent; one can see it as being split into a weak Labrador low and a weak Lapponian low. By contrast, the subtropical oceanic highs of the northern Atlantic and Pacific are fully developed. The major part of Asia is covered by the Indian low.

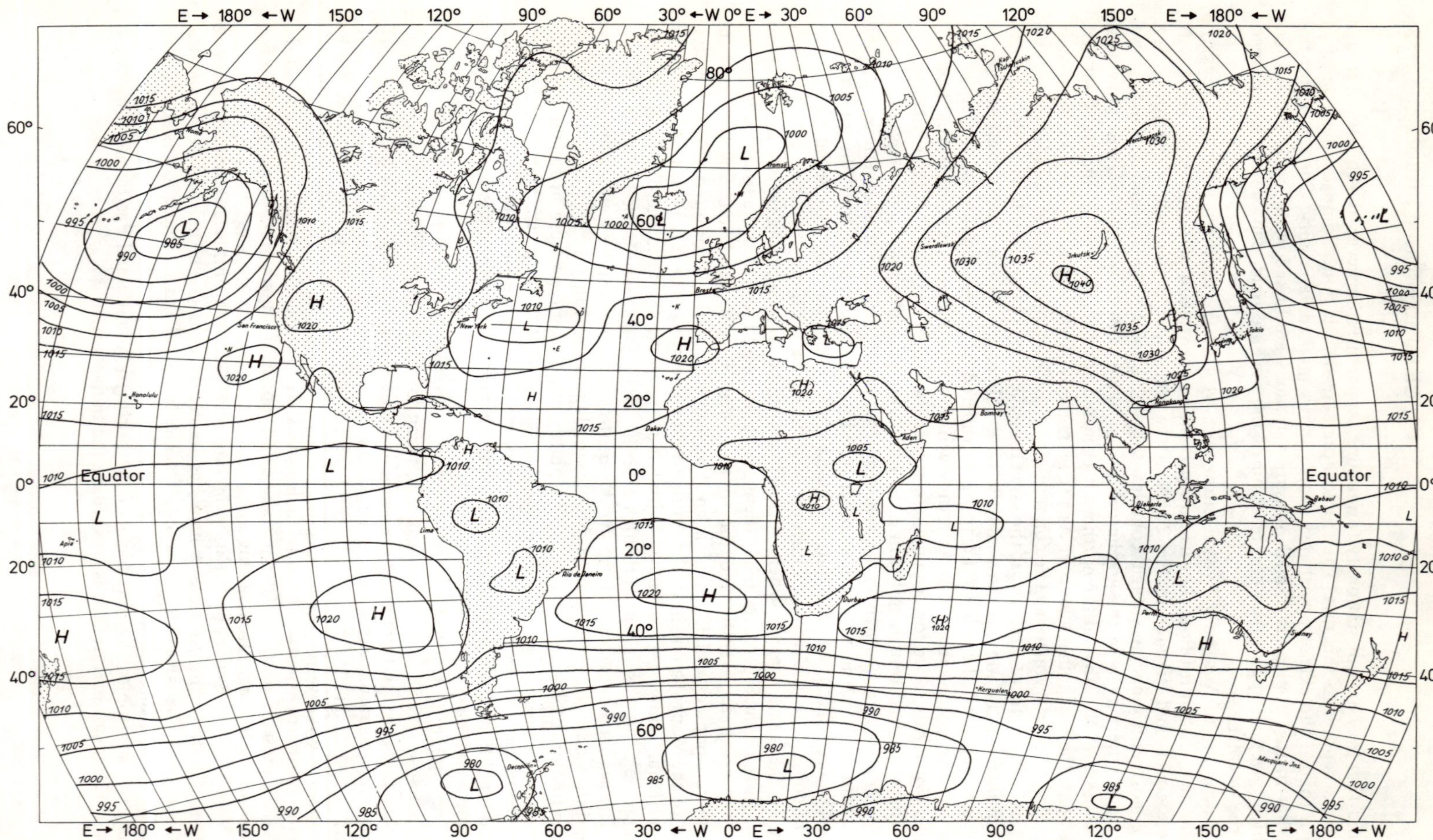

Fig. 26. Monthly atmospheric pressure in January 1958

Januar 1958

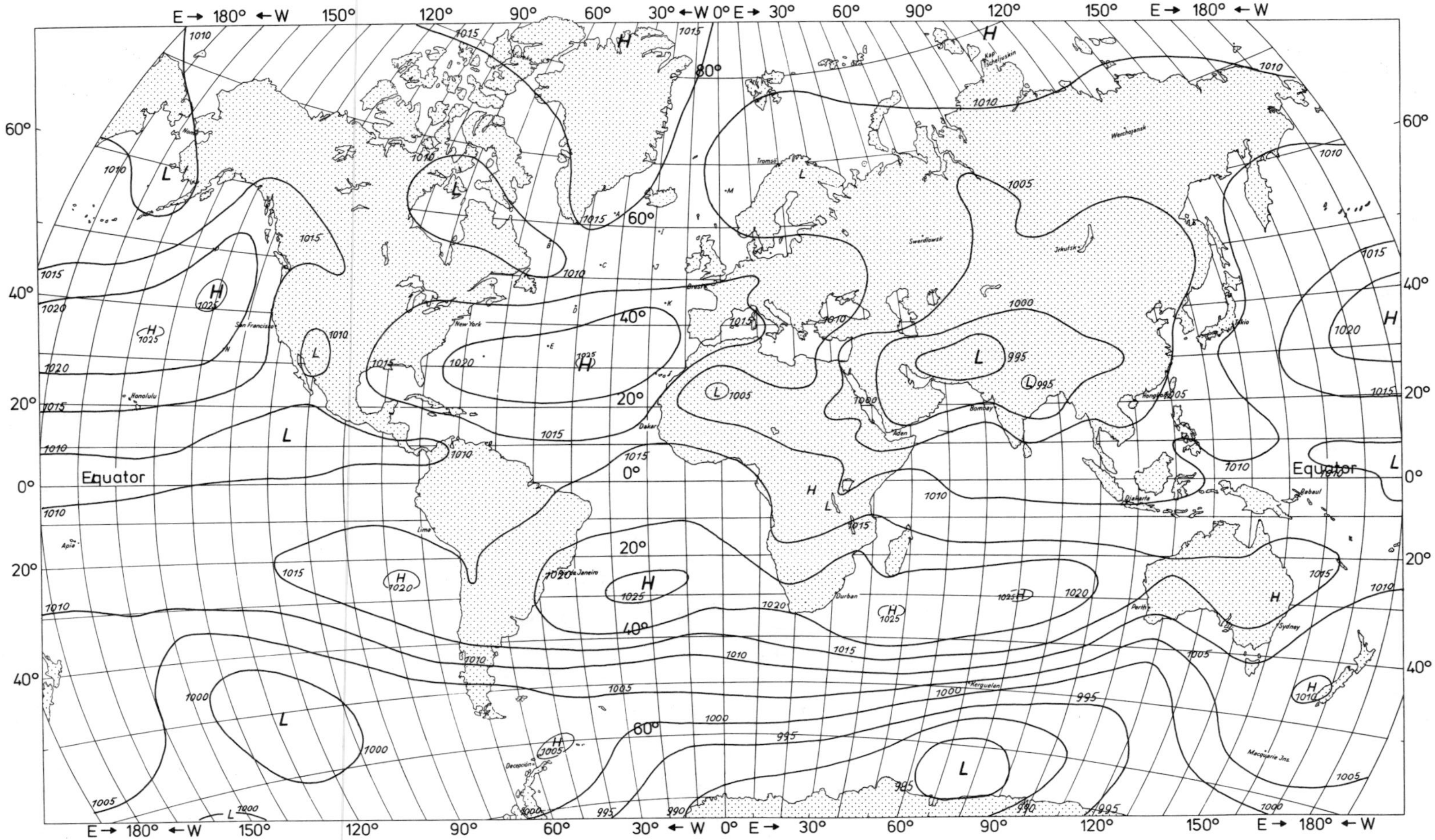

Fig. 27. Monthly atmospheric pressure in July 1958

Juli 1958

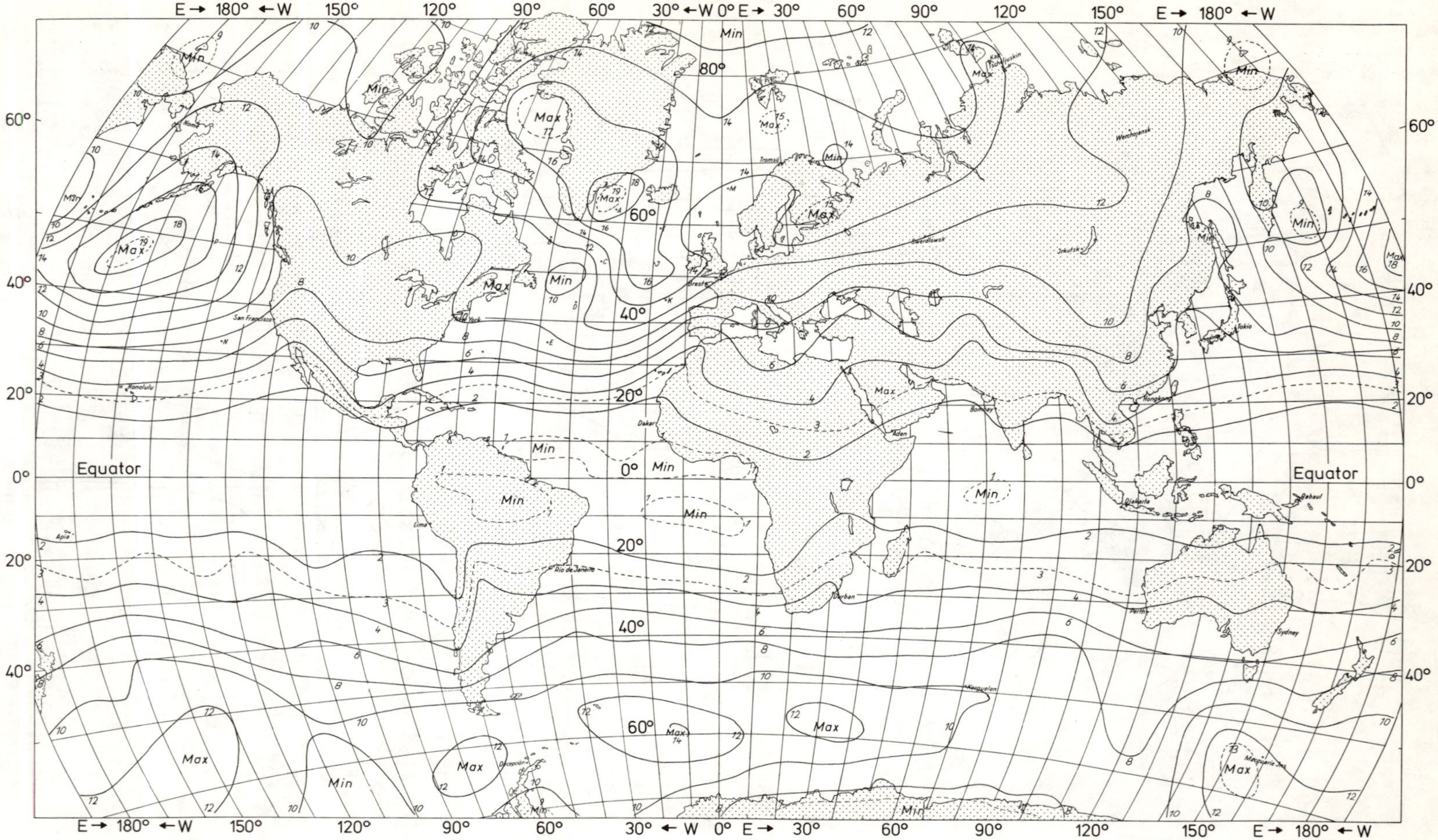

Fig. 28. Standard deviation of pressure, January–March 1958

The equatorial trough lies in the eastern Pacific at 10° N; in the Atlantic it is somewhat more to the south in the western part, and somewhat more to the north in the eastern part. Over America there is the Mexican low, over Africa the West Sahara low, from which the equatorial trough extends eastwards to the Indian low; from there it approaches the equator in the western Pacific by way of the Philippines, staying very close to it until about 150° W. The southern subtropical highs are shifted northward and lie over the southeastern Pacific, the southern Atlantic and the southern Indian Ocean. The fourth high is the winterly Australian high, replacing the high of the southwestern Pacific.

In the southern subpolar trough lies a low in a very unusual place, west of the southern tip of South America, like a shifted Bellinghausen low. The Ross Sea low, the low north of Queen Maud Land and that north of Queen Mary Land are also present, though somewhat shifted.

To arrive at a better knowledge of the behaviour of the pressure field, it is very necessary to complement the monthly mean maps of atmospheric pressure by maps indicating the standard deviation of the atmospheric pressure.

We shall therefore consider here the two maps in Fig. 28 ("Standard deviation of pressure, January–March 1958") and Fig. 29 ("Standard deviation of pressure, July–September 1958"). These underline the strong seasonal variation in the northern hemisphere, where there are maximum values up to 19 mbar in winter, contrasting with 10–11 mbar in summer. In the southern subpolar trough the values are 13–14 mbar in summer and 15–16 mbar in winter. They are arranged almost according to the zones, whilst in the northern hemisphere in subpolar latitudes they behave differently, with the variation having maximum values over the oceans, and minimum values over the continents. Near the equator there is a tendency towards lower values over the oceans. This is an indication of the fact that the forces in operation are stronger over the continents. Here, the damp and unstable movements of the air, and the connected displays of hydrodynamic forces, are strongest in the tropical zone. Correspondingly, they form the Mexican low, the West Sahara low, the Upper Nile low, the Indian low, the Amazon low, the Zambesi Highland low and the North Australian low. The maximum standard deviations on earth appear in the oceanic subpolar regions, which indicates that the conversion of energy and the displays of hydrodynamic forces connected with the circulation of the atmospheric water occur in the zones of strong westerly wind. With the passing lows and highs, this finally causes the following centres of action: the Aleutian low, the Icelandic low, the Bellinghausen Sea low, the low north of Queen Maud Land, that north of Queen Mary Land, the Ross Sea low.

In the annual mean, the equatorial trough covers about 34% of the earth's surface, and together with the adjacent belts of high pressure 74%. The major problem of weather forecasting is the changeability of the field of atmospheric pressure; this is also present, and more enigmatic, in the tropics. In subtropical latitudes the forecasts are generally more reliable because many of the variations in the pressure field – though unfortunately not all – and thus the changes in the weather can be perceived from the migrations of cyclones and anticyclones, fronts and air-mass boundaries.

1.3.5. Weather situation, general weather situation, climate situation

Finally, three concepts should be defined which form a scale like the notions of weather, general weather and climate; the definitions draw upon basic physics to explain the phenomena.

The weather situation in a given place at a given moment is the distribution of atmospheric pressure at sea level over the surrounding area for about 500 km, or 5 degrees of latitude, or more, supplemented by the distribution of air masses or fronts, if any. The weather of a place can usually be qualified physically, according to the weather situation.

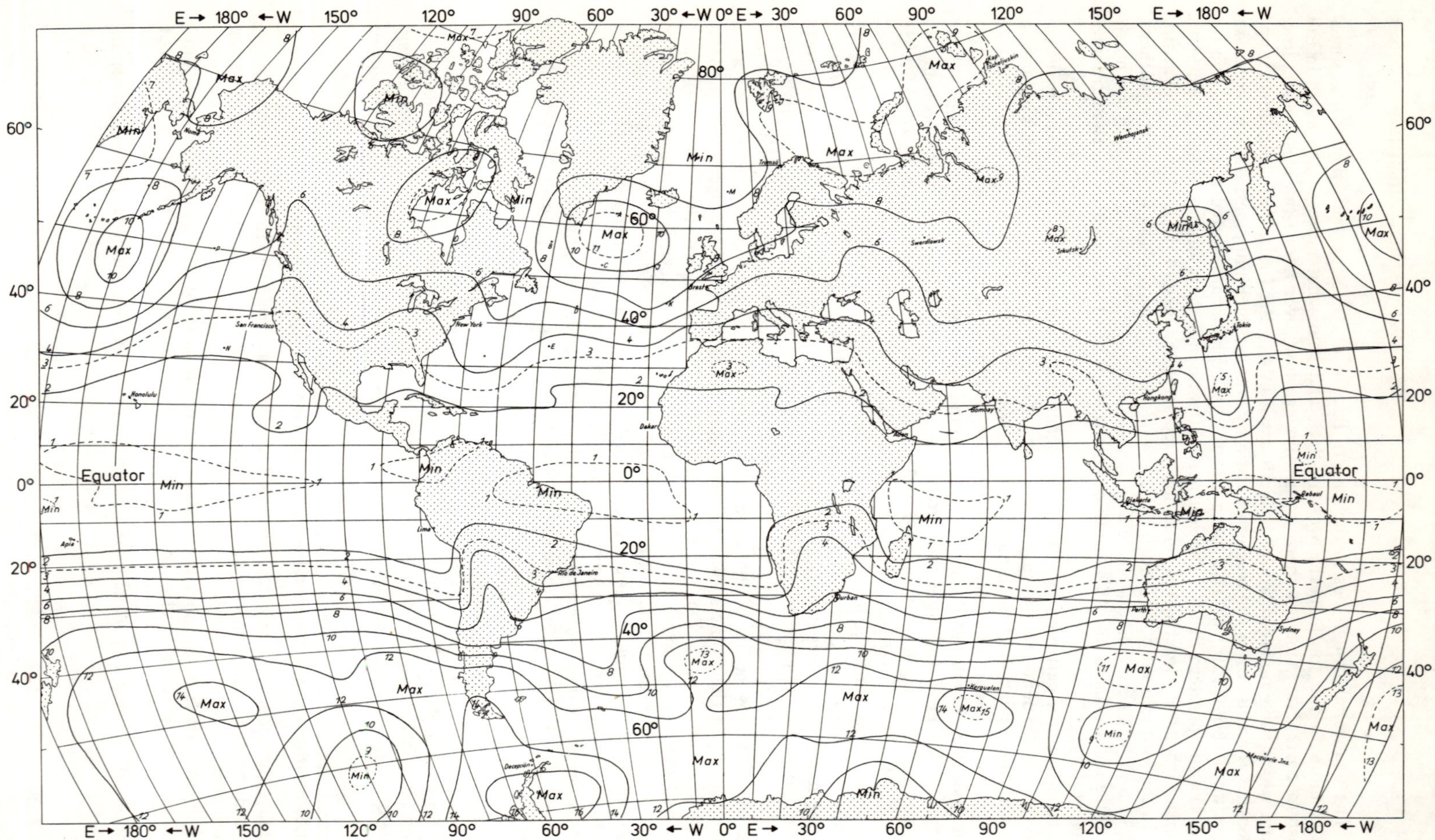

Fig. 29. Standard deviation of pressure, July–September 1958

By analogy we can say that the general weather situation of a given place or area at a given moment or during a given time-span is its mean pressure distribution and dispersion error – i.e. the variation in position and intensity of the areas of extreme pressure changes within 24 hours – over an area of about 20° of latitude measured meridionally and 40° of longitude measured zonally from the centre of the area. According to the general weather situation, the general weather of a place or area can usually be qualified physically for a given time-span.

Correspondingly we can say that the climate situation of a given place or area at a given month or season is the mean pressure distribution and dispersion error for a number of years in relation to the major atmospheric centres of action, and is especially determined by the frequency distribution of the general weather situations proper.

In accordance with the climate situation, the climate of a place or area can usually be qualified physically.

2. Bioclimatic considerations

2.1. Man between the poles and the equator

Here we shall consider the ability of man to adapt himself to the changing heat conditions on earth. There are large areas man is not able to thrive in, but nowhere where he cannot at least exist. In this respect he is alone among all the higher animals, approached only by the dog and cat. There are human settlements in the Himalayas and the Andes up to heights above 4000 m. The eskimo builds his igloo in regions of eternal frost. The bedouins put up their tents in the parching heat of the desert. In the hot damp virgin forest of the Amazon basin, naked Indians live beneath airy roofs of leaves and twigs. On 23rd January 1962, the world press reported that, at the Russian meteorological station of Vostok in the Antarctic, at 3420 m altitude, an air temperature of $-90\ ^\circ C$ had been observed, a value lower than ever before noted. Man braves the worst cold conditions on the fringes of the Antarctic continent, when air of $-20\ ^\circ C$ advances with the velocity of a storm, 30 m/s, which can knock a man down.

Apart from man, who is impelled by the desire to learn about conditions in these uninhabitable areas, the emperor penguin is to be found in this icy hell. The male bird warms the egg with the heat of its body while the hen searches for food far off in the sea before returning to relieve it. The fatty tissues of its skin and its feathers, with the warmed layers of air among them, provide sufficient protection against the bitter cold. Thus one may infer that cold can be kept out by adequate protection; but from experience of conditions in hot zones one may add: "but heat cannot". Travelling by ship through the Red Sea and the Persian Gulf has a bad reputation because of the extreme humidity of the atmosphere, which hardly permits the human body to lose the heat it receives from radiation from the sun, the sky, the land and the sea, and from food. Heat exhaustion and heat-stroke are not uncommon here.

Fig. 30. Activity of the human body and the related heat effect generated in one hour by metabolism

Food is the fuel for the human 'motor'. It supplies the body with the energy necessary to generate strength for its movements. Part of the food is kept in reserve in the form of fatty tissue to insulate the body against rapid loss of heat, and to serve as an energy supply for lean periods. Hibernating warm-blooded animals have particularly effecient fat storage; their bodies, when completely at rest, have a very slow combustion rate and survive the winter well. Similarly, a human being at rest in the protection of bed needs least food. In this state, human metabolism produces 76 watts, i.e. 76 joules are produced per second. The joule is the unit of energy as defined in physics. One joule per second is equal to one watt, which is defined as the unit of effect. The greater the movement of the body, the greater the fuel requirement. The following are the values approximately generated in the human body by metabolism in connection with the activity indicated (see also Fig. 30: "Activity of the human body and the related heat effect generated in one hour by metabolism"):

Effect generated by metabolism, in watts	Activity
76	sleeping
93	sitting
163	typing rapidly
233	walking at 4.3 km/h
349	walking at 6.1 km/h
465	working with stones
582	swimming
1279	going upstairs

The more difficult the work the body has to do in a certain time, the greater the metabolic heat – i.e. the heat produced by metabolism in the body – generated in that time. During a 22-minute boat race, 1430 watts have been measured. During an exhausting exertion lasting only 22 seconds, an effect of 4571 watts has been observed. A man can produce between 582 and 1163 watts in an hour. The maximum daily rate normally varies between 97 and 242 watts, and can rise to approximately 485 during a 6-day cycle race.

Taking as units one day and one sq m of the body surface (which varies between 0.9 and 2.4 sq m), the metabolic heat effect is about 50.5 watts per sq m. It is remarkable that among the vast range of mammals, despite the great differences in size, the metabolic effects differ relatively little. The mouse has a weight of about 18 gm and a metabolic heat effect of 57.3 watts/sq m, and the elephant, with a weight of about 3700 kg, 82.4 watts/sq m. The heating of the human body by metabolism by chemical means functions far more economically than heating processes in technology. The cells in which food combustion takes place have to be heated to 115 °C in order to produce the heat effects observed. Whilst the heating of the human body is done chemically, its cooling works only physically. The metabolic heat is brought to the skin by the blood, the mean temperature of which is 34 °C. As long as the air is colder than the skin, heat is given off by the skin. This heat loss will be larger or smaller according to the air movement. It also depends on clothing, the skin losing much more heat if unprotected, and much less if protected. Clothing slows down the heat loss; as this depends to a considerable extent on the wind, one has to protect the body more effectively in the cold season in windy coastal areas than in the often calm interiors of the continents. More fuel is used in windy areas, not only by the human body but also by the heating systems of houses. Besides cooling by contact with the air, there is cooling by irradiation, long-wave or infra-red radiation, which depends on the temperature of the object concerned. At a temperature of –30 °C, the irradiation effect is 202 watts/sq m, at 0 °C 321 watts/sq m, and at 30 °C 487 watts/sq m. The great influence of this effect can be experienced by anybody who sits down in the open air on a calm starlit night. A lack of warm clothing will cause shivering, as the heat loss by irradiation is not compensated for.

Studies made in experimental climatic chambers demonstrate still more clearly the influence of radiation. Persons clad only in swimming trunks felt quite comfortable in a room the walls of which had a temperature of 25 °C, whilst the air in it was at 10 °C. If, however, the air was heated to 25 °C and the walls kept at freezing point, the persons soon began to shiver. The metabolic heat effect was not able to compensate for the effect caused by radiative heat loss. This was the difference between the total irradiation of the body and the infra-red radiation received from the wall. At 25 °C, the heat effect of the wall was 456 watts/sq m; at 0 °C only 321 watts/sq m. At 25 °C wall temperature, the heat loss of the body by irradiation is compensated for by the metabolic heat effect combined with the infra-red radiation received from the wall. The cooling effect of irradiation is mostly ignored, although it can be unhealthy, for example sleeping on the deck of a ship at night in the tropics without protection, as practised by greenhorns.

The most effective cooling of the body, however, results from the evaporation of the perspiration that bedews the skin. In a temperate climate, man takes in 2.5–3.5 litres of water daily. Half of this water passes through the kidneys, leaving the body as urine, a quarter of it escapes from the lungs in the breath, and the remaining quarter leaves the skin as perspiration through the 2.5 million sweat glands, of which there are as many as 2000 per sq cm in certain places. If the body is not particularly active, the sweat evaporates at once, invisibly causing a loss of heat. In this way 0.75–1 litre of water can pass through the skin in the course of a day. When people are exposed to unaccustomed heat conditions, however, up to 16 litres of perspiration can pass through the skin in a day, given a sufficient supply of water. Water consumption normally rises in the tropics to 4–8 litres a day. To evaporate one litre we need 590 kilocalories, equivalent to 2470 kilojoules or 686 watt-hours, and a body from which sweat evaporates loses considerable thermal energy. Cooling by evaporation is the most effective method in the tropics. It is important for acclimatisation in foreign parts that the skin be able to perspire quickly and easily. Natives of the tropics have this facility. New arrivals at equatorial ports are particularly impressed by the fact that the hard-working natives wear shirts soaked with sweat. This is a normal sight, which one soon becomes accustomed to. This natural defence of the human body against climatic stress remains intact in Europeans only as long as both the water budget and the salt budget in their bodies are kept in balance. This means that the water lost every day by the body must be completely replaced, and also – a fact often not taken into consideration – that sufficient salt must be supplied to replace the quantities lost by perspiration. Investigations in Singapore proved that up to 18 gm of common salt per day is lost by perspiration. In one litre of sweat perspired by a European there are 2.5 gm of salt. Part of the salt brought to the surface by the sweat is re-absorbed by the skin after the evaporation of the water. Otherwise the salt loss would be greater. If the salt is not replaced, the mineral budget of the body is disturbed, causing damage. Shortage of chlorine, which is an ingredient of common salt lacking from gastric acid, causes loss of appetite and a weakening of the body's resistance to parasites. The kidneys, too, need chlorine to digest the water. Lack of sodium, the second ingredient of common salt, causes the muscles to fail. The natives of hot countries are seldom in danger of desalinating their bodies, as these are better adapted to the copious perspiration caused by the heat. One litre of their sweat contains only 0.5 gm of salt. Whether this adaptation has anything to do with the unsolved problem indicated by the question: "Why are negroes black?" cannot be said. Scientific study of the problems of acclimatisation is still in its infancy; when it matures, findings important for the future of mankind may be expected.

The world-wide distribution of the human species is doubtless primarily to be attributed to its adaptability, and was made possible only by man's learning to protect his body from cold over the course of the thousands of millenia of his evolution. About 55% of mankind today are more or less completely clothed, whilst 15% live completely unclothed. For these latter, environmental conditions are ideal when, at a mean value of

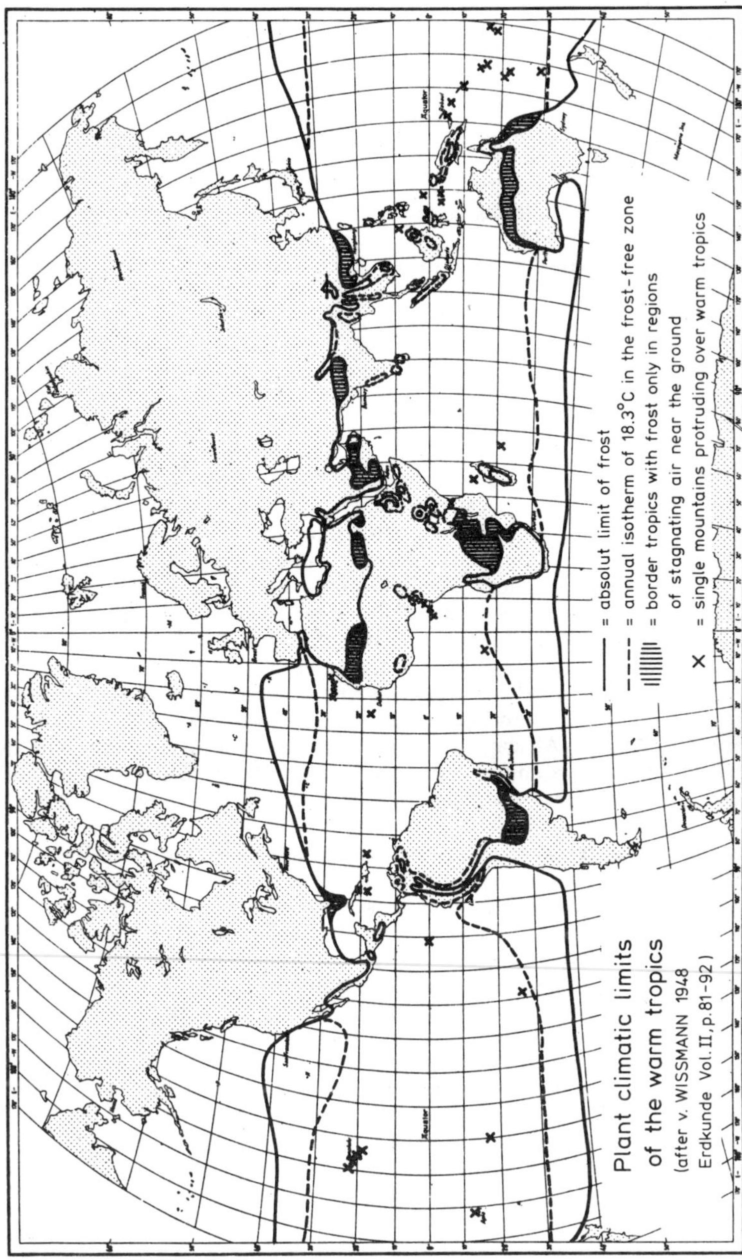

Fig. 31. Plant climatic limits of the warm tropics

atmospheric humidity, the air temperature is 29 °C. If the air temperature were to fall to -1 °C, the metabolic heat production would have to rise eightfold to keep the temperature of the body and the skin constant. This proves that a naked body is not compatible with an air temperature of -1 °C. An inadequately clothed body cannot survive a night at such a low temperature, as is proved by the number of persons killed by frost in northern India in December 1961, when an unusually powerful invasion of cold air in areas not normally affected caused a drop in air temperature to below freezing point.

The interesting question as to how close frost approaches to the equator is answered by Fig. 31 ("Plant climatic limits of the warm tropics"). A remarkable line is the annual isotherm of 18.3 °C in the frost-free zone, which indicates that there is a border region of insufficient heat for plants (for example coffee) which need large quantities of it.

Apart from the direct protection of the body by clothing, there is protection by buildings of stone or wood. If we group mankind according to the type of housing, we obtain figures similar to those relating to clothing: 55% dwell in solid houses, and about 15% live in huts without solid roofs.

It would be quite wrong to commiserate with the naked, improperly housed natives, exposed to the inclemency of the weather. Far from being "wild", they live in socially well-ordered conditions, and are perhaps more content than many people in the so-called civilised countries. This is the point at which a comment is called for on an opinion often thought-lessly expressed by peoply from the polar latitudes. In ignorance of the true conditions, they tend to interpret as laziness the apparent slowness of movement during physical labour which is instinctive in hot zones, where the physical stress on the body is considerably greater than in middle latitudes. Here a day's work generally produces more visible results than in the tropics, where a considerable amount of the work of the body is taken up in compensating for the stress, becoming visible only in sweat production. All the greater should be our admiration for the excellent workmanship shown by the huge monuments surviving from days of yore in the hot zones, often overgrown by the virgin forest, giving evidence of both physical achievement and power of thought.

In 1976, the mean global density of population was 30 per sq km, according to data published in the United Nations Statistical Yearbook. The inhabited area of the earth's surface amounts to about 135,000,000 sq km, and its population in 1976 was about 4045 million. There are considerable differences in population density between the continents. The greatest density is to be found in Europe, at 66 per sq km; the figures for the other continents are: Asia 54, Africa 14, America 14, and Australia, including Oceania, 2.

Australia provides a good example of the way man adapts to climate; the eastern part of the continent, with considerable precipitation, is the most densely inhabited, while the deserts are practically uninhabited. The same goes for other parts of the earth. In 1961 the Polish geographer *J. Staszewski* published an instructive study on density of population and climatic areas, according to *Köppen*'s classification. The 5 classes are labelled A, B, C, D and E. A brief description of them is given in the following table. A, C and D are tree climates, and are sub-divided according to the behaviour of the rainy season; Af, Cf and Df are always wet, while Aw, Cw and Dw have a dry season in winter. There is also Cs, with a dry season in summer. The B climates are sub-divided according to dryness (Bs, Bw), and the ice climates according to cold (Et, Ef). It must be noted that *Staszewski* compiled his paper before 1961, and that his data will therefore differ from those applicable to the present. As these, however, are not available, we shall compare the above-mentioned figures for population density with those of the period preceding 1961, namely Europe 54.4, Asia 31.8, Africa 6.5, America 7.8, and Australia & Oceania 1.6. Thus we can see that the populations of the continents containing developing countries have increased considerably; indeed, one can speak of a world population explosion.

Density of population and climatic areas:

A	tropical rainy climates	15.9
B	dry climates	5.2
C	humid mesothermal climates	58.6
D	humid microthermal climates	16.1
E	polar climates	0.6
Af	tropical rain forest	18.4
Aw	tropical savanna	14.4
BS	steppe	7.9
BW	desert	1.9
Cw	warm with dry winter	61.1
Cs	warm with dry summer (Mediterranean)	41.1
Cf	humid temperate	60.3
Dw	cold with dry winter	19.8
Df	cold with moist winter	15.0
ET	tundra	0.5
EF	perpetual frost	0.0

Although the values for population density contained in Fig. 32 ("Distribution of population according to climatic regions") are no longer correct, it can be seen that the human race, whose origin may be assumed to have been in a hot zone, tends to inhabit warm regions for preference. This can be shown on a small scale, for example, in Europe, by the fact that people from northern, north-western and central Europe often spend their holidays in the Mediterrean region. However, the D climates will prove more important for the future of mankind than the A climates, as *Köppen* forecast. In the A regions, there are extensive areas with virtually no human population, for example the Amazon lowlands of Brazil, whilst others are more densely populated, such as western Java or the Malabar coast of India, with densities of about 400 per sq km.

In these areas, apart from acclimatisation problems, the more important one of nutrition will be more difficult to solve than in Canada or Siberia. Nobody knows what the course of future development will actually be, but to reflect on it is a most important task for those who mould world politics. "The world has room for all" is no longer true in view of the population explosion, and it is urgently necessary to take steps to prevent a further development in the wrong direction.

2.2. Man and climate

There are over 4000 million people on earth today, and a further 200,000 are born every day. Most of them stay in or near the place of their birth all their lives, without ever visiting foreign countries. To them, the daily and seasonal variations in the weather exist only in the form in which they experience them, and they find it impossible to imagine other forms, even if they hear from travellers or the communications media of meteorological conditions markedly different from those familiar to them. People generally thrive best in the long run under the weather conditions they have been accustomed to from their youth, and find it hard to accommodate to changed climatic conditions. This kind of accommodation is properly called acclimatisation. For this it is essential that all the functions of the body are in good working order. Just how acclimatisation takes place is still not known in detail. Most people achieve it within a few years in a foreign climate, so that no physical disadvantages ensue for people in good shape to begin with. At the same time, changes such as an imaginary exchange of residence between a group of Eskimos from the north of Greenland and a group of Hindus from the Brahmaputra would definitely turn out to their detriment. A number of things would be necessary to maintain the health of the two groups on the physical plane. On the psychological level, however, great difficulties would arise because of the complete change in their environment. The children might form an exception, as children find it easier to adapt to physically and psychologically changed conditions than adults. As in all such cases, mental processes would be most important. Such groups are different in their external appearance, but perhaps even more so in the structure of their minds and personalities. The latter are of great influence when the body is tryping to acclimatise itself, and this means that it is very difficult to measure acclimatisation exactly in figures.

It must always be remembered that people's psychological constitution affects the process of acclimatisation, either negatively or positively. This is the explanation for the im-

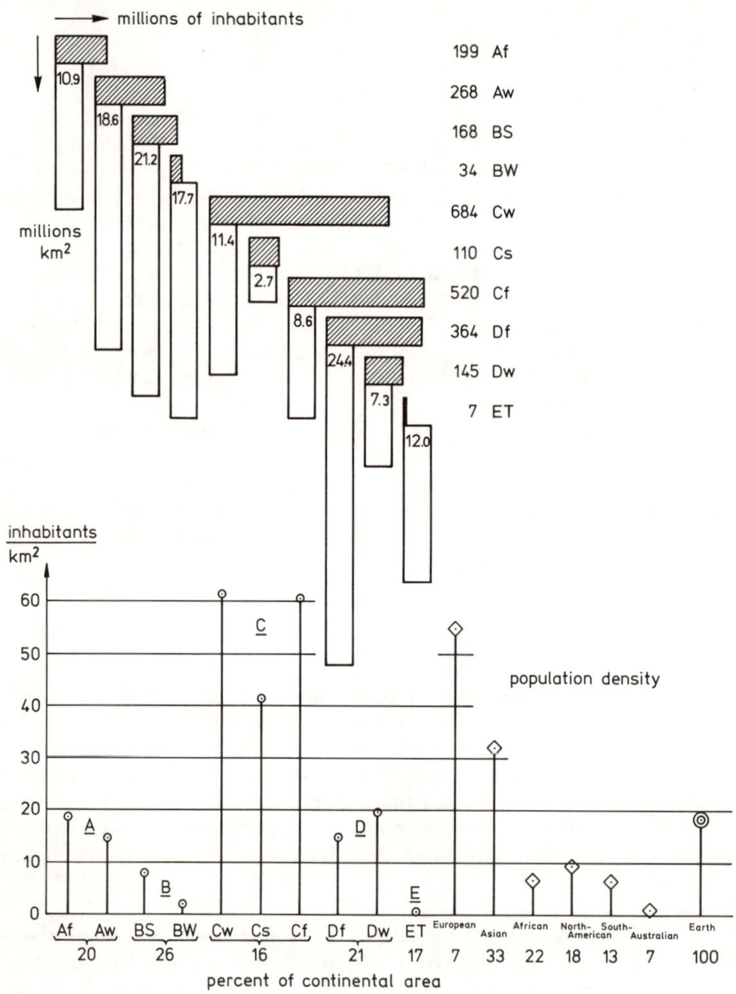

Fig. 32. Distribution of population according to climatic regions

portant principle that only morally unexceptionable persons of firm character can best serve overseas and redound to the reputation and advantage of their firm, office or country. For the maintenance of their health, it is necessary that their bodies adapt without difficulty to the often totally different climatic conditions. As a rule, this is no problem provided that they are healthy and sane. The fitness of the body can be verified by a medical examination.

It is only by experience that we can gain a correct picture of weather conditions in foreign zones. Goethe said in 1829: "What we have never seen we have not assimilated; properly speaking, it is not our concern." People from certain parts of the middle latitudes, for example Central Europe, have the advantage over the people of the tropics or subtropics, inasmuch as they experience the weather in many forms in the course of a year. This makes it easy for them to imagine foreign climates without travelling abroad.

The changeability of the weather in certain areas is – seen from the standpoint of human history – a very important environmental factor, contributing to the highest achievements of civilisation. In this respect, little impetus is to be expected from the hot tropics. The word 'tropics' suggests oppressive heat, such as can seldom be experienced in the higher latitudes, except briefly at the height of summer. The principal difference between the normal heat of summer in higher latitudes and the heat of the tropics lies in the atmospheric humidity. One cubic metre of tropical air contains twice as much water vapour as that in higher latitudes on hot summer days. This drier air is able to absorb more evaporation than that in the tropics, thus relieving the human body from heat and avoiding hyperthermia. In the tropics a great strain is put on the body. Only if he is able to adapt quickly and without difficulty to the changed circumstances can a person from a different region enjoy life there.

Like all other living creatures, man is involved in a continual exchange of matter and energy with his environment. His body needs food, water and oxygen. It transforms these substances and, after utilising them for growth and activity, returns them to the environment. It has to derive energy from organic carburets contained in the food, in order to perform work and to keep body temperature constant. Work is performed by exercising effort on the environment by means of the muscles, thus causing changes. It can be seen how one principle prevails increasingly with the higher species of animals; this is called homoeostasis by biologists. As early as 1859, *Claude Bernard* noted the constancy of the "inner milieu": "All life processes, different as they may be from one another, have the sole function of keeping constant conditions in the inner milieu." Physiology has the task of studying the organs that fulfil this principle. Psychology, too, however, has a part to play. As *Hofstätter* states in the Fischer Dictionary of Psychology (Fischers Lexikon, Vol. 6, 1957, p. 49): "When internal homoeostatic functions fail or are insufficient, there will be modifications of behaviour and experience tending to restore the disturbed equilibrium in a more or less specific way." This can happen in cases of marked changes in the environment.

2.3. The heat budget of the body

2.3.1. Metabolism

Man shares with the higher animals the tendency to maintain a constant body temperature. This is a special case of homoeostasis, called homoeothermy. It is most perfectly maintained in mammals. Warm-blooded animals generally need more energy per mass unit than cold-blooded ones, which reduce their activity in the cold season, and hence their basal turnover, sleeping throughout the winter. It is still a mystery how nature transforms the energy which animals absorb with their food without their being heated to higher temperatures.

Plants are able to synthesise organic carburets from basic organic material, a process called assimilation. In the case of green plants, this is called photosynthesis. They are able to obtain the organic matter they need for growth and survival, and are accordingly termed autotrophic. Animals, however, are heterotrophic, which means that they are not able to produce organic material directly, but require the organic carburets produced by plants.

Part of the food containing organic matter rich in energy is oxidized in the body of the animal. The solar energy bound to food by photosynthesis is set free and used to heat the interior of the body. The gradual breaking-down of the organic carburets rich in energy by digestion is termed biological oxidization. Matter capable of being oxidized is absorbed as food, transported to all parts of the body, and comes into contact with the arterial blood, which has been enriched with oxygen by the lungs. When it encounters the carburets, a very complex biological oxidization process sets in. The end-products are carbonic acid and water, together with heat, which is generated all over the body, especially in the muscles and inner organs such as the liver. Almost all the energy transformed by metabolism takes the form of heat if the body is at rest, except for a small amount which is transformed into electric energy needed for the functions of the nervous system. Energy gained by metabolism also performs the work of digestion, and other movements of the body by which work is done externally. The more energy is used for these purposes, the less is available for the maintenance of the body temperature, and the more metabolism is therefore stepped up to ensure the constancy of the body temperature.

2.3.2. Clothing

First let us present a simple model of the heat budget of the human body. Let the human body be simulated by a sphere of homogeneous material, and clothing by a covering shell. In the interior of the sphere there is a source of heat from which a uniform heat flow is given off. The heat flow density q watt/sq m is variable and depends on the distance from the source and the qualities of the materials through which the heat passes.

In a stationary state, the body, having a surface of F sq m, loses the amount of q.F watts (energy per second), which is constantly being produced by the source. In the interior of the body, the temperature is called T_i, while T_b is the temperature of the body surface, which is at the same time that of the inner side of the shell. The temperature of the outside of the shell is T_c and the air temperature T_a. Given a constant heat flow, T_i will be greater than T_b, T_b greater than T_c, and T_c greater than T_a. T_i depends essentially on the intensity of the source and on the specific heat of the material of which the sphere is made. Its thermal conductivity ${}_iL_b$ determines the temperature gradient T_i-T_b. The heat flow density is determined according to the equation: $q = {}_iL_b\,(T_i-T_b)$.

The inverse value of the thermal conductivity is called thermal resistance $1/{}_iL_b = {}_iW_b$.

When the body is covered by a shell, the thermal conductivity being ${}_bL_c$ and the thermal resistancc ${}_bW_c$, the temperature of its exterior T_c must be determined by a similar equation: $q = {}_bL_c\,(T_b-T_c)$ approximately. There is now a temperature gradient from the interior of the body to the outside of the shell of T_i-T_c, a thermal conductivity of ${}_iL_c$ and a thermal resistance of ${}_iW_c$. We thus have $(T_i-T_c) = {}_iW_c \cdot q = (T_i-T_b)+(T_b-T_c) = {}_iW_b \cdot q + {}_bW_c \cdot q$, so that ${}_iW_c = {}_iW_b + {}_bW_c$. This is a rough approximation as to be precise, the two values q in the equation are not equal but depend on the distance from the source. However, the calculations based on it are adequate to demonstrate the general principle.

The total thermal resistance is equal to the sum of the partial thermal resistances. If the thermal resistance from the interior of the body to the surface is small, the corresponding temperature gradient is relatively small at a given heat flow. We now arrange the model so that it is similar to the human body, assuming the following values:

T_i = 37.5 °C, $_iW_b$ = 0.042 K.sq m/watt, q = 60 watt/sq m. Then $(T_i - T_b)$ = 2.5 K (K = degrees Kelvin), and T_b = 35 °C. We select a thermal resistance of the shell with a value which is three times the value of the sphere $_bW_c$ = 0.125 sq m/watt; then $(T_b - T_c)$ = 7.5 K and T_c = 27.5 °C. Clothing corresponding to this shell shall be called "tropical clothing". If we now take the double thermal resistance of the shell $_bW_c$ = 0.25 K sq m/watt, then $(T_b - T_c)$ = 15.0 K and T_c = 20 °C.

To simplify the following considerations, clothing with a thermal resistance of 0.25 K m/watt is defined as a unit of clothing. If, in the model, the shell has a thermal resistance of $_bW_c$ = 0.50 K sq m/watt, then $(T_b - T_c)$ = 30.0 K and T_c = 5 °C. The corresponding clothing is called "normal clothing", and is equivalent to two units. Continuing the pro-

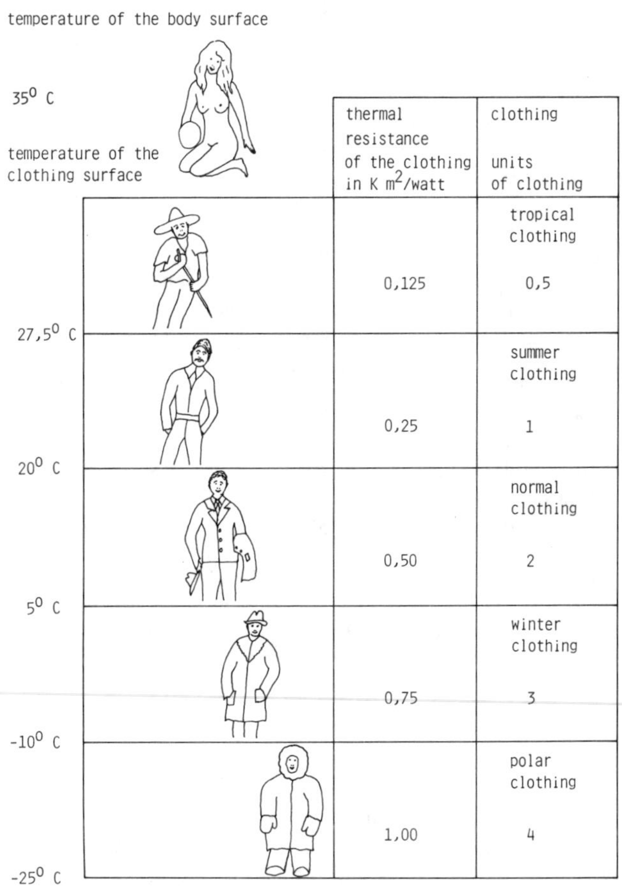

Let it be noted that the indicated values of temperature are merely a rough estimate. In general there are no fixed boundaries but fluid ones between neighbouring conditions.

Fig. 33. Insulation against loss of heat by clothing

cess, we arrive at the following values (see also Fig. 33: "Insulation against loss of heat by clothing"):

Clothing	${}_bW_c$	(T_b-T_c)	T_c	Units of clothing
none	0.00 K sq m/watt	0.0 K	35 °C	0
tropical	0.125	7.5	27.5	0.5
summer	0.25	15	20	1
normal	0.50	30	5	2
winter	0.75	45	-10	3
polar	1.00	60	-25	4
ultra-polar	1.25	75	-40	5

The temperature of the outside of the shell – if there is no shell, the figure is equal to T_b – determines the infra-red heat flow density of the body q_r. Its effective irradiation is q_r-q_e, q_e being the infra-red heat flow density of the environment, with a temperature of T_e. If T_c and T_e differ less than about 5 K, we have q_r-q_e = 6.00 (T_c-T_c) watt/sq m. The heat flow from the outside of the shell into the air by convection has a density of q_c= ${}_cL_a$ (T_c-T_a) watt(sq m. The total heat flow from the centre of the body is here split up into two flows: the convective heat flow and the infra-red heat flow. Its density is approximately $q = q_r - q_e + q_c = 6.00 (T_c - T_e) + {}_cL_a (T_c - T_a)$. To simplify the equation we take T_e as equal to T_a, giving $q = (6.00 + {}_cL_a)(T_c - T_a)$. For given values of q (for example = 60 watt/sq m), T_c and ${}_cL_a$, T_a is also a fixed value. This gives $T_c - T_a = q/(6.00 + {}_cL_a)$. The convective thermal conductivity depends decisively on the wind. An estimated value is ${}_cL_a = 26.4 \sqrt{v}$ watt/sq m·K, the wind velocity v being measured in m/s. Conductivity is small with light winds and large with strong winds. The relation is given by the following table:

Wind in m/s	0.008	0.1	0.2	0.3	0.4	0.5	0.6	0.7	0.8
${}_cL_a$ in watt/sq m·K	2.35	8.33	11.9	14.5	16.6	18.6	20.5	22.0	23.6
Wind in m/s	0.9	1.0	1.3	1.5	1.8	2.0	2.5	3.0	3.5
${}_cL_a$ in watt/sq m·K	25.0	26.4	30.2	32.4	35.5	37.2	41.7	45.5	49.2
Wind in m/s	4	5	6	7	8	9	10	15	20
${}_cL_a$ in watt/sq m·K	52.8	58.9	64.5	69.8	74.6	79.2	83.3	102	119

When the air is practically calm, the thermal conductivity is ${}_cL_a \mp 2.35$ and $q = (6.00 + 2.35)(T_c - T_a) = 8.35 (T_c - T_a)$ and $T_c - T_a = 60/8.35 = 7.2$ K. In this case the effective infra-red irradiation, at 43 watt/sq m, exceeds the convective heat loss of 17 watt/sq m. The following table demonstrates how this loss q_c increases with the wind:

At v =	$T_c - T_a$ =	and q_c =
0.1 m/s	4.2 K	35 watt/sq m
1.0	1.9	50
3.5	1.1	54
9	0.7	56

Compared with the values of the thermal conductivity of the air, ranging from 2.35 to over 125, the thermal conductivity of the human body has a very small variation, with values of 21.0 to 27.7 and a mean of 23.6 if the air temperature is over 30 °C. At 39 °C, 45.0 watt/sq m·K have been measured for a clothed person. If the air temperature is below 30 °C, its mean value is 14.7 and the values range from 12.5 to 18.5 watt/sq mK. Certainly, lower values can also exist, so that the range of thermal conductivity of the human body is from 10 to 50 watt/sq m · K.

2.3.3. Heat regulation and heat flow in the body

The essential flows of energy in the human body are treated here particularly according to Chapter V ("Bioclimatology") of the compendium *Lehrbuch der Bäder- und Klimaheilkunde* (Manual of Spa and Climatic Treatment) by *Pfleiderer* and *Büttner*, Berlin (Springer) 1940, p. 609–949.

Part of the energy set free by biological oxidization is used to do work outside the body, and is thus not able to heat the body. However, intensification of the oxidization can cause a considerable increase in heat internally. If the body is working, the amount of energy consumed by this work is less than the kinetic energy produced in the body to do this work. We can call the difference between these two kinds of energy "waste energy". In order to estimate the efficiency of the body, one has to subtract the basic effect from the metabolic effect; this gives the "pure expenditure" of the effect of working. The efficiency may be defined as the effect produced externally expressed as a percentage of the "pure expenditure". The amount depends on the constitution of the body, the state of fitness and the speed of work, and is about 15–20%, sometimes 35%. This means that, to achieve a working effect of 100 watt/man, a very fit man has to increase his metabolic effect only by about 285 watts (35% of which is 100), while a person quite out of training requires 665 (of which 100 is 15%). The heat flow outwards will be increased – assuming body temperature does not change – by about 185 or 565 watt/man respectively, as 100 watt/man are lost because of the working effect. The heat burden at the same working effect can in the case of an untrained person be three times that for a trained one. This example shows that the amount of metabolic effect indicated in Fig. 30 must be regarded as a rough approximation only. A man with a mass of 100 kg, in going upstairs 10 m, does work against gravity, which exercises an acceleration of 10 m/s, amounting to 10,000 watt·seconds. If he does it in 50 seconds, his effect is 200 watt. A well-trained sportsman has to increase his metabolic effect about 570 watt (obviously only for a short time). If he moves faster, then his effect is greater. Untrained people cannot keep up such effects, and need more time for the same work. If they take 100 seconds to go upstairs 10 m, the effect is 100 watt; for this, their metabolic effect has to be increased by about 665 watt.

Let us now return to the generation of heat in the interior of the body by biological oxidization. The heat enters the blood stream, by which most of it is transported. This "convection" performed by the blood extends into the skin. (Convection is the transfer of energy by the smallest particles of a flow of matter.)

In the outer layer of the skin, which is 0.1 to 1 mm thick, the heat flow no longer takes place by convection but by physical thermal conduction in the tissue of the epidermis, which has no blood vessels. The intensity of the blood flow and the exchange in the middle layers depends greatly upon regulation performed by the nervous system and by local stimuli on the blood vessels. At low air temperatures, the passage of heat can be much reduced by an almost complete interruption of the blood flow. This is less so with fatty skin than less fatty skin. The fat of the cutis is a thick pad which acts like a dead insulating layer if the blood flow is interrupted. A very thick horny layer performs the same function.

Generally the core of the body is warmer than the skin. The heat flow is stronger the stronger the blood flow and thus the total thermal conductivity of the skin and the decrease of temperature in the skin. The blood flow counteracts the decrease of temperature in the skin. In the cold, the blood flow is small and the decrease in temperature from core to skin is large. In a comfortably warm environment, the blood flow is strong and the decrease in temperature small. The heat flow tends to be constant; this is the principle of physical heat regulation. In the mean, the heat flow is in a stationary state equal to the metabolic effect. The heat flow density, that is, the flow through one sq metre, is 1.34 kWh/sq m day = 53 Watt/sq m in an adult at rest and fasting. Normally the heat flows outwards. In the bloodless epidermis the uniform heat flow is split up into three flows:

1. heat flow by infra-red irradiation;
2. heat flow by heat conduction through the outer boundary layer and from there by heat convection through the turbulent air into the free atmosphere of the environment;
3. heat flow by evaporation.

With regard to 1.: in calm air, 43% of the heat is lost by radiation = about 38 watt/sq m in an adult. This happens at a mean difference in temperature between skin and environment of 6 K. On clear calm nights, the loss by radiation can amount to 200 watt/sq m or more. In a hot environment, the flow can reverse its direction.

With regard to 2.: at low temperatures and with strong wind, this type of heat loss predominates. It can be regulated both instinctively and consciously, in two ways: first, the surface emitting heat can be considerably reduced by crouching and bending; secondly, the skin temperature can be regulated, which is much more effective. It varies even under normal conditions between 20° and 34 °C, in some special cases even between -6° and 45 °C; this is caused by the circulation of the blood, which is regulated by the cooling or heating of the skin.

With regard to 3.: we need approximately 700 watt-hours to evaporate one litre of water. Heat is always lost when sweat evaporates. This loss is very small when the skin is cold; the heat flow density reaches about 7 watts/sq m; the upper limit is about 230 watt/sq m and a maximum amount of water evaporation of 500 gm per hour.

Part of the water, and also of the heat, is transported through the lungs. The mean value for heat is 10–20 watt/person; in the tropics it is less, while in polar regions and in high mountains 20–30 watt/person can be reached. It can often be neglected. However, it can sometimes be of psychological importance; when returning from a prolonged stay in the tropics, the increased heat loss caused by drawing deep breaths can cause the sensation of being relieved of an oppressive heat burden.

2.3.4. The heat budget

The first principle of the regulation of the human "heat engine" is that of the conservation of the core temperature. There are many variants of a balanced heat budget. The second principle is the conservation of the inner heat generation which, however, undergoes changes by way of chemical heat regulation, particularly under the influence of heat or cold, work, or fever. The third is that of the conservation of the mean skin temperature, which ranges between 20° and 35 °C. The heat regulation is established chiefly by ascertaining more or less precisely the three quantities: core temperature, heat generation and skin temperature.

Active heat regulation is performed by heat generation, the blood flow through the skin and thus the total thermal conductivity of the skin and the mucous membranes of the respiratory organs and the activity of the sweat glands, which can cause the relative humidity of the skin to rise to 100%, and finally to some extent the attitude of the body as well as the effective surface of the body and the ventilation of the lungs. Passive heat regulation or compensation is performed by skin temperature, the relative humidity of the skin, and the temperature of the mucous membranes. These are in turn strictly governed by the active regulators, the core temperature and the environmental conditions.

Clothing can best be classified as an improved layer of insulation between the core and the air as it were a second skin, bringing a reduction in total thermal conductivity.

The final heating (heat intake +S) or cooling (heat loss –S) of the human body results essentially from the collective action of three factors:

1. M = the metabolic heat effect. It always has a positive quantitiy.
2. R = the radiation effect. It influences the body or its clothing directly in the form of a radiative heating effect (effective radiation +R) or a radiative cooling effect (effective irradiation −R).
3. C = the convective effect. This is the effect of the heat flow caused by the exchange of air between the proximity of the skin and the wider environment, causing either convective heating (convective heat intake -C) or convective cooling (convective heat loss +C).

The equation of the heat budget is S = M + R - C.

If S = 0, i.e. if the body is in heat balance, as is normally the case, then the metabolic effect is reduced by a value M, which is equal to the sum of the negative effective radiation flow and the convective heat flow M = - R + C.

Then the core temperature of the body is constant. This means that no heat is stored in the interior of the body, but all metabolic heat is at once transported outwards. M is always positive, so that at least one of the two quantities R and C must be negative.

The following cases can happen (see Fig. 34: "The human heat budget", which illustrates the conditions graphically):

Case 1: The metabolic heat is transported outwards both by radiative and by convective cooling;

Case 2: The metabolic heat is transported outwards, together with radiative heating, by convective cooling;

Case 3: The metabolic heat is transported outwards, together with convective heating, by radiative cooling.

The radiation effect R is composed of the incoming global radiation effect of short waves G – the sum of direct solar radiation and sky radiation effect – and the infra-red radiation of the environment E, and finally the infra-red radiation of the human body itself B. G and E are received by the body, whilst B is given off by it. Thus R = G + E - B.

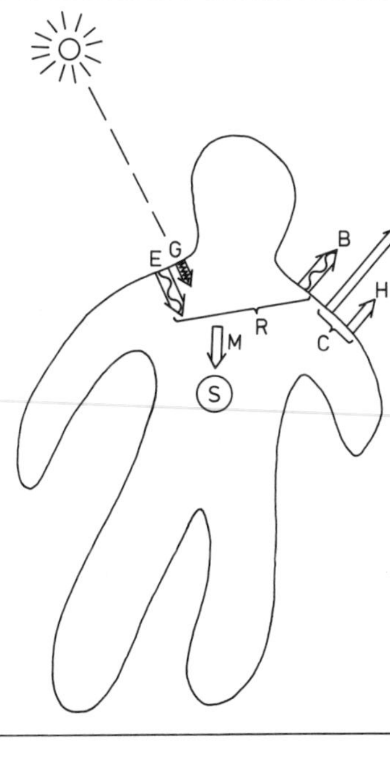

Fig. 34. The human heat budget

G depends on the altitude of the sun, and the angle of incidence formed by the radiation with the surface of the body under consideration. The radiation intensity – one could also say the radiation flow density – is, for a horizontal surface with an altitude of the sun of 90°: 1.12 kW/sq m, at 60° 0.95 kW/sq m, at 30° 0.50 kW/sq m, and at 10° 0.13 kW/sq m. (1 kW = 1000 watt.)

The global radiation is variable in time and space. The course of its values at different latitudes in the course of a year is shown in the following table which, together with the two following ones, is taken from the book *Strahlungsklima* (Radiation Climate) by *R. Schulze*, 1970. Values for UV radiation are sometimes of use for the medical consultant, the chemist, or the technical engineer.

All three tables apply to cloudless days.

The last two tables contain the values of UV-A radiation at 350 nm and UV-B radiation at 307.5 nm in a band of 10 Ångström units.

Table: Monthly and yearly means of the diurnal sums of the global radiation in kWh/sq m per day, after *R. Schulze* 1970, p. 73, Tab. 9.

	Jan	Feb	Mar	Apr	Mai	Jun	Jul	Aug	Sep	Oct	Nov	Dec	Year
North Pole	–	–	–	3.1	6.9	8.9	7.9	4.9	0.80	–	–	–	2.70
80°	–	–	0.77	3.4	6.9	8.8	7.8	4.8	1.55	0.13	–	–	2.85
70°	–	0.51	2.0	4.5	7.1	8.5	7.8	5.5	2.9	0.95	0.06	–	3.32
60°	0.55	1.53	3.4	5.7	7.7	8.8	8.2	6.5	4.3	2.2	0.81	0.34	4.17
50°	1.66	2.8	4.7	6.7	8.4	9.1	8.8	7.4	5.5	3.4	1.97	1.35	5.16
40°	3.0	4.2	5.9	7.5	8.8	9.3	9.0	8.1	6.5	4.8	3.4	2.6	6.09
30°	4.4	5.6	6.9	8.1	9.0	9.2	9.1	8.4	7.4	6.1	4.7	4.1	6.92
20°	5.8	6.7	7.8	8.5	8.8	8.9	8.8	8.6	8.0	7.1	6.0	5.5	7.54
10°	7.1	7.7	8.3	8.5	8.4	8.3	8.3	8.4	8.3	7.9	7.2	6.8	7.93
Equator	8.1	8.5	8.6	8.3	7.8	7.5	7.6	8.0	8.4	8.4	8.2	7.9	8.11
10°	8.9	8.8	8.4	7.7	6.9	6.4	6.5	7.2	8.1	8.6	8.8	8.8	7.92
20°	9.4	9.0	8.1	6.9	5.7	5.1	5.4	6.3	7.5	8.6	9.2	9.5	7.56
30°	9.6	8.8	7.4	5.8	4.4	3.8	4.1	5.2	6.7	8.2	9.3	9.8	6.92
40°	9.6	8.3	6.5	4.6	3.1	2.5	2.7	3.9	5.6	7.5	9.1	9.9	6.11
50°	9.3	7.6	5.4	3.3	1.84	1.25	1.49	2.6	4.5	6.6	8.7	9.7	5.19
60°	8.7	6.6	4.1	2.0	0.72	0.31	0.47	1.36	3.1	5.6	8.0	9.3	4.17
70°	8.2	5.5	2.8	0.84	–	–	–	0.38	1.78	4.3	7.2	9.1	3.33
80°	8.2	4.7	1.42	0.099	–	–	–	–	0.62	3.2	7.0	9.3	2.87
South Pole	8.1	4.6	0.60	–	–	–	–	–	–	2.9	7.0	9.4	2.72

Table: Monthly and yearly sums of UV radiation at 350 nm in Ws/sq cm per month or year, after *R. Schulze* 1970, p. 94, Tab. 12

	Jan	Feb	Mar	Apr	Mai	Jun	Jul	Aug	Sep	Oct	Nov	Dec	Year
North Pole	–	–	–	17	41	52	46	27	–	–	–	–	183
80°	–	–	3	19	42	52	47	28	8	–	–	–	179
70°	–	2	11	26	43	51	47	32	16	5	–	–	233
60°	3	8	20	34	46	53	49	39	24	12	4	2	294
50°	9	16	28	40	50	55	53	44	32	20	11	7	365
40°	19	25	35	45	53	56	54	49	39	29	20	15	439
30°	26	33	41	49	54	55	54	51	44	36	28	24	495
20°	34	40	47	51	53	53	53	52	48	42	36	32	541
10°	42	46	50	51	50	50	50	50	50	47	43	40	569
Equator	49	50	52	50	47	45	46	48	50	50	49	47	583
10°	53	53	50	46	41	38	39	43	48	52	53	53	569
20°	56	54	49	42	34	31	32	38	45	51	55	56	543
30°	58	53	44	35	26	22	24	31	40	49	56	58	496
40°	57	50	39	27	18	13	15	23	34	45	55	59	435
50°	55	46	32	19	10	7	8	15	27	40	52	58	369
60°	52	40	24	11	5	1	2	7	18	33	47	56	296
70°	50	35	15	4	–	–	–	1	10	26	43	54	238
80°	49	27	8	–	–	–	–	–	3	19	41	56	203
South Pole	48	26	–	–	–	–	–	–	–	16	41	56	187

Table: Monthly and yearly sums of UV radiation at 307.5 nm in Ws/sq cm per month or year, after *R. Schulze* 1970, p. 96, Tab. 13

	Jan	Feb	Mar	Apr	Mai	Jun	Jul	Aug	Sep	Oct	Nov	Dec	Year
North Pole	–	–	–	0.06	0.3	0.8	0.4	0.08	–	–	–	–	1.6
80°	–	–	0.01	0.3	0.8	1.2	1.0	0.4	0.06	–	–	–	3.8
70°	–	0.01	0.1	0.8	1.7	2.2	2.0	1.0	0.3	0.05	–	–	8.2
60°	0.01	0.07	0.4	1.7	2.8	3.5	3.2	1.9	0.9	0.2	0.02	–	14.7
50°	0.09	0.3	1.2	2.8	4.0	4.5	4.3	3.1	1.8	0.7	0.2	0.06	23.1
40°	0.4	1.0	2.3	3.8	4.8	5.2	5.0	4.1	2.9	1.3	0.6	0.3	31.7
30°	1.1	2.0	3.3	4.6	5.3	5.5	5.4	4.8	3.9	2.4	1.4	1.0	40.7
20°	2.2	3.1	4.3	5.0	5.4	5.5	5.4	5.2	4.6	3.5	2.5	2.0	48.7
10°	3.5	4.2	5.0	5.1	5.1	4.9	5.0	5.2	5.0	4.4	3.7	3.3	54.4
Equator	4.6	4.9	5.2	4.8	4.5	4.2	4.2	4.8	5.0	5.0	4.7	4.5	56.4
10°	5.4	5.3	5.2	4.2	3.6	3.1	3.3	4.1	4.7	5.2	5.3	5.4	54.8
20°	5.8	5.3	4.7	3.3	2.4	1.9	2.1	3.1	4.1	5.1	5.5	5.9	49.2
30°	5.9	5.0	4.0	2.1	1.3	0.9	1.0	2.0	3.2	4.6	5.4	6.0	41.4
40°	5.4	4.3	2.9	1.2	0.5	0.3	0.4	1.0	2.0	3.7	4.8	5.7	32.2
50°	4.6	3.3	1.8	0.5	0.1	0.07	0.08	0.4	1.1	2.7	4.0	4.9	23.6
60°	3.5	2.1	0.8	0.1	0.02	–	0.01	0.07	0.4	1.5	2.8	3.8	15.1
70°	2.2	1.1	0.3	0.01	–	–	–	0.05	0.09	0.7	1.7	2.5	8.7
80°	1.1	0.4	0.06	–	–	–	–	–	0.01	0.2	0.7	1.3	3.8
South Pole	0.4	0.1	–	–	–	–	–	–	–	0.04	0.3	0.8	1.6

30–40% of the global radiation is reflected by the human skin if it is fair, only 15–30% if it is swarthy. A very high value has been measured in the Californian desert, of 165 W/person.

Infra-red radiation is made up of several components. In a closed room, there is radiation coming from the walls, the ceiling and the floor; in the open, the radiation from the ground and that emitted by the atmospheric water vapour. A body with a temperature T_eK has a radiation intensity E_e, according to *Stefan-Boltzmann*'s law $E_e = \& T_e^4$, where $\& = 5.764.10^{-8}$ W/sq m·K^4 is a radiation constant. The following table contains corresponding values of E_e in W/sq m and T_e transformed into °C.

T_e	E_e	T_e	E_e	T_e	E_e	T_e	E_e	T_e	E_e	T_e	E_e
-70	98	-65	108	-60	119	-55	131	-50	143	-45	156
-40	171	-35	185	-30	202	-25	218	-20	237	-15	256
-10	276	- 5	298	0	321	5	345	10	371	15	398
20	426	25	456	30	487	35	520	40	554	45	591
50	629	55	669	60	710	65	755	70	800	75	847
80	890	85	949	90	1003	95	1059	100	1118		

In a closed room, the walls of which have a temperature of $T_e = 20$ °C, this being also the temperature of the air and of other objects in the environment, the effective radiation is $R = E-B = 426-456 = -30$ w/sq m if the infra-red irradiation of the person is calculated for a temperature of the surface of his clothing of 25 °C = T_b. If the man had no clothing, his temperature would be $T_b = 35$ °C and $R = -100$ W/sq m. For a person with a surface of 1.5 sq m, the heat flow would be 150 W/person. Such a flow could only be sustained if the metabolic effect considerably exceeded the basic one. The clothed person has a cooling of only 45 W/person; there needs to be additional convection cooling to take the generated heat outwards and prevent part of it from being stored in the interior.

Let us now consider the components of the convective heat effect C. It is composed of the effect caused by the flow of sensible heat D – called enthalpy, to be precise – and the flow of latent heat H released into the atmosphere by the evaporation of water. This gives $C = D + H$. The surface temperature of the body is T_b. Let us suppose that the air in the immediate vicinity of the body has this temperature, and a vapour pressure of e_b. Mostly the surrounding air is cooler than the body ($T_a < T_b$), so that heat flows from the body into the air. If m kg of air are exchanged through 1 sq m of a control surface near the body, within an hour the air flowing to the free environment brings about an enthalpy of $D_b = m\ c_p\ T_b$ and an amount of latent heat $H_b = m\ L\ s_b$. c_p is the specific heat of air at a constant atmospheric pressure, $c_p = 0.28$ W/kg K. L is the heat of vaporisation of water L = 700 Wh/kg; s_b is the specific humidity of the air $s_b = 0.622\ e_b/P$ (kg H_2O/kg air); e_b is the vapour pressure and P the atmospheric pressure. The air mass m conveys altogether energy of C_b outwards through one sq m in an hour; this means that $C_b = D_b + H_b = m\ c_p(T_b + L\ s_b/c_p) = mc_p\ \theta_b$. θ_b is equivalent temperature thus: $\theta_b = T_b + L\ s_b c_p = T_b + 0.622\ .700\ e_b/P.0.28 = T_b + 1555\ e_b/P$.

From the outside, an equivalent air mass m kg is introduced through one sq m of the control surface. This brings energy $C_a = m\ c_p\ (T_a + L\ s_a/c_p)$ per sq m per second near to the body. Similarly to C_b, we have for $C_a = D_a + H_a = m\ c_p\theta_a$, θ_a representing the equivalent temperature of the free air of the environment. The resulting convective heat effect is $C = C_b - C_a = m\ c_p(\theta_b - \theta_a)$, if both the mass exchange m and the values of temperature and humidity are constant. C being a heat flow, m is as it were a mass flow or, more precisely, a mass exchange per hour through one sq m. It corresponds to a ventilation factor dm/dt mentioned in the following observation by *Robitsch* 1939, in the *Hann-Süring* handbook of meteorology, p. 137–8:

"The most important factor in the formula of evaporation is that of ventilation dm/dt. By definition, dm is an air mass which in the time dt passes through the control surface by exchange, increasing or decreasing its vapour pressure from a value e_a to e_b by evaporation or condensation or sublimation. It is not important whether a larger or a smaller air mass is involved in the exchange of water vapour, the change in vapour pressure being correspondingly larger or smaller. The ventilation factor being an ideal quantity of a complex nature, it is not necessary to analyse it further physically, for the processes of exchange are so complicated by the aerodynamic conditions on the evaporating surface – resulting from the shape and the quality and direction of the air movement – that an analysis promises little success. Implicitly, the ventilation factor contains the influences of the molecular exchange of water vapour as well as of the water mass transported through the surface of evaporation by its own convection."

A convective heat flow C from the body to the free environment exists as long as the equivalent temperature near the body θ_b – determined by the skin temperature T_b and the vapour pressure of the air near the skin e_b – is greater than the equivalent temperature of the free air of the environment θ_a – determined by the air temperature T_a and its vapour pressure e_a. This heat flow C has two components: $C = D + H$, the flow of sensible heat D and the flow of latent heat H, that is, the heat bound to the water vapour – H. We thushave $D = D_b\text{-}D_a = m\ c_p(T_b\text{-}T_a)$ and $H = H_b\text{-}H_a = m\ c_p(e_b\text{-}e_a)\ 1555/P$.

If the air temperature is equal to that of the body, then there will not be a flow of sensible heat away from the body. If the air temperature is higher, then even heat flows into the body, the more the warmer the air. To avoid the danger of death by heat, the body has to find another way of bringing about heat flow into the environment. This is made possible by evaporation. The skin starts to perspire visibly, and starts to cool considerably when evaporation sets in. This is possible when the air of the environment is sufficiently dry to absorb a large quantity of water vapour. Evaporation already takes effect at lower air temperatures; this is termed 'perspiratio insensibilis', or imperceptible perspiration through the skin.

At air temperatures above 35 °C, evaporation is the only way in which the body can dissipate heat, provided that the air is able to hold the evaporating water vapour. Heat dissipation becomes more difficult when the air is already very humid, as in the wet tropics, which are more oppressive to Europeans than the dry tropics. If the heat cannot be got rid of, the resulting hyperthermy can lead to death. Compare the notorious "Black Hole of Calcutta" (1756), the incident in which the potentate Siraj-Uddaula of Bengal, after the capture of the town, locked 146 English prisoners in a cell at Fort William intended for only a few prisoners. There were only a handful of survivors.

2.3.5. How the human body reacts to thermal stress

(after *Lee, Douglas, H.K.* 1958, UNESCO Arid Zone Research X, p. 102–125)

Under cold conditions	*Under hot conditions*
1. Thermoregulation responses	
Constriction of skin blood vessels	Dilatation of skin blood vessels
Concentration of blood	Dilution of blood
Flexion to reduce exposed body surface	Extension to increase exposed body surface
Increased muscle tone	Decreased muscle tone
Shivering	Sweating
Inclination to increased activity	Inclination to decreased activity

2. Consequential disturbances	
Increased urine volume	Decreased urine volume
	Mobilization of tissue fluid
	Thirst and dehydration
Danger of inadequate blood supply to skin of fingers, toes, and exposed parts, leading to frostbite	Difficulty in maintaining blood supply to brain, leading to dizziness, nausea, fainting (heat exhaustion)
	Difficulty in maintaining chloride balance, leading to heat cramps
Discomfort leading to neuroses	Discomfort leading to neuroses
Increased hunger	Decreased appetite
3. Failure of regulation	
Falling body temperature	Rising body temperature
Drowsiness	Disordered cell functions
	Heat regulating centre impaired
Cessation of heart beat and respiration	Failure of nervous regulation terminating in cessation of breathing

2.3.6. The need for cooling in the tropics

The air temperature in the middle latitudes being lower than in the tropics, the heat conditions peculiar to the latter are seldom experienced in the former. Either the heat of summer is not as high as that of the warm deserts or, where the air temperature is comparable to that in the wet tropics, the humidity is well below that to be regularly encountered near the equator, as far as water vapour content is concerned. For example, if on a summer's day in Hamburg (Germany) the air temperature is 30 °C and the relative humidity 50% – there being about 15 gm of water vapour in a cubic metre of air – the atmosphere is considered very close. In the wet tropics, for example at Georgetown (Guyana), Monrovia (Liberia), Colombo (Sri Lanka), Singapore, etc., 30 °C is approximately the mean daily maximum air temperature. The relative humidity is 70%, and a cubic metre of air contains 21 gm of water vapour.

As we have already seen, the equivalent temperature of the air is the temperature reached when the total content of water vapour in it is condensed and the heat of condensation used to heat the air at constant atmospheric pressure. In the case of Hamburg, the equivalent temperature is 62 °C, in the tropics 75 °C. If this is multiplied by 0.28 (= the value of the specific heat of air at a constant pressure in Wh/kg·K), we have the so-called "heat content" of the air, expressed in Wh/kg. One can say that the equivalent temperature is the temperature of dry air with the same heat content as the humid air.

Under atmospheric conditions with an equivalent temperature of about 55 °C, the limit of comfortable conditions, with average relative humidity, is reached. Even a small increase in temperature over this creates the feeling of a heat burden. A fixed limit cannot be laid down because of the differing susceptibility of different individuals. The air at Hamburg in the example given would need to be cooled by 7 °C to approach the comfort limit, that in the tropics by 20 °C. In Hamburg, heat amounting to 7x0.28 = 2.0 Wh would need to be removed from each kg of air in order to attain an equivalent temperature of 55 °C, whereas in the tropics the amount of heat would be 20x0.28 = 5.6 Wh. In an enclosed space, air normally has to be replaced at the rate of 280 litres a minute, or 16.8 cubic m per hour per person, i.e. 21.4 kg of air. This requires a cooling effect of 42.8 watt in Hamburg, and 119.8 watt in the tropics, under the conditions mentioned.

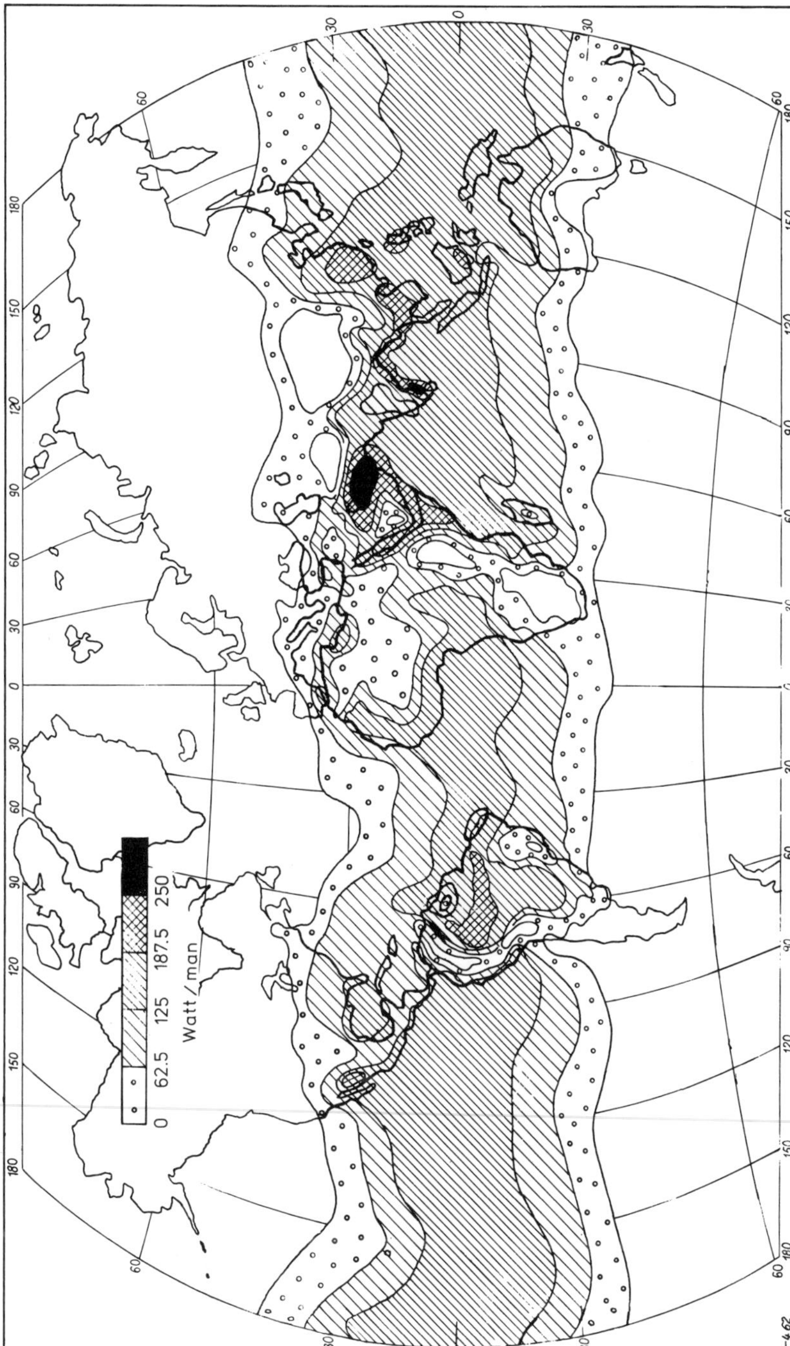

Fig. 35. World map of the "cooling effect" in the hottest month

The map in Fig. 35 ("World map of the 'cooling effect' in the hottest month") shows the cooling effect that must be supplied for one person in the hottest months in order to cool the air to 55 °C equivalent temperature. Over wide areas of the tropical seas, 100 to 150 watts have to be supplied per person to do this. Values exceeding 175 watts are to be encountered in the Gulf of California, the lowlands of the Amazon, the Red Sea, on the coast of Somalia, in the Persian Gulf and the Gulf of Oman, where 250 watts can even be exceeded, on the coasts of Bengal and of the China Sea.

3. The Climates of the earth

3.1. Climate classification

At the end of section 2.1., "Man between the Poles and the Equator", climatic areas were mentioned as classified by the famous German climatologist *Wladimir Köppen* (1846–1940). His system has been used all over the world. His first paper, "A tentative classification of climates, especially according to their relations to flora" was published in German in the "Geographical Journal" in 1900. In 1918 he published a more differentiated system under the title "Classification of climates according to temperature, precipitation and the course of the year". In 1923 his famous comprehensive work, *Outlines of Climatology* was published by Walter de Gruyter in Berlin and Leipzig, a second edition following in 1931. At that time, he began, together with his younger colleague *Geiger* and 35 contributors from all over the world, to edit a compendium of climatology which, owing to the Second World War, was unfortunately never completed.

Some years ago, *Landsberg* and his team published *Climates of the World*, a substantial compendium in 15 volumes, catering to a long-felt need on the part of specialists. The quantity of climatological data recorded has multiplied over the last fifty years. In 1927 a comprehensive collection of such data began to be published in Washington. These "World Weather Records" were of great value. In 1962, the World Meteorological Organization published climatological normals of temperature and amount of precipitation for well over 1000 stations. In the 1971 edition, normals of total duration of sunshine, and number of days with precipitation equal to or greater than 1.0 mm, were added for most of the stations. "Tables of Temperature, Relative Humidity and Precipitation for the World" have been published by the Meteorological Office in London since 1958. These show directly the behaviour of the elements of climate mentioned in the course of the year. Today we can confine ourselves to the essential part of these data if we wish to present a relatively simple classification of climates for public use and global survey.

Climatic tables permit climatic singularities to be expressed more clearly than complicated coding, the meaning of which has always to be looked up before use.

Köppen successfully adduced geographical and botanical facts in order to classify climates in his system. Up to his death at 94, he was engaged in climatological studies, striving continually to improve on his system. It is possible that he would have approved of the classification proposed by the American climatologist *Glenn T. Trewartha* as a development of his own system. We call this the *Köppen-Trewartha* system. In my opinion, *Trewartha* has simplified *Köppen*'s system effectively. There are hardly any difficulties involved in learing and applying it in its present form.

In seeking a simple global description of climates, we shall reject the inclusion of all details, particularly under consideration of the possibility existing today of having at hand complete series of normals of the main climatic elements. A full description can be found in *Trewartha*'s *Introduction to Climate* (*McGraw-Hill* Book Co., 1968). Here we have set limits to the classification, defining its mean features only as far as is necessary for our purposes. Like *Köppen* in 1931, *Trewartha* arranges the climates of the earth in six classes – after the first six letters of the alphabet – in the order of generally declining temperature at sea-level from the equator to the poles (with the exception of the dry climates B), thus:

Class of climates	*Description*	
A tropical climates	over 17 °C throughout the year	—— frost limit
C subtropical climates	8–12 months over 9 °C	
D temperate climates	4–7 months over 9 °C	
E subarctic climates	1–3 months over 9 °C	—— forest limit
F polar climates	no month over 9 °C	—— dryness limit
B dry climates	evaporation exceeds precipitation	

About a quarter of the continental areas of the earth have dry climates (B). We include the climate of a place among the class of dry climates if the mean amount of precipitation measured in mm is smaller than a dryness limit RB. This is, according to *Trewartha*, given by a simple formula RB = 20(t–10 + 0.3 PS), in which t is the average annual temperature in °C and PS the summer percentage of precipitation observed.

(NB: In the northern hemisphere, winter is the low sun period and includes the months October to March, summer the high sun period, including the months April to September. For the southern hemisphere the two seasons are reversed.)

If the amount of annual precipitation R observed at a given place is not as much as half of RB, the climate is called a desert climate BW, and the value RW = RB/2 = 10(t–10 + 0.3PS) mm desert limit. If R is less than RB but more than RW, we call the climate a steppe climate BS.

Here are some examples:

station	t	precipitation summer	R mm	PS	RW	RB	climate formula
Port Said	22	6	63	10	149	297	BM
Alexandria	21	7	169	4	122	244	BS
Zaragoza	15	183	339	54	212	424	BS
Madrid	14	171	436	39	157	314	Cs

The climate of Madrid is not a dry climate!

The steppe climate BS is semiarid; the desert climate BW is arid. There is insufficient water for the growth of forests. Special types of B climates are the marine desert climates BM; M indicates high air humidity. If the air is at the same time relatively cool and fog or wet mist is frequent, *Köppen* called this climate garua climate: Bn, which is a subtype of the marine desert climates BM.

The climate classes A, C, D and E, on the other hand, are "tree climates", with sufficient precipitation and warmth to permit the growth of standard kinds of tree. A has four and C three sub-divisions, according to the character of the rainy saison: Ar is a tropical rain climate, Am is a tropical monsoonal rain climate with only a small dry season, Aw a tropical climate with summer rain, As a tropical climate with winter rain; Cr is a subtropical rain climate, Cw a subtropical climate with summer rain, Cs a subtropical climate with winter rain.

The letter r indicates rain or rainy, i.e. with no dry season; m indicates monsoonal, i.e. with a small dry season; s stands for dry in summer, w for dry in winter. In the case of A, a month is called rainy when the precipitation amounts to at least 60 mm, and a climate is termed Ar when more than 9 months are rainy. If less than 10 months are rainy, the climate will be termed Am when the amount of annual precipitation R observed at a given place is equal to or greater than R = (100–rn) 25 mm; rn is the mean monthly precipitation of the driest month.

In the case of C, we call it rainy: Cr when the difference between the driest and the wettest month is less than is required for s and w, or the driest month of summer has more than 29 mm precipitation. A climate is termed Cs when the annual precipitation is below 890 mm, and there is at least three times as much rain in winter as in summer; and we call a climate Cw when there is at least ten times as much rain in summer as in winter.

The A, B and C climates are distinguished by precipitation, the D, E and F by air temperature. The temperate climates D are divided into two types according to the mean temperature of the coldest month, which is above or equal to 0 °C in the case of the temperate oceanic climate DO, and below 0 °C in that of the temperate continental climate DC, with the exception of inner continental regions where 0 °C and more are allowed.

The subarctic or boreal climates E are not subdivided by *Trewartha*. Here, however, his method of dividing the D climates into two types shall also be applied to the E climates. The subarctic climates E are thus divided into two types, according to the mean temperature of the coldest month, which is above -10 °C in the case of the subarctic oceanic climate EO, and below -10 °C in that of the subarctic continental climate EC.

The polar climates F are beyond the limits of forests because of lack of warmth. This limit coincides with a line on the poleward side of which no month has a mean air temperature of as much as 10 °C. The climate classes F are subdivided into two types according to cold: with a mean temperature of the warmest month between 0 °C and 10 °C we have a tundra climate FT, and with every month below 0 °C we have an ice climate FI.

There are thus 16 principal climates:

Ar tropical rain climate
Am tropical monsoonal rain climate
Aw tropical summer rain climate
As tropical winter rain climate

BS steppe climate
BW desert climate
BM marine desert climate

Cr subtropical rain climate
Cw subtropical summer rain climate
Cs subtropical winter rain climate

DO temperate oceanic climate
DC temperate continental climate

EO subarctic oceanic climate
EC subarctic continental climate

FT tundra climate
FI ice climate

In the preceding classification, hill and mountain climates are not considered. These can be expressed by the climatic notations for neighbouring low lands of corresponding moisture conditions, and indicated by G for hills/low mountains and H for high mountains, which cancel the restriction on thermal margins set for flat lands, for which increasing heights the air temperature decreases. G applies to heights from about 500 m to 2500 m, and H to heights above 2500 m. This definition and those in the preceding sentences are the only ones that differ from those given by *Trewartha*. The designations would be "mountain climate" and "high mountain climate" respectively. If necessary, thermal conditions can be expressed by an appropriate term from the thermal standard scale here proposed for international use:

Universal Thermal Scale

Thermal conditions		are termed	and coded
35 °C to	...	severely hot	i
28 °C to	34 °C	very hot	h
23 °C to	27 °C	hot	a
18 °C to	22 °C	warm	b
10 °C to	17 °C	mild	l
0 °C to	9 °C	cool	k
- 9 °C to	- 1 °C	cold	o
-24 °C to	-10 °C	very cold	c
-39 °C to	-25 °C	severely cold	d
... to	-40 °C	excessively cold	e

If the climate of a place is characterised by the corresponding letters, it can be expressed according to *Köppen*'s system in a simple climatic formula as follows:

All we need are the monthly and annual mean values of temperature and precipitation of the place.

First of all the desert limit of precipitation has to be calculated:
RW = 10(t-10) + 300 Rs/R, where t is the mean annual temperature, R is the mean annual precipitation, and Rs is the mean precipitation in summer (April–September in the northern hemisphere, and October–March in the southern).
The climate is BW if the mean precipitation R is less than RW.
The climate is BM if the mean precipitation R is less than RW and the place is near the coast and has a high air humidity.
The climate is BS if R is less than 2 RW.
The class is not B if R is equal to or greater than 2 RW.
Secondly one has to ask how many months are over 17 °C. If all months are, the climate is A. Then one has to ask how many months have more than 59 mm precipitation.
The climate is Ar if more than 9 months do; otherwise one has to calculate the value $R' = 25(100\text{–rn})$, where rn is the mean monthly precipitation of the driest month.
The climate is Am if R is equal to or greater than R′; otherwise
the climate is Aw if winter is drier than summer, and
the climate is As if summer is drier than winter.
If not all months, or none, are above 17 °C, we have to ask how many months are over 9 °C.
The class of climate is F if there are no months over 9 °C,
the class of climate is E if there are 1–3 months over 9 °C,
the class of climate is D if there are 4–7 months over 9 °C, and
the class of climate is C if there are more than 7 months over 9 °C.
The climate is Cr if the class is C and the driest month of summer has more than 29 mm precipitation.
The climate is Cs if the driest month is in summer with less than 30 mm and its precipitation is exceeded at least three times by the wettest month in winter, and R is less than 890 mm; otherwise
the climate is Cr.
The climate is Cw if the driest month is in winter and its precipitation is exceeded at least ten times by the wettest month in summer; otherwise
the climate is Cr.
The foregoing procedure can also be applied to characterise D climates by means of r, s and w. s and w can also be applied without restriction to B, E and F climates.
The next question is how many months there are below 0 °C:
the climate is FI if the class is F and all months are below 0 °C; otherwise *the climate is FT.*

The climate is DO if the class is D and no month is below 0 °C; otherwise *the climate is DC.*
The climate is EO if the class is E and no month is below -9 °C; otherwise *the climate is EC.*

The forgoing simple formula of a given place is prefixed by a G if the height of the place is between 500 m and below 2500 m, by an H if it is 2500 m or more. In addition to this all formulae in our climatic tables are expanded by two code letters of the thermal standard scale indicating the warmth of summer and the cold of winter corresponding to the maximum and minimum of the mean monthly air temperature. Some stations with extreme seasonal thermal conditions follow as examples:

a) low land (below 500 m)

place/country or region, climate,	climatic formula warmest month	coldest month
Atbara/Sudan desert climate	BWih severely hot	very hot
Canton Island/Pacific Ocean tropical winter rain climate	Ashh very hot	very hot
Kayes/Mali steppe climate	BSia severely hot	hot
Lima/Peru marine desert climate	BMbl warm	mild
Alahabad/India subtropical summer rain climate	Cwil severely hot	mild
Sian/China temperate continental climate	DChk very hot	cool
Mukden/China temperate continental climate	DCac hot	very cold
Jakutsk/USSR subarctic continental climate	ECbe warm	excessively cold
Ushuaia/Argentina tundra climate	FTkk cool	cool

b) mountain and high mountain

place/country or region, climate,	climatic formula warmest month	coldest month
Quito/Ecuador tropical summer rain high mountain climate	HAwll mild	mild
La Paz/Bolivia tropical summer rain high mountain climate	HAwlk mild	cool
Zacatecas/Mexico steppe high mountain climate	HBSll mild	mild
Lhasa/Tibet steppe high mountain climate	HBSlo mild	cold
Ulan Batar/Mongolia steppe mountain climate	GBSld mild	severely cold
Vostok/Antarctic ice high mountain climate	HFIde severely cold	excessively cold

3.2 The Climate of the tropics

Meteorologically speaking, the tropics are the zone between 30° latitude north and south of the equator, for polewards of these latitudes there normally lie the axes of the oceanic subtropical anticyclones or high pressure belts. The trade winds originate at the equatorial margins of these, and blow into the equatorial low-pressure belt or trough. The falling pressure gradient into this weakens more and more as the equator is approached, and the wind becomes weaker and weaker, until a calm often ensues. This belt of calms near the equator is known as the doldrums. Meteorologists call it the "intertropical convergence zone".

The tropics do not fall within fixed limits, but form a more or less extensive transitional zone. Thermally, the tropics span the part of the earth which receives more heat from the sun than it loses by irradiation. The input of heat is transferred to the polar caps, and used to keep warm the air there which, without this heat from the tropics, would be very much colder, as the irradiation in higher latitudes is by no means compensated for by solar radiation.

In the flat lands of the tropics the air temperature does not as a rule sink to freezing point. Frosts may occur only in certain higher regions, and in depressions in the marginal tropics, when pockets of cold air are generated during nights with strong irradiation and a lack of wind.

The vertical decrease in temperature amounts to about 5.5 °C for every 1000 m of height. Near the equator, freezing point lies as high as 4700–5000 m in the free atmosphere. Alexander von Humboldt discovered that in Mexico at 19° north the limit of eternal snow was at 4600 m, i.e. only 200 m lower than near the equator at Quito. There, in the Andes, the seasonal variation of the snow limit is only about 60–70 m, while in Mexico it is about 800 m.

The great heat which the tropics receive causes a great deal of water to evaporate. In most areas of the marginal tropics, the loss of water is not compensated for by rain, which would return water in a liquid state to the ground. Only close to the equator does the rain exceed the evaporation. Where there is a surplus of rain, the salinity at the surface of the sea is lower than in regions of prevailing evaporation, as is to be seen in Fig. 36, representing the zonal distribution of sea-surface salinity and of the differences

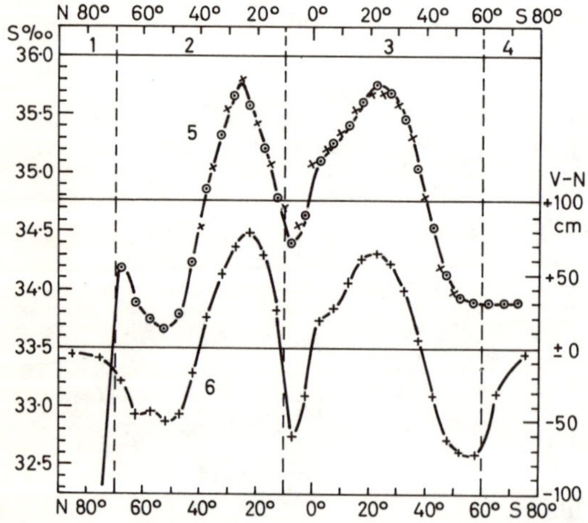

Fig. 36. Zonal distribution of sea surface salinity and of the surplus of evaporation on the oceans

(V–N) in the ocean (after *Wüst, G.* 1954, Archiv für Meteorologie A 1, p. 317). The difference (V–N) is the annual surplus of evaporation. N is the annual amount of precipitation, and V the annual amount of evaporation. It can be seen that evaporation predominates beyond the marginal tropics.

In the course of a year, the total evaporation from the sea is 336 Tt, and that from the land 70 Tt (1 Tt = 10^{12} tonnes). 296 Tt are precipitated over the sea, 110 Tt over the land. Consequently 40 Tt of water are transferred from the sea to the land in gaseous form in the course of a year. Only 70 Tt evaporate here, however, so that 40 Tt run off, completing the cycle. The distribution of precipitation is very heterogeneous, both over the sea and over the land. A quarter of the continental area of the earth has less than 25 cm annual precipitation. (The figures in this section are rough guesses, exact data not being available for the moment.)

Figure 36 shows that the area of tropical surplus of rain is located between the equator and 10° latitude north, that is, in the intertropical convergence zone, represented in the pressure field by the equatorial trough. The highest annual rainfall of each continent is to be found near this trough: 9000 mm at Andagoya (Colombia, S. America), 10,000 mm on the western slopes of the Cameroun Mts. (Africa), and 11,000 mm at Cherrapunji (Assam, Asia). The last amount is attained in spite of a dry season of 4 months in this part of the Indian subcontinent.

Let us now consider the constantly wet regions of the tropics, characterised by high air humidity, great heat, and a small seasonal variation in temperature. The conditions obtaining at Balboa (Panama), height 9 m, index p and P; Freetown (Sierra Leone), height 19 m, index f and F; and Saigon (Vietnam), height 9 m, index s and S are shown in a hygrothermal diagram as examples in Figure 37. The small letters on the left show the mean daily minima of temperature combined with the corresponding value of vapour pressure, while the capitals on the right give the mean daily maxima of temperature and the corresponding vapour pressure. In spite of the great distance between the places men-

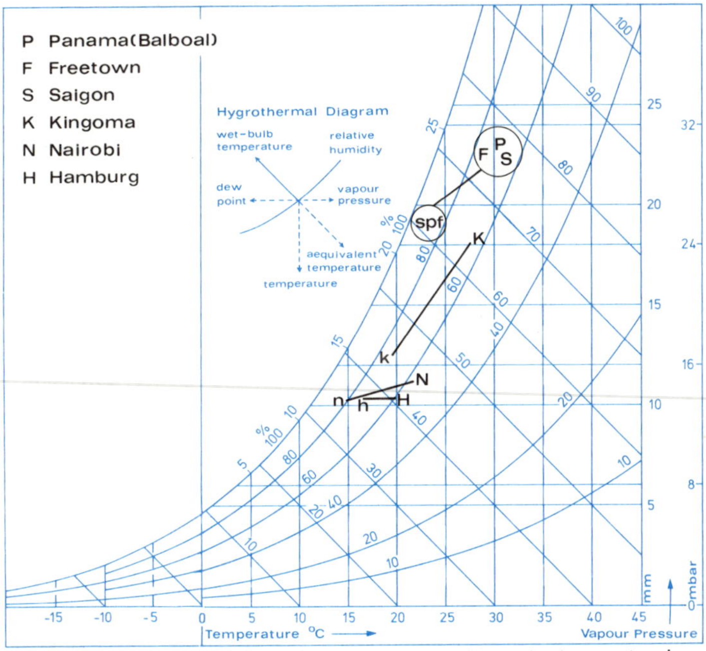

Fig. 37. Samples of atmospheric humidity and temperature conditions in the wet tropics

tioned, the atmospheric conditions of temperature and humidity are very similar, and hold good for many other places in equatorial coastal regions. At sunrise there are very high values of relative humidity – over 90% – and temperature minima of 22–24 °C. With a rapid and powerful rise in temperature to 29–32 °C in the early afternoon, the relative humidity decreases to about 70%, but the vapour pressure rises from about 19 mm to 22–23 mm and the equivalent temperature from 60 to 75 °C. These conditions are very far from being comfortable. In order to escape from the heat in these latitudes, one has, if air-conditioned rooms are not available, to go to higher places. How altitude operates in the tropics is shown by the examples of Kigoma (Tanzania), height 884 m, index k and K, and Nairobi (Kenya), height 1791 m, index n and N. (For comparison, Hamburg (Germany), height 22 m, index h and H, July, is added.) These regions are not among the constantly wet regions of the tropics, but have marked dry seasons. One can see how, with increasing altitude, and the corresponding decrease in temperature, conditions approach those of summer in temperature zones, and offer a climate pleasant to many people.

In the dry tropics and the subtropics, the atmosphere contains less than 20 kg of water vapour above one sq metre, in deserts often less than 10 kg, in the humid tropics on the other hand 40–55 kg, and in India during the monsoon even more than 60 kg. As was already mentioned, a quarter of the continental area of the earth has less than 25 cm annual precipitation. 12% of this area consists of desert. Where deserts are in the tropics, the air is heated from the ground, but no water enters. Solar radiation reaches the ground with little diminution. Here we have a maximum input of solar energy to the ground. The energy received is used to heat the ground, which in turn warms the air to such an extent that these regions have the highest air temperatures on earth.

In Death Valley (California, USA), on July 10th, 1913, the air temperature rose to 57 °C. On September 15th, 1933, the highest air temperature of 58 °C was measured in El Azizia/Lybia in Africa. Even higher values may occur in places in the dry zones of the earth where conditions are favourable for the heating of the air. The upper limit for the temperature of the free atmosphere may be expected to be about 60 °C. The town of Lugh Ferrandi in Somalia has the highest observed mean annual air temperature: 30 °C.

In direct contrast to the strong heating of the air above the ground is the situation over the open sea. Here the temperature of the sea surface and of the adjacent air differs only by a few tenths of a degree near the equator. the bulk of the incoming solar radiation energy is used to evaporate water. Much of the energy is also expended in this way on the land surface of the humid tropics. For this reason, the highest temperatures are recorded some distance from the equator, where seasonal variations in temperature are considerable. The hot season does not continue all the year round, but takes the form of a hot summer, which can often cause many people great heat stress. At the time of the maximum altitude of the sun, the air is often warmer than the human skin, at about 35 °C. If the skin is dry, it is heated by contact with the air, the more so the more strongly the wind blows. For protection against the burning winds of the desert, the natives wear the burnous, a loose white wrap which simultaneously affords protection against solar radiation.

Since the skin is not inanimate but a living organism – the most extensive organ of the human body, incidentally – it provides a good defence against over-heating. Even when it is dry, water is evaporating invisibly from its outer surface. When the air is hot and dry, evaporation provides cooling; this not only dissipates the heat generated inside the body by metabolism but also that supplied by contact with the heated air.

Even in the sauna (Finnish steam bath), an air temperature of 100 °C can be tolerated for an hour without injury to health, providing the relative humidity is below 5%. Yet, if the latter were 95% and the temperature only 45 °C, one would suffer a heat collapse. This does not happen in the open air. At temperatures of over 40 °C, the relative humidity is normally below 50%; in deserts it sinks in extreme cases below 10%.

At low humidity levels, a large quantity of water evaporating from the skin can enter the air. A naked man walking in the desert loses one litre of liquid per hour. Losses of water of up to 16 litres a day have been measured in persons doing physical work in hot dry climates. The loss of water will be less where clothing is worn; but the cooling effect will be reduced. If the water transpired by the body is not replaced several times a day, danger to life can arise even within one day, as the body may be unable to keep its temperature constant because of lack of water. At a body temperature of about 43 °C human beings lose consciousness, and die at 44–45 °C.

At Baghdad (Iraq), height 33 m, index b and B, the mean daily maximum of temperature in July is 43 °C, and the relative humidity 11%. During the night, the temperature falls to 24 °C and the relative humidity increases to 34%. The great variation of temperature in summer, at 20 °C, contrasts with 10 °C in winter, as can be seen from Fig. 38. The vapour pressure is less in winter than in summer. The relative humidity, by contrast, is greatly increased, for the air temperature is relatively low. The same applies to conditions at Alice Springs (Australia), height 570 m, index a and A, and Cairo (Egypt), height 115 m, index c and C. Here, too, the left-hand letters show the mean daily minima of temperature and the corresponding vapour pressure values, the capitals on the right the corresponding maxima and co-ordinated vapour pressure values. The dotted line indicates conditions in January, the unbroken line those in July. Cairo has the least rainfall. The annual mean is 22 mm and the mean number of days with 1 or more mm is 5. The figures for Baghdad are 151 mm and 23 days, those for Alice Springs 250 mm and 31 days. We may also consider Khartoum (Sudan), height 384 m, index d and D, and Fort Lamy (Chad), height 294 m, index e and E. Their values are: Khartoum 164 mm and 19 days, Fort Lamy 646 mm and 58 days. Figure 39 shown the conditions for January and July in a hygrothermal diagram. The conditions in winter can be compared with those at places far from the equator as to the small amount of vapour pressure, but differ considerably from them in their higher temperature and greater interdiurnal variations. Khartoum is often reached in summer by the intertropical convergence zone, which causes 150–200 mm of rain in the southern Sudan in August. Annual rain of 1000 mm and more in the catchment area of the Nile causes the extensive marshes on the White Nile, the Sudd Region, to persist. Only to the north of Khartoum is the extreme desert climate of the Nubian and Libyan deserts to be found. South of Khartoum, summer rain falls, mostly as heavy showers, often accompanied by thunderstorms, particulary in the afternoon and evening. Especially at the beginning of the rainy season, the strong wind-squalls which accompany such thundery showers cause a special form of sand-storm, known as habubs. At Khartoum and Atbara, there are an average of 18–20 habubs a year. Like Khartoum, Lake Chad is also beyond the boundary of the desert region; in summer, temperature and humidity conditions are found here such as continually obtain in the regions of tropical rain climate (see Fig. 37).

3.3. Marine dry regions

If we look at the map of global rainfall distribution over the oceans, the widespread scarcity of rain in the tropics is striking, being interrupted only by a wetter belt near the equator. Over a number of years, the mean shows rain less than 5% of the time over vast areas. Polewards of about 30° latitude, the frequency of rainfall increases rapidly to more than 25% at 40–50° latitude.

With the changing of the seasons, the equatorial belt of maximum rainfall activity shifts to and fro, indicating the oscillation of the intertropical convergence zone. The trade winds of both hemispheres converge in it. The zone of transition between the wind systems, called the doldrums by mariners, often has light variable winds accompanied by heavy rain and squalls. In the Atlantic, the convergence zone extends in north-

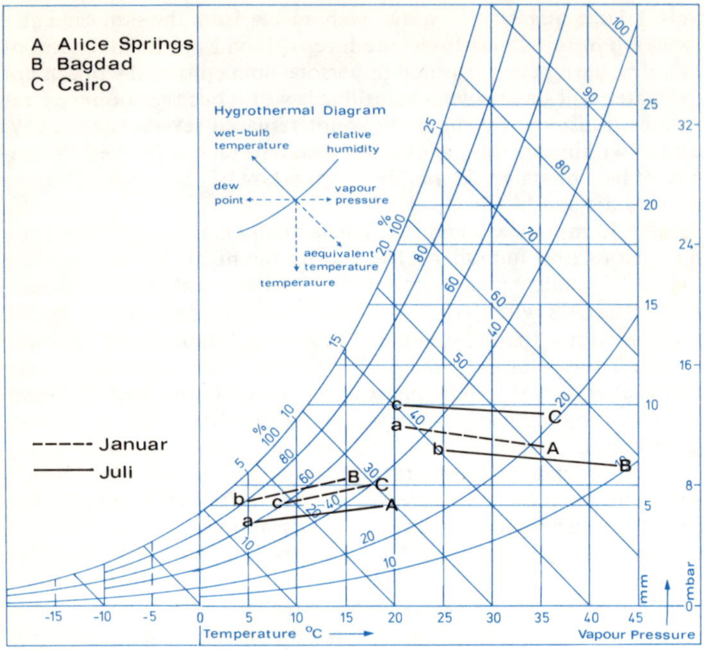

Fig. 38. Samples of atmospheric humidity and temperature conditions in dry tropics

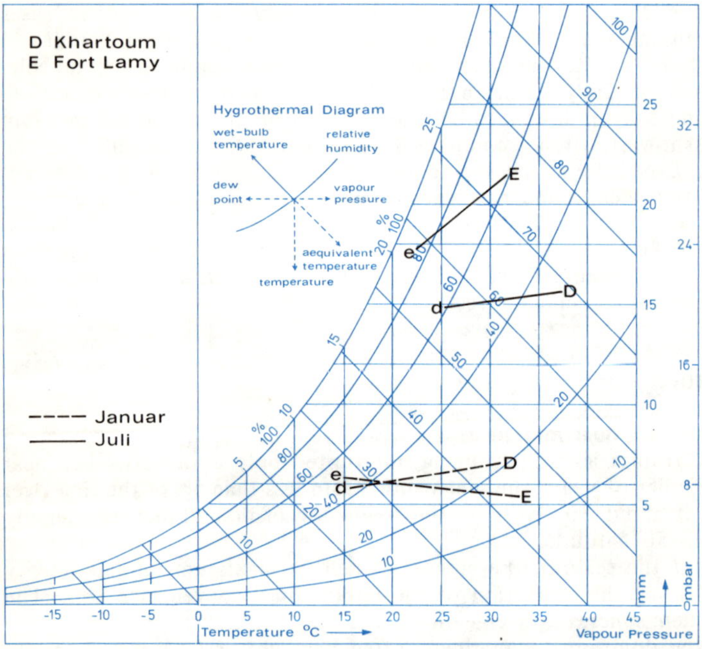

Fig. 39. Samples of atmospheric humidity and temperature conditions in desert and steppe regions

ern winter from the mouth of the Amazon to Upper Guinea, and in northern summer from Surinam to Senegal. At about 30° west longitude it reaches its southernmost position in March at 1° latitude north. Its oscillation amounts to 13° of latitude. Southward of Central America, at 90° longitude west, it oscillates between 3° and 9° latitude north, i.e. only 6° of latitude. Compared with the oscillation of the declination of the sun from 23.5° S to 23.5° N, i.e. 47° of latitude, the annual oscillation of the convergence zone

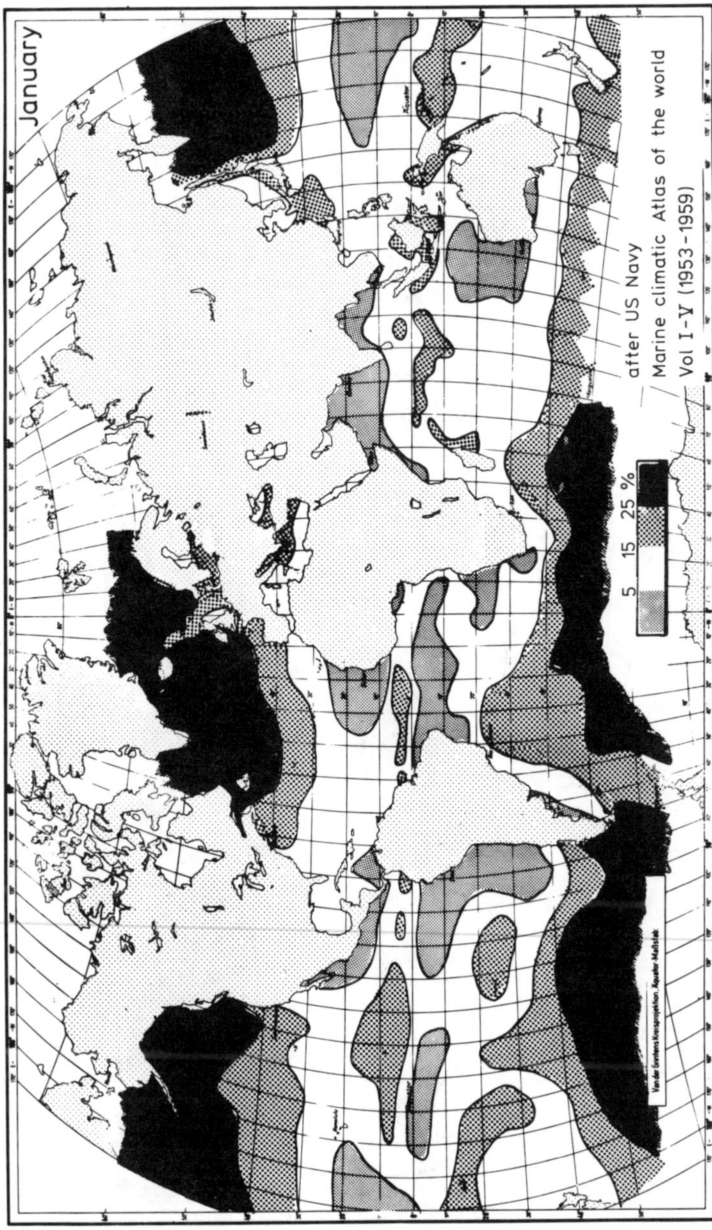

Fig. 40. Mean frequency of precipitation over the oceans in January

is small in these marine regions. Greater oscillations occur in the Indian Ocean, where in the northern summer summer monsoon rains are caused, which penetrate far into the continent of Asia.

The position in the marine dry regions corresponds to the oscillations of the intertropical convergence zone. On the map for January (Fig. 40), one can see an extensive area with a rain frequency below 5%: the Pacific coast of Central America, parts of the south-

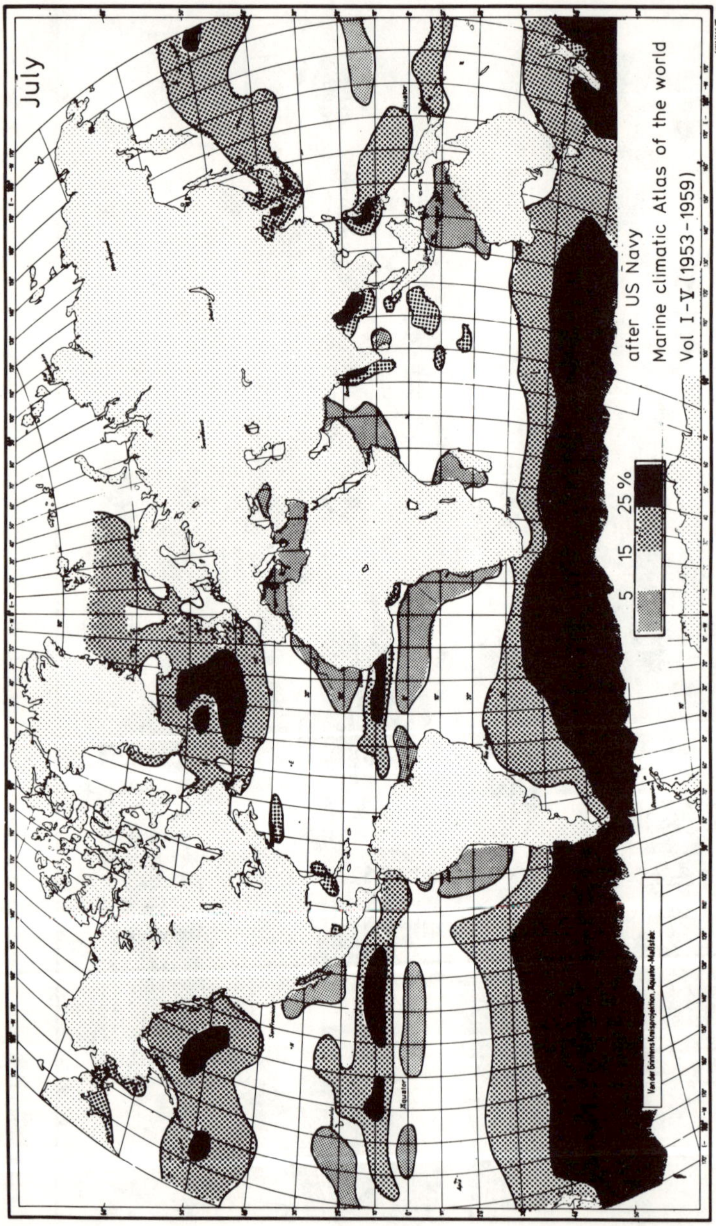

Fig. 41. Mean frequency of precipitation over the oceans in July

ern Caribbean, nearly the whole region along the coast of West Africa north of the equator and westwards to the Cape Verde Islands, the Red Sea, the Arabian Sea, the Gulf of Bengal and the region from the Marianas to Hawaii. Regions with such low frequency in the southern hemisphere are the western coast of South America from Ecuador to central Chile, with a broad westward band just south of the equator, the eastern coast of Argentina, and northern Brazil, from where a band runs to the African coast, nearly all of which falls within the region, together with a wide belt on the coast of western Australia, and the Great Australian Bight.

On the July map (Fig. 41), the dry regions of the northern hemisphere on the coast of America and Africa are farther north, and smaller than in January. They now also include the Mediterranean and the Black Sea. In the Indian Ocean, they are reduced to a small area east of Sri Lanka. The central Pacific dry region, from the Marianas to Hawaii, is also smaller than in January. In the southern hemisphere, the dry regions are smaller and further north than in January. Between Africa and Madagascar there is another dry region with under 5% rain frequency. Within these seasonal oscillations of the dry regions, some marine zones remain poor in rain for the whole year. They almost coincide with the regions having a high evaporation surplus.

The process of water evaporation, which is so important for all the meteorological processes of the globe, is most effective in the dry regions of the oceans, while it is absent in the neighbouring desert regions of the continents. A characteristic of these deserts is the lack of clouds. However, where cold sea-water wells up on the coast, hazy weather results, with the formation of a closed sheet of cloud (called "Garua" on the coast of Peru, where this phenomenon can best be observed). This is to be found in South America over the coastal desert of Peru and the Atacama Desert, the Namib Desert (bordering on to the Kalahari further inland) in South Africa, and, in much weaker form, on part of the western and southern coast of Australia reached by the wide dry region of the interior. Similar conditions are found on the coast of California where it is met by the dry region of North America.

The northern desert belt of the Old World, extending from the coast of Mauretania to Mongolia, includes the Sahara, Arabia, the Lut Desert in Iran, and the Aralo-Caspian and central Asian deserts. Where air heated above hot deserts is transported in a short distance to the sea in places where lack of clouds cause it to receive solar radiation almost unreduced, the evaporation on the surface of the sea is the most effective, and the air is most abundantly supplied with water vapour. This happens in the marine regions around the Arabian peninsula. The Red Sea, the Gulf of Aden, the Persian Gulf and the Gulf of Oman are notorious because of the great heat. Aboard ship, this heat is further aggravated by the heating of the body of the ship itself.

4. Climatic Descriptions of the Countries of the World

4.0. Introductory remarks

4.0.1. General

Any human action needs knowledge that transcends scientific laws and data. Inasfar as this knowledge refers to general weather conditions at different places on earth, they will be treated here in principle for the various countries of the earth in such a way that their characteristic features become obvious.

For rapid reference, the various forms of general weather in the different seasons are explained and special weather phenomena are mentioned, as described in the introduction. According to the mean heat conditions of the climate concerned, tips for clothing, and also an account of heat stress, are given for every month in the climatic tables, indicated by a code letter. For the thoughtful reader who has time and leisure and is eager to know more about the weather and the general circulation of the atmosphere, the predominant general weather situations are described; he can then estimate the mean behaviour of the climatic elements of temperature and precipitation, in the light of their mean values for the months and the year. These form the basis of the climate classification explained in detail in the preceding chapter.

One aim of climatology would be achieved if classification were carried out for every place on earth; then the "meteorological environment" of each place would be known and described. In most cases, however, the user – especially the tourist – is too much of a layman to have sufficient imaginative power to conceive the mean and regular course of the general weather at any given place. The attempt is therefore made here to analyse this course in greater detail. In the beginning there is always the weather as observed by our senses, and it remains paramount if we apply climatology to practical human requirements – leaving aside purely scientific considerations here.

4.0.2. Arrangement of the descriptions

This chapter is arranged according to the continents, as follows: 1. Europe, 2. Asia, 3. Africa, 4. North America, 5. South America, 6. Australia and Oceania.

In each of the six sub-chapters (4.1.–4.6.), the predominant general weather situations are described and hygrothermal diagrams of selected stations are given for parts of the continents. These diagrams show mean daily extreme values of air temperature and humidity (vapour pressure) as described on pages 29–30. The dotted curve on the left indicates the mean daily minima of temperature and the corresponding maxima of vapour pressure, the values for which have mostly had to be estimated. The unbroken line on the right indicates the corresponding temperature maxima and vapour pressure minima. If the two elements of temperature and vapour pressure have a marked annual course, the conditions are shown for every month by a corresponding letter. "J" indicates conditions for January. In the mean, the hourly values of the elements of many days in January are placed between the "J" of the dotted curve and the "J" of the unbroken curve. As the months June and July also begin with J, they are assigned the letters U and L; furthermore, G is assigned to August and R to March. All the other months are designated by their initials.

Each sub-chapter is divided into sections, of which there are 140 altogether (4.1.1.–4.6.140), according to the number of countries or territories for which climatic descriptions are given. This number depended on the data available, and also on the variety of the climatic conditions of the continents concerned.

For each country or territory, the climatic descriptions and tables are given for every climatic area to be found in it. This means that there are as many parts of the first sub-section as there are climatic areas. These areas are not separated by fixed boundaries, but by fluid ones. The second sub-section indicates which of the predominant general weather situations, and which hygrothermal diagrams, apply to the country or region.

In addition, local winds will be mentioned if their names characterise certain features of the climate. Most of them are taken from the excellent book *The Winds of the Earth and their Names* (in Germany) by *Heinz Schamp*, published by Franz Steiner, Wiesbaden 1964, in the series "Geographical Knowledge". This can be recommended to those interested for study. Another book deserving of attention, and from which some names of winds are quoted, is the *Meteorological Glossary* published by the Meteorological Office, London.

For each area, four types of information are given:

1. A description of the climats according to the climate classification.
2. General heat conditions, characterised for summer and winter (in equatorial regions for high and low sun periods) according to the sensation of warmth experienced by people from temperate latitudes, and divided into severely hot, very hot, hot, warm, mild, cool, cold, very cold, severely cold, and excessively cold.
3. Predominant form of general weather in the various seasons, normally characterised as follows:
   dry and sunny; dry and cloudy; dry and overcast; some precipitation and cloudy; some precipitation and overcast; variable and cloudy; variable and overcast; rainy and overcast.
   Especially in temperate climates, the following forms often alternate in rapid cycles: dry and sunny, dry and cloudy, rainy and overcast, showers and variable cloud, few showers and cloudy, dry and cloudy or sunny. For brevity's sake, we shall term these series of forms of general weather "variable general weather", and indicate them under 3. with the notion "variable".
4. Special weather phenomena: if the frequency of their occurrence is noteworthy, thunderstorms, fog, wind or storm, and snow cover will be mentioned.

By looking at the corresponding world map, it will often be possible to form a picture of the conditions of the area in question, and to compare them with conditions in the reader's home environment (which is strongly to be recommended). In addition to Fig. 1, which shows the world distribution of thunderstorms, Fig. 42 shows the "Global Distribution of Fog", and Figs. 43–45 show global maps of the probability of snow cover of 2.5 cm or more at the end of the months of September, January and May. Fig. 42 has been re-drawn according to a map published in the German geographical handbook *General Climate Geography* (Fig. 37, p. 152) by *Blüthgen* 1966, and Figs. 43–45 are drawn according to the maps published by *R.R. Dickson* and *J. Posey* in the monthly *Weather Review* of the US Weather Bureau, No. 95 (1967), p. 347–353.

To continue with the contents of the first subsection, there will follow the climatic tables for all stations situated in the area. Each table is organised according to this schedule: at the top of each table are given the name of the station and its geographical coordinates, latitude, longitude and height. The names of the stations are given as provided by the countries to the World Meteorological Organisation, and published in its publication WMO No. 117, TP 52.

Several names are given as well in English spelling in the added "List of Stations in alphabetical Order" (see Annex I). – Below the name is the climatic formula (e.g. see at the end of sub-chapter 3.1., p. 85) and its table-number consisting of the number of the country or territory and the place of the station in it.

Finally there are added two world maps fig. 46 and 47 showing the sea surface temperature for the months of extreme values in most areas: February and August. The maps are re-drawn according to a map of A. Defant.

In the first three data lines indicated by tx, t, and tn, mean values are given: tx of the monthly and yearly temperature maxima, t of monthly and yearly temperature, tn of monthly and yearly temperature minima. Mean monthly values of temperature extremes are of greater value than absolute extremes, which can be expected to occur very seldom. The fourth line, called "clothing", indicates what kind of clothing is to be recommended, in accordance with the considerations put forward under section 2.3.2 "Clothing", and with regard to the mean monthly heat conditions.

p stands for polar clothing: fur lining provides very high insulation against loss of heat in the icy cold;

w stands for winter clothing: thick closely-knit woollen materials provide sufficient insulation against loss of heat at temperatures below and around freezing point;

n stands for normal clothing: normally used in spring and autumn in temperate zones;

s stands for summer clothing: provides little insulation against heat loss;

t stands for tropical clothing: very light and thin clothing for great heat. In cases of great sultriness, underclothes – best made from natual fibres – should be changed, according to the heat conditions, several times a day in some regions. It is furthermore recommended to wear:

1. a light hat as protection from solar radiation, and
2. light woollen clothing as protection against loss of heat at night, if this is to be expected.

The fifth line, called "stress", indicates in which months great heat stress is to be expected, as a result of sultriness or great heat. Readers are strongly advised not only to note the data for the month in question, but always to consider those for the months before and after. Given the variability of general weather conditions, this is valid not only for stress and clothing indications, but for all monthly data in the tables. In the stress line,

o stands for "occassionally": stress occurs only for a short period during the day-time, generally early in the afternoon;

f stands for "frequently": stress occurs frequently, often for several hours, in the daytime;

ff stands for "very frequently": stress occurs very frequently, more often than not also during the night, or permanently both day and night, generally with a maximum in the afternoon.

The indications for clothing and stress cannot be regarded as valid for any and every person. Human sensitivity to heat varies considerably from one individual to the next. Under given heat conditions, one man will feel quite comfortable, another would prefer it hotter, while a third is already suffering from heat stress. But the assumption is made that the majority of people respond similarly to the indications.

The sixth line – designated "p" – contains mean amounts of precipitation in millimetres (mm), and the seventh (dp) the mean number of days with precipitation equal to or greater than 1.0 mm. If there is a limit other than 1.0 mm, it is put just before dp. The eighth line – marked "sunshine" – contains the mean monthly and yearly values for the total duration of sunshine in hours, and the last line – designated "%" – the percental ratio of the duration given in line 8 to the duration on cloudless days.

Sources of the data:

tx and tn: MOL for all data (except a few German data)
t, p, dp, and sunshine: mostly CLINO, some MOL, a few WWR and NMS.
clothing and stress: the author's definitions.

CLINO stands for *Climatological Normals*, published by the World Meteorological Organization, Geneva (Switzerland), 1971 in its publication WMO No. 117, TP 52.
MOL stands for Meteorological Office, London, which since 1958 has published *Tables of Temperature, Relative Humidity and Precipitation for the World*, Parts I–VI. The newly-edited Part III, Europe and the Azores, 1972, also contains sunshine data.
WWR stands for *World Weather Records 1951–60*, Washington 1966.
NMS stands for individual publications of the National Meteorological Services.

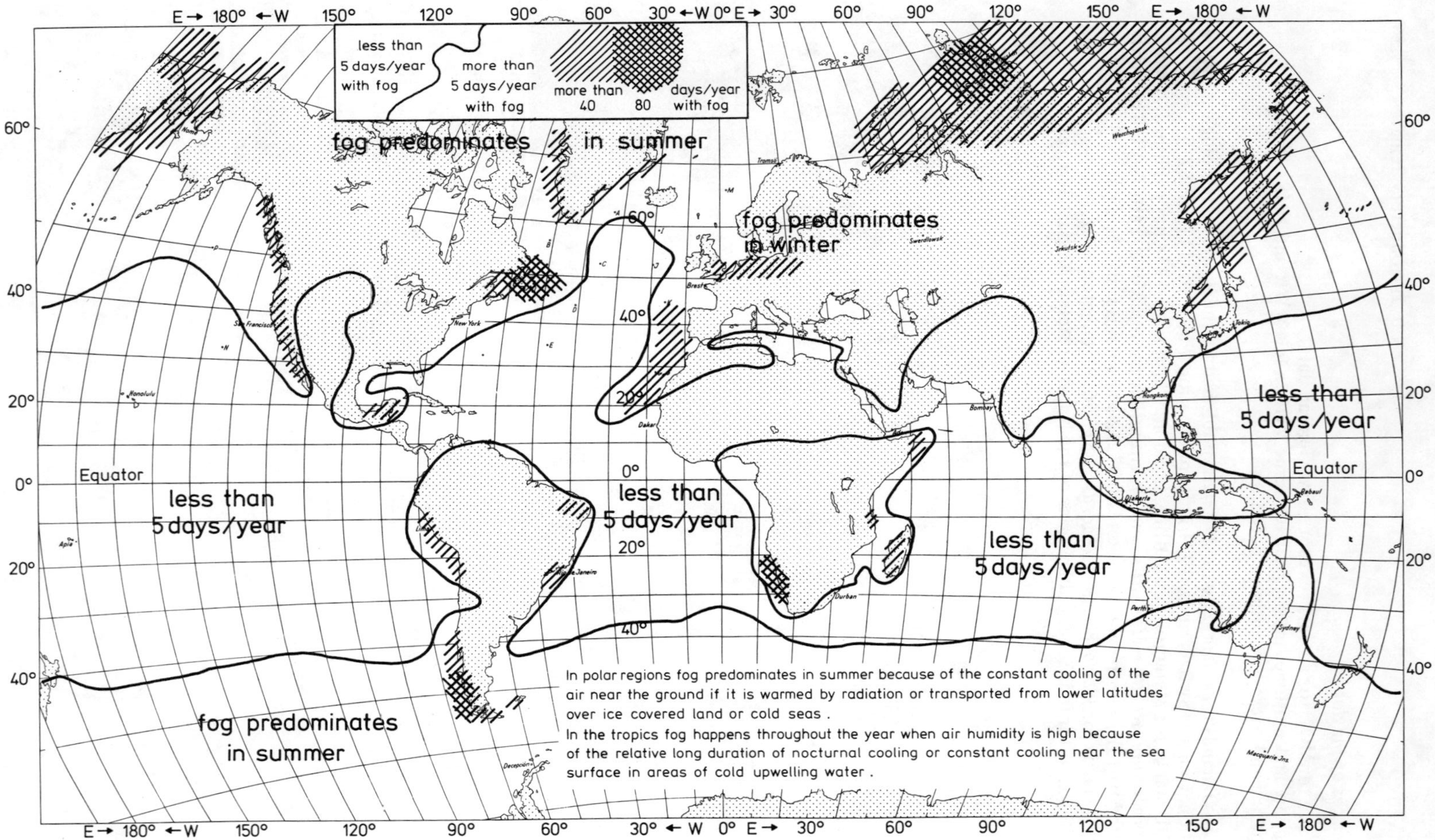

Fig. 42. Global distribution of fog

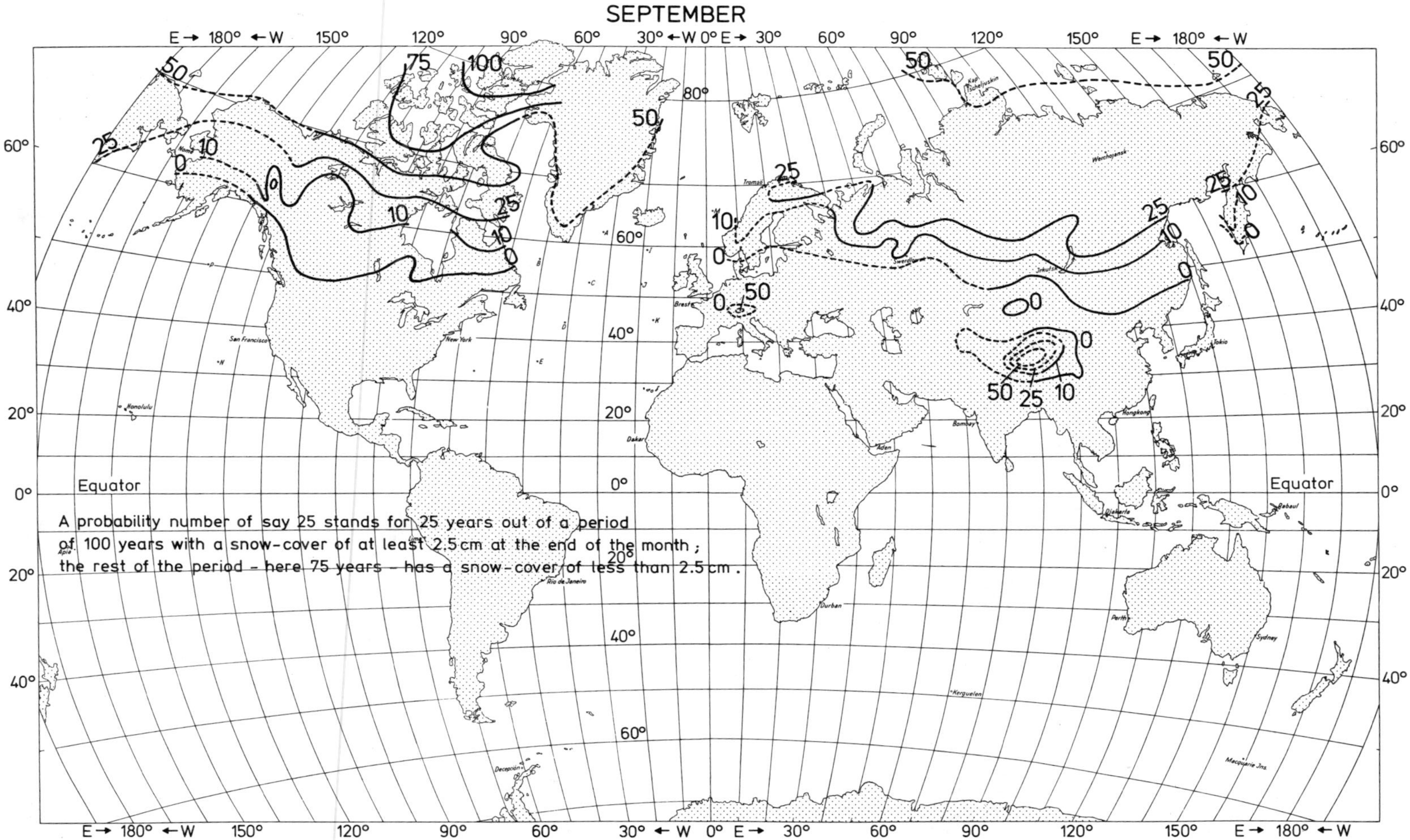

Fig. 43. Snow cover probability in September

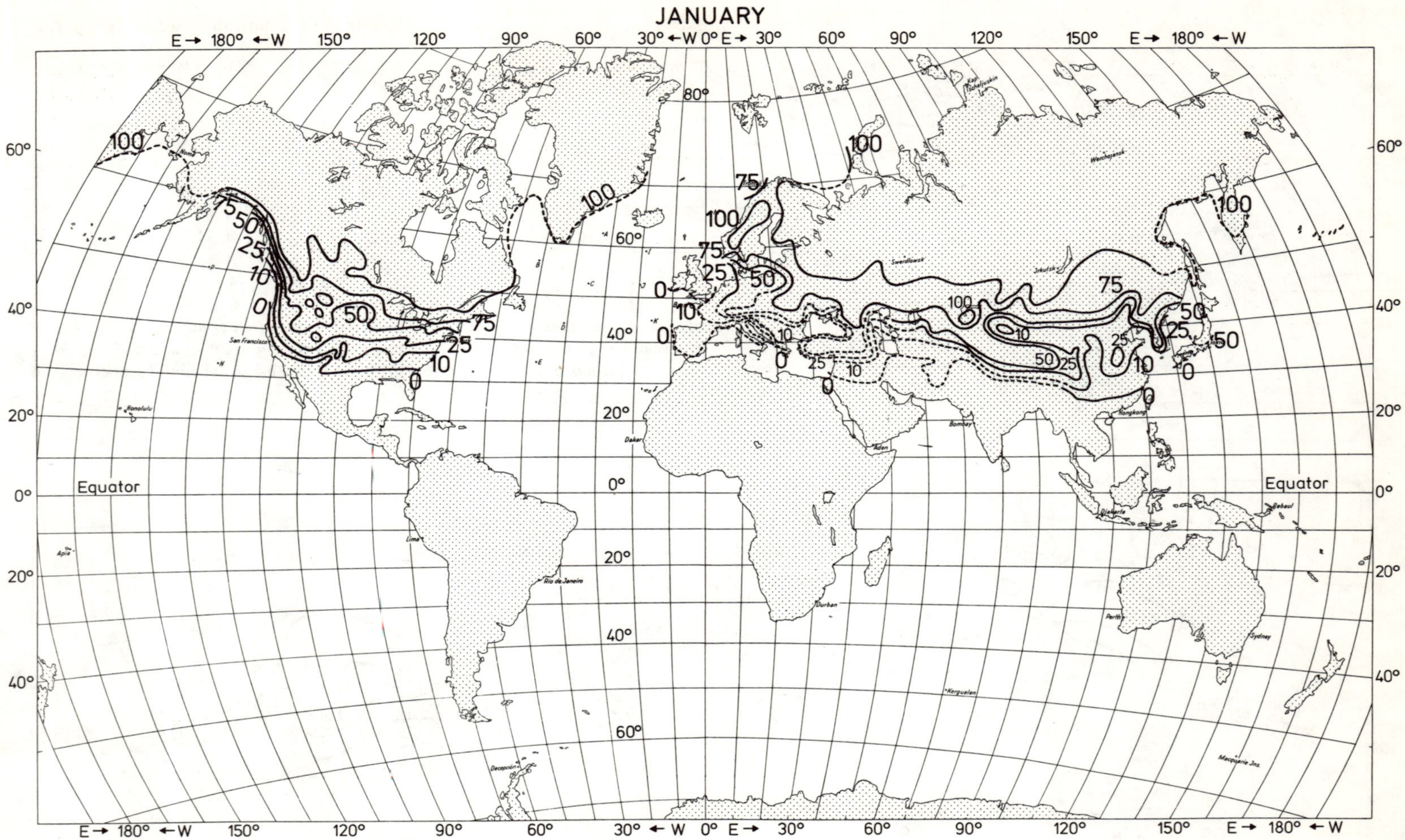

Fig. 44. Snow cover probability in January

MAY

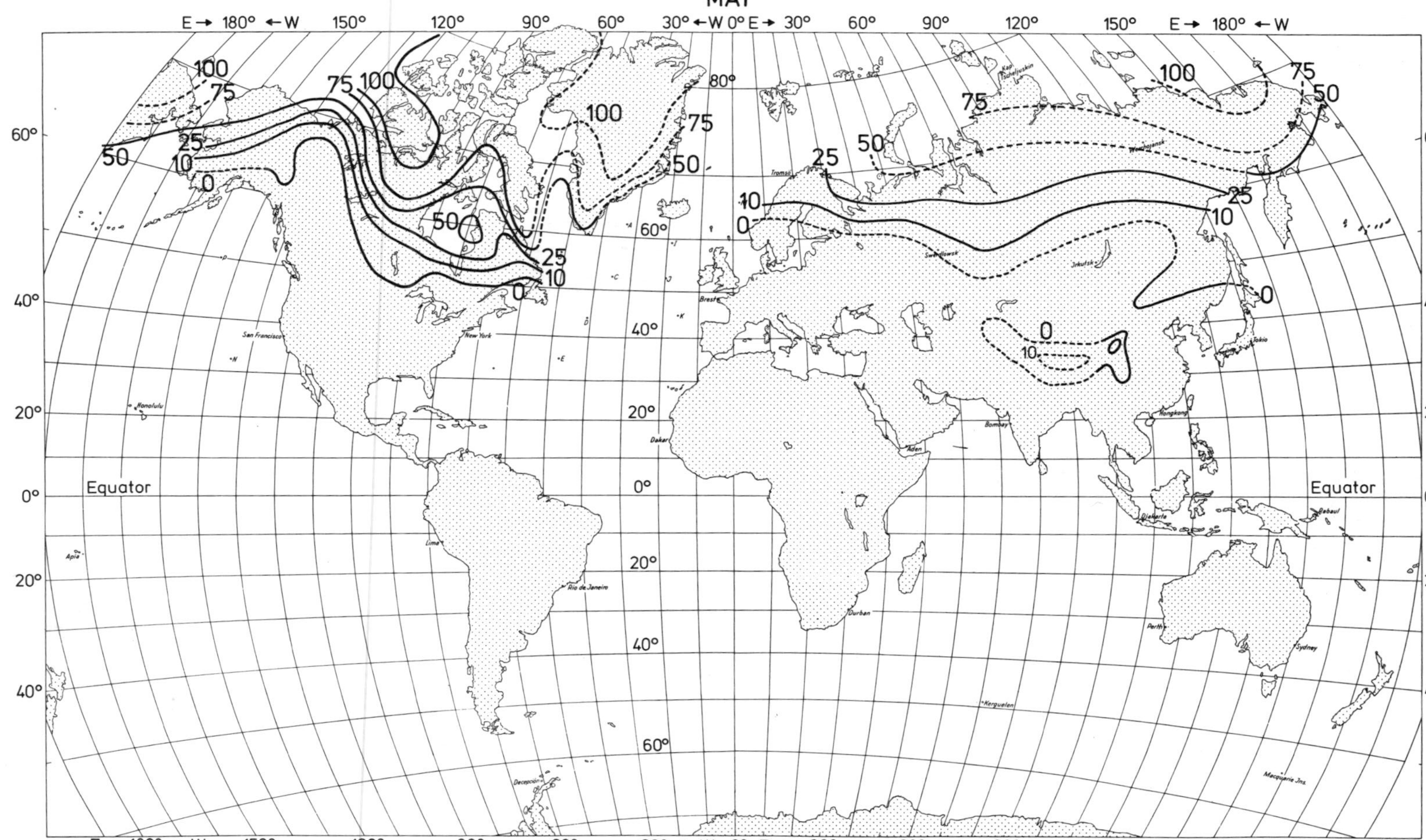

Fig. 45. Snow cover probability in May

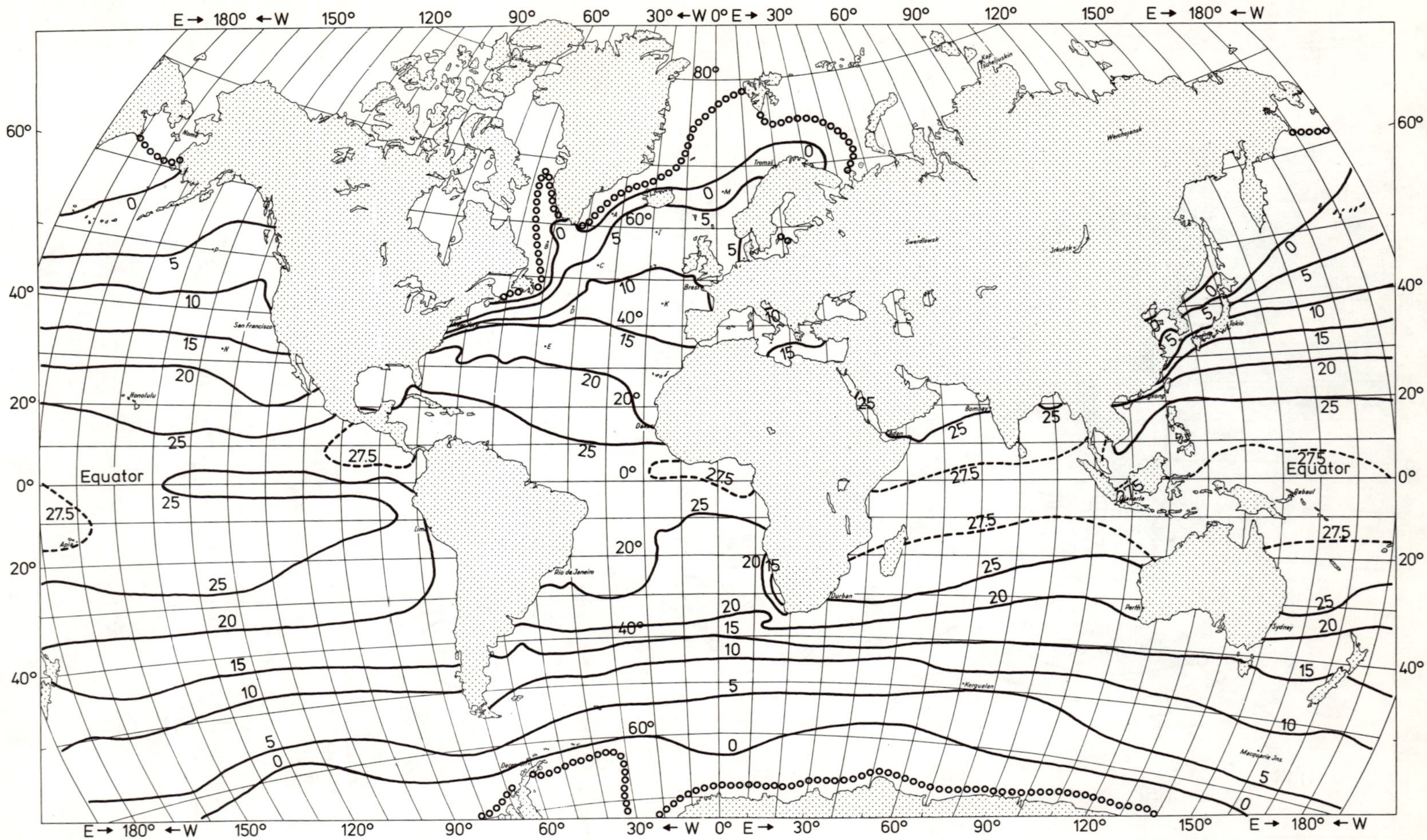

Fig. 46. Mean sea surface temperature in February

—— isotherm (figures indicate °C and are on the side towards higher temperature)

oooo pack-ice boundary zone

(after: DEFANT 1961, Phys. Oceanography Vol I, plate 3A)

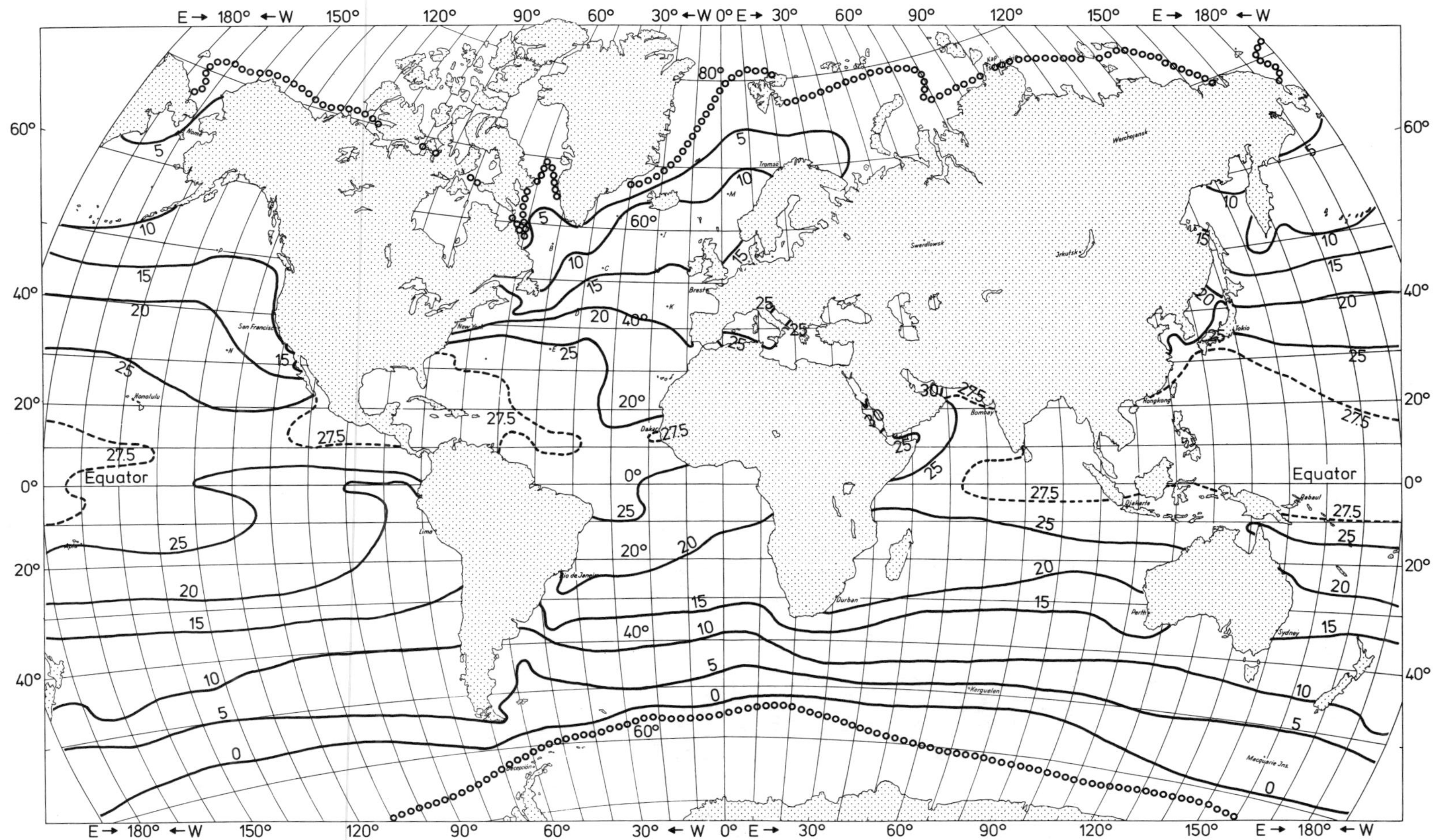

Fig. 47. Mean sea surface temperature in August

—— isotherm (figures indicate °C and are on the side towards higher temperature)
oooo pack-ice boundary zone

(after: DEFANT 1961, Phys. Oceanography Vol I, plate 3B)

4.1. Europe (including Asian USSR)

4.1.0.1. Predominant general weather situations

Iceland: In summer, the Icelandic depression is generally weakly developed, so that periods of sinking air often last longer than those of rising air. In winter, periods of depressions tracking east or north-east last for weeks. They are replaced now and then by short periods of sinking air influenced by an anticyclone over Greenland, or at times by a wedge of the Azores anticyclone.

Northern, Central and Eastern Europe: All the year round, unsettled weather is caused by depressions or secondary depressions normally migrating from west to east and by following wedges or anticyclones. Sometimes, however, a region of low or high pressure can be stationary over a wide area, resulting in uniform weather there for some days or even weeks.

Southern Europe (Portugal, Spain, Gibraltar, Southern France, Italy, Malta, Albania, Greece): In summer, a wedge of high pressure over Southern Europe causes sinking air over the Mediterranean region. At times, secondary depressions from depressions tracking farther north reach the region, and now and then part of the area will be influenced for several days by a depression having its origin there. In winter, depressions or secondary depressions often traverse the region from west to east. Wedges following the depressions now and then develop into stationary anticyclones, which determine the weather for some time.

4.1.0.2. Hygrothermal diagrams

There follow hygrothermal diagrams for the stations

Oslo, Stockholm, Helsinki, Kew (London), Dublin, Reijkjavik, Köbenhavn, De Bilt, Bruxelles (Uccle), Berne, Paris, Madrid, Lisboa, Berlin, Wien, Praha, Warszawa, Budapest, Beograd, Bucuresti, Sofia, Roma, Athinai, Moskva.

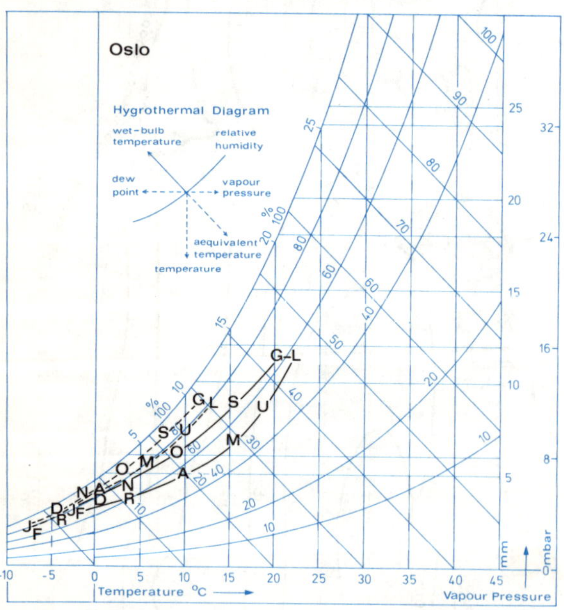

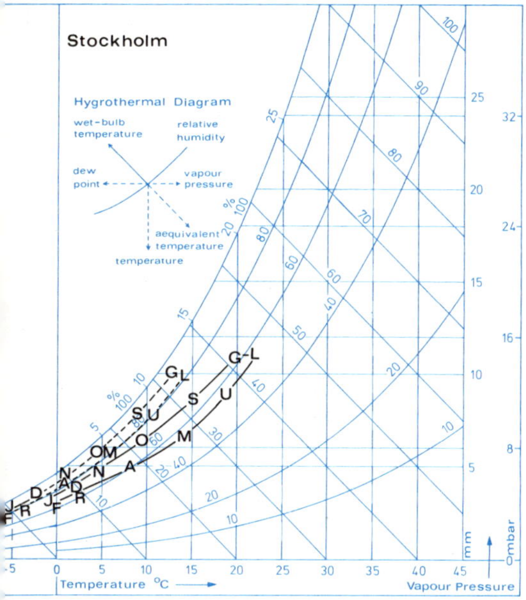
Stockholm
Hygrothermal Diagram
wet-bulb temperature
relative humidity
dew point
vapour pressure
aequivalent temperature
temperature
Temperature °C
Vapour Pressure
mm
mbar

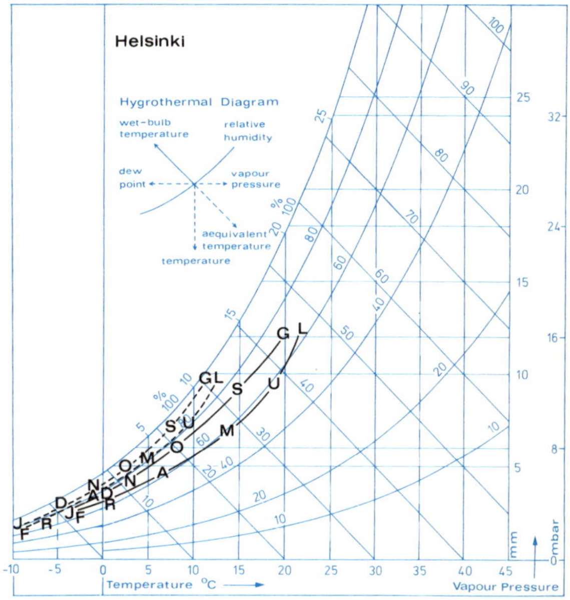
Helsinki
Hygrothermal Diagram
wet-bulb temperature
relative humidity
dew point
vapour pressure
aequivalent temperature
temperature
Temperature °C
Vapour Pressure
mm
mbar

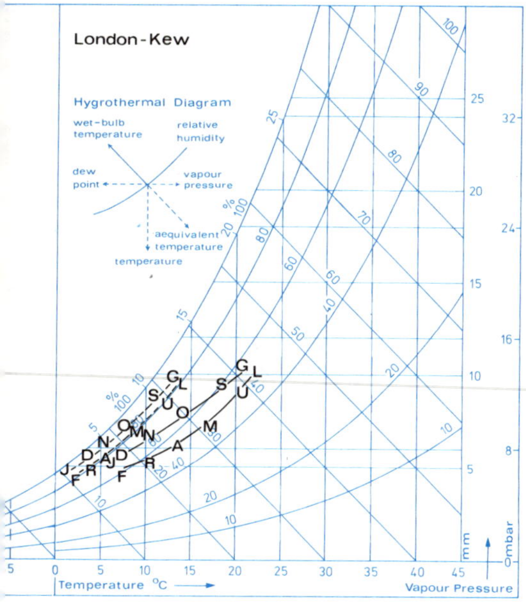
London-Kew
Hygrothermal Diagram
wet-bulb temperature
relative humidity
dew point
vapour pressure
aequivalent temperature
temperature
Temperature °C
Vapour Pressure
mm
mbar

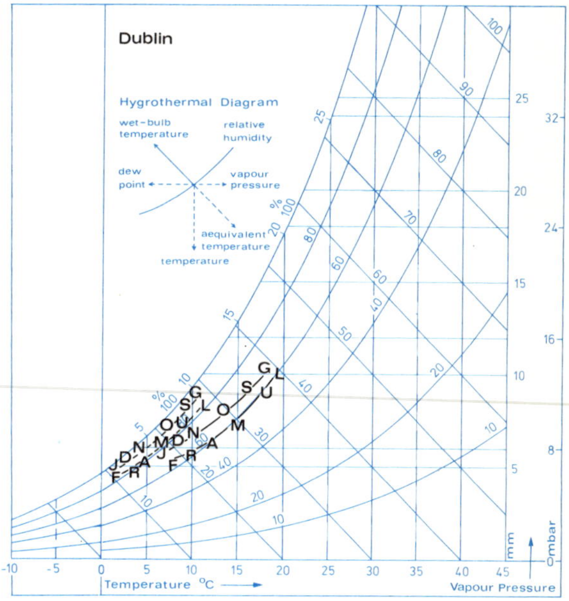
Dublin
Hygrothermal Diagram
wet-bulb temperature
relative humidity
dew point
vapour pressure
aequivalent temperature
temperature
Temperature °C
Vapour Pressure
mm
mbar

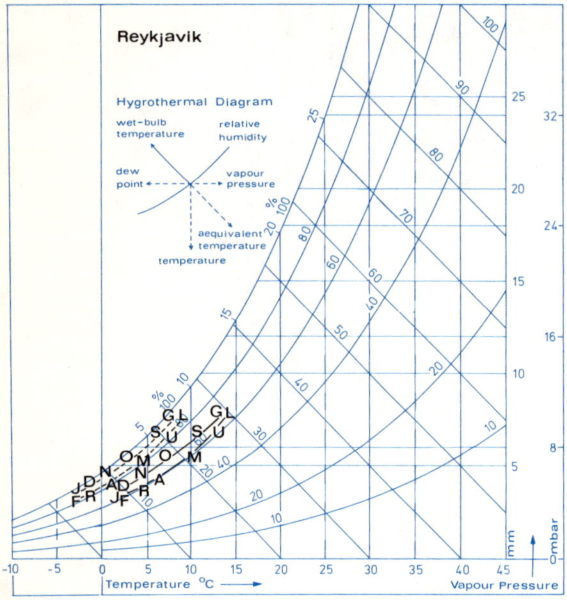
Reykjavik
Hygrothermal Diagram
wet-bulb temperature
relative humidity
dew point
vapour pressure
aequivalent temperature
temperature
Temperature °C
Vapour Pressure
mm
mbar

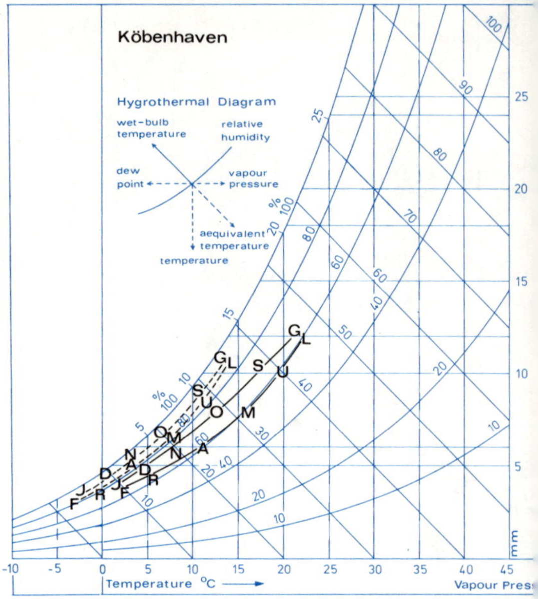
Köbenhaven
Hygrothermal Diagram
wet-bulb temperature
relative humidity
dew point
vapour pressure
aequivalent temperature
temperature
Temperature °C
mm

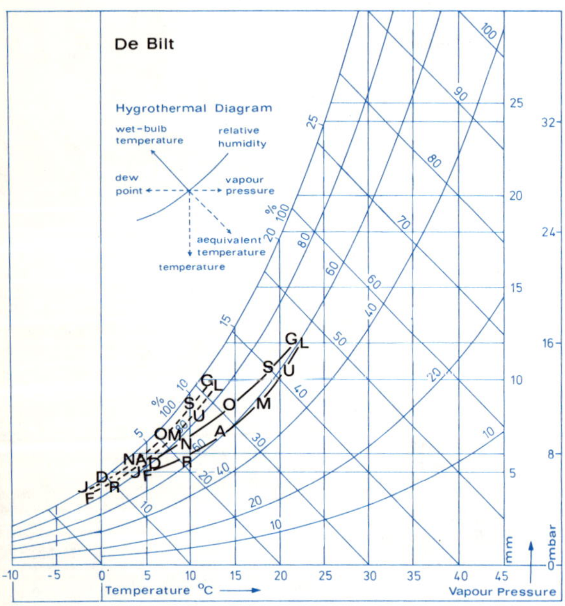
De Bilt
Hygrothermal Diagram
wet-bulb temperature
relative humidity
dew point
vapour pressure
aequivalent temperature
temperature
Temperature °C
Vapour Pressure
mm
mbar

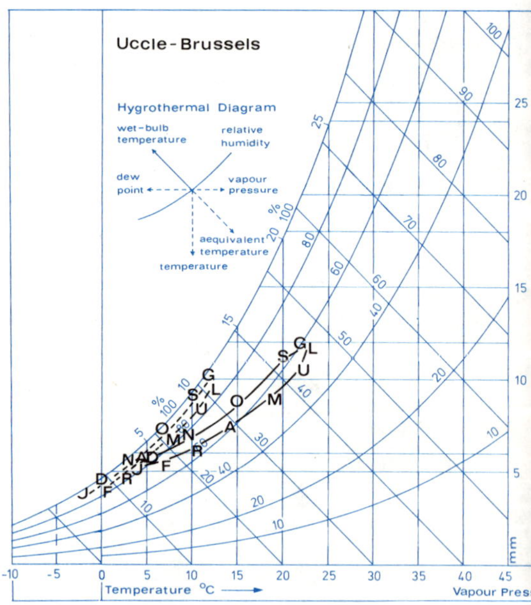
Uccle-Brussels
Hygrothermal Diagram
wet-bulb temperature
relative humidity
dew point
vapour pressure
aequivalent temperature
temperature
Temperature °C
mm

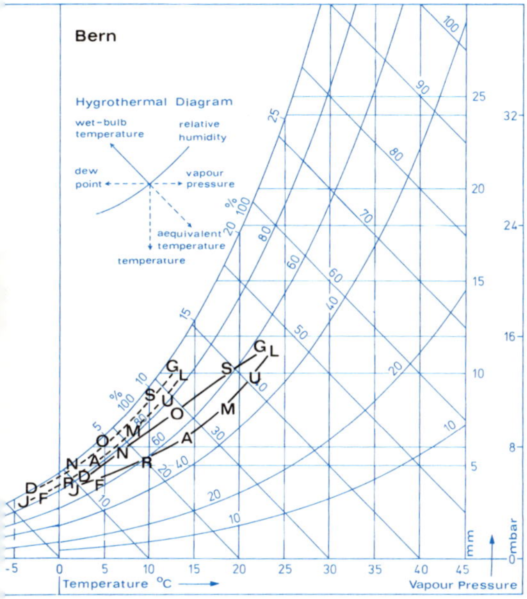
Bern
Hygrothermal Diagram
wet-bulb temperature
relative humidity
dew point
vapour pressure
aequivalent temperature
temperature
Temperature °C
Vapour Pressure
mm
mbar

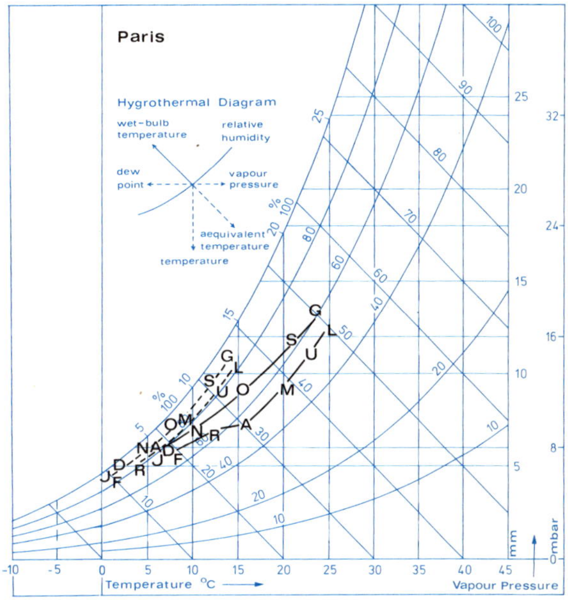
Paris
Hygrothermal Diagram
wet-bulb temperature
relative humidity
dew point
vapour pressure
aequivalent temperature
temperature
Temperature °C
Vapour Pressure
mm
mbar

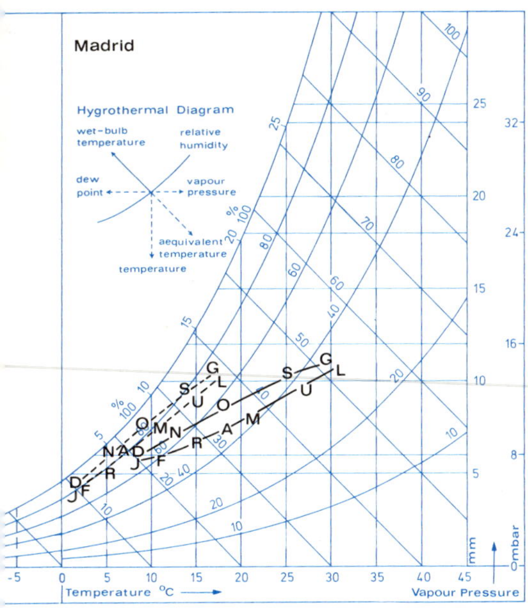
Madrid
Hygrothermal Diagram
wet-bulb temperature
relative humidity
dew point
vapour pressure
aequivalent temperature
temperature
Temperature °C
Vapour Pressure
mm
mbar

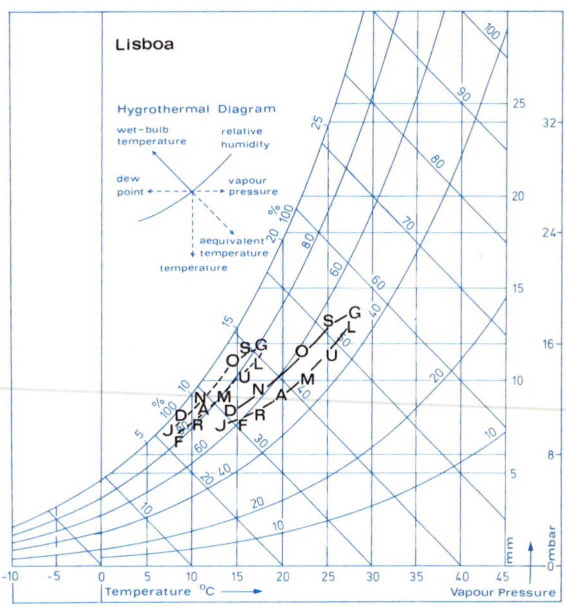
Lisboa
Hygrothermal Diagram
wet-bulb temperature
relative humidity
dew point
vapour pressure
aequivalent temperature
temperature
Temperature °C
Vapour Pressure
mm
mbar

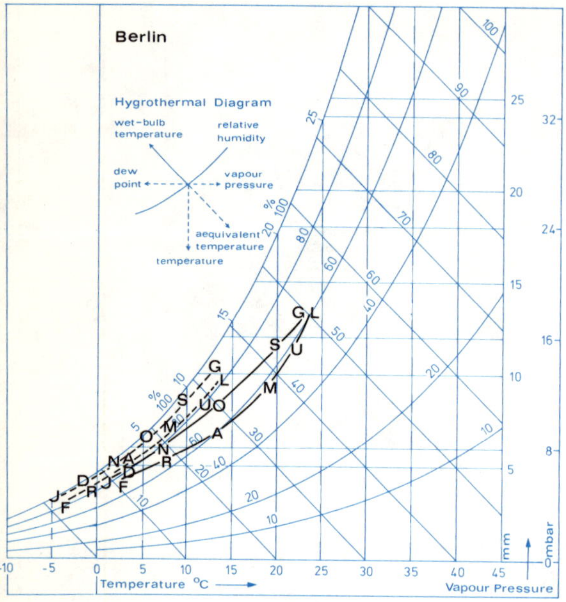
Berlin
Hygrothermal Diagram
wet-bulb temperature
relative humidity
dew point
vapour pressure
aequivalent temperature
temperature
Temperature °C
Vapour Pressure
mm
mbar

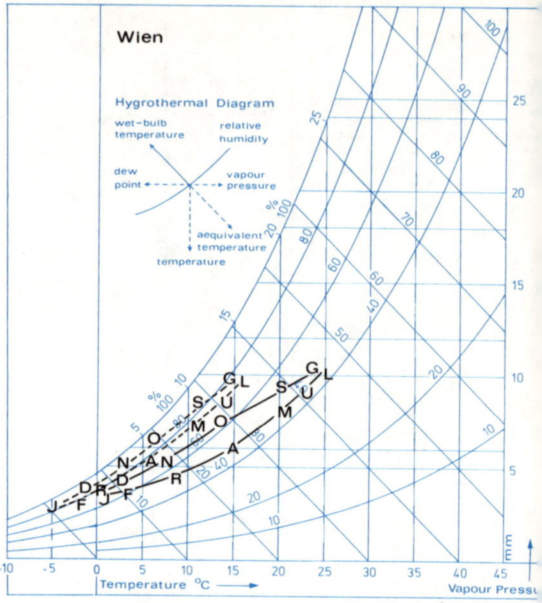
Wien
Hygrothermal Diagram
wet-bulb temperature
relative humidity
dew point
vapour pressure
aequivalent temperature
temperature
Temperature °C
mm

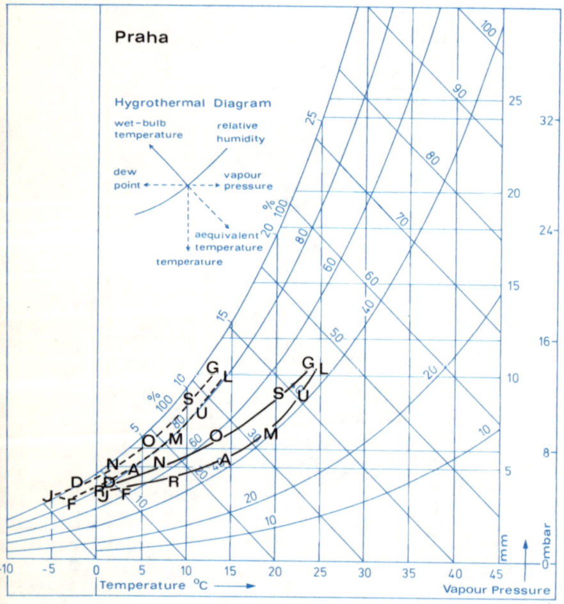
Praha
Hygrothermal Diagram
wet-bulb temperature
relative humidity
dew point
vapour pressure
aequivalent temperature
temperature
Temperature °C
Vapour Pressure
mm
mbar

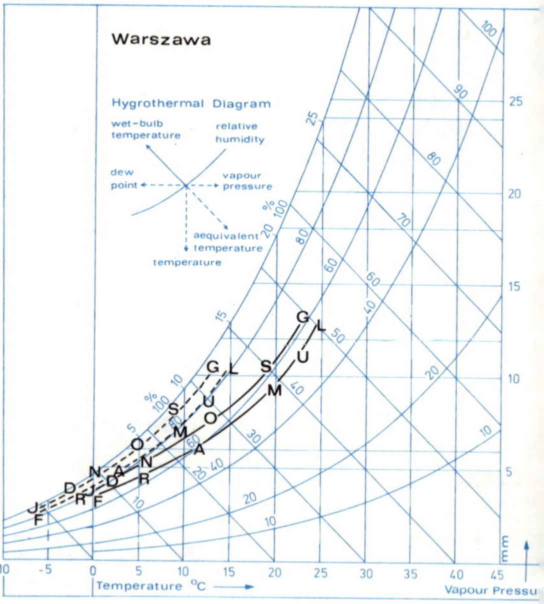
Warszawa
Hygrothermal Diagram
wet-bulb temperature
relative humidity
dew point
vapour pressure
aequivalent temperature
temperature
Temperature °C
mm

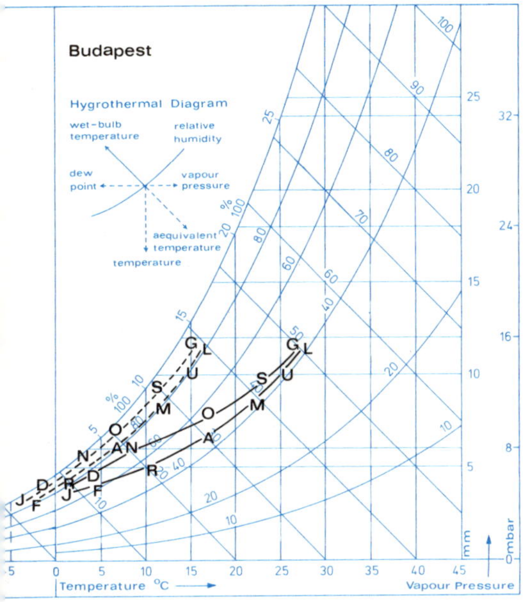
Budapest
Hygrothermal Diagram
wet-bulb temperature
relative humidity
dew point
vapour pressure
aequivalent temperature
temperature
Temperature °C
Vapour Pressure
mm
mbar

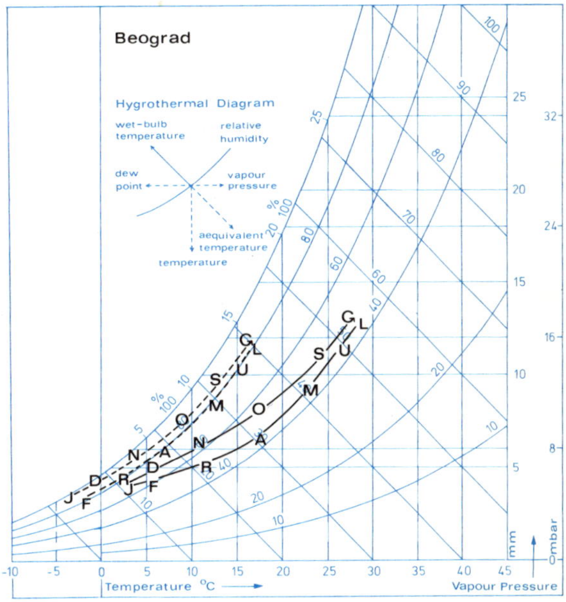
Beograd
Hygrothermal Diagram
wet-bulb temperature
relative humidity
dew point
vapour pressure
aequivalent temperature
temperature
Temperature °C
Vapour Pressure
mm
mbar

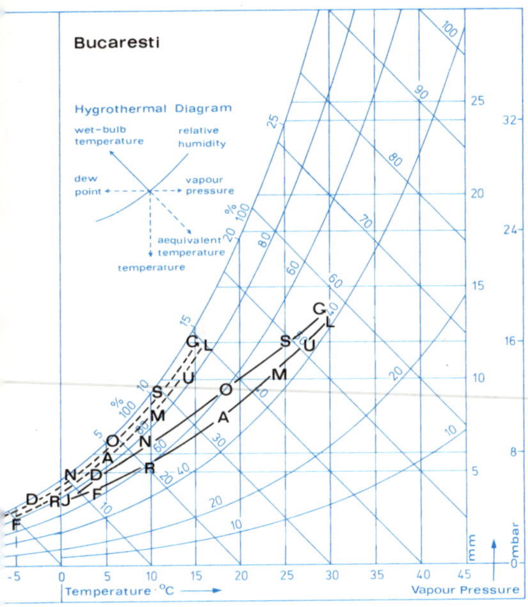
Bucaresti
Hygrothermal Diagram
wet-bulb temperature
relative humidity
dew point
vapour pressure
aequivalent temperature
temperature
Temperature °C
Vapour Pressure
mm
mbar

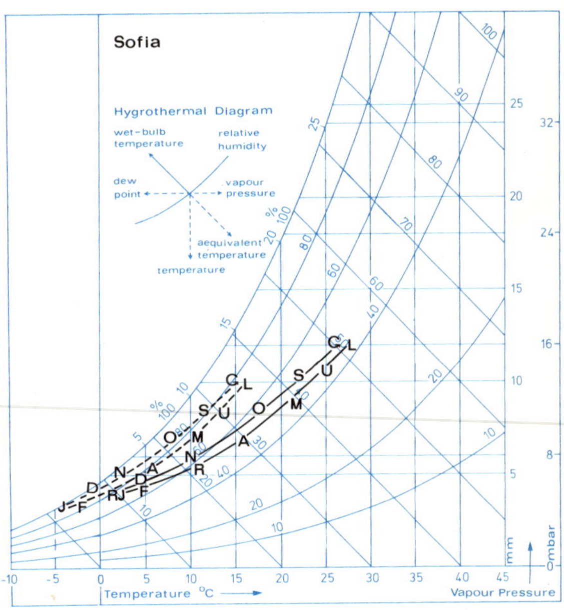
Sofia
Hygrothermal Diagram
wet-bulb temperature
relative humidity
dew point
vapour pressure
aequivalent temperature
temperature
Temperature °C
Vapour Pressure
mm
mbar

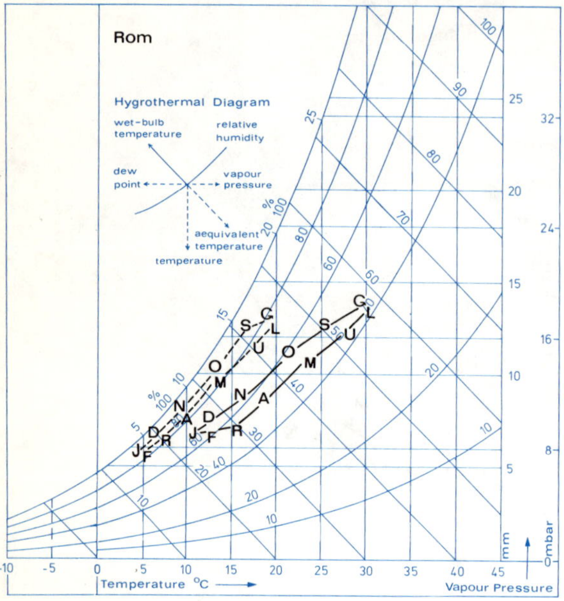
Rom
Hygrothermal Diagram
wet-bulb temperature
relative humidity
dew point
vapour pressure
aequivalent temperature
temperature
Temperature °C
Vapour Pressure
mm
mbar

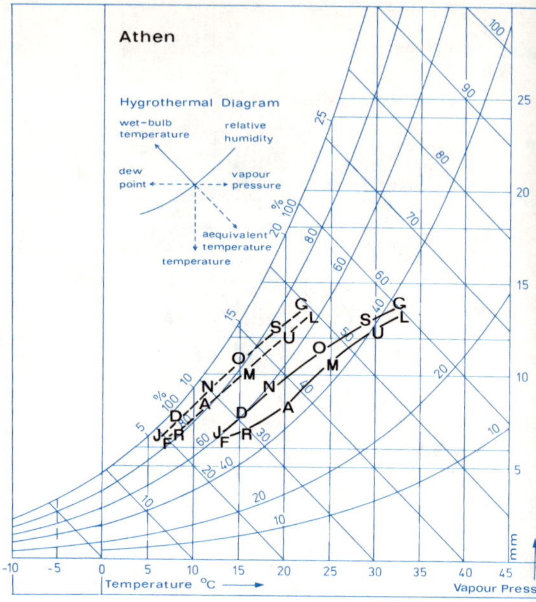
Athen
Hygrothermal Diagram
wet-bulb temperature
relative humidity
dew point
vapour pressure
aequivalent temperature
temperature
Temperature °C
Vapour Press
mm

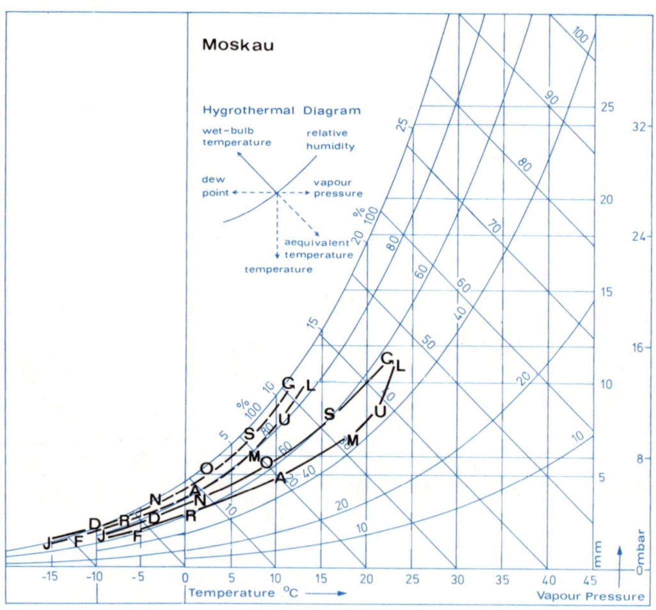
Moskau
Hygrothermal Diagram
wet-bulb temperature
relative humidity
dew point
vapour pressure
aequivalent temperature
temperature
Temperature °C
Vapour Pressure
mm
mbar

4.1.1. Norway

4.1.1.1. Climatic descriptions and tables

1. Climatic area: Islands in the Norwegian Sea
Svalbard (Spitsbergen and Björnöya) and Jan Mayen

1.1. Tundra Climate FT
1.2. Summer brief and cool, winter cold.
1.3. Throughout the year variable general weather predominates.
1.4. Wind: often stormy. Fog: prevailing in summer. Snow cover most of the year.

Isfjord Radio/Spitsbergen

latitude: 78°04'n longitude: 13°38'e height: 9 m

FTkc	001-1	Jan	Feb	Mar	Apr	May	Jun	Jul	Aug	Sep	Oct	Nov	Dec	Year
	tx	1	1	1	2	4	8	11	10	7	5	3	2	12 °C
	t	-10	-10	-12	-8	-3	2	5	4	1	2	5	8	-4 °C
	tn	-21	-23	-23	-21	-11	-3	1	0	-5	-12	-17	-20	-26 °C
	clothing stress	p	p	p	w	w	w	n	w	w	w	w	w	–
	p	22	31	31	16	21	26	32	36	36	44	23	36	354 mm
0.1 mm	dp	13	12	12	10	10	9	11	14	14	13	14	14	146 days
	sunshine	–	–	70	217	245	155	151	151	69	9	–	–	1067 hours
	%	–	–	20	36	33	22	20	20	16	40	–	–	23 %

Björnöya

latitude: 74°31'n longitude: 19°10'e height: 14 m

FTko	001-2	Jan	Feb	Mar	Apr	May	Jun	Jul	Aug	Sep	Oct	Nov	Dec	Year
	tx	3	2	3	3	6	10	12	12	9	7	4	3	14 °C
	t	-5	-6	-7	-5	-1	2	4	5	3	1	-2	-4	-1 °C
	tn	-17	-20	-21	-17	-9	-3	0	0	-2	-7	-11	-15	-23 °C
	clothing stress	w	w	w	w	w	w	w	n	w	w	w	w	–
	p	32	27	30	22	21	28	25	39	47	49	28	27	365 mm
0.1 mm	dp	21	18	20	17	15	14	13	17	21	22	20	21	219 days

Jan Mayen

latitude: 70°57'n longitude: 8°40'w height: 9 m

FTko	001-3	Jan	Feb	Mar	Apr	May	Jun	Jul	Aug	Sep	Oct	Nov	Dec	Year
	tx	4	3	3	4	6	10	12	12	10	8	6	4	13 °C
	t	-4	-5	-5	-3	0	2	5	6	4	1	-1	-3	0 °C
	tn	-14	-16	-16	-13	-8	-2	0	1	-2	-7	-10	-13	-18 °C
	clothing stress	w	w	w	w	w	w	n	n	w	w	w	w	–
	p	79	54	63	58	23	28	36	61	83	93	82	75	735 mm
0.1 mm	dp	22	20	21	19	15	14	16	18	19	21	21	22	228 days

2. *Climatic area:* Provinces (Fylkers) north of 62° northern latitude
2.1. Subarctic Oceanic Climate EO and
Subarctic Oceanic Mountain Climate GEO
2.2. Summer cool in the north, otherwise moderately mild, winter cold.
2.3. Variable general weather prevails throughout the year, on the whole rich in precipitation and clouds.
2.4. Winds: frequent gales, over high ground severe gales frequent.
Fog: frequent in the mountains, at times in coastal areas.
Snow cover: at higher levels throughout the winter.

Vardö

latitude: 70°22'n longitude: 31°06'e height: 15 m

EOlo	001-4	Jan	Feb	Mar	Apr	May	Jun	Jul	Aug	Sep	Oct	Nov	Dec	Year
	tx	3	2	3	6	11	17	19	18	14	9	6	4	21 °C
	t	-4	-5	-4	-1	3	6	9	10	7	2	-1	-3	2 °C
	tn	-14	-14	-13	-9	-4	0	4	4	0	-5	-9	-12	-16 °C
	clothing stress	w	w	w	w	w	n	n	n	n	w	w	w	– –
	p	32	23	28	23	31	33	40	49	52	52	38	31	432 mm
0.1 mm	dp	13	12	13	10	10	9	7	10	13	14	11	12	134 days

Tromsö

latitude: 69°42'n longitude: 19°01'e height: 24 m

EOlo	001-5	Jan	Feb	Mar	Apr	May	Jun	Jul	Aug	Sep	Oct	Nov	Dec	Year
	tx	5	4	5	9	14	22	24	21	16	11	7	6	26 °C
	t	-3	-3	-2	1	5	9	12	11	8	4	0	-1	3 °C
	tn	-12	-12	-11	-8	-3	1	5	4	0	-4	-8	-10	-14 °C
	clothing stress	w	w	w	w	n	n	n	n	n	w	w	w	– –
	p	118	94	113	75	65	57	56	83	115	131	97	115	1119 mm
0.1 mm	dp	12	11	11	12	10	13	13	12	14	14	11	12	144 days
	sunshine	2	35	101	176	172	202	226	176	97	52	7	–	1246 hours
	%	21	17	28	37	25	28	30	31	24	18	6	–	27 %

Bodö

latitude: 67°17'n longitude: 14°25'e height: 13 m

EOlo	001-6	Jan	Feb	Mar	Apr	May	Jun	Jul	Aug	Sep	Oct	Nov	Dec	Year
	tx	6	5	6	10	16	22	25	23	18	12	8	7	26 °C
	t	-2	-2	-1	2	6	10	14	13	9	5	2	0	5 °C
	tn	-12	-11	-11	-7	-2	2	6	5	1	-3	-6	-9	-13 °C
	clothing stress	w	w	w	w	n	n	n	n	n	n	w	w	– –
	p	93	71	74	75	52	72	70	87	123	131	95	107	1050 mm
0.1 mm	dp	14	15	14	12	9	12	11	14	16	17	14	15	163 days
	sunshine	7	45	121	160	177	198	186	181	99	58	21	–	1253 hours
	%	6	20	34	34	29	27	26	34	25	20	13	–	27 %

Orlandet

latitude: 63°42′n longitude: 9°37′e height: 7 m

EOlo	001-7	Jan	Feb	Mar	Apr	May	Jun	Jul	Aug	Sep	Oct	Nov	Dec	Year
	t	-1	-1	1	4	8	11	14	13	10	7	4	1	6 °C
	clothing stress	w	w	w	w	n	n	n	n	n	n	w	w	– –
	p	92	83	81	76	53	59	65	68	111	125	95	91	999 mm
	dp	14	13	13	15	12	12	10	10	16	17	14	14	160 days

Trondheim

latitude: 63°25′n longitude: 10°27′e height: 115 m

EOlo	001-8	Jan	Feb	Mar	Apr	May	Jun	Jul	Aug	Sep	Oct	Nov	Dec	Year
	tx	7	6	9	13	20	24	27	24	20	14	10	8	27 °C
	t	-3	-3	-1	3	8	11	14	13	9	5	2	-1	5 °C
	tn	-17	-15	-13	-6	-1	3	6	5	1	-4	-8	-13	-19 °C
	clothing stress	w	w	w	w	n	n	n	n	n	n	w	w	– –
	p	68	67	67	60	48	66	70	78	92	98	67	76	857 mm
	dp	12	11	11	12	10	13	13	12	14	14	11	12	144 days
	sunshine	19	63	138	162	193	181	182	178	116	72	31	11	1347 hours
	%	11	26	38	36	34	29	30	35	30	23	15	8	29

3. *Climatic area:* Coastal areas south of 63° northern latitude and bordering on western slopes of the mountains

3.1. Temperate Oceanic Climate DO and
Temperate Oceanic Mountain Climate GDO

3.2. Summer mild, winter cool, in the mountains cold increasing with height.

3.3. Variable general weather prevails throughout the year, on the whole rich in precipitation; long dry periods are rare.

3.4. Wind: frequent gales, in winter often severe. Fog: frequent in the mountains, at times in coastal areas. Snow cover in the lowlands not persistent.

Bergen

latitude: 60°24′n longitude: 5°19′e height: 44 m

DOlk	001-9	Jan	Feb	Mar	Apr	May	Jun	Jul	Aug	Sep.	Oct	Nov	Dec	Year
	tx	9	8	12	15	22	24	26	25	21	16	12	10	28 °C
	t	2	1	3	6	10	13	15	15	12	8	6	3	8 °C
	tn	-7	-7	-5	-2	2	5	8	8	5	1	-2	-5	-8 °C
	clothing stress	w	w	w	n	n	n	n	n	n	n	n	w	– –
	p	179	139	109	140	83	126	141	167	228	236	207	203	1958 mm
	dp	16	14	12	14	11	13	15	15	17	19	16	18	180 days
	sunshine	22	54	127	142	191	175	167	132	103	69	29	13	1224 hours
	%	11	21	35	32	36	31	30	27	26	22	13	7	27 %

Stavanger

		latitude: 58°53'n				longitude: 5°28'e				height: 8 m				
DOlk	001-10	Jan	Feb	Mar	Apr	May	Jun	Jul	Aug	Sep	Oct	Nov	Dec	Year
	tx	8	7	11	15	21	24	25	25	20	15	11	9	27 °C
	t	1	0	2	6	10	12	15	15	12	8	5	3	7 °C
	tn	-8	-7	-5	-1	2	6	8	8	6	2	-1	-4	-9 °C
	clothing	w	w	w	n	n	n	n	n	n	n	n	w	–
	stress													–
	p	88	59	38	55	48	69	89	108	122	122	112	105	1015 mm
	dp	14	12	9	11	9	10	11	13	15	17	15	16	152 days
	sunshine	48	79	140	168	226	222	197	159	141	80	45	33	1538 hours
	%	22	30	38	39	43	40	36	33	36	25	19	17	34 %

4. *Climatic Area:* Provinces (Fylkers) Aust Agdar*, Telemark*, Buskerud, Akershus*, Oslo, Östfold*, Hedmark, Opland (except northern mountains, which have GDO Climate).
   * except coastal areas, with DO Climate.

4.1. Temperate Continental Climate DC and Temperate Continental Mountain Climate GDC.
4.2. Summer mild, winter cold increasing with height in mountains.
4.3. Variable general weather prevails throughout the year, yet dry periods and also sunny days are not lacking.
4.4. Wind: occasional gales, possible at all seasons.
Fog: at times. Snow cover in winter.

Oslo

		latitude: 59°56'n				longitude: 10°44'e				height: 96 m				
DClo	001-11	Jan	Feb	Mar	Apr	May	Jun	Jul	Aug	Sep	Oct	Nov	Dec	Year
	tx	6	7	11	17	24	27	28	26	21	16	10	7	30 °C
	t	-5	-4	0	5	11	15	17	16	11	6	1	-2	6 °C
	tn	-16	-16	-12	-5	0	4	8	7	1	-3	-8	-14	-18 °C
	clothing	w	w	w	n	n	n	s	n	n	n	w	w	–
	stress							o						–
	p	49	35	26	44	44	71	84	96	83	76	69	63	740 mm
	dp	10	9	5	8	8	10	12	12	11	11	12	12	120 days
	sunshine	45	83	152	182	233	244	219	138	183	112	41	25	1657 hours
	%	22	32	42	42	44	43	39	28	47	35	18	13	37

4.1.1.2. Predominant general weather situations are described in 4.1.0.1. under Northern, Central and Eastern Europe. A hygrothermal diagram of Oslo is to be found in 4.1.0.2.
Wind names: Elvegust, also called Sno, is a katabatic wind in the inner parts of the Norwegian Fjords. (Katabatic wind is a local cold wind caused by the gravitation of cold air falling off high ground.) It produces spraying foam – called havrög – by its down-gusts.

4.1.2. Sweden

4.1.2.1. Climatic descriptions and tables

1. Climatic area: northern part of the country: on the Baltic Sea to the province (Län) of Västerbotten, in the mountainous uplands and in the west southward to Värmland

1.1. Subarctic continental Climate EC and
Subarctic Continental Mountain Climate GEC

1.2. Summer moderately mild, winter very cold

1.3. Variable weather prevails throughout the year. Yet dry periods and sunny days are not lacking.

1.4. Wind: especially in coastal areas frequent gales, prevailing in winter. Snow cover: in winter.

Karesuando

latitude: 68°27'n longitude: 22°30'e height: 327 m

EClc	002-1	Jan	Feb	Mar	Apr	May	Jun	Jul	Aug	Sep	Oct	Nov	Dec	Year
	tx	2	0	5	9	16	24	27	23	17	9	4	2	28 °C
	t	-14	-14	-10	-4	3	10	14	11	5	-1	-7	-11	-2 °C
	tn	-33	-33	-31	-22	-8	0	4	0	-5	-16	-25	-30	-37 °C
	clothing	p	p	p	w	w	n	n	n	n	w	w	p	–
	stress													–
	p	19	18	17	19	26	46	63	57	41	26	26	22	380 mm
	dp	6	6	5	5	6	9	9	10	8	6	7	7	84 days
	sunshine	2	46	137	201	225	272	287	155	134	87	22	0	1568 hours
	%	2	21	38	42	35	38	39	28	33	30	15	–	34 %

Haparanda

latitude: 65°50'n longitude: 24°09'e height: 7 m

EClc	002-2	Jan	Feb	Mar	Apr	May	Jun	Jul	Aug	Sep	Oct	Nov	Dec	Year
	tx	2	1	6	10	19	24	28	24	18	11	6	3	28 °C
	t	-11	-11	-7	-1	6	12	16	14	8	2	-3	-7	2 °C
	tn	-30	-28	-27	-17	-4	2	6	3	-3	-10	-17	-24	-32 °C
	clothing	p	p	w	w	n	n	n	n	n	w	w	w	–
	stress													–
	p	40	36	24	34	30	41	54	71	65	53	58	46	552 mm
	dp	9	8	5	8	6	7	8	8	9	10	12	11	100 days
	sunshine	24	50	151	209	269	299	308	214	154	98	34	8	1818 hours
	%	17	21	42	45	45	42	47	41	39	32	19	8	39 %

Stensele

latitude: 65°04'n longitude: 17°10'e height: 330 m

EClc	002-3	Jan	Feb	Mar	Apr	May	Jun	Jul	Aug	Sep	Oct	Nov	Dec	Year
	tx	3	3	7	11	19	24	26	23	18	11	5	3	27 °C
	t	-12	-11	-7	0	6	11	14	12	7	1	-4	-8	1 °C
	tn	-32	-31	-28	-19	-6	-1	2	0	-4	-12	-20	-27	-35 °C
	clothing	p	p	w	w	n	n	n	n	n	w	w	w	–
	stress													–
	p	30	23	21	25	33	57	80	67	47	37	38	36	494 mm
	dp	8	7	5	6	6	9	11	10	9	8	9	9	96 days

Östersund

latitude: 63° 11′n longitude: 14° 30′e height: 366 m

EOlo	002-4	Jan	Feb	Mar	Apr	May	Jun	Jul	Aug	Sep	Oct	Nov	Dec	Year
	tx	5	4	8	14	21	26	28	25	19	13	7	5	28 °C
	t	-8	-7	-4	1	7	11	15	13	9	4	-1	-4	3 °C
	tn	-26	-23	-20	-11	-3	1	5	4	-1	-7	-12	-19	-28 °C
	clothing	w	w	w	w	n	n	n	n	n	w	w	w	–
	stress													–
	p	34	23	23	29	31	69	77	74	51	43	42	36	532 mm
	dp	9	6	6	6	6	10	11	10	9	8	9	8	99 days
	sunshine	24	56	125	168	250	221	247	228	126	76	32	15	1568 hours
	%	13	23	34	38	45	36	41	45	32	25	16	10	34 %

2. *Climatic area:* southern part of the country: except the area near Skagerrak, Kattegat, and Öresund, extending north to southern Värmland; on the coast the regions on the Baltic Sea to Västernorrland

2.1. Temperate Continental Climate DC

2.2. Summer mild, winter cold.

2.3. Variable general weather prevails throughout the year. Yet dry periods and sunny days are not lacking.

2.4. Wind: especially in coastal areas gales frequent, prevailing in winter. Snow cover in winter. Fog: at times, especially in winter.

Harnösand

latitude: 62° 38′n longitude: 17° 57′e height: 8 m

DClo	002-5	Jan	Feb	Mar	Apr	May	Jun	Jul	Aug	Sep	Oct	Nov	Dec	Year
	tx	6	6	10	15	21	25	27	26	21	15	9	6	28 °C
	t	-6	-6	-3	2	8	13	16	15	10	5	1	-3	4 °C
	tn	-22	-21	-19	-10	-2	2	7	4	-1	-5	-11	-17	-24 °C
	clothing	w	w	w	w	n	n	n	n	n	n	w	w	–
	stress													–
	p	62	37	32	46	34	52	58	78	68	63	87	80	697 mm
	dp	11	8	7	7	5	7	7	8	8	9	11	12	100 days
	sunshine	58	70	147	204	266	261	273	239	148	102	59	27	1854 hours
	%	31	28	41	46	48	43	46	48	38	33	28	17	41 %

Karlstadt

latitude: 59° 22′n longitude: 13° 28′e height: 47 m

DClo	002-6	Jan	Feb	Mar	Apr	May	Jun	Jul	Aug	Sep	Oct	Nov	Dec	Year
	tx	6	6	11	17	23	26	28	26	21	16	10	7	29 °C
	t	-4	-4	-1	4	10	14	17	16	12	6	2	-1	6 °C
	tn	-20	-19	-14	-7	-1	4	8	6	1	-4	-9	-15	-22 °C
	clothing	w	w	w	w	n	n	s	n	n	n	w	w	–
	stress							o						–
	p	40	25	22	39	35	48	60	78	68	62	67	50	594 mm
	dp	8	7	5	7	6	8	9	10	9	10	11	10	100 days
	sunshine	49	70	141	189	272	299	290	237	162	84	44	25	1862 hours
	%	23	27	39	43	52	54	52	49	42	26	19	13	41 %

Stockholm

latitude: 59° 21′n longitude: 17° 57′e height: 11 m

DCbo	002-7	Jan	Feb	Mar	Apr	May	Jun	Jul	Aug	Sep	Oct	Nov	Dec	Year
	tx	5	5	10	16	22	26	28	26	21	15	9	7	29 °C
	t	-3	-3	-1	4	10	15	18	17	12	7	3	0	7 °C
	tn	-14	-13	-11	-5	0	6	10	9	4	-1	-6	-9	-16 °C
	clothing	w	w	w	w	n	n	s	s	n	n	w	w	–
	stress							o	o					–
	p	43	30	26	31	34	45	61	76	60	48	53	48	555 mm
	dp	10	7	6	7	7	8	9	10	9	9	10	11	102 days
	sunshine	41	76	151	208	292	318	295	248	174	103	41	26	1973 hours
	%	19	29	41	48	55	57	53	51	45	32	18	14	44 %

Jönköping

latitude: 57° 46′n longitude: 14° 11′e height: 99 m

DClo	002-8	Jan	Feb	Mar	Apr	May	Jun	Jul	Aug	Sep	Oct	Nov	Dec	Year
	tx	6	7	12	18	24	27	28	27	22	17	10	8	29 °C
	t	-3	-3	-1	4	9	14	16	15	11	7	3	1	6 °C
	tn	-17	-18	-15	-8	-3	2	5	4	-1	-5	-8	-13	-21 °C
	clothing	w	w	w	w	n	n	n	n	n	n	w	w	–
	stress													–
	p	34	24	24	31	40	55	71	68	57	49	41	42	536 mm
	dp	9	7	6	7	8	8	11	10	10	9	10	9	104 days
	sunshine	29	54	116	152	227	233	218	192	125	76	29	19	1470 hours
	%	13	20	32	35	44	43	40	40	32	24	12	9	33 %

Göteborg

latitude: 57° 42′n longitude: 11° 58′e height: 31 m

DClo	002-9	Jan	Feb	Mar	Apr	May	Jun	Jul	Aug	Sep	Oct	Nov	Dec	Year
	tx	6	6	10	16	23	26	28	26	21	16	10	8	28 °C
	t	-1	-1	1	6	11	14	17	16	13	9	4	2	8 °C
	tn	-12	-12	-9	-3	2	6	10	9	3	-1	-4	-9	-14 °C
	clothing	w	w	w	n	n	n	s	n	n	n	w	w	–
	stress							o						–
	p	51	34	29	39	34	54	86	84	75	65	62	57	670 mm
	dp	10	8	6	8	7	9	11	11	11	11	11	12	114 days
	sunshine	48	76	151	201	274	286	285	245	178	108	47	29	1928 hours
	%	21	29	41	47	53	53	53	51	46	33	20	14	43 %

Visby

latitude: 57° 40′n longitude: 18° 21′e height: 47 m

DClo	002-10	Jan	Feb	Mar	Apr	May	Jun	Jul	Aug	Sep	Oct	Nov	Dec	Year
	tx	6	5	9	15	22	25	27	26	21	15	10	7	28 °C
	t	-1	-1	0	4	9	14	17	17	13	8	4	2	7 °C
	tn	-10	-10	-8	-4	0	5	10	9	5	0	-3	-6	-12 °C
	clothing	w	w	w	w	n	n	s	s	n	n	w	w	–
	stress							o	o					–
	p	53	42	29	31	30	32	52	56	51	52	48	53	529 mm
	dp	11	9	7	6	6	6	8	8	9	10	10	12	100 days
	sunshine	28	56	128	208	292	317	302	255	186	108	30	22	1932 hours
	%	12	21	35	48	57	59	56	54	48	33	12	11	43 %

3. Climatic area: Provinces of Halland, Malmöhus, Kristianstad
3.1. Temperate Oceanic Climate CO
3.2. Summer mild, winter cool.
3.3. Variable general weather prevails throughout the year. Yet dry periods and sunny days are not lacking.
3.4. Wind: gales frequent. Fog: at times, especially in winter. Snow cover variable in winter.

4.1.2.2. Predominant general weather situations are described in 4.1.0.1. under Northern, Central and Western Europe. A hygrothermal diagram of Stockholm is to be found in 4.1.0.2.

4.1.3. Finland

4.1.3.1. Climatic descriptions and tables

1. Climatic area: Lappi, Oulu, eastern part of Vaasa, Kuopio, Mikkeli, Häme, small northern part of Kymi
1.1. Subarctic Continental Climate EC, in the southern part: Subarctic Oceanic Climate EO
1.2. Summer mild, winter very cold or cold.
1.3. Variable general weather prevails; occasionally there are periods of dry sunny days.
1.4. Wind: gales at times. Snow cover in winter.

Sodankylä

latitude: 67°22'n longitude: 26°39'e height: 180 m

EClc	003-1	Jan	Feb	Mar	Apr	May	Jun	Jul	Aug	Sep	Oct	Nov	Dec	Year
	tx	1	0	5	10	19	25	28	24	17	10	4	2	28 °C
	t	-14	-13	-9	-2	5	11	15	12	6	0	-6	-10	0 °C
	tn	-37	-36	-34	-24	-9	-2	2	-2	-6	-17	-26	-32	-39 °C
	clothing	p	p	w	w	n	n	n	n	n	w	w	p	–
	stress													–
	p	27	26	20	31	31	56	74	71	57	43	39	31	508 mm
	dp	8	8	6	8	7	10	10	11	10	9	10	10	107 days
	sunshine	8	51	137	198	223	265	279	197	101	63	19	–	1540 hours
	%	7	23	38	42	36	37	39	37	25	21	12	–	33 %

Kajaani

latitude: 64°17'n longitude: 27°41'e height: 136 m

EClc	003-2	Jan	Feb	Mar	Apr	May	Jun	Jul	Aug	Sep	Oct	Nov	Dec	Year
	tx	2	1	6	13	21	26	28	25	18	11	6	3	29 °C
	t	-11	-10	-7	0	7	13	16	14	8	2	-3	-7	2 °C
	tn	-28	-28	-26	-16	-5	0	5	2	-3	-9	-16	-24	-32 °C
	clothing	p	p	w	w	n	n	n	n	n	w	w	w	–
	stress													–
	p	31	22	23	33	36	66	71	70	61	51	40	32	536 mm
	dp	10	8	6	8	8	11	11	11	11	11	11	10	116 days
	sunshine	17	57	149	180	256	262	278	215	112	56	18	6	1606 hours
	%	10	24	41	40	45	41	45	42	28	18	9	5	35 %

Luonetjärvi

EOlo	003-3	Jan	Feb	Mar	Apr	May	Jun	Jul	Aug	Sep	Oct	Nov	Dec	Year
		latitude: 62°24′n						longitude: 25°40′e			height: 145 m			
	t	-9	-9	-5	1	8	13	16	14	9	3	-2	-6	3 °C
	clothing stress	w	w	w	w	n	n	n	n	n	w	w	w	–
	p	40	27	26	35	41	54	81	85	71	63	53	43	619 mm
	dp	11	8	6	10	8	9	11	11	11	10	11	11	117 days
	sunshine	21	62	152	178	242	256	262	232	120	55	24	12	1615 hours
	%	11	25	42	40	44	43	45	46	31	13	11	8	36 %

2. *Climatic area:* Provinces: western part of Vaasa, Turku-Pori, Uusimaa, Kymi (without small northern part), Ahvenanmaa (Åland Islands)
2.1. Temperate Continental Climate DC
2.2. Summer mild, winter cold.
2.3. Variable general weather prevails; occasionally there are periods of dry and sunny days.
2.4. Wind: gales at times. Snow cover in winter.

Vaasa

DClo	003-4	Jan	Feb	Mar	Apr	May	Jun	Jul	Aug	Sep	Oct	Nov	Dec	Year
		latitude: 63°03′n						longitude: 21°46′e			height: 8 m			
	tx	4	3	7	14	20	25	28	25	19	12	8	5	29 °C
	t	-7	-8	-5	1	8	13	16	15	10	4	0	-4	4 °C
	tn	-24	-22	-20	-11	-2	2	7	5	0	-6	-12	-17	-28 °C
	clothing stress	w	w	w	w	n	n	n	n	n	w	w	w	–
	p	32	21	18	30	31	47	62	63	65	58	49	37	513 mm
	dp	9	6	6	7	6	8	8	9	10	10	11	10	100 days
	sunshine	24	65	134	191	304	311	308	238	152	69	47	12	1855 hours
	%	13	26	37	43	54	51	52	47	39	22	23	8	41 %

Jokioinen

DClo	003-5	Jan	Feb	Mar	Apr	May	Jun	Jul	Aug	Sep	Oct	Nov	Dec	Year
		latitude: 60°49′n						longitude: 23°29′e			height: 103 m			
	t	-7	-8	-5	2	9	14	16	15	10	4	0	-4	4 °C
	clothing stress	w	w	w	w	n	n	n	n	n	w	w	w	–
	p	35	27	25	33	39	42	66	74	61	61	51	41	555 mm
	dp	10	8	6	8	8	8	10	10	11	11	11	11	112 days
	sunshine	36	76	158	192	232	264	244	212	140	62	29	22	1666 hours
	%	18	30	43	44	43	46	43	43	36	20	13	12	37 %

Turku

latitude: 60°31′n longitude: 22°16′e height: 54 m

DClo 003-6	Jan	Feb	Mar	Apr	May	Jun	Jul	Aug	Sep	Oct	Nov	Dec	Year
tx	4	4	7	16	23	26	28	26	20	14	9	6	29 °C
t	-6	-7	-4	2	9	14	17	16	11	5	1	-3	5 °C
tn	-24	-23	-21	-10	-4	2	6	4	-2	-6	-12	-18	-27 °C
clothing	w	w	w	w	n	n	s	n	n	n	w	w	–
stress							o						–
p	44	28	24	35	31	46	64	78	65	64	58	51	588 mm
dp	10	7	6	8	6	8	9	10	10	10	11	11	106 days
sunshine	42	79	169	200	256	282	270	236	156	75	36	25	1827 hours
%	21	31	46	46	48	49	48	48	40	24	16	14	40 %

Helsinki

latitude: 60°19′n longitude: 24°58′e height: 58 m

DClo 003-7	Jan	Feb	Mar	Apr	May	Jun	Jul	Aug	Sep	Oct	Nov	Dec	Year
tx	3	3	8	15	22	26	28	26	20	14	8	3	29 °C
t	-7	-7	-4	2	9	14	17	16	10	5	1	-3	4 °C
tn	-21	-20	-17	-8	-1	4	8	7	1	-5	-9	-15	-24 °C
clothing	w	w	w	w	n	n	s	n	n	n	w	w	–
stress							o						–
p	49	34	32	41	38	47	68	71	70	72	61	58	641 mm
dp	10	9	7	8	7	8	10	10	9	11	11	11	111 days
sunshine	39	72	152	204	270	294	295	251	152	80	33	22	1864 hours
%	19	28	42	47	51	52	52	51	39	25	19	12	41 %

4.1.3.2. Predominant general weather situations are described in 4.1.0.1. under Northern, Central and Western Europe. A hygrothermal diagram of Helsinki is to be found in 4.1.0.2.

4.1.4. United Kingdom of Great Britain and Northern Ireland

4.1.4.1. Climatic description and tables

1. Temperate Oceanic Climate DO and Temperate Oceanic Mountain Climate GDO
2. Summer mild, winter cool.
3. Variable general weather prevails; now and then periods of dry and also sunny days occur.
4. Wind: frequent gales in coastal areas also severe. Fog: maximum of occurrence in the east and southeast of England, especially in winter.

Lerwick

latitude: 60°08′n longitude: 1°11′w height: 82 m

DOlk 004-1	Jan	Feb	Mar	Apr	May	Jun	Jul	Aug	Sep	Oct	Nov	Dec	Year
tx	9	9	10	12	14	17	18	17	16	14	11	10	19 °C
t	3	5	4	5	8	10	12	12	11	8	6	4	7 °C
tn	-4	-4	-3	-2	1	3	7	6	4	1	-1	-3	-6 °C
clothing	w	w	w	n	n	n	n	n	n	n	n	w	–
stress													–
p	122	96	78	76	58	63	79	80	79	118	127	133	1129 mm
dp	20	16	14	15	11	10	12	12	14	18	19	20	181 days
sunshine	25	51	90	132	165	158	125	117	105	67	33	14	1082 hours
%	12	20	25	30	31	28	22	24	27	21	15	8	24 %

Stornoway

latitude: 58° 13′n longitude: 6° 20′w height: 3 m

DOlk	004-2	Jan	Feb	Mar	Apr	May	Jun	Jul	Aug	Sep	Oct	Nov	Dec	Year
	tx	11	10	12	13	18	20	20	20	18	16	12	12	22 °C
	t	4	4	6	7	9	12	13	13	12	9	7	6	9 °C
	tn	-4	-4	-3	-1	1	4	6	6	4	1	-1	-3	-6 °C
	clothing	w	w	n	n	n	n	n	n	n	n	n	n	–
	stress													–
	p	107	75	63	65	52	68	87	88	97	118	111	110	1041 mm
	dp	21	16	14	15	12	12	15	15	16	20	19	20	195 days
	sunshine	35	62	108	142	195	173	128	133	111	76	45	26	1234 hours
	%	16	23	30	33	38	32	23	28	29	24	19	13	27 %

Aberdeen

latitude: 57° 12′n longitude: 2° 12′w height: 59 m

DOlk	004-3	Jan	Feb	Mar	Apr	May	Jun	Jul	Aug	Sep	Oct	Nov	Dec	Year
	tx	11	11	14	17	19	23	23	22	21	17	13	11	25 °C
	t	2	3	4	7	9	12	14	14	12	9	6	4	8 °C
	tn	-6	-5	-4	-2	0	3	5	5	2	0	-2	-4	-8 °C
	clothing	w	w	w	n	n	n	n	n	n	n	n	w	–
	stress													–
	p	77	54	52	50	62	53	92	73	65	90	91	78	837 mm
	dp	14	11	10	10	10	10	13	11	10	13	14	13	139 days
	sunshine	54	80	108	148	181	181	157	147	129	96	59	45	1385 hours
	%	24	30	30	35	35	34	29	31	33	30	24	22	31 %

Eskdalemuir

latitude: 55° 19′n longitude: 3° 12′w height: 239 m

DOlk	004-4	Jan	Feb	Mar	Apr	May	Jun	Jul	Aug	Sep	Oct	Nov	Dec	Year
	tx	9	10	14	16	21	24	24	23	20	16	12	10	26 °C
	t	1	2	4	6	9	12	13	13	11	8	5	3	7 °C
	tn	-9	-9	-7	-4	-2	2	4	3	1	-2	-5	-7	-11 °C
	clothing	w	w	w	n	n	n	n	n	n	n	n	w	–
	stress													–
	p	175	112	98	97	87	108	131	120	136	148	153	162	1527 mm
	dp	17	15	13	13	11	13	16	14	15	16	16	18	177 days
	sunshine	44	65	95	130	174	169	140	129	100	78	50	36	1210 hours
	%	18	24	26	31	35	32	27	28	26	24	20	16	27 %

Manchester Airport (tx and tn Birkenhead/Bidstone)

latitude: 53° 21′n longitude: 2° 16′w height: 77 m

DOlk	004-5	Jan	Feb	Mar	Apr	May	Jun	Jul	Aug	Sep	Oct	Nov	Dec	Year
	tx	12	12	14	17	22	25	25	24	22	18	14	12	27 °C
	t	3	3	6	8	12	14	16	16	14	10	7	5	10 °C
	tn	-4	-3	-1	1	3	7	10	9	7	3	0	-2	-5 °C
	clothing	w	w	n	n	n	n	n	n	n	n	n	n	–
	stress													–
	p	77	53	45	49	57	61	79	81	67	78	80	72	799 mm
	dp	13	11	9	10	10	11	13	12	12	12	14	14	141 days
	sunshine	34	55	99	138	177	179	154	151	116	85	43	32	1263 hours
	%	14	20	27	33	36	35	30	33	30	26	17	14	28 %

Waddington

latitude: 53° 10′n longitude: 0° 31′w height: 68 m

DOlk	004-6	Jan	Feb	Mar	Apr	May	Jun	Jul	Aug	Sep	Oct	Nov	Dec	Year
	t	3	4	6	8	11	14	16	16	14	10	6	4	9 °C
	clothing	w	w	n	n	n	n	n	n	n	n	n	w	–
	stress													–
	p	54	42	37	37	46	51	62	58	47	53	62	48	597 mm
	dp	11	9	8	9	8	9	10	10	9	9	11	10	113 days
	sunshine	56	71	110	166	206	194	186	167	142	107	60	45	1510 hours
	%	22	25	30	40	42	38	36	36	37	32	23	19	34 %

Gorleston

latitude: 52° 35′n longitude: 1° 43′e height: 2 m

DOlk	004-7	Jan	Feb	Mar	Apr	May	Jun	Jul	Aug	Sep	Oct	Nov	Dec	Year
	tx	12	12	15	18	20	24	25	25	23	19	14	12	26 °C
	t	4	4	6	8	11	14	16	16	15	11	8	5	10 °C
	tn	-4	-3	-2	1	3	6	8	10	6	2	0	-2	-5 °C
	clothing	w	w	n	n	n	n	n	n	n	n	n	n	–
	stress													–
	p	57	42	37	36	38	44	57	56	53	65	68	52	605 mm
	dp	12	9	8	9	7	8	9	9	9	11	13	11	115 days
	sunshine	54	69	125	166	209	215	210	189	152	111	58	46	1604 hours
	%	21	25	34	40	43	43	41	41	40	33	22	19	36 %

Kew (west of London)

latitude: 51° 28′n longitude: 0° 19′w height: 5 m

DObk	004-8	Jan	Feb	Mar	Apr	May	Jun	Jul	Aug	Sep	Oct	Nov	Dec	Year
	tx	12	12	16	19	23	26	27	26	23	19	14	12	29 °C
	t	4	4	7	9	12	16	18	17	15	11	7	5	10 °C
	tn	-4	-4	-2	0	3	7	9	8	5	0	-1	-3	-6 °C
	clothing	w	w	n	n	n	n	n	n	n	n	n	n	–
	stress													–
	p	53	40	37	38	46	46	56	59	50	57	64	48	594 mm
	dp	11	9	8	8	8	8	9	9	9	9	10	9	107 days
	sunshine	46	64	113	160	199	213	198	188	142	98	53	40	1514 hours
	%	18	23	31	38	41	43	39	41	37	29	20	16	34 %

Plymouth

latitude: 50° 21′n longitude: 4° 07′w height: 27 m

DOlk	004-9	Jan	Feb	Mar	Apr	May	Jun	Jul	Aug	Sep	Oct	Nov	Dec	Year
	tx	12	12	15	18	21	24	24	24	22	18	15	13	26 °C
	t	6	6	7	9	12	14	16	16	15	12	9	7	11 °C
	tn	-3	-3	-1	1	3	6	8	8	6	2	0	-2	-5 °C
	clothing	n	n	n	n	n	n	n	n	n	n	n	n	–
	stress													–
	p	105	77	73	55	65	58	71	80	82	94	115	115	990 mm
	dp	14	11	10	9	9	9	10	10	11	12	14	15	134 days
	sunshine	60	80	133	182	219	222	198	198	152	114	67	52	1677 hours
	%	23	28	36	44	46	45	40	44	40	34	25	21	37 %

Aldergrove / Northern Ireland

DOlk	004-10	Jan	Feb	Mar	Apr	May	Jun	Jul	Aug	Sep	Oct	Nov	Dec	Year
				latitude: 54°39'n				longitude: 6°13'w				height: 66 m		
	tx	11	11	15	17	21	24	23	23	21	17	13	12	26 °C
	t	4	4	6	8	10	13	15	15	13	10	7	5	9 °C
	tn	-6	-6	-4	-2	-1	3	6	5	3	0	-3	-4	-8 °C
	clothing	w	w	n	n	n	n	n	n	n	n	n	n	–
	stress													–
	p	80	52	50	48	52	68	94	77	80	83	72	90	846 mm
	dp	15	12	11	11	10	12	14	13	13	14	13	16	154 days
	sunshine	45	65	104	149	196	179	136	135	107	81	53	35	1285 hours
	%	18	24	28	35	39	35	26	29	28	25	21	15	29 %

4.1.4.2. Predominant general weather situations are described in 4.1.0.1. under Northern, Central and Eastern Europe. A hygrothermal diagram of Kew (nr. London) is to be found in 4.1.0.2.

4.1.5. Ireland

4.1.5.1. Climatic description and tables

1. Temperate Oceanic Climate DO
2. Summer mild, winter cool.
3. Variable general weather prevails; now and then periods of dry and also sunny days occur.
4. Wind: gales frequent, in coastal areas at times severe.

Malin Head

DOlk	005-1	Jan	Feb	Mar	Apr	May	Jun	Jul	Aug	Sep	Oct	Nov	Dec	Year
				latitude: 55°22'n				longitude: 7°20'w				height: 25 m		
	tx	12	11	14	15	18	21	21	22	20	17	14	12	23 °C
	t	5	5	7	8	10	12	14	14	13	10	8	6	10 °C
	tn	-2	-2	0	1	3	6	8	8	6	3	1	0	-3 °C
	clothing	n	n	n	n	n	n	n	n	n	n	n	n	–
	stress													–
	p	113	89	56	52	50	85	109	87	90	96	82	144	1053 mm
	dp	18	15	12	13	11	13	17	14	16	17	17	19	182 days
	sunshine	41	63	109	154	207	177	142	167	114	72	51	31	1328 hours
	%	17	23	30	36	41	34	27	36	30	22	20	14	29 %

Belmullet

DOlk	005-2	Jan	Feb	Mar	Apr	May	Jun	Jul	Aug	Sep	Oct	Nov	Dec	Year
				latitude: 54°14'n				longitude: 10°00'w				height: 10 m		
	tx	11	12	14	16	20	20	20	20	19	17	13	12	22 °C
	t	6	6	7	9	11	13	14	15	13	11	8	7	10 °C
	tn	-4	-2	-1	0	3	6	7	6	6	3	0	-2	-4 °C
	clothing	n	n	n	n	n	n	n	n	n	n	n	n	–
	stress													–
	p	129	82	73	65	63	73	88	89	106	120	126	141	1155 mm
	dp	18	13	16	15	14	12	14	17	16	18	20	22	195 days
	sunshine	58	71	105	156	216	181	142	158	118	91	58	40	1393 hours
	%	24	26	29	37	44	35	27	34	31	28	23	17	31 %

Clones

DOlk	005-3	Jan	Feb	Mar	Apr	May	Jun	Jul	Aug	Sep	Oct	Nov	Dec	Year
				latitude: 54°11'n				longitude: 7°14'w				height: 89 m		
	tx	13	12	15	17	21	23	23	22	21	18	14	11	25 °C
	t	4	4	6	8	10	13	14	14	12	9	6	5	9 °C
	tn	-6	-6	-3	-1	0	3	6	5	3	0	-4	-4	-7 °C
	clothing	w	w	n	n	n	n	n	n	n	n	n	n	–
	stress													–
	p	92	63	60	51	60	74	91	85	87	89	83	100	935 mm
	dp	13	12	11	10	12	13	15	14	15	13	13	19	160 days
	sunshine	56	70	98	149	188	164	127	119	98	75	52	37	1233 hours
	%	23	25	27	35	38	32	25	26	26	23	20	16	27 %

Claremorris

DOlk	005-4	Jan	Feb	Mar	Apr	May	Jun	Jul	Aug	Sep	Oct	Nov	Dec	Year
				latitude: 53°43'n				longitude: 8°59'w				height: 69 m		
	tx	12	12	16	17	22	23	23	22	20	18	14	12	25 °C
	t	4	5	6	8	10	13	14	14	12	10	7	5	9 °C
	tn	-6	-5	-3	-2	0	3	5	4	3	0	-3	-4	-8 °C
	clothing	w	w	n	n	n	n	n	n	n	n	n	n	–
	stress													–
	p	121	83	75	62	67	78	98	91	108	111	110	131	1135 mm
	dp	17	13	14	12	12	12	15	15	17	14	16	21	178 days
	sunshine	55	73	104	148	194	162	128	139	100	78	59	39	1279 hours
	%	22	26	28	35	39	32	25	30	26	24	23	17	28 %

Mullingar

DOlk	005-5	Jan	Feb	Mar	Apr	May	Jun	Jul	Aug	Sep	Oct	Nov	Dec	Year
				latitude: 53°31'n				longitude: 7°21'w				height: 110 m		
	tx	12	12	16	18	22	24	24	23	21	18	14	12	26 °C
	t	4	4	6	8	11	14	15	15	13	10	6	5	9 °C
	tn	-5	-4	-3	-1	1	4	6	6	4	0	-2	-4	-6 °C
	clothing	w	w	n	n	n	n	n	n	n	n	n	n	–
	stress													–
	p	91	66	60	55	62	72	92	81	94	92	84	101	950 mm
	dp	14	10	12	10	11	11	14	14	14	13	13	19	155 days
	sunshine	58	73	107	151	189	174	139	145	110	87	62	45	1340 hours
	%	23	26	29	36	38	34	27	31	29	26	24	19	30 %

Dublin Airport

DOlk	005-6	Jan	Feb	Mar	Apr	May	Jun	Jul	Aug	Sep	Oct	Nov	Dec	Year
				latitude: 53°26'n				longitude: 6°15'w				height: 81 m		
	tx	13	13	16	18	21	24	24	24	22	19	15	13	26 °C
	t	5	5	6	8	11	14	15	15	13	10	7	6	10 °C
	tn	-6	-5	-4	-2	0	4	6	5	2	-1	-3	-4	-7 °C
	clothing	n	n	n	n	n	n	n	n	n	n	n	n	–
	stress													–
	p	70	52	49	45	58	55	70	72	73	70	68	76	758 mm
	dp	11	10	10	8	10	10	12	13	12	10	12	15	133 days
	sunshine	63	78	112	166	209	200	164	162	128	102	65	54	1503 hours
	%	25	28	31	40	42	39	32	35	33	31	25	23	33 %

Shannon Airport

DOlk	005-7	Jan	Feb	Mar	Apr	May	Jun	Jul	Aug	Sep	Oct	Nov	Dec	Year
latitude: 52°41′n	longitude: 8°55′w	height: 7 m												
	tx	12	13	16	18	22	25	24	24	21	18	15	13	26 °C
	t	5	6	7	9	12	14	16	16	14	11	8	6	10 °C
	tn	-6	-5	-3	-1	0	4	6	5	4	-1	-3	-4	-7 °C
	clothing	n	n	n	n	n	n	n	n	n	n	n	n	–
	stress													–
	p	95	65	55	53	62	60	84	77	88	90	93	108	930 mm
	dp	15	11	13	10	11	12	14	14	15	14	14	19	162 days
	sunshine	57	73	104	154	205	181	152	158	111	86	62	43	1386 hours
	%	23	26	28	37	42	36	30	34	29	26	24	18	31 %

Kilkenny

DOlk	005-8	Jan	Feb	Mar	Apr	May	Jun	Jul	Aug	Sep	Oct	Nov	Dec	Year
latitude: 52°40′n	longitude: 7°16′w	height: 64 m												
	tx	11	13	15	18	21	24	23	23	22	18	14	13	25 °C
	t	4	5	7	8	11	14	15	15	13	10	7	5	9 °C
	tn	-8	-5	-4	-2	0	3	5	3	1	-2	-5	-6	-9 °C
	clothing	w	n	n	n	n	n	n	n	n	n	n	n	–
	stress													–
	p	90	63	64	49	63	59	71	68	84	80	85	94	870 mm
	dp	13	10	12	12	13	11	13	12	12	12	12	14	146 days
	sunshine	60	70	104	144	194	176	144	156	120	91	72	58	1389 hours
	%	20	25	28	34	40	35	28	34	31	27	27	24	31 %

Valentia Observatory

DOlk	005-9	Jan	Feb	Mar	Apr	May	Jun	Jul	Aug	Sep	Oct	Nov	Dec	Year
latitude: 51°56′n	longitude: 10°15′w	height: 14 m												
	tx	12	13	15	17	20	23	22	23	20	18	14	13	25 °C
	t	7	7	8	9	12	14	15	15	14	12	9	8	11 °C
	tn	-2	-2	-1	1	3	6	8	7	6	2	0	-1	-3 °C
	clothing	n	n	n	n	n	n	n	n	n	n	n	n	–
	stress													–
	p	164	107	103	74	86	81	107	95	122	140	151	168	1398 mm
	dp	20	15	14	13	13	13	15	15	16	17	18	21	190 days
	sunshine	49	70	109	155	202	177	145	151	115	88	59	41	1361 hours
	%	19	25	30	37	42	35	29	33	30	26	22	17	30 %

Roche's Point

DOlk	005-10	Jan	Feb	Mar	Apr	May	Jun	Jul	Aug	Sep	Oct	Nov	Dec	Year
latitude: 51°48′n	longitude: 8°15′w	height: 41 m												
	tx	12	12	14	16	18	21	22	21	20	17	14	13	23 °C
	t	6	6	8	9	11	14	15	16	14	11	9	7	11 °C
	tn	-2	0	0	2	4	7	9	8	7	4	2	0	-2 °C
	clothing	n	n	n	n	n	n	n	n	n	n	n	n	–
	stress													–
	p	99	64	77	52	65	55	71	64	84	87	91	101	910 mm
	dp	13	11	12	11	12	11	11	11	11	12	13	13	141 days
	sunshine	58	64	106	153	196	191	163	157	126	98	73	60	1445 hours
	%	23	23	29	37	40	38	32	34	33	29	27	25	32 %

4.1.5.2. Predominant general weather situations are described in 4.1.0.1 under Northern, Central and Eastern Europe. A hygrothermal diagram of Dublin is to be found in 4.1.0.2.

4.1.6. Iceland

4.1.6.1. Climatic description and tables

1. Subarctic Oceanic Climate EO
2. Summer mild, winter cool.
3. Variable weather prevails, yet periods of dry and sunny days are also possible.
4. Wind: frequent gales. Fog: occasionally, especially in summer. Snow cover: in winter.

Akureyri

latitude: 65°41'n longitude: 18°05'w height: 5 m

EOlo	006-1	Jan	Feb	Mar	Apr	May	Jun	Jul	Aug	Sep	Oct	Nov	Dec	Year
	tx	10	9	11	12	18	21	21	20	17	14	12	10	23 °C
	t	-2	-2	0	2	6	9	11	10	8	4	1	0	4 °C
	tn	-15	-14	-14	-9	-4	1	3	2	-2	-7	-12	-13	-18 °C
	clothing	w	w	w	w	n	n	n	n	n	w	w	w	–
	stress													–
	p	45	42	42	32	15	22	35	39	46	57	45	54	474 mm
0.1 mm	dp	14	13	12	11	8	9	11	11	12	14	14	16	145 days
	sunshine	6	32	76	105	172	172	147	113	75	51	13	0	962 hours
	%	4	14	21	23	29	25	23	22	19	17	7	–	21 %

Holar i Hornafirdi

latitude: 64°18'n longitude: 15°12'w height: 17 m

EOlk	006-2	Jan	Feb	Mar	Apr	May	Jun	Jul	Aug	Sep	Oct	Nov	Dec	Year
	tx	6	8	9	11	14	17	18	16	14	11	10	8	19 °C
	t	0	0	2	3	6	9	11	10	8	5	3	1	5 °C
	tn	-9	-10	-9	-6	-3	2	4	3	1	-4	-7	-9	-12 °C
	clothing	w	w	w	w	n	n	n	n	n	n	w	w	–
	stress													–
	p	191	115	132	108	90	83	93	116	162	170	187	185	1632 mm
0.1 mm	dp	18	14	16	15	13	13	15	16	15	18	17	18	188 days
	sunshine	32	72	116	133	181	144	149	140	110	79	48	19	1223 hours
	%	19	30	32	29	32	23	24	27	28	26	24	15	27 %

Reykjavik

latitude: 64°08'n longitude: 21°56'w height: 16 m

EOlk	006-3	Jan	Feb	Mar	Apr	May	Jun	Jul	Aug	Sep	Oct	Nov	Dec	Year
	tx	8	8	10	11	15	17	18	17	15	12	10	8	19 °C
	t	0	0	2	3	7	10	11	11	9	5	3	1	5 °C
	tn	-11	-10	-10	-6	-2	3	5	4	0	-4	-8	-10	-13 °C
	clothing	w	w	w	w	n	n	n	n	n	n	w	w	–
	stress													–
	p	90	65	65	53	42	41	48	66	72	97	85	81	805 mm
0.1 mm	dp	20	17	18	18	16	15	15	16	19	21	18	20	213 days
	sunshine	21	57	106	138	185	189	178	159	105	71	32	8	1249 hours
	%	13	23	29	31	32	30	29	31	27	23	16	6	27 %

4.1.6.2. Predominant general weather situations are described in 4.1.0.1. under Iceland. A hygrothermal diagram of Reykjavik is to be found in 4.1.0.2.

4.1.7. Greenland

4.1.7.1. Climatic description and tables

1. Tundra Climate FT, in the interior Ice Mountain Climate GFI or Ice High Mountain Climate HFI
2. Summer cool, winter very cold, in the north severely cold.
3. Variable weather prevails, yet periods of dry and sunny days also occur.
4. Wind: gales frequent, often severe. Fog: frequent in summer in coastal areas. Snow cover in the interior all the year.

Peary Land

latitude: 82° 10'n longitude: 30° 30'w height: 9 m

FTkd	007-1	Jan	Feb	Mar	Apr	May	Jun	Jul	Aug	Sep	Oct	Nov	Dec	Year
	tx	-16	-16	-14	-9	7	13	16	17	6	-11	-12	-17	17 °C
	t	-31	-29	-27	-22	-8	3	7	4	-5	-19	-22	-28	-15 °C
	tn	-39	-41	-39	-35	-18	-4	1	-4	-16	-28	-35	-37	-42 °C
	clothing	p	p	p	p	w	w	n	w	w	p	p	p	–
	stress													–
	p	2	3	2	1	2	3	3	3	13	15	3	2	53 mm
	dp	0	1	1	0	0	1	1	2	2	3	1	0	12 days

Nord

latitude: 81° 36'n longitude: 16° 40'w height: 35 m

FTkd	007-2	Jan	Feb	Mar	Apr	May	Jun	Jul	Aug	Sep	Oct	Nov	Dec	Year
	t	-29	-30	-31	-24	-11	0	4	1	-8	-18	-24	-27	-16 °C
	clothing	p	p	p	p	p	w	w	w	w	p	p	p	–
	stress													–

Danmarkshavn

latitude: 76° 46'n longitude: 18° 46'w height: 17 m

FTkc	007-3	Jan	Feb	Mar	Apr	May	Jun	Jul	Aug	Sep	Oct	Nov	Dec	Year
	tx	-5	-18	-7	-6	4	9	14	12	7	-1	-9	-6	15 °C
	t	-22	-23	-23	-17	-7	1	4	2	-4	-13	-19	-21	-12 °C
	tn	-35	-37	-37	-32	-22	-8	-2	-6	-13	-23	-33	-34	-38 °C
	clothing	p	p	p	p	w	w	w	w	w	p	p	p	–
	stress													–
	p	30	18	18	2	5	5	0	15	8	8	25	18	152 mm
	dp	7	5	6	1	3	1	0	2	2	2	9	5	43 days

Thule

latitude: 76° 31'n longitude: 68° 50'w height: 77 m

FTkd	007-4	Jan	Feb	Mar	Apr	May	Jun	Jul	Aug	Sep	Oct	Nov	Dec	Year
	tx	-6	-6	-6	-2	4	13	14	12	6	4	2	-6	14 °C
	t	-24	-25	-26	-17	-5	3	6	4	-2	-20	-18	-23	-12 °C
	tn	-36	-36	-36	-31	-21	-5	-1	-3	-14	-21	-31	-34	-38 °C
	clothing	p	p	p	p	w	w	n	w	w	p	p	p	–
	stress													–
	p	8	9	5	3	5	6	17	14	15	17	13	6	117 mm

Kap Tobin (tx and tn Scoresbysund)

latitude: 70°25′n longitude: 21°58′w height: 41 m

FTkc	007-5	Jan	Feb	Mar	Apr	May	Jun	Jul	Aug	Sep	Oct	Nov	Dec	Year
	tx	-1	-1	0	3	6	12	14	11	9	3	2	0	14 °C
	t	-16	-16	-16	-12	-4	0	2	3	0	-5	-10	-13	-7 °C
	tn	-31	-32	-32	-26	-13	-4	-1	-3	-6	-17	-23	-29	-35 °C
	clothing	p	p	p	p	w	w	w	w	w	w	p	p	–
	stress													–
	p	46	36	23	36	10	20	38	18	43	36	28	48	381 mm
	dp	5	5	3	4	2	3	4	4	5	3	4	5	47 days

Egedesminde (tx and tn Jakobshavn)

latitude: 68°42′n longitude: 52°52′w height: 47 m

FTkc	007-6	Jan	Feb	Mar	Apr	May	Jun	Jul	Aug	Sep	Oct	Nov	Dec	Year
	tx	2	1	3	6	11	15	17	16	12	8	5	4	17 °C
	t	-14	-16	-16	-9	-1	4	6	6	3	-2	-6	-11	-5 °C
	tn	-32	-33	-32	-25	-13	-2	2	-1	-7	-14	-20	-25	-36 °C
	clothing	p	p	p	w	w	w	n	n	w	w	w	p	–
	stress													–
	p	10	10	13	13	15	20	30	36	33	23	18	13	234 mm
0.1 mm	dp	8	7	9	8	8	8	8	9	11	11	11	10	108 days

Angmagsalik

latitude: 65°37′n longitude: 37°39′w height: 35 m

FTko	007-7	Jan	Feb	Mar	Apr	May	Jun	Jul	Aug	Sep	Oct	Nov	Dec	Year
	tx	3	2	6	9	12	16	18	17	13	8	5	4	19 °C
	t	-7	-8	-6	-3	2	6	7	7	4	0	-3	-5	0 °C
	tn	-22	-23	-22	-19	-10	-3	-2	-2	-4	-10	-16	-19	-26 °C
	clothing	w	w	w	w	w	n	n	n	w	w	w	w	–
	stress													–
	p	74	61	66	53	51	46	38	53	84	119	76	69	790 mm
0.1 mm	dp	11	9	9	8	7	7	6	7	8	10	9	9	100 days

Prins Christian Sund (p and dp Torgilsbu)

latitude: 60°03′n longitude: 43°12′w height: 76 m

FTko	007-8	Jan	Feb	Mar	Apr	May	Jun	Jul	Aug	Sep	Oct	Nov	Dec	Year
	tx	3	4	3	7	9	17	17	19	14	12	5	5	19 °C
	t	-4	-4	-3	0	2	5	7	7	5	2	0	-3	1 °C
	tn	-13	-14	-13	-9	-5	-1	2	2	-1	-3	-7	-10	-17 °C
	clothing	w	w	w	w	w	n	n	n	n	w	w	w	–
	stress													–
	p	150	224	124	119	236	160	71	96	239	193	155	165	1933 mm
0.1 mm	dp	12	13	12	9	15	11	8	11	12	11	11	12	137 days

4.1.7.2. Predominant general weather situations are described in 4.4.0.1. under Alaska, Canada and Greenland. A hygrothermal diagram is not presented.

4.1.8. Denmark and Faroe Islands

4.1.8.1. Climatic descriptions and tables

1. Climatic area: Faroe Islands
1.1. Subarctic Oceanic Climate EO
1.2. Summer mild, winter cool.
1.3. Variable weather prevails, often rain.
1.4. Wind: gales frequent, often severe. Fog: frequent. Snow cover in winter variable.

Thorshavn

latitude: 62°01'n longitude: 6°45'w height: 24 m

EOlk	008-1	Jan	Feb	Mar	Apr	May	Jun	Jul	Aug	Sep	Oct	Nov	Dec	Year
	tx	10	10	10	11	14	16	17	17	15	13	11	10	18 °C
	t	4	4	5	5	7	9	11	11	10	8	6	5	7 °C
	tn	-4	-5	-4	-3	0	3	6	6	3	0	-1	-3	-6 °C
	clothing	w	w	n	n	n	n	n	n	n	n	n	n	–
	stress													–
	p	149	136	114	106	67	74	79	96	132	157	157	167	1433 mm
0.1 mm	dp	25	22	23	22	16	16	18	20	21	24	24	26	257 days
	sunshine	15	41	76	113	137	140	110	102	82	58	23	7	903 hours
	%	8	16	21	25	25	24	19	20	21	19	11	4	20 %

2. Climatic area: Denmark
2.1. Temperate Oceanic Climate DO
2.2. Summer mild, winter cool.
2.3. Variable weather prevails, yet there are also periods of dry and sunny days.
2.4. Wind: gales frequent, sometimes also in summer. Fog: prevails in winter.

Aalborg (tx and tn, sunshine Tylstrup)

latitude: 57°06'n longitude: 9°52'e height: 3 m

DOlo	008-2	Jan	Feb	Mar	Apr	May	Jun	Jul	Aug	Sep	Oct	Nov	Dec	Year
	tx	7	7	11	17	24	26	28	27	22	17	11	9	29 °C
	t	0	-1	2	6	11	14	16	16	13	9	5	2	8 °C
	tn	-12	-12	-10	-4	-1	3	6	6	2	-2	-4	-9	-15 °C
	clothing	w	w	w	n	n	n	n	n	n	n	n	w	–
	stress													–
	p	45	31	26	31	33	42	70	72	69	58	55	44	576 mm
	dp	15	11	10	12	9	11	13	14	15	15	17	16	158 days
	sunshine	40	72	134	190	266	273	264	229	165	102	46	28	1809 hours
	%	17	27	37	44	52	51	49	48	43	31	19	13	40 %

København

latitude: 55°38'n longitude: 12°40'e height: 5 m

DObk	008-3	Jan	Feb	Mar	Apr	May	Jun	Jul	Aug	Sep	Oct	Nov	Dec	Year
	tx	7	8	11	18	24	26	27	26	22	17	11	9	28 °C
	t	0	0	2	7	12	16	18	17	14	9	5	3	8 °C
	tn	-8	-8	-6	-2	2	6	10	9	5	1	-2	-5	-10 °C
	clothing	w	w	w	n	n	n	s	s	n	n	n	w	–
	stress							o	o					–
	p	49	39	32	38	42	47	71	66	62	59	48	49	602 mm
0.1 mm	dp	17	14	12	13	11	13	14	14	15	16	16	17	172 days
	sunshine	36	55	118	161	245	245	239	205	157	87	34	19	1601 hours
	%	15	20	32	38	49	47	45	44	41	27	14	9	36 %

Dueodde (tx and tn, sunshine Sandvig)

latitude: 55°00′w longitude: 15°05′e height: 6 m

DOlk	008-4	Jan	Feb	Mar	Apr	May	Jun	Jul	Aug	Sep	Oct	Nov	Dec	Year
	tx	7	6	10	15	21	24	26	25	22	16	11	8	27 °C
	t	0	0	1	5	9	14	17	17	14	10	6	3	8 °C
	tn	-7	-7	-5	-1	2	7	11	12	8	3	-1	-3	-9 °C
	clothing	w	w	w	n	n	n	s	s	n	n	n	w	–
	stress							o	o					–
	p	48	33	29	31	32	42	57	58	61	60	54	48	553 mm
0.1 mm	dp	15	13	11	10	9	9	10	11	12	14	15	16	144 days
	sunshine	38	55	121	189	271	284	257	224	181	98	36	26	1780 hours
	%	16	20	33	45	54	55	49	48	47	30	14	12	40 %

4.1.8.2. Predominant general weather situations are described in 4.1.0.1. under Northern, Central, and Western Europe. A hygrothermal diagram of København is to be found in 4.1.0.2.

4.1.9. Netherlands

4.1.9.1. Climatic description and tables

1. Temperate Oceanic Climate DO
2. Summer mild, winter cool.
3. Variable weather prevails, now and then periods of dry and also sunny days appear.
4. Wind: gales frequent, in coastal areas at times severe. Fog: occasionally, particularly in winter.

Eelde

latitude: 53°08′n longitude: 6°35′e height: 4 m

DOlk	009-1	Jan	Feb	Mar	Apr	May	Jun	Jul	Aug	Sep	Oct	Nov	Dec	Year
	tx	10	11	16	20	24	28	28	27	26	19	13	11	30 °C
	t	1	1	4	8	12	15	16	16	14	10	6	3	9 °C
	tn	-9	-10	-6	-2	0	4	6	6	4	0	-4	-7	-13 °C
	clothing	w	w	w	n	n	n	n	n	n	n	n	w	–
	stress													–
	p	69	50	44	45	53	58	86	95	75	57	71	64	767 mm
0.1 mm	dp	21	19	17	16	15	15	18	19	18	20	22	22	222 days
	sunshine	51	66	107	176	212	204	181	166	141	81	52	29	1465 hours
	%	20	24	29	42	43	40	35	36	37	24	20	12	33 %

Den Helder

latitude: 52°58′n longitude: 4°45′e height: 4 m

DOlk	009-2	Jan	Feb	Mar	Apr	May	Jun	Jul	Aug	Sep	Oct	Nov	Dec	Year
	tx	9	9	14	17	22	25	27	26	24	18	13	10	28 °C
	t	3	2	5	8	11	15	17	17	15	11	7	4	10 °C
	tn	-6	-6	-4	1	4	8	10	10	8	3	-1	-4	-8 °C
	clothing	w	w	n	n	n	n	s	s	n	n	n	w	–
	stress							o	o					–
	p	65	46	39	40	37	37	64	71	74	89	79	65	706 mm
0.1 mm	dp	21	18	17	15	13	12	15	15	18	20	21	22	207 days
	sunshine	54	75	126	179	227	235	214	204	149	101	48	39	1651 hours
	%	21	27	34	43	46	46	42	44	39	30	18	16	37 %

De Bilt

latitude: 52°06'n longitude: 5°11'e height: 0 m

DOlk	009-3	Jan	Feb	Mar	Apr	May	Jun	Jul	Aug	Sep	Oct	Nov	Dec	Year
	tx	10	12	18	21	26	29	30	29	26	20	14	12	32 °C
	t	2	2	5	8	12	16	17	17	14	10	6	3	9 °C
	tn	-9	-8	-5	-1	1	4	7	7	4	0	-3	-7	-12 °C
	clothing	w	w	n	n	n	n	s	s	n	n	n	w	–
	stress							o	o					–
	p	69	52	44	49	52	57	78	89	71	72	70	64	766 mm
	dp	13	10	9	10	9	10	11	11	11	12	12	12	131 days
	sunshine	56	69	127	163	211	222	198	185	145	101	50	41	1568 hours
	%	22	25	35	39	43	44	39	41	38	30	19	17	35 %

Zuid Limburg

latitude: 50°55'n longitude: 5°46'e height: 115 m

DOlk	009-4	Jan	Feb	Mar	Apr	May	Jun	Jul	Aug	Sep	Oct	Nov	Dec	Year
	tx	12	13	19	23	26	29	31	29	27	22	15	12	32 °C
	t	2	2	6	9	13	16	18	17	15	10	6	3	10 °C
	tn	-8	-8	-4	-1	1	5	8	8	5	0	-3	-6	-11 °C
	clothing	w	w	n	n	n	n	s	s	n	n	n	w	–
	stress							o	o					–
	p	63	56	46	48	64	66	85	90	73	49	62	61	763 mm
0.1 mm	dp	20	18	16	13	15	15	16	17	16	17	19	20	202 days
	sunshine	44	65	109	164	202	201	181	176	145	103	53	38	1481 hours
	%	17	23	30	40	42	41	36	39	38	31	20	15	33 %

4.1.9.2. Predominant general weather situations are described in 4.1.0.1. under Northern, Central and East Europe. A hygrothermal diagram of De Bilt is to be found in 4.1.0.2.

4.1.10. Belgium

4.1.10.1. Climatic description and tables

1. Temperate Oceanic Climate DO
2. Summer mild, winter cool.
3. Variable weather prevails, now and then periods of dry and also sunny days appear.
4. Wind: frequent gales, in coastal areas at times severe. Fog: occasionally, particularly in winter.

Oostende

latitude: 51°14'n longitude: 2°55'e height: 10 m

DOlk	010-1	Jan	Feb	Mar	Apr	May	Jun	Jul	Aug	Sep	Oct	Nov	Dec	Year
	tx	10	12	15	20	23	26	27	27	25	21	15	12	30 °C
	t	3	4	6	9	12	15	16	17	15	12	7	4	10 °C
	tn	-6	-3	-2	2	5	8	10	10	8	4	-1	-4	-8 °C
	clothing	w	w	n	n	n	n	n	s	n	n	n	w	–
	stress								o					–
	p	41	38	31	38	34	38	62	58	56	68	74	60	598 mm
0.1 mm	dp	13	12	11	12	11	10	12	13	10	13	15	16	148 days

Uccle (Brussels)

DObk	010-2	Jan	Feb	Mar	Apr	May	Jun	Jul	Aug	Sep	Oct	Nov	Dec	Year
				latitude: 50°48'n				longitude: 4°21'e				height: 104 m		
	tx	10	14	18	22	26	30	31	30	28	22	16	12	33 °C
	t	2	3	6	9	14	16	18	17	15	10	6	3	10 °C
	tn	-8	-6	-4	-1	1	6	7	7	5	2	-3	-7	-10 °C
	clothing	w	w	n	n	n	n	s	s	n	n	n	w	–
	stress							o	o					–
	p	73	59	52	54	57	57	79	75	68	72	71	68	785 mm
	dp	14	11	10	11	10	10	11	10	11	11	11	13	133 days
	sunshine	52	71	132	161	202	212	198	189	153	108	55	40	1573 hours
	%	20	25	36	39	42	43	40	42	40	32	20	16	35 %

4.1.10.2. Predominant general weather situations are described in 4.1.0.1. under Northern, Central and East Europe. A hygrothermal diagram of Uccle is to be found in 4.1.0.2.

4.1.11. Luxembourg

4.1.11.1. Climatic description and table

1. Temperate Oceanic Climate DO
2. Summer mild, winter cool.
3. Variable weather prevails, now and then periods of dry and also sunny days appear.
4. Wind: gales at times. Fog: occasionally, particularly in winter.

Luxembourg

DOlk	011-1	Jan	Feb	Mar	Apr	May	Jun	Jul	Aug	Sep	Oct	Nov	Dec	Year
				latitude: 49°37'n				longitude: 6°03'e				height: 330 m		
	tx	9	12	18	23	26	29	31	30	26	20	13	10	32 °C
	t	0	1	5	8	13	16	17	17	14	9	5	1	9 °C
	tn	-10	-10	-6	-2	1	6	8	7	4	-1	-3	-6	-12 °C
	clothing	w	w	n	n	n	n	s	s	n	n	n	w	–
	stress							o	o					–
	p	73	56	43	54	60	64	66	74	63	55	64	68	740 mm
0.1 mm	dp	13	10	8	9	9	10	10	10	10	10	11	12	122 days
	sunshine	45	68	143	171	201	189	199	183	143	95	28	22	1487 hours
	%	17	24	39	41	42	39	40	41	38	28	10	9	33 %

4.1.11.2. Predominant general weather situations are described in 4.1.0.1. under Northern, Central, and Eastern Europe.

4.1.12. Switzerland

4.1.12.1. Climatic description and tables

1. Temperate Oceanic Climate DO and Temperate Oceanic Mountain Climate GDO and Temperate Oceanic High Mountain Climate HDO
2. In the northern lowlands: summer mild, winter cool;
   in the southern lowlands: summer warm, winter cool.
3. Variable weather prevails; at times, periods with dry and sunny days, in the south more often than in the north.
4. Wind: gales frequent, especially in the mountains and in winter, at times severe. Fog: frequent in the lowlands in winter, when the highlands are often above the cloud cover. Snow cover: permanent above about 3 km.

Zürich

latitude: 47°23′n longitude: 8°34′e height: 569 m

GDObo	012-1	Jan	Feb	Mar	Apr	May	Jun	Jul	Aug	Sep	Oct	Nov	Dec	Year
	tx	10	13	19	24	28	30	32	31	28	22	14	11	33 °C
	t	-1	0	4	9	13	16	18	17	14	9	4	0	8 °C
	tn	-10	-9	-5	-1	2	7	9	9	5	0	-3	-7	-12 °C
	clothing	w	w	w	n	n	n	s	s	n	n	w	w	–
	stress							o	o					–
	p	75	70	66	80	107	136	143	131	108	80	76	65	1137 mm
	dp	12	10	9	11	13	13	13	13	10	10	10	10	135 days
	sunshine	47	79	149	173	207	220	238	219	166	108	51	37	1693 hours
	%	17	27	41	42	44	46	49	49	44	32	18	14	38 %

Säntis

latitude: 47°15′n longitude: 9°21′e height: 2496 m

GDOko	012-2	Jan	Feb	Mar	Apr	May	Jun	Jul	Aug	Sep	Oct	Nov	Dec	Year
	tx	0	0	2	5	10	13	16	15	13	8	4	1	16 °C
	t	-9	-9	-7	-4	0	4	6	6	4	-1	-4	-8	-2 °C
	tn	-20	-20	-17	-14	-10	-5	-3	-2	-6	-11	-14	-18	-22 °C
	clothing	w	w	w	w	w	w	n	n	w	w	w	w	–
	stress													–
	p	202	180	164	166	197	249	302	278	208	183	190	169	2488 mm
	dp	14	13	13	15	14	17	17	16	13	11	12	13	170 days
	sunshine	112	123	166	160	184	174	196	186	170	163	129	117	1880 hours
	%	41	42	45	39	39	36	41	42	45	48	46	44	42 %

Jungfraujoch

latitude: 46°33′n longitude: 7°59′e height: 3576 m

HDOoc	012-3	Jan	Feb	Mar	Apr	May	Jun	Jul	Aug	Sep	Oct	Nov	Dec	Year
	tx	-5	-5	-5	-3	1	4	6	6	4	1	-3	-4	7 °C
	tn	-26	-26	-22	-21	-16	-11	-9	-9	-12	-17	-20	-24	-29 °C
	clothing	p	p	p	p	w	w	w	w	w	w	p	p	–
	sunshine	91	100	153	164	162	182	208	197	173	140	101	92	1763 hours
	%	33	34	42	40	35	38	43	45	46	41	36	34	39 %

St. Moritz

					latitude: 46° 30′n			longitude: 9° 51′e			height: 1833 m		
GDOlo	012-4	Jan	Feb	Mar	Apr	May	Jun	Jul	Aug	Sep	Oct	Nov	Dec	Year
	tx	5	7	9	13	16	21	22	23	20	15	9	5	24 °C
	tn	-19	-18	-15	-8	-3	1	3	2	-1	-8	-12	-18	-22 °C
	clothing	w	w	w	w	n	n	n	n	n	w	w	w	–
	stress													–
	p	40	42	36	45	71	89	105	101	83	77	67	48	804 mm
0.3 mm	dp	9	9	8	8	12	13	13	12	10	10	9	9	122 days
	sunshine	101	126	163	171	176	185	216	195	166	146	104	81	1830 hours
	%	36	43	44	42	38	39	45	44	44	43	37	30	41 %

Genève

					latitude: 46° 12′n			longitude: 6° 09′e			height: 405 m		
DObk	012-5	Jan	Feb	Mar	Apr	May	Jun	Jul	Aug	Sep	Oct	Nov	Dec	Year
	tx	12	13	18	22	26	30	32	31	27	21	15	12	33 °C
	t	1	2	6	10	14	18	20	19	16	10	6	2	10 °C
	tn	-8	-7	-4	0	3	7	9	9	6	1	-2	-6	-10 °C
	clothing	w	w	n	n	n	s	s	s	n	n	n	w	–
	stress						o	o	o					–
	p	63	56	55	51	67	89	64	94	99	72	83	59	852 mm
	dp	9	8	8	8	9	10	7	9	9	8	9	9	103 days
	sunshine	54	98	167	204	236	261	287	257	189	123	61	42	1979 hours
	%	19	33	45	50	51	55	60	58	50	36	21	16	44 %

Lugano

					latitude: 46° 00′n			longitude: 8° 58′e			height: 276 m		
DObk	012-6	Jan	Feb	Mar	Apr	May	Jun	Jul	Aug	Sep	Oct	Nov	Dec	Year
	tx	13	17	21	25	28	31	32	32	29	24	17	14	33 °C
	t	2	4	8	12	15	19	21	20	17	12	7	3	12 °C
	tn	-6	-5	-2	1	5	9	11	11	8	2	-1	-5	-7 °C
	clothing	w	w	n	n	n	s	s	s	s	n	n	w	–
	stress						o	o	o	o				–
	p	63	67	98	148	215	198	185	196	159	173	147	95	1744 mm
	dp	6	6	7	10	13	12	10	10	9	9	8	7	108 days
	sunshine	117	143	171	186	191	234	268	243	189	147	110	102	2102 hours
	%	42	49	46	46	41	50	56	55	50	43	39	38	47 %

4.1.12.2. Predominant general weather situations are described in 4.1.0.1. under Northern, Central, and Eastern Europe, yet Southern Europe should be consulted, too. A hygrothermal diagram of Bern is to be found in 4.1.0.2.

4.1.13. France

4.1.13.1. Climatic descriptions and tables

1. Climatic area: France without the Mediterraenean Coast and hinterland.

1.1. Temperate Oceanic Climate DO and Temperate Oceanic Mountain Climate GDO and Temperate Oceanic High Mountain Climate HDC

1.2. Summer in the north mild, in the interior and otherwise warm, winter cool.

1.3. Variable weather prevails; now and then occur periods with dry and also sunny days, in the south more often than in the north.

1.4. Wind: gales at times, especially in coastal areas, and in the mountains.

Cherbourg

latitude: 49°39′n longitude: 1°38′w height: 12 m

DOlk 013-1	Jan	Feb	Mar	Apr	May	Jun	Jul	Aug	Sep	Oct	Nov	Dec	Year
tx	13	13	16	18	22	24	26	25	24	20	16	13	28 °C
t	6	6	8	10	12	15	16	17	16	13	10	8	11 °C
tn	-1	-1	0	3	5	8	11	11	9	5	2	0	-3 °C
clothing	n	n	n	n	n	n	n	s	n	n	n	n	–
stress								o					–
p	109	75	62	49	41	39	55	71	79	99	133	119	931 mm
dp	14	12	11	8	9	8	8	10	13	14	14	16	137 days
sunshine	56	60	143	163	204	214	233	218	151	74	45	47	1608 hours
%	21	21	39	40	43	44	47	49	40	22	16	19	36 %

Paris – Le Bourget

latitude: 48°58′n longitude: 2°27′e height: 53 m

DObk 013-2	Jan	Feb	Mar	Apr	May	Jun	Jul	Aug	Sep	Oct	Nov	Dec	Year
tx	13	15	20	24	27	31	33	31	29	24	16	13	34 °C
t	3	4	7	10	14	17	19	18	16	11	7	4	11 °C
tn	-6	-5	-1	1	4	8	11	10	7	1	-2	-4	-8 °C
clothing	w	w	n	n	n	s	s	s	n	n	n	w	–
stress						o	o	o					–
p	54	43	32	38	52	50	55	62	51	49	50	49	585 mm
dp	10	7	7	6	8	9	8	9	8	8	8	9	99 days
sunshine	49	75	144	181	219	230	224	201	158	118	53	40	1692 hours
%	18	26	39	44	46	47	46	45	42	35	19	16	38 %

Nancy

latitude: 48°41′n longitude: 6°13′e height: 217 m

DObk 013-3	Jan	Feb	Mar	Apr	May	Jun	Jul	Aug	Sep	Oct	Nov	Dec	Year
tx	11	14	20	23	27	30	32	31	28	22	15	12	33 °C
t	1	2	6	9	13	16	18	18	15	9	5	2	10 °C
tn	-11	-11	-8	-3	0	5	7	6	2	-3	-5	-9	-14 °C
clothing	w	w	n	n	n	n	s	s	n	n	n	n	–
stress							o	o					–
p	67	55	41	49	54	77	60	67	65	55	61	61	712 mm
dp	11	10	10	7	10	11	11	11	10	9	10	12	122 days
sunshine	52	84	157	164	217	221	236	214	166	119	54	43	1727 hours
%	19	29	43	40	46	46	48	48	44	35	19	17	39 %

Strasbourg

latitude: 48°35′n longitude: 7°38′e height: 154 m

DObk 013-4	Jan	Feb	Mar	Apr	May	Jun	Jul	Aug	Sep	Oct	Nov	Dec	Year
tx	12	14	20	24	27	31	32	31	28	22	15	13	34 °C
t	0	2	6	10	14	17	19	18	15	10	5	1	10 °C
tn	-10	-10	-6	-2	1	6	8	7	3	-2	-4	-8	-14 °C
clothing	w	w	n	n	n	s	s	s	n	n	n	w	–
stress						o	o	o					–
p	39	33	30	39	60	77	77	80	58	42	41	31	607 mm
dp	9	9	8	7	10	12	12	12	10	8	8	9	114 days
sunshine	50	78	157	171	214	219	229	217	166	111	49	39	1700 hours
%	18	27	43	42	45	45	47	49	44	33	18	15	38 %

Brest

DOlk	013-5	Jan	Feb	Mar	Apr	May	Jun	Jul	Aug	Sep	Oct	Nov	Dec	Year
				latitude: 48°27′n				longitude: 4°25′w				height: 103 m		
	tx	13	14	17	20	23	24	26	26	23	21	16	13	29 °C
	t	6	6	8	9	12	14	16	16	15	12	9	7	11 °C
	tn	-3	-4	0	1	3	6	8	8	6	3	1	-1	-6 °C
	clothing	n	n	n	n	n	n	n	n	n	n	n	n	–
	stress													–
	p	133	96	83	69	68	56	62	80	87	104	138	150	1126 mm
	dp	18	14	14	9	9	9	11	10	13	13	16	19	155 days
	sunshine	70	90	157	195	224	226	222	210	172	125	74	62	1827 hours
	%	26	31	43	48	47	47	45	47	45	37	27	24	41 %

Dijon

DObk	013-6	Jan	Feb	Mar	Apr	May	Jun	Jul	Aug	Sep	Oct	Nov	Dec	Year
				latitude: 47°16′n				longitude: 5°05′e				height: 227 m		
	tx	12	14	20	24	27	30	33	32	28	23	16	13	34 °C
	t	1	3	7	10	14	18	20	19	16	10	6	2	10 °C
	tn	-9	-8	-4	-1	2	7	9	9	5	0	-4	-7	-12 °C
	clothing	w	w	n	n	n	s	s	s	n	n	n	w	–
	stress						o	o	o					–
	p	64	42	42	46	64	81	58	77	72	60	76	57	739 mm
	dp	10	9	9	6	9	11	8	10	9	9	9	11	110 days
	sunshine	69	93	185	212	249	252	269	244	207	148	71	45	2044 hours
	%	25	32	50	52	53	53	56	55	55	44	25	17	46 %

Nantes

DObk	013-7	Jan	Feb	Mar	Apr	May	Jun	Jul	Aug	Sep	Oct	Nov	Dec	Year
				latitude: 47°10′n				longitude: 1°37′w				height: 27 m		
	tx	13	16	20	23	26	30	32	30	28	23	17	14	33 °C
	t	5	5	8	11	14	17	19	19	16	12	8	6	12 °C
	tn	-5	-5	-2	1	3	7	10	9	6	1	-3	-3	-7 °C
	clothing	n	n	n	n	n	s	s	s	n	n	n	n	–
	stress						o	o	o					–
	p	81	66	57	45	58	44	48	63	73	75	83	89	782 mm
	dp	12	10	11	7	9	8	8	10	11	10	12	15	123 days
	sunshine	75	93	155	192	218	230	235	221	170	132	76	58	1855 hours
	%	27	32	42	47	47	48	49	50	45	39	27	22	41 %

Bourges

DObk	013-8	Jan	Feb	Mar	Apr	May	Jun	Jul	Aug	Sep	Oct	Nov	Dec	Year
				latitude: 47°04′n				longitude: 2°23′e				height: 227 m		
	t	3	4	8	11	14	18	19	19	16	11	7	4	11 °C
	clothing	w	w	n	n	n	s	s	s	n	n	n	w	–
	stress						o	o	o					–
	p	58	50	44	46	68	60	52	63	57	58	61	58	675 mm
	dp	11	10	10	7	9	10	7	10	9	11	10	12	116 days
	sunshine	67	89	150	183	201	216	235	206	172	138	65	54	1776 hours
	%	24	31	41	45	43	45	49	47	45	41	23	20	40 %

Limoges

latitude: 45°49′n longitude: 1°17′e height: 284 m

DObk 013-9	Jan	Feb	Mar	Apr	May	Jun	Jul	Aug	Sep	Oct	Nov	Dec	Year
tx	14	16	20	24	27	31	33	32	28	24	19	14	34 °C
t	3	4	7	10	13	17	18	18	15	11	7	4	11 °C
tn	-10	-10	-6	-3	0	4	6	5	2	-3	-6	-7	-13 °C
clothing	w	w	n	n	n	s	s	s	n	n	n	w	–
stress						o	o	o					–
p	89	76	66	65	80	67	71	72	84	80	88	94	932 mm
dp	13	11	11	8	10	11	9	10	10	10	11	14	128 days
sunshine	73	94	164	188	214	237	240	225	183	142	79	64	1903 hours
%	26	32	45	46	46	50	50	51	48	42	28	24	43 %

Lyon

latitude: 45°43′n longitude: 4°57′e height: 201 m

DObk 013-10	Jan	Feb	Mar	Apr	May	Jun	Jul	Aug	Sep	Oct	Nov	Dec	Year
tx	14	16	20	23	28	32	34	34	29	24	18	14	35 °C
t	2	3	7	11	15	18	21	20	17	11	7	3	11 °C
tn	-8	-9	-4	-1	3	7	10	9	5	0	-3	-6	-11 °C
clothing	w	w	n	n	n	s	s	s	s	n	n	w	–
stress						o	o	o	o				–
p	52	46	53	56	69	85	56	89	93	77	80	57	813 mm
dp	10	8	9	6	8	10	7	9	8	9	9	10	103 days
sunshine	63	98	176	202	239	258	284	254	198	133	65	48	2018 hours
%	22	33	48	50	52	55	60	58	52	39	23	18	45 %

Bordeaux

latitude: 44°50′n longitude: 0°42′w height: 51 m

DObk 013-11	Jan	Feb	Mar	Apr	May	Jun	Jul	Aug	Sep	Oct	Nov	Dec	Year
tx	16	17	22	25	28	32	34	33	30	25	20	16	35 °C
t	5	6	9	12	15	18	20	20	17	13	8	6	12 °C
tn	-6	-6	-3	-1	3	6	8	8	6	0	-3	-5	-8 °C
clothing	n	n	n	n	n	s	s	s	s	n	n	n	–
stress						o	o	o	o				–
p	90	75	63	48	61	65	56	70	84	83	96	109	900 mm
dp	13	10	11	8	10	9	8	9	10	10	11	13	122 days
sunshine	80	101	172	203	224	251	253	238	190	148	84	64	2008 hours
%	28	34	47	50	49	54	53	54	50	43	29	23	45 %

Toulouse

latitude: 43°37′n longitude: 1°22′e height: 152 m

DObk 013-12	Jan	Feb	Mar	Apr	May	Jun	Jul	Aug	Sep	Oct	Nov	Dec	Year
tx	16	17	21	23	26	32	34	34	30	25	19	16	36 °C
t	4	5	9	11	15	19	21	21	18	13	8	3	12 °C
tn	-6	-6	-2	0	3	8	10	10	6	1	-3	-4	-9 °C
clothing	w	n	n	n	n	s	s	s	s	n	n	w	–
stress						o	o	o	o				–
p	49	46	53	50	75	61	44	54	64	45	51	67	659 mm
dp	10	8	9	7	8	9	7	6	7	8	9	11	99 days
sunshine	84	112	184	204	222	238	258	245	200	156	93	66	2062 hours
%	29	38	50	51	49	51	55	56	53	45	32	24	46 %

2. Climatic area: Mediterranean Coast of France and hinterland, Corsica.

2.1. Subtropical Winter Rain Climate Cs

2.2. Sommer hot or warm, winter cool.

2.3. Summer generally dry and sunny or cloudy; at times variable.
Winter partially variable, partially periods of dry and cloudy days.

2.4. Wind: gales infrequent in summer, frequent in winter. (See wind names under 4.1.13.2.)

Nimes

latitude: 43° 52′n longitude: 4° 24′e height: 60 m

Csak	013-13	Jan	Feb	Mar	Apr	May	Jun	Jul	Aug	Sep	Oct	Nov	Dec	Year
	tx	15	17	21	24	28	33	35	35	30	25	19	16	36 °C
	t	6	7	10	13	17	21	24	23	20	15	10	6	14 °C
	tn	-6	-6	-2	0	4	8	11	10	7	2	-1	-4	-8 °C
	clothing	n	n	n	n	s	s	t	t	s	n	n	n	–
	stress					o	o	f	f	o				–
	p	49	36	69	61	68	38	21	42	109	90	82	78	743 mm
	dp	5	5	8	5	5	6	3	4	5	7	6	7	66 days
	sunshine	143	166	200	246	279	320	361	311	240	184	139	124	2713 hours
	%	50	56	54	61	61	69	77	71	64	54	48	45	61 %

Nice

latitude: 43° 39′n longitude: 7° 12′e height: 10 m

Csak	013-14	Jan	Feb	Mar	Apr	May	Jun	Jul	Aug	Sep	Oct	Nov	Dec	Year
	tx	18	18	19	22	25	28	31	31	28	25	20	18	32 °C
	t	8	8	11	13	17	20	23	22	20	16	12	8	15 °C
	tn	0	1	2	5	8	12	15	14	13	8	4	2	-1 °C
	clothing	n	n	n	n	s	s	t	s	s	n	n	n	–
	stress					o	o	f	o	o				–
	p	68	61	73	73	68	35	20	27	77	124	129	107	862 mm
	dp	6	7	7	7	5	4	2	3	5	8	9	8	71 days
	sunshine	147	164	200	243	279	311	361	324	257	201	149	135	2771 hours
	%	51	55	54	60	61	67	77	75	68	59	51	49	62 %

Marseille

latitude: 43° 27′n longitude: 5° 13′e height: 8 m

Csak	013-15	Jan	Feb	Mar	Apr	May	Jun	Jul	Aug	Sep	Oct	Nov	Dec	Year
	tx	16	18	20	23	27	31	34	33	30	25	20	17	35 °C
	t	6	7	10	13	17	21	23	23	20	15	10	7	14 °C
	tn	-5	-5	-3	1	5	9	12	11	7	1	-2	-4	-7 °C
	clothing	n	n	n	n	s	s	t	t	s	n	n	n	–
	stress					o	o	f	f	o				–
	p	43	32	43	42	46	24	11	34	60	76	69	66	546 mm
	dp	5	4	5	4	5	3	1	3	5	7	6	7	55 days
	sunshine	137	161	210	254	290	325	369	326	254	194	147	124	2791 hours
	%	47	54	57	63	64	70	79	75	67	56	50	44	62 %

Perpignan

latitude: 42°44′n longitude: 2°52′e height: 48 m

Csak	013-16	Jan	Feb	Mar	Apr	May	Jun	Jul	Aug	Sep	Oct	Nov	Dec	Year
	tx	19	21	22	24	28	32	34	34	33	26	22	19	36 °C
	t	8	8	11	14	17	21	24	23	20	16	12	9	15 °C
	tn	-3	-3	1	4	8	12	15	15	11	5	2	-1	-4 °C
	clothing	n	n	n	n	s	s	t	t	s	n	n	n	–
	stress					o	o	f	f	o				–
	p	39	52	66	39	52	38	24	31	82	74	55	87	639 mm
0.1 mm	dp	5	4	7	4	6	5	4	4	5	6	3	7	60 days
	sunshine	163	171	217	256	263	283	320	284	226	183	157	140	2663 hours
	%	56	57	59	64	58	62	69	66	60	53	53	50	60 %

Ajaccio

latitude: 41°55′n longitude: 8°48′e height: 5 m

Csbk	013-17	Jan	Feb	Mar	Apr	May	Jun	Jul	Aug	Sep	Oct	Nov	Dec	Year
	tx	17	18	21	23	28	30	33	34	31	27	22	19	35 °C
	t	8	9	10	13	16	20	22	22	20	16	12	9	15 °C
	tn	-1	0	1	4	6	10	13	13	11	6	3	0	-2 °C
	clothing	n	n	n	n	n	s	s	s	s	n	n	n	–
	stress						o	o	o	o				–
	p	76	65	53	48	50	21	10	16	50	88	97	98	672 mm
0.1 mm	dp	9	9	7	7	7	3	1	2	4	7	9	10	75 days
	sunshine	134	138	201	241	298	339	382	343	273	206	138	117	2810 hours
	%	46	46	54	60	66	74	82	80	73	60	47	41	63 %

4.1.13.2. Predominant general weather situations are described in 4.1.0.1. under Northern, Central, and Eastern Europe as well as under Southern Europe. A hygrothermal diagram of Paris is to be found in 4.1.0.2.

Wind names: Mistral: a cold dry northerly or north-westerly wind which blows along the Mediterranean coast of France. It occurs along the coast as far as Genoa, but is most prevalent between Montpellier and Toulon. It is usually accompanied by clear weather and bright sunshine.

Bize: a cold, dry wind which blows in the winter in the mountainous regions of southern France from the North, Northeast or Northwest. The cold Northwest wind which occurs in Languedoc, near the Mediterranean coast, and differs from the Mistral in that it is accompanied by heavy clouds, has been given the name "bise noire".

4.1.14. Spain

4.1.14.1. Climatic descriptions and tables

1. Climatic area: Northern Spain from Galicia to Catalonia with exception of the dry valley of the Ebro

1.1. Subtropical Rain Climate Cr, on the Atlantic Ocean with cool summer Crb, in the mountains GCr, in the Pyrenees partly HCr.

1.2. Summer warm to hot, winter mild.

1.3. Variable weather prevails; repeatedly, especially in summer, dry sunny periods.

1.4. Wind: occasional gales, particularly in winter. Fog: now and then, especially in coastal areas and in valleys.

Santander

latitude: 43°28'n longitude: 3°49'w height: 64 m

Crbk	014-1	Jan	Feb	Mar	Apr	May	Jun	Jul	Aug	Sep	Oct	Nov	Dec	Year
	tx	18	19	23	24	25	26	28	28	29	25	21	18	32 °C
	t	9	9	12	12	14	17	19	19	18	15	12	11	14 °C
	tn	2	1	3	6	7	11	13	14	11	8	4	3	0 °C
	clothing	n	n	n	n	n	s	s	s	s	n	n	n	–
	stress						o	o	o	o				–
	p	124	91	83	75	72	62	64	89	119	158	117	154	1208 mm
	dp	16	14	13	14	15	13	12	14	14	14	16	16	173 days
	sunshine	83	110	150	175	194	192	206	196	156	137	96	72	1767 hours
	%	29	37	41	43	43	42	44	45	41	40	33	36	40 %

La Coruña

latitude: 43°22'n longitude: 8°25'w height: 65 m

Crbl	014-2	Jan	Feb	Mar	Apr	May	Jun	Jul	Aug	Sep	Oct	Nov	Dec	Year
	tx	17	18	21	22	23	26	28	28	26	24	20	17	30 °C
	t	10	10	12	12	14	16	18	19	18	15	12	10	14 °C
	tn	2	2	4	4	6	9	12	12	10	7	4	3	0 °C
	clothing	n	n	n	n	n	n	s	s	s	n	n	n	–
	stress							o	o	o				–
	p	121	80	95	70	60	46	29	47	71	92	125	139	975 mm
	dp	14	12	13	9	9	7	5	6	9	10	14	15	123 days
	sunshine	97	128	146	192	217	233	266	245	191	156	113	82	2066 hours
	%	34	43	40	48	48	50	57	56	51	45	39	29	46 %

Barcelona

latitude: 41°24'n longitude: 2°09'e height: 95 m

Crak	014-3	Jan	Feb	Mar	Apr	May	Jun	Jul	Aug	Sep	Oct	Nov	Dec	Year
	tx	18	18	21	24	27	30	32	32	29	25	21	18	33 °C
	t	9	10	12	15	18	22	24	24	22	18	14	10	16 °C
	tn	1	2	4	7	10	14	18	17	14	9	6	3	0 °C
	clothing	n	n	n	n	s	t	t	t	t	s	n	n	–
	stress					o	o	f	f	o	o			–
	p	30	40	53	45	54	40	30	47	79	77	54	49	598 mm
	dp	3	3	5	5	5	4	2	4	5	6	4	5	51 days
	sunshine	146	164	174	213	252	278	313	329	202	175	150	132	2538 hours
	%	49	55	47	53	56	61	68	76	54	51	51	46	57 %

2. *Climatic area:* parts of the valley of the Ebro and other rivers, of the high tablelands especially in the east, the lee-sides of the mountains of High Andalusia, and the bordering Mediterranean coast in the southeast of Spain.

2.1. Steppe Climate BS

2.2. Summer hot, winter mild.

2.3. Summer generally dry and sunny; at times variable. Winter: dry and cloudy periods prevail; sometimes periods of days with variable weather occur.

2.4. Wind: occasional gales, particularly in winter.

Zaragoza

latitude: 41°39'n longitude: 0°53'w height: 233 m

BSak	014-4	Jan	Feb	Mar	Apr	May	Jun	Jul	Aug	Sep	Oct	Nov	Dec	Year
	tx	17	19	24	26	30	35	37	37	33	26	21	16	38 °C
	t	6	8	11	14	17	21	23	24	21	15	10	7	15 °C
	tn	-4	-3	1	4	6	11	13	13	10	5	0	-2	-5 °C
	clothing	n	n	n	n	s	s	t	t	s	n	n	n	–
	stress					o	o	f	f	o				–
	p	16	16	30	31	48	37	17	19	31	34	28	32	339 mm
	dp	4	4	5	5	6	5	3	3	4	5	4	5	53 days
	sunshine	132	168	199	225	280	315	360	328	236	195	159	125	2722 hours
	%	45	56	54	56	62	69	78	76	63	56	54	44	61 %

Alicante

latitude: 38°22'n longitude: 0°30'w height: 82 m

BSal	014-5	Jan	Feb	Mar	Apr	May	Jun	Jul	Aug	Sep	Oct	Nov	Dec	Year
	tx	23	24	26	27	30	33	36	36	34	29	26	22	37 °C
	t	11	12	14	16	19	22	25	26	24	19	15	12	18 °C
	tn	1	1	4	6	9	13	17	17	14	9	5	3	0 °C
	clothing	n	n	n	n	s	s	t	t	t	s	n	n	–
	stress					o	o	f	f	o	o			–
	p	28	21	20	39	29	14	5	12	47	48	35	30	328 mm
	dp	4	4	4	5	4	2	1	1	4	5	4	4	42 days
	sunshine	183	192	214	254	300	327	358	328	248	215	185	171	2978 hours
	%	60	63	58	64	68	74	79	77	66	62	61	58	67 %

Almeria

latitude: 36°50'n longitude: 2°28'w height: 7 m

BSal	014-6	Jan	Feb	Mar	Apr	May	Jun	Jul	Aug	Sep	Oct	Nov	Dec	Year
	tx	19	21	23	25	29	32	35	34	33	27	23	20	36 °C
	t	11	12	14	16	18	22	25	25	23	19	16	13	18 °C
	tn	4	4	6	9	11	15	17	18	16	11	8	5	3 °C
	clothing	n	n	n	n	s	s	t	t	t	s	n	n	–
	stress					o	o	f	f	o	o			–
	p	28	19	21	28	15	5	0	5	16	26	28	35	226 mm
	dp	4	3	3	4	2	1	0	0	2	3	3	4	29 days
	sunshine	189	191	224	253	306	333	361	337	261	227	183	169	3034 hours
	%	61	62	61	64	70	76	81	80	70	65	60	56	68 %

3. Climatic area: greater part of Spain except Northern Spain (see 1. climatic area) and parts of Eastern Spain (see 2. climatic area).

3.1. Subtropical Winter Rain Climate Cs and Subtropical Winter Rain Mountain Climate GCs

3.2. Summer hot, winter mild, in the north cool.

3.3. Summer: generally dry and sunny, sometimes showers and cloudy.
Winter: partially variable, partially sunny periods.

3.4. Wind: occasional gales, particularly in winter; in summer thunderstorms accompanying the showers.

Valladolid

latitude: 41°40′n longitude: 4°43′w height: 715 m

GCsbk	014-7	Jan	Feb	Mar	Apr	May	Jun	Jul	Aug	Sep	Oct	Nov	Dec	Year
	tx	14	17	21	25	28	34	36	35	32	25	18	14	37 °C
	t	4	5	9	10	14	18	21	21	18	13	8	4	12 °C
	tn	-6	-5	-3	0	2	6	9	9	6	1	-3	-5	-8 °C
	clothing	w	n	n	n	n	s	s	s	s	n	n	w	–
	stress						o	o	o	o				–
	p	36	28	46	36	43	34	14	14	30	38	44	46	409 mm
	dp	6	5	7	6	7	5	2	2	4	5	6	7	62 days
	sunshine	107	144	181	232	265	317	368	341	243	194	132	87	2611 hours
	%	36	48	49	58	59	70	80	79	65	56	45	31	58 %

Madrid

latitude: 40°24′n longitude: 3°41′w height: 657 m

GCsak	014-8	Jan	Feb	Mar	Apr	May	Jun	Jul	Aug	Sep	Oct	Nov	Dec	Year
	tx	14	17	22	25	28	33	36	35	31	24	18	14	37 °C
	t	5	6	10	13	16	21	24	24	20	14	9	6	14 °C
	tn	-4	-3	0	2	4	9	12	12	8	3	0	-2	-6 °C
	clothing	n	n	n	n	n	s	t	t	s	n	n	n	–
	stress						o	f	f	o				–
	p	38	34	45	44	44	27	11	14	31	53	47	48	436 mm
	dp	6	4	7	6	6	4	1	1	4	5	6	6	56 days
	sunshine	157	176	188	235	280	334	383	353	255	207	164	139	2871 hours
	%	53	58	51	59	63	74	83	82	68	60	55	48	64 %

Badajoz

latitude: 38°53′n longitude: 6°49′w height: 185 m

Csak	014-9	Jan	Feb	Mar	Apr	May	Jun	Jul	Aug	Sep	Oct	Nov	Dec	Year
	tx	18	21	24	28	32	38	41	39	37	30	23	18	42 °C
	t	9	10	13	15	18	23	26	26	23	18	13	9	17 °C
	tn	-1	1	3	5	7	11	14	14	12	6	2	-1	-3 °C
	clothing	n	n	n	n	s	t	t	t	t	s	n	n	–
	stress					o	o	f	f	o	o			–
	p	61	47	68	42	37	18	3	4	25	48	61	60	474 mm
	dp	8	6	9	6	5	2	1	1	3	5	7	7	60 days
	sunshine	151	180	196	254	295	347	394	361	258	215	168	143	2962 hours
	%	50	59	53	64	67	78	81	85	69	62	56	49	66 %

Sevilla

latitude: 37°22′n longitude: 6°00′w height: 13 m

Csal	014-10	Jan	Feb	Mar	Apr	May	Jun	Jul	Aug	Sep	Oct	Nov	Dec	Year
	tx	20	23	27	30	34	39	43	42	38	33	25	20	44 °C
	t	10	12	14	16	19	23	26	26	24	19	14	11	18 °C
	tn	0	1	4	7	9	12	16	16	14	9	5	1	-1 °C
	clothing	n	n	n	n	s	t	t	t	t	s	n	n	–
	stress					o	o	f	f	o	o			–
	p	73	59	90	51	36	9	1	5	25	66	68	76	559 mm
	dp	6	6	8	6	4	1	0	0	2	5	6	7	51 days
	sunshine	164	174	187	237	286	328	370	339	246	207	170	158	2862 hours
	%	54	57	51	59	65	74	82	80	66	59	56	53	64 %

4. Climatic area: Balearic Islands
4.1. Subtropical Winter Rain Climate Cs
4.2. Summer hot, winter mild.
4.3. Summer: generally dry and sunny, sometimes showers and cloudy.
Winter: partially variable, partially sunny periods.
4.4. Wind: occasional gales.

Mahon/Menorca

latitude: 39° 52′n longitude: 4° 16′e height: 59 m

Csal 014-11	Jan	Feb	Mar	Apr	May	Jun	Jul	Aug	Sep	Oct	Nov	Dec	Year
tx	18	18	20	22	26	31	33	33	30	26	22	18	34 °C
t	10	10	12	14	15	21	24	24	22	18	14	12	17 °C
tn	3	3	4	7	9	13	17	17	15	10	7	4	2 °C
clothing	n	n	n	n	n	s	t	t	s	s	n	n	–
stress						o	f	f	o	o			–
p	60	44	48	34	30	21	4	22	72	133	92	77	637 mm
dp	6	4	7	6	6	4	1	1	4	5	6	6	56 days
sunshine	126	147	183	209	276	305	354	305	213	167	139	107	2531 hours
%	42	48	50	52	62	68	78	72	57	48	47	37	56 %

Palma de Mallorca

latitude: 39° 36′n longitude: 2° 42′e height: 45 m

Csal 014-12	Jan	Feb	Mar	Apr	May	Jun	Jul	Aug	Sep	Oct	Nov	Dec	Year
tx	18	19	21	23	27	31	33	33	31	27	22	19	34 °C
t	10	10	12	14	17	21	24	24	23	18	14	12	17 °C
tn	2	2	3	6	9	13	16	16	14	9	6	3	0 °C
clothing	n	n	n	n	s	s	t	t	t	s	n	n	–
stress					o	o	f	f	f	o			–
p	39	33	36	28	14	19	5	26	61	74	60	52	447 mm
dp	6	5	5	4	3	2	1	2	4	7	7	7	53 days
sunshine	153	174	204	229	297	310	356	332	239	195	168	141	2798 hours
%	51	57	55	58	67	69	78	78	64	56	56	48	63 %

Ibiza

latitude: 38° 52′n longitude: 1° 22′e height: 7 m

Csal 014-13	Jan	Feb	Mar	Apr	May	Jun	Jul	Aug	Sep	Oct	Nov	Dec	Year
tx	19	19	21	23	26	30	31	32	31	27	22	20	33 °C
t	11	10	13	15	18	22	24	25	23	19	15	12	17 °C
tn	2	1	3	6	9	14	17	18	14	10	6	3	0 °C
clothing	n	n	n	n	s	s	t	t	t	s	n	n	–
stress					o	o	f	f	f	o			–
p	42	35	25	22	13	22	5	12	34	85	102	47	444 mm
dp	6	6	6	5	3	3	1	1	4	8	10	8	61 days
sunshine	170	170	210	237	311	318	353	334	236	197	162	162	2860 hours
%	56	56	57	60	70	71	78	79	63	51	54	55	64 %

5. Climatic area: Canary Islands

5.1. Steppe Climate BS, in the easternmost islands Desert Climate BW, in the other islands often Subtropical Winter Rain Climate Cs on the weather-sides (north-eastern sides).

5.2. Summer hot, winter warm.

5.3. Summer: usually dry and sunny, sometimes cloudy, occasionally rain. Winter: prevailingly dry and cloudy, frequent showers, especially on the weather-sides of the islands.

5.4. Wind: trade winds predominate, at times gales occur, particularly in winter.

Tenerife

				latitude: 28°29′n				longitude: 16°20′w			height: 632 m		
GCsbl	014-14	Jan	Feb	Mar	Apr	May	Jun	Jul	Aug	Sep	Oct	Nov	Dec	Year
	t	12	12	13	13	15	16	18	19	19	18	15	12	15 °C
	clothing	n	n	n	n	n	s	s	s	s	s	n	n	–
	stress						o	o	o	o	o			–
	p	112	106	93	44	29	7	6	8	9	83	141	140	777 mm

Santa Cruz de Tenerife

				latitude: 28°27′n				longitude: 16°15′w			height: 46 m			
BSal	014-15	Jan	Feb	Mar	Apr	May	Jun	Jul	Aug	Sep	Oct	Nov	Dec	Year
	tx	23	24	26	26	28	31	34	35	32	30	27	24	37 °C
	t	17	18	18	19	20	22	24	25	24	23	20	18	21 °C
	tn	12	12	12	13	14	16	18	19	18	17	14	13	11 °C
	clothing	s	s	s	s	s	s	t	t	t	t	s	s	–
	stress	o	o	o	o	o	o	f	f	f	o	o	o	–
	p	36	39	27	13	6	0	0	0	3	31	45	51	252 mm
	dp	5	4	3	2	1	1	0	0	1	3	6	6	32 days
	sunshine	171	183	222	241	294	314	340	324	250	205	177	171	2892 hours
	%	52	58	60	63	70	75	79	79	67	57	55	53	65 %

Las Palmas – Gando / Gran Canaria

				latitude: 27°56′n				longitude: 15°25′w			height: 25 m			
BSal	014-16	Jan	Feb	Mar	Apr	May	Jun	Jul	Aug	Sep	Oct	Nov	Dec	Year
	tx	24	24	26	26	26	27	28	29	29	29	28	26	32 °C
	t	17	18	18	18	20	21	23	24	23	22	20	18	20 °C
	tn	12	12	12	13	14	17	18	19	18	17	15	13	11 °C
	clothing	s	s	s	s	s	s	t	t	t	s	s	s	–
	stress	o	o	o	o	o	o	f	f	f	o	o	o	–
	p	22	21	12	8	3	1	1	0	4	15	47	36	170 mm

4.1.14.2. Predominant general weather situations are described in 4.1.0.1. under Southern Europe and for Canary Islands under 4.3.0.1. under North Africa. A hygrothermal diagram of Madrid is to be found in 4.1.0.2.

Wind names: Embata is a wind contrary to the trade winds on the lee-sides of the Canary Islands.

Leste is a dry easterly wind laden with desert dust in the Canary Islands.

Levanter is a strong wind from the northeast which blows on the east coast of Spain, especially from February to May and October to December.
Leveche is a hot, dry southwest wind which occurs in Spain in front of an advancing depression.
Solano is a hot, dusty, moist, south-easterly or easterly wind which occurs on the east coast of Spain and in the strait of Gibraltar.

4.1.15. Gibraltar

4.1.15.1. Climatic description and table

1. Subtropical Winter Rain Climate Cs
2. Summer hot, winter mild.
3. Summer: generally dry and sunny; winter on the whole variable, frequent periods of dry and sunny days.
4. Wind: occasional gales.

Gibraltar North Front

latitude: 36°09'n longitude: 5°21'w height: 3 m

Csal	015-1	Jan	Feb	Mar	Apr	May	Jun	Jul	Aug	Sep	Oct	Nov	Dec	Year
	tx	20	21	24	24	28	30	34	34	31	28	24	21	35 °C
	t	13	13	15	17	19	22	24	24	23	20	16	14	18 °C
	tn	6	6	8	9	12	15	17	17	16	12	9	7	4 °C
	clothing	n	n	n	s	s	s	t	t	t	s	n	n	–
	stress				o	o	o	f	f	f	o			–
	p	154	101	119	62	27	5	1	3	10	64	144	125	815 mm
	dp	10	8	11	6	4	1	0	1	2	5	9	9	66 days
	sunshine	171	161	194	248	305	325	348	320	265	213	142	144	2836 hours
	%	55	52	52	63	70	74	78	76	71	61	46	48	64 %

4.1.15.2. Predominant general weather situations are described in 4.1.0.1. under Southern Europe.

4.1.16. Portugal

4.1.16.1. Climatic descriptions and tables

1. Climatic area: Portugal, northern part
1.1. Subtropical Rain Climate Cr
1.2. Summer warm, winter mild.
1.3. Variable weather prevails; dry sunny periods especially frequent in summer.
1.4. Wind: occasional gales, particularly in winter. Fog: now and then, especially in coastal areas and in valleys.

Bragança

latitude: 41°49′n longitude: 6°46′w height: 692 m

GCrbk	016-1	Jan	Feb	Mar	Apr	May	Jun	Jul	Aug	Sep	Oct	Nov	Dec	Year
	tx	14	16	20	24	26	32	35	34	31	25	18	14	36 °C
	t	4	6	8	10	13	17	20	20	17	12	8	4	12 °C
	tn	-6	-4	-2	0	2	5	8	8	5	1	-2	-4	-7 °C
	clothing	w	n	n	n	n	s	s	s	s	n	n	w	–
	stress						o	o	o	o				–
	p	149	104	133	73	69	42	15	16	39	79	110	144	973 mm
	dp	10	8	12	8	8	5	2	2	5	8	9	11	88 days
	sunshine	127	157	187	243	266	334	389	347	257	204	152	113	2776 hours
	%	43	52	51	61	59	73	84	81	68	59	51	40	62 %

O Porto

latitude: 41°14′n longitude: 8°41′w height: 73 m

Crbk	016-2	Jan	Feb	Mar	Apr	May	Jun	Jul	Aug	Sep	Oct	Nov	Dec	Year
	tx	18	20	23	26	28	32	34	33	31	28	22	18	36 °C
	t	9	10	12	14	15	18	20	20	19	16	12	10	14 °C
	tn	-1	-1	2	4	6	9	11	11	9	5	2	0	-2 °C
	clothing	n	n	n	n	n	s	s	s	s	n	n	n	–
	stress						o	o	o	o				–
	p	159	112	147	86	87	41	20	26	51	105	148	168	1150 mm
	dp	13	10	14	9	9	5	3	4	6	9	12	13	107 days
	sunshine	162	162	182	252	274	296	336	309	237	199	149	132	2690 hours
	%	55	54	49	63	61	65	73	72	63	58	50	46	60 %

Penhas Douradas

latitude: 40°25′n longitude: 7°33′w height: 1288 m

GCrlk	016-3	Jan	Feb	Mar	Apr	May	Jun	Jul	Aug	Sep	Oct	Nov	Dec	Year
	t	2	3	5	7	9	14	17	17	14	10	6	3	9 °C
	clothing	w	w	n	n	n	n	s	s	n	n	n	n	–
	stress							o	o					–
	p	276	190	238	143	147	67	25	27	82	154	262	305	1916 mm
	dp	13	10	14	11	11	6	3	3	6	11	12	13	113 days
	sunshine	126	152	167	218	245	324	375	335	241	194	140	123	2640 hours
	%	42	50	45	55	55	72	82	78	64	56	47	43	59 %

2. Climatic area: Portugal, southern part

2.1. Subtropical Winter Rain Climate Cs

2.2. Summer hot, winter mild.

2.3. Summer: generally dry and sunny, occasionally showers and cloudy; winter: variable, frequent dry and sunny.

2.4. Wind: occasional gales. Fog: at times, particularly near the coast.

Lisboa

					latitude: 38°46′n			longitude: 9°08′w			height: 110 m			
Csbl	016-4	Jan	Feb	Mar	Apr	May	Jun	Jul	Aug	Sep	Oct	Nov	Dec	Year
	tx	18	20	23	26	29	33	36	35	33	29	22	18	37 °C
	t	11	12	14	16	17	20	22	22	21	18	14	12	17 °C
	tn	3	4	6	8	9	12	14	15	14	10	7	4	2 °C
	clothing	n	n	n	n	s	s	s	s	s	s	n	n	–
	stress					o	o	o	o	o	o			–
	p	111	76	109	54	44	16	3	4	33	62	93	103	708 mm
	dp	11	8	11	7	7	2	1	1	4	7	9	11	79 days
	sunshine	161	182	206	265	301	330	378	357	279	231	174	159	3023 hours
	%	53	60	56	67	68	74	83	84	74	66	57	54	68 %

Beja

					latitude: 38°10′n			longitude: 7°52′w			height: 247 m			
Csak	016-5	Jan	Feb	Mar	Apr	May	Jun	Jul	Aug	Sep	Oct	Nov	Dec	Year
	t	9	10	12	15	17	21	24	24	22	18	13	10	16 °C
	clothing	n	n	n	n	s	s	t	t	s	s	n	n	–
	stress					o	o	o	o	o	o			–
	p	72	53	90	51	38	15	2	2	21	51	69	85	549 mm
	dp	9	7	10	7	6	3	0	1	3	6	8	9	69 days
	sunshine	162	179	199	251	291	338	378	352	274	222	168	164	2977 hours
	%	53	59	54	63	66	76	84	83	73	64	55	55	67 %

Sagres

					latitude: 36°59′n			longitude: 8°57′w			height: 41 m			
Csal	016-6	Jan	Feb	Mar	Apr	May	Jun	Jul	Aug	Sep	Oct	Nov	Dec	Year
	tx	18	18	20	22	24	27	29	28	27	25	21	19	30 °C
	t	13	13	14	15	17	18	19	19	19	18	16	13	16 °C
	tn	6	6	9	9	11	13	14	14	14	12	10	7	4 °C
	clothing	n	n	n	n	n	s	s	s	s	s	n	n	–
	stress						o	o	o	o	o			–
	p	63	47	61	38	24	5	2	1	14	46	60	56	417 mm
0.1 mm	dp	12	10	12	8	6	3	2	2	5	8	11	12	91 days
	sunshine	155	177	207	257	299	318	339	326	256	224	172	166	2896 hours
	%	50	58	56	65	68	72	76	77	68	64	56	55	65 %

3. Climatic area: Azores

3.1. Subtropical Rain Climate Cr

3.2. Summer hot or warm, winter mild.

3.3. Variable weather prevails; repeatedly especially in summer dry sunny periods.

3.4. Wind: especially in winter repeatedly storm, sometimes severe; in summer now and then tropical storms pass the islands.

Horta/Faial-Azores

latitude: 38° 31′n longitude: 28° 38′w height: 62 m

Cral	016-7	Jan	Feb	Mar	Apr	May	Jun	Jul	Aug	Sep	Oct	Nov	Dec	Year
	tx	19	19	19	21	23	25	28	29	28	25	22	20	29 °C
	t	14	14	14	15	17	19	22	23	22	19	17	15	18 °C
	tn	8	8	8	10	11	14	16	17	16	14	11	10	7 °C
	clothing	n	n	n	n	s	s	s	t	s	s	s	n	–
	stress					o	o	o	f	o	o	o		–
	p	125	106	122	68	70	48	32	44	81	110	109	113	1028 mm
	dp	15	13	16	10	10	7	5	7	10	12	13	13	131 days
	sunshine	91	100	122	153	180	177	235	235	193	144	106	87	1823 hours
	%	30	33	33	39	41	40	52	55	52	41	35	30	41 %

Ponta Delgada/Sao Miguel (Azores)

latitude: 37° 45′n longitude: 25° 40′w height: 36 m

Crbl	016-8	Jan	Feb	Mar	Apr	May	Jun	Jul	Aug	Sep	Oct	Nov	Dec	Year
	tx	20	20	20	21	23	25	27	28	27	25	23	21	28 °C
	t	14	14	14	15	16	19	21	22	21	19	17	15	17 °C
	tn	7	8	8	9	10	12	14	15	14	12	10	9	6 °C
	clothing	n	n	n	n	n	s	s	s	s	s	s	n	–
	stress						o	o	o	o	o	o		–
	p	120	100	105	67	62	42	27	29	81	103	120	102	958 mm
	dp	14	12	14	10	9	7	4	6	7	12	13	14	124 days
	sunshine	80	89	119	139	158	163	197	216	174	139	94	83	1651 hours
	%	26	29	32	35	36	37	44	51	47	40	31	28	37 %

4. Climatic area: Madeira Islands

4.1. Subtropical Winter Rain Climate Cs

4.2. Summer hot or warm, winter mild.

4.3. Summer generally dry and sunny; scattered showers. Winter variable, repeatedly dry and sunny.

4.4. Wind: stormy at times, particularly in winter.

Funchal – Madeira

latitude: 32° 41′n longitude: 16° 46′w height: 58 m

Csbl	016-9	Jan	Feb	Mar	Apr	May	Jun	Jul	Aug	Sep	Oct	Nov	Dec	Year
	tx	21	21	22	22	23	25	26	29	28	27	24	22	31 °C
	t	16	16	16	17	18	20	21	22	22	21	19	17	18 °C
	tn	9	9	9	10	12	13	16	17	16	14	12	11	8 °C
	clothing	n	n	n	s	s	s	s	s	s	s	s	s	–
	stress				o	o	o	o	o	o	o	o	o	–
	p	84	85	71	44	21	5	2	2	29	82	97	94	616 mm
	dp	7	6	7	4	2	1	0	1	2	7	7	7	51 days
	sunshine	165	158	205	219	210	194	240	251	227	202	162	167	2400 hours
	%	52	51	55	56	49	45	55	60	61	57	51	54	54 %

5. Climatic area: Cape Verde Islands

5.1. Marine Desert Climate BM

5.2. Wet season (July to November) hot; dry season (December – June) warm or hot.

5.3. July to November: frequent showers and clouds; now and then periods of dry and sunny days. December to June generally dry and cloudy or sunny.

5.4. Wind: sporadic gales. Fog: at times.

Mindelo – São Vicente – Cabo Verde

latitude: 16°53′n longitude: 25°00′w height: 2 m

BMab	016-10	Jan	Feb	Mar	Apr	May	Jun	Jul	Aug	Sep	Oct	Nov	Dec	Year
	tx	26	25	26	26	26	27	29	29	28	28	28	27	30 °C
	t	21	21	22	22	23	24	25	26	26	26	25	23	24 °C
	tn	17	16	16	16	17	19	18	21	21	20	18	17	14 °C
	clothing	s	s	s	s	t	t	t	t	t	t	t	t	–
	stress	o	o	o	f	f	f	ff	ff	ff	ff	ff	f	–
	p	2	4	1	0	0	0	2	20	39	30	18	9	125 mm
0.1 mm	dp	1	1	0	0	0	0	1	4	5	2	2	2	18 days
	sunshine	229	237	276	291	307	276	248	226	225	239	225	202	2981 hours
	%	66	72	74	77	77	70	61	57	61	65	66	59	67 %

Sal – Cabo Verde

latitude: 16°44′n longitude: 22°57′w height: 55 m

BMab	016-11	Jan	Feb	Mar	Apr	May	Jun	Jul	Aug	Sep	Oct	Nov	Dec	Year
	t	21	21	21	21	22	23	24	26	27	26	24	22	23 °C
	clothing	s	s	s	s	s	t	t	t	t	t	t	s	–
	stress	f	f	f	f	f	ff	ff	ff	ff	ff	ff	f	–
	p	2	2	1	0	0	0	18	20	27	15	10	13	108 mm
	dp	1	0	0	0	0	0	0	1	2	2	1	1	8 days
	sunshine	188	205	245	241	247	219	173	162	219	227	209	179	2514 hours
	%	54	63	66	64	62	56	43	43	60	62	61	52	57 %

Praia – São Tiago – Cabo Verde

latitude: 14°54′n longitude: 23°31′w height: 27 m

BMab	016-12	Jan	Feb	Mar	Apr	May	Jun	Jul	Aug	Sep	Oct	Nov	Dec	Year
	tx	28	28	29	30	30	31	31	31	31	31	31	28	32 °C
	t	23	22	23	23	24	25	26	26	27	27	26	24	25 °C
	tn	18	18	18	19	20	21	22	23	23	23	22	20	17 °C
	clothing	t	s	t	t	t	t	t	t	t	t	t	t	–
	stress	f	f	f	ff	ff	ff	ff	ff	ff	ff	ff	ff	–
	p	2	4	0	0	0	0	9	38	86	84	35	12	271 mm
0.1 mm	dp	1	0	0	0	0	0	2	7	7	4	2	1	24 days
	sunshine	223	237	279	285	307	279	217	202	216	245	234	205	2928 hours
	%	63	72	75	76	77	71	54	52	59	67	68	59	66 %

4.1.16.2. Predominant general weather situations are described in 4.1.0.1. under Southern Europe, and for the Cape Verde Islands in 4.3.0.1. under West Africa. A hygrothermal diagram of Lisboa is to be found in 4.1.0.2.

Wind name: Leste is a dry easterly wind with desert dust in the Madeira Islands.

4.1.17. Federal Republic of Germany and West Berlin

4.1.17.1. Climatic descriptions and tables

1. Climatic area: West Berlin

1.1. Temperate Continental Climate DC

1.2. Summer warm, winter cool.

1.3. Generally variable weather; occasionally periods of dry and cloudy, sometimes sunny days.

1.4. Wind: occasional gales. Snow cover: now and then.

Berlin

latitude: 52° 28′n longitude: 13° 24′e height: 50 m

DCbk	017-1	Jan	Feb	Mar	Apr	May	Jun	Jul	Aug	Sep	Oct	Nov	Dec	Year
	tx	9	10	16	22	28	30	32	31	28	21	13	10	33 °C
	t	0	0	4	9	14	18	19	19	15	10	5	1	10 °C
	tn	-12	-12	-7	-2	2	6	9	8	4	-1	-4	-9	-15 °C
	clothing	w	w	w	n	n	s	s	s	n	n	n	w	–
	stress						o	o	o					–
	p	41	37	30	39	44	60	67	65	45	45	44	39	556 mm
	dp	10	8	7	9	8	9	10	9	8	9	9	9	105 days
	sunshine	56	78	151	193	239	244	242	212	194	123	50	36	1818 hours
	%	22	28	41	46	49	48	48	46	51	37	19	15	40 %

*2. Climatic area: Federal Republic of G*ermany

2.1. Temperate Oceanic Climate DO and Temperate Oceanic Mountain Climate GDO, locally Temperate Oceanic High Mountain Climate HDO.

2.2. Summer mild, winter cool in the lowlands, in higher regions cold.

2.3. Variable weather prevails; now and then periods of dry and also sunny days occur.

2.4. Wind: occasionally storm, in coastal areas and at higher levels more frequent than in the interior. Snow cover in lowlands now and then. Fog: occasionally, especially in winter.

Schleswig

latitude: 54° 32′n longitude: 9° 33′e height: 48 m

DOlk	017-2	Jan	Feb	Mar	Apr	May	Jun	Jul	Aug	Sep	Oct	Nov	Dec	Year
	tx	7	8	13	18	24	26	27	26	22	17	11	9	29 °C
	t	0	0	2	7	11	15	17	16	13	9	5	2	8 °C
	tn	-10	-9	-6	-2	1	5	8	8	4	-1	-5	-9	-12 °C
	clothing	w	w	w	n	n	n	n	n	n	n	n	w	–
	stress													–
	p	74	60	46	58	61	63	95	104	87	89	73	70	880 mm
	dp	14	11	9	10	9	10	12	13	12	12	12	13	136 days
	sunshine	49	66	128	191	249	248	238	187	175	102	46	32	1711 hours
	%	20	24	35	45	50	48	46	40	46	31	18	14	38 %

Helgoland

latitude: 54° 11′n longitude: 7° 53′e height: 57 m

DOlk	017-3	Jan	Feb	Mar	Apr	May	Jun	Jul	Aug	Sep	Oct	Nov	Dec	Year
	tx	8	6	10	14	21	23	25	24	20	16	12	9	25 °C
	t	2	2	3	6	10	13	16	16	14	10	6	3	8 °C
	tn	-6	-5	-3	1	4	8	11	12	10	4	1	-3	-8 °C
	clothing	w	w	w	n	n	n	n	n	n	n	n	w	–
	stress													–
	p	58	44	38	40	36	41	79	83	85	100	71	56	731 mm
	dp	11	10	10	9	8	8	9	13	11	13	13	14	129 days
	sunshine	48	68	139	185	230	237	214	217	150	93	44	33	1658 hours
	%	19	24	37	43	46	46	41	46	39	28	17	14	36 %

Hamburg

latitude: 53°38'n longitude: 10°00'e height: 14 m

DOlk	017-4	Jan	Feb	Mar	Apr	May	Jun	Jul	Aug	Sep	Oct	Nov	Dec	Year
	tx	9	10	16	21	26	29	31	30	26	19	13	10	32 °C
	t	0	0	3	8	12	16	17	17	14	9	5	2	9 °C
	tn	-11	-11	-7	-2	1	5	8	7	4	-1	-4	-8	-14 °C
	clothing	w	w	w	n	n	n	s	s	n	n	n	w	–
	stress							o	o					–
	p	57	48	39	52	53	64	84	83	63	59	59	59	720 mm
	dp	11	10	8	10	10	10	12	12	11	11	11	12	127 days
	sunshine	51	64	131	186	230	222	220	183	171	100	44	28	1630 hours
	%	21	23	36	44	47	43	43	40	45	30	17	12	36 %

Emden

latitude: 53°22'n longitude: 7°13'e height: 1 m

DOlk	017-5	Jan	Feb	Mar	Apr	May	Jun	Jul	Aug	Sep	Oct	Nov	Dec	Year
	tx	8	9	14	20	26	27	28	27	24	18	12	9	30 °C
	t	1	1	4	8	12	15	17	17	14	10	6	3	9 °C
	tn	-9	-7	-4	-1	2	6	9	9	5	0	-4	-7	-12 °C
	clothing	w	w	w	n	n	n	s	s	n	n	n	w	–
	stress							o	o					–
	p	57	47	39	43	51	59	92	88	68	68	69	57	737 mm
	dp	12	10	8	10	8	10	12	12	12	11	12	11	129 days
	sunshine	50	66	116	192	234	217	202	183	160	100	54	34	1608 hours
	%	20	23	31	45	47	42	39	39	41	30	20	14	35 %

Hannover

latitude: 52°28'n longitude: 9°42'e height: 55 m

DOlk	017-6	Jan	Feb	Mar	Apr	May	Jun	Jul	Aug	Sep	Oct	Nov	Dec	Year
	tx	10	11	17	22	27	29	31	30	27	20	14	11	32 °C
	t	0	1	4	8	13	16	17	17	14	9	5	2	9 °C
	tn	-12	-11	-7	-2	1	5	8	7	3	-2	-4	-10	-16 °C
	clothing	w	w	w	n	n	n	s	s	n	n	n	w	–
	stress							o	o					–
	p	50	48	39	47	53	66	83	72	52	57	55	50	672 mm
	dp	10	10	8	10	10	10	11	11	9	10	10	9	117 days
	sunshine	46	69	120	184	227	214	206	188	165	105	52	34	1609 hours
	%	18	24	32	44	46	42	40	41	43	31	19	14	35 %

Essen

latitude: 51°24'n longitude: 6°57'e height: 129 m

DObk	017-7	Jan	Feb	Mar	Apr	May	Jun	Jul	Aug	Sep	Oct	Nov	Dec	Year
	tx	10	12	17	21	26	29	30	29	26	20	15	11	31 °C
	t	2	2	5	9	13	16	18	17	15	10	6	3	10 °C
	tn	-9	-8	-4	-1	2	6	8	8	5	0	-3	-7	-12 °C
	clothing	w	w	n	n	n	n	s	s	n	n	n	n	–
	stress							o	o					–
	p	73	63	47	61	63	75	86	90	66	67	72	66	829 mm
	dp	13	11	10	11	10	11	12	12	11	11	12	13	138 days
	sunshine	40	61	119	173	212	204	179	171	150	98	53	34	1494 hours
	%	15	22	32	42	44	41	36	38	39	29	20	14	33 %

Kassel

latitude: 51°19′n longitude: 9°29′e height: 163 m

DObk	017-8	Jan	Feb	Mar	Apr	May	Jun	Jul	Aug	Sep	Oct	Nov	Dec	Year
	tx	9	10	17	22	26	29	30	30	27	20	14	10	32 °C
	t	0	1	5	9	13	16	18	17	14	9	5	1	9 °C
	tn	-12	-12	-7	-2	1	5	8	7	3	-1	-4	-9	-15 °C
	clothing	w	w	n	n	n	n	s	s	n	n	n	w	–
	stress							o	o					–
	p	46	43	32	47	58	66	72	66	52	53	49	46	630 mm
	dp	11	10	8	10	9	10	10	10	9	10	10	10	118 days
	sunshine	48	73	137	188	221	213	203	181	150	103	51	28	1596 hours
	%	18	26	37	45	46	43	40	40	39	31	19	11	36 %

Frankfurt am Main

latitude: 50°03′n longitude: 8°35′e height: 113 m

DObk	017-9	Jan	Feb	Mar	Apr	May	Jun	Jul	Aug	Sep	Oct	Nov	Dec	Year
	tx	10	12	19	24	29	31	33	32	28	21	14	10	34 °C
	t	1	2	6	10	14	18	19	18	15	10	6	3	10 °C
	tn	-10	-8	-4	0	3	7	10	9	5	0	-2	-7	-12 °C
	clothing	w	w	n	n	n	s	s	s	n	n	n	w	–
	stress						o	o	o					–
	p	55	46	40	36	54	70	63	78	58	54	46	64	663 mm
	dp	9	9	9	9	9	9	10	10	9	9	9	11	111 days
	sunshine	45	63	129	154	206	227	228	204	146	94	36	31	1563 hours
	%	17	22	35	37	43	46	46	45	38	28	13	12	35 %

Nürnberg

latitude: 49°30′n longitude: 11°05′e height: 318 m

DObo	017-10	Jan	Feb	Mar	Apr	May	Jun	Jul	Aug	Sep	Oct	Nov	Dec	Year
	tx	9	12	17	22	27	30	31	31	27	22	14	10	33 °C
	t	-1	0	4	8	13	17	18	17	14	8	4	0	8 °C
	tn	-14	-14	-8	-4	-1	5	7	6	1	-3	-6	-12	-18 °C
	clothing	w	w	w	n	n	s	s	s	n	n	w	w	–
	stress						o	o	o					–
	p	43	39	35	40	55	71	90	75	46	46	41	42	623 mm
	dp	10	10	8	9	10	11	12	11	9	9	9	10	115 days
	sunshine	55	81	153	189	231	221	229	214	175	125	54	39	1766 hours
	%	21	28	42	46	49	45	46	48	46	37	20	15	39 %

Stuttgart

latitude: 48°50′n longitude: 9°12′e height: 315 m

DObk	017-11	Jan	Feb	Mar	Apr	May	Jun	Jul	Aug	Sep	Oct	Nov	Dec	Year
	tx	11	13	19	23	27	30	32	31	29	22	16	12	33 °C
	t	0	1	5	10	14	17	19	18	15	10	5	1	10 °C
	tn	-12	-10	-6	-2	2	7	9	8	4	-1	-4	-9	-15 °C
	clothing	w	w	n	n	n	s	s	s	n	n	n	w	–
	stress						o	o	o					–
	p	48	42	38	51	74	94	79	79	62	48	48	40	703 mm
	dp	9	9	7	9	10	12	11	10	10	9	9	8	114 days
	sunshine	70	90	151	181	225	204	236	218	178	136	66	60	1814 hours
	%	26	31	41	44	48	42	48	49	47	40	24	23	40 %

München

latitude: 48°08′n longitude: 11°42′e height: 528 m

GDObo	017-12	Jan	Feb	Mar	Apr	May	Jun	Jul	Aug	Sep	Oct	Nov	Dec	Year
	tx	11	12	18	23	26	29	31	31	27	22	15	12	32 °C
	t	-2	-1	3	8	12	16	18	17	14	8	5	-1	8 °C
	tn	-16	-15	-9	-4	1	5	8	7	3	-2	-6	-12	-18 °C
	clothing	w	w	w	n	n	n	s	s	n	n	n	w	–
	stress							o	o					–
	p	59	55	51	62	107	125	140	104	87	67	57	50	964 mm
	dp	11	10	9	10	12	14	13	12	10	9	8	10	128 days
	sunshine	56	72	142	173	217	201	226	211	176	130	54	41	1699 hours
	%	21	25	39	42	46	42	46	47	46	38	19	16	38 %

Freiburg im Breisgau

latitude: 48°01′n longitude: 7°50′e height: 259 m

DObk	017-13	Jan	Feb	Mar	Apr	May	Jun	Jul	Aug	Sep	Oct	Nov	Dec	Year
	tx	12	13	19	24	28	30	32	32	28	22	16	13	35 °C
	t	1	3	6	10	14	17	19	19	15	10	5	2	10 °C
	tn	-10	-10	-6	-1	2	6	9	8	4	-1	-5	-9	-14 °C
	clothing	w	w	n	n	n	s	s	s	n	n	n	w	–
	stress						o	o	o					–
	p	61	53	52	68	79	117	106	100	98	67	69	52	922 mm
	dp	10	9	11	13	12	12	12	12	10	10	10	11	131 days
	sunshine	53	82	153	173	228	233	247	234	174	122	62	47	1808 hours
	%	19	28	42	42	48	48	51	53	46	36	22	18	40 %

Friedrichshafen

latitude: 47°39′n longitude: 9°29′e height: 407 m

DObk	017-14	Jan	Feb	Mar	Apr	May	Jun	Jul	Aug	Sep	Oct	Nov	Dec	Year
	tx	9	10	16	21	26	28	29	28	24	19	14	10	30 °C
	t	-1	0	4	9	13	17	18	18	14	9	4	0	9 °C
	tn	-11	-10	-6	-2	1	6	8	7	4	-1	-5	-10	-14 °C
	clothing	w	w	w	n	n	s	s	s	n	n	w	w	–
	stress						o	o	o					–
	p	63	56	53	60	95	112	137	113	93	66	59	54	961 mm
	dp	10	10	8	10	11	13	13	12	10	9	10	10	126 days
	sunshine	51	69	158	186	232	216	239	225	177	107	57	39	1756 hours
	%	19	24	43	46	49	45	49	51	47	32	20	15	39 %

Zugspitze

latitude: 47°25′n longitude: 10°59′e height: 2962 m

HDOkc	017-15	Jan	Feb	Mar	Apr	May	Jun	Jul	Aug	Sep	Oct	Nov	Dec	Year
	tx	-2	-2	0	2	7	11	13	12	10	6	1	1	14 °C
	t	-12	-12	-10	-7	-2	0	2	2	1	-3	-7	-10	-5 °C
	tn	-23	-23	-20	-18	-12	-8	-5	-6	-9	-14	-17	-21	-26 °C
	clothing	p	p	p	w	w	w	w	w	w	w	w	p	–
	stress													–
	p	203	165	160	178	168	201	210	144	115	129	121	153	1948 mm
	dp	14	15	14	16	16	18	18	17	13	11	12	14	178 days
	sunshine	116	120	164	163	170	136	167	171	178	176	136	118	1815 hours
	%	42	41	45	40	36	28	35	39	47	52	48	45	41 %

4.1.17.2. Predominant general weather situations are described in 4.1.0.1. under Northern, Central, and Eastern Europe. A hygrothermal diagram of Berlin is to be found in 4.1.0.2.

Wind name: Föhn: a warm, dry wind which blows down the slopes on the leeward side of a ridge of mountains. The name originated in the Alps where the föhn is very prevalent, especially on the northern slopes; the wind also occurs on the eastern side of the Rocky Mountains, where it is known as the Chinook. Owing to its warmth and dryness it melts the snow very quickly and causes a considerable rise in the temperature. It is frequently accompanied by lenticular clouds.

4.1.18. German Democratic Republic

4.1.18.1. Climatic description and tables

1. Except for the Harz Mountains and the western parts of the Districts of Rostock and Schwerin with Temperate Oceanic Climate DO, there is Temperate Continental Climate DC and Temperate Continental Mountain Climate GDC
2. Summer mild, partly warm, winter cool, in higher regions cold.
3. Variable weather prevails; now and then periods of dry and cloudy or sunny days occur.
4. Wind: occasional gales, more frequently in coastal areas and at higher levels than in the lowlands. Fog: predominant in winter. Snow cover: in lowlands in the northwestern part occasionally, in other parts more frequently.

Warnemünde

				latitude: 54°11'n			longitude: 12°05'e			height: 10 m			
DObk	018-1	Jan	Feb	Mar	Apr	May	Jun	Jul	Aug	Sep	Oct	Nov	Dec	Year
	tx	8	9	15	20	27	28	30	28	25	19	12	9	32 °C
	t	0	0	3	7	12	15	18	17	14	10	5	2	8 °C
	tn	-12	-10	-7	-3	0	4	7	7	3	-2	-6	-9	-15 °C
	clothing	w	w	w	n	n	n	s	s	n	n	n	w	–
	stress							o	o					–
	p	44	32	33	40	49	54	78	64	60	52	46	43	595 mm
	dp	11	8	9	9	8	9	11	11	9	10	10	11	116 days
	sunshine	37	62	108	168	229	225	217	186	159	93	48	22	1554 hours
	%	15	22	30	40	46	44	42	40	41	28	19	10	35 %

Greifswald

				latitude: 54°06'n			longitude: 13°27'e			height: 2 m			
DCbo	018-2	Jan	Feb	Mar	Apr	May	Jun	Jul	Aug	Sep	Oct	Nov	Dec	Year
	tx	9	9	15	20	25	28	29	28	25	19	12	10	30 °C
	t	-1	-1	2	7	12	16	18	18	14	9	4	1	8 °C
	tn	-11	-12	-7	-3	1	5	8	8	4	0	-4	-8	-15 °C
	clothing	w	w	w	n	n	n	s	s	n	n	w	w	–
	stress							o	o					–
	p	40	33	30	39	45	55	69	55	59	51	36	41	553 mm
	dp	10	8	9	9	8	9	10	11	8	10	9	11	112 days
	sunshine	50	66	139	184	260	264	250	222	184	116	51	37	1823 hours
	%	20	24	38	44	52	51	48	48	48	35	20	16	41 %

Neustrelitz

DClo	018-3	Jan	Feb	Mar	Apr	May	Jun	Jul	Aug	Sep	Oct	Nov	Dec	Year
		latitude: 53°21'n				longitude: 13°05'e				height: 70 m				
	tx	9	10	16	21	27	30	30	29	26	19	12	10	32 °C
	t	-1	-1	2	7	12	16	17	17	13	8	4	1	8 °C
	tn	-14	-15	-9	-5	-1	3	6	5	1	-3	-6	-9	-18 °C
	clothing	w	w	w	n	n	n	s	s	n	n	w	w	–
	stress							o	o					–
	p	37	34	29	44	56	66	70	65	53	52	38	37	581 mm
	dp	11	8	10	9	9	9	11	11	9	9	9	11	116 days
	sunshine	58	73	122	174	234	272	225	186	185	100	43	29	1701 hours
	%	23	26	33	41	48	53	44	40	48	30	17	12	38 %

Magdeburg

DCbo	018-4	Jan	Feb	Mar	Apr	May	Jun	Jul	Aug	Sep	Oct	Nov	Dec	Year
		latitude: 52°06'n				longitude: 11°34'e				height: 85 m				
	tx	10	11	17	23	28	30	32	31	28	21	14	11	33 °C
	t	-1	0	4	8	13	16	18	17	14	9	4	1	9 °C
	tn	-13	-12	-7	-2	1	6	8	8	3	-1	-4	-10	-16 °C
	clothing	w	w	w	n	n	n	s	s	n	n	w	w	–
	stress							o	o					–
	p	35	33	28	34	49	61	64	57	38	43	40	33	515 mm
	dp	9	8	8	8	8	9	10	9	8	8	7	10	103 days
	sunshine	56	73	142	178	221	231	212	200	174	103	50	39	1679 hours
	%	22	26	39	43	45	46	42	44	46	31	19	16	37 %

Brocken

GDOlo	018-5	Jan	Feb	Mar	Apr	May	Jun	Jul	Aug	Sep	Oct	Nov	Dec	Year
		latitude: 51°48'n				longitude: 10°37'e				height: 1152 m				
	tx	6	5	10	13	18	21	22	21	19	14	10	7	23 °C
	t	-5	-5	-2	1	6	9	11	11	8	4	0	-3	3 °C
	tn	-15	-15	-12	-8	-4	0	3	3	0	-4	-8	-12	-18 °C
	clothing	w	w	w	w	n	n	n	n	n	w	w	w	–
	stress													–
	p	158	126	94	105	96	115	143	117	105	122	115	126	1422 mm
	dp	18	14	15	15	13	14	15	15	14	15	16	18	182 days
	sunshine	57	68	122	141	186	195	170	163	135	86	51	50	1424 hours
	%	22	24	33	34	38	39	34	36	35	26	19	21	32 %

Wernigerode (tx and tn, sunshine Quedlinburg)

DOlk	018-6	Jan	Feb	Mar	Apr	May	Jun	Jul	Aug	Sep	Oct	Nov	Dec	Year
		latitude: 51°51'n				longitude: 10°46'e				height: 240 m				
	tx	10	12	17	22	27	29	31	30	26	21	14	11	32 °C
	t	0	0	3	8	12	16	17	17	14	9	5	2	8 °C
	tn	-14	-12	-7	-3	1	5	8	7	2	-2	-7	-11	-17 °C
	clothing	w	w	w	n	n	n	s	s	n	n	n	w	–
	stress							o	o					–
	p	49	41	36	46	60	69	73	60	43	51	50	49	627 mm
	sunshine	53	64	99	141	202	204	198	180	135	93	60	46	1475 hours
	%	20	23	27	34	42	41	40	40	35	28	22	20	33 %

Leipzig

latitude: 51°24′n longitude: 12°24′e height: 137 m

DCbo	018-7	Jan	Feb	Mar	Apr	May	Jun	Jul	Aug	Sep	Oct	Nov	Dec	Year
	tx	10	12	17	23	27	30	32	32	28	22	14	11	33 °C
	t	-1	0	3	8	13	16	18	18	14	9	4	1	9 °C
	tn	-13	-12	-7	-2	2	7	10	8	4	-1	-4	-9	-16 °C
	clothing	w	w	w	n	n	n	s	s	n	n	w	w	–
	stress							o	o					–
	p	37	36	34	38	47	67	73	59	38	46	39	45	559 mm
	dp	10	8	9	9	9	9	10	10	8	8	8	10	109 days
	sunshine	56	69	139	177	223	231	223	211	174	121	56	46	1726 hours
	%	22	24	38	43	46	46	44	46	46	36	21	19	38 %

Goerlitz

latitude: 51°10′n longitude: 14°57′e height: 238 m

DCbo	018-8	Jan	Feb	Mar	Apr	May	Jun	Jul	Aug	Sep	Oct	Nov	Dec	Year
	tx	8	9	15	20	26	28	30	29	25	19	13	9	31 °C
	t	-2	-2	2	8	12	16	18	17	14	8	4	0	8 °C
	tn	-12	-11	-7	-2	2	6	9	8	4	-1	-6	-11	-16 °C
	clothing	w	w	w	n	n	n	s	s	n	n	w	w	–
	stress							o	o					–
	p	47	42	44	48	65	71	98	76	52	54	46	45	688 mm
	dp	11	9	9	9	10	11	12	11	9	9	9	11	119 days

Dresden

latitude: 51°08′n longitude: 13°46′e height: 129 m

DCbo	018-9	Jan	Feb	Mar	Apr	May	Jun	Jul	Aug	Sep	Oct	Nov	Dec	Year
	tx	10	11	17	22	27	31	32	31	28	21	14	10	33 °C
	t	0	-1	4	9	14	17	19	18	14	9	5	2	9 °C
	tn	-14	-13	-9	-2	1	6	9	8	3	-2	-4	-11	-17 °C
	clothing	w	w	w	n	n	s	s	s	n	n	n	w	–
	stress							o	o					–
	p	42	40	42	48	66	69	120	67	49	53	41	43	680 mm
	dp	10	8	9	10	10	10	11	10	9	8	8	10	111 days
	sunshine	59	72	123	163	210	211	211	206	162	117	56	50	1640 hours
	%	23	26	34	39	43	42	42	45	42	35	21	21	37 %

Erfurt

latitude: 50°59′n longitude: 10°58′e height: 316 m

DClo	018-10	Jan	Feb	Mar	Apr	May	Jun	Jul	Aug	Sep	Oct	Nov	Dec	Year
	tx	10	11	17	23	27	29	31	31	27	21	14	11	32 °C
	t	-2	-1	3	8	12	16	17	16	13	8	4	0	8 °C
	tn	-14	-14	-9	-3	1	5	8	7	3	-2	-6	-12	-18 °C
	clothing	w	w	w	n	n	n	s	n	n	n	w	w	–
	stress							o						–
	p	33	31	28	34	58	67	71	55	46	45	34	30	532 mm
	dp	8	7	8	9	9	9	11	10	8	8	7	8	103 days
	sunshine	54	75	135	168	215	231	226	211	167	113	55	46	1696 hours
	%	21	26	37	41	45	47	45	47	44	34	20	19	38 %

Fichtelberg

GDClo	018-11	Jan	Feb	Mar	Apr	May	Jun	Jul	Aug	Sep	Oct	Nov	Dec	Year
	latitude: 50°26′n longitude: 12°57′e height: 1215 m													
	tx	3	5	9	13	19	22	24	23	19	15	10	5	25 °C
	t	-6	-5	-2	2	6	10	12	11	8	4	-1	-4	3 °C
	tn	-16	-15	-12	-8	-4	0	3	3	0	-6	-11	-14	-19 °C
	clothing	w	w	w	w	n	n	n	n	n	w	w	w	–
	stress													–
	p	94	92	79	81	94	101	141	102	87	89	74	77	1109 mm
	dp	15	12	14	15	14	13	14	14	12	12	12	14	162 days
	sunshine	53	70	90	132	214	225	217	211	141	102	57	43	1555 hours
	%	20	25	25	32	45	46	44	47	37	30	21	17	35 %

4.1.18.2. Predominant general weather situations are described in 4.1.0.1. under Northern, Central, and Eastern Europe.

4.1.19. Austria

4.1.19.1. Climatic description and tables

1. Temperate Continental Climate DC and Temperate Continental Mountain Climate GDC, locally Temperate Continental High Mountain Climate HDC
2. Summer warm; winter cool, at higher levels cold.
3. Variable weather prevails throughout the year; periods of dry and cloudy or sunny days occur frequently.
4. Wind: occasional gales, at higher levels often severe. Fog: predominant in winter especially in the valleys; at the same time higher levels often in bright sunshine. Snow cover: at higher levels usually for many weeks.

Wien

DCbo	019-1	Jan	Feb	Mar	Apr	May	Jun	Jul	Aug	Sep	Oct	Nov	Dec	Year
	latitude: 48°15′n longitude: 16°22′e height: 212 m													
	tx	9	12	18	23	27	29	31	30	27	21	14	11	32 °C
	t	-1	0	5	10	15	18	20	19	16	10	5	1	10 °C
	tn	-12	-10	-6	0	4	9	11	10	6	0	-3	-8	-14 °C
	clothing	w	w	n	n	n	s	s	s	n	n	n	w	–
	stress						o	o	o					–
	p	40	43	45	45	70	67	83	72	41	56	53	45	660 mm
	dp	8	7	8	8	8	9	9	9	7	8	8	8	96 days
	sunshine	57	84	138	184	235	249	266	250	199	129	55	45	1891 hours
	%	21	29	38	45	50	52	55	56	52	38	20	17	42 %

St. Pölten (tx and tn Krems an der Donau)

DCbo	019-2	Jan	Feb	Mar	Apr	May	Jun	Jul	Aug	Sep	Oct	Nov	Dec	Year
	latitude: 48°12′n longitude: 15°38′e height: 282 m													
	tx	9	13	19	25	28	30	33	31	28	22	15	12	34 °C
	t	-2	-1	4	9	14	17	19	18	14	9	4	0	9 °C
	tn	-12	-11	-6	-1	2	8	10	9	4	0	-4	-8	-14 °C
	clothing	w	w	w	n	n	s	s	s	n	n	w	w	–
	stress						o	o	o					–
	p	32	37	40	49	89	102	108	92	58	53	45	36	741 mm
	dp	9	8	7	8	10	11	11	11	8	8	8	8	107 days

Salzburg

latitude: 47°48′n longitude: 13°00′e height: 430 m

DCbo	019-3	Jan	Feb	Mar	Apr	May	Jun	Jul	Aug	Sep	Oct	Nov	Dec	Year
	tx	10	12	19	24	27	30	32	31	28	22	16	11	33 °C
	t	-2	-1	4	8	13	16	18	17	14	8	3	-1	8 °C
	tn	-16	-16	-9	-3	1	5	8	7	3	-2	-6	-13	-20 °C
	clothing	w	w	w	n	n	n	s	s	n	n	w	w	–
	stress							o	o					–
	p	73	70	70	89	127	167	191	163	111	82	70	65	1278 mm
	dp	13	12	12	14	15	16	16	15	12	11	12	12	159 days
	sunshine	74	97	141	158	190	208	219	218	180	135	70	64	1754 hours
	%	27	33	38	39	40	43	45	49	47	40	25	24	39 %

Innsbruck

latitude: 47°16′n longitude: 11°24′e height: 582 m

GDCbo	019-4	Jan	Feb	Mar	Apr	May	Jun	Jul	Aug	Sep	Oct	Nov	Dec	Year
	tx	10	13	20	24	28	31	33	31	28	22	16	11	33 °C
	t	-3	0	5	9	14	17	18	17	15	9	3	-1	9 °C
	tn	-16	-13	-7	-2	2	6	8	7	3	-2	-6	-13	-18 °C
	clothing	w	w	n	n	n	s	s	s	n	n	w	w	–
	stress						o	o	o					–
	p	57	52	43	55	77	114	140	113	84	71	57	48	911 mm
	dp	8	8	7	9	11	14	14	13	10	8	8	8	118 days
	sunshine	73	105	158	166	189	190	210	199	177	145	86	68	1766 hours
	%	26	36	43	41	40	40	43	45	47	43	31	26	39 %

Sonnblick

latitude: 47°03′n longitude: 12°57′e height: 3107 m

HDCkc	019-5	Jan	Feb	Mar	Apr	May	Jun	Jul	Aug	Sep	Oct	Nov	Dec	Year
	tx	-3	-3	-2	0	4	7	10	10	7	4	0	-2	11 °C
	t	-13	-13	-11	-8	-4	-1	2	1	0	-4	-8	-11	-6 °C
	tn	-26	-25	-22	-19	-14	-9	-7	-7	-10	-15	-19	-23	-28 °C
	clothing	p	p	p	w	w	w	w	w	w	w	w	p	–
	stress													–
	p	115	108	112	153	136	142	154	134	104	118	108	111	1495 mm
	dp	13	12	13	16	15	17	18	16	12	12	12	13	170 days
	sunshine	109	118	147	139	148	150	167	165	161	147	112	108	1671 hours
	%	39	41	40	34	32	31	35	37	43	43	40	41	37 %

Graz

latitude: 46°59′n longitude: 15°27′e height: 342 m

DCbo	019-6	Jan	Feb	Mar	Apr	May	Jun	Jul	Aug	Sep	Oct	Nov	Dec	Year
	tx	9	12	19	23	27	29	31	31	28	21	14	10	32 °C
	t	-4	-2	3	9	14	17	19	18	15	9	3	-1	8 °C
	tn	-13	-12	-7	-1	3	7	9	8	4	-1	-5	-10	-16 °C
	clothing	w	w	w	n	n	s	s	s	n	n	w	w	–
	stress						o	o	o					–
	p	31	35	34	52	93	126	114	91	80	79	57	48	840 mm
	dp	6	6	6	8	11	11	11	10	8	8	8	7	102 days
	sunshine	75	107	151	178	210	234	259	240	188	133	70	61	1906 hours
	%	27	37	41	44	45	49	54	54	50	39	25	23	43 %

Klagenfurt

		latitude: 46°39'n						longitude: 14°21'e				height: 452 m	
DCbo	019-7	Jan	Feb	Mar	Apr	May	Jun	Jul	Aug	Sep	Oct	Nov	Dec	Year
	tx	6	11	18	22	27	29	31	30	28	21	13	8	32 °C
	t	-5	-3	3	9	13	17	19	18	14	8	2	-2	8 °C
	tn	-17	-16	-8	-3	1	6	8	7	3	-2	-6	-13	-20 °C
	clothing	w	w	w	n	n	s	s	s	n	n	w	w	–
	stress						o	o	o					–
	p	39	42	39	69	88	124	122	102	87	87	73	54	926 mm
	dp	6	6	6	9	11	12	11	10	8	8	7	7	100 days
	sunshine	74	116	165	188	210	231	256	241	183	118	54	42	1878 hours
	%	27	40	45	46	45	49	53	55	48	35	19	16	42 %

4.1.19.2. Predominant general weather situations are described in 4.1.0.1. under Northern, Central, and Eastern Europe. A hygrothermal diagram of Wien is to be found in 4.1.0.2.
Wind name: Föhn (see 4.1.17.2.).

4.1.20. Czechoslovakia

4.1.20.1. Climatic description and tables

1. Temperate Continental Climate DC and
   Temperate Continental Mountain Climate GDC
2. Summer warm, winter cool or cold.
3. Variable weather prevails, now and then interrupted by periods of dry and cloudy or sunny days.
4. Wind: occasional gales. Snow cover: at higher levels usually for many weeks.

Praha

		latitude: 50°06'n						longitude: 14°17'e				height: 374 m	
DCbo	020-1	Jan	Feb	Mar	Apr	May	Jun	Jul	Aug	Sep	Oct	Nov	Dec	Year
	tx	10	11	18	22	28	31	33	32	29	22	14	10	34 °C
	t	-3	-2	2	8	13	17	18	18	14	8	3	0	8 °C
	tn	-13	-12	-8	-2	2	7	9	8	4	-2	-5	-10	-16 °C
	clothing	w	w	w	n	n	n	s	s	n	n	w	w	–
	stress							o	o					–
	p	23	24	23	32	61	67	82	66	36	42	26	26	508 mm
	dp	7	6	6	7	10	10	11	9	7	8	6	7	95 days
	sunshine	57	81	143	177	234	245	243	234	182	121	49	41	1807 hours
	%	22	28	39	43	49	50	49	52	48	36	18	16	40 %

Cheb

		latitude: 50°05'n						longitude: 12°24'e				height: 474 m	
DClo	020-2	Jan	Feb	Mar	Apr	May	Jun	Jul	Aug	Sep	Oct	Nov	Dec	Year
	tx	7	10	16	21	26	29	31	30	26	20	12	8	31 °C
	t	-3	-2	2	7	12	15	17	16	12	7	2	-1	7 °C
	tn	-15	-17	-10	-4	-1	3	6	5	-1	-4	-7	-12	-19 °C
	clothing	w	w	w	n	n	n	s	n	n	n	w	w	–
	stress							o						–
	p	38	36	32	39	53	67	84	64	46	45	37	37	578 mm
	dp	10	10	8	8	9	11	11	10	9	10	9	9	113 days
	sunshine	39	65	133	168	215	222	216	202	165	100	44	29	1598 hours
	%	15	23	36	41	45	45	44	45	43	30	16	12	36 %

Ostrava

latitude: 49°47′n longitude: 18°16′e height: 257 m

DCbo	020-3	Jan	Feb	Mar	Apr	May	Jun	Jul	Aug	Sep	Oct	Nov	Dec	Year
	t	-3	-1	3	8	13	16	18	17	14	8	4	0	8 °C
	clothing	w	w	w	n	n	n	s	s	n	n	w	w	–
	stress							o	o					–
	p	30	29	35	39	79	91	99	92	59	51	39	32	675 mm
	dp	9	9	8	10	10	11	12	11	8	8	8	8	112 days
	sunshine	55	67	123	171	218	225	238	231	176	117	55	45	1721 hours
	%	21	23	34	41	46	46	48	51	46	35	20	18	38 %

Brno

latitude: 49°09′n longitude: 16°42′e height: 238 m

DCbo	020-4	Jan	Feb	Mar	Apr	May	Jun	Jul	Aug	Sep	Oct	Nov	Dec	Year
	tx	8	11	18	23	28	30	32	31	28	22	14	10	33 °C
	t	-3	-1	4	9	14	18	19	18	15	9	4	0	9 °C
	tn	-14	-13	-8	-3	2	6	9	8	3	-2	-5	-10	-16 °C
	clothing	w	w	w	n	n	s	s	s	n	n	w	w	–
	stress						o	o	o					–
	p	26	24	21	33	55	81	73	67	37	41	39	30	527 mm
	dp	7	6	5	7	8	9	10	9	6	7	7	8	88 days
	sunshine	54	76	145	187	240	251	259	246	193	126	53	39	1869 hours
	%	20	26	39	45	51	52	53	55	51	37	19	15	42 %

Poprad

latitude: 49°04′n longitude: 20°15′e height: 707 m

GDClo	020-5	Jan	Feb	Mar	Apr	May	Jun	Jul	Aug	Sep	Oct	Nov	Dec	Year
	t	-6	-4	0	6	10	14	16	15	12	6	1	-3	6 °C
	clothing	w	w	w	n	n	n	n	n	n	n	w	w	–
	stress													–
	p	28	25	34	34	68	85	90	78	43	42	48	33	608 mm
	dp	7	6	7	7	10	11	11	10	6	6	8	8	97 days
	sunshine	94	112	169	189	231	231	255	241	208	156	85	81	2052 hours
	%	35	39	46	46	49	48	52	54	55	46	31	32	46 %

Sliac

latitude: 48°38′n longitude: 19°09′e height: 318 m

DCbo	020-6	Jan	Feb	Mar	Apr	May	Jun	Jul	Aug	Sep	Oct	Nov	Dec	Year
	t	-4	-2	2	8	13	17	18	18	14	8	3	0	8 °C
	clothing	w	w	w	n	n	s	s	s	n	n	w	w	–
	stress						o	o	o					–
	p	45	46	45	40	77	86	83	66	48	57	69	55	717 mm
	dp	7	8	7	7	10	10	9	9	6	8	10	9	102 days
	sunshine	66	88	152	190	235	249	275	253	194	128	56	44	1930 hours
	%	24	31	41	46	50	51	56	57	51	38	20	17	43 %

4.1.20.2. Predominant general weather situations are described in 4.1.0.1. under Northern, Central, and Eastern Europe. A hygrothermal diagram of Praha is to be found in 4.1.0.2.

4.1.21. Poland

4.1.21.1. Climatic description and tables

1. Temperate Continental Climate DC and
   Temperate Continental Mountain Climate GDC
2. Summer mild in the north, warm in the south, winter moderately cold.
3. Variable weather prevails, now and then interrupted by periods of dry and cloudy or sunny days.
4. Wind: occasional gales. Fog: at times in winter, particularly in the north. Snow cover: often of long duration.

Gdansk

latitude: 54° 23′n longitude: 18° 36′e height: 12 m

DCbo	021-1	Jan	Feb	Mar	Apr	May	Jun	Jul	Aug	Sep	Oct	Nov	Dec	Year
	tx	8	8	12	17	24	27	28	27	25	19	12	9	30 °C
	t	-1	-2	1	6	11	15	18	17	13	9	4	1	8 °C
	tn	-12	-13	-7	-3	2	6	10	10	5	1	-4	-7	-15 °C
	clothing	w	w	w	n	n	n	s	s	n	n	w	w	–
	stress							o	o					–
	p	31	24	19	32	33	58	73	70	53	38	29	39	499 mm
	dp	9	7	5	7	7	8	10	9	8	8	7	9	95 days
	sunshine	39	70	134	163	244	259	236	225	174	105	45	32	1726 hours
	%	16	25	37	39	49	50	46	49	45	32	18	14	38 %

Koszalin

latitude: 54° 12′n longitude: 16° 11′e height: 34 m

DClo	021-2	Jan	Feb	Mar	Apr	May	Jun	Jul	Aug	Sep	Oct	Nov	Dec	Year
	t	-1	-2	1	6	10	15	17	16	13	9	4	2	8 °C
	clothing	w	w	w	n	n	n	s	n	n	n	w	w	–
	stress							o						–
	p	54	33	27	41	45	65	97	97	72	71	43	51	697 mm
	dp	12	8	6	8	9	9	12	12	12	10	9	11	117 days

Suwalki

latitude: 54° 06′n longitude: 22° 57′e height: 170 m

DCbo	021-3	Jan	Feb	Mar	Apr	May	Jun	Jul	Aug	Sep	Oct	Nov	Dec	Year
	tx	4	4	11	19	27	29	29	28	25	18	10	6	31 °C
	t	-4	-6	-2	5	12	16	18	17	12	7	1	-2	6 °C
	tn	-22	-20	-15	-5	0	3	6	6	1	-3	-10	-15	-24 °C
	clothing	w	w	w	n	n	n	s	s	n	n	w	w	–
	stress							o	o					–
	p	32	27	26	39	36	70	83	84	57	48	36	42	580 mm
	dp	9	7	6	9	8	9	12	10	10	7	9	11	107 days
	sunshine	24	60	146	139	226	250	218	207	162	85	29	22	1568 hours
	%	10	22	40	33	46	49	42	45	42	26	11	10	35 %

Szczecin

DCbo	021-4	Jan	Feb	Mar	Apr	May	Jun	Jul	Aug	Sep	Oct	Nov	Dec	Year
	latitude: 53°24′n longitude: 14°37′e height: 7 m													
	t	0	-2	2	7	12	16	18	17	13	9	4	2	8 °C
	clothing	w	w	w	n	n	n	s	s	n	n	w	w	–
	stress							o	o					–
	p	33	26	26	31	38	65	83	57	46	35	30	41	511 mm
	dp	9	7	6	7	7	9	11	11	9	9	8	10	103 days
	sunshine	38	63	139	170	232	214	210	187	174	101	43	24	1595 hours
	%	15	23	38	41	47	42	41	41	45	31	17	10	35 %

Bialystok

DCbo	021-5	Jan	Feb	Mar	Apr	May	Jun	Jul	Aug	Sep	Oct	Nov	Dec	Year
	latitude: 53°07′n longitude: 23°11′e height: 141 m													
	tx	4	5	12	20	28	30	30	29	26	20	11	7	32 °C
	t	-4	-5	-1	6	12	17	18	17	12	7	2	-1	7 °C
	tn	-20	-20	-13	-5	1	4	7	7	1	-3	-9	-14	-23 °C
	clothing	w	w	w	n	n	s	s	s	n	n	w	w	–
	stress						o	o	o					–
	p	33	36	28	39	42	67	70	69	53	44	35	44	560 mm
	dp	9	10	6	8	8	8	11	10	9	7	8	10	105 days
	sunshine	26	50	144	162	207	244	232	215	161	94	29	20	1584 hours
	%	10	18	39	39	42	48	45	47	42	28	11	8	35 %

Torun

DCbo	021-6	Jan	Feb	Mar	Apr	May	Jun	Jul	Aug	Sep	Oct	Nov	Dec	Year
	latitude: 53°03′n longitude: 18°35′e height: 70 m													
	t	-2	-3	1	7	12	17	18	17	13	8	3	0	8 °C
	clothing	w	w	w	n	n	s	s	s	n	n	w	w	–
	stress						o	o	o					–
	p	29	24	18	30	32	56	110	56	36	26	25	42	484 mm
	dp	8	7	5	7	7	8	13	7	8	7	8	10	95 days

Poznan

DCbo	021-7	Jan	Feb	Mar	Apr	May	Jun	Jul	Aug	Sep	Oct	Nov	Dec	Year
	latitude: 52°25′n longitude: 16°51′e height: 29 m													
	tx	8	9	16	21	28	31	32	30	28	21	13	9	33 °C
	t	-2	-2	2	7	13	17	18	18	13	8	3	1	8 °C
	tn	-15	-16	-8	-3	1	4	9	8	2	-1	-6	-10	-18 °C
	clothing	w	w	w	n	n	s	s	s	n	n	w	w	–
	stress						o	o	o					–
	p	29	27	29	34	38	62	85	55	40	29	27	37	492 mm
	dp	8	7	5	8	8	8	11	8	8	7	7	10	95 days
	sunshine	41	65	133	157	223	220	208	204	177	107	40	28	1603 hours
	%	16	23	36	38	46	44	41	45	46	32	15	12	36 %

Warszawa

latitude: 52°11′n longitude: 20°58′e height: 107 m

DCbo	021-8	Jan	Feb	Mar	Apr	May	Jun	Jul	Aug	Sep	Oct	Nov	Dec	Year
	tx	7	8	14	21	28	30	31	30	28	21	13	8	32 °C
	t	-2	-3	1	7	13	17	19	18	13	8	3	0	8 °C
	tn	-17	-16	-9	-3	3	6	10	9	3	-1	-6	-12	-19 °C
	clothing	w	w	w	n	n	s	s	s	n	n	w	w	–
	stress						o	o	o					–
	p	25	28	20	32	40	60	79	47	41	31	31	37	471 mm
	dp	8	7	5	7	7	8	11	8	8	6	8	8	93 days
	sunshine	21	51	131	138	192	198	203	208	171	106	36	10	1465 hours
	%	8	18	36	33	39	39	40	46	45	32	14	4	33 %

Zielona Góra

latitude: 51°56′n longitude: 15°31′e height: 174 m

DCbo	021-9	Jan	Feb	Mar	Apr	May	Jun	Jul	Aug	Sep	Oct	Nov	Dec	Year
	t	-1	-2	2	7	13	17	18	17	14	9	4	1	8 °C
	clothing	w	w	w	n	n	s	s	s	n	n	w	w	–
	stress						o	o	o					–
	p	41	33	35	45	49	62	94	70	46	40	35	45	595 mm
	dp	11	10	6	9	9	10	11	10	7	8	8	10	108 days

Wroclaw

latitude: 51°06′n longitude: 16°53′e height: 124 m

DCbo	021-10	Jan	Feb	Mar	Apr	May	Jun	Jul	Aug	Sep	Oct	Nov	Dec	Year
	tx	8	10	16	22	27	30	31	30	28	22	15	11	33 °C
	t	-1	-2	2	8	13	17	18	18	14	9	4	1	8 °C
	tn	-18	-17	-9	-4	0	4	8	7	2	-2	-6	-11	-20 °C
	clothing	w	w	w	n	n	s	s	s	n	n	w	w	–
	stress						o	o	o					–
	p	24	23	27	42	50	61	100	57	41	33	28	33	519 mm
	dp	6	6	6	8	8	8	10	8	8	7	7	8	91 days
	sunshine	48	67	124	146	200	204	202	199	156	112	48	38	1544 hours
	%	18	24	34	35	41	41	40	44	41	34	18	15	34 %

Czestochowa

latitude: 50°48′n longitude: 19°06′e height: 263 m

DCbo	021-11	Jan	Feb	Mar	Apr	May	Jun	Jul	Aug	Sep	Oct	Nov	Dec	Year
	t	-2	-3	1	7	12	17	18	17	13	8	3	1	8 °C
	clothing	w	w	w	n	n	s	s	s	n	n	w	w	–
	stress						o	o	o					–
	p	41	32	35	44	53	79	102	67	55	32	39	43	622 mm
	dp	11	9	8	9	9	10	12	9	8	6	9	11	112 days

Šniežka

GDCko	021-12	Jan	Feb	Mar	Apr	May	Jun	Jul	Aug	Sep	Oct	Nov	Dec	Year
		latitude: 50°44'n					longitude: 15°44'e				height: 1603 m			
	tx	3	4	5	8	14	17	18	18	16	12	8	5	20 °C
	t	-8	-7	-3	-2	3	7	9	9	6	2	-2	-4	-3 °C
	tn	-19	-19	-15	-12	-7	-2	1	1	-4	-8	-12	-16	-22 °C
	clothing	w	w	w	w	w	n	n	n	n	w	w	w	–
	stress													–
	p	122	94	102	115	107	127	182	140	106	126	100	119	1440 mm
0.1 mm	dp	23	19	19	20	19	18	22	19	18	22	21	22	242 days
	sunshine	66	88	117	119	176	182	161	160	138	104	69	63	1443 hours
	%	25	31	32	29	37	37	32	35	36	31	26	25	32 %

Zamość

DCbo	021-13	Jan	Feb	Mar	Apr	May	Jun	Jul	Aug	Sep	Oct	Nov	Dec	Year
		latitude: 50°43'n					longitude: 23°16'e				height: 219 m			
	t	-3	-4	0	7	13	17	18	18	13	8	3	0	8 °C
	clothing	w	w	w	n	n	s	s	s	n	n	s	s	–
	stress						o	o	o					–
	p	26	34	29	32	68	72	89	76	51	44	38	38	597 mm
	dp	9	8	8	7	9	9	10	9	8	8	8	10	102 days

Sandomierz

DCbo	021-14	Jan	Feb	Mar	Apr	May	Jun	Jul	Aug	Sep	Oct	Nov	Dec	Year
		latitude: 50°41'n					longitude: 21°44'e				height: 200 m			
	t	-3	-3	1	7	13	17	19	18	15	8	3	0	8 °C
	clothing	w	w	w	n	n	s	s	s	n	n	w	w	–
	stress						o	o	o					–
	p	29	26	33	38	59	80	96	65	50	36	35	37	584 mm
	dp	8	6	7	7	8	10	10	8	7	6	7	9	95 days

Klodzko

DClo	021-15	Jan	Feb	Mar	Apr	May	Jun	Jul	Aug	Sep	Oct	Nov	Dec	Year
		latitude: 50°26'n					longitude: 16°39'e				height: 320 m			
	t	-2	-3	1	6	12	15	17	16	12	8	3	1	7 °C
	clothing	w	w	w	n	n	n	s	n	n	n	w	w	–
	stress							o						–
	p	23	27	29	42	64	71	119	74	41	42	32	28	592 mm
	dp	6	6	7	8	9	9	12	8	8	6	6	6	91 days

Krakow

		latitude: 50°05'n					longitude: 20°01'e				height: 213 m		
DCbo	021-16	Jan	Feb	Mar	Apr	May	Jun	Jul	Aug	Sep	Oct	Nov	Dec	Year
	tx	8	10	17	21	27	30	31	30	27	22	15	11	32 °C
	t	-3	-3	1	7	13	17	18	18	13	8	3	1	8 °C
	tn	-16	-15	-8	-2	3	7	10	9	3	-1	-6	-10	-18 °C
	clothing	w	w	w	n	n	s	s	s	n	n	w	w	–
	stress						o	o	o					–
	p	33	30	31	45	62	90	95	78	46	34	34	35	613 mm
	dp	9	8	8	8	9	11	12	10	8	6	8	9	106 days
	sunshine	43	55	113	147	189	199	220	210	151	111	48	36	1522 hours
	%	16	19	31	36	40	41	44	47	40	33	18	14	34 %

Przemysl

		latitude: 49°48'n					longitude: 22°46'e				height: 239 m		
DCbo	021-17	Jan	Feb	Mar	Apr	May	Jun	Jul	Aug	Sep	Oct	Nov	Dec	Year
	tx	9	10	17	22	27	29	32	30	27	23	16	11	33 °C
	t	-2	-3	1	8	13	17	19	18	13	8	3	1	8 °C
	tn	-18	-18	-11	-4	1	5	9	7	2	-2	-8	-12	-21 °C
	clothing	w	w	w	n	n	s	s	s	n	n	w	w	–
	stress						o	o	o					–
	p	34	30	27	41	59	90	89	74	58	49	44	48	643 mm
	dp	9	8	8	8	9	10	10	9	8	6	9	10	103 days
	sunshine													hours

Zakopane

		latitude: 49°17'n					longitude: 19°57'e				height: 833 m		
GDCbo	021-18	Jan	Feb	Mar	Apr	May	Jun	Jul	Aug	Sep	Oct	Nov	Dec	Year
	tx	9	10	14	17	23	25	27	26	24	21	14	11	28 °C
	t	-5	-5	0	4	10	13	15	18	11	6	1	-2	5 °C
	tn	-22	-23	-15	-8	-2	1	5	3	-1	-6	-12	-17	-25 °C
	clothing	w	w	w	w	n	n	n	s	n	n	w	w	–
	stress								o					
	p	41	45	58	76	91	173	200	152	99	83	54	49	1121 mm
0.1 mm	dp	16	16	15	17	16	19	20	18	13	15	13	16	194 days
	sunshine													hours

4.1.21.2. Predominant general weather situations are described in 4.1.0.1. under Northern, Central, and Eastern Europe. A hygrothermal diagram of Warszawa is to be found in 4.1.0.2.

4.1.22. Hungaria

4.1.22.1. Climatic description and tables

1. Temperate Continental Climate DC and
   Temperate Mountain Climate GDC
2. Summer warm, winter cool or cold.
3. Variable weather prevails, now and then interrupted by a period of dry and cloudy or sunny days.
4. Wind: occasionally gales; snow cover: at higher levels often for many weeks.

Miskolc

latitude: 48°08′n longitude: 20°48′e height: 120 m

DCbo	022-1	Jan	Feb	Mar	Apr	May	Jun	Jul	Aug	Sep	Oct	Nov	Dec	Year
	tx	7	10	18	24	28	31	34	34	29	24	15	10	35 °C
	t	-3	1	4	10	16	19	21	20	16	10	4	0	10 °C
	tn	-16	-14	-8	-2	3	7	10	9	3	-2	-6	-11	-18 °C
	clothing	w	w	w	n	n	s	s	s	n	n	w	w	–
	stress						o	o	o					–
	p	31	31	28	39	70	84	66	66	40	49	56	40	600 mm
0.1 mm	dp	10	10	9	10	12	13	10	10	8	10	13	12	127 days
	sunshine	59	78	139	184	250	258	295	260	199	132	59	38	1951 hours
	%	22	27	38	45	53	54	61	58	52	39	21	15	44 %

Magyaróvár (tx and tn, sunshine Györ)

latitude: 47°53′n longitude: 17°16′e height: 130 m

DCbo	022-2	Jan	Feb	Mar	Apr	May	Jun	Jul	Aug	Sep	Oct	Nov	Dec	Year
	tx	9	12	19	24	28	31	33	33	29	24	16	12	34 °C
	t	-2	0	5	11	16	19	21	20	16	10	5	1	10 °C
	tn	-14	-11	-6	0	4	9	11	10	5	-1	-4	-10	-16 °C
	clothing	w	w	n	n	n	s	s	s	n	n	n	w	–
	stress						o	o	o					–
	p	33	36	40	38	63	67	80	68	35	56	53	46	615 mm
0.1 mm	dp	12	11	10	10	12	11	11	10	8	10	12	12	129 days
	sunshine	77	92	120	181	253	244	267	275	203	143	59	62	1976 hours
	%	28	32	33	44	54	51	55	62	54	42	21	24	44 %

Budapest

latitude: 47°31′n longitude: 19°02′e height: 130 m

DCbo	022-3	Jan	Feb	Mar	Apr	May	Jun	Jul	Aug	Sep	Oct	Nov	Dec	Year
	tx	8	12	19	24	29	32	34	34	30	25	15	11	35 °C
	t	-1	1	6	12	17	20	22	21	17	11	6	2	11 °C
	tn	-12	-10	-5	1	5	10	12	11	6	1	-3	-8	-14 °C
	clothing	w	w	n	n	s	s	s	s	s	n	n	w	–
	stress					o	o	o	o	o				–
	p	42	44	39	45	72	76	54	51	34	56	69	48	630 mm
0.1 mm	dp	12	12	11	11	13	13	10	9	7	10	14	13	136 days
	sunshine	58	85	140	196	250	275	309	283	213	145	60	43	2057 hours
	%	21	29	38	48	53	57	64	64	56	43	21	16	46 %

Debrecen

latitude: 47°23′n longitude: 21°37′e height: 128 m

DCbo	022-4	Jan	Feb	Mar	Apr	May	Jun	Jul	Aug	Sep	Oct	Nov	Dec	Year
	tx	8	12	19	25	29	32	35	34	30	24	17	12	36 °C
	t	-3	-1	4	11	16	20	22	21	16	10	4	0	10 °C
	tn	-16	-13	-8	-2	3	8	10	9	4	-2	-6	-12	-18 °C
	clothing	w	w	w	n	n	s	s	s	n	n	w	w	–
	stress						o	o	o					–
	p	35	36	30	36	61	80	59	64	41	49	53	40	584 mm
0.1 mm	dp	13	11	11	11	13	13	11	10	8	10	12	13	136 days
	sunshine	59	85	151	198	256	278	309	279	214	150	68	46	2093 hours
	%	21	29	41	48	55	58	64	63	56	44	24	17	47 %

Szeged

latitude: 46°15′n longitude: 20°09′e height: 97 m

DCao	022-5	Jan	Feb	Mar	Apr	May	Jun	Jul	Aug	Sep	Oct	Nov	Dec	Year
	tx	9	12	20	25	29	32	35	35	31	26	17	12	36 °C
	t	-1	1	6	12	17	21	23	22	18	12	6	2	12 °C
	tn	-16	-13	-6	-1	4	9	11	11	6	0	-4	-16	-18 °C
	clothing	w	w	n	n	s	s	s	s	s	n	n	w	–
	stress					o	o	o	o	o				–
	p	34	38	35	41	63	63	51	47	42	46	59	39	558 mm
0.1 mm	dp	12	11	10	11	13	11	8	8	7	9	12	12	124 days
	sunshine	63	94	147	199	259	283	322	298	255	162	77	56	2215 hours
	%	23	32	40	49	56	60	67	68	67	48	27	21	49 %

Pécs

latitude: 46°03′n longitude: 18°14′e height: 124 m

DCao	022-6	Jan	Feb	Mar	Apr	May	Jun	Jul	Aug	Sep	Oct	Nov	Dec	Year
	tx	11	14	20	25	28	31	34	34	30	25	18	13	35 °C
	t	-1	1	6	12	17	21	23	22	18	12	6	2	12 °C
	tn	-13	-12	-5	-1	4	9	10	10	5	0	-4	-9	-16 °C
	clothing	w	w	n	n	s	s	s	s	s	n	n	w	–
	stress					o	o	o	o	o				–
	p	41	46	41	58	66	69	64	55	47	64	71	45	667 mm
0.1 mm	dp	11	10	10	10	13	11	10	9	7	10	12	12	125 days
	sunshine	67	96	141	189	246	274	311	289	210	150	69	54	2096 hours
	%	24	33	38	47	53	58	65	66	56	44	24	20	47 %

4.1.22.2. Predominant general weather situations are described in 4.1.0.1. under Northern, Central, and Eastern Europe. A hygrothermal diagram of Budapest is to be found in 4.1.0.2.

4.1.23. Yugoslavia

4.1.23.1. Climatic descriptions and tables

1. Climatic area: Yugoslavia, except the Adriatic Coast and hinterland

1.1. Temperate Continental Climate DC and
Temperate Continental Mountain Climate GDC

1.2. Summer warm, partly hot; winter cool, cold in the mountains.

1.3. Variable weather prevails, now and then periods of dry and cloudy or sunny days occur.

1.4. Wind: occasional gales. Snow cover: at higher levels for many weeks.

Zagreb

latitude: 45°49′n longitude: 15°59′e height: 163 m

DCbk	023-1	Jan	Feb	Mar	Apr	May	Jun	Jul	Aug	Sep	Oct	Nov	Dec	Year
	tx	11	14	20	24	28	31	33	32	29	23	17	13	34 °C
	t	0	2	7	12	16	20	22	21	18	12	7	2	12 °C
	tn	-10	-10	-3	2	6	10	12	12	7	2	-2	-6	-13 °C
	clothing	w	w	n	n	n	s	s	s	s	n	n	w	–
	stress						o	o	o	o				–
	p	56	54	47	59	86	95	79	74	70	88	89	67	864 mm
0.1 mm	dp	14	11	10	12	14	14	12	10	10	13	15	14	149 days
	sunshine	54	87	137	176	211	241	280	266	193	128	61	45	1879 hours
	%	19	30	37	43	46	51	59	61	51	38	21	17	42 %

Beograd

				latitude: 44°48′n				longitude: 20°28e				height: 132 m	
DCak	023-2	Jan	Feb	Mar	Apr	May	Jun	Jul	Aug	Sep	Oct	Nov	Dec	Year
	tx	13	15	22	26	30	33	35	36	32	27	20	16	37 °C
	t	0	2	6	12	17	20	23	22	18	12	7	2	12 °C
	tn	-12	-10	-5	0	5	10	12	11	7	2	-3	-8	-15 °C
	clothing	w	w	n	n	s	s	s	s	s	n	n	w	–
	stress					o	o	o	o	o				–
	p	48	46	46	54	75	96	60	55	50	55	61	55	701 mm
0.1 mm	dp	14	13	12	13	14	13	9	9	8	11	14	14	144 days
	sunshine	68	94	150	192	237	272	308	286	230	164	85	66	2152 hours
	%	24	32	41	47	52	58	65	65	61	48	29	24	48 %

Sarajevo

				latitude: 43°52′n				longitude: 18°26′e				height: 637 m	
GDCbo	023-3	Jan	Feb	Mar	Apr	May	Jun	Jul	Aug	Sep	Oct	Nov	Dec	Year
	tx	13	15	20	23	28	30	33	34	30	23	18	15	35 °C
	t	-1	1	5	10	14	17	20	20	16	10	5	2	10 °C
	tn	-14	-13	-8	-2	1	7	9	8	3	0	-6	-9	-15 °C
	clothing	w	w	n	n	n	s	s	s	n	n	n	w	–
	stress						o	o	o					–
	p	71	69	50	59	84	86	68	62	71	84	98	87	889 mm
0.1 mm	dp	16	14	13	13	16	14	12	8	9	12	15	15	157 days
	sunshine	57	91	136	161	196	230	278	266	199	135	70	44	1863 hours
	%	20	31	37	40	43	50	59	61	53	39	24	16	42 %

Skoplje

				latitude: 41°59′n				longitude: 21°28′e				height: 241 m	
DCak	023-4	Jan	Feb	Mar	Apr	May	Jun	Jul	Aug	Sep	Oct	Nov	Dec	Year
	tx	12	18	22	26	30	34	36	37	32	26	18	16	38 °C
	t	1	3	6	12	17	22	24	24	19	12	7	3	12 °C
	tn	-11	-10	-7	-1	3	8	10	10	3	-1	-5	-8	-15 °C
	clothing	w	w	n	n	s	s	t	t	s	n	n	w	–
	stress					o	o	f	f	o				–
	p	46	41	38	34	52	49	35	37	42	58	71	43	546 mm
0.1 mm	dp	11	8	9	8	12	8	7	4	7	9	12	9	104 days
	sunshine	63	107	136	191	221	265	318	306	221	145	74	60	2107 hours
	%	21	36	37	48	49	58	69	71	59	42	25	21	47 %

2. *Climatic area:* Adriatic Coast and hinterland

2.1. Subtropical Rain Climate Cr

2.2. Summer hot, winter mild or cool, in the mountains cold.

2.3. Variable weather prevails; now and then periods of dry and sunny or cloudy days occur.

2.4. Wind: occasional gales, at times severe. Snow cover: at times.

Split

latitude: 43° 31′n longitude: 16° 26′e height: 128 m

Crak	023-5	Jan	Feb	Mar	Apr	May	Jun	Jul	Aug	Sep	Oct	Nov	Dec	Year
	tx	15	16	19	23	28	32	35	35	30	25	19	16	36 °C
	t	8	8	10	14	19	23	26	25	25	17	12	10	16 °C
	tn	-1	-1	2	6	10	14	16	16	13	8	4	1	-4 °C
	clothing	n	n	n	n	s	t	t	t	t	s	n	n	–
	stress					o	f	f	f	f	o			–
	p	76	74	53	62	60	53	40	32	55	71	110	130	816 mm
0.1 mm	dp	12	10	10	8	10	9	6	4	7	10	12	14	113 days
	sunshine	115	136	189	230	260	316	367	340	245	188	129	106	2621 hours
	%	40	46	51	57	57	68	78	78	65	55	44	38	59 %

Titograd

latitude: 42° 26′n longitude: 19° 17′e height: 53 m

Crak	023-6	Jan	Feb	Mar	Apr	May	Jun	Jul	Aug	Sep	Oct	Nov	Dec	Year
	tx	15	18	21	26	31	34	38	38	33	27	20	17	39 °C
	t	6	6	10	14	19	24	26	26	22	15	11	8	15 °C
	tn	-4	-4	0	4	8	13	16	16	12	6	0	-2	-5 °C
	clothing	n	n	n	n	s	t	t	t	s	n	n	n	–
	stress					o	o	f	f	o				–
	p	179	195	135	98	105	60	40	63	113	202	213	229	1632 mm
0.1 mm	dp	13	14	11	10	11	6	5	4	7	11	16	15	123 days
	sunshine	100	112	175	210	236	304	357	332	243	181	117	114	2482 hours
	%	34	38	47	52	52	66	77	77	65	52	40	39	56 %

4.1.23.2. Predominant general weather situations are described in 4.1.0.1. under Northern, Central, and Eastern Europe. A hygrothermal diagram of Beograd is to be found in 4.1.0.2.

Wind names: Bora is a cold and dry, often gusty wind blowing on the coast of Dalmatia. Maestro is a north-westerly wind which blows in summer in the Adriatic when pressure is low over the Balkan Peninsular. It is a fresh breeze accompanied by fine weather and light clouds.

4.1.24. Albania

4.1.24.1. Climatic description and tables

1. Subtropical Rain Climate Cr and Subtropical Rain Mountain Climate GCr.
2. Summer hot, winter cool, in the mountains cold.
3. Variable weather prevails, now and then periods of dry and sunny days occur.
4. Wind: occasional gales. Snow cover: only temporarily at low levels.

Shkoder

latitude: 42° 06′n longitude: 19° 32′e height: 43 m

Crak	024-1	Jan	Feb	Mar	Apr	May	Jun	Jul	Aug	Sep	Oct	Nov	Dec	Year
	tx	14	16	20	24	29	33	32	37	32	26	20	18	37 °C
	t	5	7	10	15	19	23	26	26	22	16	12	7	16 °C
	tn	-4	-4	0	3	7	12	15	15	11	6	2	-2	-6 °C
	clothing	n	n	n	n	s	t	t	t	s	n	n	n	–
	stress					o	o	f	f	o				–
	p	216	173	149	101	114	58	30	44	128	262	241	227	1741 mm
0.1 mm	dp	12	12	14	12	11	7	5	5	6	10	18	16	128 days
	sunshine	122	123	169	203	247	301	354	334	264	204	97	90	2508 hours
	%	42	41	46	51	55	66	76	77	70	59	33	32	56 %

Tirana

Crak	024-2	latitude: 41°20'n Jan	Feb	Mar	longitude: 19°47'e Apr	May	Jun	Jul	height: 89 m Aug	Sep	Oct	Nov	Dec	Year
	tx	17	18	22	26	30	34	36	37	32	29	23	19	38 °C
	t	7	8	11	14	18	22	25	25	22	17	13	9	16 °C
	tn	-5	-5	-1	2	6	11	13	13	9	4	1	-3	-6 °C
	clothing	n	n	n	n	s	s	t	t	s	n	n	n	–
	stress					o	o	f	f	o				–
	p	132	120	100	87	99	60	28	39	73	157	152	142	1189 mm
0.1 mm	dp	13	13	14	13	12	7	5	4	6	9	16	16	128 days
	sunshine	131	122	161	202	250	298	354	335	269	208	103	93	2526 hours
	%	44	41	44	51	56	66	77	78	72	60	35	33	57 %

Vlone

Crak	024-3	latitude: 40°28'n Jan	Feb	Mar	longitude: 19°29'e Apr	May	Jun	Jul	height: 3 m Aug	Sep	Oct	Nov	Dec	Year
	tx	18	19	23	26	31	33	34	36	32	29	25	21	36 °C
	t	9	10	12	15	19	22	25	25	22	18	15	11	17 °C
	tn	-1	-1	2	6	9	14	16	16	12	9	5	1	-3 °C
	clothing	n	n	n	n	s	s	t	t	s	s	n	n	–
	stress					o	o	f	f	o	o			–
	p	148	102	73	60	49	29	9	25	65	133	167	170	1028 mm
0.1 mm	dp	13	12	14	11	9	6	3	3	5	10	17	17	120 days
	sunshine	132	143	173	226	274	317	369	344	279	211	117	100	2685 hours
	%	44	47	47	57	61	70	80	80	74	61	39	35	60 %

4.1.24.2. Predominant general weather situations are described in 4.1.0.1. under Southern Europe.

4.1.25. Romania

4.1.25.1. Climatic descriptions and tables

1. Climatic area: Romania, except Walachia and Dobrudja

1.1. Temperate Continental Climate DC and
Temperate Continental Mountain Climate GDC

1.2. Summer warm, winter moderately cold.

1.3. Variable weather prevails, frequent periods of dry and sunny days occur.

1.4. Wind: occasional gales. Snow cover: occasionally, at higher levels of longer duration.

Iasi

DCbo	025-1	latitude: 47°10'n Jan	Feb	Mar	longitude: 27°35'e Apr	May	Jun	Jul	height: 104 m Aug	Sep	Oct	Nov	Dec	Year
	tx	8	11	19	25	30	32	35	34	30	25	17	12	36 °C
	t	-4	-2	2	10	16	20	22	21	16	10	4	-1	9 °C
	tn	-19	-18	-10	-2	3	8	11	11	9	3	-3	-8	-15 °C
	clothing	w	w	w	n	n	s	s	s	n	n	w	w	–
	stress						o	o	o					–
	p	32	30	21	37	48	73	62	64	36	35	39	29	506 mm
	dp	6	6	5	6	7	9	7	7	5	6	6	6	75 days
	sunshine	72	75	140	175	229	266	299	278	205	147	62	60	2008 hours
	%	26	26	38	43	49	56	62	63	54	43	22	23	45 %

Bistrita

latitude: 47°07'n longitude: 24°30'e height: 366 m

DCbo	025-2	Jan	Feb	Mar	Apr	May	Jun	Jul	Aug	Sep	Oct	Nov	Dec	Year
	t	-4	-2	3	9	14	18	20	19	14	9	4	-2	8 °C
	clothing	w	w	w	n	n	s	s	s	n	n	w	w	–
	stress						o	o	o					–
	p	35	42	35	56	74	90	84	71	50	51	51	41	679 mm
	dp	8	8	7	8	11	12	10	10	7	8	8	8	106 days

Cluj

latitude: 46°47'n longitude: 23°42'e height: 314 m

DCbo	025-3	Jan	Feb	Mar	Apr	May	Jun	Jul	Aug	Sep	Oct	Nov	Dec	Year
	tx	7	10	20	24	28	30	34	32	29	24	16	12	34 °C
	t	-3	-3	3	9	14	18	20	19	15	9	4	0	9 °C
	tn	-19	-18	-10	-3	2	6	9	8	1	-3	-8	-15	-22 °C
	clothing	w	w	w	n	n	s	s	s	n	n	w	w	–
	stress						o	o	o					–
	p	27	36	20	49	72	73	64	71	34	30	31	31	538 mm
	dp	6	6	5	7	10	10	8	9	5	5	7	6	84 days
	sunshine	64	88	170	176	222	233	286	271	209	163	80	54	2016 hours
	%	23	30	46	43	48	49	59	61	55	48	28	20	45 %

Arad

latitude: 46°10'n longitude: 21°19'e height: 116 m

DCbo	025-4	Jan	Feb	Mar	Apr	May	Jun	Jul	Aug	Sep	Oct	Nov	Dec	Year
	tx	10	13	21	26	30	32	35	35	31	26	18	14	36 °C
	t	-2	0	6	11	16	19	22	21	18	11	6	2	11 °C
	tn	-16	-14	-7	-2	2	7	9	9	4	-2	-5	-11	-20 °C
	clothing	w	w	n	n	n	s	s	s	s	n	n	w	–
	stress						o	o	o	o				–
	p	36	34	35	46	64	68	54	43	41	44	52	41	558 mm
0.1 mm	dp	11	11	11	10	12	11	9	8	7	9	12	12	123 days
	sunshine													hours

Sibiu

latitude: 45°48'n longitude: 24°09'e height: 452 m

DClo	025-5	Jan	Feb	Mar	Apr	May	Jun	Jul	Aug	Sep	Oct	Nov	Dec	Year
	tx	9	11	19	23	28	30	33	33	29	24	17	12	34 °C
	t	-4	-1	3	8	12	15	17	16	13	8	4	0	8 °C
	tn	-20	-19	-11	-3	2	6	9	8	2	-3	-8	-15	-24 °C
	clothing	w	w	w	n	n	n	s	n	n	n	w	w	–
	stress							o						–
	p	29	30	28	59	77	111	82	73	46	46	33	30	643 mm
	dp	6	6	6	8	11	11	9	8	6	7	6	6	92 days
	sunshine	69	89	151	167	200	216	263	255	212	162	84	67	1935 hours
	%	25	30	42	41	43	46	55	58	56	47	27	25	43 %

2. Climatic area: Walachia and Dobrudja
2.1. Temperate Continental Climate DC
2.2. Summer hot, winter cool or moderately cold.
2.3. Variable weather prevails, frequent periods of dry and sunny days.
2.4. Wind: Occasional gales. Snow cover: generally short periods only.

Sulina

				latitude: 45°09'n			longitude: 29°40'e			height: 9 m			
DCao	025-6	Jan	Feb	Mar	Apr	May	Jun	Jul	Aug	Sep	Oct	Nov	Dec	Year
	tx	11	12	17	21	27	30	33	31	28	23	18	13	33 °C
	t	-1	0	4	10	16	20	23	20	18	12	7	2	11 °C
	tn	-13	-11	-6	1	7	12	14	13	7	1	-4	-10	-16 °C
	clothing	w	w	w	n	n	s	s	s	s	n	n	w	–
	stress						o	o	o	o				–
	p	30	25	18	28	31	40	22	37	23	35	34	29	352 mm
	dp	4	4	3	4	4	4	3	3	3	4	5	5	48 days
	sunshine													hours

Bucaresti

				latitude: 44°25'n			longitude: 26°06'e			height: 82 m			
DCao	025-7	Jan	Feb	Mar	Apr	May	Jun	Jul	Aug	Sep	Oct	Nov	Dec	Year
	tx	9	12	21	26	31	34	36	36	32	27	19	13	37 °C
	t	-3	-1	5	12	17	21	23	23	18	12	6	0	11 °C
	tn	-18	-15	-9	-1	4	9	11	10	4	-1	-6	-12	-20 °C
	clothing	w	w	n	n	s	s	s	s	s	n	n	w	–
	stress					o	o	o	o	o				–
	p	43	36	35	47	69	87	55	49	30	44	43	41	578 mm
	dp	7	6	6	7	8	9	7	4	4	5	7	6	76 days
	sunshine	65	85	150	191	242	272	326	301	233	161	73	60	2159 hours
	%	23	29	41	47	53	58	69	69	62	47	25	22	48 %

4.1.25.2. Predominant general weather situations are described in 4.1.0.1. under Northern, Central, and Eastern Europe. A hygrothermal diagram of Bucaresti is to be found in 4.1.0.2.

4.1.26. Bulgaria

4.1.26.1. Climatic description and tables

1. Generally Temperate Continental Climate DC and Temperate Continental Mountain Climate GDC
2. Summer warm, in the lowlands of Maritza and Tundja as well as in the north hot, winter cool.
3. Variable weather prevails, frequent periods of dry and sunny days.
4. Wind: occasional gales. Snow cover: at higher levels of longer duration.

Lom (tx and tn: Vidim)

latitude: 43°49'n longitude: 23°14'e height: 33 m

DCao	026-1	Jan	Feb	Mar	Apr	May	Jun	Jul	Aug	Sep	Oct	Nov	Dec	Year
	tx	10	13	20	27	31	35	38	36	32	28	18	12	38 °C
	t	-2	0	6	13	18	22	24	23	18	12	6	1	12 °C
	tn	-15	-14	-7	-2	5	9	12	9	4	0	-5	-13	-22 °C
	clothing	w	w	n	n	s	s	t	s	s	n	n	w	–
	stress					o	o	f	o	o				–
	p	40	32	34	59	69	71	40	36	35	50	54	45	555 mm

Kolarovgrad

latitude: 43°16'n longitude: 26°56'e height: 198 m

DCbo	026-2	Jan	Feb	Mar	Apr	May	Jun	Jul	Aug	Sep	Oct	Nov	Dec	Year
	t	-1	1	4	10	16	19	22	22	18	12	6	2	11 °C
	clothing	w	w	w	n	n	s	s	s	s	n	n	w	–
	stress						o	o	o	o				–
	p	40	35	36	52	62	77	56	41	26	49	55	59	587 mm

Vratza

latitude: 43°12'n longitude: 23°32'e height: 348 m

DCbo	026-3	Jan	Feb	Mar	Apr	May	Jun	Jul	Aug	Sep	Oct	Nov	Dec	Year
	t	-2	0	5	12	16	20	22	22	18	12	6	1	11 °C
	clothing	w	w	n	n	n	s	s	s	s	n	n	w	–
	stress						o	o	o	o				–
	p	49	39	49	71	114	106	79	57	57	71	61	54	807 mm

Sofia

latitude: 42°49'n longitude: 23°23'e height: 588 m

GDCbo	026-4	Jan	Feb	Mar	Apr	May	Jun	Jul	Aug	Sep	Oct	Nov	Dec	Year
	tx	9	12	20	24	27	31	33	32	29	25	18	12	34 °C
	t	-2	1	5	10	15	19	21	21	17	11	6	1	10 °C
	tn	-11	-12	-7	-1	4	8	11	10	5	-1	-4	-11	-16 °C
	clothing	w	w	n	n	n	s	s	s	s	n	n	w	–
	stress						o	o	o	o				–
	p	42	31	37	55	71	90	59	43	42	55	52	44	622 mm
0.2 mm	dp	9	10	10	12	13	12	10	9	7	11	10	12	125 days
	sunshine	53	89	128	179	226	268	320	296	222	145	78	42	2046 hours
	%	18	30	35	45	50	58	69	68	59	42	27	15	46 %

Karnobat

latitude: 42°39'n longitude: 26°59'e height: 195 m

DCbk	026-5	Jan	Feb	Mar	Apr	May	Jun	Jul	Aug	Sep	Oct	Nov	Dec	Year
	t	0	2	5	10	15	19	22	22	18	13	7	3	11 °C
	clothing	w	w	n	n	n	s	s	s	s	n	n	w	–
	stress						o	o	o	o				–
	p	36	30	28	47	66	71	43	33	33	44	56	53	540 mm

4.1.26.2. Predominant general weather situations are described in 4.1.0.1. under Northern, Central, and Eastern Europe. A hygrothermal diagram of Sofia is to be found in 4.1.0.2.

4.1.27. Italy

4.1.27.1. Climatic descriptions and tables

1. Climatic area: Alps and bordering lowlands beyond the Po river

1.1. Temperate Oceanic Climate DO and Temperate Oceanic Mountain Climate GDO, locally Temperate Oceanic High Mountain Climate HDO

1.2. Summer hot (higher up warm), winter cool (higher up colder).

1.3. Variable weather prevails, too, in summer, yet periods of dry and cloudy or sunny days occur frequently.

1.4. Wind: at higher levels frequent gales, at low levels occasionally. Snow cover: only at higher levels for several months.

Bolzano

latitude: 46°30′n longitude: 11°21′e height: 271 m

DObk	027-1	Jan	Feb	Mar	Apr	May	Jun	Jul	Aug	Sep	Oct	Nov	Dec	Year
	tx	12	16	22	26	29	32	34	33	30	25	17	12	35 °C
	t	0	4	9	13	17	20	22	21	18	13	6	1	12 °C
	tn	-10	-6	-2	2	5	9	11	10	7	0	-5	-8	-11 °C
	clothing	w	w	n	n	n	s	s	s	s	n	n	w	–
	stress						o	o	o	o	o			–
	p	18	29	30	60	69	93	100	113	71	75	80	43	781 mm
	dp	3	4	5	7	8	11	10	10	6	7	8	5	84 days
	sunshine	111	125	158	171	193	210	234	206	178	154	81	88	1913 hours
	%	40	43	43	42	41	44	49	47	47	45	29	33	43 %

Udine

latitude: 46°04′n longitude: 13°14′e height: 116 m

DOak	027-2	Jan	Feb	Mar	Apr	May	Jun	Jul	Aug	Sep	Oct	Nov	Dec	Year
	tx	11	13	19	24	28	31	33	33	29	25	17	13	34 °C
	t	4	4	8	12	17	20	23	22	19	14	9	5	13 °C
	tn	-6	-4	-2	3	6	10	12	12	8	4	-1	-3	-7 °C
	clothing	w	w	n	n	s	s	s	s	s	n	n	n	–
	stress					o	o	o	o	o				–
	p	79	84	89	109	129	170	129	92	129	153	136	122	1421 mm
	dp	7	6	9	10	12	13	10	9	8	9	9	9	111 days
	sunshine	111	110	146	171	215	228	269	240	187	171	79	99	2026 hours
	%	40	38	40	42	46	48	56	55	49	50	28	37	45 %

Venezia

DOak	027-3	Jan	Feb	Mar	Apr	May	Jun	Jul	Aug	Sep	Oct	Nov	Dec	Year
latitude: 45°30′n	longitude: 12°09′e	height: 2 m												
	tx	10	13	17	22	26	30	32	31	28	23	17	12	32 °C
	t	2	4	8	12	17	21	23	23	19	13	8	4	13 °C
	tn	-4	-3	0	5	9	13	15	15	11	6	1	-2	-5 °C
	clothing	w	w	n	n	s	s	t	t	s	n	n	w	–
	stress					o	o	f	f	o				–
	p	50	54	62	70	82	84	67	67	66	95	90	67	854 mm
	dp	7	6	7	8	9	8	6	5	6	8	8	7	85 days
	sunshine	94	102	156	180	241	255	304	262	199	158	72	81	2104 hours
	%	33	35	42	44	52	54	64	60	53	46	25	30	47 %

Milano

DOak	027-4	Jan	Feb	Mar	Apr	May	Jun	Jul	Aug	Sep	Oct	Nov	Dec	Year
latitude: 45°26′n	longitude: 9°17′e	height: 103 m												
	tx	11	14	20	24	29	32	34	33	29	23	15	12	34 °C
	t	1	3	8	12	17	21	23	22	18	12	7	3	12 °C
	tn	-5	-3	1	5	8	12	15	14	11	6	1	-4	-6 °C
	clothing	w	w	n	n	s	s	s	s	s	n	n	w	–
	stress					o	o	o	o	o				–
	p	52	49	65	70	85	89	55	71	72	114	101	80	903 mm
	dp	7	7	7	8	8	7	6	6	6	7	7	8	84 days
	sunshine	61	94	160	188	220	249	291	249	185	116	46	47	1906 hours
	%	22	32	43	46	48	53	61	57	49	34	16	17	43 %

Verona

DOak	027-5	Jan	Feb	Mar	Apr	May	Jun	Jul	Aug	Sep	Oct	Nov	Dec	Year
latitude: 45°24′n	longitude: 10°53′e	height: 74 m												
	tx	12	14	19	24	28	32	34	32	28	23	16	13	34 °C
	t	1	3	8	12	17	21	23	22	19	13	7	3	12 °C
	tn	-6	-4	0	4	8	10	13	13	10	5	0	-4	-7 °C
	clothing	w	w	n	n	s	s	t	s	s	n	n	w	–
	stress					o	o	o	o	o				–
	p	37	43	50	72	72	64	70	59	65	72	77	74	755 mm
	dp	6	6	7	8	9	7	6	6	5	7	8	8	83 days
	sunshine	94	102	156	180	241	255	304	262	199	158	72	81	2104 hours
	%	33	35	42	44	52	54	64	60	53	46	25	30	47 %

Torino

DOak	027-6	Jan	Feb	Mar	Apr	May	Jun	Jul	Aug	Sep	Oct	Nov	Dec	Year
latitude: 45°04′n	longitude: 7°40′e	height: 238 m												
	tx	11	16	22	27	31	35	36	34	30	25	16	13	37 °C
	t	2	4	9	13	18	21	24	23	19	13	7	3	13 °C
	tn	-6	-5	-1	3	7	11	14	13	10	4	0	-5	-8 °C
	clothing	w	w	n	n	s	s	s	s	s	n	n	w	–
	stress					o	o	f	f	o				–
	p	23	40	57	109	76	112	72	45	62	93	96	60	845 mm
	dp	3	4	6	8	8	10	6	6	5	7	7	5	75 days
	sunshine	120	118	172	181	213	321	260	224	164	135	84	104	1998 hours
	%	42	40	47	45	46	69	55	51	43	39	29	38	45 %

2. Climatic area: Central Italy and Trieste
2.1. Subtropical Rain Climate Cr and Subtropical Rain Mountain Climate GCr
2.2. Summer hot, winter cool in the lowlands.
2.3. Variable weather prevails, periods of dry and sunny, in winter cloudy days occur frequently.
2.4. Wind: occasional gales.

Trieste

latitude: 45°39'n longitude: 13°45'e height: 20 m

Crak	027-7	Jan	Feb	Mar	Apr	May	Jun	Jul	Aug	Sep	Oct	Nov	Dec	Year
	tx	12	13	17	23	27	30	32	31	28	23	18	14	33 °C
	t	5	5	8	13	17	22	24	23	20	15	10	7	14 °C
	tn	-3	-2	1	6	9	13	15	15	12	8	2	0	-5 °C
	clothing	n	n	n	n	s	s	t	t	s	n	n	n	–
	stress					o	o	f	f	o				–
	p	66	71	52	69	72	81	84	78	90	109	100	89	961 mm
	dp	7	7	7	7	9	9	8	6	6	10	9	9	94 days
	sunshine	108	109	145	185	237	260	301	272	204	178	81	85	2165 hours
	%	38	37	39	46	51	55	63	62	54	52	28	31	48 %

Genova

latitude: 44°25'n longitude: 8°55'e height: 21 m

Crak	027-8	Jan	Feb	Mar	Apr	May	Jun	Jul	Aug	Sep	Oct	Nov	Dec	Year
	tx	16	16	19	23	27	30	31	31	29	26	20	17	32 °C
	t	8	9	11	14	18	21	24	23	21	17	12	10	16 °C
	tn	-1	1	3	6	10	14	16	17	14	9	5	2	-2 °C
	clothing	n	n	n	n	s	s	t	t	s	n	n	n	–
	stress					o	o	f	f	o	o			–
	p	79	108	103	112	66	68	40	51	126	193	190	134	1270 mm
	dp	9	7	7	8	7	5	4	5	6	9	10	8	82 days
	sunshine	130	123	175	194	235	258	303	268	201	179	105	117	2288 hours
	%	45	42	48	48	51	55	64	61	53	52	36	43	51 %

Firenze

latitude: 43°46'n longitude: 11°15'e height: 51 m

Crak	027-9	Jan	Feb	Mar	Apr	May	Jun	Jul	Aug	Sep	Oct	Nov	Dec	Year
	tx	15	17	20	25	29	33	35	36	31	26	20	17	37 °C
	t	5	7	10	14	17	22	24	23	20	16	10	7	15 °C
	tn	-5	-4	-1	3	6	10	14	13	9	4	0	-2	-6 °C
	clothing	n	n	n	n	s	t	t	t	s	n	n	n	–
	stress					o	o	f	f	o				–
	p	64	65	61	61	70	56	35	47	70	94	110	93	825 mm
	dp	9	8	8	8	8	7	3	4	6	8	10	10	88 days
	sunshine	134	134	153	205	300	305	346	322	219	177	97	96	2488 hours
	%	47	45	42	51	66	66	74	74	58	52	33	35	56 %

Pisa

latitude: 43°41'n longitude: 10°24'e height: 11 m

Crak 027-10	Jan	Feb	Mar	Apr	May	Jun	Jul	Aug	Sep	Oct	Nov	Dec	Year
tx	16	17	20	23	28	31	33	33	30	26	20	17	34 °C
t	7	8	10	13	17	21	24	23	21	16	12	8	15 °C
tn	-4	-3	-1	3	6	11	13	13	10	4	0	-2	-5 °C
clothing	n	n	n	n	s	s	t	t	s	n	n	n	–
stress					o	o	f	f	o				–
p	83	81	67	66	74	50	24	33	81	148	119	109	935 mm
dp	9	8	8	8	8	5	2	3	6	10	11	10	88 days
sunshine	127	121	148	204	244	274	331	283	215	174	99	94	2314 hours
%	44	41	40	51	53	59	70	65	57	51	34	34	52 %

Ancona

latitude: 43°37'n longitude: 13°31'e height: 105 m

Crak 027-11	Jan	Feb	Mar	Apr	May	Jun	Jul	Aug	Sep	Oct	Nov	Dec	Year
tx	15	17	20	22	25	30	32	32	29	24	21	17	34 °C
t	6	7	9	14	17	21	24	24	21	16	11	8	15 °C
tn	-2	0	2	6	9	13	16	17	14	10	5	0	-3 °C
clothing	n	n	n	n	s	s	t	t	s	n	n	n	–
stress					o	o	f	f	o				–
p	77	52	54	42	48	50	34	34	69	64	71	79	674 mm
dp	8	7	8	6	6	5	4	4	5	8	9	9	79 days
sunshine	70	91	151	204	259	268	319	286	194	146	75	72	2135 hours
%	24	31	41	51	57	58	68	66	51	42	26	26	48 %

Pescara

latitude: 42°36'n longitude: 14°12'e height: 16 m

Crak 027-12	Jan	Feb	Mar	Apr	May	Jun	Jul	Aug	Sep	Oct	Nov	Dec	Year
tx	17	19	21	22	26	30	33	34	29	25	21	19	35 °C
t	6	7	9	13	16	21	23	23	19	15	11	8	14 °C
tn	-2	-1	0	3	6	12	14	14	11	7	3	0	-3 °C
clothing	n	n	n	n	n	s	t	t	s	n	n	n	–
stress						o	f	f	o				–
p	69	44	53	46	47	38	30	35	48	96	75	77	658 mm
dp	8	5	7	7	6	4	3	4	6	8	9	8	75 days
sunshine	97	96	163	202	253	256	312	285	232	171	114	93	2274 hours
%	33	32	44	51	56	56	67	66	62	50	39	32	51 %

3. Climatic area: Southern Italy

3.1. Subtropical Winter Rain Climate Cs and
Subtropical Winter Rain Mountain Climate GCs

3.2. Summer hot, winter mild.

3.3. Summer generally dry and sunny or cloudy, now and then variable or scattered showers with thunderstorm. Winter partly variable, periods with showers or rain, partly periods of dry and cloudy or sunny days.

3.4. Wind: occasional gales.

Roma

latitude: 41°48′n longitude: 12°14′e height: 125 m

Csak 027-13	Jan	Feb	Mar	Apr	May	Jun	Jul	Aug	Sep	Oct	Nov	Dec	Year
tx	16	17	20	24	29	33	35	34	30	26	21	17	36 °C
t	8	9	11	14	18	22	24	24	22	17	13	10	16 °C
tn	-1	0	2	5	8	13	16	16	13	7	3	1	2 °C
clothing	n	n	n	n	s	s	t	t	s	s	n	n	–
stress					o	o	f	f	o	o			–
p	83	73	52	50	48	18	9	18	70	110	113	105	749 mm
dp	9	9	7	6	6	3	1	2	5	8	10	10	76 days
sunshine	131	124	179	212	263	275	334	307	243	189	123	111	2491 hours
%	45	41	49	53	58	60	72	71	65	55	42	39	56 %

Napoli

latitude: 40°53′n longitude: 14°18′e height: 68 m

Csak 027-14	Jan	Feb	Mar	Apr	May	Jun	Jul	Aug	Sep	Oct	Nov	Dec	Year
tx	16	18	20	24	28	32	33	34	30	27	22	18	35 °C
t	8	9	11	14	17	21	24	24	21	17	13	10	16 °C
tn	-2	0	0	4	7	11	14	15	12	7	2	0	-3 °C
clothing	n	n	n	n	s	s	t	t	s	s	n	n	–
stress					o	o	f	f	o	o			–
p	122	88	63	59	51	21	16	27	92	118	129	136	922 mm
dp	11	11	8	8	6	3	2	3	6	9	12	12	91 days
sunshine	120	117	167	200	251	269	320	304	237	188	124	99	2396 hours
%	40	39	45	50	56	59	70	71	63	54	42	34	54 %

Brindisi

latitude: 40°40′n longitude: 17°57′e height: 22 m

Csak 027-15	Jan	Feb	Mar	Apr	May	Jun	Jul	Aug	Sep	Oct	Nov	Dec	Year
tx	17	17	20	23	29	32	35	35	31	27	23	19	36 °C
t	9	10	11	14	17	22	24	24	22	18	14	11	16 °C
tn	1	2	3	6	9	14	16	17	14	10	6	3	0 °C
clothing	n	n	n	n	s	s	t	t	s	s	n	n	–
stress					o	o	f	f	o	o			–
p	79	62	43	35	36	13	13	24	38	76	88	74	581 mm
dp	11	6	6	6	5	2	1	2	4	7	10	9	69 days
sunshine	127	126	176	217	279	298	347	321	250	202	139	109	2591 hours
%	43	42	48	54	62	66	76	75	67	58	47	38	58 %

Alghero/Sardinia

latitude: 40°38′n longitude: 8°17′e height: 40 m

Csal 027-16	Jan	Feb	Mar	Apr	May	Jun	Jul	Aug	Sep	Oct	Nov	Dec	Year
tx	17	17	20	24	29	32	35	34	32	28	24	18	36 °C
t	10	10	11	13	16	21	23	24	21	18	14	11	16 °C
tn	0	1	2	4	7	11	13	14	13	8	5	2	0 °C
clothing	n	n	n	n	n	s	t	t	s	s	n	n	–
stress						o	f	f	o	o			–
p	60	42	45	44	27	17	2	9	42	77	81	86	529 mm
dp	9	7	6	7	4	2	0	1	4	8	9	12	69 days
sunshine	143	135	197	233	299	304	355	326	244	210	130	111	2687 hours
%	48	45	53	58	67	67	77	76	65	61	44	39	60 %

Taranto

Csak	027-17	Jan	Feb	Mar	Apr	May	Jun	Jul	Aug	Sep	Oct	Nov	Dec	Year
	latitude: 40°28′n longitude: 17°14′e height: 41 m													
	tx	18	18	20	22	28	32	35	34	30	26	21	18	36 °C
	t	9	10	11	14	18	23	26	26	22	18	14	11	16 °C
	tn	1	2	3	6	10	14	18	18	15	11	6	3	-1 °C
	clothing	n	n	n	n	s	t	t	t	s	s	n	n	–
	stress					o	o	f	f	o	o			–
	p	54	38	47	35	32	18	16	14	26	51	71	47	450 mm
	dp	8	5	6	5	5	3	2	2	3	6	7	6	58 days
	sunshine	126	131	156	221	284	316	341	327	246	197	140	110	2595 hours
	%	42	43	42	55	63	70	74	76	66	57	47	38	58 %

Cagliari/Sardinia

Csak	027-18	Jan	Feb	Mar	Apr	May	Jun	Jul	Aug	Sep	Oct	Nov	Dec	Year
	latitude: 39°15′n longitude: 9°03′e height: 18 m													
	tx	19	19	21	24	28	32	35	35	31	27	23	20	36 °C
	t	9	10	12	14	17	21	24	24	22	18	14	12	16 °C
	tn	2	3	4	7	10	13	17	18	15	10	7	4	1 °C
	clothing	n	n	n	n	s	s	t	t	s	s	n	n	–
	stress					o	o	f	f	o	o			–
	p	55	39	38	30	37	13	3	10	26	77	45	59	432 mm
	dp	8	7	6	6	5	2	0	1	3	8	7	10	63 days
	sunshine	128	122	187	208	269	274	327	304	229	193	122	105	2468 hours
	%	43	41	51	52	61	61	72	71	61	55	40	36	55 %

Messina/Sicily

Csal	027-19	Jan	Feb	Mar	Apr	May	Jun	Jul	Aug	Sep	Oct	Nov	Dec	Year
	latitude: 38°12′n longitude: 15°33′e height: 54 m													
	tx	18	18	21	22	28	31	34	34	31	27	23	19	35 °C
	t	11	11	13	15	18	23	26	26	23	20	16	13	18 °C
	tn	4	5	6	9	12	16	19	20	17	13	10	6	3 °C
	clothing	n	n	n	n	s	t	t	t	t	s	n	n	–
	stress					o	f	f	f	f	o			–
	p	146	100	92	52	51	15	19	18	69	144	148	120	974 mm
	dp	14	10	9	7	6	2	2	3	5	11	12	12	93 days
	sunshine	112	125	170	219	246	280	332	304	238	185	136	101	2448 hours
	%	37	41	46	55	56	63	73	72	64	53	45	34	55 %

Trapani/Sicily

Csal	027-20	Jan	Feb	Mar	Apr	May	Jun	Jul	Aug	Sep	Oct	Nov	Dec	Year
	latitude: 37°54′n longitude: 12°32′e height: 79 m													
	tx	18	18	21	24	28	32	34	34	31	28	24	19	36 °C
	t	10	11	12	14	18	22	24	25	23	19	16	12	17 °C
	tn	4	5	6	8	11	14	18	19	16	12	8	6	3 °C
	clothing	n	n	n	n	s	s	t	t	t	s	n	n	–
	stress					o	o	f	f	f	o			–
	p	71	52	47	34	18	9	2	8	31	83	84	77	516 mm
	dp	12	8	6	5	3	1	1	1	3	7	9	12	68 days
	sunshine	134	129	200	235	286	299	356	320	262	217	163	129	2730 hours
	%	44	42	54	59	65	67	79	76	70	62	53	43	61 %

Catania/Sicily

latitude: 38°28′n longitude: 15°04′e height: 16 m

Csal	020-21	Jan	Feb	Mar	Apr	May	Jun	Jul	Aug	Sep	Oct	Nov	Dec	Year
	tx	19	20	23	24	30	33	37	37	33	28	24	20	39 °C
	t	10	11	12	14	18	22	25	25	23	18	15	12	17 °C
	tn	3	4	5	8	11	16	20	20	17	12	8	5	2 °C
	clothing	n	n	n	n	s	s	t	t	t	s	n	n	–
	stress					o	o	f	f	f	o			–
	p	100	41	65	36	27	5	2	10	39	133	107	78	643 mm
	dp	8	5	6	5	4	1	1	1	4	8	8	8	59 days
	sunshine	123	128	172	199	241	287	319	281	247	186	170	139	2493 hours
	%	41	42	47	50	55	64	70	66	66	53	56	47	56 %

4.1.27.2. Predominant general weather situations are described in 4.1.0.1. under Southern Europe. A hygrothermal diagram of Roma is to be found in 4.1.0.2.
Wind names: The Scirocco is a hot, to begin with very dry south to south-easterly wind blowing from the hot deserts of North Africa and the Near East, which often reaches the countries on the Mediterranean as a wind laden with moisture.
Maestro is a north-westerly wind which blows in summer in the Adriatic when pressure is low over the Balkan Peninsular. It is a fresh breeze accompanied by fine weather and light clouds.

4.1.28. Malta

4.1.28.1. Climatic description and table

1. Subtropical Winter Rain Climate Cs
2. Summer hot, winter mild.
3. Summer generally dry and sunny. Winter partly variable, periods with showers or rain, partly periods of dry and sunny or cloudy days.
4. Wind: occasional gales.

Luga/Malta

latitude: 35°51′n longitude: 14°29′e height: 74 m

Csal	028-1	Jan	Feb	Mar	Apr	May	Jun	Jul	Aug	Sep	Oct	Nov	Dec	Year
	tx	18	20	22	25	29	33	35	36	32	28	24	21	37 °C
	t	12	12	14	16	19	23	25	26	24	21	17	14	19 °C
	tn	6	5	6	8	11	16	19	19	17	13	9	7	4 °C
	clothing	n	n	n	n	s	t	t	t	t	s	s	n	–
	stress					o	f	f	f	f	o	o		–
	p	92	48	46	36	14	2	2	7	41	157	114	83	640 mm
	dp	12	8	6	5	2	0	0	1	4	9	11	10	68 days
	sunshine	173	190	230	253	312	346	385	360	276	230	183	167	3105 hours
	%	56	62	62	64	72	79	86	86	74	65	59	55	70 %

4.1.28.2. Predominant general weather situations are described in 4.1.0.1. under Southern Europe.

4.1.29. Greece

4.1.29.1. Climatic description and tables

1. Subtropical Winter Rain Climate Cs and
   Subtropical Winter Rain Mountain Climate GCs
2. Summer hot, winter mild.
3. Summer: generally dry and sunny or cloudy. Winter: variable weather prevails, now and then periods of dry and cloudy or sunny days occur.
4. Wind: occasional gales. Snow cover: frequent at higher mountain levels during the winter.

Thessaloniki

latitude: 40°31′n longitude: 22°58′e height: 7 m

Csak	029-1	Jan	Feb	Mar	Apr	May	Jun	Jul	Aug	Sep	Oct	Nov	Dec	Year
	tx	16	18	22	26	31	35	38	37	33	28	22	18	38 °C
	t	6	7	10	14	20	25	27	27	22	17	12	8	16 °C
	tn	-4	-3	0	4	10	14	17	17	12	7	2	-2	-6 °C
	clothing	n	n	n	n	s	t	t	t	s	s	n	n	–
	stress					o	f	f	f	o	o			–
	p	44	34	35	36	40	33	20	14	28	55	56	54	449 mm
0.1 mm	dp	11	8	9	9	10	7	4	3	5	8	11	11	96 days

Larissa

latitude: 39°38′n longitude: 22°25′e height: 74 m

Cshk	029-2	Jan	Feb	Mar	Apr	May	Jun	Jul	Aug	Sep	Oct	Nov	Dec	Year
	tx	17	20	23	27	33	38	40	40	36	30	23	19	41 °C
	t	6	7	9	14	20	25	28	28	22	16	11	7	16 °C
	tn	-6	-5	-3	0	5	10	14	14	8	4	-1	-4	-8 °C
	clothing	n	n	n	n	s	t	t	t	s	n	n	n	–
	stress					o	f	f	f	o				–
	p	51	40	49	35	45	30	15	13	31	88	64	61	522 mm
0.1 mm	dp	10	8	8	7	9	6	3	2	4	9	10	10	86 days

Kerkyra-Corfu

latitude: 39°37′n longitude: 19°55′e height: 2 m

Csal	029-3	Jan	Feb	Mar	Apr	May	Jun	Jul	Aug	Sep	Oct	Nov	Dec	Year
	tx	18	19	21	24	28	33	36	36	32	28	23	20	37 °C
	t	10	10	12	15	19	24	27	27	23	18	14	11	18 °C
	tn	0	1	2	5	9	13	16	16	13	8	4	1	-1 °C
	clothing	n	n	n	n	s	t	t	t	t	s	n	n	–
	stress					o	f	f	f	f	o			–
	p	221	140	102	59	48	11	8	12	109	177	228	204	1319 mm
0.1 mm	dp	15	14	12	10	6	3	1	2	6	11	15	17	112 days

Patrai

latitude: 38°15'n longitude: 21°44'e height: 3 m

Csal	029-4	Jan	Feb	Mar	Apr	May	Jun	Jul	Aug	Sep	Oct	Nov	Dec	Year
	tx	18	20	23	27	29	33	36	36	32	29	25	21	37 °C
	t	10	10	12	16	20	24	26	26	23	19	14	11	18 °C
	tn	-1	0	1	4	8	12	14	14	11	7	3	1	-2 °C
	clothing	n	n	n	n	s	t	t	t	t	s	n	n	–
	stress					o	f	f	f	f	o			–
	p	123	87	72	50	27	13	1	6	27	82	113	148	749 mm
0.1 mm	dp	14	11	11	8	6	3	0	1	4	8	12	14	92 days

Athinai

latitude: 37°58'n longitude: 23°43'e height: 107 m

Cshl	029-5	Jan	Feb	Mar	Apr	May	Jun	Jul	Aug	Sep	Oct	Nov	Dec	Year
	tx	18	19	22	26	31	36	39	38	34	30	24	20	40 °C
	t	9	10	11	15	20	25	28	27	24	19	15	11	18 °C
	tn	1	1	3	6	12	16	19	19	15	10	6	2	-1 °C
	clothing	n	n	n	n	s	t	t	t	t	s	n	n	–
	stress					o	f	f	f	f	o			–
	p	62	36	38	23	23	14	6	7	15	51	56	71	402 mm
0.1 mm	dp	16	11	11	9	8	4	2	3	4	8	12	15	103 days
	sunshine	121	146	179	229	277	322	369	358	283	209	144	119	2758 hours
	%	40	48	48	58	63	73	82	85	76	60	47	40	62 %

Kalamai

latitude: 37°02'n longitude: 22°06'e height: 10 m

Csal	029-6	Jan	Feb	Mar	Apr	May	Jun	Jul	Aug	Sep	Oct	Nov	Dec	Year
	t	11	12	13	16	20	24	27	27	24	20	16	13	19 °C
	clothing	n	n	n	n	s	t	t	t	t	s	n	n	–
	stress					o	f	f	f	f	o			–
	p	138	83	93	48	37	14	1	6	34	93	143	162	852 mm

Heraklion

latitude: 35°20'n longitude: 25°11'e height: 48 m

Csal	029-7	Jan	Feb	Mar	Apr	May	Jun	Jul	Aug	Sep	Oct	Nov	Dec	Year
	tx	21	23	26	30	32	34	35	35	33	31	27	22	37 °C
	t	12	13	14	17	20	24	26	26	24	20	17	14	19 °C
	tn	4	4	5	8	10	14	18	19	14	12	9	6	3 °C
	clothing	n	n	n	s	s	t	t	t	t	s	s	n	–
	stress				o	o	f	f	f	f	o	o		–
	p	92	33	45	25	11	3	1	1	17	56	56	86	426 mm
0.1 mm	dp	14	9	10	6	4	1	0	0	2	6	11	14	77 days

4.1.29.2. Predominant general weather situations are described in 4.1.0.1. under Southern Europe. A hygrothermal diagram of Athinai is to be found in 4.1.0.2.

Wind name: Gregale is a strong north-easterly wind which blows in the Ionian Sea and neighbouring parts of the central Mediterranean mainly during the winter. It usually lasts two or three days and frequently reaches gale force. The wind is cold but not especially dry: some rain usually falls.

4.1.30. Union of Soviet Socialist Republics

4.1.30.1. Climatic descriptions and tables

1. *Climatic area:* Islands and littoral of the Arctic Ocean, except the European part
1.1. Tundra Climate FT
1.2. Summer brief and cool, winter very cold.
1.3. Variable weather prevails, now and then periods of dry and cloudy days occur, in summer sunny days, too.
1.4. Wind: frequent gales. Fog: frequently in summer. Snow cover: most of the year.

Mys Čeljuskin

latitude: 77°43′n longitude: 104°17′e height: 13 m

FTkd	030-1	Jan	Feb	Mar	Apr	May	Jun	Jul	Aug	Sep	Oct	Nov	Dec	Year
	t	-27	-27	-28	-21	-10	-1	2	1	-2	-10	-20	-25	-14 °C
	clothing stress	p	p	p	p	p	w	w	w	w	p	p	p	–
	p	12	14	18	21	24	25	27	28	24	21	16	14	244 mm

Ostrov Kotel'nyi

latitude: 76°00′n longitude: 137°54′e height: 10 m

FTkd	030-2	Jan	Feb	Mar	Apr	May	Jun	Jul	Aug	Sep	Oct	Nov	Dec	Year
	t	-30	-30	-27	-21	-9	0	2	2	-2	-10	-21	-26	-14 °C
	clothing stress	p	p	p	p	w	w	w	w	w	p	p	p	–
	p	8	8	6	9	10	11	20	19	11	11	10	8	131 mm

Ostrov Dikson

latitude: 73°30′n longitude: 80°14′e height: 20 m

FTkd	030-3	Jan	Feb	Mar	Apr	May	Jun	Jul	Aug	Sep	Oct	Nov	Dec	Year
	tx	-5	-8	-8	-3	1	9	17	15	9	2	-3	-4	18 °C
	t	-24	-24	-25	-16	-7	1	5	5	2	-7	-17	-21	-11 °C
	tn	-41	-40	-38	-33	-21	-8	-2	-1	-5	-21	-33	-38	-43 °C
	clothing stress	p	p	p	p	w	w	n	n	w	w	p	p	–
	p	20	13	17	9	11	23	32	46	42	21	14	18	266 mm
	dp	2	1	1	1	2	4	5	8	9	4	3	2	42 days

Ostrov Vrangelya

latitude: 70° 58′n longitude: 178° 33′w height: 3 m

FTkd	030-4	Jan	Feb	Mar	Apr	May	Jun	Jul	Aug	Sep	Oct	Nov	Dec	Year
	tx	-9	-11	-6	-3	3	9	13	12	5	1	-4	-6	13 °C
	t	-22	-26	-23	-17	-14	0	3	2	-2	-9	-17	-21	-12 °C
	tn	-38	-38	-37	-29	-22	-7	-3	-3	-10	-20	-28	-33	-41 °C
	clothing stress	p	p	p	p	p	w	w	w	w	w	p	p	–
	p	5	5	5	5	5	10	15	23	13	10	3	5	104 mm
	dp	2	2	3	2	3	3	5	6	4	5	3	2	40 days

Ostrov Četyreh-stolbovoj

latitude: 70° 38′n longitude: 162° 24′e height: 6 m

FTkd	030-5	Jan	Feb	Mar	Apr	May	Jun	Jul	Aug	Sep	Oct	Nov	Dec	Year
	t	-28	-29	-26	-18	-7	1	2	2	0	-9	-20	-26	-13 °C
	clothing stress	p	p	p	p	w	w	w	w	w	w	p	p	–
	p	9	10	8	8	10	13	21	19	16	15	14	12	155 mm

Mys Šmidta

latitude: 68° 55′n longitude: 179° 29′w height: 7 m

FTkd	030-6	Jan	Feb	Mar	Apr	May	Jun	Jul	Aug	Sep	Oct	Nov	Dec	Year
	t	-25	-27	-25	-18	-7	2	4	3	0	-8	-16	-24	-12 °C
	clothing stress	p	p	p	p	w	w	w	w	w	w	p	p	–
	p	23	21	17	17	19	25	33	42	30	25	20	24	296 mm

Mys Uzlen

latitude: 66° 10′n longitude: 169° 50′w height: 7 m

FTkc	030-7	Jan	Feb	Mar	Apr	May	Jun	Jul	Aug	Sep	Oct	Nov	Dec	Year
	tx	-7	-4	0	1	4	11	17	14	10	6	2	1	17 °C
	t	-21	-22	-20	-13	-4	2	6	5	3	-2	-9	-18	-8 °C
	tn	-36	-37	-36	-27	-19	-4	-1	-2	-3	-18	-27	-34	-41 °C
	clothing stress	p	p	p	p	w	w	n	n	w	w	w	p	–
	p	26	27	24	23	25	27	36	57	45	36	35	33	394 mm

2. Climatic area: Northern part of th East European lowlands
2.1. Subarctic Continental Climate EC
2.2. Summer cool, winter very cold.
2.3. Variable weather predominates all the year round; nevertheless there are also dry and in summer also sunny days.
2.4. Wind: frequent gales. Fog: frequently in summer. Snow cover: all the winter.

Murmansk

latitude: 68°58'n longitude: 33°03'e height: 46 m

EClc	030-8	Jan	Feb	Mar	Apr	May	Jun	Jul	Aug	Sep	Oct	Nov	Dec	Year
	tx	2	0	3	10	16	25	25	24	17	9	6	2	27 °C
	t	-11	-11	-8	-1	4	10	13	11	7	1	-4	-8	0 °C
	tn	-28	-28	-25	-13	-5	-1	4	3	-2	-10	-16	-24	-30 °C
	clothing	p	p	w	w	w	n	n	n	n	w	w	w	–
	stress													–
	p	19	16	18	19	25	40	54	60	44	30	28	23	376 mm
0.1 mm	dp	19	16	17	15	13	16	16	20	17	20	19	18	206 days
	sunshine	0	17	130	176	189	194	167	141	95	30	3	0	1142 hours
	%	–	8	36	37	29	27	22	25	24	10	2	–	25 %

Nar'jan-Mar

latitude: 67°39'n longitude: 53°01'e height: 7 m

EClc	030-9	Jan	Feb	Mar	Apr	May	Jun	Jul	Aug	Sep	Oct	Nov	Dec	Year
	tx	-2	-3	1	7	14	24	29	25	17	9	2	0	30 °C
	t	-16	-17	-15	-6	1	8	12	11	6	-1	-8	-13	-3 °C
	tn	-40	-36	-33	-26	-14	-3	2	0	-3	-18	-26	-35	-42 °C
	clothing	p	p	p	w	w	n	n	n	n	w	w	p	–
	stress													–
	p	20	14	17	21	24	38	41	59	52	39	31	22	378 mm
0.1 mm	dp	19	18	17	15	16	15	13	18	19	21	20	20	211 days
	sunshine	1	59	123	144	171	133	225	162	94	63	5	0	1180 hours
	%	1	26	34	31	27	18	30	30	24	21	3	–	25 %

Archangel'sk

latitude: 64°35'n longitude: 40°30'e height: 13 m

EClc	030-10	Jan	Feb	Mar	Apr	May	Jun	Jul	Aug	Sep	Oct	Nov	Dec	Year
	tx	-2	-1	4	14	20	26	28	27	21	11	5	1	29 °C
	t	-12	-12	-8	0	6	13	16	14	8	2	-3	-9	1 °C
	tn	-35	-30	-27	-16	-6	-1	2	1	-2	-10	-20	-30	-35 °C
	clothing	p	p	w	w	n	n	n	n	n	w	w	w	–
	stress													–
	p	33	28	28	28	39	59	63	57	66	55	44	39	539 mm
0.1 mm	dp	21	17	18	12	11	14	15	16	18	22	22	24	210 days
	sunshine	13	69	125	178	231	305	278	211	103	39	5	1	1558 hours
	%	8	29	35	39	40	47	45	41	26	13	3	1	34 %

Syktyvkar

ECIc	030-11	Jan	Feb	Mar	Apr	May	Jun	Jul	Aug	Sep	Oct	Nov	Dec	Year
		latitude: 61°40'n					longitude: 50°51'e				height: 96 m			
	tx	-2	-2	5	17	24	28	31	28	22	13	4	0	31 °C
	t	-14	-14	-8	1	7	14	17	14	8	1	-6	-12	1 °C
	tn	-39	-34	-27	-15	-4	0	5	2	-3	-14	-24	-32	-41 °C
	clothing stress	p	p	w	w	n	n	n	n	n	w	w	p	–
	p	24	19	23	30	48	55	68	57	60	49	30	29	492 mm
0.1 mm	dp	20	16	17	11	12	14	13	14	16	21	22	19	195 days

3. Climatic area: Siberia, except the southern Siberian Lowlands

3.1. Subarctic Continental Climate EC

3.2. Summer: cool near the Arctic Ocean, otherwise mild, in the southern part warm. Winter: in the western and southern parts very cold, otherwise severely cold and in the centre (Yakut area) excessively cold.

3.3. Variable weather prevails, in summer now and then dry and sunny days occur; in winter periods of cloudless days are not infrequent.

3.4. Wind: occasional gales. Fog: frequent in summer. Snow cover: all the winter.

Hatanga

ECId	030-12	Jan	Feb	Mar	Apr	May	Jun	Jul	Aug	Sep	Oct	Nov	Dec	Year
		latitude: 71°59'n					longitude: 102°28'e				height: 24 m			
	t	-32	-31	-29	-18	-6	6	12	9	2	-11	-26	-29	-13 °C
	clothing stress	p	p	p	p	w	n	n	n	w	p	p	p	–
	p	17	13	12	12	16	27	38	48	39	31	24	19	296 mm

Bulun

ECIe	030-13	Jan	Feb	Mar	Apr	May	Jun	Jul	Aug	Sep	Oct	Nov	Dec	Year
		latitude: 70°45'n					longitude: 127°47'e				height: 35 m			
	tx	-19	-18	-8	0	10	23	26	24	13	1	-12	-17	27 °C
	t	-41	-37	-28	-17	-6	7	11	8	1	-12	-31	-37	-15 °C
	tn	-53	-52	-45	-35	-19	-4	1	-1	-26	-31	-46	-52	-56 °C
	clothing stress	p	p	p	p	w	n	n	n	w	p	p	p	–
	p	8	5	8	8	13	25	38	41	38	23	10	8	223 mm
0.1 mm	dp	12	10	11	8	7	10	12	12	13	18	13	12	138 days

Čokurdah

ECId	030-14	Jan	Feb	Mar	Apr	May	Jun	Jul	Aug	Sep	Oct	Nov	Dec	Year
		latitude: 70°37'n					longitude: 147°53'e				height: 48 m			
	t	-36	-34	-29	-19	-5	6	10	8	1	-12	-25	-31	-14 °C
	clothing stress	p	p	p	p	w	n	n	n	w	p	p	p	–
	p	7	8	6	8	10	13	22	24	15	12	11	9	145 mm

Olenek

ECle	030-15	Jan	Feb	Mar	Apr	May	Jun	Jul	Aug	Sep	Oct	Nov	Dec	Year
		latitude: 68° 30′n					longitude: 112° 26′e				height: 127 m			
	t	-40	-37	-27	-13	-1	11	15	10	3	-11	-31	-37	-13 °C
	clothing stress	p	p	p	p	w	n	n	n	w	p	p	p	–
	p	10	7	7	10	20	38	54	55	26	22	13	13	275 mm

Verhojansk

ECle	030-16	Jan	Feb	Mar	Apr	May	Jun	Jul	Aug	Sep	Oct	Nov	Dec	Year
		latitude: 67° 33′n					longitude: 135° 23′e				height: 137 m			
	tx	-30	-22	-9	2	16	27	29	25	16	2	-14	-28	29 °C
	t	-47	-43	-30	-14	3	13	16	11	3	-14	-36	-44	-15 °C
	tn	-62	-59	-53	-37	-20	-1	2	-3	-13	-36	-52	-59	-62 °C
	clothing stress	p	p	p	p	w	n	n	n	w	p	p	p	–
	p	7	5	5	4	5	25	33	30	13	11	10	7	155 mm
0.1 mm	dp	8	7	4	4	5	7	8	8	7	7	8	8	81 days

Ilirnej

ECld	030-17	Jan	Feb	Mar	Apr	May	Jun	Jul	Aug	Sep	Oct	Nov	Dec	Year
		latitude: 67° 20′n					longitude: 168° 14′e				height: 426 m			
	t	-30	-33	-28	-18	-4	7	11	8	1	-13	-25	-31	-13 °C
	clothing stress	p	p	p	p	w	n	n	n	w	p	p	p	–
	p	12	9	7	6	7	21	48	40	18	13	14	11	206 mm

Salehard

ECIc	030-18	Jan	Feb	Mar	Apr	May	Jun	Jul	Aug	Sep	Oct	Nov	Dec	Year
		latitude: 66° 32′n					longitude: 66° 32′e				height: 35 m			
	tx	-8	-5	-1	4	9	18	25	22	14	5	-1	-6	26 °C
	t	-22	-22	-19	-8	-1	9	14	12	6	-3	-14	-21	-6 °C
	tn	-43	-39	-37	-27	-19	-3	4	2	-4	-18	-34	-39	-46 °C
	clothing stress	p	p	p	w	w	n	n	n	n	w	p	p	–
	p	24	20	24	32	39	51	57	57	54	46	31	29	464 mm
0.1 mm	dp	7	7	7	6	8	8	10	10	11	10	9	8	101 days

Šelagoncy

ECle	030-19	Jan	Feb	Mar	Apr	May	Jun	Jul	Aug	Sep	Oct	Nov	Dec	Year
		latitude: 66° 17′n					longitude: 114° 17′e				height: 325 m			
	t	-41	-36	-26	-12	1	12	15	10	3	-11	-30	-39	-13 °C
	clothing stress	p	p	p	p	w	n	n	n	w	p	p	p	–
	p	8	6	7	11	24	36	51	62	23	22	16	12	278 mm

Turuhansk

ECld	030-20	Jan	Feb	Mar	Apr	May	Jun	Jul	Aug	Sep	Oct	Nov	Dec	Year
	latitude: 65°47′n longitude: 87°57′e height: 32 m													
	tx	-11	-7	-1	6	13	21	28	24	17	4	-3	-7	28 °C
	t	-26	-23	-17	-7	1	11	16	13	6	-4	-20	-25	-6 °C
	tn	-52	-43	-39	-32	-17	-2	4	2	-6	-26	-42	-49	-54 °C
	clothing stress	p	p	p	w	w	n	n	n	n	w	p	p	–
	p	23	16	20	24	32	55	67	74	67	55	32	31	496 mm
0.1 mm	dp	11	11	11	10	11	12	12	14	16	18	15	13	154 days

Zyrjanka

ECld	030-21	Jan	Feb	Mar	Apr	May	Jun	Jul	Aug	Sep	Oct	Nov	Dec	Year
	latitude: 65°44′n longitude: 150°54′e height: 43 m													
	t	-39	-34	-26	-13	2	13	16	12	4	-11	-27	-35	-12 °C
	clothing stress	p	p	p	p	w	n	n	n	w	p	p	p	–
	p	14	13	8	8	8	28	52	47	21	19	19	17	254 mm

Tura

ECld	030-22	Jan	Feb	Mar	Apr	May	Jun	Jul	Aug	Sep	Oct	Nov	Dec	Year
	latitude: 64°10′n longitude: 100°04′e height: 140 m													
	t	-36	-31	-20	-7	4	13	16	12	5	-6	-26	-34	-9 °C
	clothing stress	p	p	p	w	w	n	n	n	n	w	p	p	–
	p	10	10	9	12	22	48	62	48	34	24	20	16	307 mm

Viljusk

ECbd	030-23	Jan	Feb	Mar	Apr	May	Jun	Jul	Aug	Sep	Oct	Nov	Dec	Year
	latitude: 63°46′n longitude: 121°37′e height: 107 m													
	tx	-16	-12	-1	6	18	28	30	28	19	8	-8	-16	31 °C
	t	-38	-32	-31	-7	5	15	18	14	6	-8	-27	-36	-9 °C
	tn	-54	-48	-42	-31	-14	1	4	0	-7	-29	-43	-53	-56 °C
	clothing stress	p	p	p	w	n	n	s	n	n	w	p	p	–
	p	9	7	6	11	16	33	43	36	24	18	13	10	226 mm

Ojmjakon

GECle	030-24	Jan	Feb	Mar	Apr	May	Jun	Jul	Aug	Sep	Oct	Nov	Dec	Year
	latitude: 63°16′n longitude: 143°09′e height: 726 m													
	t	-47	-43	-34	-15	1	12	15	11	2	-16	-36	-44	-16 °C
	clothing stress	p	p	p	p	w	n	n	n	w	p	p	p	–
	p	7	6	5	4	10	32	40	37	20	12	11	9	193 mm

Sejmčan

latitude: 62°55′n longitude: 152°25′e height: 207 m

ECld	030-25	Jan	Feb	Mar	Apr	May	Jun	Jul	Aug	Sep	Oct	Nov	Dec	Year
	t	-39	-34	-26	-12	3	14	16	12	4	-11	-27	-36	-11 °C
	clothing	p	p	p	p	w	n	n	n	w	p	p	p	–
	stress													–
	p	28	21	11	8	11	34	41	39	25	20	31	27	296 mm

Jakutsk

latitude: 62°05′n longitude: 125°45′e height: 103 m

ECbe	030-26	Jan	Feb	Mar	Apr	May	Jun	Jul	Aug	Sep	Oct	Nov	Dec	Year
	tx	-28	-17	-4	6	21	29	32	28	20	7	-9	-22	32 °C
	t	-43	-37	-23	-7	7	16	20	16	6	-8	-28	-40	-10 °C
	tn	-57	-52	-42	-28	-11	3	6	2	-7	-27	-45	-54	-58 °C
	clothing	p	p	p	w	n	n	s	n	n	w	p	p	–
	stress							o						–
	p	7	6	5	7	16	31	43	38	22	16	13	9	213 mm
0.1 mm	dp	7	7	4	5	6	8	9	9	10	10	10	7	92 days

Podkamennaja Tunguska

latitude: 61°36′n longitude: 90°00′e height: 60 m

EClc	030-27	Jan	Feb	Mar	Apr	May	Jun	Jul	Aug	Sep	Oct	Nov	Dec	Year
	t	-24	-21	-14	-2	5	14	17	14	8	-2	-16	-24	-4 °C
	clothing	p	p	p	w	n	n	s	n	n	w	p	p	–
	stress							o						–
	p	28	17	18	27	48	66	70	73	56	52	40	30	527 mm

Erbogačen

latitude: 61°16′n longitude: 108°01′e height: 287 m

ECld	030-28	Jan	Feb	Mar	Apr	May	Jun	Jul	Aug	Sep	Oct	Nov	Dec	Year
	t	-31	-27	-17	-5	5	14	17	13	6	-5	-22	-29	-7 °C
	clothing	p	p	p	w	n	n	s	n	n	w	p	p	–
	stress							o						–
	p	14	10	8	16	24	42	60	49	30	32	21	17	323 mm

Surgut

latitude: 61°15′n longitude: 73°30′e height: 43 m

EClc	030-29	Jan	Feb	Mar	Apr	May	Jun	Jul	Aug	Sep	Oct	Nov	Dec	Year
	tx	-7	-4	2	9	19	25	28	26	19	10	1	-3	28 °C
	t	-21	-19	-14	-2	5	14	17	14	8	-1	-13	-20	-3 °C
	tn	-44	-38	-34	-23	-12	-1	6	3	-3	-18	-34	-42	-47 °C
	clothing	p	p	p	w	n	n	s	n	n	w	p	p	–
	stress							o						–
	p	19	17	19	23	44	62	66	68	62	47	30	27	484 mm
0.1 mm	dp	14	10	11	9	14	14	12	15	16	19	15	17	166 days

Hanty-Mansijsk

latitude: 60° 58′n longitude: 69° 04′e height: 40 m

ECbc	030-30	Jan	Feb	Mar	Apr	May	Jun	Jul	Aug	Sep	Oct	Nov	Dec	Year
	t	-19	-17	-11	0	7	14	18	15	8	0	-11	-18	-1 °C
	clothing	p	p	p	w	n	n	s	n	n	w	p	p	–
	stress							o						–
	p	23	19	22	28	47	59	76	65	62	46	35	28	510 mm

Aldan

latitude: 58° 37′n longitude: 125° 22′e height: 682 m

GECld	030-31	Jan	Feb	Mar	Apr	May	Jun	Jul	Aug	Sep	Oct	Nov	Dec	Year
	t	-28	-25	-17	-6	4	14	17	14	6	-6	-20	-26	-6 °C
	clothing	p	p	p	w	w	n	s	n	n	w	p	p	–
	stress							o						–
	p	19	15	15	27	63	73	88	83	70	49	28	21	551 mm

Enisejsk

latitude: 58° 27′n longitude: 92° 09′e height: 78 m

ECbc	030-32	Jan	Feb	Mar	Apr	May	Jun	Jul	Aug	Sep	Oct	Nov	Dec	Year
	t	-22	-20	-11	0	8	16	18	15	8	0	-13	-21	-2 °C
	clothing	p	p	p	w	n	n	s	n	n	w	p	p	–
	stress							o						–
	p	24	17	17	19	44	57	60	60	53	42	41	36	470 mm

Kolpašev

latitude: 58° 18′n longitude: 82° 54′e height: 76 m

ECbc	030-33	Jan	Feb	Mar	Apr	May	Jun	Jul	Aug	Sep	Oct	Nov	Dec	Year
	t	-20	-18	-11	0	8	15	18	15	9	0	-12	-20	-2 °C
	clothing	p	p	p	w	n	n	s	n	n	w	p	p	–
	stress							o						–
	p	18	12	15	21	51	58	77	79	52	37	31	24	475 mm

Kirensk

latitude: 57° 46′n longitude: 108° 07′e height: 261 m

ECbd	030-34	Jan	Feb	Mar	Apr	May	Jun	Jul	Aug	Sep	Oct	Nov	Dec	Year
	tx	-7	-3	6	13	24	31	32	29	22	13	1	-4	33 °C
	t	-27	-24	-14	-2	7	15	18	15	7	-2	-17	-25	-4 °C
	tn	-51	-46	-38	-25	-9	0	4	1	-6	-22	-38	-48	-52 °C
	clothing	p	p	p	w	n	n	s	n	n	w	p	p	–
	stress							o						–
	p	19	11	7	14	27	66	67	56	36	28	25	25	381 mm
0.1 mm	dp	17	15	14	11	12	13	12	14	14	15	19	18	174 days

Tomsk

latitude: 56°30′n longitude: 84°58′e height: 122 m

ECIc	030-35		Jan	Feb	Mar	Apr	May	Jun	Jul	Aug	Sep	Oct	Nov	Dec	Year
		tx	-4	-2	4	14	25	29	30	28	23	14	3	-2	31 °C
		t	-21	-18	-11	-2	8	14	17	15	9	0	-12	-18	-2 °C
		tn	-41	-36	-33	-18	-6	2	7	3	-3	-15	-32	-40	-44 °C
		clothing	p	p	p	w	n	n	n	n	n	w	p	p	–
		stress													–
		p	28	18	20	23	41	69	66	66	41	51	46	38	506 mm
0.1 mm		dp	20	14	13	11	14	15	13	15	13	18	20	22	188 days

Krasnoyarsk

latitude: 56°00′n longitude: 92°53′e height: 194 m

ECbc	030-36		Jan	Feb	Mar	Apr	May	Jun	Jul	Aug	Sep	Oct	Nov	Dec	Year
		tx	-1	1	7	17	27	32	32	29	26	15	6	2	33 °C
		t	-17	-15	-8	2	9	16	18	15	9	2	-10	-16	0 °C
		tn	-39	-33	-26	-13	-6	2	8	6	-4	-17	-28	-37	-41 °C
		clothing	p	p	w	w	n	n	s	n	n	w	p	p	–
		stress													–
		p	12	9	10	22	38	58	83	65	47	34	25	17	419 mm
0.1 mm		dp	5	7	6	7	10	10	8	11	11	13	10	10	108 days

Bomnak

latitude: 54°43′n longitude: 128°56′e height: 357 m

ECbd	030-37		Jan	Feb	Mar	Apr	May	Jun	Jul	Aug	Sep	Oct	Nov	Dec	Year
		t	-32	-24	-14	-2	8	15	18	15	9	-3	-20	-30	-5 °C
		clothing	p	p	p	w	n	n	s	n	n	w	p	p	–
		stress							o						–
		p	4	3	9	26	45	82	113	130	90	33	23	8	566 mm

Troickij Priisk

latitude: 54°37′n longitude: 113°37′e height: 1310 m

GECld	030-38		Jan	Feb	Mar	Apr	May	Jun	Jul	Aug	Sep	Oct	Nov	Dec	Year
		t	-29	-26	-18	-7	2	10	13	10	3	-6	-20	-27	-8 °C
		clothing	p	p	p	w	w	n	n	n	w	w	p	p	–
		stress													–
		p	2	2	4	10	19	48	118	104	48	9	6	4	374 mm

Irkutsk

latitude: 52°16′n longitude: 104°21′e height: 485 m

ECbc	030-39		Jan	Feb	Mar	Apr	May	Jun	Jul	Aug	Sep	Oct	Nov	Dec	Year
		tx	-6	-4	7	17	26	30	31	28	23	17	6	-2	32 °C
		t	-21	-18	-9	2	9	15	18	15	8	1	-11	-18	-1 °C
		tn	-39	-35	-29	-17	-8	-1	3	2	-6	-19	-28	-39	-42 °C
		clothing	p	p	w	w	n	n	s	n	n	w	p	p	–
		stress							o						–
		p	12	8	9	15	29	83	102	99	49	20	17	15	458 mm
		dp	3	3	2	4	8	7	9	11	8	6	4	4	69 days
		sunshine	86	132	205	221	239	247	244	222	178	141	83	50	2046 hours
		%	34	47	56	53	49	49	48	49	47	42	31	21	46 %

Čita

		latitude: 52°01′n				longitude: 113°20′e				height: 685 m				
GECbd	030-40	Jan	Feb	Mar	Apr	May	Jun	Jul	Aug	Sep	Oct	Nov	Dec	Year
	tx	-9	-6	6	18	24	32	33	29	24	14	3	-4	34 °C
	t	-26	-21	-12	0	9	16	18	16	8	-1	-15	-23	-3 °C
	tn	-43	-40	-32	-19	-8	-1	4	1	-6	-21	-32	-41	-45 °C
	clothing	p	p	p	w	n	n	s	n	n	w	p	p	–
	stress							o						–
	p	2	2	4	7	20	51	93	90	51	12	7	4	343 mm
	dp	0	1	1	1	5	6	10	11	5	3	1	1	47 days

4. *Climatic area:* Northern coast of the Sea of Okhotsk, Kamchatka, Coast of the Gulf of Anadyr, Kommandorski Islands, Kuril Islands

4.1. Subarctic Continental Climate EC in the north and inner parts of Kamchatka, otherwise Subarctic Oceanic Climate EO.

4.2. Summer mild or cool, winter very cold in the north, in the islands cold.

4.3. Variable weather prevails.

4.4. Wind: frequent gales. Snow cover: on the continent the whole winter.

Anadyr

		latitude: 64°47′n				longitude: 177°34′e				height: 62 m				
EClc	030-41	Jan	Feb	Mar	Apr	May	Jun	Jul	Aug	Sep	Oct	Nov	Dec	Year
	tx	-3	-2	-4	1	6	16	19	17	12	4	0	-2	21 °C
	t	-22	-22	-20	-12	-3	5	11	10	4	-5	-14	-20	-7 °C
	tn	-41	-40	-36	-31	-18	-3	3	2	-6	-18	-29	-38	-42 °C
	clothing	p	p	p	p	w	n	n	n	w	w	p	p	–
	stress													–
	p	21	14	16	13	9	13	39	46	26	27	17	19	260 mm
0.1 mm	dp	12	10	11	8	7	10	12	12	13	18	13	12	138 days

Korf

		latitude: 60°21′n				longitude: 166°00′e				height: 2 m				
EClc	030-42	Jan	Feb	Mar	Apr	May	Jun	Jul	Aug	Sep	Oct	Nov	Dec	Year
	t	-12	-14	-12	-6	1	8	12	12	8	-1	-8	-13	-2 °C
	clothing	p	p	p	w	w	n	n	n	n	w	w	p	–
	stress													–
	p	34	14	19	19	15	25	50	57	45	45	29	24	376 mm

Ohotsk

		latitude: 59°22′n				longitude: 143°12′e				height: 6 m				
EClc	030-43	Jan	Feb	Mar	Apr	May	Jun	Jul	Aug	Sep	Oct	Nov	Dec	Year
	tx	-9	-8	-2	4	11	15	21	20	18	9	-3	-6	23 °C
	t	-22	-19	-14	-5	1	6	12	13	9	-2	-14	-20	-5 °C
	tn	-36	-35	-30	-22	-11	-1	5	4	-2	-16	-27	-32	-38 °C
	clothing	p	p	p	w	w	n	n	n	n	w	p	p	–
	stress													–
	p	11	6	14	17	38	44	65	55	54	39	25	10	378 mm
0.1 mm	dp	4	3	4	7	7	8	10	9	11	6	4	4	77 days

Ostrov Beringa

latitude: 55°12'n longitude: 165°59'e height: 6 m

EOlo	030-44	Jan	Feb	Mar	Apr	May	Jun	Jul	Aug	Sep	Oct	Nov	Dec	Year
	tx	1	2	3	6	11	19	21	22	17	12	4	2	22 °C
	t	-4	-4	-3	-1	2	5	9	11	9	5	0	-2	2 °C
	tn	-32	-34	-27	-21	-8	1	7	7	-1	-9	-20	-26	-38 °C
	clothing	w	w	w	w	w	n	n	n	n	n	w	w	–
	stress													–
	p	38	23	30	24	31	28	48	59	53	72	64	46	516 mm

Petropavlovsk

latitude: 52°59'n longitude: 158°39'e height: 24 m

EOlo	030-45	Jan	Feb	Mar	Apr	May	Jun	Jul	Aug	Sep	Oct	Nov	Dec	Year
	tx	-1	-1	2	7	12	18	22	22	18	12	4	1	24 °C
	t	-8	-8	-5	0	4	8	12	14	10	5	-2	-6	2 °C
	tn	-21	-21	-18	-10	-3	1	4	6	2	-4	-11	-17	-23 °C
	clothing	w	w	w	w	w	n	n	n	n	n	w	w	–
	stress													–
	p	60	30	47	44	42	45	74	74	132	83	69	119	820 mm
	dp	11	7	8	9	7	9	9	9	7	7	8	10	101 days

Simušir

latitude: 46°51'n longitude: 151°52'e height: 26 m

EOlo	030-46	Jan	Feb	Mar	Apr	May	Jun	Jul	Aug	Sep	Oct	Nov	Dec	Year
	t	-5	-5	-3	0	3	5	9	11	10	7	2	-2	3 °C
	clothing	w	w	w	w	w	n	n	n	n	n	w	w	–
	stress													–
	p	102	63	90	117	119	70	112	162	198	175	123	130	1461 mm

5. *Climatic area:* Sakhalin Island and southern Far East

5.1. Temperate Continental Climate DC and Temperate Continental Mountain Climate GDC.

5.2. Summer warm, winter very cold.

5.3. Variable weather prevails, especially in summer.

5.4. Wind: frequent gales. Fog: occasionally in summer. Snow cover: every winter, at times variable.

Nikolaevsk-na-Amure

latitude: 53°09'n longitude: 140°42'e height: 47 m

DClc	030-47	Jan	Feb	Mar	Apr	May	Jun	Jul	Aug	Sep	Oct	Nov	Dec	Year
	tx	-10	-6	2	8	15	23	27	27	22	14	2	-4	28 °C
	t	-24	-20	-13	-3	4	12	17	16	11	2	-10	-20	-2 °C
	tn	-36	-34	-30	-19	-6	1	6	6	-1	-12	-25	-34	-40 °C
	clothing	p	p	p	w	w	n	s	n	n	w	p	p	–
	stress							o						–
	p	19	20	18	32	43	41	55	72	74	55	43	31	503 mm

Alexandrovsk

latitude: 50°54'n longitude: 142°10'e height: 30 m

DClc	030-48	Jan	Feb	Mar	Apr	May	Jun	Jul	Aug	Sep	Oct	Nov	Dec	Year
	tx	-2	-2	4	11	18	22	26	27	23	18	8	1	28 °C
	t	-20	-16	-9	0	5	11	16	17	12	5	-5	-13	0 °C
	tn	-31	-29	-24	-12	-4	1	6	7	2	-7	-17	-27	-33 °C
	clothing	p	p	w	w	n	n	n	s	n	n	w	p	–
	stress								o					–
	p	21	22	22	35	46	35	55	62	91	73	45	48	555 mm
0.1 mm	dp	15	15	10	14	12	10	14	15	15	19	16	16	171 days

Blagoveščensk

latitude: 50°16'n longitude: 127°30'e height: 137 m

DCbc	030-49	Jan	Feb	Mar	Apr	May	Jun	Jul	Aug	Sep	Oct	Nov	Dec	Year
	tx	-11	-5	7	17	28	30	33	32	27	21	3	-7	33 °C
	t	-24	-19	-10	3	11	18	22	19	12	3	-12	-21	0 °C
	tn	-37	-34	-27	-11	-2	6	11	8	-1	-13	-26	-37	-38 °C
	clothing	p	p	p	w	n	s	s	s	n	w	p	p	–
	stress						o	o	o					–
	p	5	3	9	21	45	100	120	106	80	26	13	6	534 mm
	dp	1	2	2	5	6	12	10	9	7	4	3	3	64 days

Habarovsk

latitude: 48°31'n longitude: 135°10'e height: 72 m

DCbc	030-50	Jan	Feb	Mar	Apr	May	Jun	Jul	Aug	Sep	Oct	Nov	Dec	Year
	tx	-11	-6	3	14	23	27	30	31	24	18	4	-6	31 °C
	t	-22	-17	-8	3	11	18	21	20	14	5	-8	-18	2 °C
	tn	-35	-34	-25	-13	-1	7	11	12	4	-8	-24	-33	-37 °C
	clothing	p	p	w	w	n	s	s	s	n	n	w	p	–
	stress						o	o	o					–
	p	9	8	13	32	55	73	102	115	87	39	15	10	558 mm
	dp	3	2	3	4	8	11	8	9	9	4	4	4	69 days

Nevel'sk

latitude: 47°03'n longitude: 142°03'e height: 30 m

DCbc	030-51	Jan	Feb	Mar	Apr	May	Jun	Jul	Aug	Sep	Oct	Nov	Dec	Year
	tx	1	1	5	12	18	21	24	27	23	18	12	4	27 °C
	t	-10	-9	-4	2	7	11	16	18	14	8	0	-6	4 °C
	tn	-19	-18	-14	-8	-2	2	7	8	4	-3	-10	-15	-21 °C
	clothing	p	w	w	w	n	n	n	s	n	n	w	w	–
	stress								o					–
	p	33	23	28	46	58	64	91	102	91	89	79	58	762 mm
	dp	13	8	9	9	9	10	7	9	9	11	12	14	120 days

Vladivostok

latitude: 43°07′n longitude: 131°54′e height: 138 m

DCbc	030-52	Jan	Feb	Mar	Apr	May	Jun	Jul	Aug	Sep	Oct	Nov	Dec	Year
	tx	-2	3	8	16	21	26	29	29	26	20	11	4	31 °C
	t	-14	-10	-3	4	10	13	17	20	16	9	-1	-9	4 °C
	tn	-25	-22	-16	-4	2	7	12	14	7	-3	-13	-21	-26 °C
	clothing	p	p	w	w	n	n	s	s	n	n	w	w	–
	stress							o	o					–
	p	19	17	27	44	78	105	116	149	144	72	39	15	824 mm
	dp	2	2	4	5	8	10	10	9	7	5	4	3	69 days

6. *Climatic area:* Central and southern part of the East European Lowlands, the southern Siberian lowlands, large areas of the northern and northeastern Kazakh Soviet Socialist Republic
6.1. Temperate Continental Climate DC, locally steppe climate BS.
6.2. Summer warm, partly hot, winter cold in the extreme west and southwest, else very cold.
6.3. Variable weather prevails; in summer often dry and sunny.
6.4. Wind: occasional gales. Snow cover: in the northern part for many weeks, elsewhere less.

Leningrad

latitude: 59°58′n longitude: 30°18′e height: 44 m

DCbo	030-53	Jan	Feb	Mar	Apr	May	Jun	Jul	Aug	Sep	Oct	Nov	Dec	Year
	tx	2	2	6	17	24	26	28	27	24	15	9	4	29 °C
	t	-8	-8	-4	3	10	15	18	17	11	5	0	-4	5 °C
	tn	-24	-25	-19	-7	0	4	7	7	2	-4	-11	-20	-28 °C
	clothing	w	w	w	w	n	n	s	s	n	n	w	w	–
	stress							o	o					–
	p	36	32	25	34	41	54	69	77	58	52	45	36	559 mm
0.1 mm	dp	21	17	14	12	13	12	13	14	17	18	18	22	191 days
	sunshine	13	45	121	144	244	313	270	257	145	72	15	3	1642 hours
	%	6	17	33	33	46	55	48	53	37	23	7	2	36 %

Tallin

latitude: 59°25′n longitude: 24°48′e height: 44 m

DClo	030-54	Jan	Feb	Mar	Apr	May	Jun	Jul	Aug	Sep	Oct	Nov	Dec	Year
	tx	3	2	5	17	22	27	27	26	23	15	9	4	29 °C
	t	-5	-6	-3	3	9	14	17	16	12	7	1	-2	5 °C
	tn	-23	-22	-18	-7	-1	4	7	6	1	-2	-11	-18	-26 °C
	clothing	w	w	w	w	n	n	s	n	n	n	w	w	–
	stress							o						–
	p	33	26	24	32	41	49	71	68	75	65	45	39	568 mm
0.1 mm	dp	19	16	13	11	12	10	13	15	16	17	18	19	179 days
	sunshine	32	48	132	186	227	327	311	262	145	71	24	9	1774 hours
	%	15	18	36	43	43	59	56	54	37	22	10	5	39 %

Vologda

DClc	030-55	Jan	Feb	Mar	Apr	May	Jun	Jul	Aug	Sep	Oct	Nov	Dec	Year
		latitude: 59°17′n						longitude: 39°52′e				height: 118 m		
	tx	-1	0	6	17	25	26	29	27	23	16	8	2	29 °C
	t	-11	-12	-6	2	10	15	17	16	9	3	-3	-9	3 °C
	tn	-35	-32	-26	-13	-2	1	4	2	-2	-10	-19	-28	-35 °C
	clothing	p	p	w	w	n	n	s	n	n	w	w	w	–
	stress							o						–
	p	2	2	4	10	19	48	118	104	48	9	6	4	374 mm
0.1 mm	dp	19	14	16	13	14	13	14	14	16	19	18	21	191 days
	sunshine	27	59	134	187	264	271	292	241	124	55	26	13	1693 hours
	%	13	23	37	43	50	49	53	50	32	17	11	7	37 %

Kirov

DCbc	030-56	Jan	Feb	Mar	Apr	May	Jun	Jul	Aug	Sep	Oct	Nov	Dec	Year
		latitude: 58°39′n						longitude: 49°37′e				height: 164 m		
	tx	-2	-1	4	18	27	28	30	28	24	14	6	0	30 °C
	t	-14	-13	-7	2	10	16	18	16	9	2	-6	-12	2 °C
	tn	-35	-30	-23	-11	-2	2	6	3	-1	-11	-19	-29	-36 °C
	clothing	p	p	w	w	n	n	s	n	n	w	w	p	–
	stress							o						–
	p	33	24	26	28	45	58	72	69	54	56	38	35	538 mm
0.1 mm	dp	22	16	17	12	12	12	15	11	16	20	20	21	194 days
	sunshine	29	93	117	165	258	298	251	257	121	63	29	23	1704 hours
	%	13	35	32	38	49	54	46	53	31	20	12	12	38 %

Riga

DClo	030-57	Jan	Feb	Mar	Apr	May	Jun	Jul	Aug	Sep	Oct	Nov	Dec	Year
		latitude: 56°58′n						longitude: 24°04′e				height: 3 m		
	tx	3	3	8	20	24	27	29	28	25	18	12	5	30 °C
	t	-7	-6	-2	6	11	15	16	16	13	8	2	-4	6 °C
	tn	-22	-22	-18	-6	-1	2	5	5	1	-5	-11	-19	-26 °C
	clothing	w	w	w	n	n	n	n	n	n	n	w	w	–
	stress													–
	p	31	29	27	33	44	45	53	70	64	62	62	47	568 mm
0.1 mm	dp	19	18	16	13	13	11	12	16	17	19	19	21	194 days
	sunshine	32	52	132	172	225	315	312	268	167	82	38	19	1816 hours
	%	14	19	36	40	44	59	58	57	43	25	16	9	40 %

Velikie Luki

DCbo	030-58	Jan	Feb	Mar	Apr	May	Jun	Jul	Aug	Sep	Oct	Nov	Dec	Year
		latitude: 56°23′n						longitude: 30°36′e				height: 98 m		
	tx	2	2	7	20	26	27	29	28	25	17	11	3	30 °C
	t	-8	-8	-4	4	12	16	19	16	11	5	0	-5	4 °C
	tn	-30	-28	-22	-8	-1	2	5	3	-2	-8	-17	-24	-32 °C
	clothing	w	w	w	w	n	n	s	n	n	n	w	w	–
	stress							o						–
	p	22	19	15	30	37	47	60	67	50	44	27	22	440 mm
0.1 mm	dp	19	15	14	11	16	12	14	13	13	15	16	19	177 days
	sunshine	47	85	133	154	260	307	276	244	170	83	33	23	1816 hours
	%	20	31	36	36	51	58	52	52	44	25	13	11	40 %

Kazan

		latitude: 55°47′n					longitude: 49°11′e				height: 64 m		
DCbc	030-59	Jan	Feb	Mar	Apr	May	Jun	Jul	Aug	Sep	Oct	Nov	Dec	Year
	tx	0	0	4	18	28	29	31	30	26	16	7	1	32 °C
	t	-13	-13	-7	4	12	18	19	18	11	4	-5	-10	3 °C
	tn	-33	-29	-23	-10	0	3	7	5	0	-9	-16	-27	-34 °C
	clothing	p	p	w	w	n	s	s	s	n	w	w	p	–
	stress						o	o	o					–
	p	16	17	18	25	46	50	66	63	45	45	21	23	435 mm
0.1 mm	dp	18	13	13	10	11	13	13	11	11	17	16	18	164 days
	sunshine	61	118	151	216	285	307	271	295	155	89	35	38	2021 hours
	%	26	43	41	51	56	58	51	63	40	27	14	17	45 %

Moskva

		latitude: 55°45′n					longitude: 37°34′e				height: 156 m		
DCbc	030-60	Jan	Feb	Mar	Apr	May	Jun	Jul	Aug	Sep	Oct	Nov	Dec	Year
	tx	0	2	6	20	27	26	29	28	25	18	10	3	30 °C
	t	-10	-10	-4	5	12	17	19	17	11	4	-2	-7	4 °C
	tn	-29	-26	-19	-8	1	4	7	6	0	-8	-14	-23	-30 °C
	clothing	p	p	w	n	n	s	s	s	n	w	w	w	–
	stress						o	o	o					–
	p	31	28	33	35	52	67	74	74	58	51	36	36	575 mm
0.1 mm	dp	18	15	15	13	13	12	15	14	13	15	15	23	181 days
	sunshine	30	58	113	161	242	256	258	218	136	73	32	20	1597 hours
	%	13	21	31	38	48	49	49	46	35	22	13	9	35 %

Kaunas

		latitude: 54°53′n					longitude: 23°53′e				height: 75 m		
DCbo	030-61	Jan	Feb	Mar	Apr	May	Jun	Jul	Aug	Sep	Oct	Nov	Dec	Year
	tx	3	3	8	21	25	28	30	30	27	18	12	4	30 °C
	t	-5	-5	-1	6	13	16	18	17	13	7	2	-2	6 °C
	tn	-22	-22	-17	-4	0	5	7	5	0	-4	-9	-18	-25 °C
	clothing	w	w	w	n	n	n	s	s	n	n	w	w	–
	stress							o						–
	p	33	35	29	40	49	69	98	92	57	48	39	36	625 mm
0.1 mm	dp	19	16	15	12	15	12	11	14	14	16	16	19	179 days
	sunshine	43	61	143	151	221	301	311	263	189	101	46	33	1863 hours
	%	18	22	39	36	44	58	60	56	49	31	18	15	41 %

Kaliningrad

		latitude: 54°43′n					longitude: 20°30′e				height: 21 m		
DCbo	030-62	Jan	Feb	Mar	Apr	May	Jun	Jul	Aug	Sep	Oct	Nov	Dec	Year
	tx	6	6	13	21	27	29	31	30	25	19	11	7	32 °C
	t	-3	-2	1	6	12	15	18	16	13	8	3	-1	7 °C
	tn	-15	-14	-10	-3	1	5	9	8	4	-2	-8	-13	-20 °C
	clothing	w	w	w	n	n	n	s	n	n	n	w	w	–
	stress							o						–
	p	46	35	37	39	45	62	81	97	76	69	61	59	707 mm
	dp	11	9	9	8	8	9	10	12	11	11	12	13	123 days
	sunshine	40	56	93	156	257	252	257	211	153	93	36	22	1607 hours
	%	16	20	25	37	52	49	49	45	40	28	14	10	36 %

Minsk

latitude: 53°52′n longitude: 27°32′e height: 234 m

DCbo	030-63	Jan	Feb	Mar	Apr	May	Jun	Jul	Aug	Sep	Oct	Nov	Dec	Year
	tx	0	2	8	21	26	29	29	29	26	18	12	3	30 °C
	t	-8	-7	-3	5	12	16	18	16	12	6	0	-5	5 °C
	tn	-27	-23	-20	-5	1	5	7	6	0	-5	-12	-20	-29 °C
	clothing	w	w	w	n	n	n	s	n	n	n	w	w	–
	stress							o						–
	p	34	30	26	42	59	72	83	81	62	47	40	30	606 mm
0.1 mm	dp	19	17	15	13	15	12	12	13	11	15	18	21	181 days
	sunshine	61	61	127	148	240	296	292	242	194	91	29	30	1811 hours
	%	25	22	35	35	49	58	57	52	51	28	11	13	40 %

Brest

latitude: 52°07′n longitude: 23°41′e height: 144 m

DCbo	030-64	Jan	Feb	Mar	Apr	May	Jun	Jul	Aug	Sep	Oct	Nov	Dec	Year
	tx	3	5	11	24	27	30	31	30	28	22	14	4	31 °C
	t	-6	-5	-1	8	14	17	19	17	14	9	3	-4	7 °C
	tn	-22	-21	-14	-3	2	5	8	7	2	-3	-8	-18	-24 °C
	clothing	w	w	w	n	n	n	s	n	n	n	w	w	–
	stress							o						–
	p	37	46	36	42	62	64	68	72	39	41	54	42	603 mm
0.1 mm	dp	20	19	14	13	15	12	12	12	11	13	19	18	178 days
	sunshine	50	61	139	160	236	290	287	251	185	119	48	40	1866 hours
	%	20	22	38	38	49	58	57	55	48	36	18	17	42 %

Orenburg

latitude: 51°45′n longitude: 55°06′e height: 109 m

DCbc	030-65	Jan	Feb	Mar	Apr	May	Jun	Jul	Aug	Sep	Oct	Nov	Dec	Year
	tx	-1	-0	6	24	30	34	35	34	30	19	8	2	37 °C
	t	-15	-14	-8	6	15	20	22	20	14	5	-5	-12	4 °C
	tn	-34	-31	-26	-10	0	4	9	6	-2	-10	-14	-28	-34 °C
	clothing	p	p	w	n	n	s	s	s	n	n	w	p	–
	stress						o	o	o					–
	p	21	20	26	22	38	37	43	30	30	41	25	25	358 mm
0.1 mm	dp	15	11	12	7	10	12	9	9	9	14	14	14	136 days
	sunshine	89	119	154	241	289	286	326	299	214	130	85	71	2303 hours
	%	35	42	42	58	60	57	65	66	56	39	32	29	51 %

Kiev

latitude: 50°24′n longitude: 30°27′e height: 179 m

DCbo	030-66	Jan	Feb	Mar	Apr	May	Jun	Jul	Aug	Sep	Oct	Nov	Dec	Year
	tx	3	5	11	24	29	31	32	32	27	22	14	6	33 °C
	t	-6	-5	0	8	15	19	20	19	14	8	1	-3	7 °C
	tn	-20	-20	-13	-2	4	8	11	9	4	-3	-8	-16	-23 °C
	clothing	w	w	w	n	n	s	s	s	n	n	w	w	–
	stress						o	o	o					–
	p	43	39	35	46	56	66	70	72	47	47	53	41	615 mm
0.1 mm	dp	18	18	16	11	13	11	13	12	8	10	15	19	164 days
	sunshine	42	64	112	162	257	273	287	252	189	123	51	31	1843 hours
	%	16	22	31	39	54	55	58	56	50	37	19	12	41 %

Alexandrov Gay

latitude: 50°09′n longitude: 48°33′e height: 25 m

DCac	030-67	Jan	Feb	Mar	Apr	May	Jun	Jul	Aug	Sep	Oct	Nov	Dec	Year
	tx	1	1	9	24	32	35	37	36	30	21	11	3	38 °C
	t	-11	-8	-4	8	17	20	23	22	15	7	0	-7	6 °C
	tn	-30	-28	-23	-6	2	6	10	8	0	-8	-13	-25	-31 °C
	clothing	p	w	w	n	n	s	t	s	n	n	w	w	–
	stress						o	o	o					–
	p	32	27	18	16	29	35	23	24	25	26	33	30	318 mm
0.1 mm	dp	15	11	13	6	7	9	6	6	7	10	11	13	114 days
	sunshine	77	124	137	248	312	320	335	249	215	160	80	73	2330 hours
	%	29	43	37	60	65	65	68	55	57	48	29	29	52 %

Harkov

latitude: 49°56′n longitude: 36°17′e height: 152 m

DCbo	030-68	Jan	Feb	Mar	Apr	May	Jun	Jul	Aug	Sep	Oct	Nov	Dec	Year
	tx	3	4	11	24	30	31	33	34	29	22	14	9	34 °C
	t	-7	-6	-2	8	15	19	21	20	14	7	1	-4	7 °C
	tn	-24	-23	-16	-3	3	7	10	8	1	-5	-10	-20	-26 °C
	clothing	w	w	w	n	n	s	s	s	n	n	w	w	–
	stress						o	o	o					–
	p	36	33	32	33	50	60	75	48	34	42	39	37	519 mm
0.1 mm	dp	18	14	13	10	10	9	10	9	7	8	14	20	142 days
	sunshine	39	63	118	187	275	274	309	252	215	149	57	25	1963 hours
	%	15	22	32	45	58	56	62	56	57	44	21	10	44 %

Lvov

latitude: 49°49′n longitude: 23°57′e height: 325 m

DCbo	030-69	Jan	Feb	Mar	Apr	May	Jun	Jul	Aug	Sep	Oct	Nov	Dec	Year
	tx	4	5	13	23	27	29	30	29	26	22	16	3	31 °C
	t	-5	-4	0	7	13	16	18	17	13	8	3	-2	7 °C
	tn	-21	-19	-15	-3	2	6	8	7	3	-2	-7	-16	-22 °C
	clothing	w	w	w	n	n	n	s	s	n	n	w	w	–
	stress							o	o					–
	p	33	33	30	47	66	77	96	91	47	52	46	37	655 mm
0.1 mm	dp	18	19	17	11	16	14	14	13	10	11	15	20	178 days
	sunshine	58	67	116	167	238	239	253	227	191	139	52	45	1792 hours
	%	22	23	32	40	50	49	51	50	50	41	19	18	40 %

Rostov-Na-Donu

latitude: 47°15′n longitude: 39°49′e height: 77 m

DCao	030-70	Jan	Feb	Mar	Apr	May	Jun	Jul	Aug	Sep	Oct	Nov	Dec	Year
	tx	6	10	15	25	30	34	35	35	31	23	16	11	36 °C
	t	-5	-5	0	9	17	21	24	22	16	9	2	-3	9 °C
	tn	-18	-19	-12	0	6	8	12	10	3	-4	-7	-17	-22 °C
	clothing	w	w	w	n	s	s	t	s	n	n	w	w	–
	stress					o	o	o	o					–
	p	38	41	32	39	36	58	49	37	32	44	40	37	483 mm
0.1 mm	dp	19	13	12	9	9	10	7	7	6	7	12	18	129 days
	sunshine	47	73	124	218	280	278	313	276	249	179	75	28	2140 hours
	%	17	25	34	53	60	58	65	62	66	53	27	11	48 %

Odessa

DCbo	030-71	Jan	Feb	Mar	Apr	May	Jun	Jul	Aug	Sep	Oct	Nov	Dec	Year
				latitude: 46°29′n				longitude: 30°38′e				height: 64 m		
	tx	8	10	13	22	26	30	32	32	27	22	17	11	34 °C
	t	-2	-2	2	8	15	20	22	22	17	11	5	0	10 °C
	tn	-15	-14	-7	-1	6	12	13	12	8	1	-4	-10	-16 °C
	clothing	w	w	w	n	n	s	s	s	s	n	n	w	–
	stress						o	o	o	o				–
	p	28	26	20	27	34	45	34	37	29	35	43	31	389 mm
0.1 mm	dp	14	14	11	10	9	9	7	7	6	6	11	15	119 days
	sunshine	51	66	139	208	293	291	320	303	226	178	59	43	2177 hours
	%	18	23	38	51	63	61	67	69	60	52	21	16	49 %

Simferopol

DCbk	030-72	Jan	Feb	Mar	Apr	May	Jun	Jul	Aug	Sep	Oct	Nov	Dec	Year
				latitude: 45°01′n				longitude: 33°59′e				height: 205 m		
	tx	13	15	20	25	29	32	33	33	30	26	22	17	34 °C
	t	0	0	3	9	15	19	22	21	16	10	6	2	10 °C
	tn	-14	-13	-8	-2	5	8	12	10	5	0	-4	-11	-16 °C
	clothing	w	w	w	n	n	s	s	s	n	n	n	w	–
	stress						o	o	o					–
	p	44	40	35	36	49	75	58	34	31	41	44	41	528 mm
0.1 mm	dp	15	15	12	9	9	9	8	6	7	6	13	16	125 days
	sunshine	69	110	156	220	307	321	343	313	256	233	120	65	2513 hours
	%	24	37	42	54	67	69	72	72	68	68	42	24	56 %

Novo-Pjattigorsk

GDCbo	030-73	Jan	Feb	Mar	Apr	May	Jun	Jul	Aug	Sep	Oct	Nov	Dec	Year
				latitude: 44°03′n				longitude: 43°02′e				height: 532 m		
	tx	10	14	16	24	28	31	34	32	30	23	18	14	34 °C
	t	-4	-3	0	9	15	19	22	21	16	9	2	-2	9 °C
	tn	-18	-18	-14	2	4	6	10	10	2	-6	-8	-15	-22 °C
	clothing	w	w	w	n	n	s	s	s	n	n	w	w	–
	stress						o	o	o					–
	p	13	15	25	37	65	77	70	50	51	39	25	15	482 mm
0.1 mm	dp	15	12	12	13	14	14	11	10	9	9	12	14	145 days
	sunshine	65	110	112	177	241	253	258	251	189	168	73	50	1947 hours
	%	23	37	30	44	53	54	55	58	50	49	25	18	44 %

Tobolsk

DCbc	030-74	Jan	Feb	Mar	Apr	May	Jun	Jul	Aug	Sep	Oct	Nov	Dec	Year
				latitude: 58°09′n				longitude: 68°11′e				height: 44 m		
	tx	-6	-2	4	16	26	29	30	28	23	13	2	-2	31 °C
	t	-18	-17	-11	1	10	16	18	16	9	2	-9	-17	-1 °C
	tn	-38	-32	-27	-15	-5	2	7	6	-2	-13	-28	-36	-40 °C
	clothing	p	p	p	w	n	n	s	n	n	w	w	p	–
	stress							o						–
	p	19	16	21	26	42	60	73	63	54	39	36	26	475 mm
0.1 mm	dp	13	10	10	8	12	12	12	14	14	14	15	16	150 days

Sverdlovsk

latitude: 56°48′n longitude: 60°38′e height: 237 m

DCbc	030-75	Jan	Feb	Mar	Apr	May	Jun	Jul	Aug	Sep	Oct	Nov	Dec	Year
	t	-15	-13	-8	3	10	16	18	16	9	2	-7	-13	2 °C
	clothing	p	p	w	w	n	n	s	n	n	w	w	p	–
	stress							o						–
	p	15	17	17	22	40	59	80	82	49	29	25	27	462 mm

Omsk

latitude: 54°56′n longitude: 73°24′e height: 94 m

DCbc	030-76	Jan	Feb	Mar	Apr	May	Jun	Jul	Aug	Sep	Oct	Nov	Dec	Year
	tx	-5	-3	2	17	28	31	32	29	25	16	3	-1	33 °C
	t	-19	-18	-11	2	11	17	19	16	10	2	-9	-16	0 °C
	tn	-39	-35	-29	-17	-4	3	8	4	-2	-18	-29	-38	-41 °C
	clothing	p	p	p	w	n	s	s	n	n	w	w	p	–
	stress							o						–
	p	8	6	9	18	30	53	72	46	33	23	15	12	325 mm
0.1 mm	dp	12	8	7	6	10	12	12	12	11	12	12	14	128 days

Minusinsk

latitude: 53°42′n longitude: 91°42′e height: 251 m

DCbc	030-77	Jan	Feb	Mar	Apr	May	Jun	Jul	Aug	Sep	Oct	Nov	Dec	Year
	t	-21	-19	-10	3	10	17	19	17	10	2	-10	-18	0 °C
	clothing	p	p	p	w	n	n	s	n	n	w	p	p	–
	stress							o						–
	p	9	9	10	15	32	54	66	43	33	16	17	13	316 mm

Barnaul

latitude: 53°20′n longitude: 83°42′e height: 196 m

DCbc	030-78	Jan	Feb	Mar	Apr	May	Jun	Jul	Aug	Sep	Oct	Nov	Dec	Year
	t	-18	-16	-9	3	12	18	20	17	11	3	-9	-15	1 °C
	clothing	p	p	w	w	n	s	s	s	n	w	w	p	–
	stress						o	o	o					–
	p	19	18	20	26	38	52	79	52	39	40	40	31	454 mm

Kustanaj

latitude: 53°13′n longitude: 63°37′e height: 171 m

DCbc	030-79	Jan	Feb	Mar	Apr	May	Jun	Jul	Aug	Sep	Oct	Nov	Dec	Year
	t	-18	-17	-11	4	13	19	20	18	12	3	-7	-15	2 °C
	clothing	p	p	p	w	n	s	s	s	n	w	w	p	–
	stress						o	o	o					–
	p	10	9	9	18	26	35	46	34	25	28	15	13	268 mm

Semipalatinsk

latitude: 50°21′n longitude: 80°15′e height: 206 m

DCbc	030-80	Jan	Feb	Mar	Apr	May	Jun	Jul	Aug	Sep	Oct	Nov	Dec	Year
	tx	-1	-1	4	20	31	35	36	34	29	21	8	1	37 °C
	t	-17	-16	-8	6	14	20	22	20	13	5	-7	-14	3 °C
	tn	-37	-32	-27	-13	-1	5	8	6	0	-12	-26	-31	-38 °C
	clothing	p	p	w	n	n	s	s	s	n	n	w	p	–
	stress						o	o	o					–
	p	14	15	17	19	22	30	32	23	21	22	27	22	264 mm
	dp	6	4	6	3	7	8	7	6	3	11	5	8	74 days

Karaganda

latitude: 49°48′n longitude: 73°08′e height: 555 m

GDCbc	030-81	Jan	Feb	Mar	Apr	May	Jun	Jul	Aug	Sep	Oct	Nov	Dec	Year
	t	-15	-14	-9	4	13	18	20	18	12	3	-8	-14	2 °C
	clothing	p	p	w	w	n	s	s	s	n	w	w	p	–
	stress						o	o	o					–
	p	11	11	15	22	28	41	43	28	21	24	15	14	273 mm

Uil

latitude: 49°04′n longitude: 54°41′e height: 89 m

DCac	030-82	Jan	Feb	Mar	Apr	May	Jun	Jul	Aug	Sep	Oct	Nov	Dec	Year
	tx	0	1	10	25	32	38	38	37	32	21	11	3	40 °C
	t	-12	-12	-3	9	17	22	24	22	16	6	0	-9	7 °C
	tn	-30	-30	-22	-7	2	6	11	8	0	-8	-14	-25	-32 °C
	clothing	p	p	w	n	n	s	t	s	n	n	w	w	–
	stress						o	o	o					–
	p	23	16	22	11	23	23	27	13	24	30	24	19	255 mm
0.1 mm	dp	13	8	8	5	6	8	7	5	5	9	8	10	92 days
	sunshine	102	131	165	270	315	317	332	290	268	175	113	96	2574 hours
	%	38	46	45	66	66	65	68	65	71	52	41	37	57 %

7. Climatic area: Eastern Coast of the Black Sea

7.1. Subtropical Rain Climate Cr

7.2. Summer warm or hot, winter cool.

7.3. Variable weather prevails, now and then periods of dry and cloudy or sunny days occur.

7.4. Wind: occasional gales.

Sochi

latitude: 43°35′n longitude: 39°43′e height: 31 m

Crak	030-83	Jan	Feb	Mar	Apr	May	Jun	Jul	Aug	Sep	Oct	Nov	Dec	Year
	tx	17	18	21	26	28	30	32	32	30	26	24	20	33 °C
	t	7	7	9	12	17	20	23	23	20	16	13	10	15 °C
	tn	-4	-3	-1	3	8	12	15	15	11	5	4	-2	-7 °C
	clothing	n	n	n	n	n	s	t	t	s	n	n	n	–
	stress						o	f	f	o				–
	p	201	126	130	116	93	101	60	101	106	91	143	183	1451 mm
0.1 mm	dp	17	14	15	14	12	10	7	9	9	9	12	15	143 days
	sunshine	70	125	137	169	227	238	267	277	217	142	79	21	2155 hours
	%	24	42	37	42	50	51	57	64	58	60	49	28	48 %

Sukhumi

latitude: 43°00'n longitude: 41°01'e height: 4 m

Crbk	030-84	Jan	Feb	Mar	Apr	May	Jun	Jul	Aug	Sep	Oct	Nov	Dec	Year
	tx	16	18	22	27	30	32	32	32	31	28	23	19	34 °C
	t	4	6	2	12	16	20	22	22	19	16	11	7	13 °C
	tn	-3	-2	-1	3	8	12	16	15	11	7	2	-1	-5 °C
	clothing	w	w	w	n	n	s	s	s	s	n	n	n	–
	stress						o	o	o	o				–
	p	124	114	109	117	104	99	117	99	132	107	132	145	1399 mm
0.1 mm	dp	12	11	11	12	10	9	9	7	9	9	10	13	122 days

8. *Climatic area:* The Ukrainian Coast of the Caspian Sea, large areas of the Western, Central, and Eastern Kazakh Soviet Socialist Republic, of the Uzbek SSR, the Turkmen SSR, and various valleys in the mountains of the Kirghiz SSR, the Tadzhik SSR, and some valleys in the Azorbaydzhan SSR, Armenian SSR, Georgian SSR

8.1. Steppe Climate BS and Steppe Mountain Climate GBS

8.2. Summer hot, partly very hot, winter mostly cold.

8.3. Periods of dry and cloudy or sunny days are interrupted now and then by some days with variable weather, which in winter are generally more frequent.

8.4. Wind: occasional gales. Snow cover: in the northern parts for many weeks, elsewhere less.

Turgaj

latitude: 49°38'n longitude: 63°30'e height: 123 m

BSac	030-85	Jan	Feb	Mar	Apr	May	Jun	Jul	Aug	Sep	Oct	Nov	Dec	Year
	t	-17	-16	-9	6	16	22	24	22	15	6	-5	-14	4 °C
	clothing	p	p	w	n	n	s	s	s	n	n	w	p	–
	stress						o	o	o					–
	p	12	10	9	17	16	18	27	14	9	19	12	14	177 mm

Gur'ev

latitude: 47°01'n longitude: 51°51'e height: -21 m

BSac	030-86	Jan	Feb	Mar	Apr	May	Jun	Jul	Aug	Sep	Oct	Nov	Dec	Year
	tx	3	5	14	26	34	37	39	37	33	22	14	5	39 °C
	t	-10	-8	-2	10	18	23	26	24	17	8	0	-6	8 °C
	tn	-25	-23	-16	-4	4	8	12	10	2	-7	-10	-20	-27 °C
	clothing	p	w	w	n	s	s	t	t	s	n	w	w	–
	stress					o	o	f	f	o				–
	p	14	12	12	12	16	19	20	11	8	16	10	14	164 mm
0.1 mm	dp	8	6	7	3	4	5	3	3	3	5	6	8	61 days
	sunshine	83	107	157	266	320	313	333	290	264	198	130	65	2526 hours
	%	22	32	42	74	86	87	90	78	73	53	36	17	58 %

Balhaš

latitude: 46°54'n longitude: 75°00'e height: 423 m

BSac	030-87	Jan	Feb	Mar	Apr	May	Jun	Jul	Aug	Sep	Oct	Nov	Dec	Year
	t	-15	-13	-5	8	16	22	24	22	15	6	-5	-12	5 °C
	clothing	p	p	w	n	n	s	s	s	n	n	w	p	–
	stress													–
	p	10	8	10	11	9	14	11	9	4	8	9	12	115 mm

Astrakhan

latitude: 46°16′n longitude: 48°02′e height: 18 m

BSao	030-88	Jan	Feb	Mar	Apr	May	Jun	Jul	Aug	Sep	Oct	Nov	Dec	Year
	tx	6	7	16	26	32	35	37	36	32	23	15	9	37 °C
	t	-6	-5	2	11	18	22	25	23	18	10	4	-2	10 °C
	tn	-22	-21	-13	-2	4	10	14	11	4	-4	-8	-17	-24 °C
	clothing	w	w	w	n	s	s	t	t	s	n	w	w	–
	stress					o	o	f	o	o				–
	p	16	11	14	14	16	19	10	25	22	16	16	17	196 mm
0.1 mm	dp	11	8	9	5	5	6	4	5	5	6	9	11	84 days
	sunshine	71	94	159	245	300	306	329	282	253	195	86	66	2386 hours
	%	25	32	43	60	65	65	69	64	67	57	30	25	53 %

Ašabad

latitude: 37°58′n longitude: 58°20′e height: 230 m

BShk	030-89	Jan	Feb	Mar	Apr	May	Jun	Jul	Aug	Sep	Oct	Nov	Dec	Year
	tx	16	19	27	33	38	41	41	40	37	32	26	20	42 °C
	t	2	5	9	16	23	29	31	29	24	16	8	3	16 °C
	tn	-15	-8	-4	3	9	14	17	14	7	1	-5	-8	-16 °C
	clothing	w	n	n	n	t	t	t	t	t	n	n	w	–
	stress					o	o	f	f	o				–
	p	22	21	44	38	28	6	2	1	3	11	15	19	210 mm

Kirovabad

latitude: 40°43′n longitude: 46°25′e height: 303 m

BSak	030-90	Jan	Feb	Mar	Apr	May	Jun	Jul	Aug	Sep	Oct	Nov	Dec	Year
	tx	14	17	20	26	31	35	36	35	31	27	22	19	36 °C
	t	3	4	8	12	18	22	25	25	20	15	9	5	14 °C
	tn	-7	-7	-4	2	8	11	15	15	10	3	-1	-5	-9 °C
	clothing	w	w	n	n	s	s	t	t	s	n	n	n	–
	stress					o	o	f	f	o				–
	p	12	14	17	39	38	51	26	17	25	26	11	18	294 mm
0.1 mm	dp	8	6	8	11	11	11	6	6	7	8	6	7	95 days
	sunshine	109	115	130	161	228	284	286	271	202	161	114	97	2158 hours
	%	37	38	35	40	51	63	62	63	53	46	38	34	48 %

Erivan

latitude: 40°10′n longitude: 44°30′e height: 990 m

GBSao	030-91	Jan	Feb	Mar	Apr	May	Jun	Jul	Aug	Sep	Oct	Nov	Dec	Year
	tx	5	9	17	24	28	33	35	36	32	26	17	11	36 °C
	t	-6	-3	4	12	17	22	26	26	21	14	6	0	12 °C
	tn	-19	-16	-9	-1	5	9	12	13	7	1	-4	-10	-19 °C
	clothing	w	w	w	n	n	s	t	t	s	n	n	w	–
	stress						o	f	f	o				–
	p	23	25	28	48	53	23	15	8	13	23	30	28	317 mm
0.1 mm	dp	9	8	7	11	12	6	4	2	3	5	7	8	82 days
	sunshine	89	118	169	212	283	334	359	352	300	246	144	90	2696 hours
	%	30	39	46	53	63	74	78	82	80	71	48	31	60 %

9. Climatic area: Region extending from the Caspian Sea to the Aral Sea and the Balkash Lake with the Desert of Kizil Kum

9.1. Desert Climate BW

9.2. Summer hot, in the south very hot, winter cold.

9.3. Dry and cloudy or sunny days prevail, sporadically it rains or snows.

9.4. Occasional gales. Snow cover: irregular.

Fort Ševčenko

latitude: 44° 33′n longitude: 50° 15′e height: -20 m

BWao	030-92	Jan	Feb	Mar	Apr	May	Jun	Jul	Aug	Sep	Oct	Nov	Dec	Year
	t	-3	-2	2	10	18	23	26	25	20	12	5	0	11 °C
	clothing	w	w	w	n	s	s	t	t	s	n	n	w	–
	stress					o	o	f	f	o				–
	p	7	8	8	14	11	17	15	7	14	12	9	8	130 mm

Čimbaj

latitude: 42° 57′n longitude: 59°49′e height: 66 m

BWao	030-93	Jan	Feb	Mar	Apr	May	Jun	Jul	Aug	Sep	Oct	Nov	Dec	Year
	t	-6	-4	2	12	20	24	26	24	18	10	1	-5	10 °C
	clothing	w	w	w	n	s	t	t	t	s	n	w	w	–
	stress					o	o	f	f	o				–
	p	7	14	13	12	10	6	2	2	3	7	5	8	89 mm

Krasnovodsk

latitude: 40° 02′n longitude: 52° 59′e height: 89 m

BWhk	030-94	Jan	Feb	Mar	Apr	May	Jun	Jul	Aug	Sep	Oct	Nov	Dec	Year
	tx	11	12	19	24	32	36	38	37	33	26	19	14	38 °C
	t	2	3	6	13	19	24	28	28	22	15	8	4	14 °C
	tn	-8	-6	0	5	11	16	19	19	13	7	1	-3	-10 °C
	clothing	w	w	n	n	s	t	t	t	s	n	n	w	–
	stress					o	f	f	f	o				–
	p	11	13	15	12	7	2	2	3	3	6	9	11	92 mm
0.1 mm	dp	5	4	6	5	3	1	1	1	1	2	4	6	39 days

10. Climatic area: Mountaineous regions south and southwest of Lake Balkhash, and of the Caucasian Mountains

10.1. Temperate Continental Climate DC and
Temperate Continental Mountain Climate GDC

10.2. Summer hot, winter moderately cold, cold at higher levels.

10.3. Partially variable, partially periods of dry and sunny or cloudy days.

10.4. Wind: occasional gales. Snow cover: only of longer duration at higher levels.

Alma Ata

latitude: 43°14′n longitude: 76°58′e height: 847 m

GDCao	030-95	Jan	Feb	Mar	Apr	May	Jun	Jul	Aug	Sep	Oct	Nov	Dec	Year
	tx	5	7	17	24	29	32	34	34	30	24	16	9	35 °C
	t	-7	-5	2	11	16	20	23	22	17	10	0	-5	9 °C
	tn	-25	-23	-16	-5	3	8	11	8	2	-8	-16	-20	-28 °C
	clothing	w	w	w	n	n	s	t	s	n	n	w	w	–
	stress						o	f	o					–
	p	26	32	64	89	99	59	35	23	25	46	48	35	581 mm
0.1 mm	dp	9	6	10	12	11	10	9	6	4	7	8	7	99 days

Taškent

latitude: 41°16′n longitude: 69°16′e height: 428 m

DCak	030-96	Jan	Feb	Mar	Apr	May	Jun	Jul	Aug	Sep	Oct	Nov	Dec	Year
	tx	13	17	23	29	34	37	38	37	33	28	24	17	39 °C
	t	0	3	7	14	20	25	27	25	19	13	5	1	13 °C
	tn	-19	-14	-6	-1	6	10	12	10	5	-3	-8	-11	-20 °C
	clothing	w	w	n	n	s	t	t	t	s	n	n	w	–
	stress					o	f	f	f	o				–
	p	49	51	81	58	32	12	4	3	3	23	44	57	417 mm
0.1 mm	dp	10	8	12	10	7	4	1	1	1	5	7	9	75 days

Tbilisi

latitude: 41°41′n longitude: 44°57′e height: 490 m

DCak	030-97	Jan	Feb	Mar	Apr	May	Jun	Jul	Aug	Sep	Oct	Nov	Dec	Year
	tx	14	17	21	26	30	33	36	35	32	27	21	17	36 °C
	t	1	3	6	12	18	21	25	24	20	14	8	3	13 °C
	tn	-7	-6	-2	2	7	11	15	15	10	2	-1	-5	-9 °C
	clothing	w	w	n	n	s	s	t	t	s	n	n	w	–
	stress					o	o	f	f	o				–
	p	20	21	36	43	87	69	50	37	42	46	37	20	508 mm
0.1 mm	dp	6	7	8	13	12	11	8	9	8	7	7	7	103 days
	sunshine	79	109	141	156	205	265	277	249	198	174	95	79	2027 hours
	%	21	32	38	43	55	74	74	67	55	47	26	21	46 %

Lenkoran

latitude: 38°46′n longitude: 48°52′e height: 19 m

DCak	030-98	Jan	Fcb	Mar	Apr	May	Jun	Jul	Aug	Sep	Oct	Nov	Dec	Year
	tx	11	12	17	21	26	30	32	31	28	25	17	15	33 °C
	t	2	4	7	11	17	22	24	24	20	15	9	6	13 °C
	tn	-7	-3	-2	4	9	14	16	16	12	7	2	-2	-8 °C
	clothing	w	w	n	n	n	s	t	t	s	n	n	n	–
	stress						o	f	f	o				–
	p	119	84	94	66	38	23	28	46	218	213	196	127	1253 mm
	dp	12	8	12	9	7	2	2	2	11	8	11	10	94 days

4.1.30.2. Predominant general weather situations are described in 4.1.0.1. under Northern, Central, and Eastern Europe, and in 4.2.0.1. under Northern Asia. A hygrothermal diagram of Moskva is to be found in 4.1.0.2.
Wind names: Buran is a snowstorm from the north or northeast in eastern European Russia and Siberia; here it is also called Purga, in the steppes it is called Wjuga.
Metel is a snowstorm in Central Russia.
Posjemok is high drifting snow in Russian arctic regions.
Purga: see Buran.
Ssuchowej is a hot dry duststorm in the steppes of southern Russia, blowing in summer, predominantly from the southeast.
Wjuga: see Buran.
A special phenomenon is the "débacle": the breaking up in the spring of the ice in the rivers. The term is chiefly applied to the great rivers of Russia and Siberia and of the North American Continent. The débacle lasts from two to six weeks; during this period the rivers often overflow their banks, inundating the surrounding country. In southern Russia the débacle begins about the middle of March, at latitude 55°–60° North, it begins early in April, but in the north it does not begin until May and in the extreme north of Siberia not until June. In Ontario, Canada, the débacle takes place in March and the water is free by April; in the St. Lawrence it is a little later, the river being free of ice in May.

4.2. Asia (except Asian USSR)

4.2.0.1. Predominant general weather situations

Western Near East (Turkey, Cyprus, Syria, Lebanon, Israel, Jordan)
In summer, the region is mostly dominated by a wedge of the anticyclone over North Africa, causing sinking air. Occasionally secondary depressions migrating eastward extend over the Mediterranean and cause rising air. In winter, secondary depressions repeatedly traverse the region, causing rising air in their passage. Meanwhile there is sinking air under the influence of wedges of the Azores or Asian anticyclone.

Arabia
In summer, the southern part is often influenced by the south-west monsoon, while the rest of the region has no cloud, in spite of predominantly rising air, because of lack of water vapour. In winter, the air generally sinks, under the influence of a wedge of the anticyclone over North Africa or Central Asia. Sometimes these wedges are displaced when secondary depressions transport cold air from the north into the region.

Eastern Near East (Iraq, Iran, Afghanistan, Pakistan)
In summer, the weather is mostly determined by a depression. The rising air in it does often not form clouds, because of lack of water vapour. Pakistan is frequently influenced by the Indian monsoon. In winter, a wedge of the anticyclone over Central Asia causes sinking air. It is often displaced by eastward-tracking secondary depressions belonging to depressions over Russia, which transport polar air behind their cold fronts far to the south.

India (India, Sri Lanka, Bangla Desh)
In summer, the whole region is influenced by the south-west monsoon, with its predominantly rising air, which at times is displaced for a short period in some areas by sinking air. In winter, a wedge of the continental anticyclone dominates the weather with its sinking air. The northern parts are now and then invaded by secondary depressions and their cold air, which displaces the ridge for a while. The southern parts and Sri Lanka are frequently influenced by disturbances, with rising air, which migrate from east to west in the equatorial trough of low pressure south of the Equator.

Northern Asia (Union of Soviet Socialist Republics, Mongolia)
In summer, the eastern half of the region is influenced by a depression over central Asia, and its warm and humid air fed in by the monsoon from the south. The rest of the region, especially the northern part, is traversed by eastward-tracking secondary depressions of depressions in the Arctic Sea. In winter, the central Asian anticyclone predominates, with sinking air, especially the eastern part, which, on the other hand, is frequently traversed by secondary depressions and invaded by the accompanying cold air. In the western part, eastward-tracking troughs with rising air occur, pursued by migrating wedges which at times develop into stationary anticyclones.

China
In summer, depressions coming from the south and also the west cause unsettled weather, but because of lack of water vapour they do not form clouds in some regions in the generally rising air. In winter, sinking air in the Central Asian anticyclone mostly determines the weather. Occasionally there are invasions of cold air far to the south especially on the eastern coast.

Korea and Japan
In summer, sinking air on the western slope of the North Pacific anticyclone alternates with the rising air of secondary depressions approaching from the south and south-east. In winter, the great changeability of the atmospheric pressure field determines the weather. Eastward-tracking depressions extend their troughs south and south-east, pursued by migrating wedges. Now and then anticyclones and depressions become stationary.

South-east Asia (Burma, Thailand, Viet Nam, Laos, Cambodia, Malaysia, Indonesia, Philippines)
In summer, the Equatorial Trough shifts far to the north, being in its northernmost position in August. The southwest monsoon, with its rising air, determines the weather in the regions far from the Equator. The rising movement of the air is intensified when disturbances approach from the east, and weakened when they move off to the west. Near the Equator there is often sinking air. In winter, a strong wedge of the continental anticyclone dominates the weather with its sinking air. Occasionally there are invasions of polar air behind the cold fronts of troughs coming from the north. Normally this process lasts a few days. The southern parts are frequently influenced by disturbances with rising air migrating from east to west in the Equatorial Trough south of the Equator.

Northern Indian Ocean
In summer, the Equatorial Trough degenerates into an enormous monsoon depression in the far north of the continent. In winter, it is south of the equator. There is a strong anticyclone over Central Asia, with a wedge extending to the Arabian Sea, and another anticyclone over the central North Pacific Ocean, with a wedge extending to South-east Asia. Under their influence, the air sinks and determines the weather in the region.

4.2.0.2. Hygrothermal diagrams

In the following figures, hygrothermal diagrams are given for the stations

Ankara, Damascus, Beirut, Tel Aviv, Amman, Riyadh, Baghdad, Tehran, Kabul, Lahore, Chittagong, New Delhi, Calcutta, Bombay, Colombo, Hong Kong, Seoul, Peking, Taipeh, Tokyo, Rangoon, Bangkok, Saigon, Vientiane, Phnom Penh, Singapore, Djakarta, Manila.

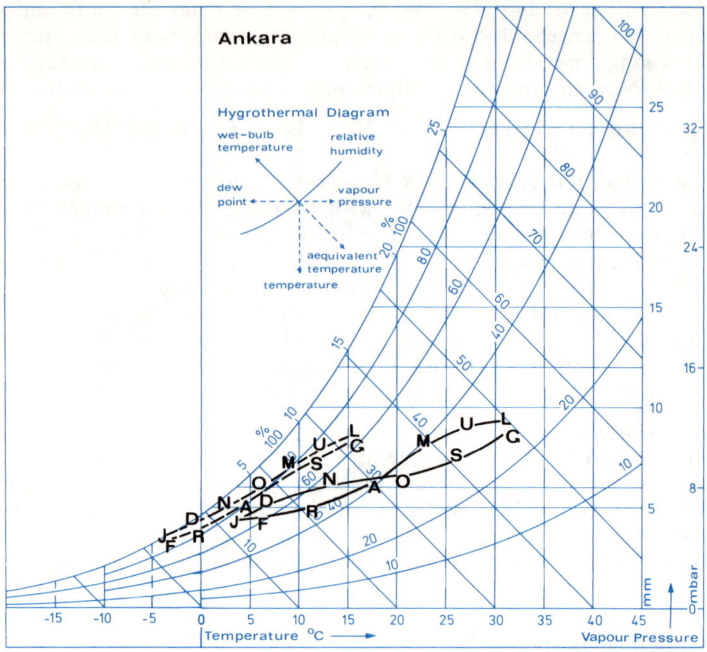
Ankara
Hygrothermal Diagram
wet-bulb temperature
relative humidity
dew point
vapour pressure
aequivalent temperature
temperature
Temperature °C
Vapour Pressure
mm
mbar

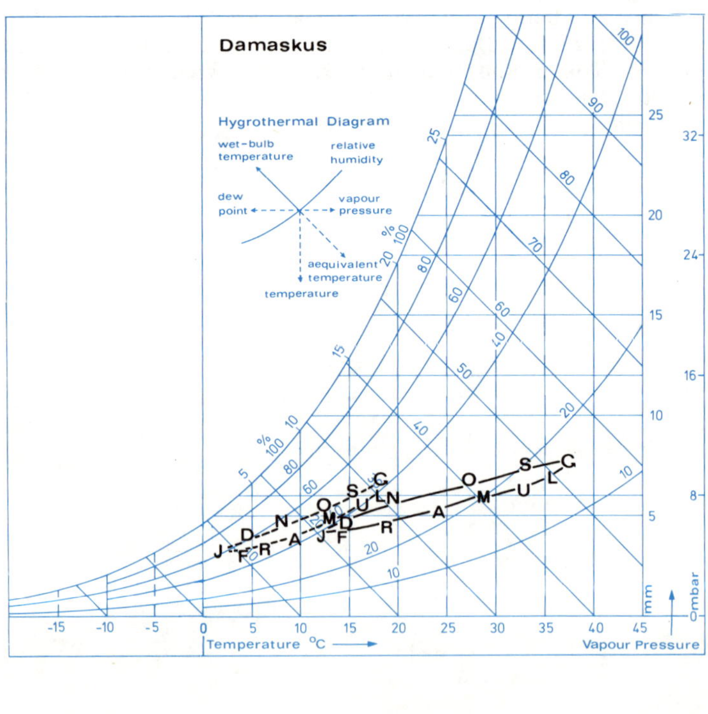
Damaskus
Hygrothermal Diagram
wet-bulb temperature
relative humidity
dew point
vapour pressure
aequivalent temperature
temperature
Temperature °C
Vapour Pressure
mm
mbar

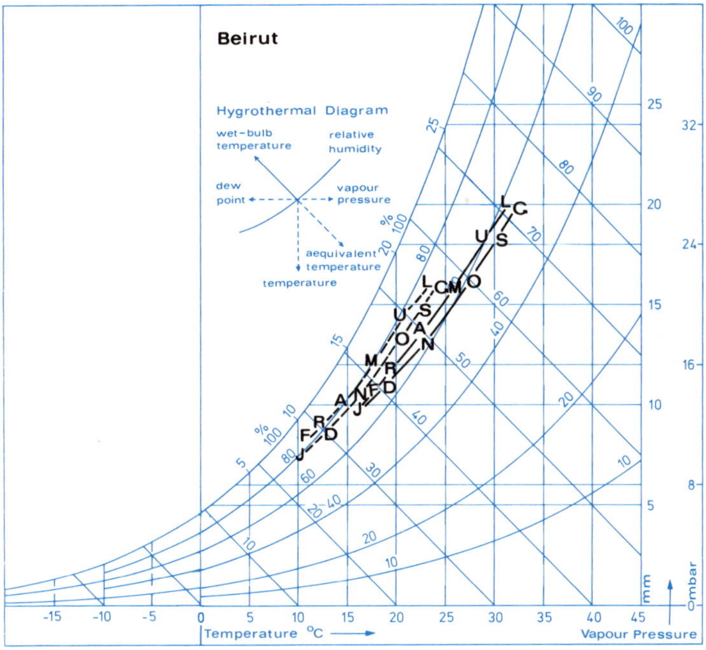
Beirut
Hygrothermal Diagram
wet-bulb temperature
relative humidity
dew point
vapour pressure
aequivalent temperature
temperature
Temperature °C
Vapour Pressure
mm
mbar

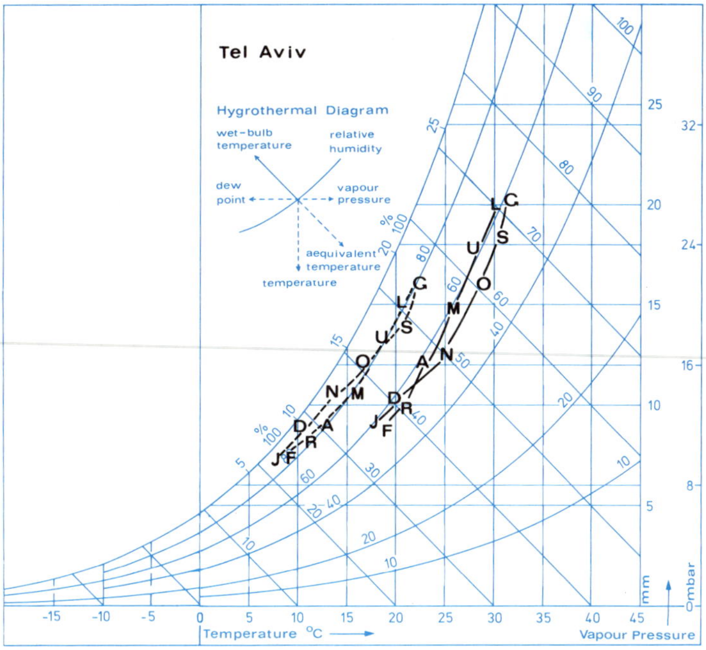
Tel Aviv
Hygrothermal Diagram
wet-bulb temperature
relative humidity
dew point
vapour pressure
aequivalent temperature
temperature
Temperature °C
Vapour Pressure
mm
mbar

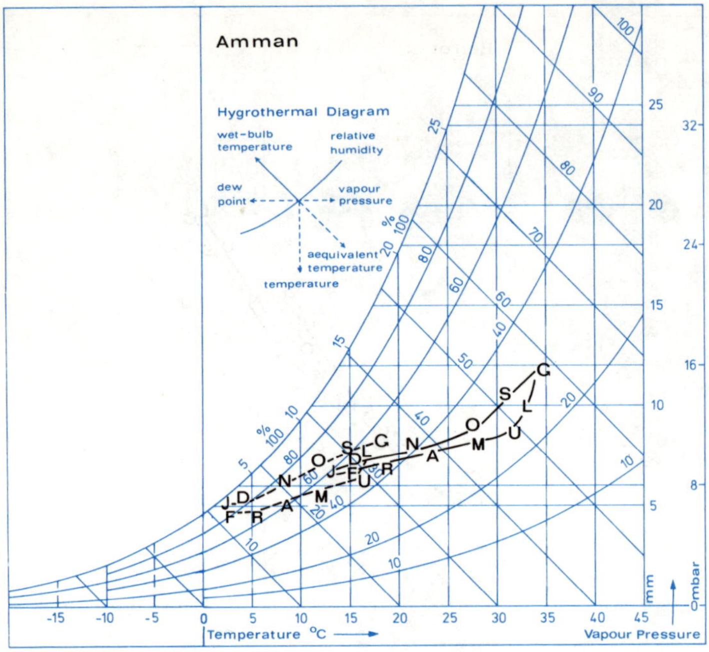
Amman
Hygrothermal Diagram
wet-bulb temperature
relative humidity
dew point
vapour pressure
aequivalent temperature
temperature
Temperature °C
Vapour Pressure
mm
mbar

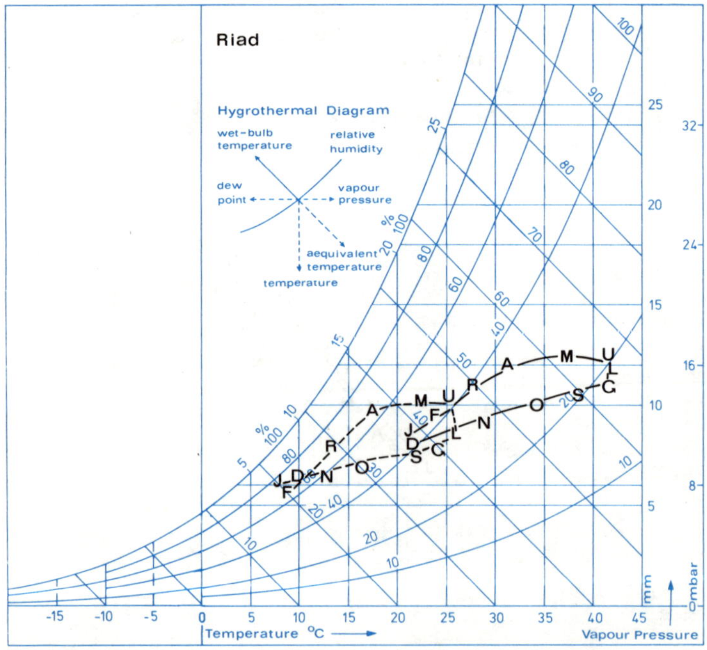
Riad
Hygrothermal Diagram
wet-bulb temperature
relative humidity
dew point
vapour pressure
aequivalent temperature
temperature
Temperature °C
Vapour Pressure
mm
mbar

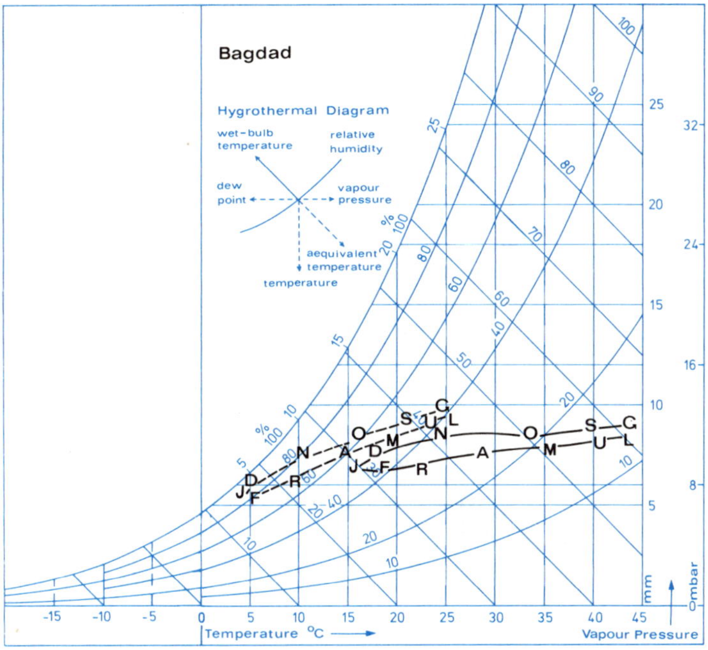
Bagdad
Hygrothermal Diagram
wet-bulb temperature
relative humidity
dew point
vapour pressure
aequivalent temperature
temperature
Temperature °C
Vapour Pressure
mm
mbar

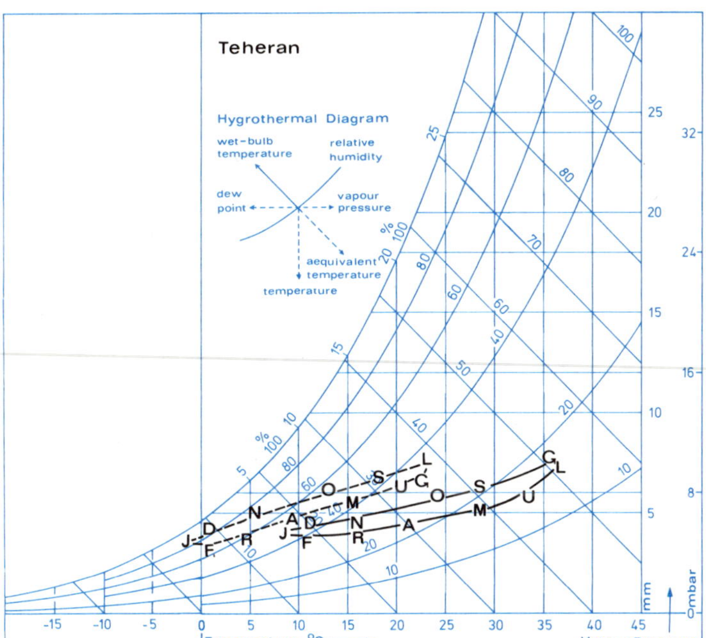
Teheran
Hygrothermal Diagram
wet-bulb temperature
relative humidity
dew point
vapour pressure
aequivalent temperature
temperature
Temperature °C
Vapour Pressure
mm
mbar

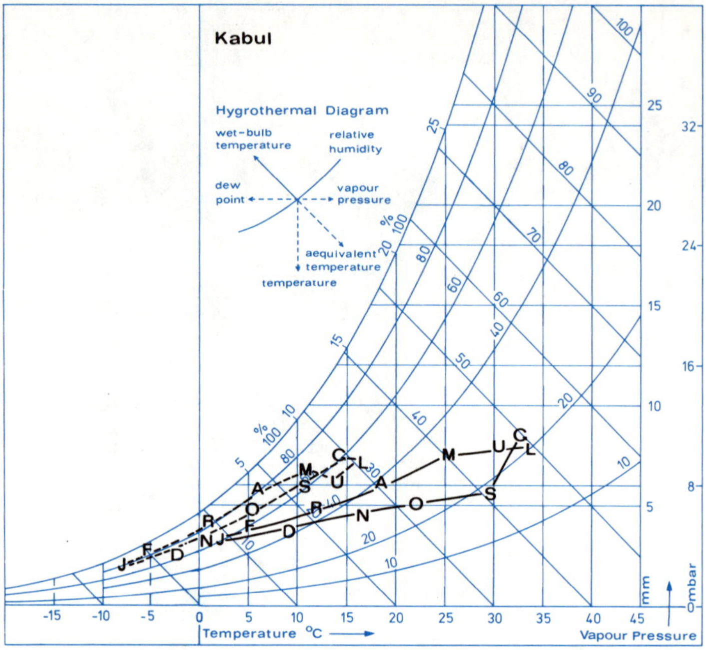
Kabul
Hygrothermal Diagram
wet-bulb temperature
relative humidity
dew point
vapour pressure
aequivalent temperature
temperature
Temperature °C
Vapour Pressure
mm
mbar

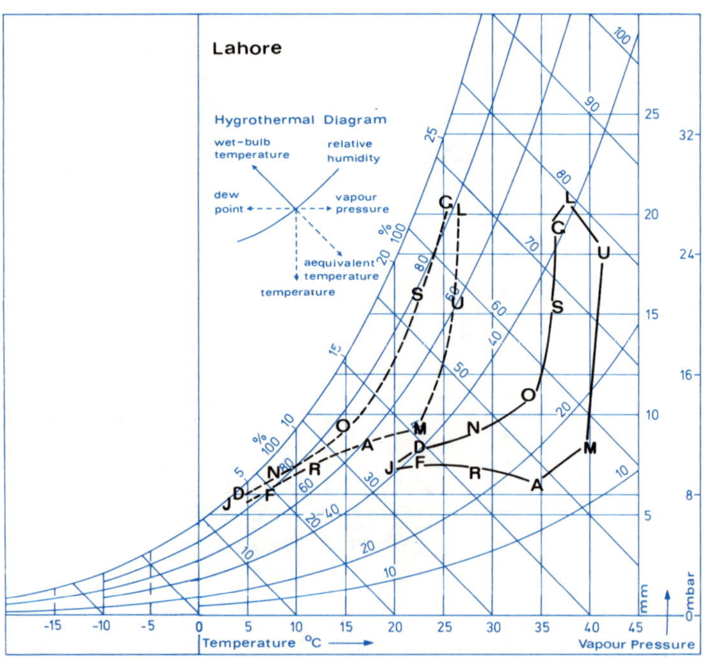
Lahore
Hygrothermal Diagram
wet-bulb temperature
relative humidity
dew point
vapour pressure
aequivalent temperature
temperature
Temperature °C
Vapour Pressure
mm
mbar

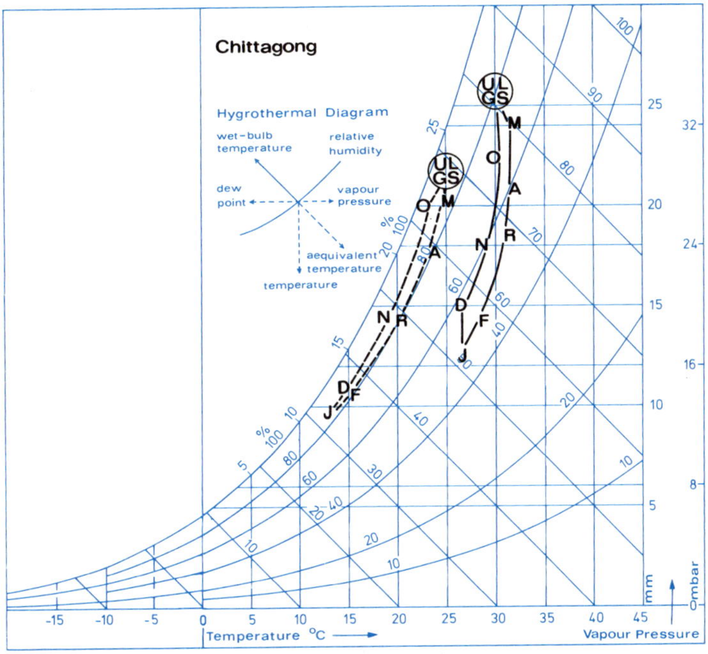
Chittagong
Hygrothermal Diagram
wet-bulb temperature
relative humidity
dew point
vapour pressure
aequivalent temperature
temperature
Temperature °C
Vapour Pressure
mm
mbar

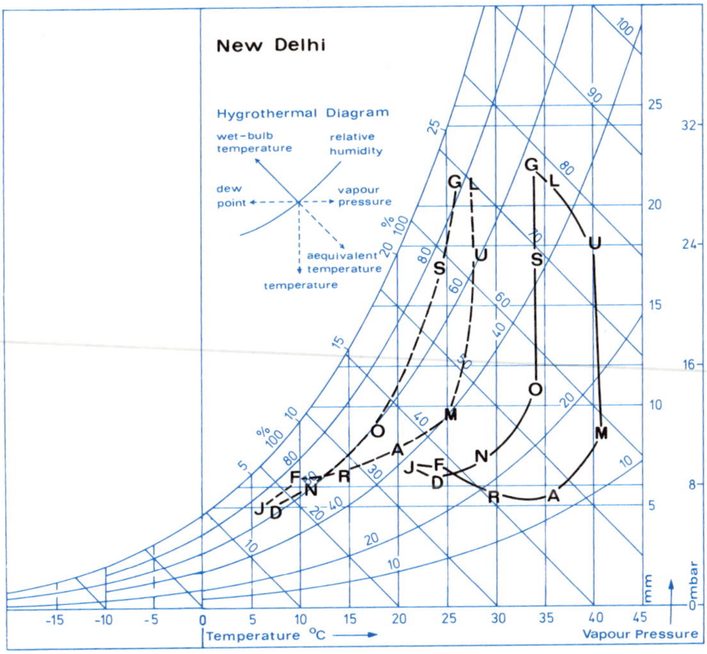
New Delhi
Hygrothermal Diagram
wet-bulb temperature
relative humidity
dew point
vapour pressure
aequivalent temperature
temperature
Temperature °C
Vapour Pressure
mm
mbar

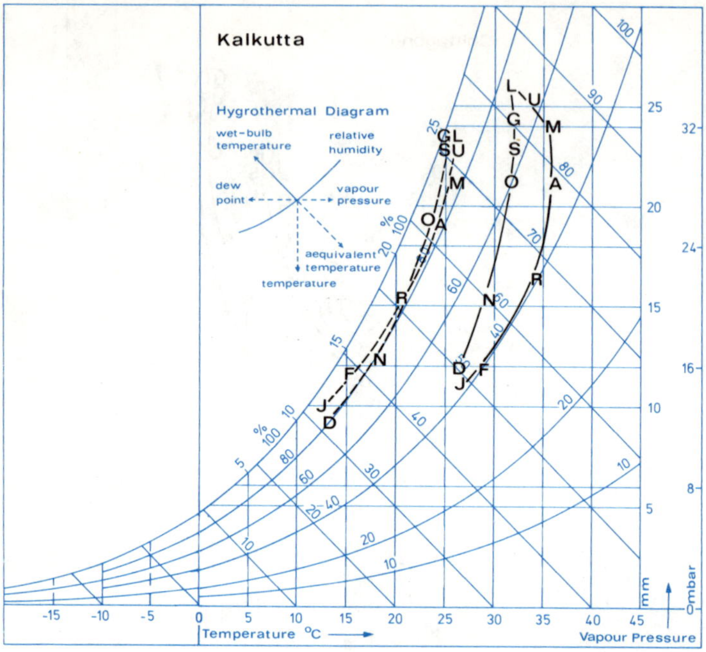
Kalkutta
Hygrothermal Diagram
wet-bulb temperature
relative humidity
dew point
vapour pressure
aequivalent temperature
temperature
Temperature °C
Vapour Pressure
mm
mbar

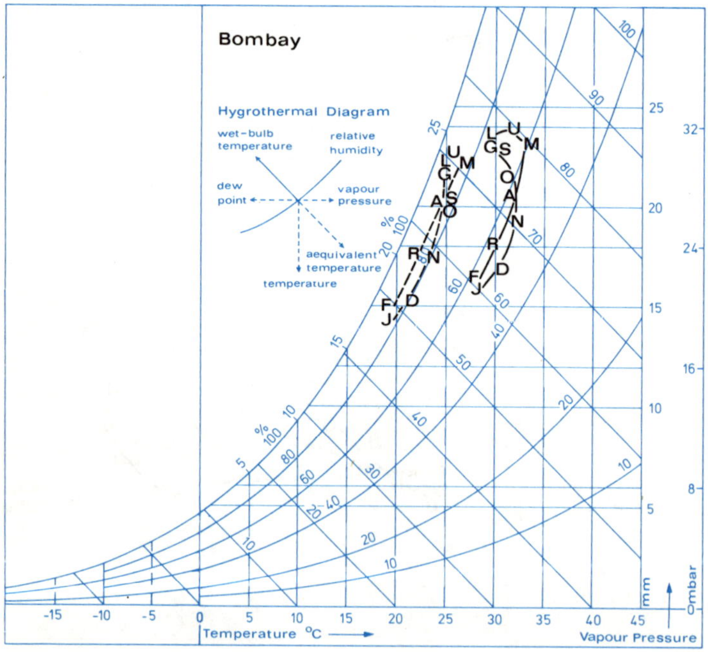
Bombay
Hygrothermal Diagram
wet-bulb temperature
relative humidity
dew point
vapour pressure
aequivalent temperature
temperature
Temperature °C
Vapour Pressure
mm
mbar

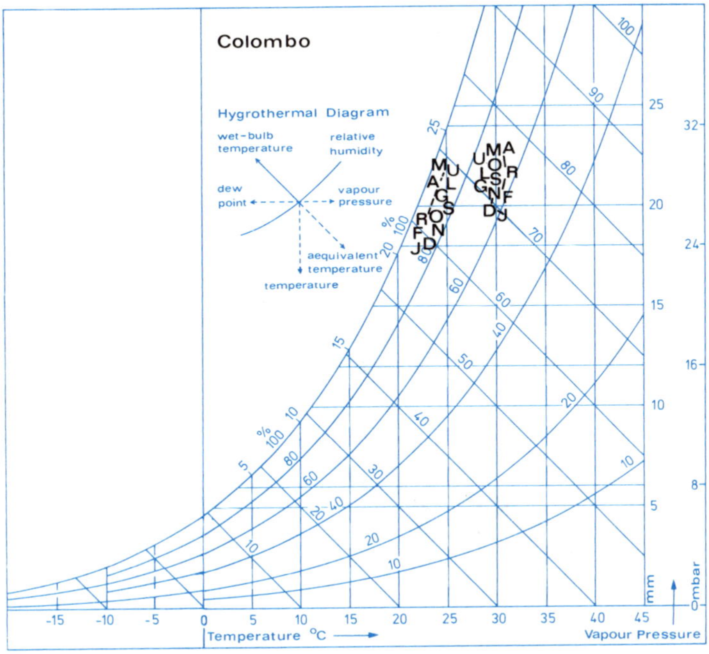
Colombo
Hygrothermal Diagram
wet-bulb temperature
relative humidity
dew point
vapour pressure
aequivalent temperature
temperature
Temperature °C
Vapour Pressure
mm
mbar

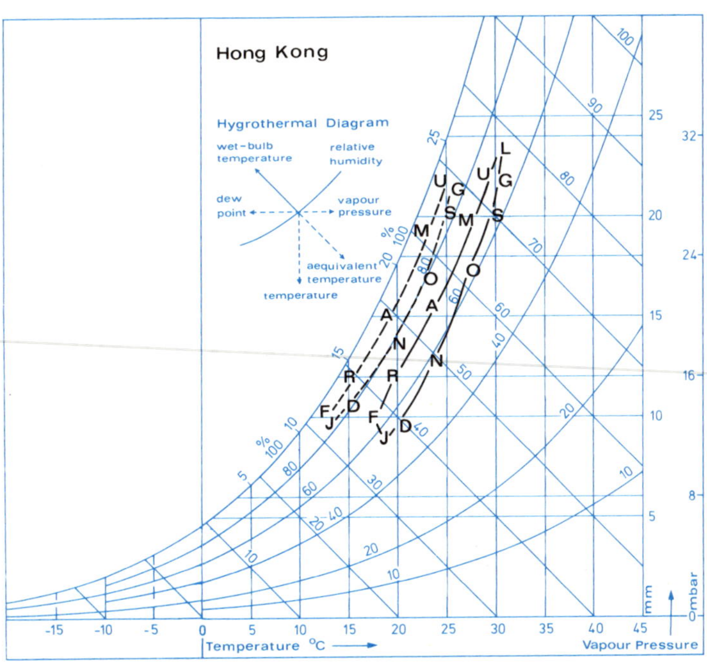
Hong Kong
Hygrothermal Diagram
wet-bulb temperature
relative humidity
dew point
vapour pressure
aequivalent temperature
temperature
Temperature °C
Vapour Pressure
mm
mbar

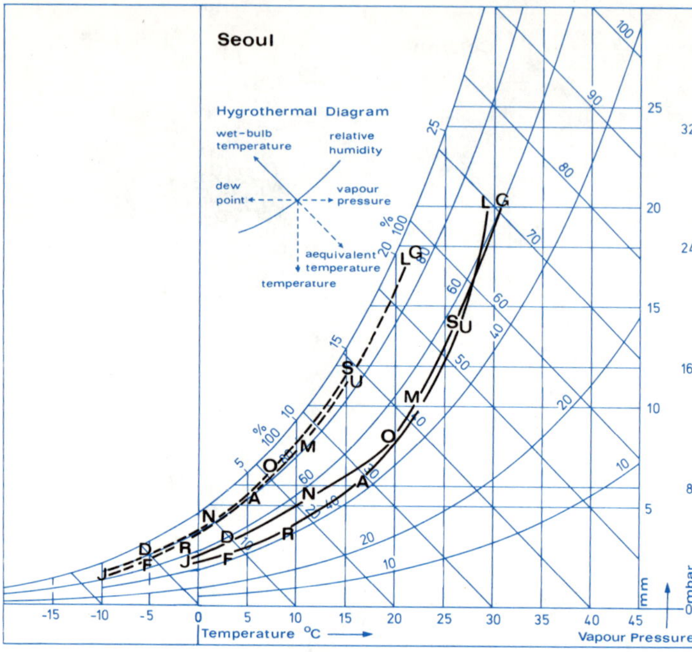
Seoul
Hygrothermal Diagram
wet-bulb temperature
relative humidity
dew point
vapour pressure
aequivalent temperature
temperature
Temperature °C
Vapour Pressure
mm
mbar

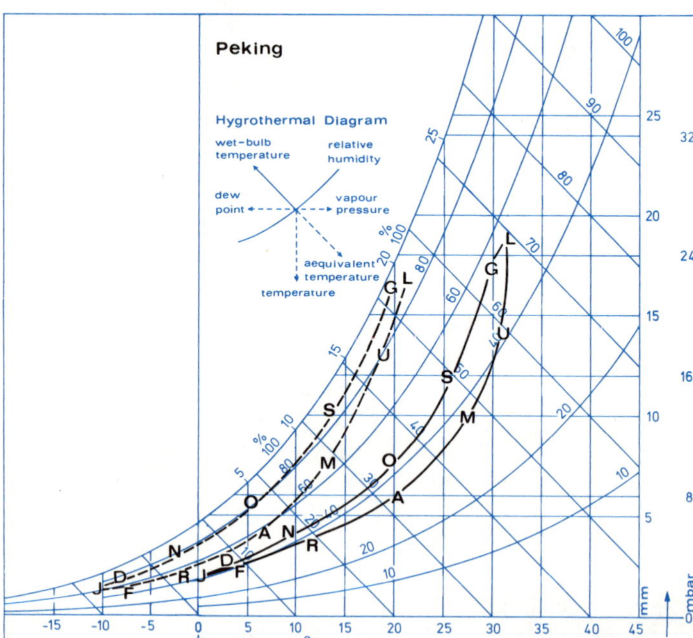
Peking
Hygrothermal Diagram
wet-bulb temperature
relative humidity
dew point
vapour pressure
aequivalent temperature
temperature
Temperature °C
Vapour Pressure
mm
mbar

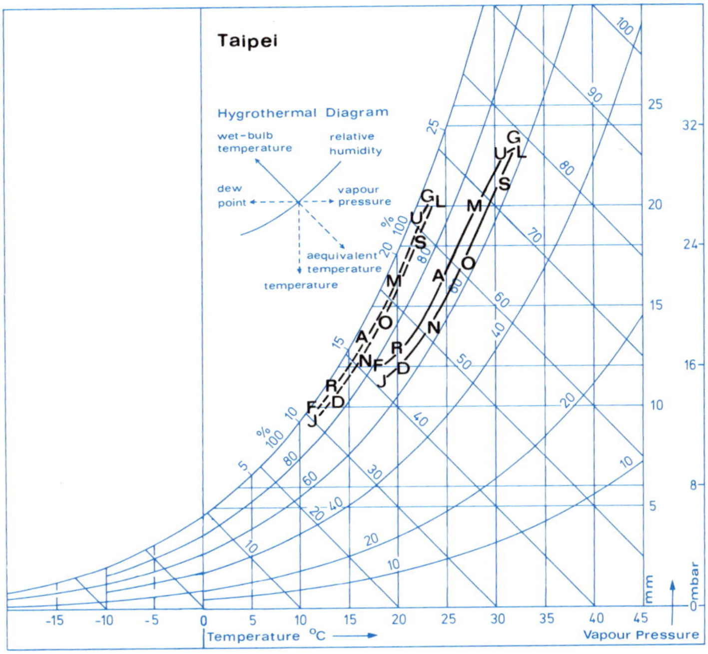
Taipei
Hygrothermal Diagram
wet-bulb temperature
relative humidity
dew point
vapour pressure
aequivalent temperature
temperature
Temperature °C
Vapour Pressure
mm
mbar

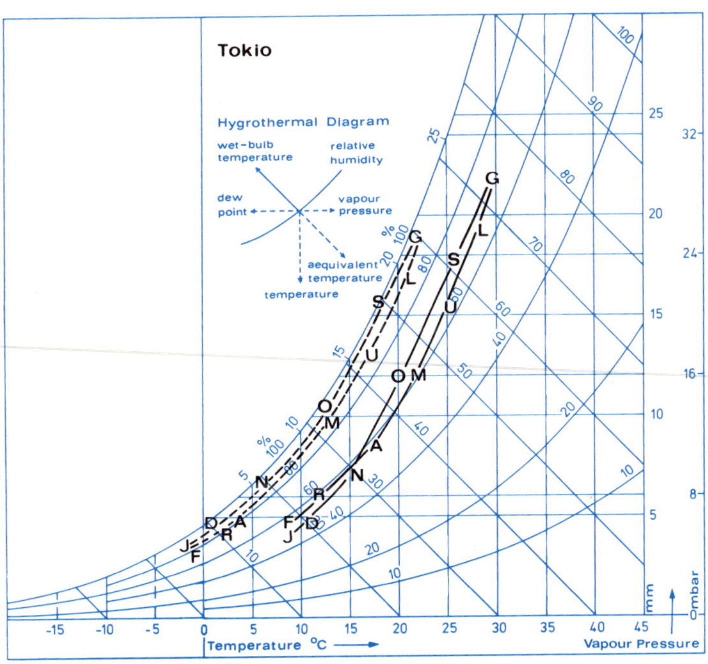
Tokio
Hygrothermal Diagram
wet-bulb temperature
relative humidity
dew point
vapour pressure
aequivalent temperature
temperature
Temperature °C
Vapour Pressure
mm
mbar

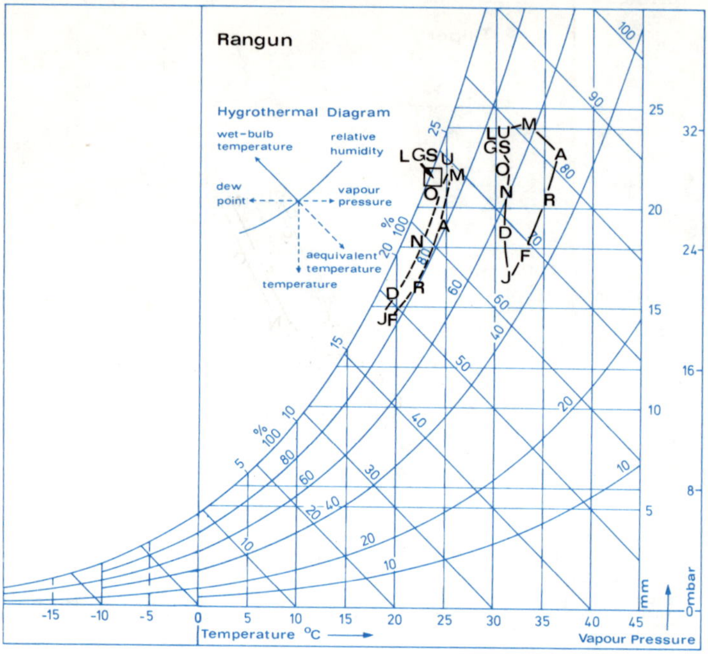
Rangun
Hygrothermal Diagram
wet-bulb temperature
relative humidity
dew point
vapour pressure
aequivalent temperature
temperature
Temperature °C
Vapour Pressure
mm
mbar

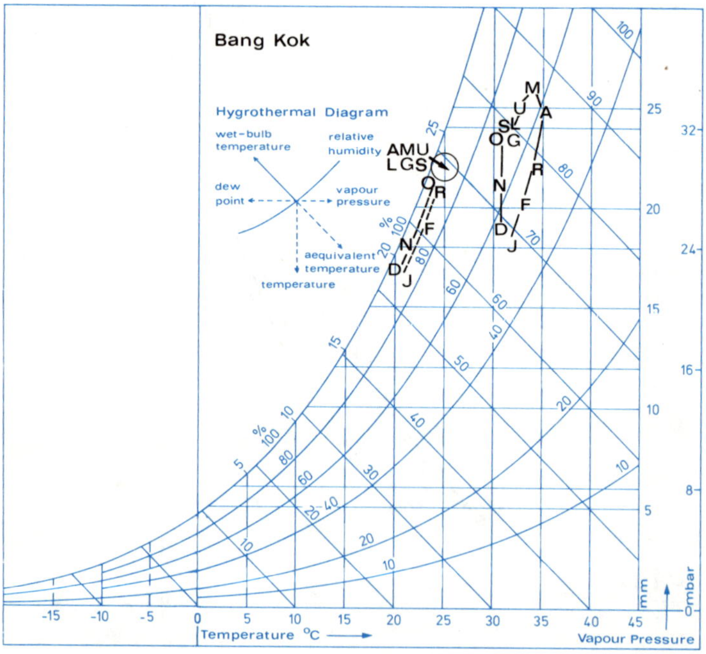
Bang Kok
Hygrothermal Diagram
wet-bulb temperature
relative humidity
dew point
vapour pressure
aequivalent temperature
temperature
Temperature °C
Vapour Pressure
mm
mbar

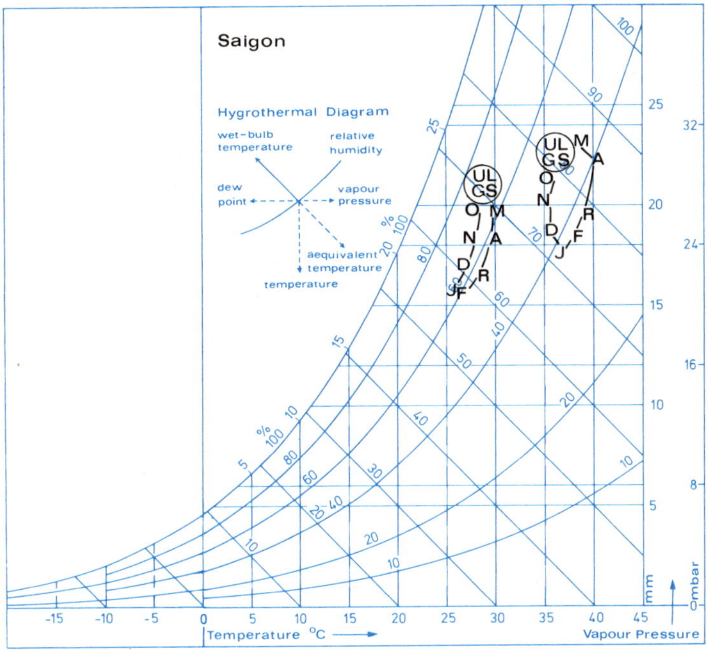
Saigon
Hygrothermal Diagram
wet-bulb temperature
relative humidity
dew point
vapour pressure
aequivalent temperature
temperature
Temperature °C
Vapour Pressure
mm
mbar

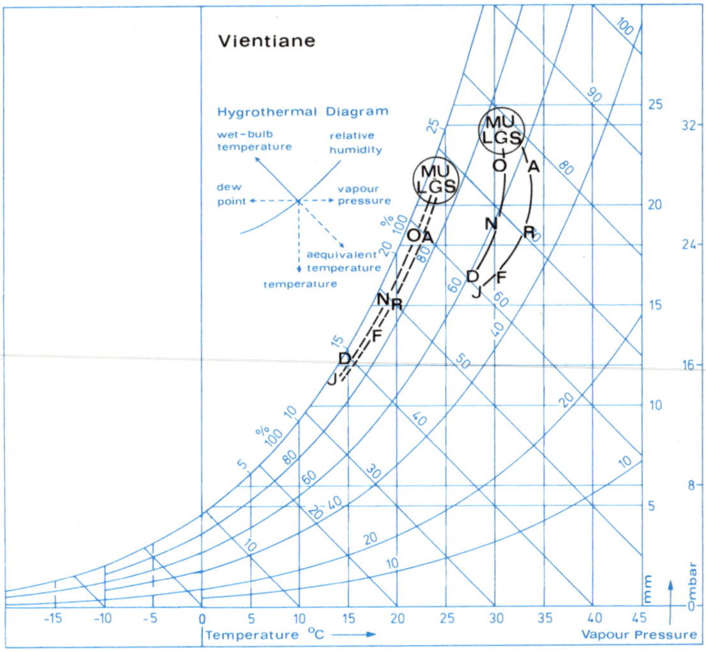
Vientiane
Hygrothermal Diagram
wet-bulb temperature
relative humidity
dew point
vapour pressure
aequivalent temperature
temperature
Temperature °C
Vapour Pressure
mm
mbar

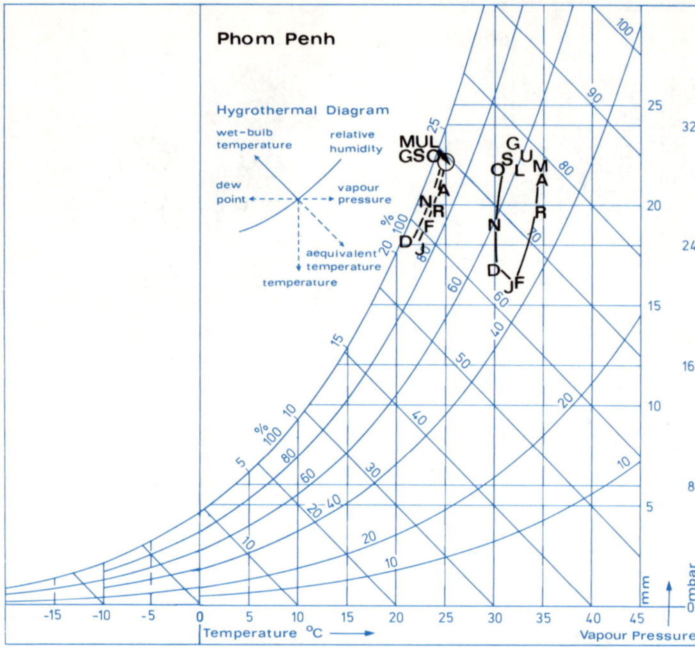
Phom Penh
Hygrothermal Diagram
wet-bulb temperature
relative humidity
dew point
vapour pressure
aequivalent temperature
temperature
Temperature °C
Vapour Pressure
mm
mbar

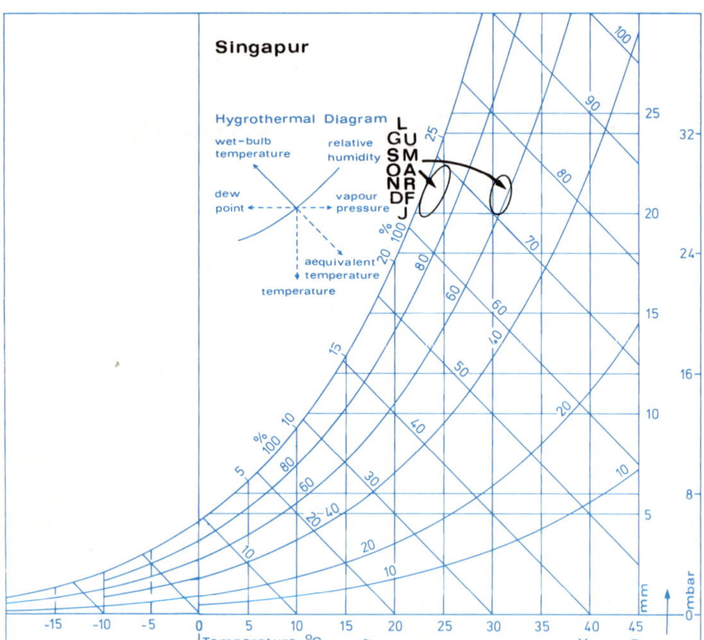
Singapur
Hygrothermal Diagram
wet-bulb temperature
relative humidity
dew point
vapour pressure
aequivalent temperature
temperature
Temperature °C
Vapour Pressure
mm
mbar

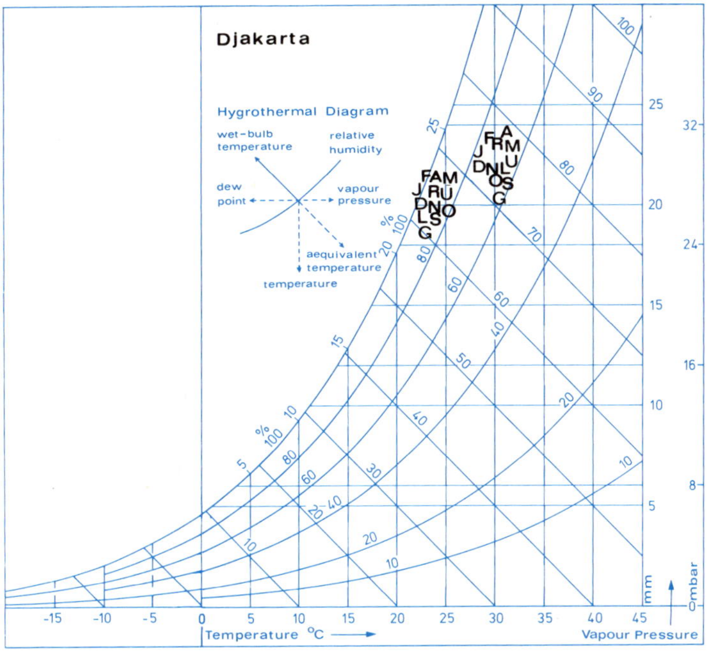
Djakarta
Hygrothermal Diagram
wet-bulb temperature
relative humidity
dew point
vapour pressure
aequivalent temperature
temperature
Temperature °C
Vapour Pressure
mm
mbar

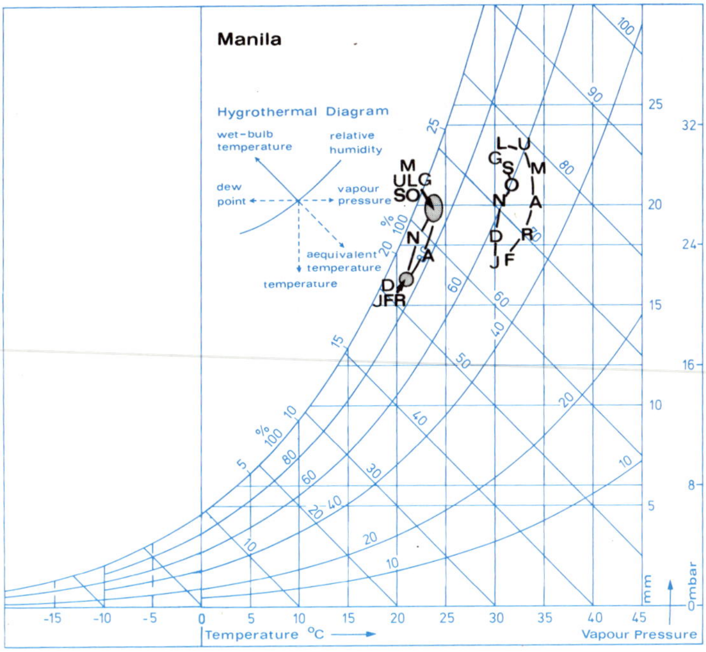
Manila
Hygrothermal Diagram
wet-bulb temperature
relative humidity
dew point
vapour pressure
aequivalent temperature
temperature
Temperature °C
Vapour Pressure
mm
mbar

4.2.31. Turkey

4.2.31.1. Climatic descriptions and tables

1. Climatic area: Eastern coast of the Black Sea
1.1. Subtropical Rain Climate Cr
1.2. Summer hot, winter cool.
1.3. Variable weather prevails; frequent periods of dry and sunny days occur.
1.4. Wind: occasional gales.

Rize

latitude: 41°02'n longitude: 40°30'e height: 4 m

Crak	031-1	Jan	Feb	Mar	Apr	May	Jun	Jul	Aug	Sep	Oct	Nov	Dec	Year
	t	7	7	8	11	16	20	22	23	20	16	12	9	14 °C
	clothing	n	n	n	n	n	s	s	t	s	n	n	n	–
	stress						o	o	f	o				–
	p	259	215	187	97	97	131	150	211	270	299	278	246	2440 mm
	dp	13	13	13	10	10	10	10	11	13	12	12	12	140 days
	sunshine	93	96	102	129	167	201	192	208	138	158	102	96	1682 hours
	%	31	32	28	32	37	44	42	48	37	46	34	33	38 %

Samsun

latitude: 41°17'n longitude: 36°20'e height: 44 m

Crak	031-2	Jan	Feb	Mar	Apr	May	Jun	Jul	Aug	Sep	Oct	Nov	Dec	Year
	tx	19	20	24	28	29	29	31	31	29	30	27	21	34 °C
	t	7	7	8	11	15	20	23	23	20	16	13	9	14 °C
	tn	-3	-2	-1	2	7	15	15	16	11	8	4	0	-4 °C
	clothing	n	n	n	n	n	s	t	t	s	n	n	n	–
	stress						o	f	f	o				–
	p	81	71	75	56	42	39	39	33	56	74	86	79	731 mm
	dp	10	10	11	9	8	6	4	4	6	8	9	9	92 days
	sunshine	87	96	112	138	211	276	313	288	195	158	123	93	2090 hours
	%	29	32	30	35	47	61	68	67	52	46	41	32	47 %

2. Climatic area: most parts of Turkey
2.1. Subtropical Winter Rain Climate Cs and
Subtropical Winter Rain Mountain Climate GCs
2.2. Summer hot, in some parts very hot, winter cool, at higher levels cold.
2.3. Summer: generally dry and sunny. Winter: variable weather prevails, though periods of dry and cloudy or sunny days also occur.
2.4. Wind: occasional gales.

Edirne

latitude: 41°40'n longitude: 26°34'e height: 48 m

Csak	031-3	Jan	Feb	Mar	Apr	May	Jun	Jul	Aug	Sep	Oct	Nov	Dec	Year
	tx	14	17	22	26	30	34	36	36	33	28	21	17	37 °C
	t	2	4	7	13	18	22	25	24	20	14	9	4	13 °C
	tn	-10	-8	-5	0	6	10	13	12	6	2	-3	-8	-13 °C
	clothing	w	w	n	n	s	s	t	t	s	n	n	w	–
	stress					o	o	f	f	o				–
	p	65	48	43	48	48	57	35	23	30	60	79	73	609 mm
	dp	8	7	6	6	7	6	4	3	3	6	7	8	72 days
	sunshine	78	111	143	195	254	291	357	335	252	177	96	78	2367 hours
	%	27	37	39	49	56	64	77	78	67	51	32	27	53 %

Kastamonu

latitude: 41°22'n longitude: 33°46'e height: 799 m

GCsbo	031-4	Jan	Feb	Mar	Apr	May	Jun	Jul	Aug	Sep	Oct	Nov	Dec	Year
	t	-1	1	4	9	14	18	20	20	16	11	5	1	10 °C
	clothing	w	w	w	n	n	s	s	s	n	n	n	w	–
	stress						o	o	o					–
	p	27	27	34	46	78	64	28	24	25	30	29	27	439 mm
	dp	7	6	7	8	11	8	4	3	4	5	6	6	74 days
	sunshine	87	119	146	186	242	288	338	326	231	195	135	84	2377 hours
	%	29	40	40	47	54	63	73	76	61	56	45	29	53 %

Istanbul

latitude: 40°58'n longitude: 29°05'e height: 40 m

Csbk	031-5	Jan	Feb	Mar	Apr	May	Jun	Jul	Aug	Sep	Oct	Nov	Dec	Year
	t	5	4	5	10	15	20	22	22	18	14	10	7	13 °C
	clothing	n	w	n	n	n	s	s	s	s	n	n	n	–
	stress						o	o	o	o				–
	p	88	80	61	37	32	28	27	22	49	62	87	96	669 mm
	dp	12	11	9	6	5	4	2	2	4	7	9	13	86 days
	sunshine	81	101	140	189	267	318	360	338	246	183	111	87	2421 hours
	%	27	34	38	47	60	70	78	79	66	53	37	30	54 %

Bursa

latitude: 40°11'n longitude: 29°04'e height: 100 m

Csak	031-6	Jan	Feb	Mar	Apr	May	Jun	Jul	Aug	Sep	Oct	Nov	Dec	Year
	tx	18	18	24	28	32	35	37	36	34	29	26	19	38 °C
	t	5	6	8	13	17	22	24	24	20	16	11	7	14 °C
	tn	-5	-5	-4	1	6	10	13	13	9	5	1	-4	-9 °C
	clothing	n	n	n	n	n	s	t	t	s	n	n	n	–
	stress						o	f	f	o				–
	p	97	86	73	56	57	33	35	18	42	57	81	90	725 mm
	dp	12	10	9	8	7	4	2	2	3	6	9	10	84 days
	sunshine	87	107	149	186	254	312	357	341	252	186	126	96	2453 hours
	%	29	35	40	47	57	69	78	80	67	54	42	33	55 %

Canakkale

latitude: 40°08'n longitude: 26°24'e height: 3 m

Csak	031-7	Jan	Feb	Mar	Apr	May	Jun	Jul	Aug	Sep	Oct	Nov	Dec	Year
	t	6	6	8	12	17	22	25	25	21	16	12	8	15 °C
	clothing	n	n	n	n	n	s	t	t	s	n	n	n	–
	stress						o	o	f	f	o			–
	p	111	73	68	36	29	25	10	11	18	51	87	88	607 mm
	dp	11	8	7	5	4	2	1	1	2	5	7	9	63 days
	sunshine	109	136	164	234	310	351	391	375	279	202	144	118	2813 hours
	%	36	45	44	59	69	78	85	88	74	58	48	41	63 %

Ankara

latitude: 39°57′n longitude: 32°53′e height: 894 m

GCsak	031-8	Jan	Feb	Mar	Apr	May	Jun	Jul	Aug	Sep	Oct	Nov	Dec	Year
	tx	11	12	21	26	29	33	36	36	33	27	22	13	37 °C
	t	0	1	5	11	16	20	23	23	18	13	7	2	12 °C
	tn	-13	-12	-9	-3	3	7	10	9	4	1	-4	-10	-17 °C
	clothing	w	w	n	n	n	s	t	t	s	n	n	w	–
	stress						o	f	f	o				–
	p	37	36	36	37	49	30	14	9	17	24	30	43	362 mm
	dp	8	8	7	7	8	5	2	1	3	4	6	8	67 days
	sunshine	96	118	174	216	282	336	384	369	291	226	159	96	2747 hours
	%	32	39	47	54	63	75	84	86	78	65	53	33	62 %

Erzurum

latitude: 39°55′n longitude: 41°16′e height: 1863 m

GCsbo	031-9	Jan	Feb	Mar	Apr	May	Jun	Jul	Aug	Sep	Oct	Nov	Dec	Year
	tx	3	4	9	18	23	27	31	31	28	22	15	7	32 °C
	t	-9	-7	-3	5	11	15	19	20	15	9	2	-6	6 °C
	tn	-24	-22	-18	-9	-1	3	7	7	1	-3	-11	-19	-26 °C
	clothing	w	w	w	n	n	n	s	s	n	n	w	w	–
	stress							o	o					–
	p	29	32	41	53	78	55	31	19	27	49	38	25	476 mm
	dp	7	7	8	9	11	8	5	3	4	6	7	6	81 days
	sunshine	105	136	158	198	257	318	363	347	285	236	156	115	2677 hours
	%	35	45	43	50	58	71	79	81	76	68	52	40	60 %

Sivas

latitude: 39°45′n longitude: 37°01′e height: 1285 m

GCsbo	031-10	Jan	Feb	Mar	Apr	May	Jun	Jul	Aug	Sep	Oct	Nov	Dec	Year
	t	-4	-3	1	8	13	17	20	20	16	11	5	-2	8 °C
	clothing	w	w	w	n	n	n	s	s	n	n	n	w	–
	stress							o	o					–
	p	44	42	42	56	59	32	8	5	16	31	40	40	415 mm
	dp	9	8	9	9	10	6	1	1	3	5	6	8	75 days
	sunshine	81	93	140	186	260	342	394	363	285	217	141	87	2588 hours
	%	27	31	38	47	58	76	86	85	76	62	47	30	58 %

Erzincan

latitude: 39°44′n longitude: 39°30′e height: 1214 m

GCsao	031-11	Jan	Feb	Mar	Apr	May	Jun	Jul	Aug	Sep	Oct	Nov	Dec	Year
	t	-3	-2	3	10	16	20	24	24	20	12	6	-1	11 °C
	clothing	w	w	w	n	n	s	t	t	s	n	n	w	–
	stress						o	f	f	o				–
	p	31	35	37	48	48	30	9	5	14	39	34	30	360 mm
	dp	7	7	8	8	9	6	2	1	2	6	6	6	65 days
	sunshine	105	117	152	198	273	333	372	357	288	217	150	115	2676 hours
	%	35	39	41	50	61	74	81	84	77	62	50	40	60 %

Kayseri

latitude: 38°47′n longitude: 35°29′n height: 1070 m

GCsao	031-12	Jan	Feb	Mar	Apr	May	Jun	Jul	Aug	Sep	Oct	Nov	Dec	Year
	t	-2	0	4	11	16	20	23	22	18	12	6	1	11 °C
	clothing	w	w	w	n	n	s	t	s	s	n	n	w	–
	stress						o	f	o	o				–
	p	37	40	38	49	45	40	7	3	11	23	29	31	353 mm
	dp	8	7	8	8	9	6	2	1	2	4	5	6	65 days
	sunshine	109	121	146	198	260	336	397	375	297	233	168	115	2755 hours
	%	36	40	40	50	59	75	87	88	79	67	55	39	62 %

Afyon

latitude: 38°45′n longitude: 30°32′e height: 1014 m

GCsbk	031-13	Jan	Feb	Mar	Apr	May	Jun	Jul	Aug	Sep	Oct	Nov	Dec	Year
	t	0	2	5	10	15	19	22	22	18	12	7	2	11 °C
	clothing	w	w	n	n	n	s	s	s	s	n	n	w	–
	stress						o	o	o	o				–
	p	50	47	49	47	61	41	25	10	20	28	36	48	462 mm
	dp	9	8	8	8	9	5	3	2	3	5	6	8	74 days
	sunshine	93	121	167	216	276	339	372	357	276	214	144	93	2668 hours
	%	31	40	45	55	62	76	82	84	74	61	48	32	60 %

Van

latitude: 38°30′n longitude: 43°23′e height: 1661 m

GCsbo	031-14	Jan	Feb	Mar	Apr	May	Jun	Jul	Aug	Sep	Oct	Nov	Dec	Year
	tx	6	8	11	17	26	29	33	33	31	24	16	9	33 °C
	t	-3	-3	0	7	13	18	22	22	17	10	4	-1	9 °C
	tn	-17	-17	-14	-7	1	5	9	9	6	-2	-6	-15	-21 °C
	clothing	w	w	w	n	n	s	s	s	n	n	w	w	–
	stress						o	o	o					–
	p	45	36	48	53	40	16	7	3	12	45	48	31	384 mm
	dp	8	7	9	9	8	3	1	1	2	6	8	6	67 days
	sunshine	130	147	161	222	282	339	360	313	234	177	150	133	2648 hours
	%	43	48	44	56	64	76	79	74	62	51	49	45	59 %

Izmir

latitude: 38°26′n longitude: 27°10′e height: 25 m

Cshk	031-15	Jan	Feb	Mar	Apr	May	Jun	Jul	Aug	Sep	Oct	Nov	Dec	Year
	tx	19	19	24	27	33	36	38	38	36	31	26	21	39 °C
	t	9	9	11	15	20	25	28	27	23	18	14	10	18 °C
	tn	-2	-2	0	3	8	13	16	16	12	7	3	-1	-4 °C
	clothing	n	n	n	n	s	t	t	t	t	s	n	n	–
	stress					o	f	f	f	f	o			–
	p	141	100	72	43	39	8	3	3	11	41	93	141	695 mm
	dp	12	9	7	6	4	1	0	0	1	4	7	11	62 days
	sunshine	124	158	195	252	307	354	388	378	312	239	168	133	3008 hours
	%	41	52	53	64	69	80	86	89	83	69	55	45	67 %

Malatya

latitude: 38°21′n longitude: 38°18′e height: 998 m

GCsao	031-16	Jan	Feb	Mar	Apr	May	Jun	Jul	Aug	Sep	Oct	Nov	Dec	Year
	t	-1	0	6	13	18	23	27	27	22	15	8	1	13 °C
	clothing	w	w	n	n	s	t	t	t	s	n	n	w	–
	stress					o	f	f	f	o				–
	p	45	46	46	53	41	16	2	2	5	32	42	39	369 mm
	dp	8	6	8	7	6	3	1	1	1	4	6	7	58 days
	sunshine	102	134	171	222	301	360	406	388	312	245	162	99	2902 hours
	%	34	44	46	56	68	81	90	91	83	70	53	34	65 %

Diyarbakir

latitude: 37°55′n longitude: 40°12′e height: 677 m

GCshk	031-17	Jan	Feb	Mar	Apr	May	Jun	Jul	Aug	Sep	Oct	Nov	Dec	Year
	t	2	4	8	14	19	26	31	30	25	17	10	4	16 °C
	clothing	w	w	n	n	s	t	t	t	t	s	n	w	–
	stress					o	f	f	f	f	o			–
	p	79	65	63	73	42	8	1	1	3	29	58	68	489 mm
	dp	11	8	9	8	6	2	0	0	1	4	7	10	66 days
	sunshine	121	150	171	222	307	378	397	378	318	251	171	124	2988 hours
	%	40	49	46	56	70	85	88	89	85	72	56	42	67 %

Konya

latitude: 37°52′n longitude: 32°30′e height: 1022 m

GCsak	031-18	Jan	Feb	Mar	Apr	May	Jun	Jul	Aug	Sep	Oct	Nov	Dec	Year
	tx	12	14	21	26	29	32	35	35	32	27	22	14	36 °C
	t	0	2	5	11	16	20	23	23	18	12	6	2	12 °C
	tn	-14	-12	-9	-3	2	6	10	9	4	-1	-7	-11	-18 °C
	clothing	w	w	n	n	n	s	t	t	s	n	n	w	–
	stress						o	f	f	o				–
	p	41	32	31	31	39	26	6	4	11	27	31	37	316 mm
	dp	7	5	6	6	6	4	1	1	2	4	4	7	52 days
	sunshine	105	144	167	222	295	342	388	375	294	239	177	112	2860 hours
	%	34	47	45	56	67	77	86	89	79	68	58	38	64 %

Isparta

latitude: 37°45′n longitude: 30°33′e height: 1043 m

GCsak	031-19	Jan	Feb	Mar	Apr	May	Jun	Jul	Aug	Sep	Oct	Nov	Dec	Year
	t	2	3	5	11	15	20	23	23	19	13	8	4	12 °C
	clothing	w	w	n	n	n	s	t	t	s	n	n	w	–
	stress						o	f	f	o				–
	p	93	76	67	51	63	38	13	13	18	37	47	99	615 mm
	dp	11	8	8	7	8	4	2	2	2	4	6	10	73 days
	sunshine	87	108	124	165	214	240	273	301	192	149	96	71	2020 hours
	%	29	35	34	42	49	54	61	71	51	43	31	24	45 %

Mugla

latitude: 37°12′n longitude: 28°21′e height: 646 m

GCsak	031-20	Jan	Feb	Mar	Apr	May	Jun	Jul	Aug	Sep	Oct	Nov	Dec	Year
	t	5	6	8	13	18	23	26	26	22	16	11	7	15 °C
	clothing	n	n	n	n	s	t	t	t	s	n	n	n	–
	stress					o	f	f	f	o				–
	p	276	170	120	60	44	21	6	9	10	64	126	273	1179 mm
	dp	15	11	9	6	5	2	1	1	1	5	7	13	76 days
	sunshine	118	146	205	243	295	357	391	369	312	239	165	121	2961 hours
	%	38	48	55	62	67	81	87	87	83	68	54	41	66 %

Urfa

latitude: 37°07′n longitude: 38°46′e height: 532 m

GCshk	031-21	Jan	Feb	Mar	Apr	May	Jun	Jul	Aug	Sep	Oct	Nov	Dec	Year
	tx	14	17	23	29	35	40	43	42	38	33	26	17	43 °C
	t	5	7	10	16	22	28	32	31	27	20	13	7	18 °C
	tn	-5	-4	-2	2	8	14	18	18	14	7	2	-3	-7 °C
	clothing	n	n	n	n	s	t	t	t	t	s	n	n	–
	stress					o	f	f	f	f	o			–
	p	106	66	69	54	24	4	0	0	1	19	39	82	464 mm
	dp	11	8	8	7	4	1	0	0	0	3	5	8	56 days
	sunshine	155	171	205	258	335	390	409	384	327	285	189	143	3251 hours
	%	51	56	55	65	76	88	91	91	87	81	62	48	73 %

Gizianter

latitude: 37°05′n longitude: 37°22′e height: 855 m

GCsak	031-22	Jan	Feb	Mar	Apr	May	Jun	Jul	Aug	Sep	Oct	Nov	Dec	Year
	t	3	4	7	13	18	24	27	27	22	15	9	4	14 °C
	clothing	w	w	n	n	s	t	t	t	s	n	n	w	–
	stress					o	f	f	f	o				–
	p	105	90	70	55	24	7	1	3	4	31	58	103	550 mm
	dp	12	9	9	7	5	1	0	0	1	4	6	10	64 days

Adana

latitude: 36°59′n longitude: 35°18′e height: 66 m

Cshk	031-23	Jan	Feb	Mar	Apr	May	Jun	Jul	Aug	Sep	Oct	Nov	Dec	Year
	tx	18	21	26	31	36	38	37	38	37	35	29	22	40 °C
	t	9	10	13	17	21	25	28	28	25	21	16	11	19 °C
	tn	-2	-1	1	4	11	14	18	19	14	10	4	0	-4 °C
	clothing	n	n	n	s	s	t	t	t	t	s	n	n	–
	stress				o	o	f	f	f	f	o			–
	p	111	93	66	45	47	18	4	5	17	42	62	102	612 mm
	dp	9	8	7	6	4	2	0	1	2	4	5	8	56 days
	sunshine	155	170	208	243	310	348	372	363	306	260	192	149	3076 hours
	%	50	56	56	62	71	79	83	86	82	74	63	50	69 %

Antalya

Cshl	031-24	Jan	Feb	Mar	Apr	May	Jun	Jul	Aug	Sep	Oct	Nov	Dec	Year
				latitude: 36°53′n				longitude: 30°42′e				height: 43 m		
	tx	18	19	23	27	33	37	41	40	37	33	28	21	41 °C
	t	10	11	13	16	20	25	28	28	25	20	16	12	19 °C
	tn	1	2	3	6	11	15	19	19	15	11	6	2	-1 °C
	clothing	n	n	n	n	s	t	t	t	t	s	n	n	–
	stress					o	f	f	f	f	o			–
	p	255	143	86	41	26	10	2	3	10	53	115	284	1028 mm
	dp	11	9	6	4	3	1	0	0	1	4	6	10	57 days
	sunshine	149	188	223	255	326	366	397	375	321	267	198	158	3223 hours
	%	48	61	60	65	74	83	89	89	86	76	65	53	70 %

3. *Climatic area:* possibly various valleys in the extreme east of the country
3.1. Steppe Climate BS and Steppe Mountain Climate GBS
3.2. Summer partly very hot or hot, winter cool or cold.
3.3. Summer: dry and sunny days prevail. Winter: dry days alternate frequently with variable weather.
3.4. Wind: occasional gales.

4.2.31.2. Predominant general weather situations are described in 4.2.0.1. under Western Near East. A hygrothermal diagram of Ankara is to be found in 4.2.0.2.
Wind name: Rageas or Ragut is a very gusty cold easterly wind which blows from Asia Minor into the Gulf of Iskenderoon (or Alexandrette).

4.2.32. Cyprus

4.2.32.1. Climatic description and table

1. Subtropical Winter Rain Climate Cs
2. Summer hot, winter mild.
3. Summer generally dry and sunny; in winter frequently variable weather.
4. Wind: occasional gales.

Nicosia

Cshl	032-1	Jan	Feb	Mar	Apr	May	Jun	Jul	Aug	Sep	Oct	Nov	Dec	Year
				latitude: 35°09′n				longitude: 33°17′e				height: 217 m		
	tx	19	20	25	32	36	39	41	41	38	34	28	21	42 °C
	t	11	11	12	17	21	26	28	28	25	21	16	12	19 °C
	tn	1	1	2	5	9	14	17	18	14	10	5	2	-1 °C
	clothing	n	n	n	s	s	t	t	t	t	s	n	n	–
	stress				o	o	f	f	ff	f	o			–
	p	74	43	37	17	21	9	1	2	7	21	32	75	339 mm
	dp	9	5	7	3	2	1	0	0	1	2	5	8	43 days
	sunshine	170	198	234	274	344	375	397	378	326	279	212	169	3356 hours
	%	55	64	63	70	79	86	89	90	87	79	68	56	75 %

4.2.32.2. Predominant general weather situations are described in 4.2.0.1. under Western Near East.

4.2.33. Syria

4.2.33.1. Climatic descriptions and tables

1. Climatic area: Northern and Western Part
1.1. Subtropical Winter Rain Climate Cs and
Subtropical Winter Rain Mountain Climate GCs
1.2. Summer very hot; winter cool.
1.3. Summer: dry and sunny, nearly all the time; winter: frequently variable weather.
1.4. Wind: occasional gales. Snow cover: in winter at higher levels.

Kamishli

latitude: 37°03′n longitude: 41°13′e height: 451 m

Cshk	033-1	Jan	Feb	Mar	Apr	May	Jun	Jul	Aug	Sep	Oct	Nov	Dec	Year
	t	7	8	11	17	22	28	32	32	27	21	14	9	19 °C
	clothing	n	n	n	s	s	t	t	t	t	s	n	n	–
	stress				o	o	f	ff	ff	f	o			–
	p	106	39	72	36	48	4	1	0	1	7	36	46	396 mm

Aleppo

latitude: 36°11′n longitude: 37°13′e height: 392 m

Cshk	033-2	Jan	Feb	Mar	Apr	May	Jun	Jul	Aug	Sep	Oct	Nov	Dec	Year
	tx	15	17	24	31	36	41	42	42	38	33	25	17	42 °C
	t	6	7	10	16	21	25	28	29	24	19	12	8	17 °C
	tn	-5	-4	-3	2	7	12	17	17	9	8	1	-3	-8 °C
	clothing	n	n	n	n	s	t	t	t	t	s	n	n	–
	stress					o	f	f	ff	f	o			–
	p	63	46	36	35	14	4	0	2	0	18	27	74	319 mm
	dp	11	10	7	4	2	0	0	0	0	4	8	10	57 days

Damascus

latitude: 33°29′n longitude: 36°14′e height: 729 m

GCshk	033-3	Jan	Feb	Mar	Apr	May	Jun	Jul	Aug	Sep	Oct	Nov	Dec	Year
	tx	17	21	26	33	34	38	39	41	37	32	27	18	41 °C
	t	8	9	12	17	21	25	27	28	24	21	14	9	18 °C
	tn	-2	-1	-1	3	8	12	13	14	11	9	3	-1	-4 °C
	clothing	n	n	n	s	s	t	t	t	t	s	n	n	–
	stress				o	o	f	f	f	f	o			–
	p	54	39	29	15	6	0	0	0	0	5	26	60	234 mm
	dp	7	6	5	3	1	0	0	0	0	2	5	8	33 days
	sunshine	161	181	211	264	313	372	403	360	309	257	195	152	3178 hours
	%	51	58	57	68	73	86	92	87	83	77	62	49	71 %

2. Climatic area: most of Eastern and Southern Part
2.1. mostly: Desert Climate BW; bordering zones in north and west: Steppe Climate BS
2.2. Summer: very hot; winter cool.
2.3. Summer: dry and sunny, nearly all the time; winter: now and then variable weather.
2.4. Wind: occasional gales.

Deir Ezzor

latitude: 35° 20′n longitude: 40° 09′e height: 203 m

BShk	033-4	Jan	Feb	Mar	Apr	May	Jun	Jul	Aug	Sep	Oct	Nov	Dec	Year
	tx	18	19	29	36	38	42	44	43	41	35	29	19	45 °C
	t	8	10	13	19	24	30	32	32	28	21	14	8	20 °C
	tn	-4	-3	-1	6	12	16	21	22	12	10	1	-3	-4 °C
	clothing	n	n	n	s	t	t	t	t	t	s	n	n	–
	stress				o	f	f	f	f	f	o			–
	p	35	33	32	20	7	2	0	0	0	3	12	33	177 mm

Palmyra

latitude: 34° 33′n longitude: 38° 18′e height: 395 m

BWhk	033-5	Jan	Feb	Mar	Apr	May	Jun	Jul	Aug	Sep	Oct	Nov	Dec	Year
	t	8	9	13	18	23	27	29	30	25	20	13	8	19 °C
	clothing	n	n	n	s	t	t	t	t	t	s	n	n	–
	stress				o	f	f	f	f	f	o			–
	p	17	19	17	13	1	2	0	0	0	1	12	19	101 mm

4.2.33.2. Predominant general weather situations are described in 4.2.0.1. under Western Near East. A hygrothermal diagram of Damascus is to be found in 4.2.0.2.

4.2.34. Lebanon

4.2.34.1. Climatic description and tables

1. Subtropical Winter Rain Climate Cs and Subtropical Winter Rain Mountain Climate GCs
2. Summer hot, winter cool, at higher levels cold.
3. Summer generally dry and sunny; in winter frequently variable weather.
4. Wind: occasional gales. Snow cover: most of the winter at higher levels.

Tripoli

latitude: 34° 27′n longitude: 35° 48′e height: 10 m

Cshl	034-1	Jan	Feb	Mar	Apr	May	Jun	Jul	Aug	Sep	Oct	Nov	Dec	Year
	t	12	13	14	17	20	24	28	26	25	22	18	14	19 °C
	clothing	n	n	n	s	s	t	t	t	t	s	s	n	–
	stress				o	o	f	f	f	f	o	o		–
	p	190	110	104	44	16	1	0	0	12	23	101	144	745 mm

Rayack

latitude: 33° 52′n longitude: 36° 00′e height: 921 m

GCsak	034-2	Jan	Feb	Mar	Apr	May	Jun	Jul	Aug	Sep	Oct	Nov	Dec	Year
	t	5	6	9	13	18	21	24	24	22	17	12	7	15 °C
	clothing	n	n	n	n	s	s	t	t	t	s	n	n	–
	stress					o	o	f	f	o	o			–
	p	159	123	76	34	16	0	0	0	1	16	66	113	603 mm
0.1 mm	dp	12	12	10	6	2	0	0	0	0	5	5	12	64 days
	sunshine	140	161	242	273	344	393	409	388	327	273	207	158	3314 hours
	%	44	52	65	70	80	90	93	93	88	77	66	51	74 %

Ksara

GCsak	034-3	Jan	Feb	Mar	Apr	May	Jun	Jul	Aug	Sep	Oct	Nov	Dec	Year
		latitude: 33°50'n						longitude: 35°53'e				height: 918 m		
	tx	17	18	24	31	32	34	36	37	36	31	26	18	38 °C
	t	6	7	10	14	19	22	24	24	22	18	13	8	15 °C
	tn	-4	-3	-1	2	6	9	12	12	9	7	2	-3	-6 °C
	clothing	n	n	n	n	s	s	t	t	s	s	n	n	–
	stress					o	o	f	f	o	o			–
	p	155	134	70	41	15	2	0	1	1	18	69	120	625 mm
0.1 mm	dp	14	14	11	6	2	0	0	0	1	5	6	13	72 days
	sunshine	112	144	198	243	304	354	412	372	327	254	180	136	3037 hours
	%	36	46	53	62	70	82	94	89	88	72	58	44	69 %

Beyrouth

Csal	034-4	Jan	Feb	Mar	Apr	May	Jun	Jul	Aug	Sep	Oct	Nov	Dec	Year
		latitude: 33°49'n						longitude: 35°29'e				height: 16 m		
	tx	21	23	27	30	33	32	33	33	32	31	28	23	35 °C
	t	14	14	15	18	21	24	26	27	26	23	19	16	20 °C
	tn	6	7	8	11	14	18	21	22	20	17	12	8	4 °C
	clothing	n	n	n	s	s	t	t	t	t	t	s	n	–
	stress				o	o	f	ff	ff	ff	f	o		–
	p	113	80	77	26	10	1	0	0	7	20	78	105	517 mm
	dp	15	12	9	5	2	0	0	0	1	4	8	12	68 days
	sunshine	149	153	195	225	307	363	369	350	276	254	198	146	2985 hours
	%	47	49	53	58	71	84	84	84	74	72	63	47	67 %

4.2.34.2. Predominant general weather situations are described in 4.2.0.1. under Western Near East. A hygrothermal diagram of Beyrouth is to be found in 4.2.0.2.

4.2.35. Israel

4.2.35.1. Climatic descriptions and tables

1. Climatic area: Northern part of the country

1.1. Subtropical Winter Rain Climate Cs and
Subtropical Winter Rain Mountain Climate GCs

1.2. Summer hot, in lowlands partly very hot; winter mild.

1.3. Summer generally dry and sunny; in winter frequently variable weather.

1.4. Wind: occasional gales.

Haifa

Cshl	035-1	Jan	Feb	Mar	Apr	May	Jun	Jul	Aug	Sep	Oct	Nov	Dec	Year
		latitude: 32°49'n						longitude: 35°00'e				height: 5 m		
	tx	23	25	31	35	38	35	33	34	34	35	32	25	39 °C
	t	14	14	16	19	22	25	27	28	26	24	20	16	21 °C
	tn	7	7	7	9	14	18	21	22	20	16	11	7	5 °C
	clothing	n	n	n	s	s	t	t	t	t	t	s	n	–
	stress				o	o	f	ff	ff	ff	f	o		–
	p	129	69	57	22	5	0	1	0	3	7	87	119	499 mm
0.1 mm	dp	11	10	6	3	1	0	0	0	1	2	5	8	46 days
	sunshine	174	175	291	300	363	372	384	378	324	307	249	208	3525 hours
	%	55	56	79	77	85	87	88	91	87	87	79	67	79 %

Lod Airport (near Java and Tel Aviv)

latitude: 32°00'n longitude: 34°54'e height: 49 m

Csal 035-2	Jan	Feb	Mar	Apr	May	Jun	Jul	Aug	Sep	Oct	Nov	Dec	Year
t	12	13	14	17	21	24	25	26	25	22	18	14	19 °C
clothing	n	n	n	s	s	t	t	t	t	t	s	n	–
stress				o	o	f	ff	ff	ff	f	o		–
p	131	93	58	15	4	0	0	0	2	18	79	29	529 mm
dp	10	8	6	3	1	0	0	0	0	2	6	9	45 days
sunshine	186	195	232	273	339	362	372	355	290	281	233	186	3304 hours
%	58	62	63	70	79	85	85	86	78	79	74	60	74 %

Jerusalem

latitude: 31°47'n longitude: 35°13'e height: 810 m

GCsak 035-3	Jan	Feb	Mar	Apr	May	Jun	Jul	Aug	Sep	Oct	Nov	Dec	Year
tx	19	21	27	32	35	36	35	35	35	33	28	21	38 °C
t	9	10	12	16	20	22	23	24	22	20	15	11	17 °C
tn	2	1	3	4	8	12	13	14	13	11	7	3	-1 °C
clothing	n	n	n	n	s	t	t	t	t	s	n	n	–
stress					o	o	f	f	o	o			–
p	128	106	85	17	4	0	0	0	1	8	61	76	492 mm
dp	9	8	7	3	1	0	0	0	0	1	5	8	42 days
sunshine	180	182	231	290	346	412	410	400	339	288	211	185	3474 hours
%	56	58	62	75	81	97	94	97	91	81	67	59	78 %

2. Climatic area: Southern part of the country

2.1. (apart from a zone bordering on area 1 with Steppe Climate BS and coastal areas with Marine Desert Climate BM)
Desert Climate BW and Desert Mountain Climate GBW

2.2. Summer very hot; winter cool or mild, in the extreme south warm.

2.3. Mostly dry and sunny all the year round; in winter scattered showers.

2.4. Wind: occasional gales.

Eilat

latitude: 29°33'n longitude: 34°57'e height: 11 m

BMhl 035-4	Jan	Feb	Mar	Apr	May	Jun	Jul	Aug	Sep	Oct	Nov	Dec	Year
tx	26	28	32	38	42	43	43	44	42	38	33	28	45 °C
t	16	18	20	24	29	32	34	34	31	27	22	17	25 °C
tn	6	7	12	13	18	22	24	25	22	18	12	7	5 °C
clothing	n	s	s	t	t	t	t	t	t	t	s	s	–
stress		o	o	f	ff	ff	ff	ff	ff	f	o	o	–
p	2	5	5	3	0	0	0	0	0	0	2	9	27 mm
dp	1	1	1	0	0	0	0	0	0	0	0	1	4 days
sunshine	230	236	251	274	320	325	348	347	291	282	245	217	3366 hours
%	71	75	68	71	67	77	81	85	78	79	76	68	76 %

4.2.35.2. Predominant general weather situations are described in 4.2.0.1. under Western Near East. A hygrothermal diagram of Tel Aviv is to be found in 4.2.0.2.

4.2.36. Jordan

4.2.36.1. Climatic descriptions and tables

1. Climatic area: Northwestern part of the country
1.1. Subtropical Winter Rain Mountain Climate GCs
1.2. Summer hot, winter mild or cool.
1.3. Summer generally dry and sunny; in winter frequently variable weather.
1.4. Wind: occasional gales.

Amman

latitude: 31°57'n longitude: 35°57'e height: 771 m

GCsak	036-1	Jan	Feb	Mar	Apr	May	Jun	Jul	Aug	Sep	Oct	Nov	Dec	Year
	tx	18	22	27	32	36	37	37	38	37	33	28	21	39 °C
	t	8	9	12	16	21	24	25	26	23	21	15	10	18 °C
	tn	0	1	1	3	8	11	14	14	13	9	5	1	-2 °C
	clothing	n	n	n	n	s	t	t	t	t	s	n	n	–
	stress					o	o	f	f	f	o			–
	p	68	59	44	13	5	0	0	0	1	4	31	48	273 mm
	dp	9	9	7	3	1	0	0	0	0	1	5	7	40 days
	sunshine	207	209	256	303	346	393	407	388	340	303	237	201	3590 hours
	%	65	67	69	78	81	92	93	94	91	85	75	64	81 %

2. Climatic area: greater part of the country
2.1. (apart from a zone bordering on area 1 with Steppe Climate BS and Steppe Mountain Climate GBS)
Desert Mountain Climate GBWh and Desert Climate BW
2.2. Summer very hot, winter mild.
2.3. Generally dry and sunny, in winter at times variable weather or scattered showers.
2.4. Wind: occasional gales and dust-storms.

H4-Station

latitude: 32°30'n longitude: 38°12'e height: 686 m

GBWhk	036-2	Jan	Feb	Mar	Apr	May	Jun	Jul	Aug	Sep	Oct	Nov	Dec	Year
	t	9	10	14	18	23	27	29	29	26	21	15	11	19 °C
	clothing	n	n	n	s	t	t	t	t	t	s	n	n	–
	stress				o	f	ff	ff	ff	ff	o			–
	p	21	16	4	14	2	0	0	0	1	9	10	6	83 mm
0.1 mm	dp	5	5	2	4	3	0	0	0	0	3	3	3	28 days
	sunshine	195	218	248	270	338	360	388	360	330	279	234	202	3420 hours
	%	61	70	67	69	79	84	89	87	89	79	74	65	77 %

Ma'an Airport

latitude: 30°10'n longitude: 35°47'e height: 1069 m

GBWak	036-3	Jan	Feb	Mar	Apr	May	Jun	Jul	Aug	Sep	Oct	Nov	Dec	Year
	t	8	10	13	16	21	25	26	27	24	20	15	10	18 °C
	clothing	n	n	n	n	s	t	t	t	t	s	n	n	–
	stress					o	f	f	ff	f	o			–
	p	9	3	3	5	0	0	0	0	0	6	3	5	34 mm
	dp	5	3	1	2	0	0	0	0	0	1	1	3	16 days
	sunshine	229	243	301	291	353	372	391	363	330	298	249	226	3646 hours
	%	71	77	81	75	83	88	91	88	89	84	78	71	82 %

4.2.36.2. Predominant general weather situations are described in 4.2.0.1. under Western Near East. A hygrothermal diagram of Amman is to be found in 4.2.0.2.

4.2.37. Arabia

4.2.37.1. Climatic descriptions and tables

1. Climatic area: nearly the whole peninsula, except coastal areas with Marine Desert Climate BM and parts of Yemen

1.1. Desert Climate BW and Desert Mountain Climate GBW

1.2. Summer severely hot or very hot, winter (low sun period) mild or warm.

1.3. Generally dry and sunny; low sun period; at times scattered showers or variable weather.

1.4. Wind: occasional gales and dust-storms.

Shuwaikh/Kuwait

latitude: 29° 20′n longitude: 47° 57′e height: 11 m

BMil	037-1	Jan	Feb	Mar	Apr	May	Jun	Jul	Aug	Sep	Oct	Nov	Dec	Year
	tx	22	23	28	36	41	43	45	45	43	38	32	24	46 °C
	t	14	16	20	26	31	35	37	37	33	27	21	15	26 °C
	tn	4	7	9	14	19	25	27	29	26	18	12	9	3 °C
	clothing	n	n	s	t	t	t	t	t	t	t	s	n	–
	stress			o	f	ff	ff	ff	ff	ff	ff	o		–
	p	15	7	8	11	3	0	0	0	0	0	25	41	111 mm
	dp	3	1	2	2	1	0	0	0	0	0	3	6	18 days
	sunshine	248	252	279	240	310	300	310	341	300	310	240	217	3347 hours
	%	76	80	75	62	73	71	72	83	81	87	75	68	75 %

Dhahran/Saudi Arabia

latitude: 26° 16′n longitude: 50° 10′e height: 25 m

BMil	037-2	Jan	Feb	Mar	Apr	May	Jun	Jul	Aug	Sep	Oct	Nov	Dec	Year
	t	17	17	22	27	32	35	36	37	34	29	24	18	28 °C
	clothing	s	s	t	t	t	t	t	t	t	t	t	s	–
	stress	o	o	f	f	ff	ff	ff	ff	ff	f	f	o	–
	p	26	15	12	2	3	0	0	0	0	0	4	24	86 mm
	dp	1	2	1	2	0	0	0	0	0	0	1	1	8 days

Riyadh/Saudi Arabia

latitude: 24°42′n longitude: 46°43′e height: 609 m

GBWil	037-3	Jan	Feb	Mar	Apr	May	Jun	Jul	Aug	Sep	Oct	Nov	Dec	Year
	tx	29	32	36	38	41	44	44	43	42	37	33	27	44 °C
	t	16	16	21	26	30	34	40	36	32	25	20	16	26 °C
	tn	0	3	5	9	17	21	21	20	18	12	9	3	-2 °C
	clothing	n	n	s	t	t	t	t	t	t	t	s	n	–
	stress			o	o	f	ff	ff	ff	ff	f	o		–
	p	24	6	6	11	15	0	0	2	0	0	11	7	82 mm

Medina/Saudi Arabia

latitude: 24°39′n longitude: 39°39′e height: 672 m

GBWil	037-4	Jan	Feb	Mar	Apr	May	Jun	Jul	Aug	Sep	Oct	Nov	Dec	Year
	t	17	20	24	29	32	35	35	36	34	29	23	19	28 °C
	clothing	n	s	t	t	t	t	t	t	t	t	t	s	–
	stress	o	o	f	ff	ff	ff	ff	ff	ff	ff	f	o	–
	p	0	3	10	4	1	0	0	0	0	0	9	10	37 mm
n	dp	0	1	2	5	1	0	0	0	0	0	5	3	17 days

Jeddah/Saudi Arabia

latitude: 21°30′n longitude: 39°12′e height: 12 m

BMha	037-5	Jan	Feb	Mar	Apr	May	Jun	Jul	Aug	Sep	Oct	Nov	Dec	Year
	tx	32	34	36	38	39	42	40	41	39	39	37	33	43 °C
	t	23	23	24	27	30	31	32	31	30	29	27	26	28 °C
	tn	13	13	14	15	18	21	22	23	22	21	18	15	11 °C
	clothing	t	t	t	t	t	t	t	t	t	t	t	t	–
	stress	f	f	f	f	ff	ff	ff	ff	ff	ff	ff	f	–
	p	17	2	13	11	13	1	0	2	0	0	6	12	77 mm

Muharraq/Bahrain

latitude: 26°16′n longitude: 50°37′e height: 2 m

BMhl	037-6	Jan	Feb	Mar	Apr	May	Jun	Jul	Aug	Sep	Oct	Nov	Dec	Year
	tx	26	29	32	37	39	40	42	42	40	36	33	28	43 °C
	t	17	18	21	26	30	32	34	34	32	29	24	19	26 °C
	tn	9	11	12	16	21	24	26	26	23	21	17	12	8 °C
	clothing	n	s	s	t	t	t	t	t	t	t	t	s	–
	stress		o	o	f	ff	ff	ff	ff	ff	ff	f	o	–
	p	16	15	11	6	1	0	0	0	0	0	9	18	76 mm
	dp	3	1	3	1	0	0	0	0	0	0	1	2	11 days

Sharjah/Oman

latitude: 25°21′n longitude: 55°23′e height: 2 m

BMhb	037-7	Jan	Feb	Mar	Apr	May	Jun	Jul	Aug	Sep	Oct	Nov	Dec	Year
	tx	27	29	36	38	39	42	44	44	42	38	34	29	45 °C
	t	18	18	22	25	28	30	33	34	31	28	24	20	26 °C
	tn	8	10	11	13	18	21	24	24	22	19	14	11	7 °C
	clothing	s	s	s	t	t	t	t	t	t	t	t	s	–
	stress	o	o	o	f	ff	ff	ff	ff	ff	ff	f	o	–
	p	34	13	9	18	1	0	2	0	0	0	18	21	115 mm
2.5 mm	dp	2	2	1	0.3	0	0	0	0	0	0	0.2	2	7 days

Muscat/Oman

latitude: 23°37′n longitude: 58°35′e height: 5 m

BMhb	037-8	Jan	Feb	Mar	Apr	May	Jun	Jul	Aug	Sep	Oct	Nov	Dec	Year
	tx	29	29	33	38	42	43	42	39	38	38	34	30	44 °C
	t	22	22	25	29	33	34	33	31	31	30	26	23	28 °C
	tn	16	16	18	21	26	29	28	26	26	24	20	17	15 °C
	clothing	s	s	t	t	t	t	t	t	t	t	t	t	–
	stress	o	o	f	ff	ff	ff	ff	ff	ff	ff	ff	f	–
	p	28	13	10	10	1	3	1	1	1	3	10	18	99 mm
	dp	2	1	1	1	0	0	0	0	0	0	1	3	8 days

Masirah Island/Oman

latitude: 20°41′n longitude: 58°54′e height: 16 m

BMhb	037-9	Jan	Feb	Mar	Apr	May	Jun	Jul	Aug	Sep	Oct	Nov	Dec	Year
	tx	28	31	35	38	41	40	36	34	34	33	32	29	41 °C
	t	22	23	25	28	31	30	27	27	26	26	25	23	27 °C
	tn	15	15	17	20	23	23	22	21	21	20	19	17	14 °C
	clothing	s	t	t	t	t	t	t	t	t	t	t	t	–
	stress	o	f	f	ff	ff	ff	ff	ff	ff	ff	f	f	–
	p	0	0	5	0	0	0	0	0	0	0	0	10	15 mm

Salalah/Oman

latitude: 17°03′n longitude: 54°06′e height: 17 m

BMha	037-10	Jan	Feb	Mar	Apr	May	Jun	Jul	Aug	Sep	Oct	Nov	Dec	Year
	tx	30	32	35	35	36	37	31	29	31	34	34	32	39 °C
	t	23	23	25	27	29	29	26	26	26	26	25	24	26 °C
	tn	14	15	17	19	23	24	23	22	21	18	17	16	14 °C
	clothing	t	t	t	t	t	t	t	t	t	t	t	t	–
	stress	f	f	f	ff	ff	ff	ff	ff	ff	ff	f	f	–
	p	0	0	0	0	0	5	28	25	3	13	0	8	81 mm
	dp	0	0	0	0	0	1	11	11	1	1	0	1	26 days

Kamaran Island/Yemen

latitude: 15°20′n longitude: 42°37′e height: 6 m

BMha	037-11	Jan	Feb	Mar	Apr	May	Jun	Jul	Aug	Sep	Oct	Nov	Dec	Year
	tx	29	30	32	34	37	38	39	38	38	37	33	30	39 °C
	t	26	26	27	28	31	32	33	33	32	31	28	26	29 °C
	tn	22	22	23	24	26	27	26	27	27	26	23	22	21 °C
	clothing	t	t	t	t	t	t	t	t	t	t	t	t	–
	stress	ff	ff	ff	ff	ff	ff	ff	ff	ff	ff	ff	ff	–
	p	5	5	3	2	1	1	13	18	2	3	10	23	86 mm
	dp	1	1	1	0	0	0	1	2	0	1	1	2	10 days

Riyan/Yemen

latitude: 14°39′n longitude: 49°23′e height: 25 m

BMha	037-12	Jan	Feb	Mar	Apr	May	Jun	Jul	Aug	Sep	Oct	Nov	Dec	Year
	tx	29	31	33	33	35	38	36	36	34	34	33	36	39 °C
	t	23	24	25	27	29	31	29	29	29	27	26	23	27 °C
	tn	16	17	18	20	23	24	22	22	23	19	18	17	16 °C
	clothing	t	t	t	t	t	t	t	t	t	t	t	t	–
	stress	f	f	f	ff	ff	ff	ff	ff	ff	ff	f	f	–
	p	8	3	15	5	1	2	2	2	0	1	18	8	64 mm
	dp	1	1	1	1	0	0	1	0	0	0	1	1	7 days

Aden/Khormaksar/Yemen

latitude: 12°50′n longitude: 45°02′e height: 3 m

BMha	037-13	Jan	Feb	Mar	Apr	May	Jun	Jul	Aug	Sep	Oct	Nov	Dec	Year
	tx	29	29	32	34	38	39	39	38	38	36	32	30	39 °C
	t	26	26	27	29	31	33	32	32	32	29	27	26	29 °C
	tn	18	20	21	21	24	27	26	24	26	20	19	19	17 °C
	clothing	t	t	t	t	t	t	t	t	t	t	t	t	–
	stress	f	ff	ff	ff	ff	ff	ff	ff	ff	ff	f	f	–
	p	7	3	6	0	1	0	3	2	7	1	3	6	39 mm

Socotra Island/Yemen

latitude: 12°38′n longitude: 53°53′e height: 42 m

BMha	037-14	Jan	Feb	Mar	Apr	May	Jun	Jul	Aug	Sep	Oct	Nov	Dec	Year
	tx	29	31	32	34	38	37	33	34	34	33	31	31	38 °C
	t	25	25	26	26	31	30	28	29	28	26	26	26	27 °C
	tn	19	19	19	21	24	24	23	24	23	20	20	19	19 °C
	clothing	t	t	t	t	t	t	t	t	t	t	t	t	–
	stress	ff	ff	ff	ff	ff	ff	ff	ff	ff	ff	ff	ff	–
	p	2	2	10	0	2	28	0	0	2	10	51	81	193 mm
	dp	1	1	0	0	0	1	0	0	0	2	6	5	16 days

Sana/Yemen

latitude: 15°23′n longitude: 44°11′e height: 2350 m

GBWbl	037-15	Jan	Feb	Mar	Apr	May	Jun	Jul	Aug	Sep	Oct	Nov	Dec	Year
	t	12	15	16	18	19	20	20	20	18	14	12	11	16 °C
	clothing	n	n	n	s	s	s	s	s	s	n	n	n	–
	stress				o	o	o	o	o	o				–
	p	0	4	21	46	46	0	20	102	3	9	0	0	251 mm
	dp	0	1	5	6	4	0	6	11	1	1	0	0	35 days
	sunshine	313	274	276	285	276	300	226	257	288	326	294	307	3422 hours
	%	89	83	74	76	69	77	56	65	78	89	86	88	77 %

2. *Climatic area:* probably parts of Asir and Jemen Mountains
2.1. Steppe Mountain Climate GBS
2.2. High sun period according to level warm or hot, low sun period mild or warm.
2.3. High sun period partly variable weather, partly dry and sunny, low sun period predominantly dry and sunny.
2.4. Wind: occasional gales and dust storms.

4.2.37.2. Predominant general weather situations are described in 4.2.0.1. under Arabia. A hygrothermal diagram of Riyadh is to be found in 4.2.0.2.
Wind names: Belat is a gusty dusty wind blowing from the north or northwest on the southeast coast of Arabia, with main frequency from mid-December to mid-March.
Khamsin: see Simoom.
Kaus is a rainy wind in the Persian Gulf blowing from the southeast.
Maskat is a dry foehn-like dusty wind which often blows in heavy gusts, falling from the mountains near Maskat.
N'aashi is a bora-like (relatively cold and gusty) wind which blows from the northeast in the Persian Gulf accompanied by cloudy weather and often bringing dust from the coast of Iran.
Shamal is a constant wind which blows all the year round from the northwest in the Persian Gulf. Great Shamal is a dusty Shamal blowing in summer. Shamal on the Red Sea is a northerly or northwesterly wind which ends a Khamsin situation by a marked cold front squall.
Sharki is a southeasterly wind of the summer season of the monsoon accompanied by overcast sky in the Persian Gulf.
Simoom is a hot, dry suffocating wind or whirlwind which occurs in the deserts of Africa and Arabia. It usually, but not invariably, carries a large quantity of sand. It does not usually last more than about 20 minutes. It is most frequent in the spring and summer.

4.2.38. Iraq

4.2.38.1. Climatic descriptions and tables

1. Climatic area: mountainous northern part and foreland

1.1. Subtropical Winter Rain Climate Cs and
Subtropical Winter Rain Mountain Climate GCs

1.2. Summer severely or very hot, winter cool.

1.3. Summer dry and sunny, in winter frequently variable weather.

1.4. Wind: occasional gales and dust-storms.

Mosul

				latitude: 36° 19′n			longitude: 43° 09′e				height: 222 m		
Cshk	038-1	Jan	Feb	Mar	Apr	May	Jun	Jul	Aug	Sep	Oct	Nov	Dec	Year
	tx	17	20	27	33	40	43	47	47	44	37	31	19	47 °C
	t	6	9	12	17	24	30	34	33	30	20	13	8	19 °C
	tn	-4	-2	-1	3	9	14	18	16	12	6	1	-3	-5 °C
	clothing	n	n	n	s	t	t	t	t	t	s	n	n	–
	stress				o	f	f	ff	ff	f	o			–
	p	70	67	65	55	20	1	0	0	0	7	43	62	390 mm
	dp	10	8	9	7	4	0	0	0	0	1	5	7	52 days
	sunshine	158	169	224	242	310	366	382	369	323	267	207	162	3179 hours
	%	51	55	61	62	71	83	85	88	86	76	67	54	71 %

Kirkuk

				latitude: 35° 28′n			longitude: 44° 24′e				height: 331 m		
Csik	038-2	Jan	Feb	Mar	Apr	May	Jun	Jul	Aug	Sep	Oct	Nov	Dec	Year
	t	9	10	13	19	26	32	35	35	30	24	16	11	22 °C
	clothing	n	n	n	s	t	t	t	t	t	t	n	n	–
	stress				o	f	ff	ff	ff	f	f			–
	p	58	66	85	47	17	0	0	0	0	3	42	62	381 mm
	dp	7	7	8	1	2	0	0	0	0	1	5	6	37 days

2. Climatic area: northern lowland

2.1. Steppe Climate BS

2.2. Summer severely or very hot, winter mild.

2.3. Summer dry and sunny, in winter frequently variable weather.

2.4. Wind: occasional gales and dust-storms.

Khanaquin

				latitude: 34° 18′n			longitude: 45° 26′e				height: 201 m		
BSil	038-3	Jan	Feb	Mar	Apr	May	Jun	Jul	Aug	Sep	Oct	Nov	Dec	Year
	t	10	12	15	21	28	33	36	36	30	25	18	12	23 °C
	clothing	n	n	n	s	t	t	t	t	t	t	s	n	–
	stress				o	f	ff	ff	ff	ff	f	o		–
	p	65	57	73	39	21	0	1	0	0	6	37	57	356 mm
	dp	6	6	8	4	2	0	0	0	0	1	4	6	37 days

3. Climatic area: greater southern part of the country
3.1. Desert Climate BW, coastal areas Marine Desert Climate BM
3.2. Summer severely or very hot, winter mild.
3.3. Summer dry and sunny, in winter occasional scattered showers or rain.
3.4. Wind: occasional gales and dust-storms.

Habbaniya

latitude: 33°22′n longitude: 43°34′e height: 45 m

BWil	038-4	Jan	Feb	Mar	Apr	May	Jun	Jul	Aug	Sep	Oct	Nov	Dec	Year
	tx	21	24	29	37	43	46	48	47	45	39	32	22	48 °C
	t	10	12	16	22	28	33	35	34	30	24	17	11	23 °C
	tn	-2	-1	2	8	14	19	22	22	17	11	4	-1	-3 °C
	clothing	n	n	n	s	t	t	t	t	t	t	s	n	–
	stress				o	f	ff	ff	ff	f	f	o		–
	p	21	17	26	17	4	0	0	0	0	2	19	20	126 mm
	dp	3	3	3	3	1	0	0	0	0	1	3	4	20 days
	sunshine	197	208	252	257	300	350	368	371	334	282	251	210	3380 hours
	%	62	67	68	66	70	81	84	89	90	80	80	68	76 %

Baghdad

latitude: 33°20′n longitude: 44°24′e height: 34 m

BWil	038-5	Jan	Feb	Mar	Apr	May	Jun	Jul	Aug	Sep	Oct	Nov	Dec	Year
	tx	20	23	29	34	40	44	47	47	44	39	30	22	48 °C
	t	10	12	16	22	28	33	35	34	31	25	17	11	23 °C
	tn	-2	1	4	10	14	21	23	22	17	12	4	1	-3 °C
	clothing	n	n	n	s	t	t	t	t	t	t	s	n	–
	stress				o	f	ff	ff	ff	f	f	o		–
	p	25	25	29	16	7	0	0	0	0	3	22	26	156 mm
	dp	4	4	4	3	1	0	0	0	0	1	3	4	23 days
	sunshine	192	204	246	256	300	348	348	354	316	273	213	194	3241 hours
	%	61	66	66	66	70	81	79	85	85	77	68	63	73 %

Rutbah

latitude: 33°02′n longitude: 40°17′e height: 615 m

GBWhk	038-6	Jan	Feb	Mar	Apr	May	Jun	Jul	Aug	Sep	Oct	Nov	Dec	Year
	tx	19	24	30	36	38	42	42	43	42	35	29	20	44 °C
	t	7	9	12	18	24	27	30	30	27	21	14	8	19 °C
	tn	-6	-3	-2	4	11	14	18	17	13	8	0	-4	-7 °C
	clothing	n	n	n	s	t	t	t	t	t	s	n	n	–
	stress				o	f	f	ff	ff	f	o			–
	p	16	13	20	20	10	1	0	0	1	5	16	21	121 mm
	dp	3	3	3	3	2	0	0	0	0	1	3	3	20 days

Kut-El-Hai

latitude: 32°10′n longitude: 46°02′e height: 15 m

BWil	038-7	Jan	Feb	Mar	Apr	May	Jun	Jul	Aug	Sep	Oct	Nov	Dec	Year
	t	11	13	17	23	31	34	36	35	32	26	18	13	24 °C
	clothing	n	n	s	t	t	t	t	t	t	t	s	n	–
	stress			o	f	ff	ff	ff	ff	ff	f	o		–
	p	22	22	26	12	7	0	0	0	0	2	19	28	138 mm
	dp	4	3	4	3	1	0	0	0	0	1	3	4	22 days

Diwaniya

latitude: 31°59'h longitude: 44°59'e height: 20 m

BWhl	038-8	Jan	Feb	Mar	Apr	May	Jun	Jul	Aug	Sep	Oct	Nov	Dec	Year
	t	10	13	16	23	30	32	34	34	32	26	18	12	23 °C
	clothing	n	n	s	t	t	t	t	t	t	t	s	n	–
	stress			o	f	ff	ff	ff	ff	ff	f	o		–
	p	17	15	19	19	9	0	0	0	0	1	15	25	121 mm
	dp	2	2	2	2	1	0	0	0	0	1	2	4	16 days

Nasiriya

latitude: 31°01'n longitude: 46°14'e height: 3 m

BWhl	038-9	Jan	Feb	Mar	Apr	May	Jun	Jul	Aug	Sep	Oct	Nov	Dec	Year
	t	12	14	17	24	30	33	34	34	32	26	19	13	24 °C
	clothing	n	n	s	t	t	t	t	t	t	t	s	n	–
	stress			o	f	ff	ff	ff	ff	ff	f	o		–
	p	17	15	19	21	5	0	0	0	0	3	21	25	123 mm
	dp	3	2	3	2	1	0	0	0	0	0	3	4	18 days

Basrah

latitude: 30°34'n longitude: 47°47'e height: 2 m

BWhl	038-10	Jan	Feb	Mar	Apr	May	Jun	Jul	Aug	Sep	Oct	Nov	Dec	Year
	tx	23	26	32	39	44	46	48	48	46	41	34	25	48 °C
	t	12	14	18	24	29	32	34	33	31	26	19	14	24 °C
	tn	-1	1	4	10	17	22	24	23	19	13	6	1	-2 °C
	clothing	n	n	s	t	t	t	t	t	t	t	s	n	–
	stress			o	f	f	ff	ff	ff	ff	f	o		–
	p	26	17	25	22	7	0	0	0	0	1	28	38	164 mm
	dp	3	3	3	3	1	0	0	0	0	0	2	4	20 days
	sunshine	224	224	264	254	306	338	323	333	306	291	229	227	3319 hours
	%	70	71	71	66	72	80	75	81	82	82	72	72	75 %

4.2.38.2. Predominant general weather situations are described in 4.2.0.1. under Eastern Near East. A hygrothermal diagram of Baghdad is to be found in 4.2.0.2.
Wind names: see Arabia 4.2.36.2. and
Reshabar or Rashaba: a name meaning "black wind" given to a strong, very gusty, northeasterly wind which blows down certain mountain ranges in southern Kurdistan. It is dry, comparatively hot in summer and cold in winter.

4.2.39. Iran

4.2.39.1. Climatic descriptions and tables

1. Climatic area: parts of the Caspian Coast and Hinterland
1.1. Subtropical Rain Climate Cr and Subtropical Rain Mountain Climate GCr
1.2. Summer hot, winter cool.
1.3. Summer: more often than not dry and sunny; winter: variable weather prevails, though with frequent periods of dry and sunny days.
1.4. Wind: occasional gales.

Pahlavi

latitude: 37°28′n longitude: 49°28′e height: -15 m

Crak	039-1	Jan	Feb	Mar	Apr	May	Jun	Jul	Aug	Sep	Oct	Nov	Dec	Year
	t	8	8	9	13	19	22	26	25	23	18	14	10	17 °C
	clothing	n	n	n	n	s	s	t	t	t	s	n	n	–
	stress					o	o	f	f	f	o			–
	p	133	125	111	75	56	57	45	131	298	326	227	173	1757 mm
	dp	11	9	11	9	6	4	4	8	12	14	11	9	108 days
	sunshine	112	96	102	135	242	249	242	226	186	108	105	108	1912 hours
	%	37	31	28	34	55	56	54	53	50	31	34	36	43 %

2. Climatic area: parts of the Caspian Coast and Hinterland, the western slopes of the mountains in the West and Southwest

2.1. Subtropical Winter Rain Climate Cs and
Subtropical Winter Rain Mountain Climate GCs

2.2. Summer hot, winter cool.

2.3. Summer: generally dry and sunny, except the Caspian Coast with occasional variable weather; winter: variable weather prevails, though with frequent periods of dry and sunny days.

2.4. Wind: occasional gales.

Babolsar

latitude: 36°43′n longitude: 52°39′e height: -21 m

Csak	039-2	Jan	Feb	Mar	Apr	May	Jun	Jul	Aug	Sep	Oct	Nov	Dec	Year
	t	8	8	10	14	19	24	26	26	23	18	14	10	17 °C
	clothing	n	n	n	n	s	t	t	t	t	s	n	n	–
	stress					o	f	f	f	f	o			–
	p	89	62	73	40	15	27	24	40	85	134	101	125	815 mm
	dp	8	7	9	7	3	3	3	5	7	8	8	8	76 days
	sunshine	158	130	136	141	201	234	232	229	165	149	150	136	2063 hours
	%	51	42	37	36	46	53	52	54	44	43	49	45	46 %

Hamadan

latitude: 35°42′n longitude: 48°43′e height: 1644 m

GCsao	039-3	Jan	Feb	Mar	Apr	May	Jun	Jul	Aug	Sep	Oct	Nov	Dec	Year
	t	-2	1	6	10	16	20	24	24	19	13	6	1	11 °C
	clothing	w	w	n	n	n	s	t	t	s	n	n	w	–
	stress						o	f	f	o				–
	p	37	40	40	79	40	1	1	0	0	16	31	28	313 mm
	dp	6	6	7	9	5	0	0	0	0	3	5	4	45 days
	sunshine	167	147	202	192	264	354	350	316	294	208	189	180	2862 hours
	%	54	48	55	49	61	81	79	75	79	59	61	59	64 %

Kermanshan

GCsak	039-4	latitude: 34° 21′n					longitude: 47°05′e				height: 1307 m			
		Jan	Feb	Mar	Apr	May	Jun	Jul	Aug	Sep	Oct	Nov	Dec	Year
	tx	13	16	21	26	32	37	40	39	36	31	24	16	41 °C
	t	2	3	7	12	17	20	25	24	20	15	9	4	13 °C
	tn	-13	-12	-8	-4	2	3	8	8	2	-2	-6	-10	-15 °C
	clothing	w	w	n	n	n	s	t	t	s	n	n	w	–
	stress						o	f	f	o				–
	p	66	58	71	56	41	0	0	0	0	10	51	61	414 mm
	dp	9	7	9	8	5	0	0	0	0	1	6	7	52 days
	sunshine	140	153	205	204	242	351	360	344	291	217	180	158	2843 hours
	%	45	49	55	52	56	81	81	82	78	62	58	52	64 %

Shiraz

GCshk	039-5	latitude: 29° 32′n					longitude: 52°38′e				height: 1491 m			
		Jan	Feb	Mar	Apr	May	Jun	Jul	Aug	Sep	Oct	Nov	Dec	Year
	t	6	8	12	16	22	26	28	27	24	18	12	8	17 °C
	clothing	n	n	n	n	s	t	t	t	t	s	n	n	–
	stress					o	f	ff	ff	f	o			–
	p	77	47	63	24	13	0	1	0	0	0	65	93	383 mm
	dp	5	5	4	5	1	0	0	0	0	0	3	4	27 days
	sunshine	223	212	267	228	332	366	360	329	318	291	240	236	3401 hours
	%	69	67	72	59	79	87	84	80	86	82	75	74	76 %

3. Climatic area: Mountainous parts in the north and northeast, the eastern slopes of the mountains in the west and southwest

3.1 Steppe climate BS and Steppe Mountain Climate GBS

3.2. Summer hot or very hot, at lower levels severely hot; winter mild or cool.

3.3. Summer generally dry and sunny; winter mostly dry and sunny or cloudy, sometimes scattered showers or variable weather.

3.4. Wind: occasional gales and dust-storms.

Tabriz

GBSak	039-6	latitude: 38° 08′n					longitude: 46° 15′e				height: 1362 m			
		Jan	Feb	Mar	Apr	May	Jun	Jul	Aug	Sep	Oct	Nov	Dec	Year
	t	1	1	5	11	16	21	24	25	21	14	6	0	12 °C
	clothing	w	w	n	n	n	s	t	t	s	n	n	w	–
	stress						o	f	f	o				–
	p	15	32	45	47	43	23	11	1	9	14	29	16	285 mm
	dp	6	6	8	9	7	3	1	1	1	5	5	4	56 days
	sunshine	136	147	167	180	254	348	366	341	297	198	186	133	2754 hours
	%	45	48	45	46	58	78	81	80	79	57	61	45	62 %

Mashhad

latitude: 36°16'n longitude: 59°38'e height: 985 m

GBSak	039-7	Jan	Feb	Mar	Apr	May	Jun	Jul	Aug	Sep	Oct	Nov	Dec	Year
	tx	16	17	24	29	33	37	37	37	34	33	25	17	39 °C
	t	2	4	7	14	18	23	26	24	20	14	7	3	14 °C
	tn	-16	-11	-6	1	8	11	13	10	6	0	-5	-8	-16 °C
	clothing	w	w	n	n	s	t	t	t	s	n	n	w	–
	stress					o	f	f	f	o				–
	p	32	31	61	51	25	4	0	0	0	6	17	20	249 mm
	dp	4	5	9	9	4	0	0	0	0	2	5	2	39 days
	sunshine	170	147	174	180	279	342	363	360	300	226	195	167	2902 hours
	%	55	48	47	46	64	71	81	86	80	64	63	55	65 %

Tehran

latitude: 35°37'n longitude: 51°40'e height: 1519 m

GBShk	039-8	Jan	Feb	Mar	Apr	May	Jun	Jul	Aug	Sep	Oct	Nov	Dec	Year
	tx	13	16	23	28	33	37	40	39	36	29	23	16	40 °C
	t	4	5	10	15	21	26	30	28	25	18	11	5	16 °C
	tn	-8	-7	-2	3	8	14	18	18	13	7	1	-4	-9 °C
	clothing	w	n	n	n	s	t	t	t	t	s	n	n	–
	stress					o	f	ff	ff	f	o			–
	p	37	23	36	31	14	2	1	1	1	5	29	27	208 mm
	dp	4	4	3	6	3	1	0	0	0	2	4	5	32 days
	sunshine	183	189	232	222	267	348	347	341	303	236	207	195	3070 hours
	%	59	61	63	57	61	80	78	81	81	67	67	64	69 %

Dezful

latitude: 32°24'n longitude: 48°23'e height: 143 m

BSil	039-9	Jan	Feb	Mar	Apr	May	Jun	Jul	Aug	Sep	Oct	Nov	Dec	Year
	tx													°C
	t	12	14	17	23	30	35	37	36	32	27	20	14	25 °C
	tn													°C
	clothing	n	n	s	t	t	t	t	t	t	t	s	n	–
	stress			o	f	ff	ff	ff	ff	ff	ff	o		–
	p	103	43	28	38	11	0	1	0	0	9	58	52	343 mm
	dp	6	5	4	4	2	0	0	0	0	1	5	5	32 days
	sunshine	155	167	177	213	257	324	307	298	285	198	174	177	2732 hours
	%	49	54	48	55	60	76	70	72	77	56	55	57	61 %

Kerman

latitude: 30°15'n longitude: 56°58'e height: 1749 m

GBSak	039-10	Jan	Feb	Mar	Apr	May	Jun	Jul	Aug	Sep	Oct	Nov	Dec	Year
	tx	22	23	26	31	38	42	43	41	38	34	29	23	43 °C
	t	6	8	12	17	21	25	27	25	22	16	10	6	16 °C
	tn	-8	-8	-6	1	6	13	13	9	5	2	-4	-8	-11 °C
	clothing	n	n	n	n	s	t	t	t	s	n	n	n	–
	stress					o	f	f	f	o				–
	p	36	22	46	22	20	2	7	0	4	3	8	32	203 mm
	dp	4	4	4	6	3	0	0	0	0	0	2	2	26 days
	sunshine	220	184	218	225	288	300	294	335	306	282	246	223	3152 hours
	%	68	59	67	58	68	71	68	81	82	79	77	70	71 %

Bushehr

latitude: 28°59′n longitude: 50°50′e height: 14 m

BShl	039-11	Jan	Feb	Mar	Apr	May	Jun	Jul	Aug	Sep	Oct	Nov	Dec	Year
	tx	23	24	31	35	38	38	39	40	38	34	31	26	41 °C
	t	14	15	19	25	28	31	33	34	31	26	21	16	24 °C
	tn	6	7	11	15	19	23	26	26	23	18	12	8	6 °C
	clothing	n	n	s	t	t	t	t	t	t	t	s	n	–
	stress			o	f	ff	ff	ff	ff	ff	f	o		–
	p	61	23	19	9	5	0	0	0	0	1	48	93	260 mm
	dp	4	3	2	2	1	0	0	0	0	0	3	3	18 days
	sunshine	223	192	220	237	288	330	316	313	297	279	252	220	3168 hours
	%	69	61	59	61	68	79	74	76	80	78	78	69	71 %

4. Climatic area: the plains of Kavir and Lut, the south of the country including most parts of the coast of the Persian Gulf

4.1. Desert Climate BW and Desert Mountain Climate GBW, coastal areas Marine Desert Climate BM

4.2. High sun period severely or very hot; low sun period mild or warm at sea level, in higher levels cool.

4.3. Generally dry and sunny; during low sun period now and then scattered showers or variable weather.

4.4. Wind: occasional gales and dust-storms.

Esfahan

latitude: 32°37′n longitude: 51°40′e height: 1598 m

GBWhk	039-12	Jan	Feb	Mar	Apr	May	Jun	Jul	Aug	Sep	Oct	Nov	Dec	Year
	tx	14	18	23	27	33	38	39	38	36	31	23	17	40 °C
	t	4	6	10	16	20	25	28	26	23	16	9	4	16 °C
	tn	-9	-8	-3	3	8	13	16	14	10	4	-2	-7	-11 °C
	clothing	w	n	n	n	s	t	t	t	t	s	n	w	–
	stress					o	f	f	f	f	o			–
	p	19	16	18	16	8	1	5	1	0	3	19	20	126 mm
	dp	3	2	2	4	2	0	0	0	0	0	1	3	17 days
	sunshine	217	206	270	249	298	363	353	344	297	254	243	220	3314 hours
	%	68	66	73	64	69	85	81	83	80	72	77	71	74 %

Seistan

latitude: 31°00′n longitude: 61°30′e height: 610 m

GBWhk	039-13	Jan	Feb	Mar	Apr	May	Jun	Jul	Aug	Sep	Oct	Nov	Dec	Year
	tx	21	26	31	36	41	43	43	42	39	36	29	22	44 °C
	t	8	11	16	22	27	31	33	32	27	21	15	9	21 °C
	tn	-4	-2	1	8	14	18	22	20	13	7	0	-4	-6 °C
	clothing	n	n	n	s	t	t	t	t	t	s	n	n	–
	stress				o	f	ff	ff	ff	f	o			–
	p	10	10	15	3	0	0	0	0	0	15	13	8	74 mm
2.5 mm	dp	2	1	2	1	0	0	0	0	0	0	0	1	7 days

Abadan

BWil	039-14	Jan	Feb	Mar	Apr	May	Jun	Jul	Aug	Sep	Oct	Nov	Dec	Year
				latitude: 30°22′n				longitude: 48°15′e			height: 3 m			
	tx	22	25	31	39	44	47	47	48	46	41	33	26	48 °C
	t	13	15	19	25	30	35	36	36	33	27	20	14	25 °C
	tn	1	3	6	12	19	22	24	23	19	14	6	2	0 °C
	clothing	n	n	s	t	t	t	t	t	t	t	s	n	–
	stress			o	f	ff	ff	ff	ff	ff	f	o		–
	p	20	15	19	15	4	0	0	0	0	1	32	41	147 mm
	dp	4	3	2	2	1	0	0	0	0	0	3	3	18 days
	sunshine	202	192	257	246	245	306	307	285	255	254	209	217	2970 hours
	%	63	61	69	64	58	72	71	69	69	71	65	69	67 %

Zahedan

GBWhk	039-15	Jan	Feb	Mar	Apr	May	Jun	Jul	Aug	Sep	Oct	Nov	Dec	Year
				latitude: 29°28′n				longitude: 60°53′e			height: 1370 m			
	t	6	12	15	19	23	27	28	26	22	18	12	7	18 °C
	clothing	n	n	n	s	t	t	t	t	s	s	n	n	–
	stress				o	f	f	f	f	o	o			–
	p	30	24	14	16	7	0	0	0	0	0	13	9	113 mm
	dp	3	3	3	3	1	0	0	0	0	0	1	2	16 days
	sunshine	232	170	229	216	291	297	313	335	300	273	222	211	3089 hours
	%	72	54	62	56	69	71	73	82	81	77	69	66	69 %

Bandar Abbas

BMhb	039-16	Jan	Feb	Mar	Apr	May	Jun	Jul	Aug	Sep	Oct	Nov	Dec	Year
				latitude: 27°11′n				longitude: 56°17′e			height: 9 m			
	tx	27	27	32	37	40	43	43	42	41	38	35	29	44 °C
	t	18	19	23	27	30	33	34	33	32	29	25	20	27 °C
	tn	9	11	13	17	22	24	27	27	24	19	15	10	8 °C
	clothing	s	s	t	t	t	t	t	t	t	t	t	s	–
	stress	o	o	f	f	ff	ff	ff	ff	ff	f	f	o	–
	p	61	30	8	5	3	0	0	0	0	0	0	41	148 mm
2.5 mm	dp	2	2	1	0	0	0	0	0	0	0	0	2	8 days
	sunshine	251	209	248	240	322	312	288	282	270	291	264	245	3222 hours
	%	69	62	66	65	83	83	74	74	74	78	75	67	73 %

Jask

BMhb	039-17	Jan	Feb	Mar	Apr	May	Jun	Jul	Aug	Sep	Oct	Nov	Dec	Year
				latitude: 25°45′n				longitude: 57°45′e			height: 4 m			
	tx	26	27	31	34	38	41	40	39	38	37	32	28	41 °C
	t	19	20	23	27	30	32	30	32	31	28	24	22	27 °C
	tn	11	13	15	19	22	26	27	27	24	21	17	14	11 °C
	clothing	s	s	t	t	t	t	t	t	t	t	t	s	–
	stress	o	o	f	ff	ff	ff	ff	ff	ff	ff	f	o	–
	p	30	23	15	5	1	1	1	1	0	5	8	30	120 mm
2.5 mm	dp	2	2	1	0	0	0	0	0	0	0	1	2	9 days

Chahbar

				latitude: 25°17′n			longitude: 60°37′e			height: 8 m			
BMhb	039-18	Jan	Feb	Mar	Apr	May	Jun	Jul	Aug	Sep	Oct	Nov	Dec	Year
	tx	27	28	31	34	37	37	37	34	35	36	33	29	39 °C
	t	19	21	23	27	29	31	30	29	28	27	24	21	26 °C
	tn	9	12	14	19	23	26	26	24	22	19	16	12	8 °C
	clothing	s	s	t	t	t	t	t	t	t	t	t	s	–
	stress	o	o	f	ff	ff	ff	ff	ff	ff	ff	f	o	–
	p	43	20	5	3	0	1	3	0	0	0	3	25	103 mm
2.5 mm	dp	2	1	0	0	0	0	0	0	0	0	0	1	5 days

4.2.39.2. Predominant general weather situations are described in 4.2.0.1. under Eastern Near East. A hygrothermal diagram of Tehran is to be found in 4.2.0.2.
Wind names: Kaus is a rainy wind in the Persian Gulf blowing from the southeast.
N'ashi is a bora-like (relatively cold and gusty) wind which blows from the northeast in the Persian Gulf, accompanied by cloudy weather and often bringing dust from the Coast of Iran.
Shamal is a constant wind which blows all the year round from the northwest in the Persian Gulf. Great Shamal is a dusty Shamal blowing in summer.
Sharki is a southeasterly wind of the summer season of the monsoon accompanied by overcast sky in the Persian Gulf.
Seistan is a stormy, often dusty, wind on the hot plains of Iran blowing in summer from the North or Northwest. It is called the "wind of 120 days".

4.2.40. Afghanistan

4.2.40.1. Climatic descriptions and tables

1. Climatic area: Northeastern Part
1.1. Subtropical Winter Rain Mountain Climate GCs and
Subtropical Winter Rain High Mountain Climate HCs
1.2. Summer warm or hot, winter cool or cold.
1.3. Summer generally dry and sunny, in winter frequently variable weather.
1.4. Wind: occasional gales and dust-storms.

Kabul Airport

				latitude: 34°33′n			longitude: 69°12′e			height: 1803 m			
GCsao	040-1	Jan	Feb	Mar	Apr	May	Jun	Jul	Aug	Sep	Oct	Nov	Dec	Year
	tx	8	12	19	26	32	35	37	37	34	28	23	17	38 °C
	t	-2	0	6	12	16	22	25	24	20	13	5	0	12 °C
	tn	-16	-14	-6	1	5	8	12	11	5	1	-4	-8	-17 °C
	clothing	w	w	n	n	n	s	t	t	s	n	n	w	–
	stress						o	f	f	o				–
	p	28	61	72	117	33	1	7	1	0	1	37	14	372 mm
	dp	5	6	7	8	6	1	1	1	0	1	2	4	42 days
	sunshine	182	168	175	202	313	347	352	346	290	289	234	187	2795 hours
	%	58	54	47	52	72	80	80	83	78	82	75	61	63 %

2. *Climatic area:* Northwestern and Southeastern part
2.1. Steppe Climate BS and Steppe Mountain Climate GBS and
Steppe High Mountain Climate HBS
2.2. Summer very hot, higher up hot; winter cool, higher up cold.
2.3. Generally dry and sunny, in winter now and then variable weather.
2.4. Wind: occasional gales and dust-storms.

Mazari-Sherif

latitude: 36°42'n longitude: 67°13'e height: 378 m

BShk	040-2	Jan	Feb	Mar	Apr	May	Jun	Jul	Aug	Sep	Oct	Nov	Dec	Year
	t	4	7	11	17	23	29	32	29	23	16	7	4	17 °C
	clothing	w	n	n	s	t	t	t	t	t	s	n	n	–
	stress				o	f	f	ff	ff	f	o			–
	p	28	32	43	45	16	0	0	0	0	2	21	27	214 mm
	dp													days
	sunshine	176	128	160	181	333	355	357	331	279	249	171	159	2879 hours
	%	57	42	43	46	76	81	80	79	75	71	56	53	65 %

Herat

latitude: 34°11'n longitude: 62°13'e height: 964 m

GBShk	040-3	Jan	Feb	Mar	Apr	May	Jun	Jul	Aug	Sep	Oct	Nov	Dec	Year
	t	4	6	10	16	22	27	29	26	23	16	8	4	16 °C
	clothing	w	n	n	n	s	t	t	t	t	n	n	w	–
	stress					o	f	ff	f	f				–
	p	21	23	73	47	19	0	0	0	0	0	17	30	231 mm

Kandahar

latitude: 31°37'n longitude: 65°48'e height: 1010 m

GBShk	040-4	Jan	Feb	Mar	Apr	May	Jun	Jul	Aug	Sep	Oct	Nov	Dec	Year
	tx	19	24	29	35	38	41	41	41	37	34	29	23	42 °C
	t	5	10	14	19	25	30	32	29	25	18	10	6	18 °C
	tn	-5	-3	1	4	8	12	15	13	7	2	-3	-7	-7 °C
	clothing	n	n	n	s	t	t	t	t	t	s	n	n	–
	stress				o	f	f	ff	f	f	o			–
	p	22	21	58	46	13	0	0	0	0	0	14	51	225 mm
2.5 mm	dp	5	4	2	1	0	0	0	0	0	0	1	2	15 days
	sunshine	209	188	245	262	324	389	373	369	332	325	269	221	3506 hours
	%	65	60	66	67	76	91	86	89	89	92	85	70	79 %

3. *Climatic area:* Southwestern part (Registan and Margo plains)
3.1. Desert Climate BW and Desert Mountain Climate GBW
3.2. High sun period very hot or hot; low sun period warm or mild, in higher levels cool.
3.3. Generally dry and sunny, during low sun period now and then scattered showers or variable weather.

4.2.40.2. Predominant general weather situations are described in 4.2.0.1. under Eastern Near East. A hygrothermal diagram of Kabul is to be found in 4.2.0.2.
Wind name: Afghanetz or "Wind of 120 days" is a hot stormy dusty wind blowing predominantly from the north in summer.

4.2.41. Pakistan

4.2.41.1. Climatic descriptions and tables

1. Climatic area: parts of the mountainous regions in the north
1.1. Subtropical Rain Mountain Climate GCr and
Subtropical Rain High Mountain Climate HCr
1.2. Summer very hot, at higher levels hot or warm; winter mild, at higher levels cool or cold.
1.3. Variable weather prevails, frequent showers in the afternoon.
1.4. Wind: occasional gales.

Murree

latitude: 33°54′n longitude: 73°24′e height: 2168 m

GCrbk	041-1	Jan	Feb	Mar	Apr	May	Jun	Jul	Aug	Sep	Oct	Nov	Dec	Year
	t	3	4	8	13	18	21	20	19	18	14	10	6	13 °C
	clothing	w	w	n	n	s	s	s	s	s	n	n	n	–
	stress					o	o	o	o	o				–
	p	116	108	155	103	61	106	360	348	133	53	21	54	1618 mm
	dp	7	7	9	7	6	6	15	16	8	3	2	4	89 days

Parachinar

latitude: 33°52′n longitude: 70°05′e height: 1729 m

GCrak	041-2	Jan	Feb	Mar	Apr	May	Jun	Jul	Aug	Sep	Oct	Nov	Dec	Year
	t	4	6	10	15	21	25	25	24	22	17	11	6	16 °C
	clothing	w	n	n	n	s	t	t	t	s	n	n	n	–
	stress					o	f	f	f	o				–
	p	82	78	136	104	64	47	122	112	54	17	13	31	860 mm
	dp	5	6	8	7	6	4	8	8	5	2	1	2	64 days

Rawalpindi

latitude: 33°36′n longitude: 73°07′e height: 510 m

GCrhl	041-3	Jan	Feb	Mar	Apr	May	Jun	Jul	Aug	Sep	Oct	Nov	Dec	Year
	t	10	12	17	23	28	32	31	29	27	23	16	11	22 °C
	clothing	n	n	s	t	t	t	t	t	t	t	s	n	–
	stress			o	f	ff	ff	ff	ff	f	f	o		–
	p	60	63	68	49	32	59	205	233	99	15	7	32	922 mm
	dp	7	6	7	6	4	7	13	10	5	2	1	3	71 days
	sunshine	186	158	211	246	326	306	282	254	312	304	264	208	3056 hours
	%	59	51	57	63	76	71	64	61	84	86	84	67	69 %

2. Climatic area: Mountainous region in the west from the North-West Frontier Province to Northern Baluchistan

2.1. Steppe Climate BS and Steppe Mountain Climate GBS and
Steppe High Mountain Climate HBS

2.2. Summer very hot at lower levels; winter mild, at higher levels cool or cold.

2.3. Predominantly dry and sunny; occasionally variable weather.

2.4. Wind: occasional gales and dust-storms.

Peshawar

latitude: 34°01'n longitude: 71°35'e height: 359 m

BShl	041-4	Jan	Feb	Mar	Apr	May	Jun	Jul	Aug	Sep	Oct	Nov	Dec	Year
	tx	22	24	31	37	44	46	46	42	39	36	29	24	46 °C
	t	11	13	17	23	29	33	33	31	29	24	18	12	23 °C
	tn	0	2	6	11	16	21	23	22	17	11	4	1	-1 °C
	clothing	n	n	s	t	t	t	t	t	t	t	s	n	–
	stress			o	f	f	ff	ff	ff	f	f	o		–
	p	39	41	65	42	40	7	39	41	14	10	10	15	363 mm
	dp	3	4	5	4	2	1	2	2	1	1	1	2	27 days
	sunshine	191	201	189	240	289	304	288	284	227	274	239	204	2930 hours
	%	61	65	51	61	67	70	65	68	61	78	76	66	66 %

Lahore City

latitude: 31°33'n longitude: 74°20'e height: 211 m

BShl	041-5	Jan	Feb	Mar	Apr	May	Jun	Jul	Aug	Sep	Oct	Nov	Dec	Year
	tx	23	27	35	41	46	46	43	40	39	38	32	27	47 °C
	t	12	15	20	27	32	34	32	31	30	26	19	14	24 °C
	tn	1	3	7	13	18	22	22	22	19	12	6	3	0 °C
	clothing	n	n	s	t	t	t	t	t	t	t	s	n	–
	stress			o	f	f	ff	ff	ff	ff	f	o		–
	p	31	23	24	16	12	38	122	123	80	9	3	11	492 mm
	dp	2	2	3	1	1	3	6	6	3	1	0	1	29 days
	sunshine	212	225	238	289	307	286	251	251	267	287	261	220	3094 hours
	%	66	72	64	74	72	67	58	61	72	81	82	70	70 %

Fort Sandeman

latitude: 31°21'n longitude: 69°27'e height: 1406 m

GBShk	041-6	Jan	Feb	Mar	Apr	May	Jun	Jul	Aug	Sep	Oct	Nov	Dec	Year
	t	6	9	14	20	26	30	30	29	26	20	14	9	20 °C
	clothing	n	n	n	s	t	t	t	t	t	s	n	n	–
	stress				o	f	ff	ff	ff	ff	o			–
	p	24	30	42	27	21	13	49	47	6	1	6	14	280 mm
	dp	2	3	4	2	2	1	4	4	0	0	0	2	27 days

Quetta

latitude: 30°15′n longitude: 66°53′e height: 1601 m

GBSak	041-7	Jan	Feb	Mar	Apr	May	Jun	Jul	Aug	Sep	Oct	Nov	Dec	Year
	tx	17	20	24	29	34	37	38	37	34	29	23	19	38 °C
	t	4	6	11	16	21	24	27	25	20	14	9	5	15 °C
	tn	-8	-7	-3	2	6	9	13	11	4	-1	-5	-8	-11 °C
	clothing	w	n	n	n	s	t	t	t	s	n	n	n	–
	stress					o	f	f	f	o				–
	p	37	43	42	12	7	1	18	4	1	1	6	23	195 mm
	dp	3	4	5	1	1	0	0	2	1	0	1	2	20 days
	sunshine	213	217	233	274	343	355	324	337	299	311	269	224	3399 hours
	%	66	69	63	71	81	84	75	82	81	87	84	71	76 %

Kalat

latitude: 29°02′n longitude: 66°35′e height: 2017 m

GBSak	041-8	Jan	Feb	Mar	Apr	May	Jun	Jul	Aug	Sep	Oct	Nov	Dec	Year
	t	3	5	9	14	18	22	24	23	19	13	8	5	14 °C
	clothing	w	n	n	n	s	s	t	t	s	n	n	n	–
	stress					o	o	f	f	o				–
	p	54	46	38	15	6	3	30	14	2	0	3	19	230 mm
	dp	5	4	3	2	1	0	2	1	0	0	0	2	21 days

3. Climatic area: Most of the lowland plains of the country

3.1. Desert Climate BW, coastal areas Marine Desert Climate BM

3.2. High sun period severely or very hot, low sun period warm or mild.

3.3. Generally dry and sunny; now and then scattered showers and cloudy or variable weather.

3.4. Wind: occasional gales and dust-storms.

Dera Ismail Khan

latitude: 31°49′n longitude: 70°55′e height: 174 m

BWil	041-9	Jan	Feb	Mar	Apr	May	Jun	Jul	Aug	Sep	Oct	Nov	Dec	Year
	t	12	15	20	26	32	35	34	33	31	26	19	13	25 °C
	clothing	n	n	s	t	t	t	t	t	t	t	s	n	–
	stress			o	f	ff	ff	ff	ff	ff	f	o		–
	p	14	18	27	20	9	9	65	36	14	2	3	6	223 mm
	dp	1	2	3	1	1	1	2	2	1	0	0	1	16 days

Multan

latitude: 30°12′n longitude: 71°26′e height: 123 m

BWil	041-10	Jan	Feb	Mar	Apr	May	Jun	Jul	Aug	Sep	Oct	Nov	Dec	Year
	tx	23	29	35	42	47	47	44	41	42	39	34	28	48 °C
	t	13	16	22	28	33	36	34	33	32	27	20	15	26 °C
	tn	1	4	8	14	21	23	24	24	21	14	7	3	1 °C
	clothing	n	n	s	t	t	t	t	t	t	t	s	n	–
	stress			o	f	ff	ff	ff	ff	ff	f	o		–
	p	7	10	13	6	8	8	45	33	20	10	2	5	167 mm
	dp	1	1	1	1	1	1	2	2	1	0	0	1	11 days

Dalbandin

latitude: 28°53′n longitude: 64°24′e height: 849 m

GBWhk	041-11	Jan	Feb	Mar	Apr	May	Jun	Jul	Aug	Sep	Oct	Nov	Dec	Year
	t	9	12	17	23	28	32	33	32	27	21	15	10	22 °C
	clothing	n	n	s	t	t	t	t	t	t	s	n	n	–
	stress			o	f	ff	ff	ff	ff	ff	o			–
	p	25	18	12	5	2	1	7	0	0	0	1	13	84 mm
	dp	2	2	1	1	0	0	1	0	0	0	0	1	9 days

Jacobabad

latitude: 28°18′n longitude: 68°28′e height: 56 m

BWil	041-12	Jan	Feb	Mar	Apr	May	Jun	Jul	Aug	Sep	Oct	Nov	Dec	Year
	tx	26	31	39	44	48	50	47	43	42	40	35	27	50 °C
	t	15	18	24	30	35	37	35	34	32	28	22	17	27 °C
	tn	3	3	9	15	21	26	27	25	21	13	7	3	2 °C
	clothing	n	s	t	t	t	t	t	t	t	t	s	n	–
	stress		o	f	ff	ff	ff	ff	ff	ff	f	o		–
	p	8	8	7	2	4	6	37	22	1	0	1	3	99 mm
	dp	1	1	1	0	0	0	2	1	0	0	0	0	7 days
	sunshine	251	237	263	280	310	284	253	288	292	301	293	251	3303 hours
	%	77	75	71	73	74	68	59	71	79	84	91	78	74 %

Panjgur

latitude: 26°58′n longitude: 64°06′e height: 968 m

GBWhl	041-13	Jan	Feb	Mar	Apr	May	Jun	Jul	Aug	Sep	Oct	Nov	Dec	Year
	t	10	13	18	22	28	31	31	30	27	22	16	12	22 °C
	clothing	n	n	s	s	t	t	t	t	t	s	n	n	–
	stress			o	o	ff	ff	ff	ff	f	o			–
	p	24	18	17	8	3	3	27	8	1	0	1	11	121 mm
	dp	2	2	1	1	1	0	2	0	0	0	0	1	11 days

Hyderabad

latitude: 25°23′n longitude: 68°25′e height: 29 m

BWhl	041-14	Jan	Feb	Mar	Apr	May	Jun	Jul	Aug	Sep	Oct	Nov	Dec	Year
	tx	29	34	39	43	47	44	42	40	42	42	38	32	48 °C
	t	17	21	26	31	34	34	32	31	31	29	24	19	28 °C
	tn	5	9	12	18	23	25	25	25	23	18	12	8	5 °C
	clothing	s	s	t	t	t	t	t	t	t	t	t	s	–
	stress	o	o	f	ff	ff	ff	ff	ff	ff	ff	f	o	–
	p	4	5	1	2	4	6	69	44	15	3	1	3	157 mm
	dp	1	0	0	0	0	0	3	2	1	0	0	0	9 days

Pasni

latitude: 25°16′n longitude: 63°29′e height: 9 m

BMhb	041-15	Jan	Feb	Mar	Apr	May	Jun	Jul	Aug	Sep	Oct	Nov	Dec	Year
	tx	27	30	35	38	41	40	37	34	38	37	33	29	42 °C
	t	19	20	23	27	30	30	30	28	28	27	24	20	25 °C
	tn	8	7	12	16	20	23	23	22	19	15	11	8	7 °C
	clothing	s	s	t	t	t	t	t	t	t	t	t	s	–
	stress	o	o	f	f	ff	ff	ff	ff	f	f	f	o	–
	p	43	32	8	7	2	6	12	3	1	0	2	12	127 mm
	dp	2	2	1	0	0	0	1	0	0	0	0	1	7 days

Karachi

BMhb	041-16	Jan	Feb	Mar	Apr	May	Jun	Jul	Aug	Sep	Oct	Nov	Dec	Year
latitude: 24°55′n longitude: 67°09′e height: 22 m														
	tx	28	30	36	37	39	38	36	33	34	37	34	30	42 °C
	t	18	20	24	28	30	31	30	29	29	28	24	20	26 °C
	tn	8	9	14	19	23	26	25	24	23	18	14	10	7 °C
	clothing	s	s	t	t	t	t	t	t	t	t	t	s	–
	stress	o	o	f	ff	ff	ff	ff	ff	ff	ff	f	o	–
	p	7	11	6	2	0	7	96	50	15	2	2	6	204 mm
	dp	1	1	0	0	0	1	4	2	1	0	0	0	10 days
	sunshine	273	262	278	297	312	235	136	148	212	289	280	269	2991 hours
	%	82	82	75	78	75	57	32	37	57	80	85	82	67 %

4.2.41.2. Predominant general weather situations are described in 4.2.0.1. under Eastern Near East. A hygrothermal diagram of Lahore is to be found in 4.2.0.2.
Wind name: Shaitan – in Baluchistan – is a dusty whirlwind which arises in summer when the ground is overheated.

4.2.42. Bangladesh

4.2.42.1. Climatic description and tables

1. Tropical Summer Rain Climate Aw and Tropical Monsoonal Rain Climate Am
2. Summer very hot, winter warm.
3. High sun period: more often than not showers in the afternoon and cloudy; low sun period: generally dry and sunny.
4. Wind: occasional gales, in coastal areas tropical cyclones.

Sylhet

Amhb	042-1	Jan	Feb	Mar	Apr	May	Jun	Jul	Aug	Sep	Oct	Nov	Dec	Year
latitude: 24°54′n longitude: 91°53′e height: 32 m														
	t	19	21	24	26	27	27	28	28	28	26	23	20	25 °C
	clothing	s	s	t	t	t	t	t	t	t	t	t	s	–
	stress	o	o	f	f	ff	ff	ff	ff	ff	f	f	o	–
	p	17	44	79	252	497	954	684	642	455	198	14	3	3839 mm
	dp	2	2	3	11	18	22	24	24	16	8	1	0	131 days

Bogra

Awhb	042-2	Jan	Feb	Mar	Apr	May	Jun	Jul	Aug	Sep	Oct	Nov	Dec	Year
latitude: 24°51′n longitude: 89°23′e height: 20 m														
	t	18	20	25	29	29	29	29	29	29	27	23	20	26 °C
	clothing	s	s	t	t	t	t	t	t	t	t	s	s	–
	stress	o	o	f	ff	ff	ff	ff	ff	ff	ff	o	o	–
	p	14	17	27	63	195	322	317	351	275	180	13	2	1776 mm
	dp	1	1	2	4	9	19	16	16	13	6	1	0	89 days

Dacca Tejgaon

latitude: 23°46′n longitude: 90°23′e height: 7 m

Awhb	042-3	Jan	Feb	Mar	Apr	May	Jun	Jul	Aug	Sep	Oct	Nov	Dec	Year
	t	18	21	26	29	29	28	28	28	29	27	23	19	26 °C
	clothing	s	s	t	t	t	t	t	t	t	t	t	s	–
	stress	o	o	f	ff	ff	ff	ff	ff	ff	ff	f	o	–
	p	15	21	54	108	253	420	405	324	256	153	27	3	2039 mm
	dp	1	1	3	6	11	16	12	16	12	7	1	0	88 days

Narayanganj

latitude: 23°37′n longitude: 90°30′e height: 8 m

Awhb	042-4	Jan	Feb	Mar	Apr	May	Jun	Jul	Aug	Sep	Oct	Nov	Dec	Year
	tx	28	32	36	37	36	34	33	33	34	33	31	28	37 °C
	t	20	22	27	29	29	29	29	29	29	28	24	21	26 °C
	tn	10	11	15	19	20	23	24	24	24	21	15	11	9 °C
	clothing	s	s	t	t	t	t	t	t	t	t	t	s	–
	stress	o	o	f	ff	ff	ff	ff	ff	ff	ff	f	o	–
	p	14	27	46	161	245	346	348	364	242	171	29	19	2012 mm
	dp	1	2	3	7	10	14	19	18	13	6	2	0	98 days

Jessore

latitude: 23°10′n longitude: 89°13′e height: 7 m

Awhb	042-5	Jan	Feb	Mar	Apr	May	Jun	Jul	Aug	Sep	Oct	Nov	Dec	Year
	t	18	21	26	30	30	29	28	29	29	27	23	19	26 °C
	clothing	s	s	t	t	t	t	t	t	t	t	t	s	–
	stress	o	o	f	ff	ff	ff	ff	ff	ff	ff	f	o	–
	p	14	23	35	88	182	275	313	307	197	136	22	16	1608 mm
	dp	1	2	2	5	9	14	18	16	12	6	2	0	86 days

Chittagong

latitude: 22°21′n longitude: 91°50′e height: 14 m

Amhb	042-6	Jan	Feb	Mar	Apr	May	Jun	Jul	Aug	Sep	Oct	Nov	Dec	Year
	tx	29	31	34	34	34	33	33	32	33	33	31	28	35 °C
	t	20	24	26	28	28	28	28	28	28	27	24	21	26 °C
	tn	9	11	14	19	20	22	23	23	23	19	15	11	9 °C
	clothing	s	t	t	t	t	t	t	t	t	t	t	s	–
	stress	o	f	f	ff	ff	ff	ff	ff	ff	ff	f	o	–
	p	10	23	58	116	285	507	642	572	344	228	56	17	2858 mm
	dp	1	1	2	5	10	15	19	15	11	7	2	1	90 days

4.2.42.2. Predominant general weather situations are described in 4.2.0.1. under India. A hygrothermal diagram of Chittagong is to be found in 4.2.0.2.

4.2.43. Kashmir

4.2.43.1. Climatic descriptions and table

1. Climatic area: the southwestern half of the country
1.1. Subtropical Rain Mountain Climate GCr and
Subtropical Rain High Mountain Climate HCr
1.2. Summer warm, partly hot; winter cool or cold.
1.3. Variable weather frequent; often dry and sunny in the morning and rainy or showers and cloudy in the afternoon and evening.
1.4. Wind: occasional gales.

Srinagar

latitude: 34°05′n longitude: 74°50′e height: 1586 m

GCrak	043-1	Jan	Feb	Mar	Apr	May	Jun	Jul	Aug	Sep	Oct	Nov	Dec	Year
	tx	10	12	21	26	31	34	36	34	32	28	20	14	36 °C
	t	1	4	8	13	18	22	25	24	20	14	8	4	13 °C
	tn	-7	-6	-1	3	7	11	13	13	7	1	-4	-7	-9 °C
	clothing	w	w	n	n	s	s	t	t	s	n	n	w	–
	stress					o	o	f	f	o				–
	p	73	72	104	78	63	36	61	63	31	28	20	36	665 mm
	dp	9	8	10	9	8	5	7	8	4	3	2	5	78 days
	sunshine	78	115	134	189	242	246	257	233	238	243	204	116	2295 hours
	%	25	37	36	48	56	57	58	56	64	69	65	38	52 %

2. Climatic area: the northeastern half of the country
2.1. Desert Mountain Climate GBW and Desert High Mountain Climate HBW (except for zones of transition with corresponding Steppe Climate)
2.2. Summer warm or mild; winter cold.
2.3. Generally dry and sunny or cloudy, partly also overcast; now and then variable weather or showers.
2.4. Wind: occasional gales.

Leh

latitude: 34°09′n longitude: 77°34′e height: 3514 m

HBWlo	043-2	Jan	Feb	Mar	Apr	May	Jun	Jul	Aug	Sep	Oct	Nov	Dec	Year
	tx	3	7	14	18	22	26	28	27	25	20	12	8	29 °C
	t	-8	-6	0	6	10	14	17	17	13	7	1	-5	6 °C
	tn	-19	-19	-13	-5	-1	2	6	6	1	-5	-10	-17	-21 °C
	clothing	w	w	w	n	n	n	s	s	n	n	w	w	–
	stress							o	o					–
	p	12	9	12	7	7	4	16	19	12	7	3	8	116 mm
	dp	4	3	3	2	2	1	3	4	2	1	1	2	28 days

4.2.43.2. Predominant general weather situations are described in 4.2.0.1. under India.

4.2.44. India

4.2.44.1. Climatic descriptions and tables

1. Climatic area: westernmost region from Great Desert or Thar to the Gulf of Cutch
1.1. Desert Climate BW, coastal areas Marine Desert Climate BM
1.2. Summer severely or very hot; winter warm or mild.
1.3. High sun period: especially in July and August, frequent showers, particularly in the afternoon, otherwise dry and sunny; low sun period: generally dry and sunny.
1.4. Wind: occasional gales and dust-storms.

Bikaner

latitude: 28°00'n longitude: 73°18'e height: 224 m

BWil	044-1	Jan	Feb	Mar	Apr	May	Jun	Jul	Aug	Sep	Oct	Nov	Dec	Year
	tx	28	32	38	42	46	46	43	40	40	39	34	29	46 °C
	t	14	17	24	30	35	36	33	31	31	27	20	15	26 °C
	tn	3	5	10	16	21	24	24	24	23	16	8	4	2 °C
	clothing	n	s	t	t	t	t	t	t	t	t	s	n	–
	stress		o	f	f	ff	ff	ff	ff	ff	f	o		–
	p	6	7	6	5	7	27	87	104	45	6	3	2	305 mm
	dp	1	1	1	1	1	3	6	7	2	0	0	0	25 days

Jodhpur

latitude: 26°18'n longitude: 73°01'e height: 224 m

BWhl	044-2	Jan	Feb	Mar	Apr	May	Jun	Jul	Aug	Sep	Oct	Nov	Dec	Year
	tx	29	33	39	43	46	43	41	38	39	39	36	31	46 °C
	t	17	20	25	30	34	34	31	29	29	28	23	19	27 °C
	tn	4	7	12	18	23	24	24	23	22	16	11	7	3 °C
	clothing	s	s	t	t	t	t	t	t	t	t	t	s	–
	stress	o	o	f	ff	ff	ff	ff	ff	ff	f	f	o	–
	p	8	5	2	2	6	31	122	146	47	7	3	1	380 mm
	dp	1	1	0	0	1	3	8	9	3	1	0	0	27 days
	sunshine	276	269	279	300	326	288	208	189	250	307	297	285	3274 hours
	%	83	84	75	78	78	70	49	47	68	86	91	87	74 %

Dwarka

latitude: 22°22'n longitude: 69°05'e height: 11 m

BMhb	044-3	Jan	Feb	Mar	Apr	May	Jun	Jul	Aug	Sep	Oct	Nov	Dec	Year
	tx	29	31	34	33	33	33	32	31	31	36	34	31	37 °C
	t	18	21	26	30	34	31	26	25	26	25	22	19	25 °C
	tn	12	13	17	22	25	26	25	24	24	21	17	13	11 °C
	clothing	s	s	t	t	t	t	t	t	t	t	s	s	–
	stress	o	o	f	ff	ff	ff	ff	ff	ff	ff	o	o	–
	p	2	2	1	0	0	49	220	77	42	5	19	2	419 mm
	dp	0	0	0	0	0	3	10	6	3	1	0	0	25 days

2. Climatic area: region ranging from the Punjab in the north to Hyderabad and Eastern Bombay in the south, bordering the western desert region

2.1. Steppe Climate BS

2.2. High sun period: severely or very hot; low sun period: mild, warm in the south.

2.3. High sun period: in the morning often dry and sunny, in the afternoon frequent showers. Low sun period: generally dry and sunny or cloudy.

2.4. Wind: occasional gales, at times dust-storms.

Amritsar

latitude: 31°38'n longitude: 74°52'e height: 234 m

BShl	044-4	Jan	Feb	Mar	Apr	May	Jun	Jul	Aug	Sep	Oct	Nov	Dec	Year
	tx													°C
	t	12	14	20	25	30	33	31	30	29	24	18	13	23 °C
	tn													°C
	clothing	n	n	s	t	t	t	t	t	t	t	s	n	–
	stress			o	f	ff	ff	ff	ff	ff	f	o		–
	p	37	11	25	8	11	32	169	168	106	54	10	11	646 mm
	dp	4	2	3	2	2	3	10	11	5	2	1	2	46 days
	sunshine	208	216	241	288	312	278	229	242	247	288	265	220	3034 hours
	%	65	69	65	74	73	65	53	59	66	81	84	70	68 %

Ludhiana

latitude: 30°56'n longitude: 75°52'e height: 247 m

BShl	044-5	Jan	Feb	Mar	Apr	May	Jun	Jul	Aug	Sep	Oct	Nov	Dec	Year
	t	13	16	21	27	33	34	31	30	30	26	20	15	25 °C
	clothing	n	n	s	t	t	t	t	t	t	t	s	n	–
	stress			o	f	ff	ff	ff	ff	ff	f	o		–
	p	35	35	29	11	9	54	191	173	136	35	3	14	725 mm
	dp	4	3	3	2	2	4	10	10	6	1	0	2	46 days

New Delhi

latitude: 28°35'n longitude: 77°12'e height: 216 m

BShl	044-6	Jan	Feb	Mar	Apr	May	Jun	Jul	Aug	Sep	Oct	Nov	Dec	Year
	tx	26	29	37	42	44	44	40	37	37	37	32	27	45 °C
	t	14	17	23	29	34	34	31	30	29	26	20	16	25 °C
	tn	2	4	9	15	21	22	23	23	20	13	7	3	2 °C
	clothing	n	s	t	t	t	t	t	t	t	t	s	n	–
	stress		o	f	f	ff	ff	ff	ff	ff	f	o		–
	p	25	22	17	7	8	65	211	173	150	31	1	5	715 mm
	dp	3	2	2	1	2	4	12	12	6	2	0	1	47 days
	sunshine	229	246	250	261	249	192	177	187	217	288	285	252	2833 hours
	%	71	78	68	68	59	46	41	46	59	81	89	79	64 %

Agra

latitude: 27°10'n longitude: 78°02'e height: 169 m

BSil	044-7	Jan	Feb	Mar	Apr	May	Jun	Jul	Aug	Sep	Oct	Nov	Dec	Year
	tx	27	32	38	42	46	45	46	37	37	37	33	28	46 °C
	t	15	18	24	30	34	35	31	29	29	26	21	16	26 °C
	tn	2	4	8	13	21	23	24	24	20	12	6	4	1 °C
	clothing	n	s	t	t	t	t	t	t	t	t	s	n	–
	stress		o	f	f	ff	ff	ff	ff	ff	f	o		–
	p	16	9	11	5	10	60	210	263	154	23	2	4	767 mm
	dp	2	1	2	1	1	4	12	14	8	2	0	1	48 days

Jaipur

BShl	044-8	latitude: 26°55′n			longitude: 75°50′e				height: 436 m					
		Jan	Feb	Mar	Apr	May	Jun	Jul	Aug	Sep	Oct	Nov	Dec	Year
	t	16	18	22	29	33	33	30	29	28	26	21	17	25 °C
	clothing	n	s	s	t	t	t	t	t	t	t	s	s	–
	stress		o	o	f	ff	ff	ff	ff	f	f	o	o	–
	p	11	8	9	4	14	57	196	205	82	12	4	8	610 mm
0.2 mm	dp	2	3	2	3	2	8	16	14	7	1	0	1	59 days
	sunshine	267	257	301	297	313	231	143	146	222	313	297	279	3065 hours
	%	81	81	81	77	75	56	34	36	60	87	91	86	69 %

Kota

BSib	044-9	latitude: 25°11′n			longitude: 75°51′e				height: 257 m					
		Jan	Feb	Mar	Apr	May	Jun	Jul	Aug	Sep	Oct	Nov	Dec	Year
	t	18	21	26	31	36	35	30	29	29	28	23	19	27 °C
	clothing	s	s	t	t	t	t	t	t	t	t	s	s	–
	stress	o	o	f	ff	ff	ff	ff	ff	ff	f	o	o	–
	p	9	4	5	3	7	79	311	268	135	15	6	3	845 mm
	dp	2	1	1	1	1	6	15	14	8	1	1	0	50 days

Ahmedabad

BShb	044-10	latitude: 23°04′n			longitude: 72°38′e				height: 55 m					
		Jan	Feb	Mar	Apr	May	Jun	Jul	Aug	Sep	Oct	Nov	Dec	Year
	tx	33	35	39	42	43	41	38	36	37	38	36	34	43 °C
	t	20	23	27	31	34	33	29	28	29	28	24	21	27 °C
	tn	10	12	16	21	23	24	23	23	23	21	16	12	9 °C
	clothing	s	s	t	t	t	t	t	t	t	t	t	s	–
	stress	o	o	f	ff	ff	ff	ff	ff	ff	ff	f	o	–
	p	4	0	1	2	5	81	316	213	163	13	5	1	804 mm
	dp	1	0	0	0	0	6	16	14	7	1	1	0	46 days
	sunshine	301	286	288	303	338	258	149	127	201	295	300	298	3144 hours
	%	89	89	77	80	82	63	36	32	54	82	91	90	71 %

Veraval

BShb	044-11	latitude: 20°54′n			longitude: 70°22′e				height: 8 m					
		Jan	Feb	Mar	Apr	May	Jun	Jul	Aug	Sep	Oct	Nov	Dec	Year
	tx	32	33	37	37	33	32	31	29	32	37	35	33	38 °C
	t	22	22	25	27	28	30	28	27	27	28	26	23	26 °C
	tn	11	12	14	18	23	25	24	24	23	19	17	13	10 °C
	clothing	s	s	t	t	t	t	t	t	t	t	t	t	–
	stress	o	o	f	ff	ff	ff	ff	ff	ff	f	f	f	–
	p	1	1	0	5	5	135	305	146	65	28	7	1	699 mm
	dp	0	0	0	0	0	6	16	12	6	2	0	0	44 days

Poona

		latitude: 18°32′n				longitude: 73°51′e				height: 559 m				
GBShb	044-12		Jan	Feb	Mar	Apr	May	Jun	Jul	Aug	Sep	Oct	Nov	Dec	Year
	tx	33	36	39	41	41	37	31	31	33	35	33	33	41 °C
	t	21	23	26	29	30	28	25	25	25	26	23	21	25 °C
	tn	7	8	11	16	19	21	21	20	18	14	9	7	6 °C
	clothing	s	s	t	t	t	t	t	t	t	t	s	s	–
	stress	o	o	f	f	ff	ff	ff	ff	ff	f	o	o	–
	p	2	0	3	18	35	103	187	106	127	92	37	5	715 mm
	dp	0	0	0	2	3	10	22	16	10	7	2	1	72 days
	sunshine	295	283	307	295	307	189	87	109	164	243	264	285	2828 hours
	%	85	87	82	78	76	48	21	28	45	67	78	83	66 %

Bagampet

		latitude: 17°27′n				longitude: 78°28′e				height: 545 m				
GBShb	044-13	Jan	Feb	Mar	Apr	May	Jun	Jul	Aug	Sep	Oct	Nov	Dec	Year
	tx	32	36	39	41	43	41	35	34	34	33	32	31	43 °C
	t	22	24	27	30	32	29	26	26	26	25	22	21	26 °C
	tn	12	14	18	20	22	21	21	21	21	17	13	11	10 °C
	clothing	s	t	t	t	t	t	t	t	t	t	s	s	–
	stress	o	f	ff	ff	ff	ff	ff	ff	ff	f	o	o	–
	p	2	10	13	23	30	107	165	147	163	71	25	5	761 mm
	dp	0	1	1	3	4	9	16	13	12	6	3	1	68 days
	sunshine	307	288	289	287	290	199	120	144	173	240	261	292	2890 hours
	%	88	88	77	76	72	50	30	36	47	66	77	85	65 %

3. Climatic area: Northernmost region including the Himalayas and adjacent lowland plains

3.1. Subtropical Summer Rain Climate Cw and Subtropical Summer Rain Mountain Climate GCw and Subtropical Summer Rain High Mountain Climate HCw

3.2. Summer hot, partly very or severely hot in the lowlands; winter mild in the lowlands.

3.3. High sun period: in the morning often dry and sunny, in the afternoon frequent showers, in July and August even periods of rainy days; low sun period: generally dry and sunny.

3.4. Wind: occasional gales and squalls.

Simla

		latitude: 31°06′n				longitude: 77°10′e				height: 2205 m				
GCwbk	044-14	Jan	Feb	Mar	Apr	May	Jun	Jul	Aug	Sep	Oct	Nov	Dec	Year
	tx	13	14	19	23	27	27	24	22	22	21	17	16	28 °C
	t	5	8	11	15	19	20	18	18	17	14	11	8	14 °C
	tn	-3	-3	2	5	8	11	13	13	11	7	3	-1	-4 °C
	clothing	n	n	n	n	s	s	s	s	s	n	n	n	–
	stress					o	o	o	o	o				–
	p	82	31	70	33	51	140	456	408	202	72	10	18	1574 mm
0.2 mm	dp	10	10	8	7	9	15	26	25	15	3	1	3	134 days

Mukteswar Kumaon

GCwbk	044-15	latitude: 29°28'n				longitude: 79°39'e				height: 2311 m				
		Jan	Feb	Mar	Apr	May	Jun	Jul	Aug	Sep	Oct	Nov	Dec	Year
	t	6	7	11	16	19	19	18	17	16	14	11	8	14 °C
	clothing	n	n	n	n	s	s	s	s	n	n	n	n	–
	stress					o	o	o	o					–
	p	56	56	45	34	51	143	332	285	200	79	4	23	1308 mm
	dp	5	5	5	4	6	11	22	21	14	4	1	2	101 days

Mohanbari

Cwhl	044-16	latitude: 27°29'n				longitude: 95°01'e				height: 111 m				
		Jan	Feb	Mar	Apr	May	Jun	Jul	Aug	Sep	Oct	Nov	Dec	Year
	t	15	18	21	23	25	27	28	28	28	25	20	16	23 °C
	clothing	n	s	s	t	t	t	t	t	t	t	s	n	–
	stress		o	o	f	f	ff	ff	ff	ff	f	o		–
	p	39	61	100	204	356	507	523	419	352	166	27	21	2775 mm
	dp	5	8	10	15	18	22	23	20	17	10	3	3	153 days

Darjeeling

GCwbk	044-17	latitude: 27°03'n				longitude: 88°16'e				height: 2127 m				
		Jan	Feb	Mar	Apr	May	Jun	Jul	Aug	Sep	Oct	Nov	Dec	Year
	tx	13	14	19	21	22	22	23	23	22	19	17	14	23 °C
	t	6	8	11	14	16	17	18	18	17	15	11	8	13 °C
	tn	-1	0	3	6	8	12	13	13	12	7	3	1	-1 °C
	clothing	n	n	n	n	n	s	s	s	s	n	n	n	–
	stress						o	o	o	o				–
	p	22	27	53	109	187	522	713	573	419	116	14	5	2760 mm
	dp	2	4	6	10	18	24	28	26	20	7	2	1	150 days

Gwalior

Cwil	044-18	latitude: 26°14'n				longitude: 78°15'e				height: 207 m				
		Jan	Feb	Mar	Apr	May	Jun	Jul	Aug	Sep	Oct	Nov	Dec	Year
	t	15	18	25	31	36	36	30	29	28	26	20	16	26 °C
	clothing	n	s	t	t	t	t	t	t	t	t	s	n	–
	stress		o	f	ff	ff	ff	ff	ff	ff	f	o		–
	p	22	7	9	3	10	69	283	282	167	42	3	6	903 mm

Darbhanga

Cwhl	044-19	latitude: 26°10'n				longitude: 85°54'e				height: 49 m				
		Jan	Feb	Mar	Apr	May	Jun	Jul	Aug	Sep	Oct	Nov	Dec	Year
	t	17	19	24	29	30	30	29	29	29	27	22	18	25 °C
	clothing	n	s	t	t	t	t	t	t	t	t	s	s	–
	stress		o	f	ff	ff	ff	ff	ff	ff	f	o	o	–
	p	20	14	10	20	57	190	330	291	247	70	7	2	1258 mm
	dp	2	2	1	2	4	10	16	16	12	4	1	0	70 days

Gauhati

latitude: 26°05′n longitude: 91°43′e height: 54 m

Cwhl	044-20	Jan	Feb	Mar	Apr	May	Jun	Jul	Aug	Sep	Oct	Nov	Dec	Year
	t	16	19	22	26	27	28	28	28	26	26	21	18	24 °C
	clothing	n	s	s	t	t	t	t	t	t	t	s	s	–
	stress		o	o	f	f	ff	ff	ff	ff	f	o	o	–
	p	11	18	53	171	274	292	301	263	190	90	11	5	1679 mm
	dp	1	3	5	10	17	17	17	15	12	6	2	1	106 days

Allahabad

latitude: 25°27′n longitude: 81°44′e height: 98 m

Cwil	044-21	Jan	Feb	Mar	Apr	May	Jun	Jul	Aug	Sep	Oct	Nov	Dec	Year
	tx	28	32	38	43	46	46	38	36	36	36	32	28	46 °C
	t	16	19	25	31	35	34	30	29	29	26	21	17	26 °C
	tn	9	7	12	17	23	24	24	24	23	15	9	5	4 °C
	clothing	n	s	t	t	t	t	t	t	t	t	s	n	–
	stress		o	f	ff	ff	ff	ff	ff	ff	f	o		–
	p	20	22	14	5	8	100	283	333	195	40	6	6	1032 mm
	dp	3	3	2	1	1	6	16	18	10	4	0	1	64 days
	sunshine	245	258	273	294	295	220	158	152	208	270	279	276	2928 hours
	%	74	81	73	77	71	53	38	38	56	75	85	84	66 %

Daltonganj

latitude: 24°03′n longitude: 84°04′e height: 221 m

Cwhl	044-22	Jan	Feb	Mar	Apr	May	Jun	Jul	Aug	Sep	Oct	Nov	Dec	Year
	t	17	19	24	30	34	33	29	28	28	26	20	17	26 °C
	clothing	s	s	t	t	t	t	t	t	t	t	s	s	–
	stress	o	o	f	ff	ff	ff	ff	ff	ff	f	o	o	–
	p	31	28	21	9	13	140	346	358	223	57	7	4	1237 mm
	dp	3	3	3	2	2	9	20	21	14	5	1	1	82 days

4. Climatic area: Eastern and southern parts of the country

4.1. Tropical Summer Rain Climate Aw, Tropical Monsoonal Rain Climate Am, Tropical Summer Rain Mountain Climate GAw, Tropical Monsoonal Rain Mountain Climate GAm

4.2. High sun period: very hot; low sun period: hot or warm.

4.3. High sun period: variable weather prevails, at times periods of rainy days; otherwise generally dry and sunny or cloudy in the morning, in the afternoon frequent showers. Low sun period: generally dry and sunny; now and then showers.

4.4. Wind: occasional gales, in coastal areas tropical storms.

Dhubri

latitude: 26°01′n longitude: 89°59′e height: 35 m

Amhb	044-23	Jan	Feb	Mar	Apr	May	Jun	Jul	Aug	Sep	Oct	Nov	Dec	Year
	tx	26	29	35	36	35	33	33	31	33	32	29	26	37 °C
	t	18	20	24	27	26	27	28	28	28	26	22	19	24 °C
	tn	9	9	13	18	19	22	24	24	23	19	14	11	8 °C
	clothing	s	s	t	t	t	t	t	t	t	t	s	s	–
	stress	o	o	f	ff	ff	ff	ff	ff	ff	f	o	o	–
	p	11	19	45	154	481	644	448	305	328	145	12	1	2593 mm
	dp	1	2	3	9	18	20	18	16	13	6	1	0	108 days

Patna

latitude: 25°36′n longitude: 85°06′e height: 51 m

Awhb	044-24	Jan	Feb	Mar	Apr	May	Jun	Jul	Aug	Sep	Oct	Nov	Dec	Year
	tx	26	31	38	41	43	41	37	34	35	34	31	27	43 °C
	t	18	20	26	31	33	32	30	30	29	28	23	19	26 °C
	tn	7	9	13	19	22	23	24	24	23	19	12	8	7 °C
	clothing	s	s	t	t	t	t	t	t	t	t	s	s	–
	stress	o	o	f	ff	ff	ff	ff	ff	ff	ff	o	o	–
	p	26	6	8	9	22	142	282	244	224	54	6	2	1027 mm
0.2 mm	dp	3	4	3	1	2	11	19	21	15	3	0	1	83 days

Cherrapunji

latitude: 25°15′n longitude: 91°44′e height: 1313 m

GAmbl	044-25	Jan	Feb	Mar	Apr	May	Jun	Jul	Aug	Sep	Oct	Nov	Dec	Year
	tx	19	21	24	26	26	26	26	25	26	24	23	21	28 °C
	t	12	13	17	19	19	20	20	20	20	19	16	13	17 °C
	tn	5	5	8	11	13	15	15	17	16	13	9	6	4 °C
	clothing	n	n	s	s	s	s	s	s	s	s	n	n	–
	stress				o	o	o	o	o	o	o			–
	p	20	41	179	605	1705	2875	2455	1827	1231	447	47	5	11437 mm
	dp	2	4	9	17	25	27	29	26	22	11	2	1	175 days

Silchar

latitude: 24°49′n longitude: 92°48′e height: 29 m

Amhb	044-26	Jan	Feb	Mar	Apr	May	Jun	Jul	Aug	Sep	Oct	Nov	Dec	Year
	tx	28	31	34	34	36	36	36	36	36	34	32	29	37 °C
	t	19	20	24	26	27	28	28	28	28	27	23	20	25 °C
	tn	8	9	13	17	19	22	23	23	23	18	13	9	8 °C
	clothing	s	s	t	t	t	t	t	t	t	t	t	s	–
	stress	o	o	f	f	ff	ff	ff	ff	ff	f	f	o	–
	p	15	46	131	313	493	594	547	488	378	207	44	7	3263 mm
	dp	2	4	8	14	20	23	26	24	18	10	2	1	153 days

Sagar

latitude: 23°51′n longitude: 78°45′e height: 551 m

GAwhb	044-27	Jan	Feb	Mar	Apr	May	Jun	Jul	Aug	Sep	Oct	Nov	Dec	Year
	t	18	21	26	30	34	31	26	25	26	25	22	19	25 °C
	clothing	s	s	t	t	t	t	t	t	t	t	s	s	–
	stress	o	o	f	ff	ff	ff	ff	ff	ff	ff	o	o	–
	p	30	13	10	4	8	145	463	418	228	44	24	7	1394 mm
	dp	3	2	2	1	1	10	20	20	13	3	1	1	76 days

Indore

latitude: 22°43′n longitude: 75°48′e height: 567 m

GAwhb	044-28	Jan	Feb	Mar	Apr	May	Jun	Jul	Aug	Sep	Oct	Nov	Dec	Year
	tx	30	33	38	41	42	40	34	32	33	34	32	29	43 °C
	t	18	20	24	29	32	30	26	25	25	24	20	18	24 °C
	tn	5	6	10	16	21	22	21	21	19	13	8	6	3 °C
	clothing	s	s	t	t	t	t	t	t	t	t	s	s	–
	stress	o	o	f	f	ff	ff	ff	ff	ff	f	o	o	–
	p	85	1	3	3	13	145	316	267	221	48	22	3	1127 mm
	dp	2	0	1	1	1	9	18	17	12	3	1	1	65 days

Calcutta

latitude: 22°32′n longitude: 88°20′e height: 6 m

Awhb	044-29	Jan	Feb	Mar	Apr	May	Jun	Jul	Aug	Sep	Oct	Nov	Dec	Year
	tx	29	32	37	39	39	37	34	33	34	33	31	28	40 °C
	t	20	23	28	30	31	30	29	29	29	28	24	21	27 °C
	tn	9	11	16	20	21	23	24	24	24	20	14	10	9 °C
	clothing	s	s	t	t	t	t	t	t	t	t	t	s	–
	stress	o	o	f	ff	ff	ff	ff	ff	ff	ff	f	o	–
	p	13	24	27	43	121	259	301	306	290	160	35	3	1582 mm
	dp	1	2	2	4	8	15	21	21	17	9	2	0	102 days
	sunshine	257	241	264	261	257	144	118	124	144	208	250	260	2528 hours
	%	76	75	71	69	63	36	28	31	39	58	75	78	57 %

Nagpur

latitude: 21°06′n longitude: 79°03′e height: 310 m

Awib	044-30	Jan	Feb	Mar	Apr	May	Jun	Jul	Aug	Sep	Oct	Nov	Dec	Year
	tx	32	36	40	43	46	44	35	34	35	35	33	31	46 °C
	t	21	24	28	32	36	32	28	27	27	26	22	20	27 °C
	tn	9	11	14	19	23	22	22	22	21	16	11	8	7 °C
	clothing	s	t	t	t	t	t	t	t	t	t	s	s	–
	stress	o	f	f	ff	ff	ff	ff	ff	ff	f	o	o	–
	p	14	19	22	20	13	210	407	288	173	65	17	3	1251 mm
	dp	2	2	2	3	2	12	20	16	12	4	1	0	77 days
	sunshine	288	283	291	285	301	189	100	110	160	260	283	298	2848 hours
	%	85	87	78	75	74	47	24	28	43	72	85	88	64 %

Cuttack

latitude: 20°48′n longitude: 85°56′e height: 27 m

Awhb	044-31	Jan	Feb	Mar	Apr	May	Jun	Jul	Aug	Sep	Oct	Nov	Dec	Year
	t	22	25	29	32	33	31	29	29	29	28	24	22	28 °C
	clothing	s	t	t	t	t	t	t	t	t	t	t	s	–
	stress	o	f	ff	ff	ff	ff	ff	ff	ff	f	f	o	–
	p	10	27	19	26	70	207	355	365	252	168	41	5	1545 mm
	dp	1	2	2	2	6	13	20	20	16	10	2	1	96 days

Akola

latitude: 20°42′n longitude: 77°02′e height: 282 m

Awhb	044-32	Jan	Feb	Mar	Apr	May	Jun	Jul	Aug	Sep	Oct	Nov	Dec	Year
	t	22	24	28	32	35	32	28	27	27	27	23	21	27 °C
	clothing	s	t	t	t	t	t	t	t	t	t	t	s	–
	stress	o	f	f	ff	ff	ff	ff	ff	ff	f	f	o	–
	p	9	8	7	7	11	146	261	170	178	46	28	6	877 mm
	dp	1	1	1	1	2	10	16	12	10	3	2	1	60 days

Jagdalpur

latitude: 19°05′n longitude: 82°02′e height: 553 m

GAwhb	044-33	Jan	Feb	Mar	Apr	May	Jun	Jul	Aug	Sep	Oct	Nov	Dec	Year
	tx	31	34	38	40	42	40	32	32	32	32	31	29	42 °C
	t	20	23	26	30	32	29	25	25	26	25	21	19	25 °C
	tn	7	10	13	18	20	21	21	21	20	15	9	6	6 °C
	clothing	s	t	t	t	t	t	t	t	t	t	s	s	–
	stress	o	f	f	ff	ff	ff	ff	ff	ff	f	o	o	–
	p	5	15	17	51	65	212	396	380	246	115	24	4	1530 mm
	dp	1	2	2	6	6	14	23	22	18	8	2	0	104 days

Bombay

latitude: 18°54′n longitude: 72°49′e height: 11 m

Awha	044-34	Jan	Feb	Mar	Apr	May	Jun	Jul	Aug	Sep	Oct	Nov	Dec	Year
	tx	31	32	33	34	34	34	31	31	31	34	33	32	35 °C
	t	24	25	27	29	30	29	28	27	27	28	28	26	27 °C
	tn	17	18	20	23	26	24	24	24	23	23	21	18	17 °C
	clothing	t	t	t	t	t	t	t	t	t	t	t	t	–
	stress	f	f	ff	ff	ff	ff	ff	ff	ff	ff	ff	f	–
	p	2	1	0	3	16	520	709	419	297	88	21	2	2078 mm
	dp	0	0	0	0	2	17	27	23	16	4	1	0	93 days
	sunshine	291	274	290	289	298	162	68	87	153	251	276	288	2727 hours
	%	84	84	78	77	74	41	17	22	42	69	82	84	61 %

Vishakhapatnam

latitude: 17°43′n longitude: 83°14′e height: 3 m

Awha	044-35	Jan	Feb	Mar	Apr	May	Jun	Jul	Aug	Sep	Oct	Nov	Dec	Year
	tx	29	31	33	34	37	38	36	35	34	33	31	29	39 °C
	t	23	25	28	32	32	31	29	30	29	28	26	24	28 °C
	tn	17	18	21	23	23	23	23	23	24	22	19	17	16 °C
	clothing	t	t	t	t	t	t	t	t	t	t	t	t	–
	stress	f	f	ff	ff	ff	ff	ff	ff	ff	ff	ff	f	–
	p	7	15	9	13	53	88	125	99	167	261	90	17	944 mm
	dp	1	1	1	2	3	7	12	10	12	11	4	1	65 days

Masulipatnam

latitude: 16°11′n longitude: 81°08′e height: 3 m

Awha	044-36	Jan	Feb	Mar	Apr	May	Jun	Jul	Aug	Sep	Oct	Nov	Dec	Year
	t	24	25	27	30	33	32	29	29	29	28	25	23	28 °C
	clothing	t	t	t	t	t	t	t	t	t	t	t	t	–
	stress	f	f	f	ff	ff	ff	ff	ff	ff	ff	f	f	–
	p	1	11	9	18	36	106	199	152	156	264	110	15	1077 mm
	dp	0	1	1	1	3	9	15	12	12	10	5	1	70 days

Belgaum

latitude: 15°51′n longitude: 74°32′e height: 753 m

GAwhb	044-37	Jan	Feb	Mar	Apr	May	Jun	Jul	Aug	Sep	Oct	Nov	Dec	Year
	t	22	24	26	28	27	24	22	22	23	24	23	22	24 °C
	clothing	s	t	t	t	t	t	t	t	t	t	t	s	–
	stress	o	f	f	ff	ff	f	f	f	f	f	f	o	–
	p	1	2	12	57	87	211	505	284	120	168	38	7	1492 mm
	dp	0	0	1	5	7	17	26	24	14	11	4	1	110 days

Goa-Panjim

Amha	044-38	Jan	Feb	Mar	Apr	May	Jun	Jul	Aug	Sep	Oct	Nov	Dec	Year
latitude: 15°23′n	longitude: 73°49′e	height: 48 m												
	t	24	25	27	30	32	32	29	29	29	28	25	23	28 °C
	clothing	t	t	t	t	t	t	t	t	t	t	t	t	–
	stress	f	f	f	ff	ff	ff	ff	ff	ff	ff	f	f	–
	p	2	0	1	18	87	869	923	456	253	119	36	3	2767 mm
	dp	0	0	1	5	7	17	26	24	14	11	4	1	110 days

Madras

Awha	044-39	Jan	Feb	Mar	Apr	May	Jun	Jul	Aug	Sep	Oct	Nov	Dec	Year
latitude: 13°00′n	longitude: 80°11′e	height: 16 m												
	tx	31	33	35	37	42	41	39	38	37	36	32	30	42 °C
	t	24	26	28	30	33	32	31	30	30	28	26	25	29 °C
	tn	17	17	19	23	24	24	23	23	23	22	19	18	16 °C
	clothing	t	t	t	t	t	t	t	t	t	t	t	t	–
	stress	f	f	ff	ff	ff	ff	ff	ff	ff	ff	ff	f	–
	p	24	7	15	25	52	53	83	124	118	267	308	157	1233 mm
	dp	2	1	1	2	2	7	10	11	9	12	11	6	73 days
	sunshine	271	274	304	290	282	203	157	193	197	206	196	226	2799 hours
	%	76	83	81	78	71	52	39	50	54	56	57	64	63 %

Mangalore

Amha	044-40	Jan	Feb	Mar	Apr	May	Jun	Jul	Aug	Sep	Oct	Nov	Dec	Year
latitude: 12°52′n	longitude: 74°51′e	height: 22 m												
	tx	34	34	34	34	34	32	31	30	30	32	33	34	35 °C
	t	26	27	28	29	29	27	26	26	26	27	27	27	27 °C
	tn	19	20	22	23	23	22	22	23	22	22	20	19	18 °C
	clothing	t	t	t	t	t	t	t	t	t	t	t	t	–
	stress	ff	ff	ff	ff	ff	ff	ff	ff	ff	ff	ff	ff	–
	p	5	2	9	40	233	980	1059	577	279	206	71	18	3479 mm
	dp	0	0	1	3	10	26	30	26	20	13	6	1	137 days

Bangalore

GAwab	044-41	Jan	Feb	Mar	Apr	May	Jun	Jul	Aug	Sep	Oct	Nov	Dec	Year
latitude: 12°58′n	longitude: 77°35′e	height: 921 m												
	tx	31	33	36	37	37	34	32	31	32	31	30	29	37 °C
	t	21	23	26	27	27	24	23	23	23	23	22	20	24 °C
	tn	12	13	15	18	18	18	18	18	17	17	13	12	12 °C
	clothing	s	t	t	t	t	t	t	t	t	t	s	s	–
	stress	o	f	f	f	f	f	f	f	f	f	o	o	–
	p	3	10	6	46	117	80	117	147	143	185	54	16	924 mm
	dp	1	1	1	4	9	9	13	14	11	11	5	2	81 days
	sunshine	257	252	282	255	242	148	98	130	150	178	189	223	2405 hours
	%	72	76	75	68	61	38	25	33	41	48	55	63	54 %

Fort Cochin

Amha	044-42	Jan	Feb	Mar	Apr	May	Jun	Jul	Aug	Sep	Oct	Nov	Dec	Year
latitude: 9°58′n	longitude: 76°14′e	height: 3 m												
	t	27	28	28	29	28	26	26	26	26	27	27	27	27 °C
	clothing	t	t	t	t	t	t	t	t	t	t	t	t	–
	stress	f	f	ff	ff	ff	ff	ff	ff	ff	ff	f	f	–
	p	10	34	50	145	364	756	572	386	235	333	184	37	3106 mm
	dp	1	2	4	9	15	27	26	23	17	17	11	3	156 days

Pamban

Awha	044-43	Jan	Feb	Mar	Apr	May	Jun	Jul	Aug	Sep	Oct	Nov	Dec	Year
				latitude: 9°16′n				longitude: 79°18′e				height: 11 m		
	tx	29	31	33	34	34	34	33	33	33	33	31	29	35 °C
	t	26	27	28	30	30	30	29	29	29	28	27	26	28 °C
	tn	22	22	23	24	25	25	25	25	24	23	23	22	22 °C
	clothing	t	t	t	t	t	t	t	t	t	t	t	t	–
	stress	ff	ff	ff	ff	ff	ff	ff	ff	ff	ff	ff	ff	–
	p	66	19	24	68	24	5	9	12	16	175	306	196	922 mm
	dp	5	2	2	4	2	1	1	1	2	9	13	10	51 days

Trivandrum

Awha	044-44	Jan	Feb	Mar	Apr	May	Jun	Jul	Aug	Sep	Oct	Nov	Dec	Year
				latitude: 8°29′n				longitude: 76°57′e				height: 64 m		
	tx	30	31	32	33	32	31	29	29	24	24	24	24	33 °C
	t	27	27	28	28	28	26	26	26	27	27	27	27	27 °C
	tn	20	21	23	24	23	22	22	23	23	22	22	21	19 °C
	clothing	t	t	t	t	t	t	t	t	t	t	t	t	–
	stress	ff	ff	ff	ff	ff	ff	ff	ff	ff	ff	ff	ff	–
	p	19	21	44	122	249	331	211	164	123	271	207	73	1835 mm
	dp	2	2	4	9	13	22	19	15	12	14	14	6	131 days
	sunshine	264	249	260	210	195	120	133	177	195	171	171	236	2381 hours
	%	73	74	69	57	50	32	34	46	53	46	49	65	54 %

4.2.44.2. Predominant general weather situations are described in 4.2.0.1. under India. Hygrothermal diagrams of Bombay, Calcutta and New Delhi are to be found in 4.2.0.2.
Wind names: Elephanta is a south-south-easterly wind with heavy rain and thundersqualls occurring in September and October on the Malabar coast.
Kal-baishakhi is a thunderstorm at the beginning of the rainy or monsoon season which starts in May/June and ends in September/October.
Nor'-westers are heavy thunderstorms in the overheated lowland plain of the Ganges river and at the mouths of the Ganges and Brahmapootra at the beginning of the rainy or monsoon season.
Peesash is a dust whirl.

4.2.45. Sri Lanka

4.2.45.1. Climatic description and tables

1. Tropical Rain Climate Ar *and* Tropical Rain Mountain Climate GAw, locally Tropical Winter Rain Climate As
2. Very hot all the year round.
3. Generally dry and sunny in the morning, in the afternoon often showers and cloudy, yet now and then periods of rainy days.
4. Wind: occasional gales and squalls in thunderstorms.

Mannar

Asha	045-1	Jan	Feb	Mar	Apr	May	Jun	Jul	Aug	Sep	Oct	Nov	Dec	Year
				latitude: 8°59′n				longitude: 79°55′e				height: 3 m		
	t	26	27	28	29	30	29	28	28	28	28	27	26	28 °C
	clothing	t	t	t	t	t	t	t	t	t	t	t	t	–
	stress	ff	ff	ff	ff	ff	ff	ff	ff	ff	ff	ff	ff	–
	p	87	34	44	77	44	5	7	16	23	168	242	202	949 mm

Trincomalee

Arha	045-2	Jan	Feb	Mar	Apr	May	Jun	Jul	Aug	Sep	Oct	Nov	Dec	Year
				latitude: 8°35′n				longitude: 81°15′e				height: 3 m		
	tx	28	30	32	35	36	35	36	36	36	34	32	29	36 °C
	t	26	26	27	29	30	30	30	29	29	28	26	26	28 °C
	tn	21	21	22	23	23	24	23	23	23	22	22	22	21 °C
	clothing	t	t	t	t	t	t	t	t	t	t	t	t	–
	stress	ff	ff	ff	ff	ff	ff	ff	ff	ff	ff	ff	ff	–
	p	211	95	48	77	68	18	54	103	89	235	355	374	1727 mm
	dp	10	4	4	5	5	2	3	6	6	13	17	16	91 days
	sunshine	202	260	279	264	242	216	232	220	234	189	189	183	2710 hours
	%	56	78	75	71	62	57	59	57	64	51	54	51	61 %

Kandy

Araa	045-3	Jan	Feb	Mar	Apr	May	Jun	Jul	Aug	Sep	Oct	Nov	Dec	Year
				latitude: 7°27′n				longitude: 80°38′e				height: 458 m		
	t	23	24	25	26	26	25	24	24	24	25	24	23	24 °C
	clothing	t	t	t	t	t	t	t	t	t	t	t	t	–
	stress	f	f	f	f	f	f	f	f	f	f	f	f	–
	p	170	61	137	152	165	185	150	119	152	295	272	226	2084 mm
0.2 mm	dp	11	4	9	12	11	18	17	14	14	18	18	13	159 days
	sunshine	189	198	214	189	178	123	136	146	165	158	141	170	2007 hours
	%	52	59	57	51	46	33	35	38	45	43	40	47	45 %

Nuwara Eliya

GArll	045-4	Jan	Feb	Mar	Apr	May	Jun	Jul	Aug	Sep	Oct	Nov	Dec	Year
				latitude: 6°58′n				longitude: 80°46′e				height: 1881 m		
	tx	23	23	23	24	24	23	22	23	23	22	23	23	24 °C
	t	14	14	15	16	16	16	16	16	16	16	15	14	15 °C
	tn	1	1	2	4	6	10	9	9	7	7	6	1	-1 °C
	clothing	n	n	n	n	n	n	n	n	n	n	n	n	–
	stress													–
	p	170	43	109	119	175	277	300	198	226	269	241	203	2332 mm
2.5 mm	dp	13	6	11	15	18	25	25	22	20	22	22	17	216 days

Colombo

Arha	045-5	Jan	Feb	Mar	Apr	May	Jun	Jul	Aug	Sep	Oct	Nov	Dec	Year
	latitude: 6°54'n longitude: 79°52'e height: 6 m													
	tx	32	33	33	32	32	30	30	30	31	31	31	32	33 °C
	t	26	26	27	28	28	27	27	27	27	27	26	26	27 °C
	tn	19	19	21	22	22	23	23	23	23	22	21	20	18 °C
	clothing	t	t	t	t	t	t	t	t	t	t	t	t	–
	stress	ff	ff	ff	ff	ff	ff	ff	ff	ff	ff	ff	ff	–
	p	88	96	118	260	353	212	140	124	153	354	324	175	2397 mm
	dp	7	6	8	14	19	18	12	11	13	19	16	10	152 days
	sunshine	245	254	251	216	198	162	189	195	186	202	192	242	2532 hours
	%	67	75	67	59	51	43	49	51	51	54	54	67	57 %

Hombantota

Asha	045-6	Jan	Feb	Mar	Apr	May	Jun	Jul	Aug	Sep	Oct	Nov	Dec	Year
	latitude: 6°07'n longitude: 81°08'e height: 20 m													
	t	26	26	27	28	28	28	28	28	27	27	27	26	27 °C
	clothing	t	t	t	t	t	t	t	t	t	t	t	t	–
	stress	ff	ff	ff	ff	ff	ff	ff	ff	ff	ff	ff	ff	–
	p	101	58	66	109	121	55	43	42	45	126	187	121	1074 mm
	sunshine	195	209	229	207	174	111	133	130	153	167	165	158	2032 hours
	%	53	62	61	56	45	30	34	34	42	45	46	43	46 %

4.2.45.2. Predominant general weather situations are described in 4.2.0.1. under India. A hygrothermal diagram of Colombo is to be found in 4.2.0.2.

4.2.46. Mongolia

4.2.46.1. Climatic description and table

1. Northern part: Steppe Mountain Climate GBS.
   Southern part: Desert Mountain Climate GBW
2. Summer mild, winter very cold, at higher levels severely cold.
3. Summer: normally dry and sunny, in steppe repeatedly, in desert occasionally showers or rain and cloudy or overcast. Winter: mostly dry and cloudy.
4. Wind: occasional gales and dust-storms, in winter snow-storms. Snow cover: in winter.

Ulan Bator

GBSld	046-1	Jan	Feb	Mar	Apr	May	Jun	Jul	Aug	Sep	Oct	Nov	Dec	Year
	latitude: 47°56'n longitude: 106°59'e height: 1337 m													
	tx	-9	-3	9	21	26	32	31	29	25	18	7	-5	33 °C
	t	-26	-21	-13	-1	6	14	16	14	8	-1	-13	-22	-3 °C
	tn	-39	-38	-33	-19	-9	-1	4	2	-7	-19	-29	-38	-41 °C
	clothing	p	p	p	w	n	n	n	n	n	w	p	p	–
	stress													–
	p	1	1	2	5	10	28	76	51	23	5	5	2	209 mm
	dp	1	1	2	2	4	5	10	8	3	2	2	1	41 days

4.2.46.2. Predominant general weather situations are described in 4.2.0.1. under Northern Asia.

4.2.47. Nepal, Bhutan

4.2.47.1. Climatic description and table

1. Subtropical Summer Rain Climate Cw and Subtropical Summer Rain Mountain Climate GCw and Subtropical Summer Rain High Mountain Climate HCw
2. In the lowlands: summer hot, partly very hot; winter mild. At higher levels summer mild or cool, winter cold.
3. Summer: in the morning often dry and sunny, in the afternoon frequent showers, from June to October also periods of rainy days. Winter: generally dry and sunny; now and then scattered showers.
4. Wind: occasional gales, thunderstorms with squalls. Snow cover: in winter at higher levels.

Khatmandu

				latitude: 27°42'n			longitude: 85°20'e				height: 1334 m		
GCwal	047-1	Jan	Feb	Mar	Apr	May	Jun	Jul	Aug	Sep	Oct	Nov	Dec	Year
	tx	23	24	30	33	33	33	32	32	32	30	26	23	34 °C
	t	10	12	17	20	23	24	24	24	23	20	15	11	19 °C
	tn	-1	2	3	7	12	16	18	18	16	9	3	0	-1 °C
	clothing	n	n	s	s	t	t	t	t	t	s	n	n	–
	stress			o	o	f	f	f	f	f	o			–
	p	19	9	36	32	99	198	378	346	168	37	5	2	1328 mm
0.2 mm	dp	1	5	2	6	10	15	21	20	12	4	1	0	97 days
	sunshine	183	158	236	285	177	147	81	74	99	161	156	158	1915 hours
	%	56	50	64	74	42	35	19	18	27	45	48	49	43 %

4.2.47.2. Predominant general weather situations are described in 4.2.0.1. under India.

4.2.48. Islands in the Indian Ocean north of the equator

4.2.48.1. Climatic description and tables

1. Tropical Summer Rain Climate Aw or Tropical Rain Climate Ar
2. Hot or very hot all the time.
3. In the morning predominantly dry and sunny, in the afternoon – especially from May to October – frequent showers, occasionally periods of rainy days.
4. Wind: occasional gales, thunderstorms with squalls, tropical storms.

Table Island / Andaman Islands

				latitude: 14°11'n			longitude: 93°22'e				height: 27 m		
Awha	048-1	Jan	Feb	Mar	Apr	May	Jun	Jul	Aug	Sep	Oct	Nov	Dec	Year
	tx	29	31	32	33	33	31	31	31	31	31	31	30	34 °C
	t	26	26	27	29	29	28	27	27	27	28	27	26	27 °C
	tn	22	22	23	24	23	22	23	23	22	23	23	22	21 °C
	clothing	t	t	t	t	t	t	t	t	t	t	t	t	–
	stress	ff	ff	ff	ff	ff	ff	ff	ff	ff	ff	ff	ff	–
	p	20	3	5	15	221	384	343	292	287	165	145	61	1940 mm
2.5 mm	dp	1	0	0	1	11	17	18	16	16	10	7	2	99 days

Port Blair / Andaman Islands

Awha	048-2	Jan	Feb	Mar	Apr	May	Jun	Jul	Aug	Sep	Oct	Nov	Dec	Year
		latitude: 11°40′n				longitude: 92°43′e				height: 79 m				
	tx	31	31	32	33	33	31	31	30	31	31	31	30	33 °C
	t	27	26	28	29	28	27	26	27	26	27	27	26	27 °C
	tn	19	19	19	22	21	22	22	22	22	22	21	20	18 °C
	clothing	t	t	t	t	t	t	t	t	t	t	t	t	–
	stress	ff	ff	ff	ff	ff	ff	ff	ff	ff	ff	ff	ff	–
	p	40	22	16	65	396	596	472	448	439	360	217	105	3176 mm
	dp	2	2	2	7	17	25	24	23	22	19	13	8	165 days
	sunshine	276	272	289	260	160	101	119	133	133	180	225	257	2405 hours
	%	77	82	77	70	41	26	30	34	36	49	65	72	54 %

Amini Divi / Laccadive Islands

Awha	048-3	Jan	Feb	Mar	Apr	May	Jun	Jul	Aug	Sep	Oct	Nov	Dec	Year
		latitude: 11°07′n				longitude: 72°44′e				height: 4 m				
	tx	32	31	34	35	36	33	32	31	32	33	33	33	36 °C
	t	27	28	28	30	30	28	28	28	28	28	27	27	28 °C
	tn	21	22	23	24	24	23	23	23	23	22	22	21	20 °C
	clothing	t	t	t	t	t	t	t	t	t	t	t	t	–
	stress	ff	ff	ff	ff	ff	ff	ff	ff	ff	ff	ff	ff	–
	p	18	1	1	38	94	363	305	196	160	147	66	33	1422 mm
2.5 mm	dp	1	0	0	2	5	17	16	11	10	9	4	2	77 days

Minicoy / Laccadive Islands

Awha	048-4	Jan	Feb	Mar	Apr	May	Jun	Jul	Aug	Sep	Oct	Nov	Dec	Year
		latitude: 8°18′n				longitude: 73°00′e				height: 2 m				
	tx	31	31	32	34	34	32	31	31	31	32	32	31	34 °C
	t	26	27	28	29	29	28	27	27	27	27	27	26	27 °C
	tn	22	22	23	24	23	23	23	22	23	23	23	22	21 °C
	clothing	t	t	t	t	t	t	t	t	t	t	t	t	–
	stress	ff	ff	ff	ff	ff	ff	ff	ff	ff	ff	ff	ff	–
	p	35	25	16	52	200	293	212	200	144	185	141	76	1579 mm
	dp	3	2	2	4	12	21	19	16	14	13	10	5	121 days

Car Nicobar / Nicobar Islands

Arha	048-5	Jan	Feb	Mar	Apr	May	Jun	Jul	Aug	Sep	Oct	Nov	Dec	Year
		latitude: 9°15′n				longitude: 92°48′e				height: 8 m				
	tx	31	32	33	34	33	32	32	32	32	32	32	31	34 °C
	t	28	28	28	28	28	28	28	28	27	27	27	27	28 °C
	tn	22	22	23	23	23	23	23	23	23	23	23	22	21 °C
	clothing	t	t	t	t	t	t	t	t	t	t	t	t	–
	stress	ff	ff	ff	ff	ff	ff	ff	ff	ff	ff	ff	ff	–
	p	99	30	53	89	317	315	236	259	328	295	290	198	2509 mm
2.5 mm	dp	4	2	3	5	14	15	12	13	15	14	13	9	119 days

4.2.48.2. Predominant general weather situations are described in 4.2.0.1. under India.

4.2.49. Hong Kong

4.2.49.1. Climatic description and table

1. Subtropical Summer Rain Climate Cw
2. Summer very hot, winter warm.
3. In the morning predominantly dry and sunny; in the afternoon during the high sun period often showers, occasionally periods of rainy days; during the low sun period generally dry and sunny or cloudy, scattered showers or rain.
4. Wind: occasional gales, at times typhoons.

Hong Kong

				latitude: 22°18'n				longitude: 114°10'e			height: 33 m			
Cwhl	049-1	Jan	Feb	Mar	Apr	May	Jun	Jul	Aug	Sep	Oct	Nov	Dec	Year
	tx	23	23	27	29	32	32	33	33	33	31	28	24	34 °C
	t	15	16	18	22	26	28	28	28	27	25	21	17	23 °C
	tn	7	8	11	15	20	23	23	23	22	18	13	9	6 °C
	clothing	n	n	s	s	t	t	t	t	t	t	s	s	–
	stress			o	f	ff	ff	ff	ff	ff	f	o	o	–
	p	30	60	70	133	332	479	286	415	364	33	46	17	2265 mm
0.1 mm	dp	4	5	6	8	13	17	15	15	13	5	3	3	107 days
	sunshine	147	113	93	122	160	156	234	200	194	232	202	180	2030 hours
	%	43	35	25	32	39	39	56	50	53	64	61	54	46 %

4.2.49.2. Predominant general weather situations are described in 4.2.0.1. under China. A hygrothermal diagram of Hong Kong is to be found in 4.2.0.2.

4.2.50. Macao

4.2.50.1. Climatic description and table

1. Subtropical Summer Rain Climate Cw
2. Summer very hot, winter mild.
3. In the morning predominantly dry and sunny; in the afternoon during the high sun period often showers, occasionally periods with rainy days; during the low sun period generally dry and sunny or cloudy, scattered showers or rain.
4. Wind: occasional gales, at times typhoons.

Macao

				latitude: 22°12'n				longitude: 113°33'e			height: 55 m			
Cwhl	050-1	Jan	Feb	Mar	Apr	May	Jun	Jul	Aug	Sep	Oct	Nov	Dec	Year
	t	15	15	18	22	26	28	28	28	28	24	21	16	22 °C
	clothing	n	n	s	s	t	t	t	t	t	t	s	n	–
	stress			o	f	ff	ff	ff	ff	ff	f	o		–
	p	26	54	83	156	263	346	292	293	205	57	43	28	1846 mm
	dp	4	6	8	9	13	16	16	15	11	5	3	3	109 days
	sunshine	120	83	62	91	154	164	234	202	188	213	185	156	1852 hours
	%	35	26	17	24	38	41	56	50	51	59	56	47	42 %

4.2.50.2. Predominant general weather situations are described in 4.2.0.1. under China.

4.2.51. North Korea

4.2.51.1. Climatic description and tables

1. Temperate Continental Climate DC and
   Temperate Continental Mountain Climate GDC
2. Summer warm or hot, winter cold.
3. Summer: variable weather prevails, repeatedly dry and sunny; winter normally dry and sunny or cloudy, at times variable weather.
4. Wind: occasional gales. Snow cover: in winter. Fog: occasionally.

Unggi

latitude: 42° 19′n longitude: 130° 24′e height: 89 m

DCbo	051-1	Jan	Feb	Mar	Apr	May	Jun	Jul	Aug	Sep	Oct	Nov	Dec	Year
	t	-9	-7	-1	5	10	14	19	21	17	10	1	-6	6 °C
	clothing	w	w	w	n	n	n	s	s	s	n	w	w	–
	stress								o					–
	p	5	7	22	29	72	102	117	202	120	61	23	9	769 mm
	dp	1	1	3	4	9	11	12	12	8	4	3	2	70 days
	sunshine	205	212	232	222	198	168	158	177	213	220	174	180	2359 hours
	%	70	71	63	55	44	37	34	41	57	64	59	64	53 %

Sinuiju

latitude: 40° 06′n longitude: 124° 23′e height: 8 m

DCao	051-2	Jan	Feb	Mar	Apr	May	Jun	Jul	Aug	Sep	Oct	Nov	Dec	Year
	t	-9	-6	1	9	15	20	24	24	19	12	3	-6	9 °C
	clothing	w	w	w	n	n	s	t	t	s	n	w	w	–
	stress						o	f	f	o				–
	p	8	8	29	52	87	115	265	280	105	53	31	19	1052 mm
	dp	2	2	5	5	7	9	12	11	7	6	5	3	75 days
	sunshine	208	218	236	246	251	231	183	202	237	239	192	189	2630 hours
	%	70	72	64	62	56	51	40	47	63	69	64	65	59 %

Wonsan

latitude: 39° 11′n longitude: 127° 26′e height: 37 m

DCao	051-3	Jan	Feb	Mar	Apr	May	Jun	Jul	Aug	Sep	Oct	Nov	Dec	Year
	tx	8	9	17	26	31	33	35	34	30	25	19	12	36 °C
	t	-4	-2	2	10	15	19	23	23	19	13	6	-1	10 °C
	tn	-16	-14	-8	-2	4	10	15	16	8	2	-7	-13	-17 °C
	clothing	w	w	w	n	n	s	s	s	s	n	n	w	–
	stress						o	f	f	o				–
	p	29	32	46	68	86	126	273	312	178	70	61	29	1310 mm
	dp	3	4	5	5	7	9	14	13	9	5	5	5	82 days
	sunshine	205	206	232	234	236	207	174	177	198	223	192	192	2475 hours
	%	68	68	63	59	53	46	38	42	53	64	64	66	55 %

4.2.51.2. Predominant general weather situations are described in 4.2.0.1. under Korea and Japan.

4.2.52. South Korea

4.2.52.1. Climatic description and tables

1. Temperate Continental Climate DC and Temperate Continental Mountain Climate GDC, in the southern part: Subtropical Rain Climate Cr
2. Summer hot, winter cold or cool.
3. Variable weather prevails, in summer more than in winter, yet repeatedly dry and sunny.
4. Wind: frequent gales in coastal areas. Snow cover: in winter, not permanent. Fog: occasionally.

Kangnung

latitude: 37°45'n longitude: 128°54'e height: 27 m

DCao	052-1	Jan	Feb	Mar	Apr	May	Jun	Jul	Aug	Sep	Oct	Nov	Dec	Year
	tx	12	13	21	27	31	33	36	35	30	28	22	14	36 °C
	t	-1	0	5	12	17	20	24	24	20	14	9	2	12 °C
	tn	-14	-12	-7	-1	6	9	16	16	10	3	-3	-11	-14 °C
	clothing	w	w	n	n	s	s	t	t	s	n	n	w	–
	stress					o	o	f	f	o				–
	p	37	73	73	70	64	135	212	191	198	88	88	53	1282 mm
	dp	3	5	6	6	6	9	12	10	8	5	5	5	80 days
	sunshine	209	189	212	228	252	193	162	182	181	217	185	189	2398 hours
	%	69	62	57	58	57	44	36	43	48	62	61	64	54 %

Seoul

latitude: 37°34'n longitude: 126°58'e height: 86 m

DCao	052-2	Jan	Feb	Mar	Apr	May	Jun	Jul	Aug	Sep	Oct	Nov	Dec	Year
	tx	8	11	17	24	29	33	34	35	30	26	18	12	36 °C
	t	-5	-2	4	10	16	21	24	25	20	13	6	-1	11 °C
	tn	-19	-16	-9	-1	6	12	18	17	8	0	-8	-16	-20 °C
	clothing	w	w	w	n	n	s	t	t	s	n	n	w	–
	stress						o	f	ff	o				–
	p	17	21	56	68	86	169	358	224	142	49	36	32	1258 mm
	dp	3	3	6	6	7	9	14	10	7	5	5	5	80 days
	sunshine	195	197	219	238	262	220	189	219	218	244	190	176	2567 hours
	%	64	64	59	60	60	50	42	52	58	70	62	59	58 %

Pusan

latitude: 35°06'n longitude: 129°01'e height: 71 m

Crak	052-3	Jan	Feb	Mar	Apr	May	Jun	Jul	Aug	Sep	Oct	Nov	Dec	Year
	tx	13	14	18	22	26	28	32	33	29	26	21	16	33 °C
	t	2	4	7	12	17	20	24	25	22	17	11	5	14 °C
	tn	-9	-8	-3	3	8	12	17	19	13	6	-1	-7	-10 °C
	clothing	w	w	n	n	n	s	t	t	s	n	n	n	–
	stress						o	f	ff	o				–
	p	25	44	89	114	139	198	248	165	205	73	44	39	1382 mm
	dp	5	3	7	8	7	10	10	8	8	5	4	3	78 days
	sunshine	205	190	213	219	241	196	183	232	182	218	196	198	2473 hours
	%	66	62	58	56	55	45	41	55	49	62	63	65	56 %

Mokpo

latitude: 34°47′n longitude: 126°23′e height: 33 m

Crak	052-4	Jan	Feb	Mar	Apr	May	Jun	Jul	Aug	Sep	Oct	Nov	Dec	Year
	tx	13	13	18	23	27	29	33	34	31	26	21	16	34 °C
	t	1	2	6	12	17	21	25	26	22	16	10	4	13 °C
	tn	-8	-7	-3	2	8	14	18	19	13	7	-1	-6	-9 °C
	clothing	w	w	n	n	s	s	t	t	s	n	n	w	–
	stress					o	o	ff	ff	o				–
	p	37	40	59	83	102	136	183	188	156	55	44	43	1126 mm
	dp	8	6	6	6	6	8	9	9	8	5	6	7	84 days
	sunshine	137	152	199	209	233	199	182	228	190	218	168	138	2253 hours
	%	44	49	54	53	54	46	41	55	51	62	54	45	51 %

Cheju

latitude: 33°31′n longitude: 126°32′e height: 22 m

Crak	052-5	Jan	Feb	Mar	Apr	May	Jun	Jul	Aug	Sep	Oct	Nov	Dec	Year
	t	5	5	8	12	16	20	25	26	22	17	12	8	15 °C
	clothing	n	n	n	n	n	s	t	t	s	s	n	n	–
	stress						o	f	ff	o				–
	p	59	76	73	82	89	158	210	227	250	88	69	60	1441 mm
	dp	10	9	8	8	8	8	9	9	10	6	7	9	100 days
	sunshine	74	113	174	204	226	198	226	248	177	198	132	87	2058 hours
	%	23	36	47	52	52	46	51	60	48	56	42	28	46 %

4.2.52.2. Predominant general weather situations are described in 4.2.0.1. under Korea and Japan. A hygrothermal diagram of Seoul is to be found in 4.2.0.2.

4.2.53. China

4.2.53.1. Climatic descriptions and tables

1. Climatic area: South-western Inner Mongolia, northern Suiyuan, Ningsia, western Kansu, Sinkian (except Tienshan Mountains)

1.1. Desert Climate BW and Desert Mountain Climate GBW

1.2. Summer: at low levels warm or hot; winter: cool, at higher levels cold or very cold.

1.3. Summer: normally dry and sunny, occasional showers or rain. Winter: generally dry and sunny or cloudy, at times overcast.

1.4. Wind: occasional gales and dust storms.

Kashgar

latitude: 39°24′n longitude: 76°07′e height: 1309 m

GBWao	053-1	Jan	Feb	Mar	Apr	May	Jun	Jul	Aug	Sep	Oct	Nov	Dec	Year
	tx	4	13	22	29	34	36	39	37	33	27	19	9	39 °C
	t	-2	0	8	15	21	25	27	24	21	14	5	-2	13 °C
	tn	-17	-12	-4	3	7	12	15	13	7	0	-8	-15	-18 °C
	clothing	w	w	n	n	s	t	t	t	s	n	n	w	–
	stress					o	f	f	f	o				–
	p	15	3	13	5	8	5	10	8	2	2	5	8	83 mm
2.5 mm	dp	1	0	1	1	1	1	1	2	1	0	1	1	11 days

2. Climatic area: Western Inner Mongolia (except south-western part), parts of the Northern Region, Tienshan Mountains, eastern Kansu, Tsinghai, northern Sikiang, Tibet (except southernmost part)

2.1. Mostly Steppe Climate BS, Steppe Mountain Climate GBS

2.2. Summer: at low levels, warm or hot; winter: cool, at higher levels cold or very cold.

2.3. Summer: variable weather prevails; winter: generally dry and sunny or cloudy.

2.4. Wind: occasional gales and dust-storms.

Urumchi

latitude: 43°47′n longitude: 87°37′e height: 913 m

GDCac	053-2	Jan	Feb	Mar	Apr	May	Jun	Jul	Aug	Sep	Oct	Nov	Dec	Year
	tx	-1	0	11	24	31	33	34	34	28	20	12	3	35 °C
	t	-16	-14	-4	8	18	22	24	22	17	6	-6	-13	5 °C
	tn	-29	-29	-23	-8	-2	7	10	9	3	-13	-22	-28	-30 °C
	clothing	p	p	w	n	s	s	t	s	n	n	w	p	–
	stress				o	o	f	o						–
	p	9	15	15	33	25	33	16	35	15	47	22	11	276 mm
0.1 mm	dp	10	11	7	8	5	6	7	5	4	9	12	11	95 days
	sunshine	186	150	195	204	298	270	279	270	264	214	144	146	2619 hours
	%	65	51	53	51	65	58	59	62	70	62	49	53	59 %

Paotou

latitude: 40°34′n longitude: 109°50′e height: 1044 m

GBSac	053-3	Jan	Feb	Mar	Apr	May	Jun	Jul	Aug	Sep	Oct	Nov	Dec	Year
	t	-12	-9	0	7	16	21	23	21	14	5	-2	-10	6 °C
	clothing	p	w	w	n	n	s	t	s	n	n	w	p	–
	stress						o	f	o					–
	p	0	4	4	19	31	29	82	76	31	23	4	0	303 mm
0.1 mm	dp	1	4	3	5	6	7	10	10	6	4	3	2	61 days
	sunshine	226	212	257	276	285	288	282	254	270	251	183	211	2996 hours
	%	76	70	70	69	64	64	61	59	72	72	61	73	67 %

Tientsin

latitude: 39°06′n longitude: 117°10′e height: 3 m

BSao	053-4	Jan	Feb	Mar	Apr	May	Jun	Jul	Aug	Sep	Oct	Nov	Dec	Year
	t	-4	-2	5	13	20	24	27	26	21	14	5	-2	12 °C
	clothing	w	w	n	n	s	t	t	t	s	n	n	w	–
	stress					o	f	f	f	o				–
	p	3	4	9	17	28	63	177	149	45	15	13	5	528 mm
0.1 mm	dp	2	2	3	4	6	9	13	11	6	4	3	2	65 days
	sunshine	183	189	232	246	285	273	254	232	234	242	192	177	2740 hours
	%	61	62	63	62	64	61	56	54	62	70	64	60	61 %

Hsinging

latitude: 36°35′n longitude: 101°55′e height: 2244 m

GBSbo	053-5	Jan	Feb	Mar	Apr	May	Jun	Jul	Aug	Sep	Oct	Nov	Dec	Year
	t	-6	-3	3	9	13	16	18	18	13	8	0	-5	7 °C
	clothing	w	w	w	n	n	n	s	s	n	n	w	w	–
	stress							o	o					–
	p	1	2	5	18	34	46	73	92	74	27	4	2	378 mm
0.1 mm	dp	2	2	3	4	8	12	12	12	12	8	2	1	78 days
	sunshine	214	203	195	213	245	243	239	232	189	211	210	226	2621 hours
	%	69	66	53	54	56	55	53	55	51	60	68	75	59 %

Lhasa (Tibet)

latitude: 29°40′n longitude: 91°07′e height: 3685 m

HBSlo	053-6	Jan	Feb	Mar	Apr	May	Jun	Jul	Aug	Sep	Oct	Nov	Dec	Year
	tx	12	16	18	22	25	28	27	26	24	22	17	14	28 °C
	t	-2	1	5	8	12	17	16	16	14	9	4	0	8 °C
	tn	-14	-12	-18	-4	-1	4	6	6	4	-5	-9	-13	-15 °C
	clothing	w	w	n	n	n	n	n	n	n	n	w	w	–
	stress													–
	p	0	13	8	5	25	63	122	89	66	13	2	0	406 mm
2.5 mm	dp	0.1	0.4	1	0.4	3	8	13	10	7	2	0.1	0	45 days

3. *Climatic area:* North-eastern Region, Eastern Mongolia, parts of the Northern Region, Shensi, north-western Honan, Shantung

3.1. Temperate Continental Climate DC and
Temperate Continental Mountain Climate GDC

3.2. Summer hot, winter in the north very cold, in the south cold or cool.

3.3. Summer: variable weather prevails; however, dry and sunny days are not lacking; winter: at times variable weather, more often dry and sunny days or cloudy days.

3.4. Wind: occasional gales, at times typhoons in coastal area.

Harbin

latitude: 45°45′n longitude: 126°38′e height: 143 m

DCac	053-7	Jan	Feb	Mar	Apr	May	Jun	Jul	Aug	Sep	Oct	Nov	Dec	Year
	tx	-2	2	13	24	29	33	33	32	28	22	11	2	33 °C
	t	-20	-16	-6	6	14	20	23	22	14	6	-7	-17	3 °C
	tn	-34	-31	-23	-9	-1	7	12	8	1	-11	-21	-30	-35 °C
	clothing	p	p	w	n	n	s	t	s	n	n	w	p	–
	stress						o	f	o					–
	p	4	6	17	23	44	92	167	119	52	36	12	5	577 mm
	dp	1	1	3	3	7	9	10	11	7	5	2	2	61 days
	sunshine	148	220	248	234	257	258	267	254	216	214	183	177	2726 hours
	%	70	75	67	58	56	55	56	58	57	63	64	66	61 %

Ch'ang Ch'un

latitude: 43°55′n longitude: 125°18′e height: 216 m

DCac	053-8	Jan	Feb	Mar	Apr	May	Jun	Jul	Aug	Sep	Oct	Nov	Dec	Year
	tx	-1	2	14	23	31	33	33	33	30	24	15	3	34 °C
	t	-17	-13	-4	7	14	20	24	22	15	7	-4	-14	5 °C
	tn	-31	-27	-20	-8	0	8	12	11	2	-8	-19	-26	-31 °C
	clothing	p	p	w	n	n	s	t	s	n	n	w	p	–
	stress						o	f	o					–
	p	8	8	15	20	51	102	183	135	61	38	15	8	644 mm
0.1 mm	dp	6	5	6	7	11	14	16	12	11	8	6	6	108 days

Mukden

latitude: 41°48′n longitude: 123°23′e height: 43 m

DCac	053-9	Jan	Feb	Mar	Apr	May	Jun	Jul	Aug	Sep	Oct	Nov	Dec	Year
	tx	3	7	17	26	31	36	35	34	31	25	17	6	37 °C
	t	-12	-8	0	9	17	22	26	24	18	9	1	-9	8 °C
	tn	-26	-23	-16	-6	3	9	16	14	3	-4	-16	-24	-28 °C
	clothing	p	w	w	n	n	s	t	t	s	n	w	w	–
	stress						o	f	f	o				–
	p	8	8	18	28	67	84	183	170	64	36	28	15	709 mm
	dp	2	2	3	5	7	8	11	8	7	5	4	3	65 days

Peking

latitude: 39°57'n longitude: 116°19'e height: 52 m

DCao	053-10	Jan	Feb	Mar	Apr	May	Jun	Jul	Aug	Sep	Oct	Nov	Dec	Year
	t	-5	-2	5	14	20	25	26	26	20	13	4	-3	12 °C
	clothing	w	w	n	n	s	t	t	t	s	n	w	w	–
	stress					o	f	ff	f	o				–
	p	4	5	8	17	35	78	243	141	58	16	11	3	619 mm
0.1 mm	dp	3	3	3	4	6	8	13	11	7	3	3	2	66 days
	sunshine	205	198	236	240	267	261	220	223	228	245	192	192	2706 hours
	%	69	66	64	60	60	58	48	52	61	71	64	66	61 %

Dairen

latitude: 38°54'n longitude: 121°54'e height: 96 m

DCao	053-11	Jan	Feb	Mar	Apr	May	Jun	Jul	Aug	Sep	Oct	Nov	Dec	Year
	tx	7	8	15	22	27	31	32	32	28	24	17	10	33 °C
	t	-5	-4	2	9	16	20	24	24	20	14	5	-2	10 °C
	tn	-16	-14	-8	-1	7	13	17	18	12	2	-6	-13	-17 °C
	clothing	w	w	w	n	n	s	t	t	s	n	n	w	–
	stress						o	f	f	o				–
	p	10	7	15	25	46	49	155	133	66	33	25	14	578 mm
	dp	1	1	2	3	4	5	8	8	5	4	3	4	48 days
	sunshine	192	201	242	255	282	276	232	239	243	242	183	174	2762 hours
	%	64	66	65	64	64	62	51	56	65	69	60	59	62 %

Tsingtao

latitude: 36°04'n longitude: 120°19'e height: 79 m

DCao	053-12	Jan	Feb	Mar	Apr	May	Jun	Jul	Aug	Sep	Oct	Nov	Dec	Year
	tx	8	9	16	21	27	29	31	32	29	25	19	12	33 °C
	t	-1	1	4	10	16	20	24	26	21	16	8	1	12 °C
	tn	-17	-9	-6	2	8	13	18	18	12	6	-3	-9	-18 °C
	clothing	w	w	w	n	n	s	t	t	s	n	n	w	–
	stress						o	f	f	o				–
	p	10	10	20	38	41	84	155	147	84	33	20	15	657 mm
2.5 mm	dp	4	3	4	6	7	8	13	11	8	4	4	4	76 days
	sunshine	183	175	211	225	248	216	186	220	213	229	186	170	2463 hours
	%	60	57	57	57	57	49	42	52	57	65	60	56	55 %

Sian

latitude: 34°15'n longitude: 108°35'e height: 412 m

DChk	053-13	Jan	Feb	Mar	Apr	May	Jun	Jul	Aug	Sep	Oct	Nov	Dec	Year
	tx	11	15	27	33	37	39	41	38	34	27	20	12	42 °C
	t	0	2	9	15	21	26	28	26	20	14	7	1	14 °C
	tn	-11	-8	-1	4	11	17	20	19	12	5	-2	-8	-13 °C
	clothing	w	w	n	n	s	t	t	t	s	n	n	w	–
	stress					o	f	ff	ff	o				–
	p	5	11	23	41	58	49	89	97	116	57	24	8	578 mm
0.1 mm	dp	3	5	7	8	8	8	11	11	12	8	6	4	91 days
	sunshine	108	107	121	138	170	186	217	198	132	136	111	112	1738 hours
	%	34	35	33	35	39	43	49	47	35	39	36	37	39 %

4. *Climatic area:* Honan (except north-western part), eastern Hupe, Hunan, Kuangsi, Kiangsi and Eastern Region (except Shantung)
4.1. Subtropical Rain Climate Cr and Subtropical Rain Mountain Climate GCr
4.2. Summer hot, in the south very hot; winter cool, in the south mild.
4.3. Variable weather prevails with more rainy days in summer than in winter; however, dry and sunny days are not lacking.
4.4. Wind: occasional gales, typhoons at times in coastal areas.

Nanking

latitude: 32°04′n longitude: 118°47′e height: 62 m

Crhk 053-14	Jan	Feb	Mar	Apr	May	Jun	Jul	Aug	Sep	Oct	Nov	Dec	Year
tx	13	18	27	29	33	36	38	37	34	29	24	18	38 °C
t	2	4	9	15	20	25	28	28	23	18	11	5	16 °C
tn	-9	-6	-2	3	10	16	21	21	13	7	-1	-6	-11 °C
clothing	w	w	n	n	s	t	t	t	t	s	n	n	–
stress					o	f	ff	ff	f	o			–
p	41	51	76	102	81	183	206	117	94	51	41	30	1073 mm
dp	8	7	6	9	7	8	8	8	5	3	4	8	81 days

Shanghai

latitude: 31°12′n longitude: 121°26′e height: 7 m

Crak 053-15	Jan	Feb	Mar	Apr	May	Jun	Jul	Aug	Sep	Oct	Nov	Dec	Year
tx	17	17	24	29	32	35	37	36	33	29	24	19	38 °C
t	3	4	8	14	19	23	27	27	23	18	12	6	15 °C
tn	-7	-5	-2	3	8	14	19	19	13	6	-1	-6	-8 °C
clothing	w	w	n	n	s	t	t	t	t	s	n	n	–
stress					o	f	ff	ff	f	o			–
p	49	62	85	91	96	177	148	139	132	74	53	38	1143 mm
dp	6	9	9	9	9	11	9	9	11	4	6	6	98 days
sunshine	124	105	136	144	167	141	214	232	159	174	141	140	1877 hours
%	39	34	37	37	39	33	49	56	43	49	44	45	42 %

Hankow

latitude: 30°35′n longitude: 114°17′e height: 37 m

Crhk 053-16	Jan	Feb	Mar	Apr	May	Jun	Jul	Aug	Sep	Oct	Nov	Dec	Year
tx	16	18	25	31	34	36	37	37	36	29	23	18	38 °C
t	4	6	10	17	21	27	30	30	25	19	13	7	17 °C
tn	-4	-3	-1	5	11	18	21	21	15	10	1	-3	-5 °C
clothing	w	n	n	n	s	t	t	t	t	s	n	n	–
stress					o	f	ff	ff	f	o			–
p	46	48	96	152	165	244	180	96	71	81	48	28	1255 mm
dp	6	6	10	7	11	10	9	6	5	5	5	5	85 days

Chiu-Chiang

latitude: 29°45′n longitude: 116°08′e height: 46 m

Crhk 053-17	Jan	Feb	Mar	Apr	May	Jun	Jul	Aug	Sep	Oct	Nov	Dec	Year
tx	18	16	24	32	34	36	39	39	36	31	24	20	39 °C
t	5	6	10	17	22	26	30	29	24	19	13	7	17 °C
tn	-4	-3	1	6	11	17	22	22	16	9	3	-2	-6 °C
clothing	n	n	n	n	s	t	t	t	t	s	n	n	–
stress					o	f	ff	ff	f	o			–
p	64	84	150	180	175	244	142	132	89	96	69	43	1468 mm
dp	9	10	10	13	14	11	7	7	7	5	7	9	109 days

Pei-Yü-Shan

latitude: 28°53'n longitude: 122°16'e height: 82 m

Crak	053-18	Jan	Feb	Mar	Apr	May	Jun	Jul	Aug	Sep	Oct	Nov	Dec	Year
	tx	16	15	17	21	25	27	32	32	30	27	23	20	32 °C
	t	6	6	9	14	18	22	27	27	24	20	14	11	16 °C
	tn	-2	-1	2	7	12	18	22	23	18	13	7	2	-2 °C
	clothing	n	n	n	n	s	s	t	t	t	s	n	n	–
	stress					o	f	ff	ff	ff	o			–
	p	51	79	112	114	127	152	69	104	165	71	56	53	1153 mm
	dp	9	10	12	12	11	13	7	7	9	7	7	8	110 days

Nanchang

latitude: 28°40'n longitude: 115°58'e height: 49 m

Crhk	053-19	Jan	Feb	Mar	Apr	May	Jun	Jul	Aug	Sep	Oct	Nov	Dec	Year
	t	6	7	10	16	21	25	29	30	25	19	14	7	17 °C
	clothing	n	n	n	n	s	t	t	t	t	s	n	n	–
	stress					o	ff	ff	ff	ff	o			–
	p	55	108	192	250	289	295	258	111	109	57	70	70	1864 mm
0.1 mm	dp	9	12	14	16	16	14	10	9	10	9	9	10	137 days
	sunshine	121	79	93	90	108	126	195	276	279	236	186	158	1947 hours
	%	37	25	25	23	26	30	46	67	75	66	58	49	44 %

Ch'ang-Sha

latitude: 28°11'n longitude: 112°47'e height: 60 m

Crhk	053-20	Jan	Feb	Mar	Apr	May	Jun	Jul	Aug	Sep	Oct	Nov	Dec	Year
	tx	18	19	28	31	35	36	38	39	36	30	25	20	39 °C
	t	4	6	11	17	23	26	30	30	26	19	16	7	18 °C
	tn	-6	-2	1	8	14	20	22	22	16	10	3	-3	-6 °C
	clothing	w	n	n	s	s	t	t	t	t	s	n	n	–
	stress				o	o	ff	ff	ff	f	o			–
	p	48	94	135	145	208	221	112	109	69	76	69	38	1324 mm
	dp	8	11	12	13	13	12	6	6	6	7	8	7	109 days

Hengchow

latitude: 26°56'n longitude: 112°25'e height: 64 m

Crhk	053-21	Jan	Feb	Mar	Apr	May	Jun	Jul	Aug	Sep	Oct	Nov	Dec	Year
	tx	16	17	28	33	35	36	38	41	37	33	26	22	41 °C
	t	4	7	12	16	24	26	30	30	26	19	13	8	18 °C
	tn	-4	1	2	6	14	18	23	23	18	7	4	-1	-5 °C
	clothing	n	n	n	s	t	t	t	t	t	s	n	n	–
	stress				o	f	ff	ff	ff	ff	o			–
	p	69	132	94	193	168	272	84	145	64	84	119	51	1475 mm
	dp	11	15	14	17	12	15	7	7	6	8	9	9	130 days

Kweilin

latitude: 25°15′n longitude: 110°10′e height: 167 m

Crhk	053-22	Jan	Feb	Mar	Apr	May	Jun	Jul	Aug	Sep	Oct	Nov	Dec	Year
	t	9	10	13	19	24	24	28	28	26	22	16	11	19 °C
	clothing	n	n	n	s	t	t	t	t	t	s	n	n	–
	stress				o	f	ff	ff	ff	ff	o			–
	p	51	79	161	223	359	370	236	200	101	87	53	47	1967 mm
0.1 mm	dp	12	14	19	19	20	19	19	18	8	8	11	11	176 days
	sunshine	81	40	59	105	118	150	186	217	207	149	126	136	1573 hours
	%	24	13	16	27	28	36	44	54	56	41	38	41	35 %

Wu-Ch'iu-Hsü

latitude: 25°00′n longitude: 119°27′e height: 63 m

Cral	053-23	Jan	Feb	Mar	Apr	May	Jun	Jul	Aug	Sep	Oct	Nov	Dec	Year
	tx	19	18	21	24	28	30	32	32	32	28	23	21	32 °C
	t	11	10	12	17	21	25	27	27	26	22	19	14	19 °C
	tn	5	5	6	11	16	19	23	23	21	17	12	8	4 °C
	clothing	n	n	n	n	s	t	t	t	t	s	s	n	–
	stress					o	ff	ff	ff	ff	f	o		–
	p	30	56	89	109	135	155	94	99	76	28	20	23	914 mm
	dp	4	6	9	10	8	8	6	6	3	1	2	2	65 days

Amoy

latitude: 24°26′n longitude: 118°04′e height: 5 m

Crhl	053-24	Jan	Feb	Mar	Apr	May	Jun	Jul	Aug	Sep	Oct	Nov	Dec	Year
	tx	22	22	25	28	31	33	35	34	34	32	28	24	35 °C
	t	14	13	16	17	24	27	29	29	28	25	21	17	22 °C
	tn	6	7	8	12	17	20	24	24	22	17	13	8	6 °C
	clothing	n	n	n	s	t	t	t	t	t	t	s	n	–
	stress				o	f	ff	ff	ff	ff	f	o		–
	p	33	76	89	124	158	178	132	168	109	48	30	33	1178 mm
	dp	4	6	7	10	8	10	9	7	6	1	2	5	75 days

Pakhoi

latitude: 21°28′n longitude: 109°05′e height: 4 m

Crhl	053-25	Jan	Feb	Mar	Apr	May	Jun	Jul	Aug	Sep	Oct	Nov	Dec	Year
	t	16	15	18	24	27	28	29	28	28	26	21	17	23 °C
	clothing	n	n	s	t	t	t	t	t	t	t	s	n	–
	stress			o	f	ff	ff	ff	ff	ff	f	o		–
	p	30	41	69	89	180	269	500	483	254	91	43	41	2090 mm
	dp	7	9	12	10	11	14	17	17	13	7	5	6	127 days

5. *Climatic area:* Southernmost Tibet, southern Sikiang, South-western Region, western Hupe, Kuangtung (except Hainan)

5.1. Subtropical Summer Rain Climate Cw and Subtropical Summer Rain Mountain Climate GCw and Subtropical Summer Rain High Mountain Climate HCw (Hainan Tropical instead of Subtropical)

5.2. In lowlands summer very hot, winter mild (Hainan warm).

5.3. High sun period: variable weather prevails, often dry and sunny in the morning; low sun period: generally dry and sunny or cloudy, scattered showers or rain.

5.4. Wind: occasional gales, typhoons at times in coastal areas.

Changtu

latitude: 31°11′n longitude: 96°59′e height: 3200 m

HCwlo	053-26	Jan	Feb	Mar	Apr	May	Jun	Jul	Aug	Sep	Oct	Nov	Dec	Year
	t	-2	1	5	8	12	14	16	15	13	9	3	1	8 °C
	clothing	w	w	n	n	n	n	n	n	n	n	w	w	–
	stress													–
	p	2	4	11	23	79	88	135	99	81	29	1	4	556 mm
0.1 mm	dp	3	3	6	9	16	19	23	20	19	10	1	2	131 days
	sunshine	183	181	186	186	226	180	155	202	177	202	216	192	2285 hours
	%	57	58	50	48	53	42	36	49	48	57	68	61	51 %

I-Ch'ang

latitude: 30°43′n longitude: 111°13′e height: 113 m

Cwhk	053-27	Jan	Feb	Mar	Apr	May	Jun	Jul	Aug	Sep	Oct	Nov	Dec	Year
	tx	16	19	27	33	36	38	40	41	37	31	27	21	41 °C
	t	6	7	11	18	22	26	29	29	24	19	13	8	18 °C
	tn	-2	-2	3	7	11	17	21	21	16	9	4	-1	-3 °C
	clothing	n	n	n	s	s	t	t	t	t	s	n	n	–
	stress				o	o	f	ff	ff	f	o			–
	p	23	30	48	99	124	158	208	178	102	74	33	18	1095 mm
	dp	6	7	10	12	13	12	14	12	10	11	7	5	118 days

Chengtu

latitude: 30°40′n longitude: 104°04′e height: 498 m

Cwak	053-28	Jan	Feb	Mar	Apr	May	Jun	Jul	Aug	Sep	Oct	Nov	Dec	Year
	tx	13	16	23	27	32	33	33	33	28	23	19	13	34 °C
	t	6	8	13	18	22	24	26	26	22	18	12	8	17 °C
	tn	1	1	3	8	12	17	20	19	16	11	6	1	0 °C
	clothing	n	n	n	s	s	t	t	t	s	s	n	n	–
	stress					o	f	ff	ff	f	o			–
	p	7	15	25	56	96	122	304	303	139	53	18	8	1146 mm
	dp	2	4	5	9	10	11	14	12	14	10	4	1	96 days
	sunshine	31	62	81	102	133	111	102	170	78	71	60	56	1058 hours
	%	10	20	22	26	31	26	24	41	21	20	19	18	24 %

Chunking

latitude: 29°33′n longitude: 106°33′e height: 230 m

Cwhk	053-29	Jan	Feb	Mar	Apr	May	Jun	Jul	Aug	Sep	Oct	Nov	Dec	Year
	tx	14	19	27	32	33	36	39	39	36	27	23	17	41 °C
	t	7	10	15	19	23	25	29	30	25	19	14	11	19 °C
	tn	2	3	6	11	16	18	21	21	17	13	8	4	1 °C
	clothing	n	n	n	s	t	t	t	t	t	s	n	n	–
	stress				o	f	ff	ff	ff	f	o			–
	p	15	20	38	99	142	180	142	122	150	112	48	20	1088 mm
	dp	4	4	7	9	13	11	6	6	11	13	9	6	99 days

Yatung / Tibet

latitude: 27° 29′n longitude: 88°55′e height: 2989 m

HCwlk	053-30	Jan	Feb	Mar	Apr	May	Jun	Jul	Aug	Sep	Oct	Nov	Dec	Year
	tx	13	14	16	19	20	22	22	22	21	19	16	13	23 °C
	t	0	2	4	8	11	13	15	14	13	9	4	1	8 °C
	tn	-11	-10	-7	-3	1	6	8	7	4	-3	-7	-9	-12 °C
	clothing	w	w	w	n	n	n	n	n	n	n	w	w	–
	stress													–
	p	15	48	64	99	107	119	130	117	102	53	18	5	876 mm
2.5 mm	dp	2	4	6	11	11	13	15	16	11	3	1	1	94 days

Kunming

latitude: 25°02′n longitude: 102°43′e height: 1893 m

GCwbl	053-31	Jan	Feb	Mar	Apr	May	Jun	Jul	Aug	Sep	Oct	Nov	Dec	Year
	tx	21	23	26	28	31	28	29	29	28	26	24	21	31 °C
	t	10	11	14	18	20	20	20	20	18	16	12	10	16 °C
	tn	-2	-1	3	8	10	12	13	14	10	7	3	1	-2 °C
	clothing	n	n	n	s	s	s	s	s	s	n	n	n	–
	stress				o	o	o	f	f	o				–
	p	3	18	21	31	99	192	214	220	161	95	31	11	1096 mm
	dp	2	4	3	5	11	15	19	20	14	12	7	3	115 days
	sunshine	251	237	245	237	195	105	102	130	117	167	180	202	2169 hours
	%	75	74	66	62	47	26	24	32	32	46	55	61	49 %

T'eng-Chung

latitude: 25°00′n longitude: 98°40′e height: 1635 m

GCwbk	053-32	Jan	Feb	Mar	Apr	May	Jun	Jul	Aug	Sep	Oct	Nov	Dec	Year
	tx	19	21	24	26	28	28	29	28	27	26	22	19	29 °C
	t	8	10	13	16	18	20	20	21	20	17	16	9	16 °C
	tn	-2	-1	2	6	9	14	14	16	14	7	2	-2	-3 °C
	clothing	n	n	n	n	s	s	s	s	s	s	n	n	–
	stress					o	o	o	o	o	o			–
	p	13	38	36	69	128	236	312	282	163	158	41	23	1499 mm
	dp	3	7	7	11	15	22	26	24	18	13	5	3	154 days

Meng-Tzu

latitude: 23° 30′n longitude: 103° 30′e height: 1251 m

GCwal	053-33	Jan	Feb	Mar	Apr	May	Jun	Jul	Aug	Sep	Oct	Nov	Dec	Year
	tx	26	28	32	34	36	33	33	33	32	29	28	27	36 °C
	t	14	16	19	22	24	24	24	24	23	20	16	14	20 °C
	tn	4	5	7	13	14	16	17	17	15	10	7	3	3 °C
	clothing	n	n	s	s	t	t	t	t	s	s	n	n	–
	stress				o	f	f	f	f	o	o			–
	p	8	18	28	41	127	132	196	198	96	51	56	15	966 mm
0.25 mm	dp	2	5	5	8	12	15	18	20	12	10	7	2	116 days

Canton

Cwhl	053-34	Jan	Feb	Mar	Apr	May	Jun	Jul	Aug	Sep	Oct	Nov	Dec	Year
		latitude: 23°00'n					longitude: 113°13'e				height: 18 m			
	tx	26	27	28	32	33	34	36	36	34	32	29	27	36 °C
	t	14	14	17	22	26	27	28	28	27	24	20	16	22 °C
	tn	3	6	8	14	20	22	24	23	21	16	9	5	2 °C
	clothing	n	n	s	s	t	t	t	t	t	t	s	n	–
	stress			o	o	ff	ff	ff	ff	ff	f	o		–
	p	27	65	101	185	256	292	264	249	149	49	51	34	1722 mm
	dp	5	8	11	12	15	17	16	13	10	4	3	5	119 days
	sunshine	124	79	84	81	155	153	214	214	204	211	192	158	1868 hours
	%	37	25	23	21	38	38	51	53	55	58	58	47	47 %

Lungchow

Cwhl	053-35	Jan	Feb	Mar	Apr	May	Jun	Jul	Aug	Sep	Oct	Nov	Dec	Year
		latitude: 22°22'n					longitude: 106°45'e				height: 266 m			
	tx	27	29	33	36	37	37	37	37	36	34	31	30	38 °C
	t	14	15	19	23	27	29	29	29	28	24	19	18	23 °C
	tn	6	7	11	14	19	22	23	22	20	14	9	6	3 °C
	clothing	n	n	s	t	t	t	t	t	t	t	s	s	–
	stress			o	f	ff	ff	ff	ff	ff	f	o	o	–
	p	20	36	48	81	178	216	229	231	142	64	30	20	1295 mm
	dp	5	9	5	8	11	12	16	14	9	6	3	4	102 days

Yulin – Hainan

Awhb	053-36	Jan	Feb	Mar	Apr	May	Jun	Jul	Aug	Sep	Oct	Nov	Dec	Year
		latitude: 18°14'n					longitude: 109°32'e				height: 2 m			
	t	21	22	24	27	28	28	28	28	27	26	23	21	26 °C
	clothing	s	s	t	t	t	t	t	t	t	t	t	s	–
	stress	o	o	f	ff	ff	ff	ff	ff	ff	ff	f	o	–
	p	11	7	21	28	150	197	149	189	293	190	54	43	1332 mm
0.1 mm	dp	4	3	3	4	10	14	14	16	19	12	7	7	113 days
	sunshine	211	203	211	210	229	255	285	236	204	242	210	195	2691 hours
	%	61	62	57	56	57	64	70	60	55	66	62	57	61 %

4.2.53.2. Predominant general weather situations are described in 4.2.0.1. under China. A hygrothermal diagram of Peking is to be found in 4.2.0.2.
Wind names: Karaburan is a north-easterly dust storm blowing in summer in the Tarim Desert of Sinkiang.
Yellow wind is a westerly or north-westerly wind charged with yellow loess dust – which blows particularly in winter.

4.2.54. Taiwan

4.2.54.1. Climatic descriptions and tables

1. Climatic area: Smaller eastern part of the island
1.1. Subtropical Rain Climate Cr and Subtropical Rain Mountain Climate GCr
1.2. Summer very hot, winter mild in low levels.
1.3. All the year through variable weather prevails; however, often dry and sunny in the morning.
1.4. Wind: occasional gales, at times typhoons.

Taipei

latitude: 25°02′n longitude: 121°31′e height: 9 m

Crhl	054-1	Jan	Feb	Mar	Apr	May	Jun	Jul	Aug	Sep	Oct	Nov	Dec	Year
	tx	27	27	29	32	33	35	36	36	34	33	31	28	36 °C
	t	15	15	18	21	24	27	28	28	27	23	20	17	22 °C
	tn	7	7	9	12	16	19	22	22	19	15	10	8	5 °C
	clothing	n	n	s	s	t	t	t	t	t	t	s	n	–
	stress			o	o	f	ff	ff	ff	ff	f	o		–
	p	91	147	164	182	205	322	269	266	189	117	71	77	2100 mm
	dp	10	12	13	12	12	13	11	11	10	10	10	10	132 days
	sunshine	87	76	90	111	140	168	226	217	192	143	108	93	1650 hours
	%	26	24	24	29	34	41	54	54	52	40	33	28	37 %

Hwalien

latitude: 23°58′n longitude: 121°37′e height: 19 m

Crhl	054-2	Jan	Feb	Mar	Apr	May	Jun	Jul	Aug	Sep	Oct	Nov	Dec	Year
	tx	26	27	28	29	31	32	33	33	33	31	29	28	34 °C
	t	17	18	20	22	22	26	28	27	26	24	22	19	23 °C
	tn	10	11	12	14	18	20	22	22	20	16	13	11	9 °C
	clothing	s	s	s	s	t	t	t	t	t	t	s	s	–
	stress	o	o	o	o	f	ff	ff	ff	ff	f	o	o	–
	p	59	84	112	106	180	210	203	244	362	255	174	89	2078 mm
	dp	10	12	13	12	16	13	8	10	12	12	11	10	138 days
	sunshine	68	62	81	90	140	183	264	236	177	143	102	93	1637 hours
	%	20	19	22	24	34	45	63	59	48	40	31	28	37 %

2. *Climatic area:* greater part of the country, except eastern section and southern part
2.1. Subtropical Summer Rain Climate Cw and Subtropical Summer Rain Mountain Climate GCw, southern part: Tropical Monsoonal Rain Climate Am and Tropical Monsoonal Rain Mountain Climate GAm
2.2. Summer very hot, winter mild in low levels.
2.3. High sun period: variable weather prevails, often dry and sunny in the morning; low sun period: generally dry and sunny, now and then scattered showers or rain.
2.4. Wind: occasional gales, at times typhoons.

Taichung

latitude: 24°09′n longitude: 120°41′e height: 85 m

Cwhl	054-3	Jan	Feb	Mar	Apr	May	Jun	Jul	Aug	Sep	Oct	Nov	Dec	Year
	t	16	16	19	22	26	27	28	28	27	24	21	18	23 °C
	clothing	n	n	s	s	t	t	t	t	t	t	s	s	–
	stress			o	f	f	ff	ff	ff	ff	f	o	o	–
	p	31	72	101	135	211	438	282	337	133	20	14	32	1803 mm
0.1 mm	dp	8	10	12	11	13	16	15	18	9	4	5	7	128 days
	sunshine	174	144	158	168	186	210	242	217	231	232	198	186	2346 hours
	%	52	45	42	44	45	51	58	54	63	64	60	56	53 %

Penghu

				latitude: 23° 32′n				longitude: 119°33′e				height: 11 m		
Cwhl	054-4	Jan	Feb	Mar	Apr	May	Jun	Jul	Aug	Sep	Oct	Nov	Dec	Year
	t	16	16	19	22	26	28	28	28	28	25	22	18	23 °C
	clothing	n	n	s	s	t	t	t	t	t	t	s	s	–
	stress			o	f	f	ff	ff	ff	ff	f	o	o	–
	p	23	44	65	76	98	189	195	178	128	47	19	28	1090 mm
	dp	3	5	6	7	8	10	9	8	5	2	2	4	68 days

Tainan

				latitude: 23°00′n				longitude: 120°13′e				height: 14 m		
Cwhl	054-5	Jan	Feb	Mar	Apr	May	Jun	Jul	Aug	Sep	Oct	Nov	Dec	Year
	t	17	18	21	24	27	28	28	28	28	25	22	19	24 °C
	clothing	s	s	s	t	t	t	t	t	t	t	s	s	–
	stress	o	o	f	f	ff	ff	ff	ff	ff	f	o	o	–
	p	14	27	50	82	182	445	508	431	181	34	17	22	1991 mm
0.1 mm	dp	5	6	7	7	10	15	17	19	11	4	3	4	108 days
	sunshine	195	184	198	210	229	228	236	214	243	245	213	198	2593 hours
	%	58	57	53	55	56	56	57	53	66	68	64	59	58 %

Hengchu

				latitude: 22°00′n				longitude: 120°45′e				height: 24 m		
Amhb	054-6	Jan	Feb	Mar	Apr	May	Jun	Jul	Aug	Sep	Oct	Nov	Dec	Year
	tx	29	30	32	33	34	33	33	33	33	32	31	29	34 °C
	t	21	21	23	25	27	28	28	28	27	26	24	22	25 °C
	tn	13	14	14	17	21	22	23	23	22	19	17	14	12 °C
	clothing	s	s	t	t	t	t	t	t	t	t	t	s	–
	stress	o	o	f	f	ff	ff	ff	ff	ff	ff	f	o	–
	p	17	26	21	53	151	466	584	538	357	137	87	25	2462 mm
	dp	3	4	3	4	8	16	17	17	14	7	4	4	102 days
	sunshine	177	175	205	204	208	228	229	192	201	208	189	170	2386 hours
	%	52	54	55	54	51	56	55	48	54	57	57	51	54 %

4.2.54.2. Predominant general weather situations are described in 4.2.0.1. under South-Eastern Asia. A hygrothermal diagram of Taipei is to be found in 4.2.0.2.

4.2.55. Japan

4.2.55.1. Climatic descriptions and tables

1. Climatic area: Hokkaido and northern part of Hondo

1.1. Temperate Continental Climate DC and
Temperate Continental Mountain Climate GDC

1.2. Summer warm, winter cold.

1.3. Variable weather prevails; however, periods of dry and sunny days also occur.

1.4. Wind: occasional gales. Snow cover: in winter for several months in Hokkaido.

Wakkanai

latitude: 45°25′n longitude: 141°41′e height: 3 m

DCbo 055-1	Jan	Feb	Mar	Apr	May	Jun	Jul	Aug	Sep	Oct	Nov	Dec	Year
tx	3	3	7	14	19	22	26	28	25	19	14	6	28 °C
t	-6	-6	-2	4	8	12	17	20	17	11	3	-3	6 °C
tn	-14	-14	-10	-5	1	5	9	13	7	1	-6	-12	-21 °C
clothing	w	w	w	w	n	n	s	s	s	n	w	w	–
stress							o	o	o				–
p	94	62	65	63	78	70	112	105	152	129	120	112	1161 mm
dp	20	15	13	9	10	8	9	8	11	13	17	20	153 days
sunshine	42	79	146	184	203	177	164	180	188	157	73	38	1630 hours
%	15	27	40	45	44	38	34	41	50	46	25	14	36 %

Asahikawa

latitude: 43°46′n longitude: 142°42′e height: 113 m

DCbo 055-2	Jan	Feb	Mar	Apr	May	Jun	Jul	Aug	Sep	Oct	Nov	Dec	Year
t	-9	-8	-3	4	11	16	20	21	15	9	1	-5	6 °C
clothing	w	w	w	w	n	n	s	s	n	n	w	w	–
stress							o	o					–
p	82	61	56	61	78	75	125	144	136	109	118	101	1144 mm
dp	18	14	13	10	11	10	10	10	13	14	17	20	159 days
sunshine	74	95	154	187	200	202	186	188	158	139	77	55	1715 hours
%	26	32	42	46	44	44	40	43	42	40	26	20	38 %

Nemuro

latitude: 43°20′n longitude: 145°35′e height: 26 m

DCbo 055-3	Jan	Feb	Mar	Apr	May	Jun	Jul	Aug	Sep	Oct	Nov	Dec	Year
tx	4	4	7	14	19	22	26	27	24	19	15	9	28 °C
t	-5	-6	-2	3	7	10	14	18	16	11	5	-1	6 °C
tn	-15	-16	-13	-5	-1	3	6	10	7	1	-5	-11	-17 °C
clothing	w	w	w	w	n	n	n	s	n	n	n	w	–
stress								o					–
p	49	40	77	77	99	97	104	106	152	124	92	63	1081 mm
dp	8	6	8	9	10	9	9	9	10	10	10	8	104 days
sunshine	154	168	192	194	182	152	132	147	158	174	151	151	1955 hours
%	53	57	52	48	40	33	28	34	42	51	52	54	44 %

Sapporo

latitude: 43°03′n longitude: 141°20′e height: 18 m

DCbo 055-4	Jan	Feb	Mar	Apr	May	Jun	Jul	Aug	Sep	Oct	Nov	Dec	Year
tx	5	5	9	19	26	28	31	32	28	23	17	9	32 °C
t	-6	-5	-1	6	11	16	20	21	18	12	4	-1	8 °C
tn	-19	-20	-15	-7	-1	4	9	11	4	-2	-8	-17	-22 °C
clothing	w	w	w	n	n	n	s	s	s	n	w	w	–
stress							o	o	o				–
p	111	83	67	66	59	67	100	107	145	113	112	104	1136 mm
dp	17	14	12	10	8	9	9	10	11	12	13	15	139 days
sunshine	99	112	158	199	213	205	190	201	173	166	112	92	1920 hours
%	34	38	43	50	47	45	41	46	46	48	38	33	43 %

Akita

DCao	055-5	Jan	Feb	Mar	Apr	May	Jun	Jul	Aug	Sep	Oct	Nov	Dec	Year
	latitude: 39°43′n longitude: 140°06′e height: 10 m													
	tx	7	7	15	22	26	29	32	33	30	24	19	13	34 °C
	t	-1	-1	2	8	13	18	22	24	19	13	7	2	11 °C
	tn	-12	-13	-8	-3	2	8	14	16	8	2	-2	-8	-14 °C
	clothing	w	w	w	n	n	s	s	t	s	n	n	w	–
	stress						o	o	f	o				–
	p	123	102	107	128	119	138	190	164	205	176	179	158	1789 mm
	dp	22	18	15	12	11	11	12	9	14	15	18	22	177 days
	sunshine	48	67	135	189	205	196	172	223	171	160	107	45	1712 hours
	%	16	22	37	48	46	44	38	52	46	46	36	15	38 %

2. Climatic area: Northern Hondo transition zone to Central Hondo

2.1. Temperate Oceanic Climate DO and Temperate Oceanic Mountain Climate GDO

2.2. Summer hot to very hot, winter cool.

2.3. Variable weather prevails; however, periods of dry and sunny days also occur.

2.4. Wind: occasional gales, at times typhoons.

Miyako

DObk	055-6	Jan	Feb	Mar	Apr	May	Jun	Jul	Aug	Sep	Oct	Nov	Dec	Year
	latitude: 39°39′n longitude: 141°58′e height: 47 m													
	tx	13	12	19	25	28	28	33	34	30	25	21	17	34 °C
	t	0	0	3	9	13	16	20	22	19	13	7	2	10 °C
	tn	-12	-10	-8	-3	1	8	12	15	8	2	-3	-9	-13 °C
	clothing	w	w	w	n	n	n	s	s	s	n	n	w	–
	stress							o	o	o				–
	p	57	73	88	93	91	129	132	132	203	158	70	62	1288 mm
	dp	5	6	8	7	8	10	10	10	11	9	6	5	96 days
	sunshine	180	170	198	209	211	174	164	190	147	162	163	161	2130 hours
	%	60	56	54	53	47	39	36	45	39	47	54	55	48 %

Sendai

DOak	055-7	Jan	Feb	Mar	Apr	May	Jun	Jul	Aug	Sep	Oct	Nov	Dec	Year
	latitude: 38°16′n longitude: 140°54′e height: 40 m													
	t	0	1	4.	9	14	18	22	24	20	14	8	3	11 °C
	clothing	w	w	w	n	n	s	s	t	s	n	n	w	–
	stress						o	o	f	o				–
	p	37	44	62	95	100	155	167	136	191	133	61	50	1232 mm
	dp	6	5	7	9	9	12	12	11	11	9	6	8	104 days
	sunshine	151	155	192	203	206	153	134	163	129	149	142	131	1908 hours
	%	50	51	52	51	47	34	30	38	34	43	47	44	43 %

Niigata

latitude: 37°55′n longitude: 139°03′e height: 4 m

DOak	055-8	Jan	Feb	Mar	Apr	May	Jun	Jul	Aug	Sep	Oct	Nov	Dec	Year
	tx	10	11	18	24	27	30	34	35	33	26	21	14	35 °C
	t	2	2	5	10	15	20	24	26	21	16	10	5	13 °C
	tn	-6	-6	-3	1	6	11	17	18	12	6	1	-3	-6 °C
	clothing	w	w	n	n	n	s	t	t	s	n	n	n	–
	stress						o	f	ff	o				–
	p	194	126	121	104	95	127	193	107	177	165	171	264	1841 mm
	dp	23	18	16	12	10	10	11	8	13	15	18	23	176 days
	sunshine	70	82	140	201	220	202	207	255	170	145	110	59	1861 hours
	%	23	27	38	51	50	46	46	60	45	42	36	20	42 %

3. *Climatic area:* Japan, south of 37° latitude north (except southernmost group of Tyukyu Islands with Tropical Rain Climate Ar)

3.1. Subtropical Rain Climate Cr and Subtropical Rain Mountain Climate GCr

3.2. Summer hot or very hot; winter cool, in the south mild or warm.

3.3. Variable weather prevails; however, periods of dry and sunny days also occur.

3.4. Wind: occasional gales and typhoons.

Kanazawa

latitude: 36°33′n longitude: 136°39′e height: 29 m

Crak	055-9	Jan	Feb	Mar	Apr	May	Jun	Jul	Aug	Sep	Oct	Nov	Dec	Year
	t	2	2	6	11	16	20	24	26	22	16	10	6	14 °C
	clothing	w	w	n	n	n	s	t	t	s	n	n	n	–
	stress						o	f	ff	o				–
	p	309	191	173	164	135	170	223	154	248	217	225	353	2559 mm
	dp	25	20	17	13	11	12	12	9	13	14	17	24	185 days
	sunshine	62	84	143	195	213	186	190	241	159	151	123	71	1818 hours
	%	20	27	39	50	49	42	42	57	43	43	40	24	41 %

Tokyo

latitude: 35°41′n longitude: 139°46′e height: 6 m

Crak	055-10	Jan	Feb	Mar	Apr	May	Jun	Jul	Aug	Sep	Oct	Nov	Dec	Year
	tx	16	16	20	24	27	31	33	33	32	26	22	18	34 °C
	t	4	4	8	13	18	21	25	26	23	17	11	6	15 °C
	tn	-6	-5	-3	2	6	12	17	18	13	6	1	-4	-6 °C
	clothing	w	w	n	n	s	s	t	t	s	s	n	n	–
	stress					o	o	f	ff	o	o			–
	p	48	73	101	135	131	182	146	147	217	220	101	61	1563 mm
	dp	5	6	9	10	10	11	9	9	12	11	8	5	104 days
	sunshine	186	166	176	180	193	150	181	204	136	136	144	169	2021 hours
	%	60	54	48	46	44	34	41	49	36	39	47	56	45 %

Yonago

latitude: 35°26′n longitude: 133°21′e height: 8 m

Crak	055-11	Jan	Feb	Mar	Apr	May	Jun	Jul	Aug	Sep	Oct	Nov	Dec	Year
	t	4	4	7	12	16	20	25	26	22	16	11	6	14 °C
	clothing	w	w	n	n	n	s	t	t	s	n	n	n	–
	stress						o	f	ff	o				–
	p	154	152	122	110	127	170	212	127	267	156	116	141	1858 mm
	dp	20	17	15	11	10	12	12	8	14	12	13	18	159 days
	sunshine	84	96	147	206	223	190	203	257	163	166	137	91	1960 hours
	%	27	31	40	52	51	44	46	61	44	47	44	30	44 %

Nagoya

Crak	055-12	latitude: 35°10′n					longitude: 136°58′e				height: 56 m		
	Jan	Feb	Mar	Apr	May	Jun	Jul	Aug	Sep	Oct	Nov	Dec	Year
t	3	4	7	13	18	22	26	27	23	16	11	6	14 °C
clothing	w	w	n	n	s	s	t	t	s	n	n	n	–
stress					o	f	ff	ff	o				–
p	49	64	100	137	145	204	178	155	212	160	86	57	1546 mm
dp	6	6	9	10	10	12	11	9	12	9	7	6	107 days
sunshine	178	181	204	200	215	178	203	238	166	172	175	167	2272 hours
%	57	59	55	51	49	41	46	57	44	49	56	55	51 %

Osaka

Crhk	055-13	latitude: 34°39′n					longitude: 135°32′e				height: 8 m		
	Jan	Feb	Mar	Apr	May	Jun	Jul	Aug	Sep	Oct	Nov	Dec	Year
tx	14	15	19	24	28	31	34	35	33	27	23	17	36 °C
t	4	5	8	14	18	22	27	28	24	17	12	7	16 °C
tn	-3	-4	-2	2	7	12	18	20	14	7	2	-2	-4 °C
clothing	w	n	n	n	s	s	t	t	t	s	n	n	–
stress					o	f	ff	ff	f	o			–
p	43	58	96	127	122	193	177	118	171	122	81	52	1359 mm
dp	6	7	10	9	9	11	9	7	11	9	7	6	100 days
sunshine	150	144	175	196	209	181	213	242	175	166	156	146	2153 hours
%	48	47	47	50	48	42	48	58	47	47	50	48	48 %

Hiroshima

Crak	055-14	latitude: 34°22′n					longitude: 132°26′e				height: 30 m		
	Jan	Feb	Mar	Apr	May	Jun	Jul	Aug	Sep	Oct	Nov	Dec	Year
tx	15	15	19	23	27	29	34	34	32	27	22	17	35 °C
t	4	5	8	13	17	21	25	27	23	17	12	1	15 °C
tn	-4	-4	-3	1	6	12	17	19	13	6	1	-3	-6 °C
clothing	w	n	n	n	s	s	t	t	s	s	n	n	–
stress					o	o	f	ff	o	o			–
p	45	70	106	158	154	249	250	116	216	115	67	51	1596 mm
dp	6	7	10	10	9	11	11	7	11	7	6	6	100 days
sunshine	146	153	188	205	222	191	211	248	178	190	177	155	2264 hours
%	47	49	51	52	51	44	48	59	48	54	57	51	51 %

Kochi

Crak	055-15	latitude: 33°34′n					longitude: 133°33′e				height: 2 m		
	Jan	Feb	Mar	Apr	May	Jun	Jul	Aug	Sep	Oct	Nov	Dec	Year
t	5	6	10	14	18	22	26	26	24	18	13	8	16 °C
clothing	n	n	n	n	s	s	t	t	t	s	n	n	–
stress					o	f	ff	ff	f	o			–
p	55	97	177	261	279	344	369	344	350	184	108	80	2646 mm
dp	5	7	10	11	11	13	11	13	13	9	7	5	117 days
sunshine	194	179	200	196	201	163	195	222	178	184	182	191	2285 hours
%	62	58	54	50	47	38	44	53	48	52	58	62	51 %

Fukuoka

latitude: 33°55′n longitude: 130°23′e height: 4 m

Crak	055-16	Jan	Feb	Mar	Apr	May	Jun	Jul	Aug	Sep	Oct	Nov	Dec	Year
	t	5	6	9	14	18	22	26	27	23	17	12	8	15 °C
	clothing	n	n	n	n	s	s	t	t	t	s	n	n	–
	stress					o	f	ff	ff	f	o			–
	p	69	83	98	129	127	270	253	171	244	102	80	78	1703 mm
	dp	10	9	11	10	9	11	11	9	11	7	8	10	116 days
	sunshine	110	124	171	188	205	177	205	238	169	189	160	121	2057 hours
	%	35	40	46	48	47	41	47	57	45	54	51	39	46 %

Nagasaki

latitude: 32°44′n longitude: 129°52′e height: 27 m

Crak	055-17	Jan	Feb	Mar	Apr	May	Jun	Jul	Aug	Sep	Oct	Nov	Dec	Year
	tx	17	17	21	24	27	30	33	33	32	27	23	19	34 °C
	t	6	8	10	15	18	22	26	27	24	19	14	9	16 °C
	tn	-3	-3	0	4	9	14	18	21	15	9	3	-1	-4 °C
	clothing	n	n	n	n	s	s	t	t	t	s	n	n	–
	stress					o	f	ff	ff	f	o			–
	p	70	83	116	182	207	294	288	189	253	106	83	87	1957 mm
	dp	11	9	10	11	11	13	12	9	11	7	7	10	119 days
	sunshine	114	132	177	188	202	172	210	252	186	200	170	128	2131 hours
	%	36	42	48	48	47	40	48	61	50	57	54	41	48 %

Kagoshima

latitude: 31°34′n longitude: 130°33′e height: 5 m

Crak	055-18	Jan	Feb	Mar	Apr	May	Jun	Jul	Aug	Sep	Oct	Nov	Dec	Year
	tx	19	19	23	26	29	31	34	34	33	29	25	21	34 °C
	t	7	8	11	15	19	23	27	27	24	19	14	9	17 °C
	tn	-4	-4	-2	2	7	13	19	20	13	5	1	-2	-4 °C
	clothing	n	n	n	n	s	t	t	t	t	s	n	n	–
	stress					o	f	ff	ff	f	o			–
	p	75	116	149	228	249	454	343	220	213	120	90	79	2337 mm
	dp	10	10	12	12	11	14	13	11	10	7	7	9	126 days
	sunshine	153	147	178	179	190	159	222	244	201	198	180	171	2222 hours
	%	48	47	48	46	45	37	51	59	54	56	57	55	50 %

Hachijojima

latitude: 33°06′n longitude: 139°47′e height: 81 m

Cral	055-19	Jan	Feb	Mar	Apr	May	Jun	Jul	Aug	Sep	Oct	Nov	Dec	Year
	tx	20	19	19	22	25	26	31	31	31	28	24	22	31 °C
	t	10	10	12	16	19	22	25	26	25	21	17	13	18 °C
	tn	2	2	3	7	12	16	19	21	18	14	9	4	2 °C
	clothing	n	n	n	n	s	s	t	t	t	s	n	n	–
	stress					o	f	ff	ff	ff	o			–
	p	192	207	254	239	267	323	180	238	365	515	370	193	3343 mm
	dp	16	15	15	13	14	15	10	13	16	18	16	14	177 days
	sunshine	108	108	142	166	171	121	196	235	181	132	113	118	1797 hours
	%	34	35	38	43	40	28	45	57	49	37	36	38	40 %

Torishima

Cral	055-20	Jan	Feb	Mar	Apr	May	Jun	Jul	Aug	Sep	Oct	Nov	Dec	Year
	latitude: 30° 29′n longitude: 140° 18′e height: 83 m													
	t	15	14	16	18	20	23	26	27	27	24	20	17	21 °C
	clothing	n	n	n	s	s	s	t	t	t	t	s	n	–
	stress				o	o	f	ff	ff	ff	f	o		–
	p	100	110	92	159	261	249	108	114	155	286	191	131	1954 mm
	dp	15	12	11	10	14	14	7	9	10	14	14	10	140 days

Naze

Crhl	055-21	Jan	Feb	Mar	Apr	May	Jun	Jul	Aug	Sep	Oct	Nov	Dec	Year
	latitude: 28° 23′n longitude: 129° 30′e height: 4 m													
	tx	24	24	25	29	31	34	34	34	33	31	28	26	35 °C
	t	14	15	16	19	22	25	28	28	26	23	20	16	21 °C
	tn	7	7	7	9	13	17	22	22	18	14	11	7	6 °C
	clothing	n	n	n	s	s	t	t	t	t	s	s	n	–
	stress				o	o	f	ff	ff	ff	o	o		–
	p	163	184	220	221	362	443	231	281	287	247	224	160	3033 mm
	dp	17	16	15	14	16	17	12	15	14	13	12	13	173 days
	sunshine	80	78	106	125	138	136	236	213	187	141	103	93	1633 hours
	%	24	25	29	32	33	33	55	52	50	39	32	29	37 %

Mawashi/Naha

Crhl	055-22	Jan	Feb	Mar	Apr	May	Jun	Jul	Aug	Sep	Oct	Nov	Dec	Year
	latitude: 26° 14′n longitude: 127° 41′e height: 36 m													
	tx	25	25	26	28	30	33	34	33	33	31	29	26	34 °C
	t	16	16	18	20	23	26	28	27	27	24	21	18	22 °C
	tn	8	8	9	13	15	19	23	23	21	17	14	10	7 °C
	clothing	n	n	s	s	t	t	t	t	t	t	s	s	–
	stress			o	o	f	ff	ff	ff	ff	f	o	o	–
	p	121	137	168	165	246	329	180	270	165	184	174	166	2305 mm
	dp	13	13	13	11	13	14	10	14	11	10	10	11	143 days
	sunshine	91	87	120	132	166	179	252	223	220	115	120	90	1795 hours
	%	27	27	32	34	40	43	60	55	59	32	37	28	42 %

Minamidaitojima

Crhl	055-23	Jan	Feb	Mar	Apr	May	Jun	Jul	Aug	Sep	Oct	Nov	Dec	Year
	latitude: 25° 50′n longitude: 131° 14′e height: 16 m													
	t	17	18	19	22	24	26	28	28	28	25	22	20	23 °C
	clothing	s	s	s	s	t	t	t	t	t	t	s	s	–
	stress	o	o	o	f	f	ff	ff	ff	ff	f	o	o	–
	p	76	98	85	123	177	241	103	187	129	189	166	104	1676 mm
	dp	9	9	9	9	11	11	10	13	11	13	11	9	124 days
	sunshine	122	126	141	145	165	194	270	238	222	168	121	114	2026 hours
	%	37	39	38	38	40	47	64	59	60	47	37	35	46 %

Ishigakijima

Arhb	055-24	latitude: 24° 20′n				longitude: 124° 10′e				height: 7 m				
		Jan	Feb	Mar	Apr	May	Jun	Jul	Aug	Sep	Oct	Nov	Dec	Year
	t	18	18	20	22	25	28	29	28	28	25	22	20	24 °C
	clothing	s	s	s	s	t	t	t	t	t	t	s	s	–
	stress	o	o	o	f	f	ff	ff	ff	ff	f	o	o	–
	p	140	118	157	170	240	230	174	215	233	171	197	183	2196 mm
	dp	13	12	11	10	12	12	11	14	13	12	13	14	146 days
	sunshine	99	101	124	148	190	212	275	251	230	182	135	103	2050 hours
	%	30	32	33	39	46	52	66	62	62	51	41	31	46 %

4.2.55.2. Predominant general weather situations are described in 4.2.0.1. under Korea and Japan. A hygrothermal diagram of Tokyo is to be found in 4.2.0.2.
Wind names: Bai-u are weak easterly winds accompanied by abundant rain causing a maximum of rainfall, often in the second half of June.
Kogarashi is a snow-carrying wind which blows from the north or north-west.
Nowaki is a rainy wind from the north in autumn.
Soyakaze is a mild rainy wind in spring.

4.2.56. Burma

4.2.56.1 Climatic description and tables

1. (except small northernmost part with Subtropical Summer Rain Climate Cw and Subtropical Summer Rain Mountain Climate GCw) Tropical Summer Rain Climate Aw and Tropical Monsoonal Rain Climate Am
2. In the north, summer very hot, winter warm; otherwise very hot thorough the year.
3. High sun period: variable weather prevails; generally dry and sunny or cloudy in the morning; in the afternoon frequent showers; periods of rainy days, especially from June to September. Low sun period: generally dry and sunny, at times scattered showers or rain.
4. Wind: occasional gales.

Myitkyina

Cwhl	056-1	latitude: 25° 22′n				longitude: 97° 24′e				height: 147 m				
		Jan	Feb	Mar	Apr	May	Jun	Jul	Aug	Sep	Oct	Nov	Dec	Year
	t	17	19	23	25	28	28	28	28	27	25	21	17	24 °C
	clothing	s	s	t	t	t	t	t	t	t	t	s	s	–
	stress	o	o	f	ff	ff	ff	ff	ff	ff	f	o	o	–
	p	10	23	23	46	160	481	476	433	256	183	39	12	2142 mm
	dp	1	3	3	5	10	19	21	19	13	9	2	1	106 days

Lashio

latitude: 22°58'n longitude: 97°51'e height: 855 m

GCwal	056-2	Jan	Feb	Mar	Apr	May	Jun	Jul	Aug	Sep	Oct	Nov	Dec	Year
	tx	25	29	32	35	34	31	31	31	31	30	28	25	36 °C
	t	16	17	22	24	25	25	25	25	24	23	19	16	22 °C
	tn	4	6	9	14	16	19	20	19	18	14	9	6	4 °C
	clothing	s	s	s	t	t	t	t	t	t	t	s	s	–
	stress	o	o	o	f	f	ff	ff	ff	ff	f	o	o	–
	p	8	8	15	56	175	249	305	325	198	145	69	23	1574 mm
	dp	1	1	1	5	12	15	27	19	14	10	5	1	111 days

Mandalay

latitude: 21°59'n longitude: 96°06'e height: 76 m

Awhb	056-3	Jan	Feb	Mar	Apr	May	Jun	Jul	Aug	Sep	Oct	Nov	Dec	Year
	tx	32	35	40	42	41	38	37	36	37	36	33	31	42 °C
	t	21	24	28	32	32	30	30	30	29	28	25	22	28 °C
	tn	11	12	16	21	22	23	24	24	23	21	16	11	10 °C
	clothing	s	t	t	t	t	t	t	t	t	t	t	s	–
	stress	o	f	ff	ff	ff	ff	ff	ff	ff	ff	f	o	–
	p	1	5	5	36	149	151	72	102	148	127	64	11	871 mm
	dp	0	0	1	3	8	7	6	8	9	7	3	1	53 days

Akyab

latitude: 20°08'n longitude: 92°53'e height: 5 m

Amhb	056-4	Jan	Feb	Mar	Apr	May	Jun	Jul	Aug	Sep	Oct	Nov	Dec	Year
	tx	29	32	34	35	34	32	32	31	32	33	32	29	36 °C
	t	21	23	26	28	29	28	27	27	28	28	25	22	26 °C
	tn	12	12	16	21	22	23	23	23	23	22	18	13	12 °C
	clothing	s	t	t	t	t	t	t	t	t	t	t	s	–
	stress	o	f	f	ff	ff	ff	ff	ff	ff	ff	f	o	–
	p	2	4	10	50	391	1151	1400	1134	577	286	130	19	5154 mm
	dp	0	0	1	2	11	25	28	27	19	9	4	1	127 days

Rangoon

latitude: 16°46'n longitude: 96°10'e height: 23 m

Amha	056-5	Jan	Feb	Mar	Apr	May	Jun	Jul	Aug	Sep	Oct	Nov	Dec	Year
	tx	34	36	38	39	38	33	32	32	32	33	33	33	39 °C
	t	25	26	29	30	29	27	27	27	27	27	27	25	27 °C
	tn	16	16	19	23	23	23	23	23	23	23	20	17	15 °C
	clothing	t	t	t	t	t	t	t	t	t	t	t	t	–
	stress	f	f	ff	ff	ff	ff	ff	ff	ff	ff	ff	f	–
	p	3	6	8	50	308	481	580	529	393	181	68	11	2618 mm
	dp	0	0	1	2	14	23	26	25	20	10	3	1	125 days
	sunshine	301	266	301	306	217	105	84	93	138	186	192	260	2448 hours
	%	86	81	81	81	54	27	21	24	38	51	56	75	55 %

Diamond Island

Amha	056-6	Jan	Feb	Mar	Apr	May	Jun	Jul	Aug	Sep	Oct	Nov	Dec	Year
				latitude: 15°51'n				longitude: 94°19'e				height: 13 m		
	tx	31	31	31	33	31	32	30	31	31	32	32	31	33 °C
	t	26	26	27	28	29	27	27	27	27	27	27	26	27 °C
	tn	21	21	23	24	23	22	23	23	23	23	23	21	20 °C
	clothing	t	t	t	t	t	t	t	t	t	t	t	t	–
	stress	ff	ff	ff	ff	ff	ff	ff	ff	ff	ff	ff	ff	–
	p	3	3	5	33	284	640	701	648	439	213	127	20	3117 mm
2.5 mm	dp	0	0	0	1	11	21	23	22	17	11	5	1	113 days

Mergui

Amha	056-7	Jan	Feb	Mar	Apr	May	Jun	Jul	Aug	Sep	Oct	Nov	Dec	Year
				latitude: 12°26'n				longitude: 98°36'e				height: 20 m		
	tx	33	34	35	35	34	33	32	32	32	32	33	32	36 °C
	t	26	27	27	28	27	27	27	27	27	27	26	26	27 °C
	tn	17	18	21	22	22	22	22	22	22	21	19	17	16 °C
	clothing	t	t	t	t	t	t	t	t	t	t	t	t	–
	stress	f	ff	ff	ff	ff	ff	ff	ff	ff	ff	ff	f	–
	p	25	53	79	124	424	762	836	762	632	307	96	20	4120 mm
2.5 mm	dp	1	3	5	6	18	25	26	26	23	16	6	2	157 days

Victoria Point

Amha	056-8	Jan	Feb	Mar	Apr	May	Jun	Jul	Aug	Sep	Oct	Nov	Dec	Year
				latitude: 9°58'n				longitude: 98°35'e				height: 47 m		
	t	27	28	28	29	27	26	26	26	26	26	26	26	27 °C
	clothing	t	t	t	t	t	t	t	t	t	t	t	t	–
	stress	f	ff	ff	ff	ff	ff	ff	ff	ff	ff	ff	f	–
	p	10	17	55	130	503	720	731	663	711	447	161	58	4206 mm
	dp	1	1	4	8	20	23	24	22	22	18	11	4	158 days

4.2.56.2. Predominant general weather situations are described in 4.2.0.1. under South-east Asia. A hygrothermal diagram of Rangoon is to be found in 4.2.0.2.

4.2.57. Thailand

4.2.57.1. Climatic description and tables

1. a) greatest part of the country: Tropical Summer Rain Climate Aw and Tropical Monsoonal Rain Climate Am
   b) Southern part of the country: Tropical Rain Climate Ar
2. High sun period: very hot in the lowlands; low sun period: hot in the north, very hot in the south in the lowlands.
3. High sun period: in the morning often dry and sunny, in the afternoon often showers; many rainy days. Low sun period: generally dry and sunny, at times scattered showers, however in the south frequent showers or rain.
4. Wind: occasional gales, squalls in thunderstorms.

Chiang Mai

latitude: 18°47′n longitude: 98°59′e height: 312 m

Awhb	057-1	Jan	Feb	Mar	Apr	May	Jun	Jul	Aug	Sep	Oct	Nov	Dec	Year
	tx	32	34	37	39	38	34	34	34	34	33	33	31	39 °C
	t	21	23	23	29	29	28	27	27	27	26	24	22	26 °C
	tn	9	11	13	18	21	22	22	23	22	18	15	11	9 °C
	clothing	s	t	t	t	t	t	t	t	t	t	t	s	–
	stress	f	f	f	ff	ff	ff	ff	ff	ff	ff	f	f	–
	p	7	12	15	49	144	146	188	231	289	126	39	10	1254 mm
	dp	1	1	2	5	12	16	18	21	18	10	4	1	108 days
	sunshine	270	266	285	264	236	170	147	137	166	226	247	270	2684 hours
	%	78	87	76	70	58	43	36	35	45	62	73	79	60 %

Udon Thani

latitude: 17°26′n longitude: 102°46′e height: 178 m

Awhb	057-2	Jan	Feb	Mar	Apr	May	Jun	Jul	Aug	Sep	Oct	Nov	Dec	Year
	t	22	25	28	30	30	29	28	28	28	27	25	22	27 °C
	clothing	s	t	t	t	t	t	t	t	t	t	t	s	–
	stress	f	ff	ff	ff	ff	ff	ff	ff	ff	ff	f	f	–
	p	6	10	40	72	172	205	251	313	310	132	26	2	1539 mm
	dp	1	3	5	7	15	16	17	18	17	7	2	0	126 days

Phitsanulok

latitude: 16°50′n longitude: 100°16′e height: 50 m

Awha	057-3	Jan	Feb	Mar	Apr	May	Jun	Jul	Aug	Sep	Oct	Nov	Dec	Year
	t	25	27	29	31	30	29	28	28	28	28	27	24	28 °C
	clothing	t	t	t	t	t	t	t	t	t	t	t	t	–
	stress	f	ff	ff	ff	ff	ff	ff	ff	ff	ff	ff	f	–
	p	7	24	38	75	160	179	206	226	275	133	37	2	1362 mm
	dp	1	2	3	5	12	15	16	18	18	10	3	0	103 days
	sunshine	264	250	262	281	271	195	160	156	152	221	254	264	2731 hours
	%	76	76	70	75	68	49	40	40	41	61	75	76	62 %

Nakhon Sawan

latitude: 15°48′n longitude: 100°10′e height: 28 m

Awha	057-4	Jan	Feb	Mar	Apr	May	Jun	Jul	Aug	Sep	Oct	Nov	Dec	Year
	t	25	28	30	31	31	29	29	28	28	28	26	24	28 °C
	clothing	t	t	t	t	t	t	t	t	t	t	t	t	–
	stress	f	ff	ff	ff	ff	ff	ff	ff	ff	ff	ff	f	–
	p	5	33	26	65	137	141	148	179	274	151	26	3	1188 mm
	dp	1	2	2	4	10	12	13	14	15	9	2	0	84 days
	sunshine	263	248	246	251	235	167	146	138	140	198	242	271	2545 hours
	%	74	75	66	67	59	43	36	35	38	54	71	78	57 %

Nakhon Ratchasima

latitude: 14°58′n longitude: 102°07′e height: 181 m

Awhb	057-5	Jan	Feb	Mar	Apr	May	Jun	Jul	Aug	Sep	Oct	Nov	Dec	Year
	tx	36	39	41	41	38	37	36	34	36	35	35	34	41 °C
	t	23	26	29	30	30	29	28	28	27	26	24	22	27 °C
	tn	10	13	18	20	22	22	22	22	22	19	14	10	9 °C
	clothing	t	t	t	t	t	t	t	t	t	t	t	s	–
	stress	f	f	ff	ff	ff	ff	ff	ff	ff	ff	f	f	–
	p	7	33	45	83	157	111	132	139	244	171	37	3	1162 mm
	dp	1	3	5	8	15	13	14	15	18	12	3	1	108 days
	sunshine	284	244	244	234	246	203	204	180	161	223	255	261	2739 hours
	%	81	74	65	63	62	52	51	46	44	61	74	75	62 %

Bangkok

latitude: 13°44′n longitude: 100°30′e height: 12 m

Awha	057-6	Jan	Feb	Mar	Apr	May	Jun	Jul	Aug	Sep	Oct	Nov	Dec	Year
	tx	36	37	38	38	38	36	36	36	35	35	34	34	39 °C
	t	26	28	29	30	30	29	28	28	28	28	27	26	28 °C
	tn	15	18	20	22	23	23	22	23	22	22	19	16	14 °C
	clothing	t	t	t	t	t	t	t	t	t	t	t	t	–
	stress	f	ff	ff	ff	ff	ff	ff	ff	ff	ff	ff	f	–
	p	9	29	34	89	166	171	178	191	306	255	57	7	1492 mm
	dp	2	3	3	6	14	16	18	18	21	16	6	1	124 days
	sunshine	288	251	271	244	234	180	149	157	138	200	256	285	2655 hours
	%	81	76	73	65	59	46	37	40	38	54	74	81	60 %

Aranyaprathet

latitude: 13°42′n longitude: 102°35′e height: 44 m

Awha	057-7	Jan	Feb	Mar	Apr	May	Jun	Jul	Aug	Sep	Oct	Nov	Dec	Year
	t	25	28	30	30	30	29	28	28	28	27	26	24	28 °C
	clothing	t	t	t	t	t	t	t	t	t	t	t	t	–
	stress	f	ff	ff	ff	ff	ff	ff	ff	ff	ff	ff	f	–
	p	7	31	67	113	177	184	219	190	270	197	57	12	1524 mm
	dp	1	3	6	10	17	19	20	20	21	15	6	1	139 days

Chanthaburi

latitude: 12°37′n longitude: 102°07′e height: 5 m

Amha	057-8	Jan	Feb	Mar	Apr	May	Jun	Jul	Aug	Sep	Oct	Nov	Dec	Year
	t	26	27	28	28	28	28	27	27	27	27	26	25	27 °C
	clothing	t	t	t	t	t	t	t	t	t	t	t	t	–
	stress	f	ff	ff	ff	ff	ff	ff	ff	ff	ff	f	f	–
	p	21	38	67	142	352	508	557	501	558	268	81	6	3098 mm
	dp	2	4	7	10	19	24	24	23	23	16	7	1	160 days
	sunshine	279	240	235	236	188	157	113	133	105	177	252	293	2408 hours
	%	78	72	63	63	48	41	28	34	29	48	73	83	54 %

Brachuap Khirikhan

latitude: 11°48′n longitude: 99°48′e height: 5 m

Awha	057-9	Jan	Feb	Mar	Apr	May	Jun	Jul	Aug	Sep	Oct	Nov	Dec	Year
	t	24	26	27	29	29	28	28	28	28	27	26	25	27 °C
	clothing	t	t	t	t	t	t	t	t	t	t	t	t	–
	stress	f	f	ff	ff	ff	ff	ff	ff	ff	ff	f	f	–
	p	33	42	48	81	114	92	97	90	111	253	172	31	1164 mm
	dp	2	3	4	6	10	14	15	15	14	14	9	2	108 days
	sunshine	280	258	237	231	215	162	108	138	128	176	231	290	2454 hours
	%	78	78	63	62	55	42	27	36	35	48	66	82	55 %

Chumphon

latitude: 10°27′n longitude: 99°15′e height: 3 m

Arha	057-10	Jan	Feb	Mar	Apr	May	Jun	Jul	Aug	Sep	Oct	Nov	Dec	Year
	t	25	26	27	28	28	28	27	27	27	27	26	24	27 °C
	clothing	t	t	t	t	t	t	t	t	t	t	t	t	–
	stress	f	ff	ff	ff	ff	ff	ff	ff	ff	ff	ff	f	–
	p	68	70	78	122	161	161	192	172	172	318	327	192	2033 mm
	dp	5	5	5	9	14	17	18	18	17	18	13	7	146 days
	sunshine	236	251	256	198	205	180	170	181	161	149	168	196	2351 hours
	%	66	75	68	53	52	47	43	47	44	40	48	55	53 %

Phuket

latitude: 7°58′n longitude: 98°24′e height: 3 m

Amha	057-11	Jan	Feb	Mar	Apr	May	Jun	Jul	Aug	Sep	Oct	Nov	Dec	Year
	tx	33	34	34	34	33	33	32	33	32	32	32	32	34 °C
	t	27	28	29	29	28	28	28	28	28	28	27	27	28 °C
	tn	21	22	22	23	23	23	23	22	22	23	22	22	21 °C
	clothing	t	t	t	t	t	t	t	t	t	t	t	t	–
	stress	ff	ff	ff	ff	ff	ff	ff	ff	ff	ff	ff	ff	–
	p	36	38	74	127	297	264	216	244	328	315	193	79	2211 mm
	dp	4	3	6	15	19	19	17	17	19	19	14	8	160 days

Songkhla

latitude: 7°11′n longitude: 100°37′e height: 10 m

Amha	057-12	Jan	Feb	Mar	Apr	May	Jun	Jul	Aug	Sep	Oct	Nov	Dec	Year
	t	27	27	28	28	29	28	28	28	28	27	27	26	28 °C
	clothing	t	t	t	t	t	t	t	t	t	t	t	t	–
	stress	ff	ff	ff	ff	ff	ff	ff	ff	ff	ff	ff	ff	–
	p	157	59	58	91	119	101	94	95	105	316	576	439	2208 mm
	dp	14	7	7	10	14	12	12	13	9	22	23	20	163 days

4.2.57.2. Predominant general weather situations are described in 4.2.0.1. under South-east Asia. A hygrothermal diagram of Bangkok is to be found in 4.2.0.2.

Wind name: Kite or junk winds are fresh south-westerly on-shore winds in the Gulf of Siam.

4.2.58. Viet Nam

4.2.58.1. Climatic description and tables

1. Tropical Summer Rain Climate Aw and Tropical Monsoonal Rain Climate Am (except the northernmost part of the country with Subtropical Summer Rain Climate Cw)
2. High sun period: very hot; low sun period: warm in the north, hot in the south.
3. High sun period: often dry and sunny in the morning, in the afternoon showers; frequent rainy days. Low sun period: normally dry and sunny.
4. Wind: occasional gales, squalls in thunderstorms.

Hanoi

latitude: 21°02'n longitude: 105°52'e height: 16 m

Cwhl	058-1	Jan	Feb	Mar	Apr	May	Jun	Jul	Aug	Sep	Oct	Nov	Dec	Year
	tx	28	29	32	34	37	38	37	36	35	33	31	29	39 °C
	t	17	18	20	24	28	30	29	29	28	25	22	19	24 °C
	tn	9	9	12	16	19	22	23	23	21	18	12	10	8 °C
	clothing	s	s	s	t	t	t	t	t	t	t	s	s	–
	stress		o	o	f	ff	ff	ff	ff	ff	f	o		–
	p	18	28	38	81	196	239	323	343	254	99	43	20	1682 mm
	dp	7	13	15	14	15	14	15	16	14	9	7	7	146 days

Hoàng-Sa (Pattle) / Paracel Islands

latitude: 16°33'n longitude: 111°37'e height: 6 m

Awha	058-2	Jan	Feb	Mar	Apr	May	Jun	Jul	Aug	Sep	Oct	Nov	Dec	Year
	t	24	24	26	28	29	29	29	29	28	27	26	24	27 °C
	clothing	t	t	t	t	t	t	t	t	t	t	t	t	–
	stress	f	f	f	ff	ff	ff	ff	ff	ff	ff	ff	ff	–
	p	30	21	28	52	57	135	106	202	228	302	130	57	1348 mm
	dp	5	3	3	4	5	7	5	9	12	14	11	10	88 days

Dà-Nang (Tourane)

latitude: 16°02'n longitude: 108°11'e height: 7 m

Amhb	058-3	Jan	Feb	Mar	Apr	May	Jun	Jul	Aug	Sep	Oct	Nov	Dec	Year
	t	21	23	24	26	28	29	29	29	27	26	24	22	26 °C
	clothing	s	s	t	t	t	t	t	t	t	t	t	s	–
	stress	f	f	f	f	ff	ff	ff	ff	ff	ff	f	f	–
	p	91	42	39	43	38	130	60	87	283	533	417	206	1970 mm
	dp	10	4	3	3	6	6	4	7	13	20	17	15	108 days

Nha-Trang

latitude: 12°15'n longitude: 109°12'e height: 6 m

Awha	058-4	Jan	Feb	Mar	Apr	May	Jun	Jul	Aug	Sep	Oct	Nov	Dec	Year
	t	24	24	26	27	28	28	28	28	27	26	25	24	26 °C
	clothing	t	t	t	t	t	t	t	t	t	t	t	t	–
	stress	f	f	f	ff	ff	ff	ff	ff	ff	ff	ff	ff	–
	p	28	12	16	33	53	49	50	52	127	259	269	148	1095 mm
	dp	6	3	2	3	5	5	4	6	13	15	14	10	87 days

Saigon

Awha	058-5	Jan	Feb	Mar	Apr	May	Jun	Jul	Aug	Sep	Oct	Nov	Dec	Year
		latitude: 10°49′n					longitude: 106°40′e					height: 10 m		
	tx	34	35	36	37	36	34	33	33	33	33	33	33	37 °C
	t	26	26	28	29	28	27	27	27	27	26	26	26	27 °C
	tn	17	19	21	22	23	22	22	22	22	21	20	18	17 °C
	clothing	t	t	t	t	t	t	t	t	t	t	t	t	–
	stress	f	ff	ff	ff	ff	ff	ff	ff	ff	ff	ff	f	–
	p	6	13	12	65	196	285	242	277	292	259	122	37	1808 mm
	dp	1	1	2	4	15	19	18	21	20	18	9	4	133 days
	sunshine	143	164	158	165	124	105	121	140	156	115	111	121	1623 hours
	%	40	49	42	44	32	27	31	36	43	31	34	37	37 %

4.2.58.2. Predominant general weather situations are described in 4.2.0.1. under Southeast Asia. A hygrothermal diagram of Saigon is to be found in 4.2.0.2.
Wind name: Crachin is special weather with wet fog in the Gulf of Tongking during the winter monsoon season with northwesterly wind, especially in January and February.

4.2.59. Laos

4.2.59.1. Climatic description and tables

1. Tropical Summer Rain Climate Aw and
   Tropical Summer Rain Mountain Climate GAw
2. Most of the year very hot, low sun period partly warm.
3. High sun period: often dry and sunny in the morning, in the afternoon showers; frequent rainy days. Low sun period: normally dry and sunny.
4. Wind: occasional gales, squalls in thunderstorms.

Luang Prabang

Awhb	059-1	Jan	Feb	Mar	Apr	May	Jun	Jul	Aug	Sep	Oct	Nov	Dec	Year
		latitude: 19°53′n					longitude: 102°08′e					height: 304 m		
	tx	33	35	38	39	39	37	36	36	36	35	33	31	40 °C
	t	20	23	26	28	28	28	27	27	27	26	23	20	24 °C
	tn	9	10	12	17	20	21	22	21	19	16	13	11	8 °C
	clothing	s	t	t	t	t	t	t	t	t	t	t	s	–
	stress	f	f	f	ff	ff	ff	ff	ff	ff	f	f	f	–
	p	17	20	37	78	128	157	204	261	146	63	16	10	1137 mm
0.1 mm	dp	2	2	4	8	13	13	17	19	12	6	3	1	100 days
	sunshine	155	168	186	150	155	150	124	124	180	217	150	124	1883 hours
	%	45	52	50	40	38	38	30	31	49	60	45	37	42 %

Vientiane

Awhb	059-2	Jan	Feb	Mar	Apr	May	Jun	Jul	Aug	Sep	Oct	Nov	Dec	Year
		latitude: 17°57′n					longitude: 102°34′e					height: 170 m		
	tx	32	35	37	38	37	34	33	34	33	33	32	32	38 °C
	t	22	24	27	29	28	28	28	27	27	26	24	21	26 °C
	tn	9	12	15	19	22	22	22	23	22	17	13	11	8 °C
	clothing	t	t	t	t	t	t	t	t	t	t	t	s	–
	stress	f	f	f	ff	ff	ff	ff	ff	ff	ff	f	f	–
	p	15	14	25	78	225	261	260	354	385	49	16	1	1683 mm
0.1 mm	dp	1	2	4	7	15	17	18	18	16	7	2	1	108 days
	sunshine	248	224	217	240	217	150	155	155	240	248	240	248	2582 hours
	%	72	69	58	64	54	38	38	39	65	68	71	72	58 %

Seno

Awhb	059-3	Jan	Feb	Mar	Apr	May	Jun	Jul	Aug	Sep	Oct	Nov	Dec	Year
		latitude: 16°40'n				longitude: 105°00'e				height: 184 m				
	t	21	24	27	29	28	28	27	27	27	25	27	21	26 °C
	clothing	s	t	t	t	t	t	t	t	t	t	t	s	–
	stress	o	f	f	ff	ff	ff	ff	ff	ff	ff	ff	f	–
	p	6	18	21	85	180	252	242	342	278	59	3	0	1468 mm
	dp													days
	sunshine	310	252	279	270	217	180	186	155	150	248	300	310	2857 hours
	%	89	77	75	72	54	46	46	39	41	68	88	90	64 %

Paksé

Awha	059-4	Jan	Feb	Mar	Apr	May	Jun	Jul	Aug	Sep	Oct	Nov	Dec	Year
		latitude: 15°07'n				longitude: 105°45'e				height: 93 m				
	t	25	27	29	30	29	29	28	27	27	27	26	24	27 °C
	clothing	t	t	t	t	t	t	t	t	t	t	t	t	–
	stress	f	f	ff	ff	ff	ff	ff	ff	ff	ff	f	f	–
	p	2	8	32	38	140	264	246	418	296	84	13	1	1542 mm

4.2.59.2. Predominant general weather situations are described in 4.2.0.1. under South-east Asia. A hygrothermal diagram of Vientiane is to be found in 4.2.0.2.

4.2.60. Khmer (Cambodia)

4.2.60.1. Climatic description and tables

1. Tropical Summer Rain Climate Aw and
   Tropical Summer Rain Mountain Climate GAw
2. Most of the year very hot.
3. High sun period: often dry and sunny in the morning, in the afternoon showers; frequent rainy days. Low sun period: normally dry and sunny.
4. Wind: occasional gales, squalls in thunderstorms.

Stung Treng

Awha	060-1	Jan	Feb	Mar	Apr	May	Jun	Jul	Aug	Sep	Oct	Nov	Dec	Year
		latitude: 13°31'n				longitude: 105°58'e				height: 54 m				
	t	24	26	29	30	28	28	27	27	26	26	25	24	27 °C
	clothing	t	t	t	t	t	t	t	t	t	t	t	t	–
	stress	f	f	ff	ff	ff	ff	ff	ff	ff	ff	ff	f	–
	p	1	11	18	73	213	259	254	344	299	183	68	6	1729 mm
	dp	0	1	2	5	14	15	17	20	19	12	6	1	112 days

Siemréap (Angkor)

Awha	060-2	Jan	Feb	Mar	Apr	May	Jun	Jul	Aug	Sep	Oct	Nov	Dec	Year
		latitude: 13°22'n				longitude: 103°51'e				height: 15 m				
	t	24	26	28	29	28	28	27	27	27	26	25	24	27 °C
	clothing	t	t	t	t	t	t	t	t	t	t	t	t	–
	stress	f	f	ff	ff	ff	ff	ff	ff	ff	ff	ff	f	–
	p	0	8	22	46	101	144	181	219	218	263	73	6	1281 mm
	dp	0	1	2	5	10	15	15	17	18	16	7	1	106 days

Kompong Cham

Awha	060-3	latitude: 12°00'n						longitude: 105°27'e				height: 16 m	
		Jan	Feb	Mar	Apr	May	Jun	Jul	Aug	Sep	Oct	Nov	Dec	Year
	t	26	27	29	29	28	28	27	27	27	27	26	25	27 °C
	clothing	t	t	t	t	t	t	t	t	t	t	t	t	–
	stress	f	f	ff	ff	ff	ff	ff	ff	ff	ff	ff	f	–
	p	2	6	29	77	246	228	219	274	253	246	112	6	1698 mm

Pochentong (Phom Penh)

Awha	060-4	latitude: 11°33'n						longitude: 104°51'e				height: 10 m	
		Jan	Feb	Mar	Apr	May	Jun	Jul	Aug	Sep	Oct	Nov	Dec	Year
	t	26	28	29	29	29	28	28	28	27	27	27	25	28 °C
	clothing	t	t	t	t	t	t	t	t	t	t	t	t	–
	stress	f	f	ff	ff	ff	ff	ff	ff	ff	ff	ff	f	–
	p	9	8	28	73	146	129	129	147	231	250	134	36	1320 mm
	dp	1	1	2	5	12	11	13	14	16	16	8	2	102 days
	sunshine	285	258	266	248	228	194	179	178	155	201	232	270	2694 hours
	%	80	78	71	67	58	51	45	46	42	55	67	76	61 %

Kampot

Awha	060-5	latitude: 10°37'n						longitude: 104°13'e				height: 5 m	
		Jan	Feb	Mar	Apr	May	Jun	Jul	Aug	Sep	Oct	Nov	Dec	Year
	t	26	27	28	28	28	27	27	27	27	27	26	26	27 °C
	clothing	t	t	t	t	t	t	t	t	t	t	t	t	–
	stress	f	ff	ff	ff	ff	ff	ff	ff	ff	ff	ff	ff	–
	p	15	33	89	141	168	226	248	390	267	280	164	45	2066 mm
	dp	3	3	6	9	12	16	16	20	18	16	10	3	132 days

4.2.60.2. Predominant general weather situations are described in 4.2.0.1. under Southeast Asia.

4.2.61. Malaysia

4.2.61.1. Climatic description and tables

1. Tropical Rain Climate Ar and Tropical Rain Mountain Climate GAr
2. Very hot all the time.
3. Predominantly dry and sunny in the morning, showers in the afternoon, frequent rainy days.
4. Wind: occasional gales, squalls in thunderstorms.

Kota Bharu

Arha	061-1	latitude: 6°10'n						longitude: 102°17'e				height: 9 m	
		Jan	Feb	Mar	Apr	May	Jun	Jul	Aug	Sep	Oct	Nov	Dec	Year
	t	26	26	27	28	28	28	27	27	27	27	26	26	27 °C
	clothing	t	t	t	t	t	t	t	t	t	t	t	t	–
	stress	ff	ff	ff	ff	ff	ff	ff	ff	ff	ff	ff	ff	–
	p	211	73	112	87	142	145	155	150	191	326	617	546	2755 mm
	dp	9	6	6	6	8	9	11	12	14	17	21	19	138 days
	sunshine	190	229	262	258	231	209	215	222	194	176	146	153	2485 hours
	%	52	68	70	70	60	56	56	58	53	47	41	42	56 %

Sandakan (Sarawak)

latitude: 5°54′n longitude: 118°04′e height: 46 m

Araa	061-2	Jan	Feb	Mar	Apr	May	Jun	Jul	Aug	Sep	Oct	Nov	Dec	Year
	tx	31	31	32	33	33	33	33	33	33	33	32	32	33 °C
	t	26	26	27	27	27	27	27	27	27	27	27	26	27 °C
	tn	22	22	23	23	23	23	23	23	23	23	23	23	21 °C
	clothing	t	t	t	t	t	t	t	t	t	t	t	t	–
	stress	ff	ff	ff	ff	ff	ff	ff	ff	ff	ff	ff	ff	–
	p	431	236	198	101	132	200	185	239	242	271	340	465	3040 mm
	dp	19	13	13	11	10	14	14	12	15	18	19	20	178 days
	sunshine	158	172	223	264	239	210	208	226	210	205	156	177	2448 hours
	%	43	51	59	72	62	56	54	59	58	55	44	48	55 %

Penang

latitude: 5°25′n longitude: 100°19′e height: 5 m

Arha	061-3	Jan	Feb	Mar	Apr	May	Jun	Jul	Aug	Sep	Oct	Nov	Dec	Year
	tx	34	34	34	34	34	34	33	33	33	33	33	33	35 °C
	t	28	28	28	28	28	28	28	27	27	27	27	27	28 °C
	tn	21	22	22	22	22	22	22	22	22	22	22	22	21 °C
	Clothing	t	t	t	t	t	t	t	t	t	t	t	t	–
	stress	ff	ff	ff	ff	ff	ff	ff	ff	ff	ff	ff	ff	–
	p	94	79	142	188	272	196	190	295	401	429	302	147	2735 mm
0.25 mm	dp	8	7	11	14	16	12	12	15	18	21	19	11	164 days
	sunshine	239	229	236	225	202	204	208	192	162	170	168	202	2436 hours
	%	65	68	63	61	53	55	54	50	44	46	47	55	55 %

Kuala Trengganu

latitude: 5°20′n longitude: 103°05′e height: 32 m

Arha	061-4	Jan	Feb	Mar	Apr	May	Jun	Jul	Aug	Sep	Oct	Nov	Dec	Year
	tx	30	31	32	33	34	33	33	33	33	32	32	30	34 °C
	t	25	26	27	27	28	27	27	27	27	27	26	26	27 °C
	tn	19	19	21	22	22	21	21	21	21	22	22	21	19 °C
	clothing	t	t	t	t	t	t	t	t	t	t	t	t	–
	stress	ff	ff	ff	ff	ff	ff	ff	ff	ff	ff	ff	ff	–
	p	292	163	160	155	135	109	117	147	190	279	610	554	2911 mm
0.25 mm	dp	19	13	12	10	12	10	12	12	17	20	22	23	182 days
	sunshine	180	203	260	246	226	192	217	211	201	180	144	152	2412 hours
	%	49	60	69	67	59	51	56	55	55	48	40	41	54 %

Cameron Highlands

latitude: 4°28′n longitude: 101°23′e height: 1449 m

GArbb	061-5	Jan	Feb	Mar	Apr	May	Jun	Jul	Aug	Sep	Oct	Nov	Dec	Year
	tx	24	24	25	25	25	25	25	24	24	24	24	24	26 °C
	t	18	18	18	18	18	18	18	18	18	18	18	18	18 °C
	tn	8	8	8	11	11	10	9	10	11	11	11	9	7 °C
	clothing	s	s	s	s	s	s	s	s	s	s	s	s	–
	stress	o	o	o	o	o	o	o	o	o	o	o	o	–
	p	168	132	216	292	246	140	122	163	262	340	330	229	2640 mm
0.25 mm	dp	17	14	19	23	22	16	15	19	22	26	24	21	238 days
	sunshine	164	143	174	147	146	147	155	136	120	112	102	115	1664 hours
	%	44	42	46	40	38	40	40	36	33	30	29	31	38 %

Kuala Lumpur

latitude: 3°07′n longitude: 101°42′e height: 38 m

Arha	061-6	Jan	Feb	Mar	Apr	May	Jun	Jul	Aug	Sep	Oct	Nov	Dec	Year
	tx	34	35	35	35	35	34	34	34	34	34	34	34	36 °C
	t	27	27	27	27	28	28	27	27	27	27	27	27	27 °C
	tn	20	21	21	22	22	22	21	21	21	21	22	21	19 °C
	clothing	t	t	t	t	t	t	t	t	t	t	t	t	–
	stress	ff	ff	ff	ff	ff	ff	ff	ff	ff	ff	ff	ff	–
	p	157	209	277	285	207	121	117	157	206	251	289	223	2499 mm
	dp	10	9	17	16	13	10	10	12	13	18	19	16	163 days
	sunshine	189	209	203	188	197	200	202	188	170	164	148	165	2223 hours
	%	51	62	54	51	52	54	53	50	47	44	41	45	50 %

Malacca

latitude: 2°12′n longitude: 102°16′e height: 45 m

Araa	061-7	Jan	Feb	Mar	Apr	May	Jun	Jul	Aug	Sep	Oct	Nov	Dec	Year
	tx	33	33	34	32	33	33	33	33	33	32	33	33	33 °C
	t	27	27	27	27	27	27	27	27	27	27	27	27	27 °C
	tn	21	21	21	21	21	21	21	21	21	21	21	21	21 °C
	clothing	t	t	t	t	t	t	t	t	t	t	t	t	–
	stress	ff	ff	ff	ff	ff	ff	ff	ff	ff	ff	ff	ff	–
	p	99	94	124	188	173	201	198	262	224	257	221	165	2206 mm
0.25 mm	dp	9	8	10	13	12	12	12	15	14	16	17	14	152 days
	sunshine	189	203	232	201	220	204	205	195	183	189	168	174	2364 hours
	%	51	60	62	55	58	55	54	52	50	50	47	47	53 %

Kuching (Sarawak)

latitude: 1°29′n longitude: 110°20′e height: 26 m

Arha	061-8	Jan	Feb	Mar	Apr	May	Jun	Jul	Aug	Sep	Oct	Nov	Dec	Year
	tx	32	33	33	34	34	34	35	34	35	34	33	33	36 °C
	t	26	26	27	28	28	28	27	27	27	27	27	27	27 °C
	tn	21	21	21	21	22	21	21	21	21	21	21	21	19 °C
	clothing	t	t	t	t	t	t	t	t	t	t	t	t	–
	stress	ff	ff	ff	ff	ff	ff	ff	ff	ff	ff	ff	ff	–
	p	610	511	328	279	262	180	196	234	218	267	358	462	3905 mm
0.1 mm	dp	24	21	22	20	21	15	18	19	20	24	26	25	255 days
	sunshine	121	124	155	165	183	183	177	167	144	146	135	136	1836 hours
	%	32	36	41	45	48	50	47	44	40	39	37	36	41 %

4.2.61.2. Predominant general weather situations are described in 4.2.0.1. under Southeast Asia. (Hygrothermal diagram: see Singapore).

Wind names: Pibut are short sharp squalls which occur on the east coast of Malacca before the southwest monsoon starts.

Sumatra is a squall which occurs in the Malacca Straits, blowing from the southwest. It is accompanied by a characteristic cloud-formation, a heavy bank or arch of cumulo-nimbus which rises to a great height. These squalls are more frequent at night, and between April to October. They are generally accompanied by thunder and lightning and torrential rain.

4.2.62. Singapore

4.2.62.1. Climatic description and table

1. Tropical Rain Climate Ar
2. Permanently very hot.
3. Predominantly dry and sunny in the morning, showers in the afternoon, frequent rainy days.
4. Wind: occasional gales, squalls in thunderstorms.

Singapore

		latitude: 1°21′n				longitude: 103°54′e				height: 10 m				
Arha	062-1	Jan	Feb	Mar	Apr	May	Jun	Jul	Aug	Sep	Oct	Nov	Dec	Year
	tx	32	33	33	33	33	33	33	33	33	33	33	32	34 °C
	t	26	27	27	28	28	28	27	27	27	27	27	26	27 °C
	tn	21	21	22	22	23	22	22	22	22	22	22	22	21 °C
	clothing	t	t	t	t	t	t	t	t	t	t	t	t	–
	stress	ff	ff	ff	ff	ff	ff	ff	ff	ff	ff	ff	ff	–
	p	285	164	154	160	131	177	163	200	122	184	236	306	2282 mm
	dp	15	11	11	11	12	11	11	13	11	13	16	15	150 days
	sunshine	156	178	189	178	184	182	196	185	166	162	140	142	2058 hours
	%	42	52	50	49	49	50	52	49	46	43	39	38	46 %

4.2.62.2. Predominant general weather situations are described in 4.2.0.1. under Southeast Asia. A hygrothermal diagram of Singapore is given in 4.2.0.2.

4.2.63. Indonesia

4.2.63.1. Climatic descriptions and tables

1. Climatic area: Indonesia except southern Celebes, eastern Java, southwestern New Guinea and islands between them

1.1. Tropical Rain Climate Ar and Tropical Rain Mountain Climate GAr

1.2. Very hot all the time.

1.3. Predominantly dry and sunny in the morning, showers in the afternoon, frequent rainy days.

1.4. Wind: occasional gales, squalls in thunderstorms.

Takengön – Sumatra

		latitude: 4°40′n				longitude: 96°50′e				height: 1186 m				
GArbb	063-1	Jan	Feb	Mar	Apr	May	Jun	Jul	Aug	Sep	Oct	Nov	Dec	Year
	tx	28	29	29	29	29	29	29	29	29	28	28	28	28 °C
	t	21	21	21	22	22	21	21	21	21	21	21	21	21 °C
	tn	13	12	13	14	13	12	12	12	12	13	13	13	10 °C
	clothing	s	s	s	s	s	s	s	s	s	s	s	s	–
	stress	o	o	o	o	o	o	o	o	o	o	o	o	–
	p	167	119	183	165	119	58	56	84	142	196	221	224	1734 mm
0.1 mm	dp	12	9	13	14	12	8	7	10	13	17	17	16	148 days

Medan/Polonia – Sumatra

		latitude: 3°34′n					longitude: 98°41′e				height: 25 m		
Araa	063-2	Jan	Feb	Mar	Apr	May	Jun	Jul	Aug	Sep	Oct	Nov	Dec	Year
	tx	32	33	33	34	34	34	34	34	33	32	32	32	35 °C
	t	26	26	26	26	27	27	26	26	26	26	26	26	26 °C
	tn	19	20	20	21	21	21	20	21	21	21	21	20	19 °C
	clothing	t	t	t	t	t	t	t	t	t	t	t	t	–
	stress	ff	ff	ff	ff	ff	ff	ff	ff	ff	ff	ff	ff	–
	p	138	94	110	151	245	150	173	196	207	256	247	158	2125 mm
	dp	13	8	10	13	14	11	13	14	16	19	19	16	166 days

Tarakan – Borneo

		latitude: 3°19′n					longitude: 117°36′e				height: 12 m		
Araa	063-3	Jan	Feb	Mar	Apr	May	Jun	Jul	Aug	Sep	Oct	Nov	Dec	Year
	tx	32	32	32	32	32	32	32	32	32	33	32	32	33 °C
	t	26	27	27	27	27	27	27	27	27	27	27	27	27 °C
	tn	22	22	22	22	22	22	22	22	22	22	22	22	22 °C
	clothing	t	t	t	t	t	t	t	t	t	t	t	t	–
	stress	ff	ff	ff	ff	ff	ff	ff	ff	ff	ff	ff	ff	–
	p	277	259	356	353	343	320	262	315	295	363	386	340	3869 mm
0.1 mm	dp	17	15	19	19	19	18	15	16	16	19	20	20	213 days

Menado – Celebes

		latitude: 1°30′n					longitude: 124°50′e				height: 2 m		
Arha	063-4	Jan	Feb	Mar	Apr	May	Jun	Jul	Aug	Sep	Oct	Nov	Dec	Year
	tx	32	32	32	32	32	32	33	34	34	33	33	32	34 °C
	t	26	26	26	27	27	27	27	28	27	27	27	27	27 °C
	tn	21	21	21	22	22	22	21	21	21	21	21	22	19 °C
	clothing	t	t	t	t	t	t	t	t	t	t	t	t	–
	stress	ff	ff	ff	ff	ff	ff	ff	ff	ff	ff	ff	ff	–
	p	465	358	305	198	160	163	119	96	86	122	218	371	2661 mm
0.5 mm	dp	21	18	16	13	13	13	9	9	8	10	14	19	163 days
	sunshine	167	175	180	231	214	219	242	264	237	217	216	208	2569 hours
	%	45	51	48	63	57	60	64	70	65	58	60	56	58 %

Pontianak – Borneo

		latitude: 0°01′s					longitude: 109°20′				height: 3 m		
Arha	063-5	Jan	Feb	Mar	Apr	May	Jun	Jul	Aug	Sep	Oct	Nov	Dec	Year
	tx	33	34	34	34	34	34	34	34	34	33	33	33	35 °C
	t	27	28	28	28	28	28	28	28	28	28	28	27	28 °C
	tn	22	22	22	23	23	23	22	22	22	23	23	23	21 °C
	clothing	t	t	t	t	t	t	t	t	t	t	t	t	–
	stress	ff	ff	ff	ff	ff	ff	ff	ff	ff	ff	ff	ff	–
	p	274	208	241	277	282	224	165	203	229	366	389	323	3181 mm
	dp	17	13	15	16	16	13	10	12	13	19	21	19	183 days

Padang – Sumatra

Araa	063-6	Jan	Feb	Mar	Apr	May	Jun	Jul	Aug	Sep	Oct	Nov	Dec	Year
				latitude: 0°56′s				longitude: 100°22′e			height: 7 m			
	tx	32	33	33	32	33	32	32	32	32	32	32	32	33 °C
	t	27	27	27	27	27	27	27	27	27	27	27	27	27 °C
	tn	22	22	22	22	23	22	22	22	22	22	22	22	21 °C
	clothing	t	t	t	t	t	t	t	t	t	t	t	t	–
	stress	ff	ff	ff	ff	ff	ff	ff	ff	ff	ff	ff	ff	–
	p	351	259	307	363	315	307	277	348	152	495	518	480	4173 mm
0.5 mm	dp	16	13	15	17	14	12	12	14	16	20	21	20	190 days
	sunshine	208	223	223	225	248	231	229	226	207	202	186	192	2600 hours
	%	55	65	59	62	66	64	61	60	57	54	51	51	59 %

Mankwari/rendani – West Irian

Araa	063-7	Jan	Feb	Mar	Apr	May	Jun	Jul	Aug	Sep	Oct	Nov	Dec	Year
				latitude: 0°53′s				longitude: 134°03′e			height: 3 m			
	t	26	26	26	26	26	26	26	26	26	27	27	27	26 °C
	clothing	t	t	t	t	t	t	t	t	t	t	t	t	–
	stress	ff	ff	ff	ff	ff	ff	ff	ff	ff	ff	ff	ff	–
	p	295	255	268	334	259	163	216	128	144	84	185	266	2597 mm
	dp	17	13	14	15	13	12	11	9	10	11	11	14	150 days
	sunshine	133	178	174	156	177	156	202	208	192	180	207	164	2126 hours
	%	35	52	46	43	47	43	54	55	53	48	57	43	48 %

Jefman – West Irian

Arha	063-8	Jan	Feb	Mar	Apr	May	Jun	Jul	Aug	Sep	Oct	Nov	Dec	Year
				latitude: 0°56′s				longitude: 131°07′e			height: 3 m			
	t	28	28	28	28	28	27	27	26	27	28	28	28	27 °C
	clothing	t	t	t	t	t	t	t	t	t	t	t	t	–
	stress	ff	ff	ff	ff	ff	ff	ff	ff	ff	ff	ff	ff	–
	p	157	130	164	206	402	284	453	265	182	186	188	164	2781 mm
	dp	14	15	15	17	19	17	22	16	17	16	14	18	200 days

Biak/Mokmer – West Irian

Araa	063-9	Jan	Feb	Mar	Apr	May	Jun	Jul	Aug	Sep	Oct	Nov	Dec	Year
				latitude: 1°11′s				longitude: 136°07′e			height: 11 m			
	t	27	27	27	27	27	27	27	27	27	27	27	27	27 °C
	clothing	t	t	t	t	t	t	t	t	t	t	t	t	–
	stress	ff	ff	ff	ff	ff	ff	ff	ff	ff	ff	ff	ff	–
	p	216	215	203	157	237	222	329	261	188	154	190	280	2652 mm
	dp	18	19	18	15	19	20	20	20	19	14	16	20	219 days

Balikpapan/Sepinggan – Borneo

Araa	063-10	Jan	Feb	Mar	Apr	May	Jun	Jul	Aug	Sep	Oct	Nov	Dec	Year
				latitude: 1°16′s				longitude: 116°54′e			height: 3 m			
	tx	32	32	32	31	32	31	29	29	31	32	32	31	33 °C
	t	27	26	27	27	27	27	26	26	27	27	27	27	27 °C
	tn	21	21	21	21	22	21	21	22	21	22	22	22	20 °C
	clothing	t	t	t	t	t	t	t	t	t	t	t	t	–
	stress	ff	ff	ff	ff	ff	ff	ff	ff	ff	ff	ff	ff	–
	p	187	172	249	196	223	265	244	230	221	140	177	235	2540 mm
	dp	15	15	18	17	17	17	17	14	15	15	14	19	193 days

Sentani – West Irian

latitude: 2°34′s longitude: 140°29′e height: 99 m

Araa	063-11	Jan	Feb	Mar	Apr	May	Jun	Jul	Aug	Sep	Oct	Nov	Dec	Year
	t	26	26	26	26	26	26	26	26	26	26	26	26	26 °C
	clothing	t	t	t	t	t	t	t	t	t	t	t	t	–
	stress	ff	ff	ff	ff	ff	ff	ff	ff	ff	ff	ff	ff	–
	p	232	239	238	183	83	75	70	108	126	84	179	210	1827 mm
	dp	16	16	17	15	11	10	10	11	10	12	14	16	158 days

Tandjoengpandan – Borneo

latitude: 2°45′s longitude: 107°39′e height: 3 m

Araa	063-12	Jan	Feb	Mar	Apr	May	Jun	Jul	Aug	Sep	Oct	Nov	Dec	Year
	tx	30	30	31	32	32	32	32	33	33	33	32	31	33 °C
	t	26	26	26	27	27	27	27	27	27	27	26	26	27 °C
	tn	22	22	22	22	22	22	21	21	21	21	22	22	20 °C
	clothing	t	t	t	t	t	t	t	t	t	t	t	t	–
	stress	ff	ff	ff	ff	ff	ff	ff	ff	ff	ff	ff	ff	–
	p	277	165	193	267	257	190	170	142	163	274	371	404	2873 mm
0.1 mm	dp	17	11	14	17	17	14	11	10	10	17	21	22	181 days

Kaimana/Utarom – West Irian

latitude: 3°40′s longitude: 133°45′e height: 3 m

Araa	063-13	Jan	Feb	Mar	Apr	May	Jun	Jul	Aug	Sep	Oct	Nov	Dec	Year
	t	27	27	27	27	27	26	25	25	26	26	27	27	26 °C
	clothing	t	t	t	t	t	t	t	t	t	t	t	t	–
	stress	ff	ff	ff	ff	ff	ff	ff	ff	ff	ff	ff	ff	–
	p	196	287	139	254	281	195	177	164	146	167	189	276	2471 mm
	dp	10	9	10	12	11	10	10	8	7	6	10	10	113 days

Ambon – Moluccas

latitude: 3°42′s longitude: 128°10′e height: 3 m

Arha	063-14	Jan	Feb	Mar	Apr	May	Jun	Jul	Aug	Sep	Oct	Nov	Dec	Year
	tx	33	33	33	32	31	29	29	29	30	32	33	33	34 °C
	t	28	28	28	27	27	26	26	26	26	27	28	28	27 °C
	tn	23	23	23	23	22	22	21	21	22	22	22	23	21 °C
	clothing	t	t	t	t	t	t	t	t	t	t	t	t	–
	stress	ff	ff	ff	ff	ff	ff	ff	ff	ff	ff	ff	ff	–
	p	127	119	135	279	516	638	602	401	241	155	114	132	3459 mm
	dp	12	13	12	12	19	21	22	21	16	10	10	13	181 days
	sunshine	192	186	211	177	158	120	115	112	150	192	219	202	2034 hours
	%	50	54	56	49	43	34	31	30	41	51	59	53	46 %

Djakarta – Java

latitude: 6°11′s longitude: 106°50′e height: 8 m

Araa	063-15	Jan	Feb	Mar	Apr	May	Jun	Jul	Aug	Sep	Oct	Nov	Dec	Year
	tx	31	31	32	32	32	32	32	32	33	33	33	32	33 °C
	t	26	26	27	27	27	27	27	27	27	27	27	27	27 °C
	tn	22	22	22	23	22	22	21	21	22	22	22	22	21 °C
	clothing	t	t	t	t	t	t	t	t	t	t	t	t	–
	stress	ff	ff	ff	ff	ff	ff	ff	ff	ff	ff	ff	ff	–
	p	335	241	201	141	116	97	61	50	78	91	151	193	1755 mm
	dp	19	17	16	11	9	7	6	5	6	8	12	14	130 days
	sunshine	140	147	192	213	220	219	229	239	225	208	180	152	2363 hours
	%	36	42	51	59	60	62	63	65	62	55	48	39	53 %

Tambora – Soembawa Island

GArab	063-16	Jan	Feb	Mar	Apr	May	Jun	Jul	Aug	Sep	Oct	Nov	Dec	Year
		latitude: 8°12′s				longitude: 117°50′e				height: 500 m				
	tx	27	27	28	28	27	27	27	28	28	28	27	27	29 °C
	t	22	22	23	23	22	22	21	21	22	23	23	22	22 °C
	tn	18	18	18	18	17	17	16	16	16	17	18	18	15 °C
	clothing	s	s	t	t	t	s	s	s	s	t	t	t	–
	stress	f	f	f	f	f	f	f	f	f	f	f	f	–
	p	932	574	564	320	140	158	71	79	71	89	292	462	3752 mm
0.1 mm	dp	24	21	20	16	10	7	5	3	3	6	17	21	153 days

Cocos Island

Araa	063-17	Jan	Feb	Mar	Apr	May	Jun	Jul	Aug	Sep	Oct	Nov	Dec	Year
		latitude: 12°05′s				longitude: 96°53′e				height: 5 m				
	tx	32	32	32	31	31	30	30	30	30	31	31	32	33 °C
	t	27	27	27	27	27	26	26	26	26	26	27	27	27 °C
	tn	23	23	23	23	23	23	22	22	22	22	23	23	21 °C
	clothing	t	t	t	t	t	t	t	t	t	t	t	t	–
	stress	ff	ff	ff	ff	ff	ff	ff	ff	ff	ff	ff	ff	–
	p	216	136	229	236	294	202	262	204	81	109	56	86	2108 mm

2. *Climatic area:* Southern part of the Archipelago comprising eastern Java, southern Celebes, southern Irian Barat and minor islands between them

2.1. Tropical Summer Rain Climate Aw and Tropical Summer Rain Mountain Climate GAw

2.2. Very hot all the time.

2.3. High sun period: dry and sunny in the morning, showers in the afternoon; rainy days now and then. Low sun period: more often than not dry and sunny.

2.4. Wind: occasional gales, squalls in thunderstorms.

Makasar/Hasanuddin – Celebes

Awha	063-18	Jan	Feb	Mar	Apr	May	Jun	Jul	Aug	Sep	Oct	Nov	Dec	Year
		latitude: 5°04′s				longitude: 119°33′e				height: 14 m				
	tx	31	31	31	32	32	32	32	32	32	33	32	31	33 °C
	t	26	26	26	27	27	27	26	27	27	28	27	26	27 °C
	tn	22	22	22	22	21	20	18	18	18	20	22	22	18 °C
	clothing	t	t	t	t	t	t	t	t	t	t	t	t	–
	stress	ff	ff	ff	ff	ff	ff	f	f	f	ff	ff	ff	–
	p	498	466	396	109	123	68	30	33	8	60	229	562	2581 mm
	dp	23	21	18	10	14	9	8	5	4	9	17	25	163 days
	sunshine	161	170	189	240	245	252	279	304	306	298	273	167	2883 hours
	%	42	49	50	67	66	71	76	82	84	79	74	43	65 %

Surabaja/Perak – Java

Awha	063-19	Jan	Feb	Mar	Apr	May	Jun	Jul	Aug	Sep	Oct	Nov	Dec	Year
		latitude: 7°13′s				longitude: 112°43′e				height: 3 m				
	t	27	26	27	27	27	27	26	26	27	28	28	27	27 °C
	clothing	t	t	t	t	t	t	t	t	t	t	t	t	–
	stress	ff	ff	ff	ff	ff	f	f	f	f	f	ff	ff	–
	p	275	245	213	111	114	59	48	32	6	31	130	244	1509 mm
	dp	17	18	15	9	9	5	5	3	1	4	11	16	113 days

Merauke/Mopah – West Irian

Awaa	063-20	Jan	Feb	Mar	Apr	May	Jun	Jul	Aug	Sep	Oct	Nov	Dec	Year
		latitude: 8°28′s				longitude: 140°23′e					height: 3 m			
	t	27	26	27	26	26	25	25	25	25	26	27	27	26 °C
	clothing	t	t	t	t	t	t	t	t	t	t	t	t	–
	stress	ff	ff	ff	ff	ff	ff	f	f	f	f	ff	ff	–
	p	204	352	202	186	83	53	37	16	8	27	62	226	1457 mm
	dp	18	17	17	13	8	10	9	5	4	5	9	13	128 days

Dili – Timor

Awaa	063-21	Jan	Feb	Mar	Apr	May	Jun	Jul	Aug	Sep	Oct	Nov	Dec	Year
		latitude: 8°35′s				longitude: 125°35′e					height: 4 m			
	tx	33	33	33	34	34	33	33	33	33	33	34	34	35 °C
	t	27	27	27	26	26	25	25	24	24	26	27	27	26 °C
	tn	23	24	23	23	22	21	20	20	20	21	23	23	18 °C
	clothing	t	t	t	t	t	t	t	t	t	t	t	t	–
	stress	ff	ff	ff	ff	ff	ff	ff	ff	ff	ff	ff	ff	–
	p	159	163	113	79	35	114	60	27	14	16	38	157	975 mm
	dp	13	13	9	8	5	5	4	2	1	2	5	12	79 days
	sunshine	202	175	227	234	260	253	275	284	288	302	280	209	2989 hours
	%	52	50	60	65	71	72	76	77	80	79	75	53	68 %

Kupang/Penfui – Timor

Awha	063-22	Jan	Feb	Mar	Apr	May	Jun	Jul	Aug	Sep	Oct	Nov	Dec	Year
		latitude: 10°10′s				longitude: 123°34′e					height: 45 m			
	tx	33	33	33	33	34	33	33	34	36	36	36	34	37 °C
	t	27	27	27	27	27	26	26	27	27	28	28	28	27 °C
	tn	22	22	22	20	19	18	18	18	18	20	22	22	17 °C
	clothing	t	t	t	t	t	t	t	t	t	t	t	t	–
	stress	ff	ff	ff	ff	ff	ff	ff	ff	ff	ff	ff	ff	–
	p	389	366	221	64	28	10	5	3	2	18	89	246	1441 mm
	dp	17	17	11	7	4	2	1	1	1	4	8	15	88 days
	sunshine	189	195	223	267	276	276	288	304	306	288	264	205	3081 hours
	%	48	56	59	75	76	80	80	83	85	75	70	52	70 %

4.2.63.2. Predominant general weather situations are described in 4.2.0.1. under Southeastern Asia. A hygrothermal diagram of Djakarta is to be found in 4.2.0.2.

Wind names: Bahorok or Bokorot is a dry warm wind blowing down from the Plateau of Kaor into the plains of northeastern Sumatra.

Gending is a dry foehn-like wind in the northern plains of Java, blowing during the low sun period.

Wam-andai is a strong, very gusty wind in western New Guinea blowing during the high sun period.

4.2.64. Philippines

4.2.64.1. Climatic description and tables

1. a) greater part of the Islands particularly on their eastern sides
   Tropical Rain Climate Ar and Tropical Rain Mountain Climate GAr
   b) often on the western sides of the islands
   Tropical Summer Rain Climate Aw and Tropical Summer Rain Mountain Climate GAw or Tropical Monsoonal Rain Climate Am and Tropical Monsoonal Rain Mountain Climate GAm
2. Very hot all the year round.
3. High sun period: dry and sunny in the morning, showers in the afternoon; rainy days now and then. Low sun period: more often than not dry and sunny days, especially on the western sides of the islands.
4. Wind: occasional gales, squalls in thunderstorms, at times typhoons.

Basco

latitude: 20° 27′n longitude: 121° 58′e height: 11 m

Arha	064-1	Jan	Feb	Mar	Apr	May	Jun	Jul	Aug	Sep	Oct	Nov	Dec	Year
	t	23	24	25	27	28	28	29	28	28	27	25	24	26 °C
	clothing	t	t	t	t	t	t	t	t	t	t	t	t	–
	stress	ff	ff	ff	ff	ff	ff	ff	ff	ff	ff	ff	ff	–
	p	221	131	151	90	189	254	279	422	403	412	354	322	3228 mm
	dp	18	12	12	8	12	13	16	18	18	19	18	19	183 days

Aparri

latitude: 18° 22′n longitude: 121° 38′e height: 4 m

Arha	064-2	Jan	Feb	Mar	Apr	May	Jun	Jul	Aug	Sep	Oct	Nov	Dec	Year
	tx	31	32	34	36	37	37	36	36	35	33	32	31	37 °C
	t	24	24	26	28	29	29	29	29	28	27	26	24	27 °C
	tn	18	18	19	21	22	23	23	23	22	21	20	18	17 °C
	clothing	t	t	t	t	t	t	t	t	t	t	t	t	–
	stress	f	f	ff	ff	ff	ff	ff	ff	ff	ff	ff	f	–
	p	146	108	51	35	106	157	165	224	307	390	386	237	2312 mm
	dp	13	9	6	4	8	9	10	11	14	18	18	18	138 days

Dagupan

latitude: 16°03′n longitude: 120° 20′e height: 2 m

Awha	064-3	Jan	Feb	Mar	Apr	May	Jun	Jul	Aug	Sep	Oct	Nov	Dec	Year
	t	26	27	28	30	30	29	28	28	28	28	27	27	28 °C
	clothing	t	t	t	t	t	t	t	t	t	t	t	t	–
	stress	ff	ff	ff	ff	ff	ff	ff	ff	ff	ff	ff	ff	–
	p	8	9	28	85	175	286	476	531	364	168	103	30	2263 mm
	dp	1	1	2	5	11	16	20	22	18	10	5	3	114 days

Manila

Awha	064-4	Jan	Feb	Mar	Apr	May	Jun	Jul	Aug	Sep	Oct	Nov	Dec	Year
		latitude: 14°31′n						longitude: 121°00′e				height: 15 m		
	tx	33	33	35	36	36	36	34	33	33	33	33	33	37 °C
	t	25	26	27	29	29	28	28	27	27	27	26	25	27 °C
	tn	17	17	18	19	21	22	22	22	22	21	19	18	16 °C
	clothing	t	t	t	t	t	t	t	t	t	t	t	t	–
	stress	f	f	f	ff	ff	ff	ff	ff	ff	ff	ff	ff	–
	p	18	7	6	24	110	236	253	480	271	201	129	56	1791 mm
	dp	3	2	1	2	6	11	15	22	18	13	9	7	109 days
	sunshine	177	198	226	258	223	162	133	133	132	158	153	152	2106 hours
	%	50	60	61	69	56	42	33	34	36	43	44	43	47 %

Legaspi

Arha	064-5	Jan	Feb	Mar	Apr	May	Jun	Jul	Aug	Sep	Oct	Nov	Dec	Year
		latitude: 13°08′n						longitude: 125°44′e				height: 19 m		
	tx	31	32	33	34	35	35	34	34	34	34	33	32	36 °C
	t	26	26	26	27	28	28	28	28	28	27	26	26	27 °C
	tn	21	21	22	22	23	23	23	23	23	22	22	21	21 °C
	clothing	t	t	t	t	t	t	t	t	t	t	t	t	–
	stress	ff	ff	ff	ff	ff	ff	ff	ff	ff	ff	ff	ff	–
	p	343	235	239	186	224	182	196	234	245	327	490	525	3426 mm
	dp	20	14	14	14	14	13	13	17	16	19	20	20	194 days

Tacloban

Arha	064-6	Jan	Feb	Mar	Apr	May	Jun	Jul	Aug	Sep	Oct	Nov	Dec	Year
		latitude: 11°15′n						longitude: 125°00′e				height: 21 m		
	t	26	26	27	28	28	28	28	28	28	28	27	26	27 °C
	clothing	t	t	t	t	t	t	t	t	t	t	t	t	–
	stress	ff	ff	ff	ff	ff	ff	ff	ff	ff	ff	ff	ff	–
	p	300	191	172	129	170	150	150	130	156	197	263	314	2322 mm
	dp	18	14	14	13	13	14	14	12	13	16	18	20	179 days

Iloilo

Amha	064-7	Jan	Feb	Mar	Apr	May	Jun	Jul	Aug	Sep	Oct	Nov	Dec	Year
		latitude: 10°42′n						longitude: 122°34′e				height: 14 m		
	tx	32	33	34	35	36	34	33	33	33	33	33	32	36 °C
	t	26	26	27	28	28	28	27	27	27	27	27	26	27 °C
	tn	21	21	22	23	23	23	23	23	23	22	22	21	21 °C
	clothing	t	t	t	t	t	t	t	t	t	t	t	t	–
	stress	ff	ff	ff	ff	ff	ff	ff	ff	ff	ff	ff	ff	–
	p	53	28	37	48	146	263	302	360	290	255	209	131	2122 mm
	dp	8	5	5	5	11	15	17	19	17	16	13	11	142 days

Cebu

Arha	064-8	Jan	Feb	Mar	Apr	May	Jun	Jul	Aug	Sep	Oct	Nov	Dec	Year
		latitude: 10°20′n						longitude: 123°54′e				height: 42 m		
	t	26	27	28	29	29	28	28	28	28	28	28	27	28 °C
	clothing	t	t	t	t	t	t	t	t	t	t	t	t	–
	stress	ff	ff	ff	ff	ff	ff	ff	ff	ff	ff	ff	ff	–
	p	101	70	55	70	133	186	207	175	211	191	160	144	1703 mm
	dp	11	8	7	6	11	14	14	13	15	16	12	12	139 days

Surigao

Arha	064-9	Jan	Feb	Mar	Apr	May	Jun	Jul	Aug	Sep	Oct	Nov	Dec	Year
		latitude: 9°48′n						longitude: 125°30′e			height: 22 m			
	tx	30	31	31	32	33	34	33	34	34	33	32	31	34 °C
	t	26	26	26	27	27	28	27	28	28	27	27	26	27 °C
	tn	21	21	21	22	23	23	23	23	23	22	22	22	21 °C
	clothing	t	t	t	t	t	t	t	t	t	t	t	t	–
	stress	ff	ff	ff	ff	ff	ff	ff	ff	ff	ff	ff	ff	–
	p	589	405	398	258	184	112	195	149	197	308	415	653	3854 mm
	dp	24	18	20	18	14	12	13	12	12	16	19	24	202 days

Zamboanga

Awha	064-10	Jan	Feb	Mar	Apr	May	Jun	Jul	Aug	Sep	Oct	Nov	Dec	Year
		latitude: 6°54′n						longitude: 122°04′e			height: 6 m			
	tx	33	33	34	34	33	34	33	34	34	33	32	31	34 °C
	t	26	26	26	27	27	28	27	28	28	27	27	26	27 °C
	tn	21	21	21	22	23	23	23	23	23	22	22	22	21 °C
	clothing	t	t	t	t	t	t	t	t	t	t	t	t	–
	stress	ff	ff	ff	ff	ff	ff	ff	ff	ff	ff	ff	ff	–
	p	51	49	44	54	96	131	120	138	139	173	135	96	1226 mm
	dp	5	5	5	6	10	12	12	12	11	11	12	9	110 days

4.2.64.2. Predominant general weather situations are described in 4.2.0.1. under Southeast Asia. A hygrothermal diagram of Manila is to be found in 4.2.0.2.

Wind names: Baguio is a local name by which the tropical cyclones experienced in the Philippines Islands are known. Cyclones or typhoons occur from July to November.

Colla is a very gusty south-southwest wind in the rainy season.

Nortadas are northerly storms occurring predominantly in winter.

4.3. Africa

4.3.0.1. Predominant general weather situations

North Africa (Morocco, Algeria, Tunisia, Libya, Egypt)
In summer, a wedge of the Azores anticyclone, extending to the eastern Mediterranean, mostly governs the weather with its sinking air. The Equatorial Trough moves from the Equator to the southern border of the Sahara till August, then going back southwards. In winter, there is often a wedge of the Azores anticyclone over North Africa. On its northern slope, the weather is unsettled, because of the influence of troughs of depressions which traverse Europe from west to east. Now and then there are stationary anticyclones over the Mediterranean, influencing the weather over extensive areas with their sinking air.

West Africa (West Africa, Niger, Mali, Mauretania, Senegal, Gambia, Chad, Upper Volta)
In summer, the weather is decisively determined by the position of the Equatorial Trough. It frequently advances as far as the southern border of the Sahara, but here sinking air of the wedge of the Azores anticyclone, which lies to the north of the region, often predominates. In winter, there is nearly always sinking air from the anticyclone over North Africa.

Sudan, Ethiopia, Somalia
In summer, rising air predominates, coming from the Equatorial Trough, which moves to the north till August, covering a wide area. Because of the lack of water, however, vapour clouds do not form in all parts. In winter, sinking air from the anticyclone over North Africa predominates.

Equatorial West Africa (Equatorial Guinea, Guinea, Sierra Leone, Gabun, Central Africa, Cameroun, Nigeria, Benin, Togo, Ghana, Ivory Coast, Liberia)
The Equatorial Trough determines the weather. It reaches its northernmost position in mid-summer and passes the northern boundaries of the countries in this region several times. Especially in some coastal regions, the influence of a wedge of the anticyclone over the South Atlantic, with its sinking air, then predominates. By mid-winter, the Equatorial Trough moves south close to the equator, so that the northern areas are influenced by the sinking air of a wedge of the Azores anticyclone or an anticyclone over North Africa.

Equatorial South Africa (Kenya, Tanzania, Uganda, Rwanda, Burundi, Zaire, Congo, Angola, Madagascar, Mozambique, Zambia, Malawi, Zimbabwe)
In summer, the Equatorial Trough is in the southern hemisphere, and its rising air influences the weather in great parts of the area. The general shower activity is intensified in front of its disturbances, which migrate from east to west, and weakened in the rear of them. In winter, there is often sinking air, from a wedge of the anticyclone over South Africa, as far as the equator. Disturbances of the Equatorial Trough, which is then in the northern hemisphere, occasionally reach the southern parts of the region.

Southern Africa (South Africa, Namibia, Botswana, Ngwane, Lesotho)
In summer, disturbances of the Equatorial Trough influence the weather at times, especially in the north of the region, with their rising air. But sinking air predominates in the region, which is covered by a ridge of high pressure between the anticyclone over the South Atlantic and the southern Indian Ocean. In winter, there is often sinking air from an anticyclone over South Africa, but troughs of depressions migrating from west to east farther south frequently traverse the region and influence the weather with their rising air.

Southern Indian Ocean

In summer (December to February), the mean position of the Equatorial Trough is from northern Madagascar to the Arafura Sea. The subtropical belt of high pressure has its mean position at 35° southern latitude. After the equinox in March, the Equatorial Trough passes the Equator to the north, yielding place to the following ridge of high pressure, which has its mean position at 30° southern latitude in midwinter. The rising of the air in the southern westerlies is intensified in winter in combination with the northward shift.

4.3.0.2. Hygrothermal diagrams

In the following figures, hygrothermal diagrams are given for the stations

Casablanca, Algiers, Tunis, Niamey, Dakar, Conakry, Tripoli, Cairo, Khartoum, Addis Ababa, Mogadishu, Nairobi, Dar-Es-Salaam, Kinshasa, Libreville, Bangui, Ndjemena, Yaounde, Lagos, Cotonou, Ougadougou, Luanda, Tananarive, Lourenço Marques, Lusaka, Salisbury, Pretoria, Cape Town.

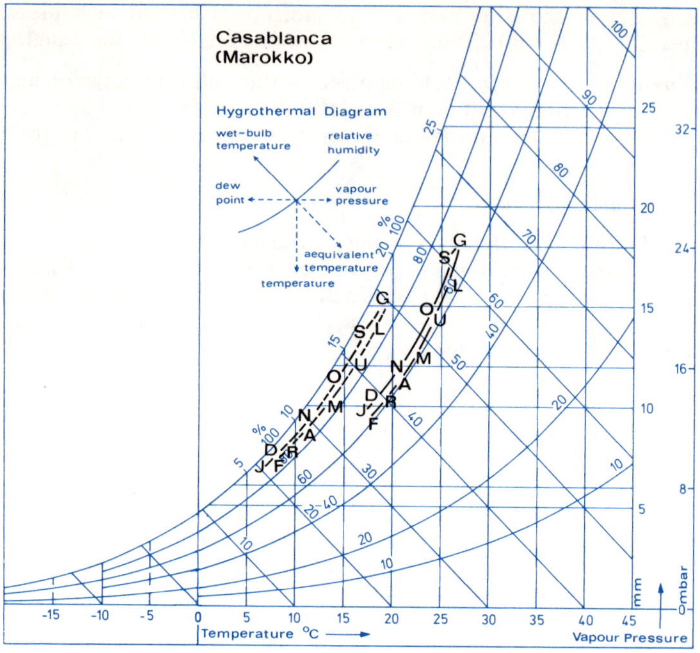
Casablanca
(Marokko)
Hygrothermal Diagram
wet-bulb temperature
relative humidity
dew point
vapour pressure
aequivalent temperature
temperature
Temperature °C
Vapour Pressure
mm
mbar

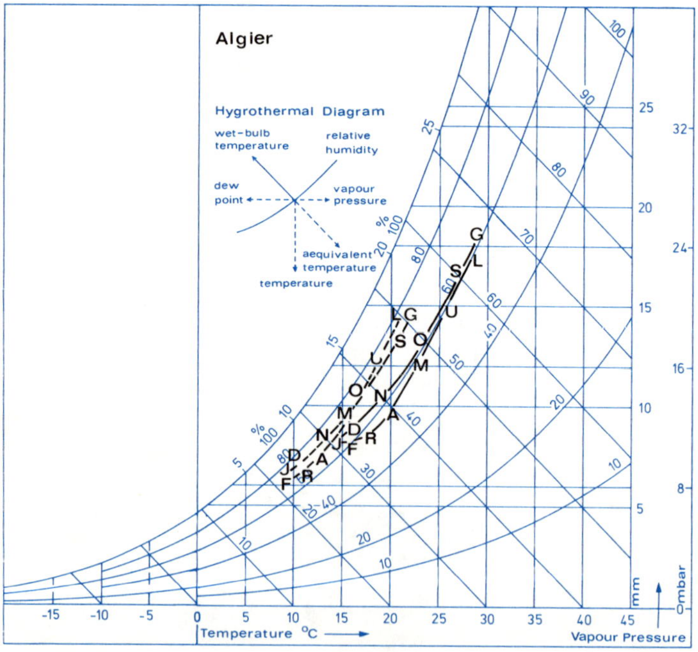
Algier
Hygrothermal Diagram
wet-bulb temperature
relative humidity
dew point
vapour pressure
aequivalent temperature
temperature
Temperature °C
Vapour Pressure
mm
mbar

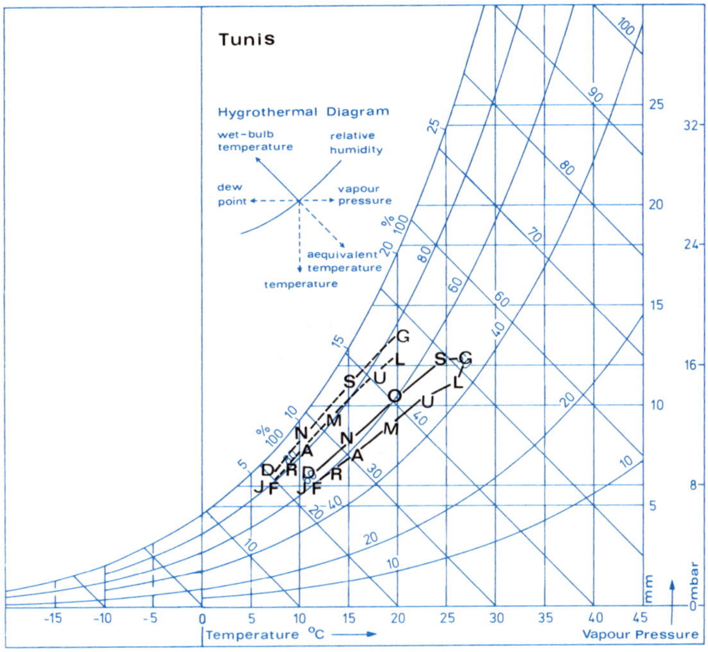
Tunis
Hygrothermal Diagram
wet-bulb temperature
relative humidity
dew point
vapour pressure
aequivalent temperature
temperature
Temperature °C
Vapour Pressure
mm
mbar

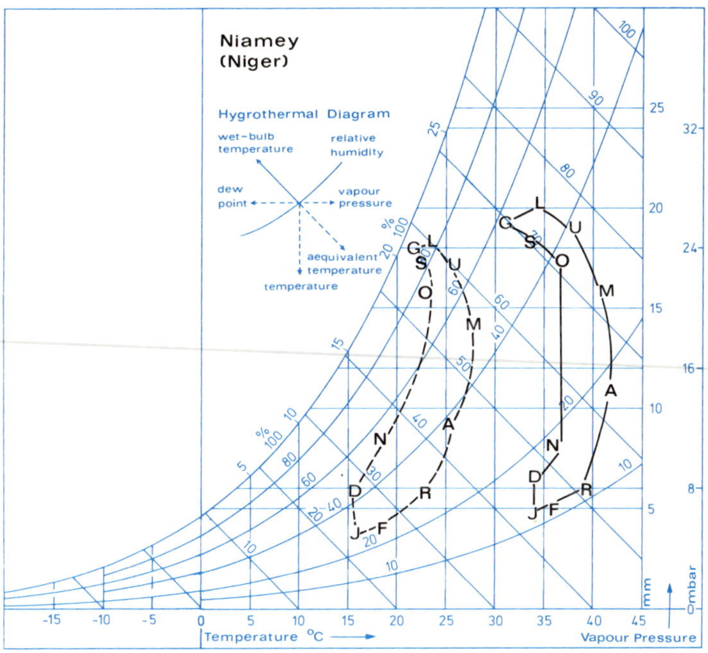
Niamey
(Niger)
Hygrothermal Diagram
wet-bulb temperature
relative humidity
dew point
vapour pressure
aequivalent temperature
temperature
Temperature °C
Vapour Pressure
mm
mbar

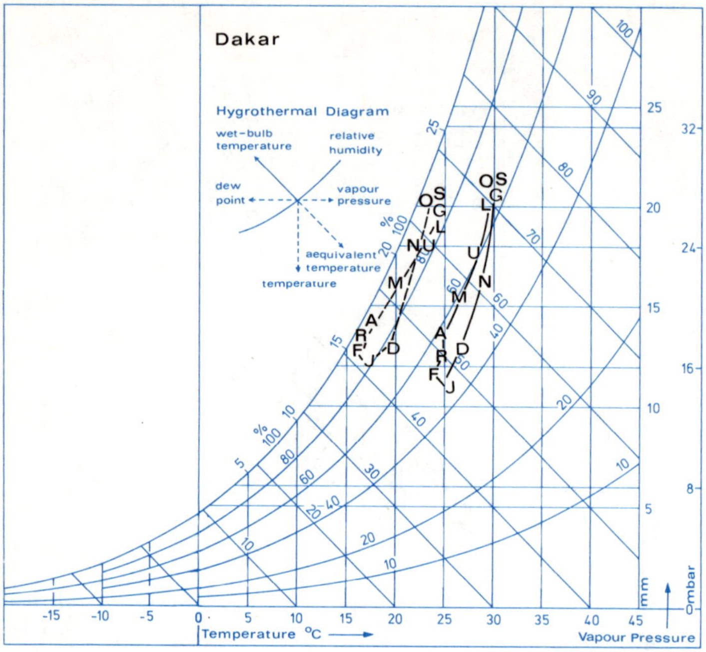
Dakar
Hygrothermal Diagram
wet-bulb temperature
relative humidity
dew point
vapour pressure
aequivalent temperature
temperature
Temperature °C
Vapour Pressure
mm
mbar

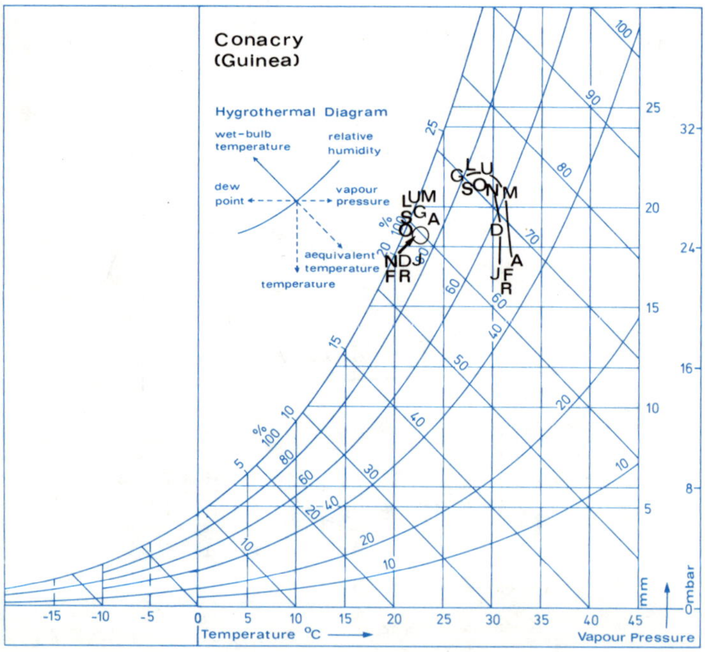
Conacry (Guinea)
Hygrothermal Diagram
wet-bulb temperature
relative humidity
dew point
vapour pressure
aequivalent temperature
temperature
Temperature °C
Vapour Pressure
mm
mbar

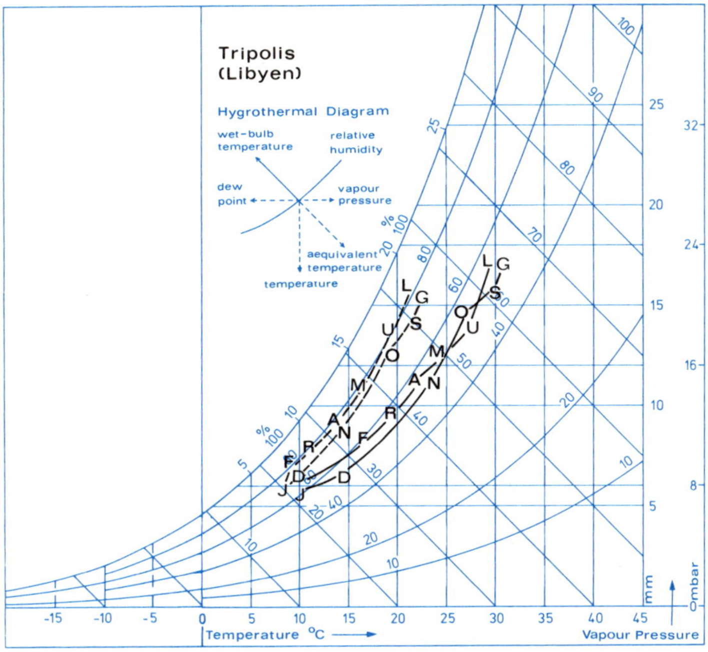
Tripolis
(Libyen)
Hygrothermal Diagram
wet-bulb temperature
relative humidity
dew point
vapour pressure
aequivalent temperature
temperature
Temperature °C
Vapour Pressure
mm
mbar

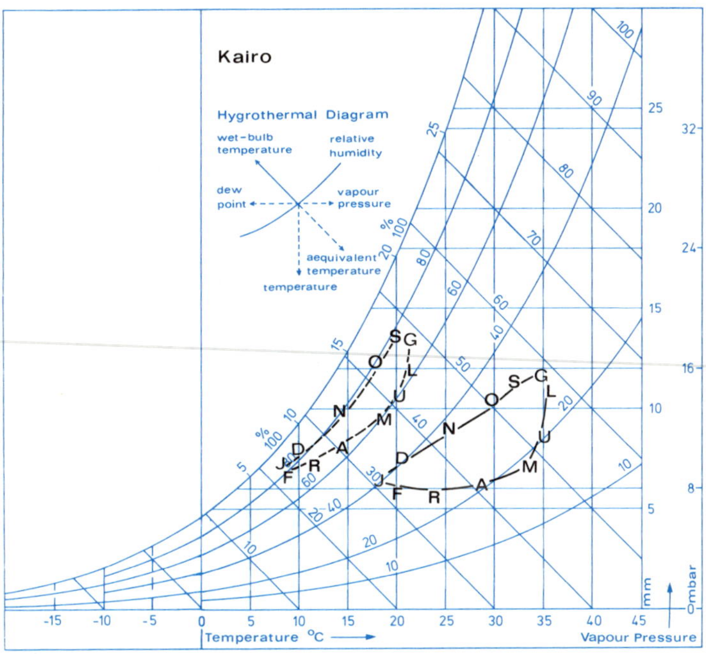
Kairo
Hygrothermal Diagram
wet-bulb temperature
relative humidity
dew point
vapour pressure
aequivalent temperature
temperature
Temperature °C
Vapour Pressure
mm
mbar

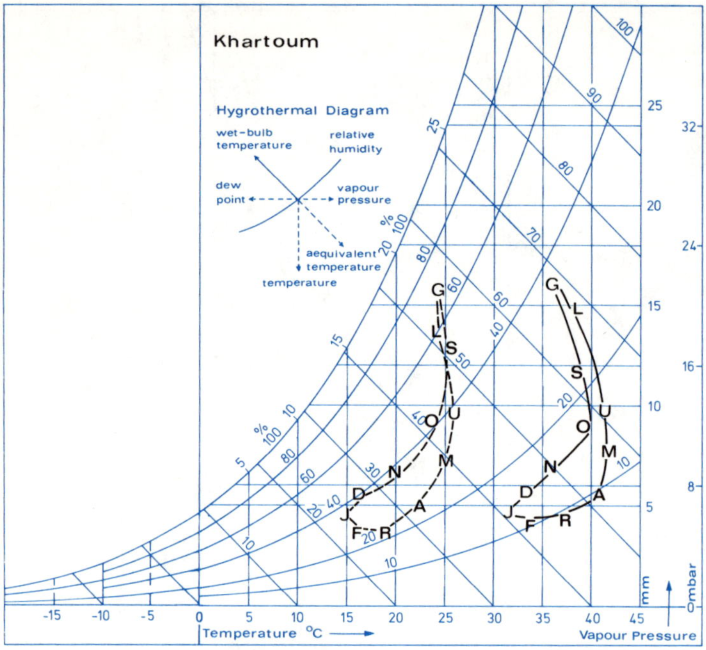
Khartoum
Hygrothermal Diagram
wet-bulb temperature
relative humidity
dew point
vapour pressure
aequivalent temperature
temperature
Temperature °C
Vapour Pressure
mm
mbar

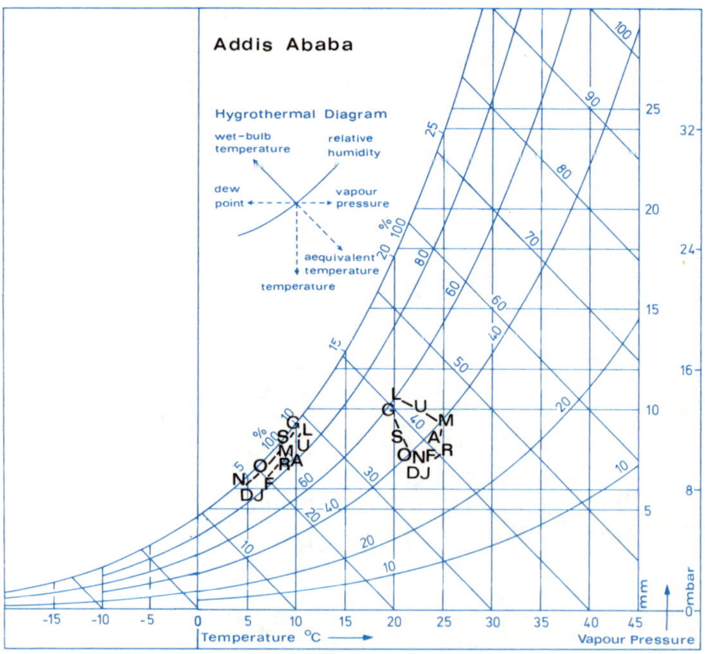
Addis Ababa
Hygrothermal Diagram
wet-bulb temperature
relative humidity
dew point
vapour pressure
aequivalent temperature
temperature
Temperature °C
Vapour Pressure
mm
mbar

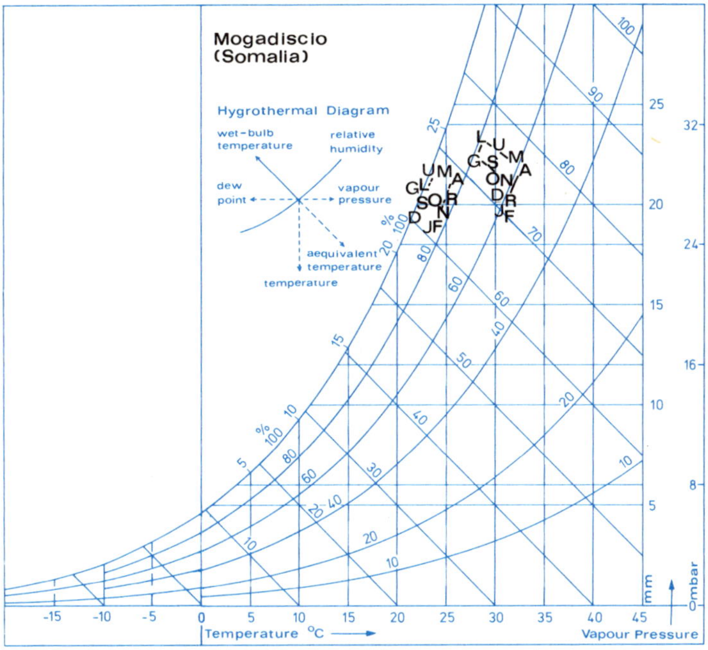
Mogadiscio
(Somalia)
Hygrothermal Diagram
wet-bulb temperature
relative humidity
dew point
vapour pressure
aequivalent temperature
temperature
Temperature °C
Vapour Pressure
mm
mbar

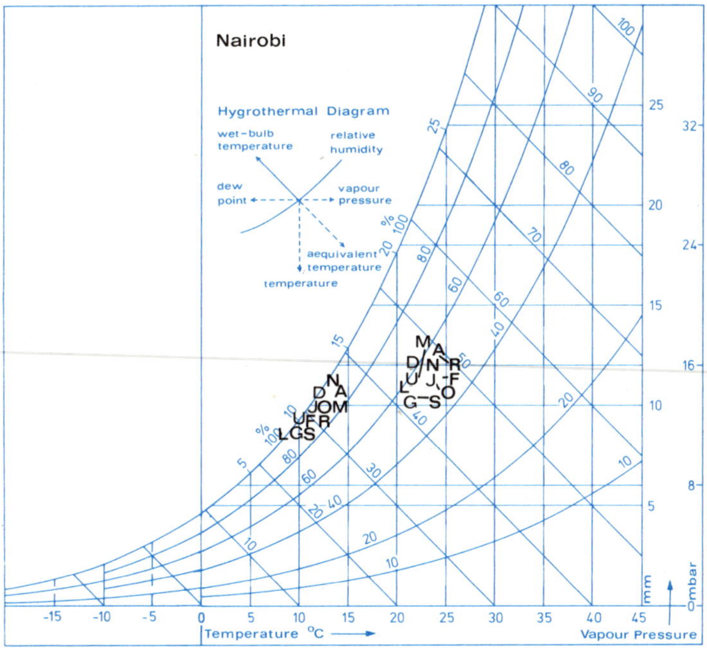
Nairobi
Hygrothermal Diagram
wet-bulb temperature
relative humidity
dew point
vapour pressure
aequivalent temperature
temperature
Temperature °C
Vapour Pressure
mm
mbar

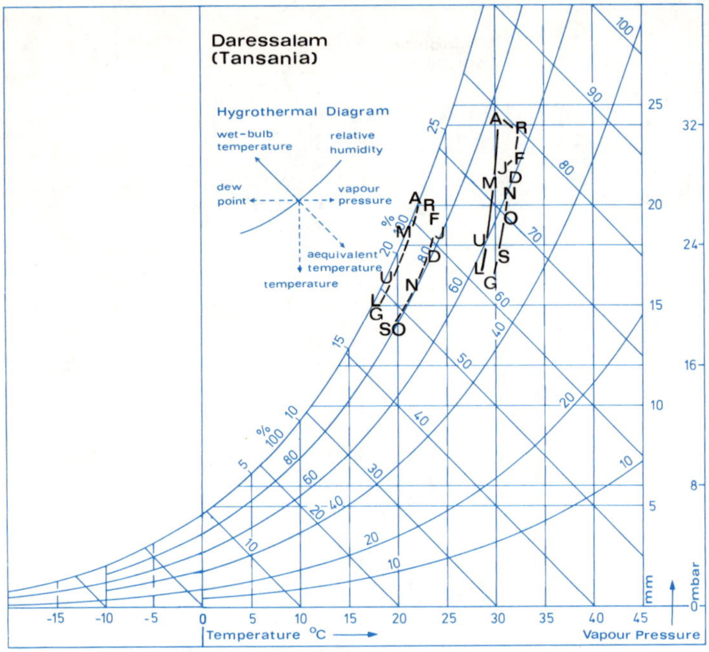
Daressalam
(Tansania)
Hygrothermal Diagram
wet-bulb temperature
relative humidity
dew point
vapour pressure
aequivalent temperature
temperature
Temperature °C
Vapour Pressure
mm
mbar

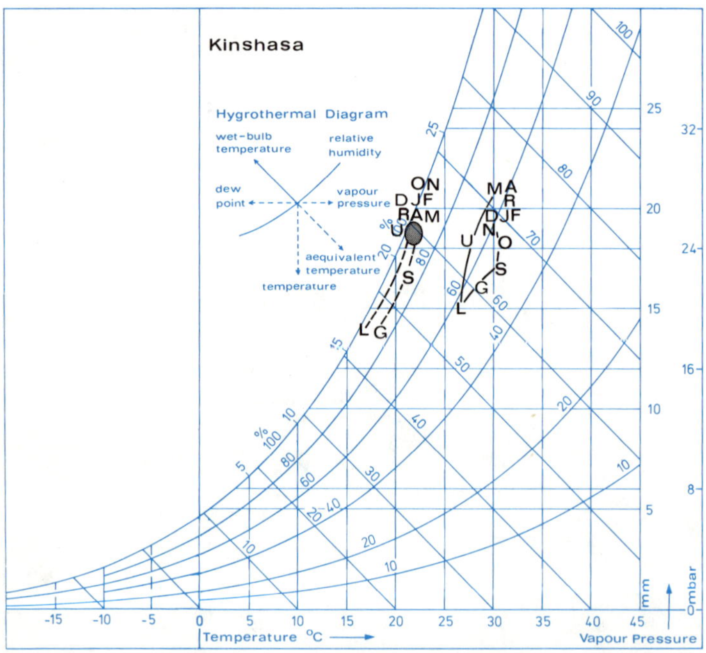
Kinshasa
Hygrothermal Diagram
wet-bulb temperature
relative humidity
dew point
vapour pressure
aequivalent temperature
temperature
Temperature °C
Vapour Pressure
mm
mbar

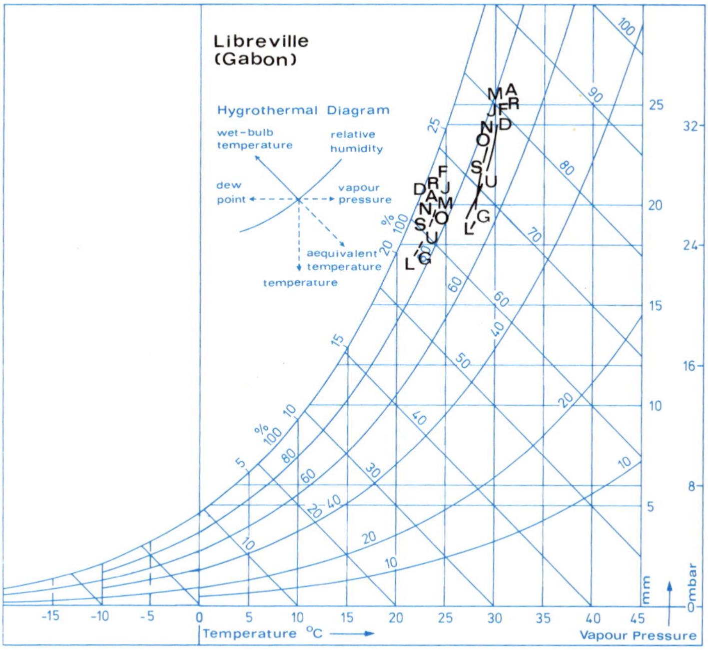
Libreville
(Gabon)
Hygrothermal Diagram
wet-bulb temperature
relative humidity
dew point
vapour pressure
aequivalent temperature
temperature
Temperature °C
Vapour Pressure
mm
mbar

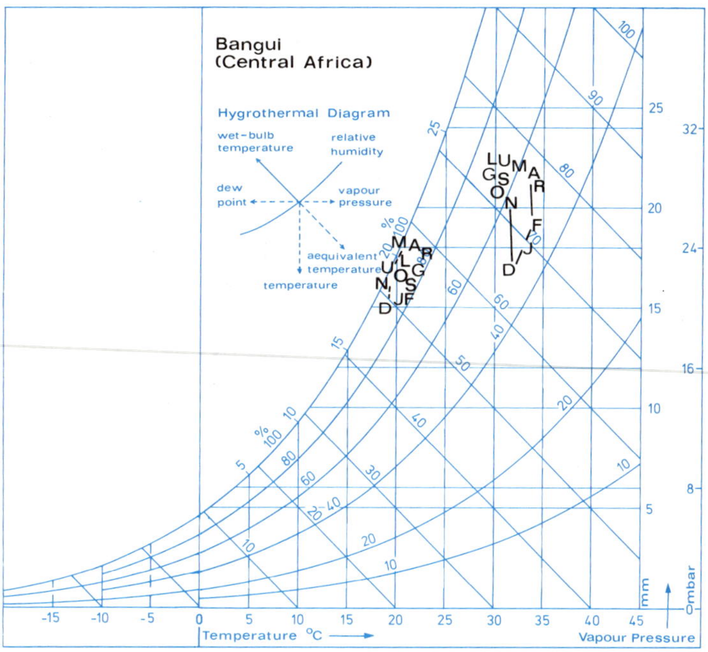
Bangui
(Central Africa)
Hygrothermal Diagram
wet-bulb temperature
relative humidity
dew point
vapour pressure
aequivalent temperature
temperature
Temperature °C
Vapour Pressure
mm
mbar

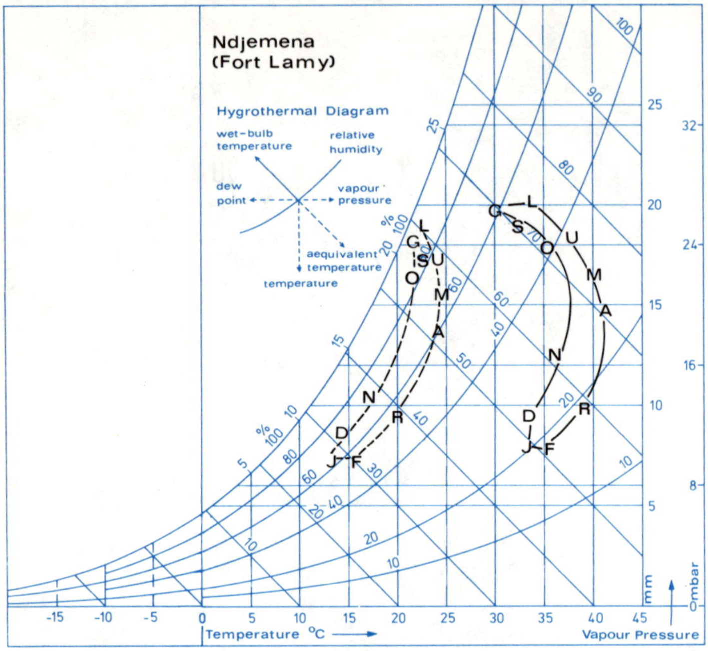
Ndjemena
(Fort Lamy)
Hygrothermal Diagram
wet-bulb temperature
relative humidity
dew point
vapour pressure
aequivalent temperature
temperature
Temperature °C
Vapour Pressure
mm
mbar

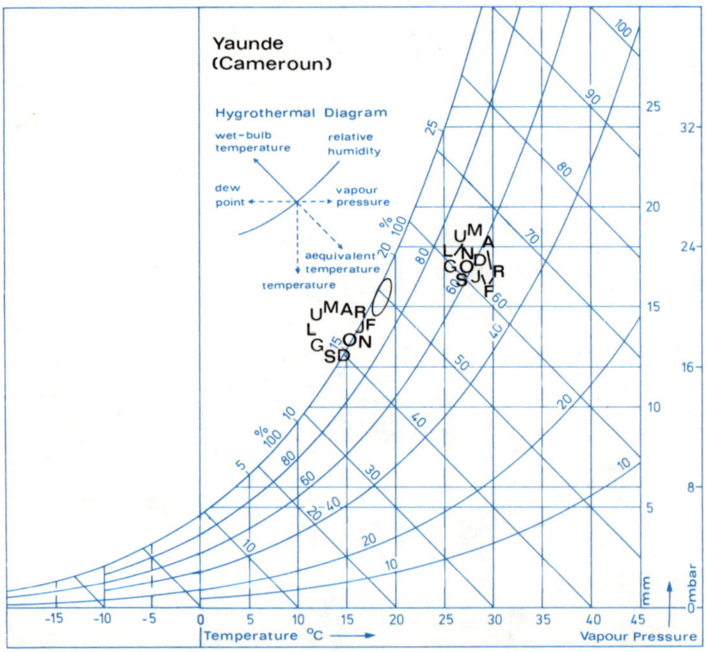
Yaunde
(Cameroun)
Hygrothermal Diagram
wet-bulb temperature
relative humidity
dew point
vapour pressure
aequivalent temperature
temperature
Temperature °C
Vapour Pressure
mm
mbar

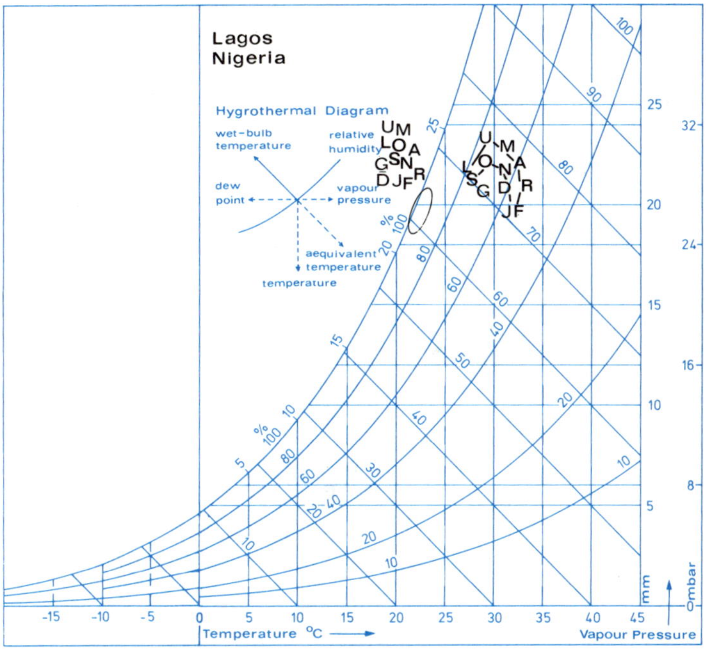
Lagos
Nigeria
Hygrothermal Diagram
wet-bulb temperature
relative humidity
dew point
vapour pressure
aequivalent temperature
temperature
Temperature °C
Vapour Pressure
mm
mbar

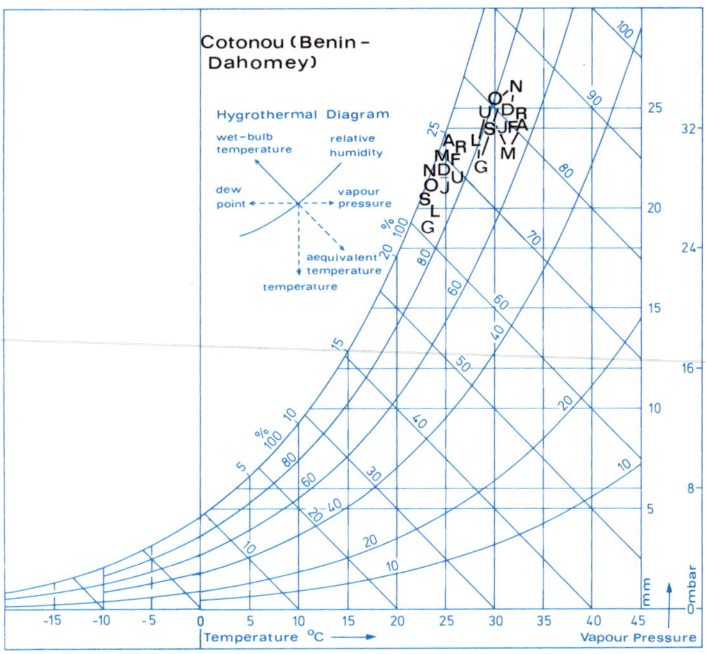
Cotonou (Benin-Dahomey)
Hygrothermal Diagram
wet-bulb temperature
relative humidity
dew point
vapour pressure
aequivalent temperature
temperature
Temperature °C
Vapour Pressure
mm
mbar

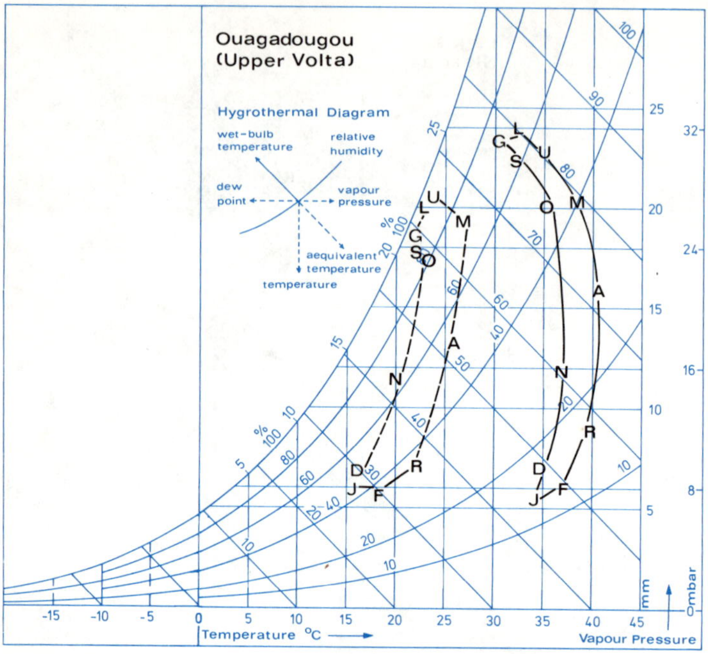
Ouagadougou
(Upper Volta)
Hygrothermal Diagram
wet-bulb temperature
relative humidity
dew point
vapour pressure
aequivalent temperature
temperature
Temperature °C
Vapour Pressure
mm
mbar

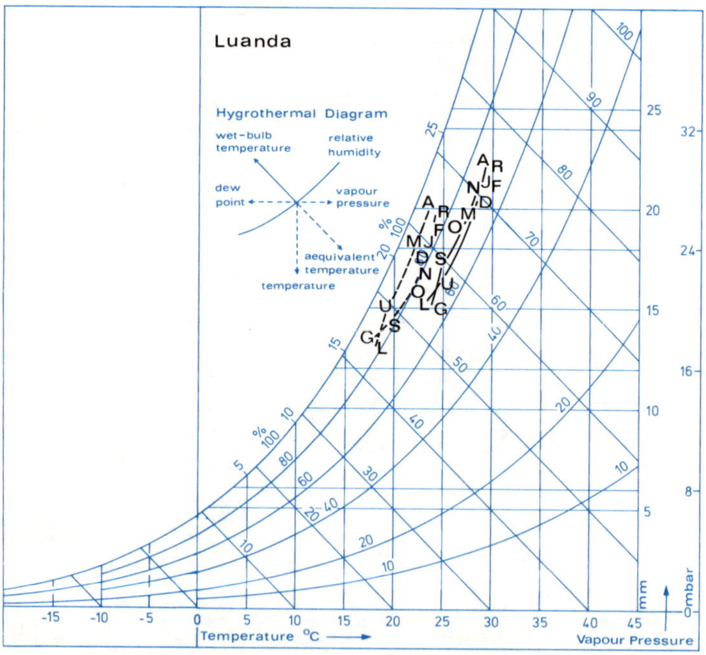
Luanda
Hygrothermal Diagram
wet-bulb temperature
relative humidity
dew point
vapour pressure
aequivalent temperature
temperature
Temperature °C
Vapour Pressure
mm
mbar

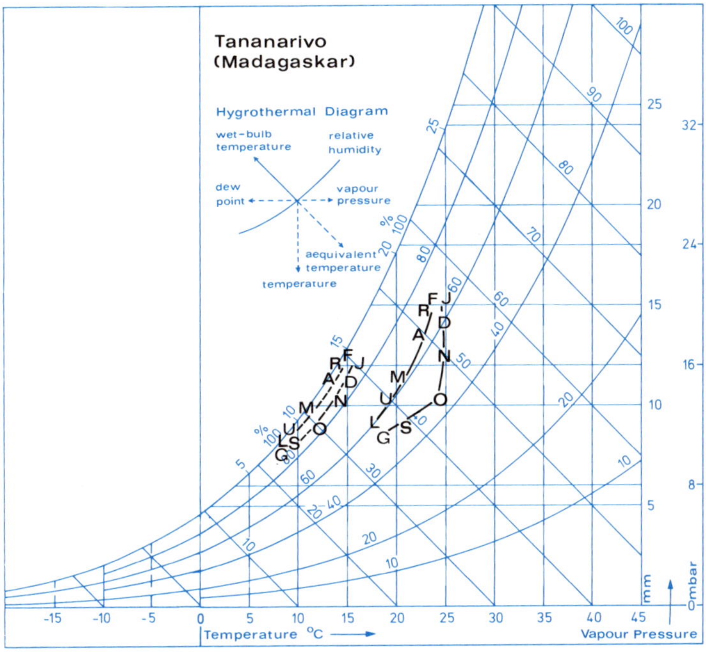
Tananarivo
(Madagaskar)
Hygrothermal Diagram
wet-bulb temperature
relative humidity
dew point
vapour pressure
aequivalent temperature
temperature
Temperature °C
Vapour Pressure
mm
mbar

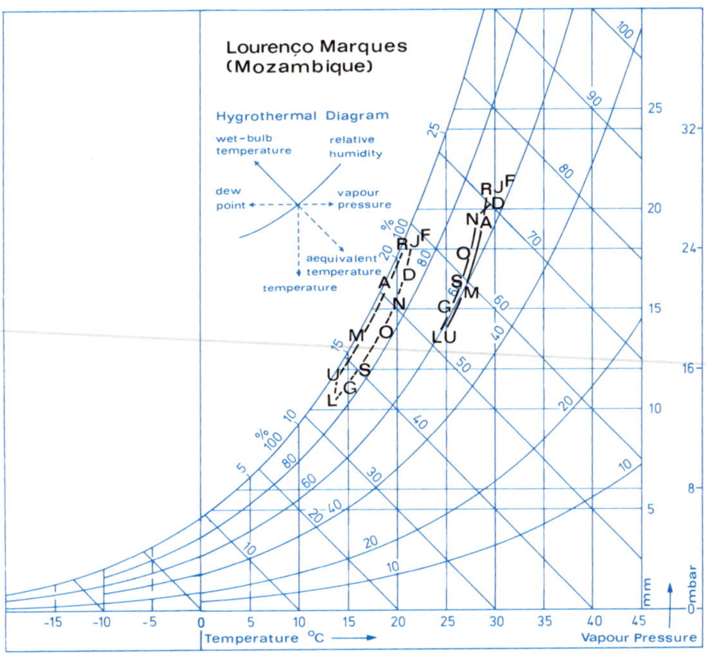
Lourenço Marques
(Mozambique)
Hygrothermal Diagram
wet-bulb temperature
relative humidity
dew point
vapour pressure
aequivalent temperature
temperature
Temperature °C
Vapour Pressure
mm
mbar

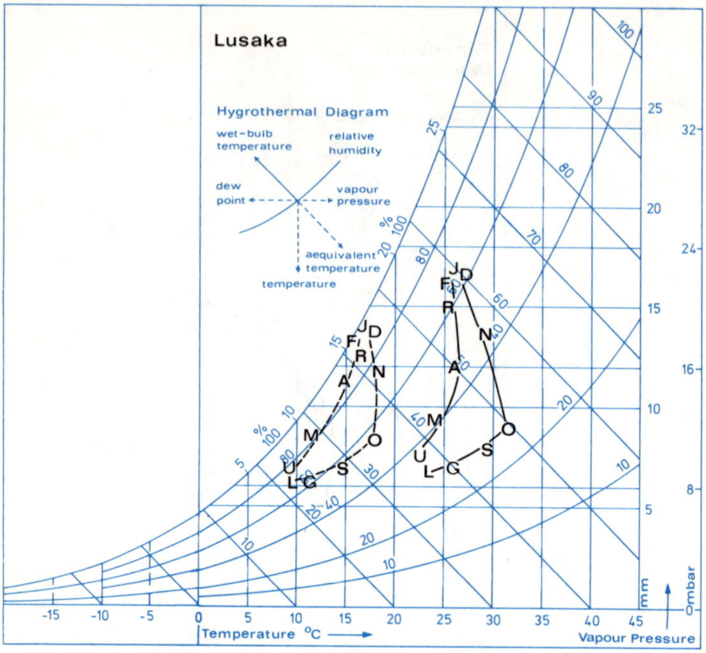
Lusaka
Hygrothermal Diagram
wet-bulb temperature
relative humidity
dew point
vapour pressure
aequivalent temperature
temperature
Temperature °C
Vapour Pressure
mm
mbar

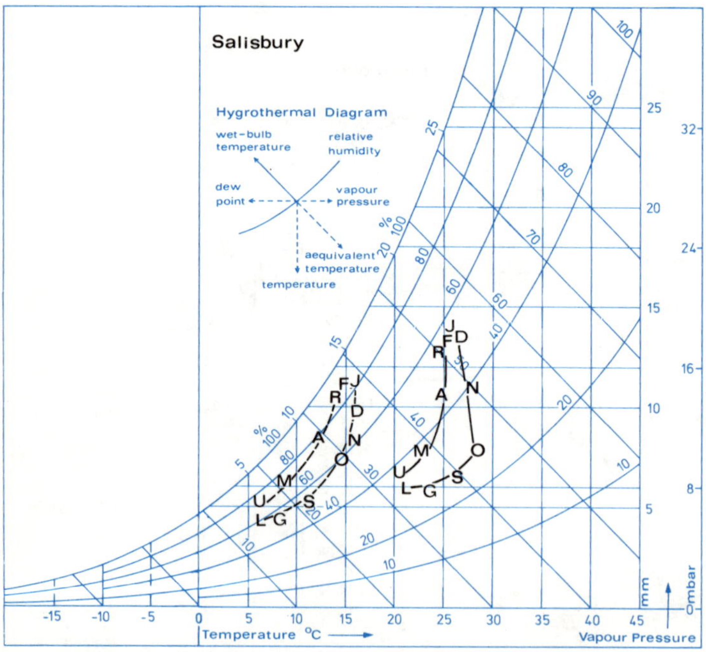
Salisbury
Hygrothermal Diagram
wet-bulb temperature
relative humidity
dew point
vapour pressure
aequivalent temperature
temperature
Temperature °C
Vapour Pressure
mm
mbar

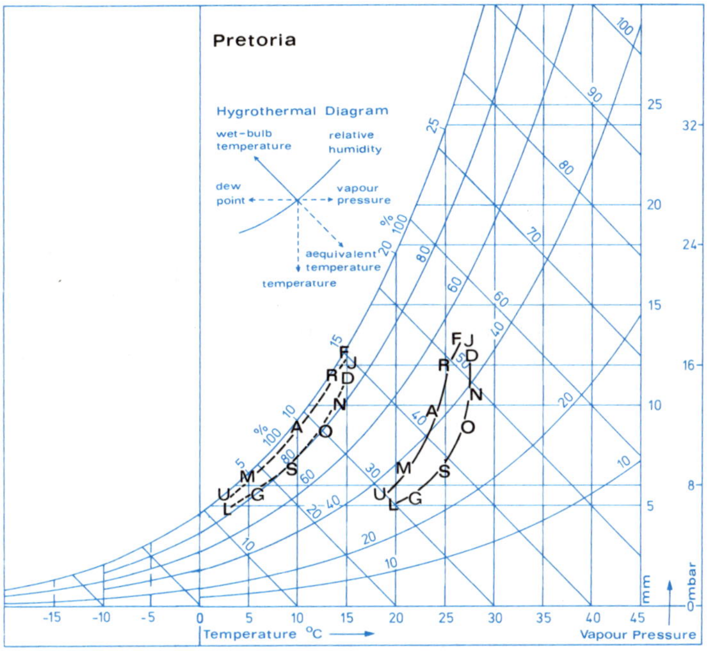
Pretoria
Hygrothermal Diagram
wet-bulb temperature
relative humidity
dew point
vapour pressure
aequivalent temperature
temperature
Temperature °C
Vapour Pressure
mm
mbar

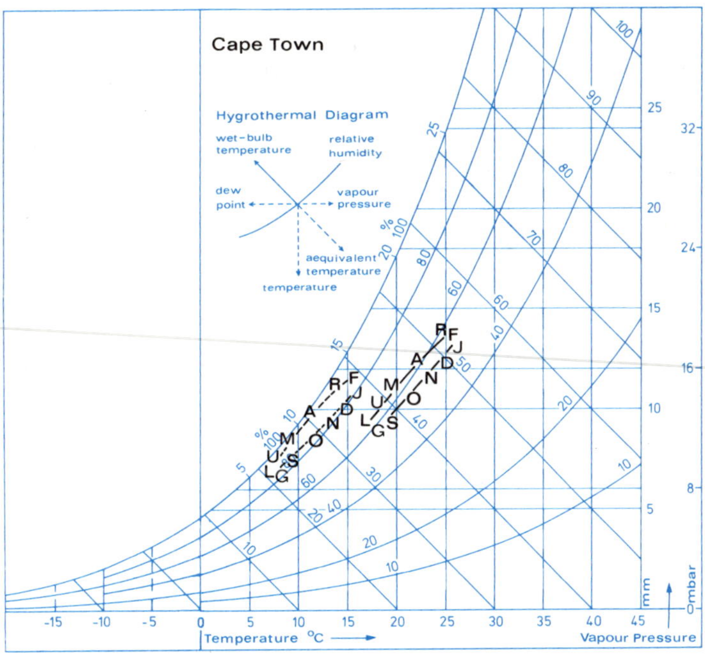
Cape Town
Hygrothermal Diagram
wet-bulb temperature
relative humidity
dew point
vapour pressure
aequivalent temperature
temperature
Temperature °C
Vapour Pressure
mm
mbar

4.3.65. West Sahara

4.3.65.1. Climatic description and tables

1. Desert Climate BW, coastal areas: Marine Desert Climate BM
2. Summer very hot, in coastal regions warm; winter mild.
3. Generally dry and sunny.
4. Wind: occasional gales. Fog: at times in coastal regions.

Smara

latitude: 26°44′n longitude: 11°26′w height: 460 m

BWhl	065-1	Jan	Feb	Mar	Apr	May	Jun	Jul	Aug	Sep	Oct	Nov	Dec	Year
	tx	27	32	35	38	35	39	47	47	44	39	34	29	48 °C
	t	15	17	20	22	22	24	27	28	26	23	20	16	22 °C
	tn	5	5	8	9	12	13	17	16	13	12	11	5	3 °C
	clothing	n	n	s	s	s	t	t	t	t	s	s	n	–
	stress			o	o	o	f	f	ff	f	o	o		–
	p	5	2	3	0	0	0	0	3	11	16	3	0	45 mm

Rio de Oro

latitude: 23°42′n longitude: 15°52′w height: 10 m

BMab	065-2	Jan	Feb	Mar	Apr	May	Jun	Jul	Aug	Sep	Oct	Nov	Dec	Year
	tx	27	28	28	29	29	30	31	33	33	35	32	28	38 °C
	t	18	18	19	19	20	21	21	22	23	22	21	19	20 °C
	tn	11	12	12	13	14	14	16	17	17	16	14	12	11 °C
	clothing	s	s	s	s	s	s	s	s	t	s	s	s	–
	stress	o	o	o	o	o	o	o	o	f	o	o	o	–
	p	2	1	2	0	0	0	0	2	10	3	3	8	32 mm
0.1 mm	dp	1	1	0	0	0	0	0	1	2	1	1	2	9 days
	sunshine	251	246	273	264	298	285	260	264	249	245	237	236	3107 hours
	%	75	77	73	69	72	70	62	66	67	68	72	71	70 %

4.3.65.2. Predominant general weather situations are described in 4.3.0.1. under North Africa.

4.3.66. Morocco

4.3.66.1. Climatic descriptions and tables

1. Climatic area: Northern and Northwestern part
1.1. Subtropical Winter Rain Climate Cs and
Subtropical Winter Rain Mountain Climate GCs
1.2. Summer hot, winter mild.
1.3. Summer generally dry and sunny; winter: frequently variable weather, at times rainy days.
1.4. Wind: occasional gales. Fog: at times.

Ceuta

latitude: 35°55'n longitude: 5°18'w height: 201 m

Csbl	066-1	Jan	Feb	Mar	Apr	May	Jun	Jul	Aug	Sep	Oct	Nov	Dec	Year
	t	12	12	13	15	17	19	22	22	21	18	14	12	16 °C
	clothing	n	n	n	n	s	s	s	s	s	s	n	n	–
	stress					o	o	o	o	o	o			–
	p	74	67	69	25	32	15	0	0	16	63	105	122	588 mm
0.1 mm	dp	9	9	8	5	3	2	0	0	2	6	10	10	64 days
	sunshine	149	150	167	219	279	306	319	298	234	183	144	140	2587 hours
	%	48	49	45	56	64	70	72	71	63	52	47	46	58 %

Fes

latitude: 34°02'n longitude: 4°59'w height: 414 m

Csak	066-2	Jan	Feb	Mar	Apr	May	Jun	Jul	Aug	Sep	Oct	Nov	Dec	Year
	tx	21	24	27	31	34	39	43	43	39	34	27	22	44 °C
	t	9	11	13	15	19	23	26	26	23	18	14	10	17 °C
	tn	-1	0	2	4	6	9	13	13	11	7	3	-1	-2 °C
	clothing	n	n	n	n	s	t	t	t	t	s	n	n	–
	stress					o	f	f	f	f	o			–
	p	83	66	85	64	27	11	1	4	13	44	56	86	541 mm
0.1 mm	dp	7	8	9	8	6	3	1	1	3	7	8	9	70 days

Rabat

latitude: 34°00'n longitude: 6°05'w height: 65 m

Csal	066-3	Jan	Feb	Mar	Apr	May	Jun	Jul	Aug	Sep	Oct	Nov	Dec	Year
	tx	22	23	27	29	31	33	36	36	34	32	28	23	34 °C
	t	12	13	15	16	17	21	23	23	22	20	16	14	18 °C
	tn	3	4	6	7	9	12	13	14	12	10	7	4	3 °C
	clothing	n	n	n	n	s	s	t	t	s	s	n	n	–
	stress					o	o	f	f	o	o			–
	p	62	62	65	44	30	9	0	1	9	54	94	93	523 mm
0.1 mm	dp	8	8	10	7	6	2	0	1	2	7	9	11	71 days

Meknès

		latitude: 33°53′n						longitude: 5°32′w				height: 549 m	
GCsal	066-4	Jan	Feb	Mar	Apr	May	Jun	Jul	Aug	Sep	Oct	Nov	Dec	Year
	t	10	11	13	15	18	22	25	26	23	19	14	11	17 °C
	clothing	n	n	n	n	s	s	t	t	t	s	n	n	–
	stress					o	f	f	f	f	o			–
	p	82	67	79	63	33	9	3	2	19	54	69	105	585 mm
	dp	8	7	8	7	5	1	0	1	2	6	7	9	61 days
	sunshine	169	181	219	242	301	315	361	337	278	234	176	154	2967 hours
	%	54	58	59	62	70	73	82	81	75	66	56	50	67 %

Casablanca

		latitude: 33°34′n						longitude: 7°40′w				height: 58 m		
Csal	066-5	Jan	Feb	Mar	Apr	May	Jun	Jul	Aug	Sep	Oct	Nov	Dec	Year
	tx	22	23	26	27	28	30	33	33	32	30	27	23	37 °C
	t	12	13	15	16	18	20	22	23	22	19	16	13	18 °C
	tn	2	3	4	7	9	12	14	15	13	10	7	4	2 °C
	clothing	n	n	n	n	s	s	s	t	s	s	n	n	–
	stress						o	o	f	o	o			–
	p	66	53	55	38	21	2	0	1	7	39	57	87	426 mm
	dp	8	6	7	5	4	1	0	0	1	5	7	9	53 days
	sunshine	170	179	225	263	293	295	309	300	267	225	172	165	2865 hours
	%	54	58	61	67	68	68	70	72	72	64	55	53	64 %

2. Climatic area: Atlas Mountains

2.1. Steppe Climate BS and Steppe Mountain Climate GBS

2.2. Summer hot or very hot, winter mild or cool.

2.3. Summer generally dry and sunny; winter mostly dry and sunny, now and then showers or rainy.

2.4. Wind: occasional gales and dust-storms.

Oudja

		latitude: 34°48′n						longitude: 1°54′w				height: 460 m		
BSak	066-6	Jan	Feb	Mar	Apr	May	Jun	Jul	Aug	Sep	Oct	Nov	Dec	Year
	t	9	10	13	15	18	22	26	26	23	18	14	10	17 °C
	clothing	n	n	n	n	s	s	t	t	t	s	n	n	–
	stress					o	f	f	f	f	o			–
	p	38	37	36	40	37	13	2	5	21	33	33	47	342 mm
0.1 mm	dp	6	7	7	7	6	4	1	2	4	6	6	8	64 days

Bouârfa

		latitude: 32°34′n						longitude: 1°58′w				height: 1310 m		
GBShk	066-7	Jan	Feb	Mar	Apr	May	Jun	Jul	Aug	Sep	Oct	Nov	Dec	Year
	tx	18	21	25	29	33	38	41	40	37	31	24	19	41 °C
	t	7	9	12	15	19	25	30	29	24	18	13	8	17 °C
	tn	-2	-3	0	2	3	9	14	15	11	6	2	-2	-4 °C
	clothing	n	n	n	n	s	t	t	t	t	s	n	n	–
	stress					o	f	ff	ff	f	o			–
	p	14	14	24	21	15	4	1	10	22	26	23	25	199 mm
0.1 mm	dp	3	2	3	4	3	1	1	2	4	4	4	4	35 days

Marrakech

BShl	066-8	Jan	Feb	Mar	Apr	May	Jun	Jul	Aug	Sep	Oct	Nov	Dec	Year
				latitude: 31°37′n				longitude: 8°02′w				height: 466 m		
	tx	24	27	31	34	37	41	46	44	41	36	30	24	46 °C
	t	11	13	16	19	21	25	29	29	25	21	16	12	20 °C
	tn	0	2	4	7	9	13	16	16	14	9	6	1	-1 °C
	clothing	n	n	n	s	s	t	t	t	t	s	n	n	–
	stress				o	o	f	f	f	f	o			–
	p	28	29	32	31	17	8	2	3	10	21	28	33	241 mm
	dp	4	4	4	4	2	1	0	1	2	3	4	4	33 days
	sunshine	221	210	248	260	292	320	253	326	281	245	208	208	3072 hours
	%	69	67	67	67	68	75	58	79	76	69	66	66	69 %

Agadir

BSal	066-9	Jan	Feb	Mar	Apr	May	Jun	Jul	Aug	Sep	Oct	Nov	Dec	Year
				latitude: 30°23′n				longitude: 9°34′w				height: 19 m		
	tx	27	29	33	34	33	32	35	38	37	35	32	27	43 °C
	t	14	15	17	18	19	21	22	23	22	20	18	15	19 °C
	tn	3	3	6	8	11	13	15	15	13	10	7	4	2 °C
	clothing	n	n	n	s	s	s	s	t	s	s	s	n	–
	stress				o	o	f	f	f	f	f	o		–
	p	48	32	24	16	5	0	0	1	6	22	30	42	226 mm
	dp	4	3	3	2	1	0	0	0	1	2	3	4	23 days
	sunshine	240	231	283	294	304	287	288	265	254	248	224	231	3149 hours
	%	74	74	76	76	72	68	67	64	68	70	70	73	71 %

Sidi Ifni

BSbl	066-10	Jan	Feb	Mar	Apr	May	Jun	Jul	Aug	Sep	Oct	Nov	Dec	Year
				latitude: 29°22′n				longitude: 10°11′w				height: 66 m		
	tx	25	28	31	32	30	31	32	31	33	31	29	26	39 °C
	t	16	16	17	18	19	20	20	21	21	21	19	16	19 °C
	tn	7	9	11	12	13	15	16	16	16	13	11	9	7 °C
	clothing	n	n	s	s	s	s	s	s	s	s	s	n	–
	stress			o	o	o	o	o	f	f	o	o		–
	p	30	22	11	11	1	1	0	1	3	13	22	35	149 mm
0.1 mm	dp	6	4	3	4	3	3	1	1	3	2	4	5	39 days

3. Climatic area: Southern and southeastern part

3.1. Desert Climate BW and Desert Mountain Climate GBW

3.2. Summer very hot, winter warm.

3.3. Generally dry and sunny; occasionally scattered showers, more frequent in winter than in summer; rainy days are a rarity.

3.4. Wind: occasional gales and dust-storms.

Quarzazate

latitude: 30° 56′n longitude: 6° 54′w height: 1136 m

GBWhK	066-11	Jan	Feb	Mar	Apr	May	Jun	Jul	Aug	Sep	Oct	Nov	Dec	Year
	tx	23	26	29	32	36	41	43	43	39	32	27	23	43 °C
	t	9	11	15	18	22	27	30	29	25	20	14	9	19 °C
	tn	-3	-2	2	5	7	13	16	16	12	7	2	-3	-4 °C
	clothing	n	n	n	s	s	t	t	t	t	s	n	n	–
	stress				o	o	f	f	f	f	o			–
	p	6	6	12	9	5	4	2	9	20	18	17	15	123 mm
	dp	1	1	2	1	1	1	1	1	2	2	2	2	17 days
	sunshine	222	235	290	308	324	322	295	283	268	261	236	228	3272 hours
	%	69	75	78	79	76	76	68	69	72	73	74	72	74 %

4.3.66.2. Predominant general weather situations are described in 4.3.0.1. under North Africa. A hygrothermal diagram of Casablanca is to be found in 4.3.0.2.
Wind names: Chergui is a hot dry wind blowing from east to southeast, especially in spring and autumn.
Gharbi are rainy winds from South to West on the Atlantic Coast, which are regularly followed by strong winds, or storm, from the northwest.
Irifi is a dry hot wind blowing from northeast in the western parts of the country.
Sahel is a strong dusty wind in the desert.

4.3.67. Algeria

4.3.67.1. Climatic descriptions and tables

1. Climatic area: Region northward of the ridge line of the coastal or Tell Atlas
1.1. Subtropical Winter Rain Climate Cs and
Subtropical Winter Rain Mountain Climate GCs
1.2. Summer hot, winter mild.
1.3. Summer: generally dry and sunny; at times shower or rain. Winter: variable weather prevails, frequently dry and sunny or cloudy.
1.4. Wind: occasional gales. Fog: at times.

Cap Carbon

latitude: 36°45′n longitude: 5°06′e height: 230 m

Csal	067-1	Jan	Feb	Mar	Apr	May	Jun	Jul	Aug	Sep	Oct	Nov	Dec	Year
	tx	21	21	24	29	31	34	37	38	35	31	26	22	39 °C
	t	12	12	14	15	18	22	24	26	24	20	16	14	18 °C
	tn	4	4	6	8	11	16	18	19	16	12	9	5	3 °C
	clothing	n	n	n	n	s	s	t	t	t	s	n	n	–
	stress					o	f	f	ff	f	o			–
	p	93	70	55	82	38	22	3	12	51	112	129	91	758 mm
0.1 mm	dp	13	11	11	8	7	4	1	2	6	9	12	13	97 days

Dar El Beida / Alger

latitude: 36°43'n longitude: 3°15'e height: 25 m

Csal	067-2	Jan	Feb	Mar	Apr	May	Jun	Jul	Aug	Sep	Oct	Nov	Dec	Year
	tx	19	22	24	28	31	33	36	37	34	29	26	21	39 °C
	t	10	11	13	15	18	22	24	25	23	19	15	12	17 °C
	tn	5	6	6	8	12	15	18	19	16	12	9	5	3 °C
	clothing	n	n	n	n	s	s	t	t	t	s	n	n	–
	stress					o	f	ff	ff	f	o			–
	p	116	76	57	65	36	14	2	4	27	84	93	117	691 mm
0.1 mm	dp	15	12	13	10	9	5	1	2	7	10	14	15	113 days
	sunshine	146	170	220	237	307	306	344	332	273	214	156	146	2851 hours
	%	47	55	59	60	70	70	77	79	73	61	51	49	64 %

Oran / Es Sénia

latitude: 35°38'n longitude: 0°37'w height: 99 m

Csal	067-3	Jan	Feb	Mar	Apr	May	Jun	Jul	Aug	Sep	Oct	Nov	Dec	Year
	tx	21	23	25	27	29	32	34	35	33	29	25	22	38 °C
	t	10	11	13	15	18	22	24	25	23	18	14	11	17 °C
	tn	6	6	8	9	12	16	19	20	17	13	8	7	4 °C
	clothing	n	n	n	n	s	s	t	t	t	s	n	n	–
	stress					o	f	ff	ff	f	o			–
	p	70	54	35	33	19	7	1	3	16	43	46	67	394 mm
0.1 mm	dp	10	9	8	7	5	3	3	2	4	7	11	10	79 days
	sunshine	161	161	236	246	299	297	350	326	267	229	162	158	2892 hours
	%	52	52	64	63	69	68	79	78	72	65	62	52	65 %

2. *Climatic area:* Atlas Mountains (except northern slopes of the Coastal Atlas and southern slopes of the Saharan Atlas)
2.1. Steppe Climate BS and Steppe Mountain Climate GBS
2.2. Summer very hot, winter mild or cool.
2.3. Predominantly dry and sunny; at times – in winter more than in summer – showers in the afternoon or occasional rain.
2.4. Wind: occasional gales, at times dust-storms.

Tébessa

latitude: 35°26'n longitude: 8°08'e height: 816 m

GBSak	067-4	Jan	Feb	Mar	Apr	May	Jun	Jul	Aug	Sep	Oct	Nov	Dec	Year
	tx	17	20	24	29	33	38	41	39	36	29	23	18	– °C
	t	6	8	10	14	18	23	26	26	22	16	11	7	16 °C
	tn	-2	-2	-1	2	6	11	13	14	10	6	1	-2	– °C
	clothing	n	n	n	n	s	s	t	t	s	n	n	n	–
	stress					o	o	f	f	o				–
	p	26	23	31	46	38	24	5	18	33	38	34	32	348 mm
0.1 mm	dp	8	7	7	6	6	4	3	3	5	5	6	7	67 days

3. Climatic area: Region southward of the ridge-line of the Saharan Atlas
3.1. Desert Climate BW and Desert Mountain Climate GBW
3.2. Summer severely or very hot, winter mild or cool.
3.3. Generally dry and sunny.
3.4. Wind: occasional gales and dust-storms.

Biskra

latitude: 34°48′n longitude: 5°44′e height: 81 m

BWhl	067-5	Jan	Feb	Mar	Apr	May	Jun	Jul	Aug	Sep	Oct	Nov	Dec	Year
	tx	21	23	26	33	37	42	46	44	41	34	27	22	47 °C
	t	11	13	16	21	25	31	34	33	29	22	16	12	22 °C
	tn	2	4	6	9	13	19	23	22	17	12	7	3	2 °C
	clothing	n	n	n	s	t	t	t	t	t	s	n	n	–
	stress				o	f	ff	ff	ff	f	o			–
	p	13	11	19	13	9	4	1	6	17	18	19	16	148 mm

Laghouat

latitude: 33°46′n longitude: 2°56′e height: 767 m

GBWhk	067-6	Jan	Feb	Mar	Apr	May	Jun	Jul	Aug	Sep	Oct	Nov	Dec	Year
	t	8	9	13	16	21	26	30	29	24	18	12	8	18 °C
	clothing	n	n	n	n	s	t	t	t	t	s	n	n	–
	stress					o	f	ff	ff	f	o			–
	p	12	11	15	16	14	11	5	8	24	22	12	19	169 mm
	sunshine	211	229	279	297	332	342	360	335	297	267	219	205	3373 hours
	%	67	74	75	76	77	79	82	80	80	76	70	67	76 %

Béchar

latitude: 31°38′n longitude: 2°15′w height: 806 m

GBWhk	067-7	Jan	Feb	Mar	Apr	May	Jun	Jul	Aug	Sep	Oct	Nov	Dec	Year
	tx	21	24	28	33	36	41	42	42	39	32	26	22	43 °C
	t	9	12	16	20	24	30	34	33	28	21	15	10	21 °C
	tn	-2	-1	3	7	10	15	21	19	14	8	2	-2	-3 °C
	clothing	n	n	n	s	t	t	t	t	t	s	n	n	–
	stress				o	f	ff	ff	ff	ff	o			–
	p	7	9	13	8	3	2	0	4	7	14	13	10	90 mm
	sunshine	245	249	307	321	347	348	363	341	300	281	240	237	3579 hours
	%	77	80	83	83	81	82	83	83	81	79	76	76	80 %

El Goléa

latitude: 30°34′n longitude: 2°54′e height: 398 m

BWhk	067-8	Jan	Feb	Mar	Apr	May	Jun	Jul	Aug	Sep	Oct	Nov	Dec	Year
	tx	23	26	32	36	41	44	46	44	42	37	29	23	46 °C
	t	9	12	16	21	26	32	34	33	30	22	15	10	22 °C
	tn	-2	1	3	8	11	18	22	22	17	9	3	-1	-3 °C
	clothing	n	n	n	s	t	t	t	t	t	s	n	n	–
	stress				o	f	ff	ff	ff	ff	o			–
	p	8	1	6	4	0	1	0	0	2	4	4	14	44 mm
0.1 mm	dp	2	1	1	2	1	1	0	0	1	2	2	3	16 days
	sunshine	260	263	322	333	360	360	391	366	324	298	255	251	3783 hours
	%	81	84	87	86	85	85	90	89	87	84	80	79	85 %

Béni Abbès

latitude: 30°08′n longitude: 2°10′w height: 498 m

BWil	067-9	Jan	Feb	Mar	Apr	May	Jun	Jul	Aug	Sep	Oct	Nov	Dec	Year
	t	11	14	18	22	27	32	36	34	30	23	17	12	23 °C
	clothing	n	n	s	s	t	t	t	t	t	t	s	n	–
	stress			o	o	f	ff	ff	ff	ff	f	o		–
	p	9	2	7	8	1	0	0	2	3	10	6	12	62 mm

Fort Flatters

latitude: 28°06′n longitude: 6°49′e height: 381 m

BWhl	067-10	Jan	Feb	Mar	Apr	May	Jun	Jul	Aug	Sep	Oct	Nov	Dec	Year
	tx	26	30	36	39	43	45	47	46	43	39	34	26	47 °C
	t	11	14	19	24	28	33	34	34	31	25	18	13	24 °C
	tn	-4	0	3	8	11	19	21	21	18	11	3	-2	-5 °C
	clothing	n	n	s	t	t	t	t	t	t	t	s	n	–
	stress			o	f	ff	ff	ff	ff	ff	f	o		mm
	p	4	2	1	3	0	0	0	0	1	1	3	4	18 mm
	sunshine	270	271	310	321	353	342	394	372	321	304	270	239	3766 hours
	%	83	86	83	83	84	82	92	91	87	85	84	74	85 %

Aoulef

latitude: 26°58′n longitude: 1°05′e height: 290 m

BWil	067-11	Jan	Feb	Mar	Apr	May	Jun	Jul	Aug	Sep	Oct	Nov	Dec	Year
	tx	27	31	35	39	44	46	48	47	44	40	34	27	48 °C
	t	14	17	21	26	30	36	38	37	34	27	20	15	26 °C
	tn	-1	3	6	11	14	23	25	24	22	14	7	2	-1 °C
	clothing	n	n	s	t	t	t	t	t	t	t	s	n	–
	stress			o	f	f	ff	ff	ff	ff	f	o		–
	p	4	0	2	0	0	0	0	1	1	1	2	5	15 mm
	dp	1	0	1	1	1	0	0	0	1	1	1	1	8 days

Ouallen

latitude: 24°36′n longitude: 1°14′e height: 347 m

BWil	067-12	Jan	Feb	Mar	Apr	May	Jun	Jul	Aug	Sep	Oct	Nov	Dec	Year
	tx	24	32	36	39	44	47	47	45	44	41	36	30	47 °C
	t	16	20	24	28	32	37	38	37	35	30	23	17	28 °C
	tn	3	7	9	13	17	24	27	25	24	16	11	5	3 °C
	clothing	n	s	t	t	t	t	t	t	t	t	s	n	–
	stress		o	f	ff	ff	ff	ff	ff	ff	f	o		–
	p	2	2	1	0	0	2	2	6	5	0	0	1	22 mm
	sunshine	301	288	347	342	347	351	363	344	318	319	288	248	3856 hours
	%	90	90	93	90	84	86	87	85	86	89	88	75	87 %

Tamanrasset

GBWhl	067-13	Jan	Feb	Mar	Apr	May	Jun	Jul	Aug	Sep	Oct	Nov	Dec	Year
				latitude: 22°47′n				longitude: 5°31′e			height: 1378 m			
	tx	24	26	29	33	36	37	37	37	36	33	29	25	38 °C
	t	12	15	18	22	26	29	29	28	27	23	18	14	22 °C
	tn	-2	1	4	7	12	17	17	18	16	11	6	1	-2 °C
	clothing	n	n	s	s	t	t	t	t	t	s	s	n	–
	stress			o	o	f	ff	ff	ff	ff	o	o		–
	p	2	0	1	5	6	12	3	18	17	3	4	4	73 mm
0.1 mm	dp	1	0	1	1	1	2	1	3	2	1	1	1	15 days
	sunshine	260	259	310	291	310	276	310	307	276	276	267	260	3400 hours
	%	77	80	83	76	76	68	75	77	75	76	80	78	77 %

4.3.67.2. Predominant general weather situations are described in 4.3.0.1. under North Africa. A hygrothermal diagram of Alger is to be found in 4.3.0.2.
Wind names: Chichili or Sheheli are winds blowing to the south or southwest, and carrying dust and sand, in the southern parts of the country.
Dshani are heat gusts with drifting dust and sand in the Sahara, occurring in the afternoon.

4.3.68. Tunisia

4.3.68.1. Climatic descriptions and tables

1. Climatic area: Northern part of the country
1.1. Subtropical Winter Rain Climate Cs and
Subtropical Winter Rain Mountain Climate GCs
1.2. Summer hot, winter mild.
1.3. Summer: generally dry and sunny; winter: frequently variable weather.
1.4. Wind: occasional gales.

Tunis (Carthag)

Csal	068-1	Jan	Feb	Mar	Apr	May	Jun	Jul	Aug	Sep	Oct	Nov	Dec	Year
				latitude: 36°50′n				longitude: 10°14′e			height: 5 m			
	tx	19	22	26	29	33	38	41	42	38	33	27	21	43 °C
	t	11	12	13	16	19	23	26	27	25	20	16	12	18 °C
	tn	2	2	4	6	8	12	16	17	13	9	6	3	2 °C
	clothing	n	n	n	n	s	t	t	t	t	s	n	n	–
	stress					o	f	f	ff	f	o			–
	p	70	47	43	42	23	11	1	11	37	56	57	70	466 mm
	dp	9	7	7	5	4	2	1	1	4	6	7	9	66 days

2. Climatic area: Southeasterly coastal area and southern Atlas
2.1. Steppe Climate BS and Steppe Mountain Climate GBS
2.2. Summer very hot, winter mild.
2.3. Generally dry and sunny, in winter frequently variable weather.
2.4. Wind: occasional gales and dust-storms.

Gabès

latitude: 33°53′n longitude: 10°06′e height: 5 m

BSal	068-2	Jan	Feb	Mar	Apr	May	Jun	Jul	Aug	Sep	Oct	Nov	Dec	Year
	tx	21	24	28	32	34	37	38	39	37	34	27	21	42 °C
	t	11	13	15	17	21	24	26	27	26	22	17	12	19 °C
	tn	3	4	6	8	12	15	19	19	17	12	8	4	2 °C
	clothing	n	n	n	s	s	t	t	t	t	s	n	n	–
	stress				o	o	f	ff	ff	ff	o			–
	p	17	17	17	17	9	2	0	1	14	41	30	19	183 mm
	dp	2	2	3	2	2	1	0	1	2	2	3	2	23 days

3. Climatic area: Southwestern part
3.1. Desert Climate BW
3.2. Summer very hot, winter mild.
3.3. Generally dry and sunny.
3.4. Wind: occasional gales and dust-storms.

4.3.68.2. Predominant general weather situations are described in 4.3.0.1. and North Africa. A hygrothermal diagram of Tunis is to be found in 4.3.0.2.
Wind name: Dshani are heat gusts with drifting dust and sand in the Sahara, occurring in the afternoon.

4.3.69. Niger

4.3.69.1. Climatic descriptions and tables

1. Climatic area: greater part of the country, except smaller part in the south
1.1. Desert Climate BW and Desert Mountain Climate GBW
1.2. Summer very hot, winter hot or warm.
1.3. Generally dry and sunny, in summer showers now and then.
1.4. Wind: occasional gales and dust-storms.

Bilma

latitude: 18°41′n longitude: 12°55′e height: 357 m

BWhl	069-1	Jan	Feb	Mar	Apr	May	Jun	Jul	Aug	Sep	Oct	Nov	Dec	Year
	tx	33	36	41	43	45	45	44	43	42	41	36	32	46 °C
	t	17	20	25	29	32	33	33	33	32	28	23	18	27 °C
	tn	2	3	8	12	17	19	20	21	18	13	7	2	1 °C
	clothing	s	s	t	t	t	t	t	t	t	t	t	s	–
	stress	o	o	f	f	ff	ff	ff	ff	ff	f	f	o	–
	p	0	0	0	0	1	1	2	11	4	2	0	0	21 mm
	dp	0	0	0	0	0	0	0	2	1	0	0	0	3 days
	sunshine	285	280	310	297	319	318	332	316	303	316	300	301	3677 hours
	%	83	86	83	79	79	80	81	80	82	87	89	88	83 %

Agadès

GBWhb	069-2	Jan	Feb	Mar	Apr	May	Jun	Jul	Aug	Sep	Oct	Nov	Dec	Year
latitude: 16°59′n longitude: 7°59′e height: 500 m														
	tx	35	38	42	44	45	44	43	41	41	41	38	34	45 °C
	t	20	23	27	31	33	33	31	30	31	29	24	21	28 °C
	tn	6	7	12	15	19	21	21	19	19	17	1	7	6 °C
	clothing	s	t	t	t	t	t	t	t	t	t	t	s	–
	stress	o	o	f	f	ff	ff	ff	ff	ff	f	f	o	–
	p	0	0	0	1	6	8	49	78	20	1	0	0	164 mm
	dp	0	0	0	0	2	3	6	10	3	0	0	0	24 days
	sunshine	294	280	279	279	288	285	298	232	282	313	300	288	3418 hours
	%	84	85	75	74	72	72	74	59	77	86	88	83	77 %

Tahoua

BWha	069-3	Jan	Feb	Mar	Apr	May	Jun	Jul	Aug	Sep	Oct	Nov	Dec	Year
latitude: 14°54′n longitude: 5°15′e height: 387 m														
	tx	37	39	42	44	44	42	39	37	33	40	38	35	44 °C
	t	24	26	30	33	33	32	29	27	28	30	28	24	29 °C
	tn	11	12	17	20	22	21	19	19	19	18	15	12	10 °C
	clothing	t	t	t	t	t	t	t	t	t	t	t	t	–
	stress	o	f	f	ff	ff	ff	ff	ff	ff	ff	f	o	–
	p	1	0	0	2	18	45	117	144	66	14	0	0	407 mm
0.1 mm	dp	0	0	0	0	2	5	8	10	7	1	0	0	33 days

N'Guigmi

BWhb	069-4	Jan	Feb	Mar	Apr	May	Jun	Jul	Aug	Sep	Oct	Nov	Dec	Year
latitude: 14°15′n longitude: 13°07′e height: 289 m														
	tx	34	36	39	41	42	41	40	38	38	38	36	32	42 °C
	t	21	23	28	30	31	31	27	28	29	28	25	21	27 °C
	tn	9	11	14	17	22	21	21	21	21	18	13	9	8 °C
	clothing	s	t	t	t	t	t	t	t	t	t	t	s	–
	stress	o	o	f	ff	ff	ff	ff	ff	ff	ff	f	o	–
	p	0	0	0	0	7	8	57	141	22	0	0	0	236 mm
	dp	0	0	0	0	1	1	4	8	2	0	0	0	16 days
	sunshine	298	285	288	285	307	282	264	229	264	313	297	307	3419 hours
	%	84	86	77	76	77	72	66	59	72	85	86	88	77 %

2. Climatic area: southern part of the country (south of 15° north)

2.1. Steppe Climate BS and Steppe Mountain Climate GBS

2.2. Summer very hot, winter hot or warm.

2.3. High sun period: generally in the morning dry and sunny, in the afternoon showers at times; low sun period: mainly dry and sunny.

2.4. Wind: occasional gales and dust-storms.

Tillabery

BSha	069-5	Jan	Feb	Mar	Apr	May	Jun	Jul	Aug	Sep	Oct	Nov	Dec	Year
		latitude: 14°12′n				longitude: 1°27′e				height: 210 m				
	t	25	28	31	33	34	32	29	28	29	30	28	25	29 °C
	clothing	t	t	t	t	t	t	t	t	t	t	t	t	–
	stress	f	ff	ff	ff	ff	ff	ff	ff	ff	ff	ff	f	–
	p	0	1	2	1	18	55	132	203	86	11	0	0	510 mm

Niamey

BSha	069-6	Jan	Feb	Mar	Apr	May	Jun	Jul	Aug	Sep	Oct	Nov	Dec	Year
		latitude: 13°29′n				longitude: 2°10′e				height: 226 m				
	tx	38	41	43	44	44	42	39	36	37	41	41	37	45 °C
	t	25	27	31	34	33	31	28	27	28	30	28	25	29 °C
	tn	11	13	16	19	21	21	19	19	20	20	14	11	10 °C
	clothing	t	t	t	t	t	t	t	t	t	t	t	t	–
	stress	f	f	ff	ff	ff	ff	ff	ff	ff	ff	ff	f	–
	p	0	0	1	3	46	79	181	206	101	21	0	0	636 mm
0.1 mm	dp	0	0	0	1	4	7	11	16	11	3	0	0	53 days
	sunshine	273	254	254	246	248	243	229	189	225	279	279	276	2996 hours
	%	77	73	68	66	63	63	57	48	61	76	81	78	68 %

Birni N'Konni

BSha	069-7	Jan	Feb	Mar	Apr	May	Jun	Jul	Aug	Sep	Oct	Nov	Dec	Year
		latitude: 13°48′n				longitude: 5°15′e				height: 274 m				
	t	24	27	31	33	33	31	28	27	28	29	27	24	28 °C
	clothing	t	t	t	t	t	t	t	t	t	t	t	t	–
	stress	f	f	ff	ff	ff	ff	ff	ff	ff	ff	f	f	–
	p	0	0	0	2	29	72	138	232	111	16	0	0	597 mm
0.1 mm	dp	0	0	0	1	4	7	11	14	9	2	0	0	48 days
	sunshine	294	271	282	267	279	279	248	208	255	301	297	301	3282 hours
	%	83	82	75	72	70	72	62	53	70	82	86	86	74 %

Maradi

BShb	069-8	Jan	Feb	Mar	Apr	May	Jun	Jul	Aug	Sep	Oct	Nov	Dec	Year
		latitude: 13°28′n				longitude: 7°05′e				height: 369 m				
	t	22	25	29	32	32	30	27	26	27	28	26	22	27 °C
	clothing	s	t	t	t	t	t	t	t	t	t	t	s	–
	stress	o	f	ff	ff	ff	ff	ff	ff	ff	ff	f	o	–
	p	0	0	0	4	32	60	164	260	110	12	0	0	642 mm

Zinder

BSha	069-9	Jan	Feb	Mar	Apr	May	Jun	Jul	Aug	Sep	Oct	Nov	Dec	Year
		latitude: 13°48′n				longitude: 9°00′e				height: 479 m				
	tx	37	39	43	43	44	42	38	36	38	39	38	35	44 °C
	t	23	27	30	32	33	31	28	26	28	30	27	23	28 °C
	tn	10	12	17	20	22	20	18	18	18	19	14	11	9 °C
	clothing	t	t	t	t	t	t	t	t	t	t	t	t	–
	stress	f	f	ff	ff	ff	ff	ff	ff	ff	ff	f	f	–
	p	0	0	0	3	27	55	153	232	71	7	0	0	549 mm
0.1 mm	dp	0	0	0	0	4	6	11	14	8	1	0	0	44 days
	sunshine	254	243	254	255	267	249	229	205	246	282	264	257	3005 hours
	%	72	74	68	68	67	64	57	52	67	77	76	73	68 %

Mainé-Soroa

BShb	069-10	Jan	Feb	Mar	Apr	May	Jun	Jul	Aug	Sep	Oct	Nov	Dec	Year
		latitude: 13° 14′n				longitude: 11° 59′e				height: 339 m				
	t	22	24	29	31	32	31	28	27	28	29	26	22	28 °C
	clothing	s	t	t	t	t	t	t	t	t	t	t	s	–
	stress	o	f	ff	ff	ff	ff	ff	ff	ff	ff	ff	o	–
	p	0	0	0	3	15	32	106	176	62	9	0	0	404 mm
0.1 mm	dp	0	0	0	0	3	5	9	12	9	2	0	0	40 days
	sunshine	291	277	276	276	288	267	239	186	243	301	291	291	3226 hours
	%	82	84	74	74	73	69	60	48	66	82	84	83	73 %

4.3.69.2. Predominant general weather situations are described in 4.3.0.1. under West Africa. A hygrothermal diagram of Niamey is to be found in 4.3.0.2.
Wind name: Harmattan is a dry and hot wind blowing from the northeast and carrying sand and laterite dust.

4.3.70. Mali

4.3.70.1. Climatic descriptions and tables

1. Climatic area: Northern part (northward of 15° north)
1.1. Desert Climate BW and Desert Mountain Climate GBW
1.2. Summer severely or very hot, winter hot or warm.
1.3. Generally dry and sunny; high sun period: showers now and then.
1.4. Wind: occasional gales and dust-storms.

Tessalit

BWhb	070-1	Jan	Feb	Mar	Apr	May	Jun	Jul	Aug	Sep	Oct	Nov	Dec	Year
		latitude: 20° 12′n				longitude: 0° 59′e				height: 494 m				
	tx	32	36	38	42	44	45	46	43	43	41	37	32	46 °C
	t	20	23	26	30	33	34	34	33	33	31	26	20	29 °C
	tn	6	8	12	14	20	24	23	21	22	19	13	7	5 °C
	clothing	s	t	t	t	t	t	t	t	t	t	t	s	–
	stress	o	f	f	ff	ff	ff	ff	ff	ff	ff	f	o	–
	p	1	0	1	0	2	7	23	55	27	1	1	0	118 mm
	dp	0	0	0	0	0	1	3	6	3	0	0	0	14 days
	sunshine	300	269	315	312	306	266	296	285	273	300	269	259	3450 hours
	%	88	83	85	82	75	66	72	72	74	83	80	76	78 %

Kidal

BWhb	070-2	Jan	Feb	Mar	Apr	May	Jun	Jul	Aug	Sep	Oct	Nov	Dec	Year
		latitude: 18° 26′n				longitude: 1° 21′e				height: 464 m				
	tx	31	36	39	43	44	45	44	42	41	41	37	33	46 °C
	t	20	24	27	31	33	34	33	31	32	31	26	21	29 °C
	tn	7	9	12	17	21	22	21	20	20	19	14	8	6 °C
	clothing	s	t	t	t	t	t	t	t	t	t	t	s	–
	stress	o	f	f	ff	ff	ff	ff	ff	ff	ff	f	o	–
	p	0	1	0	1	4	8	38	51	28	1	0	0	132 mm
0.1 mm	dp	0	0	0	0	1	2	5	7	3	0	0	0	18 days

Tombouctou (Timbuktu)

BWha	070-3	latitude: 16°46′n						longitude: 3°01′w				height: 273 m		
		Jan	Feb	Mar	Apr	May	Jun	Jul	Aug	Sep	Oct	Nov	Dec	Year
	tx	36	39	43	46	47	47	44	41	44	43	41	36	47 °C
	t	23	25	28	32	34	34	32	29	31	32	28	23	29 °C
	tn	8	9	13	17	22	23	21	21	21	19	14	8	7 °C
	clothing	t	t	t	t	t	t	t	t	t	t	t	t	–
	stress	f	f	f	ff	ff	ff	ff	ff	ff	ff	f	f	–
	p	0	0	0	1	3	19	65	95	37	5	0	0	225 mm
	dp	0	0	0	0	1	2	8	8	4	1	0	0	23 days

Gao

BWia	070-4	latitude: 16°16′n						longitude: 0°03′w				height: 260 m		
		Jan	Feb	Mar	Apr	May	Jun	Jul	Aug	Sep	Oct	Nov	Dec	Year
	tx	35	38	41	44	45	44	42	38	40	41	39	36	45 °C
	t	23	25	29	32	34	35	32	30	31	32	28	24	30 °C
	tn	9	11	14	18	22	22	21	20	21	21	14	10	8 °C
	clothing	t	t	t	t	t	t	t	t	t	t	t	t	–
	stress	f	f	f	ff	ff	ff	ff	ff	ff	ff	f	f	–
	p	0	0	1	1	6	27	75	110	36	5	0	0	261 mm
	dp	0	0	0	0	1	4	8	9	5	1	0	0	29 days
	sunshine	276	264	288	294	286	266	276	276	273	292	289	270	3341 hours
	%	79	80	77	78	72	68	68	68	74	80	85	78	75 %

Ménaka

BWia	070-5	latitude: 15°52′n						longitude: 2°13′e				height: 280 m		
		Jan	Feb	Mar	Apr	May	Jun	Jul	Aug	Sep	Oct	Nov	Dec	Year
	tx	36	38	42	44	44	44	43	39	41	42	41	36	45 °C
	t	23	26	30	33	35	34	32	30	31	32	29	24	30 °C
	tn	9	11	16	22	22	23	21	19	21	19	16	11	8 °C
	clothing	t	t	t	t	t	t	t	t	t	t	t	t	–
	stress	f	f	ff	ff	ff	ff	ff	ff	ff	ff	ff	f	–
	p	0	0	1	0	8	22	73	112	43	4	0	0	263 mm
	dp	0	0	0	0	1	4	8	10	5	1	0	0	28 days

Hombori

BWia	070-6	latitude: 15°20′n						longitude: 1°41′w				height: 288 m		
		Jan	Feb	Mar	Apr	May	Jun	Jul	Aug	Sep	Oct	Nov	Dec	Year
	tx	37	39	42	44	44	43	40	37	39	41	39	36	44 °C
	t	24	27	30	33	35	34	31	28	30	32	29	25	30 °C
	tn	12	14	17	21	24	23	21	21	21	21	17	13	11 °C
	clothing	t	t	t	t	t	t	t	t	t	t	t	t	–
	stress	f	f	ff	ff	ff	ff	ff	ff	ff	ff	ff	f	–
	p	0	0	0	1	11	43	124	191	73	11	0	0	454 mm
	dp	0	0	0	0	2	4	9	12	7	2	0	0	36 days

2. Climatic area: Central part (between 15° and 13° north)

2.1. Steppe Climate BS

2.2. Summer very hot, winter hot.

2.3. High sun period: in the morning mostly dry and sunny, in the afternoon frequent showers, sporadic rainy days. Low sun period: generally dry and sunny.

2.4. Wind: occasional gales and dust-storms.

Nioro du Sahel

latitude: 15°14′n longitude: 9°22′w height: 237 m

BSha	070-7	Jan	Feb	Mar	Apr	May	Jun	Jul	Aug	Sep	Oct	Nov	Dec	Year
	t	23	26	29	32	34	33	30	28	29	30	27	24	29 °C
	clothing	t	t	t	t	t	t	t	t	t	t	t	t	–
	stress	f	f	ff	ff	ff	ff	ff	ff	ff	ff	ff	f	–
	p	1	1	0	5	19	66	157	233	121	24	3	1	631 mm
	dp	0	0	0	1	3	5	9	12	8	2	0	0	40 days

Mopti

latitude: 14°31′n longitude: 4°06′w height: 272 m

BSha	070-8	Jan	Feb	Mar	Apr	May	Jun	Jul	Aug	Sep	Oct	Nov	Dec	Year
	tx	36	38	41	43	43	41	38	33	33	36	37	34	43 °C
	t	23	25	29	32	33	31	28	27	28	28	27	23	28 °C
	tn	10	12	14	17	22	21	21	21	21	20	17	12	9 °C
	clothing	t	t	t	t	t	t	t	t	t	t	t	t	–
	stress	f	f	f	ff	ff	ff	ff	ff	ff	ff	f	f	–
	p	1	0	0	3	19	55	154	195	104	20	1	0	552 mm
	dp	0	0	0	1	3	8	11	13	9	3	0	0	49 days
	sunshine	230	225	251	250	234	240	226	206	253	260	249	190	2814 hours
	%	65	68	67	67	59	62	56	53	69	71	72	54	63 %

Kayes

latitude: 14°26′n longitude: 11°26′w height: 47 m

BSia	070-9	Jan	Feb	Mar	Apr	May	Jun	Jul	Aug	Sep	Oct	Nov	Dec	Year
	tx	38	41	45	46	46	43	38	35	47	39	40	39	47 °C
	t	25	28	31	34	36	32	29	27	28	30	29	26	30 °C
	tn	12	15	18	21	23	20	19	20	20	19	16	13	12 °C
	clothing	t	t	t	t	t	t	t	t	t	t	t	t	–
	stress	f	f	ff	ff	ff	ff	ff	ff	ff	ff	f	f	–
	p	1	1	0	7	21	100	203	257	181	51	4	0	826 mm
	dp	0	0	0	1	2	8	10	14	12	4	1	0	52 days
	sunshine	253	233	301	286	254	219	197	187	209	241	248	201	2829 hours
	%	72	71	81	77	64	56	49	48	57	66	72	57	64 %

Segou

latitude: 13°24′n longitude: 6°09′w height: 289 m

BSha	070-10	Jan	Feb	Mar	Apr	May	Jun	Jul	Aug	Sep	Oct	Nov	Dec	Year
	tx	37	39	41	43	42	41	37	33	34	36	37	36	43 °C
	t	24	27	30	32	33	30	28	26	27	28	27	24	28 °C
	tn	12	13	17	19	22	20	19	19	20	18	16	11	11 °C
	clothing	t	t	t	t	t	t	t	t	t	t	t	t	–
	stress	f	f	ff	ff	ff	ff	ff	ff	ff	ff	f	f	–
	p	0	0	3	9	20	94	202	239	127	27	3	0	724 mm
	dp	0	0	0	1	4	10	11	14	10	2	0	0	51 days
	sunshine	286	269	296	281	271	273	251	226	247	294	283	277	3254 hours
	%	81	81	79	75	69	70	63	58	67	80	82	72	73 %

San

BSha	070-11	Jan	Feb	Mar	Apr	May	Jun	Jul	Aug	Sep	Oct	Nov	Dec	Year
				latitude: 13°20'n				longitude: 4°50'w				height: 284 m		
	t	24	26	30	32	33	30	28	27	27	29	28	24	28 °C
	clothing	t	t	t	t	t	t	t	t	t	t	t	t	–
	stress	f	f	ff	ff	ff	ff	ff	ff	ff	ff	ff	f	–
	p	1	0	3	12	43	97	185	249	121	27	5	0	743 mm
	dp	0	0	1	1	4	8	12	16	11	4	0	0	57 days

3. Climatic area: Southern part (southward of 13° north)

3.1. Tropical Summer Rain Climate Aw and
Tropical Summer Rain Mountain Climate GAw

3.2. Summer very hot, winter hot.

3.3. High sun period: generally dry and sunny in the morning, in the afternoon often showers, rainy days now and then. Low sun period: generally dry and sunny.

3.4. Wind: occasional gales and dust-storms in the dry season.

Kéniéba

Awha	070-12	Jan	Feb	Mar	Apr	May	Jun	Jul	Aug	Sep	Oct	Nov	Dec	Year
				latitude: 12°48'n				longitude: 11°21'w				height: 135 m		
	t	26	28	31	33	32	29	27	27	27	27	27	25	28 °C
	clothing	t	t	t	t	t	t	t	t	t	t	t	t	–
	stress	f	ff	ff	ff	ff	ff	ff	ff	ff	ff	ff	f	–
	p	0	1	0	7	51	171	308	382	331	141	15	1	1408 mm
	dp	0	0	0	1	5	12	15	19	17	8	1	0	79 days

Bamako

Awha	070-13	Jan	Feb	Mar	Apr	May	Jun	Jul	Aug	Sep	Oct	Nov	Dec	Year
				latitude: 12°38'n				longitude: 8°02'w				height: 332 m		
	tx	37	40	41	42	43	38	36	33	34	37	38	37	44 °C
	t	26	28	31	32	32	29	27	26	27	28	27	25	28 °C
	tn	12	15	18	21	21	19	19	19	19	18	14	12	11 °C
	clothing	t	t	t	t	t	t	t	t	t	t	t	t	–
	stress	f	f	ff	ff	ff	ff	ff	ff	ff	ff	f	f	–
	p	1	0	3	15	60	145	251	334	220	58	12	0	1099 mm
	dp	0	0	0	1	6	11	14	19	15	5	1	0	73 days
	sunshine	267	242	270	242	216	205	195	164	197	230	243	219	2690 hours
	%	75	73	72	65	55	53	49	42	54	63	70	62	61 %

Koutiala

Awha	070-14	Jan	Feb	Mar	Apr	May	Jun	Jul	Aug	Sep	Oct	Nov	Dec	Year
				latitude: 12°24'n				longitude: 5°28'w				height: 346 m		
	tx	36	39	41	42	41	38	34	33	34	36	36	36	43 °C
	t	23	26	29	31	31	29	27	26	27	28	27	24	27 °C
	tn	8	11	15	17	21	20	20	19	21	19	13	9	8 °C
	clothing	t	t	t	t	t	t	t	t	t	t	t	t	–
	stress	f	f	ff	ff	ff	ff	ff	ff	ff	ff	ff	f	–
	p	0	1	5	28	66	136	221	304	190	61	9	1	1022 mm
	dp	0	0	1	2	6	10	14	18	14	6	1	0	73 days

Bougouni

Awha	070-15	Jan	Feb	Mar	Apr	May	Jun	Jul	Aug	Sep	Oct	Nov	Dec	Year
				latitude: 11°25'n				longitude: 7°30'w				height: 351 m		
	tx	37	39	39	39	39	36	33	32	33	36	36	36	40 °C
	t	26	27	30	31	30	28	26	26	26	28	27	24	27 °C
	tn	13	17	19	21	21	19	19	19	20	19	17	13	13 °C
	clothing	t	t	t	t	t	t	t	t	t	t	t	t	–
	stress	f	f	ff	ff	ff	ff	ff	ff	ff	ff	ff	f	–
	p	0	2	11	30	94	165	292	380	255	92	13	3	1337 mm
	dp	0	0	1	3	7	11	15	18	15	9	2	0	81 days

Sikasso

Awha	070-16	Jan	Feb	Mar	Apr	May	Jun	Jul	Aug	Sep	Oct	Nov	Dec	Year
				latitude: 11°21'n				longitude: 5°41'w				height: 375 m		
	t	24	27	30	31	30	28	26	26	26	28	27	24	27 °C
	clothing	t	t	t	t	t	t	t	t	t	t	t	t	–
	stress	f	f	ff	ff	ff	ff	ff	ff	ff	ff	ff	f	–
	p	1	7	17	41	122	162	273	338	246	105	14	2	1329 mm
	dp	0	0	2	4	8	12	16	19	17	10	2	0	91 days
	sunshine	273	246	242	213	232	225	202	164	183	236	255	264	2734 hours
	%	76	74	65	57	59	59	51	42	50	64	73	74	62 %

4.3.70.2. Predominant general weather situations are described in 4.3.0.1. under West Africa.
Wind name: Harmattan is a dry hot wind blowing from the northeast and carrying sand and laterite dust.

4.3.71. Mauretania

4.3.71.1. Climatic description and tables

1. Desert Climate BW, coastal areas Marine Desert Climate BM
2. Summer very hot, coast hot in the north, winter warm in the north, hot in the south.
3. Generally dry and sunny. High sun period: showers now and then, in the south also the occasional rainy day.
4. Wind: occasional gales and dust-storms.

Fort Gouraud

BWhb	071-1	Jan	Feb	Mar	Apr	May	Jun	Jul	Aug	Sep	Oct	Nov	Dec	Year
				latitude: 22°41'n				longitude: 12°42'w				height: 297 m		
	tx	30	33	37	39	43	46	47	47	44	41	36	32	48 °C
	t	18	20	23	24	28	31	34	34	32	28	24	19	26 °C
	tn	7	8	11	11	13	17	18	19	18	16	14	8	7 °C
	clothing	s	s	t	t	t	t	t	t	t	t	t	s	–
	stress	o	o	o	f	f	ff	ff	ff	ff	f	f	o	–
	p	3	7	0	0	0	0	4	12	15	22	9	3	76 mm
0.1 mm	dp	1	1	0	0	0	0	1	2	3	1	1	0	10 days
	sunshine	264	251	310	318	326	294	288	291	261	264	249	251	3367 hours
	%	78	78	83	84	79	72	69	73	71	73	75	75	76 %

Nou Adhibou (Port Etienne)

BMab	071-2	Jan	Feb	Mar	Apr	May	Jun	Jul	Aug	Sep	Oct	Nov	Dec	Year
		latitude: 20°56'n						longitude: 17°02'w				height: 5 m		
	tx	31	34	34	38	38	39	35	35	39	38	36	32	40 °C
	t	19	19	20	20	21	23	23	24	26	24	22	19	22 °C
	tn	9	10	11	12	12	13	15	16	17	14	13	10	9 °C
	clothing	s	s	s	s	s	t	t	t	t	t	s	s	–
	stress	o	o	o	o	o	f	f	f	ff	f	o	o	–
	p	2	2	3	1	0	0	1	4	9	7	4	3	36 mm
0.1 mm	dp	0	0	0	0	0	0	0	1	1	1	1	1	6 days
	sunshine	254	257	291	306	322	303	279	288	261	257	249	248	3317 hours
	%	74	79	78	81	79	75	68	72	71	71	74	74	75 %

Tidjikja

BWhb	071-3	Jan	Feb	Mar	Apr	May	Jun	Jul	Aug	Sep	Oct	Nov	Dec	Year
		latitude: 18°34'n						longitude: 11°26'w				height: 396 m		
	tx	32	36	39	42	44	45	44	42	42	40	37	34	46 °C
	t	20	22	25	28	32	33	32	31	31	29	25	20	27 °C
	tn	6	7	11	13	18	21	19	18	19	16	11	7	6 °C
	clothing	s	s	t	t	t	t	t	t	t	t	t	t	–
	stress	o	o	f	f	ff	ff	ff	ff	ff	f	o	o	–
	p	1	4	5	0	4	8	24	58	54	10	4	3	174 mm
0.1 mm	dp	1	1	1	0	1	2	4	8	5	2	1	1	26 days
	sunshine	251	268	322	330	301	270	291	298	273	285	264	244	3399 hours
	%	73	82	86	88	75	68	71	75	74	78	78	71	77 %

Nouakshott

BMhb	071-4	Jan	Feb	Mar	Apr	May	Jun	Jul	Aug	Sep	Oct	Nov	Dec	Year
		latitude: 18°06'n						longitude: 15°57'w				height: 2 m		
	tx	34	36	39	40	43	43	39	38	41	41	38	36	44 °C
	t	21	23	25	26	27	28	28	29	30	29	26	22	26 °C
	tn	9	11	13	16	16	19	22	22	22	18	15	10	8 °C
	clothing	s	t	t	t	t	t	t	t	t	t	t	s	–
	stress	o	o	f	f	f	ff	ff	ff	ff	f	f	o	–
	p	2	1	0	0	0	2	13	64	58	11	5	16	171 mm
0.1 mm	dp	0	0	0	0	0	0	2	5	4	1	1	1	14 days
	sunshine	242	249	298	318	310	285	273	264	249	257	258	248	3250 hours
	%	70	76	80	84	77	72	67	67	68	71	76	72	73 %

Nema

BWia	071-5	Jan	Feb	Mar	Apr	May	Jun	Jul	Aug	Sep	Oct	Nov	Dec	Year
		latitude: 16°36'n						longitude: 7°16'w				height: 269 m		
	tx	36	39	41	44	45	45	43	39	41	41	40	37	46 °C
	t	24	27	30	33	36	34	32	30	31	32	30	25	30 °C
	tn	12	13	18	21	24	23	20	20	20	23	18	13	11 °C
	clothing	t	t	t	t	t	t	t	t	t	t	t	t	–
	stress	o	f	ff	ff	ff	ff	ff	ff	ff	ff	ff	f	–
	p	1	1	1	2	5	54	98	125	82	17	2	3	390 mm
0.1 mm	dp	0	0	0	0	1	4	6	9	5	2	1	1	28 days
	sunshine	276	268	304	297	288	267	270	264	261	273	273	257	3298 hours
	%	79	82	81	79	72	68	67	67	71	75	80	74	74 %

Rosso

BWhb	071-6	Jan	Feb	Mar	Apr	May	Jun	Jul	Aug	Sep	Oct	Nov	Dec	Year
		latitude: 16°30'n				longitude: 15°49'w				height: 5 m				
	t	23	24	27	28	30	30	30	29	29	29	27	22	27 °C
	clothing	t	t	t	t	t	t	t	t	t	t	t	t	–
	stress	o	f	ff	ff	ff	ff	ff	ff	ff	ff	ff	f	–
	p	2	2	0	0	2	14	49	134	81	46	3	6	340 mm

4.3.71.2. Predominant general weather situations are described in 4.3.0.1. under West Africa.
Wind name: Harmattan is a dry hot wind blowing from the northeast and carrying sand and laterite dust.

4.3.72. Senegal

4.3.72.1. Climatic descriptions and tables

1. Climatic area: Northernmost part
1.1. Desert Climate BW, coastal areas: Marine Desert Climate BM
1.2. Summer very hot, winter hot.
1.3. Generally dry and sunny; in the high sun period showers now and then.
1.4. Wind: occasional gales and dust-storms.

Podor

BWha	072-1	Jan	Feb	Mar	Apr	May	Jun	Jul	Aug	Sep	Oct	Nov	Dec	Year
		latitude: 16°38'n				longitude: 14°56'w				height: 6 m				
	t	23	25	28	30	32	32	31	30	30	30	28	23	28 °C
	clothing	t	t	t	t	t	t	t	t	t	t	t	t	–
	stress	o	o	f	ff	ff	ff	ff	ff	ff	ff	f	f	–
	p	1	3	0	0	2	28	72	109	109	36	1	4	365 mm
0.1 mm	dp	1	1	0	0	0	2	5	9	6	3	1	1	29 days

Saint-Louis

BMhb	072-2	Jan	Feb	Mar	Apr	May	Jun	Jul	Aug	Sep	Oct	Nov	Dec	Year
		latitude: 16°03'n				longitude: 16°27'w				height: 4 m				
	tx	33	36	38	34	32	32	32	34	36	36	34	34	40 °C
	t	22	22	22	22	22	26	28	28	29	28	26	23	25 °C
	tn	12	13	14	16	17	19	21	20	20	20	17	14	12 °C
	clothing	s	s	s	s	s	t	t	t	t	t	t	t	–
	stress	o	o	o	f	f	ff	ff	ff	ff	ff	f	f	–
	p	1	1	0	0	1	7	44	161	97	29	2	3	347 mm
0.1 mm	dp	0	0	0	0	0	1	5	12	9	3	1	1	33 days
	sunshine	239	229	291	306	291	252	217	229	228	248	240	232	3003 hours
	%	68	70	78	82	73	64	54	58	62	68	70	67	68 %

2. *Climatic area:* greater part of the country (except northernmost part and southern part)
2.1. Steppe Climate BS
2.2. Summer very hot, winter hot.
2.3. High sun period: predominantly dry and sunny in the morning, often showers in the afternoon. Low sun period: mostly dry and sunny.
2.4. Wind: occasional gales, showers often accompanied by thunderstorms; low sun period: occasional dust-storms.

Matan

latitude: 15° 38′n longitude: 13° 15′w height: 15 m

BSha	072-3	Jan	Feb	Mar	Apr	May	Jun	Jul	Aug	Sep	Oct	Nov	Dec	Year
	tx	36	40	42	44	46	44	40	37	37	39	39	37	46 °C
	t	23	25	28	31	34	33	30	28	28	29	27	23	28 °C
	tn	9	10	13	16	21	21	20	21	20	19	15	11	9 °C
	clothing	t	t	t	t	t	t	t	t	t	t	t	t	–
	stress	f	f	f	ff	ff	ff	ff	ff	ff	ff	f	f	–
	p	0	2	0	0	2	36	139	184	148	28	5	0	547 mm
0.1 mm	dp	0	0	0	0	1	5	8	11	9	2	1	0	37 days
	sunshine	279	257	313	312	313	285	273	229	243	279	276	270	3329 hours
	%	80	78	84	83	79	73	68	58	66	76	81	78	75 %

Linguère

latitude: 15° 32′n longitude: 15° 09′w height: 21 m

BSha	072-4	Jan	Feb	Mar	Apr	May	Jun	Jul	Aug	Sep	Oct	Nov	Dec	Year
	tx	38	41	42	44	45	44	34	36	37	41	40	38	46 °C
	t	24	26	28	30	31	31	29	29	28	28	28	24	28 °C
	tn	9	10	13	14	17	20	17	21	19	17	13	10	8 °C
	clothing	t	t	t	t	t	t	t	t	t	t	t	t	–
	stress	o	f	f	ff	ff	ff	ff	ff	ff	f	f	o	–
	p	1	2	2	0	4	31	101	209	136	45	4	2	535 mm
0.1 mm	dp	0	0	0	0	1	3	8	13	10	4	1	0	40 days
	sunshine	242	266	301	297	304	252	217	180	186	229	213	242	2928 hours
	%	69	81	81	79	76	64	54	76	51	63	62	70	66 %

Dakar/Yoff

latitude: 14° 44′n longitude: 17° 30′w height: 23 m

BShb	072-5	Jan	Feb	Mar	Apr	May	Jun	Jul	Aug	Sep	Oct	Nov	Dec	Year
	tx	34	34	37	33	33	34	34	34	34	36	35	33	38 °C
	t	21	20	21	22	23	26	27	27	28	27	26	23	24 °C
	tn	15	16	16	17	18	21	22	22	22	22	20	17	9 °C
	clothing	s	s	s	s	t	t	t	t	t	t	t	s	–
	stress	o	o	o	f	ff	ff	ff	ff	ff	ff	ff	f	–
	p	0	2	0	0	1	15	88	249	163	49	5	6	578 mm
0.1 mm	dp	0	0	0	0	1	2	9	15	13	4	1	1	46 days
	sunshine	245	240	298	303	301	252	220	183	231	254	258	239	3023 hours
	%	70	73	80	81	76	65	55	47	63	69	75	68	68 %

Kaolack

BSha	072-6	Jan	Feb	Mar	Apr	May	Jun	Jul	Aug	Sep	Oct	Nov	Dec	Year
		latitude: 14°08'n						longitude: 16°04'w				height: 6 m		
	tx	37	39	42	43	43	41	37	34	35	37	38	37	43 °C
	t	25	26	28	29	30	30	28	28	28	29	28	25	28 °C
	tn	12	13	15	17	18	22	21	21	21	19	16	13	11 °C
	clothing	t	t	t	t	t	t	t	t	t	t	t	t	–
	stress	o	o	f	ff	ff	ff	ff	ff	ff	ff	f	o	–
	p	0	1	0	0	11	65	166	124	232	75	5	1	680 mm
0.1 mm	dp	0	0	0	0	1	6	13	19	16	6	1	0	62 days
	sunshine	288	282	329	327	326	282	251	211	222	267	288	276	3348 hours
	%	82	85	88	88	82	73	63	54	61	73	84	79	75 %

3. Climatic area: southern part (east and south of Gambia)
3.1. Tropical Summer Rain Climate Aw
3.2. Very hot all the year round.
3.3. High sun period: predominantly dry in the morning, often showers in the afternoon, at times rainy days. Low sun period: generally dry and sunny.
3.4. Wind: occasional gales, showers often accompanied by thunderstorms, during the low sun period occasional dust-storms.

Tambacounda

Awha	072-7	Jan	Feb	Mar	Apr	May	Jun	Jul	Aug	Sep	Oct	Nov	Dec	Year
		latitude: 13°46'n						longitude: 13°41'w				height: 46 m		
	tx	38	41	43	44	44	42	36	34	34	37	38	38	44 °C
	t	25	27	30	32	33	30	27	26	27	28	28	24	28 °C
	tn	10	12	16	17	21	19	19	19	19	18	13	10	9 °C
	clothing	t	t	t	t	t	t	t	t	t	t	t	t	–
	stress	o	f	ff	ff	ff	ff	ff	ff	ff	ff	f	o	–
	p	0	1	0	2	20	131	196	289	231	70	2	0	942 mm
0.1 mm	dp	0	0	0	0	2	9	13	19	16	6	1	0	66 days
	sunshine	223	234	294	282	273	210	183	155	189	229	246	205	2724 hours
	%	63	71	79	76	69	54	46	40	52	62	71	58	61 %

Ziguinchor

Awha	072-8	Jan	Feb	Mar	Apr	May	Jun	Jul	Aug	Sep	Oct	Nov	Dec	Year
		latitude: 12°35'n						longitude: 16°16'w				height: 10 m		
	t	24	26	27	28	28	28	27	26	27	28	27	24	27 °C
	clothing	t	t	t	t	t	t	t	t	t	t	t	t	–
	stress	f	f	f	ff	ff	ff	ff	ff	ff	ff	ff	f	–
	p	0	1	0	0	10	125	363	532	361	146	8	1	1547 mm
0.1 mm	dp	0	0	0	0	2	11	21	25	21	11	2	0	93 days
	sunshine	229	232	298	303	285	183	146	115	162	223	249	208	2632 hours
	%	64	70	80	81	72	47	37	30	44	61	72	59	59 %

4.3.72.2. Predominant general weather situations are described in 4.3.0.1. under West Africa. A hygrothermal diagram of Dakar is to be found in 4.3.0.2.
Wind name: Harmattan is a hot dry wind blowing from the northeast and carrying sand and laterite dust.

4.3.73. Gambia

4.3.73.1. Climatic description and table

1. Tropical Summer Rain Climate Aw
2. Summer very hot, winter hot.
3. High sun period: generally dry and sunny in the morning, in the afternoon often showers, now and then rainy days. Low sun period: generally dry and sunny.
4. Wind: occasional gales, squalls in thunderstorms, in the dry season dust-storms.

Bathurst

		latitude: 13°21'n					longitude: 16°40'w				height: 26 m		
Awaa	073-1	Jan	Feb	Mar	Apr	May	Jun	Jul	Aug	Sep	Oct	Nov	Dec	Year
	tx	34	37	39	39	38	35	33	32	33	35	35	34	40 °C
	t	24	24	25	25	26	27	26	26	27	27	26	24	26 °C
	tn	9	12	14	15	16	21	21	21	19	19	14	12	9 °C
	clothing	t	t	t	t	t	t	t	t	t	t	t	t	–
	stress	o	o	f	f	f	ff	ff	ff	ff	ff	f	o	–
	p	2	2	0	0	11	78	309	329	349	104	17	1	1202 mm
0.2 mm	dp	0	0	0	0	0	7	18	25	21	8	0	0	79 days
	sunshine	273	274	316	303	322	255	186	174	180	236	252	270	3040 hours
	%	77	83	85	81	81	66	47	45	49	64	73	77	69 %

4.3.73.2. Predominant general weather situations are described in 4.3.0.1. under West Africa.
Wind name: Harmattan is a dry and hot wind blowing from the northeast and carrying sand and laterite dust.

4.3.74. Guinea-Bissau

4.3.74.1. Climatic description and tables

1. Tropical Summer Rain Climate Aw
2. Very hot all the year round.
3. High sun period: generally dry and sunny in the morning, frequent showers in the afternoon, at times rainy days. Low sun period: mostly dry and sunny.
4. Wind: occasional gales, squalls in thunderstorms, in the dry season dust-storms.

Bafata

		latitude: 12°10'n					longitude: 14°40'w				height: 43 m		
Awha	074-1	Jan	Feb	Mar	Apr	May	Jun	Jul	Aug	Sep	Oct	Nov	Dec	Year
	t	24	28	29	30	30	28	27	26	26	27	27	24	27 °C
	clothing	t	t	t	t	t	t	t	t	t	t	t	t	–
	stress	o	ff	ff	ff	ff	ff	ff	ff	ff	ff	ff	f	–
	p	0	0	0	5	34	174	314	441	367	181	20	0	1536 mm
	dp	0	0	0	1	2	12	18	22	18	12	2	0	86 days
	sunshine	245	240	304	288	298	210	170	127	165	226	246	245	2764 hours
	%	69	72	81	77	76	55	43	33	45	61	71	69	62 %

Bissau

Awaa	074-2	Jan	Feb	Mar	Apr	May	Jun	Jul	Aug	Sep	Oct	Nov	Dec	Year
		latitude: 11°52′n					longitude: 15°35′w				height: 40 m			
	t	24	26	27	27	27	27	26	26	26	27	27	25	26 °C
	clothing	t	t	t	t	t	t	t	t	t	t	t	t	–
	stress	o	f	f	f	ff	ff	ff	ff	ff	ff	ff	f	–
	p	1	1	0	1	17	175	471	682	435	195	42	2	2023 mm
	dp	0	0	0	0	2	11	22	24	21	13	2	0	95 days
	sunshine	255	227	285	283	265	188	139	105	163	215	233	232	2590 hours
	%	71	68	76	76	67	49	35	27	45	58	67	65	58 %

4.3.74.2. Predominant general weather situations are described in 4.3.0.1. under Equatorial West Africa.
Wind name: Harmattan is a hot dry wind blowing from the northeast and carrying sand and laterite dust.

4.3.75. Guinea

4.3.75.1. Climatic description and tables

1. Tropical Summer Rain Climate Aw and Tropical Monsoonal Rain Climate Am and Tropical Summer Rain Mountain Climate GAw
2. Very hot all the year round in the lowlands.
3. High sun period: often heavy showers, especially in the afternoon and at night; also frequent rainy days. Low sun period: generally dry and sunny or cloudy.
4. Wind: squalls in thunderstorms, occasional gales and dust-storms in the dry season.

Siguiri

Awha	075-1	Jan	Feb	Mar	Apr	May	Jun	Jul	Aug	Sep	Oct	Nov	Dec	Year
		latitude: 11°26′n					longitude: 9°10′w				height: 365 m			
	t	24	26	29	31	30	27	26	25	26	27	26	24	27 °C
	clothing	t	t	t	t	t	t	t	t	t	t	t	t	–
	stress	o	f	ff	ff	ff	ff	ff	ff	ff	ff	f	f	–
	p	0	2	11	32	93	186	253	346	273	117	21	2	1336 mm
0.1 mm	dp	0	0	1	4	8	13	15	18	17	11	2	0	89 days
	sunshine	288	271	291	246	251	234	183	174	189	248	246	257	2879 hours
	%	81	81	78	66	64	61	46	45	52	67	71	72	65 %

Mamou

GAwaa	075-2	Jan	Feb	Mar	Apr	May	Jun	Jul	Aug	Sep	Oct	Nov	Dec	Year
		latitude: 10°22′n					longitude: 12°05′w				height: 783 m			
	t	23	25	26	26	24	23	22	22	23	23	23	23	24 °C
	clothing	t	t	t	t	t	t	t	t	t	t	t	t	–
	stress	o	o	f	f	f	f	f	f	f	f	f	f	–
	p	6	7	39	86	164	241	328	454	378	291	79	10	2085 mm
0.1 mm	dp	1	1	4	9	14	20	23	26	24	19	7	1	149 days

Conacry

Amha	075-3	Jan	Feb	Mar	Apr	May	Jun	Jul	Aug	Sep	Oct	Nov	Dec	Year
				latitude: 9°34'n				longitude: 13°37'w			height: 26 m			
	tx	33	33	33	34	34	32	31	31	31	32	32	33	34 °C
	t	27	27	27	28	27	26	25	25	26	26	27	27	26 °C
	tn	19	21	21	21	21	21	21	21	21	21	22	21	18 °C
	clothing	t	t	t	t	t	t	t	t	t	t	t	t	–
	stress	ff	ff	ff	ff	ff	ff	ff	ff	ff	ff	ff	ff	–
	p	0	2	18	38	163	514	1480	1230	645	351	126	25	4594 mm
0.1 mm	dp	0	1	1	3	12	24	29	29	26	22	10	1	158 days
	sunshine	155	192	257	210	161	105	62	43	114	164	171	121	1756 hours
	%	43	57	69	57	41	28	16	11	31	44	49	34	40 %

Macenta

GAmha	075-4	Jan	Feb	Mar	Apr	May	Jun	Jul	Aug	Sep	Oct	Nov	Dec	Year
				latitude: 8°32'n				longitude: 9°28'w			height: 543 m			
	t	23	25	26	26	26	25	24	24	24	25	24	23	25 °C
	clothing	t	t	t	t	t	t	t	t	t	t	t	t	–
	stress	o	o	f	f	f	f	f	f	f	f	f	f	–
	p	13	57	146	166	201	270	489	553	411	265	164	56	2791 mm
0.1 mm	dp	1	5	11	13	15	18	23	25	22	18	13	3	167 days

4.3.75.2. Predominant general weather situations are described in 4.3.0.1. under Equatorial West Africa. A hygrothermal diagram of Conacry is to be found in 4.3.0.2.
Wind name: Harmattan is a hot dry wind blowing from the northeast and carrying sand and laterite dust.

4.3.76. Sierra Leone

4.3.76.1. Climatic description and tables

1. Tropical Monsoonal Rain Climate Am and
   Tropical Monsoonal Rain Mountain Climate GAm
2. Very hot throughout the year.
3. High sun period: often heavy showers, especially in the afternoon and at night; also frequent rainy days. Low sun period: generally dry and sunny or cloudy.
4. Wind: Squalls in thunderstorms, occasional gales, in winter dust-storms.

Kabala

Amha	076-1	Jan	Feb	Mar	Apr	May	Jun	Jul	Aug	Sep	Oct	Nov	Dec	Year
				latitude: 9°35'n				longitude: 11°33'w			height: 464 m			
	tx	35	37	38	37	35	34	32	32	32	33	33	34	38 °C
	t	25	27	28	27	26	25	24	24	24	25	25	24	25 °C
	tn	7	10	15	17	18	18	18	18	19	19	13	9	7 °C
	clothing	t	t	t	t	t	t	t	t	t	t	t	t	–
	stress	o	f	f	f	ff	ff	ff	ff	ff	ff	f	o	–
	p	19	16	64	106	197	315	314	405	420	357	129	31	2373 mm
0.2 mm	dp	1	2	4	7	14	20	26	28	27	23	9	1	162 days
	sunshine	251	237	232	234	217	174	136	90	150	202	210	260	2394 hours
	%	70	71	62	63	56	46	35	23	41	55	60	72	54 %

Lungi (Freetown)

Amha	076-2	latitude: 8°37′n Jan	Feb	Mar	Apr	May	longitude: 13°12′w Jun	Jul	Aug	Sep	Oct	height: 27 m Nov	Dec	Year
	tx	32	34	33	33	32	32	30	29	30	32	32	32	35 °C
	t	27	27	28	28	27	26	25	25	26	26	26	26	26 °C
	tn	19	21	22	22	21	21	21	21	21	21	21	19	18 °C
	clothing	t	t	t	t	t	t	t	t	t	t	t	t	–
	stress	ff	ff	ff	ff	ff	ff	ff	ff	ff	ff	ff	ff	–
	p	17	7	32	65	227	389	730	797	528	301	171	54	3318 mm
0.2 mm	dp	1	1	2	6	15	24	27	27	26	22	12	5	168 days
	sunshine	242	223	254	225	202	144	90	71	120	192	201	223	2187 hours
	%	67	67	68	61	52	38	23	18	33	52	57	62	49 %

Daru

Amaa	076-3	latitude: 7°59′n Jan	Feb	Mar	Apr	May	longitude: 10°51′w Jun	Jul	Aug	Sep	Oct	height: 186 m Nov	Dec	Year
	tx	34	37	37	36	35	34	32	31	33	34	34	33	37 °C
	t	25	26	27	27	27	26	25	25	26	26	26	25	26 °C
	tn	13	17	19	20	21	21	20	20	21	20	19	15	13 °C
	clothing	t	t	t	t	t	t	t	t	t	t	t	t	–
	stress	f	ff	ff	ff	ff	ff	ff	ff	ff	ff	ff	f	–
	p	12	32	106	152	251	290	285	375	408	334	206	63	2514 mm
0.2 mm	dp	1	1	7	11	16	22	24	25	25	23	15	5	175 days
	sunshine	189	184	205	174	174	141	87	71	108	167	177	180	1856 hours
	%	52	55	55	47	45	37	22	18	30	45	50	50	42 %

Bonthe

Amha	076-4	latitude: 7°32′n Jan	Feb	Mar	Apr	May	longitude: 12°30′w Jun	Jul	Aug	Sep	Oct	height: 8 m Nov	Dec	Year
	t	27	28	28	28	28	26	25	25	26	26	27	27	27 °C
	clothing	t	t	t	t	t	t	t	t	t	t	t	t	–
	stress	ff	ff	ff	ff	ff	ff	ff	ff	ff	ff	ff	ff	–
	p	17	21	49	112	267	570	882	755	616	351	179	67	3890 mm
0.2 mm	dp	1	1	5	10	20	25	27	27	26	24	17	6	189 days
	sunshine	186	170	186	156	149	105	79	46	60	140	159	177	1592 hours
	%	51	51	50	42	39	28	15	12	16	38	45	49	36 %

4.3.76.2. Predominant general weather situations are described in 4.3.0.1. under Equatorial West Africa. A hygrothermal diagram of Conary is to be found in 4.3.0.2.
Wind name: Harmattan is a hot dry wind blowing from the northeast and carrying sand and laterite dust.

4.3.77. Ascension Island

4.3.77.1. Climatic description and table

1. Marine Desert Climate BM
2. Very hot throughout the year.
3. Generally dry and sunny or cloudy; low sun period: at times overcast with rain.
4. Wind: gales rare.

Georgetown / Ascension Island

latitude: 7°56's longitude: 14°25'w height: 17 m

BMha	077-1	Jan	Feb	Mar	Apr	May	Jun	Jul	Aug	Sep	Oct	Nov	Dec	Year
	tx	31	32	32	32	32	31	30	29	29	29	29	30	33 °C
	t	26	27	27	27	28	26	26	25	24	25	25	26	26 °C
	tn	21	22	22	23	23	21	21	21	20	20	21	21	20 °C
	clothing	t	t	t	t	t	t	t	t	t	t	t	t	–
	stress	ff	ff	ff	ff	ff	ff	ff	ff	ff	ff	ff	ff	–
	p	5	10	18	28	13	13	13	10	8	8	5	3	132 mm
	dp	2	2	3	4	3	3	3	3	2	3	1	1	30 days

4.3.77.2. Predominant general weather situations are described in 4.3.0.1. under Equatorial West Africa.

4.3.78. St. Helena

4.3.78.1. Climatic description and tables

1. Partly Tropical Winter Rain Climate As, partly Marine Desert Climate BM
2. Sea level: summer hot, winter warm.
3. Variable weather prevails, especially during low sun period, although dry and sunny or cloudy days are not infrequent.
4. Wind: gales rare.

Jamestown / St. Helena

latitude: 15°55's longitude: 5°43'w height: 12 m

BMab	078-1	Jan	Feb	Mar	Apr	May	Jun	Jul	Aug	Sep	Oct	Nov	Dec	Year
	tx	28	29	29	29	29	27	26	24	24	23	24	26	27 °C
	t	24	24	24	24	22	21	20	20	20	20	21	22	22 °C
	tn	18	19	19	19	17	17	16	16	16	16	17	17	15 °C
	clothing	t	t	t	t	s	s	s	s	s	s	s	s	–
	stress	f	f	f	f	f	f	o	o	o	o	o	o	–
	p	8	10	20	10	18	18	33	10	5	2	0	3	137 mm
	dp	4	4	5	3	4	6	8	3	2	1	0	1	41 days

Hutts Gate / St. Helena

GAsbl	078-2	Jan	Feb	Mar	Apr	May	Jun	Jul	Aug	Sep	Oct	Nov	Dec	Year
				latitude: 15°57's				longitude: 5°40'w			height: 619 m			
	tx	23	23	23	23	22	21	19	18	18	19	20	21	24 °C
	t	18	19	19	18	17	17	15	14	14	14	15	16	16 °C
	tn	14	15	16	15	14	12	12	11	11	12	12	13	11 °C
	clothing	s	s	s	s	s	s	n	n	n	n	n	n	–
	stress													–
	p	53	78	107	79	71	81	109	66	56	43	30	41	814 mm
	dp	11	15	16	14	14	14	17	13	12	11	8	11	156 days

4.3.78.2. Predominant general weather situations are described in 4.3.0.1. under Equatorial West Africa.

4.3.79. Sao Thomé

4.3.79.1. Climatic descriptions and table

1. Tropical Winter Rain Climate As
2. Very hot throughout the year.
3. June to August repeatedly dry and sunny, otherwise often rain or showers in the afternoon; now and then rainy days too.
4. Wind: squalls in thunderstorms, gales rare.

São Thomé

Asha	079-1	Jan	Feb	Mar	Apr	May	Jun	Jul	Aug	Sep	Oct	Nov	Dec	Year
				latitude: 0°23'n				longitude: 6°43'e			height: 13 m			
	tx	31	32	32	32	31	30	29	29	31	30	31	31	32 °C
	t	28	26	26	26	26	25	24	24	25	25	25	26	26 °C
	tn	21	21	21	21	21	19	17	17	19	20	21	21	17 °C
	clothing	t	t	t	t	t	t	t	t	t	t	t	t	–
	stress	ff	ff	ff	ff	ff	ff	ff	ff	ff	ff	ff	ff	–
	p	82	84	132	120	113	16	0	1	18	108	100	98	872 mm
0.1 mm	dp	8	8	12	11	10	3	2	3	6	12	11	8	94 days
	sunshine	141	145	129	132	156	177	162	139	101	111	137	141	1672 hours
	%	38	42	34	36	42	49	43	37	28	30	38	38	38 %

4.3.79.2. Predominant general weather situations are described in 4.3.0.1. under Equatorial West Africa.

4.3.80. Libya

4.3.80.1. Climatic descriptions and tables

1. Climatic area: part of the coast of Cyrenaika
1.1. Subtropical Winter Rain Climate Cs
1.2. Summer hot, winter mild.
1.3. Summer: generally dry and sunny; winter: now and then days with variable weather.
1.4. Wind: occasionally storm.

Benina (sunshine Bengazi)

latitude: 32°05′n longitude: 20°16′e height: 132 m

Csal	080-1	Jan	Feb	Mar	Apr	May	Jun	Jul	Aug	Sep	Oct	Nov	Dec	Year
	tx	21	23	30	34	38	38	34	33	36	33	29	24	40 °C
	t	12	13	15	18	22	26	26	25	25	22	18	14	20 °C
	tn	6	6	7	9	12	16	18	19	17	14	11	7	5 °C
	clothing	n	n	n	s	s	t	t	t	t	s	s	n	–
	stress				o	o	f	ff	ff	ff	o	o		–
	p	67	23	15	6	3	0	0	0	0	22	28	66	231 mm
0.1 mm	dp	13	9	6	2	1	0	0	0		4	8	12	56 days
	sunshine	155	167	232	261	273	321	347	332	291	257	207	158	3001 hours
	%	49	53	63	67	64	75	80	80	78	73	65	51	67 %

2. Climatic area: coastal areas, except part of Cyrenaika
2.1. Steppe Climate BS
2.2. Summer hot, winter mild.
2.3. Generally dry and sunny, in winter now and then days with variable weather.
2.4. Wind: occasional gales and dust-storms.

Zuara

latitude: 32°55′n longitude: 12°05′e height: 3 m

BSal	080-2	Jan	Feb	Mar	Apr	May	Jun	Jul	Aug	Sep	Oct	Nov	Dec	Year
	t	12	13	15	18	20	24	26	27	26	22	17	13	20 °C
	clothing	n	n	n	s	s	t	t	t	t	s	n	n	–
	stress				o	o	f	ff	ff	ff	o			–
	p	21	13	9	23	6	1	0	0	6	38	38	61	216 mm

Idris (Tripoli)

latitude: 32°41′n longitude: 13°41′e height: 84 m

BShl	080-3	Jan	Feb	Mar	Apr	May	Jun	Jul	Aug	Sep	Oct	Nov	Dec	Year
	tx	21	23	29	35	36	39	37	36	37	34	29	23	41 °C
	t	11	13	16	19	22	27	27	28	26	22	16	13	20 °C
	tn	4	6	7	9	12	15	18	19	18	14	9	6	4 °C
	clothing	n	n	n	s	s	t	t	t	t	s	n	n	–
	stress				o	o	f	ff	ff	ff	o			–
	p	46	27	12	16	2	1	0	1	6	27	39	74	253 mm
0.1 mm	dp	11	7	5	2	3	1	0	0	2	5	7	11	55 days
	sunshine	158	172	186	225	260	285	344	335	243	214	162	146	2730 hours
	%	50	55	50	58	61	66	79	81	65	60	51	47	61 %

Misurata

		latitude: 32° 25′n					longitude: 15°06′e				height: 6 m		
BSal	080-4	Jan	Feb	Mar	Apr	May	Jun	Jul	Aug	Sep	Oct	Nov	Dec	Year
	tx	24	26	33	38	38	41	40	39	37	37	32	27	45 °C
	t	12	14	16	18	21	25	26	27	26	23	18	14	20 °C
	tn	3	4	7	8	11	13	17	18	16	14	9	4	2 °C
	clothing	n	n	n	s	s	t	t	t	t	t	s	n	–
	stress				o	o	f	ff	ff	ff	f	o		–
	p	61	22	17	15	5	0	0	0	5	27	21	76	250 mm
0.1 mm	dp	11	8	4	2	1	1	0	0	2	5	6	9	49 days

3. Climatic area: Greater part of the country, except coastal areas
3.1. Desert Climate BW and Desert Mountain Climate GBW
3.2. Summer hot or very hot, winter mild.
3.3. Generally dry and sunny, in winter now and then scattered rain.
3.4. Wind: occasional gales and dust-storms.

El Adem

		latitude: 31°51′n					longitude: 23°55′e				height: 155 m		
BWal	080-5	Jan	Feb	Mar	Apr	May	Jun	Jul	Aug	Sep	Oct	Nov	Dec	Year
	tx	22	26	30	34	38	39	37	37	37	34	29	23	41 °C
	t	12	13	15	18	21	25	26	26	24	21	17	13	19 °C
	tn	2	3	3	6	11	14	17	17	15	11	7	4	2 °C
	clothing	n	n	n	s	s	t	t	t	t	s	s	n	–
	stress				o	o	f	ff	ff	f	o	o		–
	p	15	8	13	1	5	0	0	0	1	11	7	23	84 mm
	dp	3	2	3	0	1	0	0	0	0	2	2	4	17 days

Hon

		latitude: 29°08′n					longitude: 15°57′e				height: 261 m		
BWhl	080-6	Jan	Feb	Mar	Apr	May	Jun	Jul	Aug	Sep	Oct	Nov	Dec	Year
	tx	25	29	34	41	43	46	45	43	43	38	33	27	47 °C
	t	10	13	16	21	25	28	29	29	27	22	17	12	21 °C
	tn	-3	-1	1	4	9	14	15	16	13	8	3	-1	-3 °C
	clothing	n	n	n	s	t	t	t	t	t	s	s	n	–
	stress				o	f	f	ff	ff	f	o	o		–
	p	4	3	2	3	3	1	0	0	4	9	3	5	37 mm
0.1 mm	dp	1	2	1	1	2	0	0	0	1	2	1	1	12 days

Sebha

		latitude: 27°01′n					longitude: 14°26′e				height: 440 m		
BWhl	080-7	Jan	Feb	Mar	Apr	May	Jun	Jul	Aug	Sep	Oct	Nov	Dec	Year
	tx	26	29	33	39	42	44	43	42	42	37	33	27	46 °C
	t	11	14	18	23	28	31	31	31	28	24	18	12	23 °C
	tn	2	3	6	9	14	18	20	19	18	13	6	2	-1 °C
	clothing	n	n	s	t	t	t	t	t	t	t	s	n	–
	stress			o	o	f	ff	ff	ff	ff	f	o		–
	p	3	1	2	1	3	1	0	0	0	2	5	2	18 mm

Kufra

latitude: 24°13′n longitude: 23°18′e height: 382 m

BWhl	080-8	Jan	Feb	Mar	Apr	May	Jun	Jul	Aug	Sep	Oct	Nov	Dec	Year
	tx	29	29	34	41	43	44	42	42	39	38	33	28	45 °C
	t	13	16	19	23	29	31	31	31	29	25	18	14	23 °C
	tn	1	3	6	11	14	20	20	21	18	12	8	2	1 °C
	clothing	n	n	s	t	t	t	t	t	t	t	s	n	–
	stress			o	f	f	ff	ff	ff	ff	f	o		–
	p	0	0	0	0	0	0	0	1	0	0	0	0	2 mm

4.3.80.2. Predominant general weather situations are described in 4.3.0.1. under North Africa. A hygrothermal diagram of Tripoli is to be found in 4.3.0.2.
Wind names: Gharra are winterly squalls in thunderstorms on the coast coming from the northeast.
Ghibli is a dust-storm blowing especially in springtime from the hot deserts.

4.3.81. Egypt

4.3.81.1. Climatic descriptions and tables

1. Climatic area: parts of the Mediterranean coast
1.1. Steppe Climate BS
1.2. Summer hot, winter mild.
1.3. Generally dry and sunny, in winter frequent days with showers or rain.
1.4. Wind: occasional gales and dust-storms.

Sidi Barrani

latitude: 31°38′n longitude: 25°58′e height: 23 m

BSal	081-1	Jan	Feb	Mar	Apr	May	Jun	Jul	Aug	Sep	Oct	Nov	Dec	Year
	t	13	14	15	18	20	23	25	26	24	22	19	15	20 °C
	clothing	n	n	n	s	s	t	t	t	t	s	s	n	–
	stress				o	o	f	f	f	f	o	o		–
	p	30	7	11	2	5	0	0	0	0	18	22	41	136 mm
	dp	5	2	3	1	1	0	0	0	0	2	4	8	26 days

Mersa Matruh

latitude: 31°21′n longitude: 27°13′e height: 30 m

BSal	081-2	Jan	Feb	Mar	Apr	May	Jun	Jul	Aug	Sep	Oct	Nov	Dec	Year
	tx	23	25	29	35	36	35	30	32	33	33	29	24	39 °C
	t	13	14	15	18	20	23	25	26	24	22	18	15	19 °C
	tn	6	6	7	9	12	16	18	19	18	15	11	7	4 °C
	clothing	n	n	n	s	s	t	t	t	t	s	s	n	–
	stress				o	o	f	ff	ff	ff	f	o		–
	p	20	16	8	2	2	0	0	0	1	17	21	40	127 mm
	dp	4	3	2	1	1	0	0	0	0	2	4	7	27 days

Alexandria

latitude: 31°12′n longitude: 29°57′e height: 7 m

BSal	081-3	Jan	Feb	Mar	Apr	May	Jun	Jul	Aug	Sep	Oct	Nov	Dec	Year
	tx	23	26	30	35	36	36	33	33	34	33	30	26	39 °C
	t	15	15	17	19	20	24	26	27	26	24	21	17	21 °C
	tn	7	8	9	12	15	18	21	22	20	17	12	8	6 °C
	clothing	n	n	n	s	s	t	t	t	t	t	s	n	–
	stress				o	o	ff	ff	ff	ff	f	o		–
	p	44	25	11	3	2	0	0	1	1	7	30	48	169 mm
	dp	6	5	3	1	1	0	0	0	0	2	4	7	28 days
	sunshine	220	221	276	306	332	357	369	366	327	301	243	205	3523 hours
	%	69	71	74	79	78	84	85	89	88	85	76	65	79 %

2. *Climatic area:* Greater part of the country, except parts of the Mediterranean coast

2.1. Desert Climate BW and Desert Mountain Climate GBW, coastal areas: Marine Desert Climate BM

2.2. Summer predominantly very hot, winter mild.

2.3. Dry and sunny, rarely rain.

2.4. Wind: occasional gales and dust-storms.

Salloum

latitude: 31°32′n longitude: 25°11′e height: 6 m

BMal	081-4	Jan	Feb	Mar	Apr	May	Jun	Jul	Aug	Sep	Oct	Nov	Dec	Year
	tx	23	25	30	36	38	38	35	36	37	34	30	24	43 °C
	t	14	15	16	19	21	25	26	26	25	23	20	16	21 °C
	tn	4	4	6	7	11	14	17	17	15	13	9	4	2 °C
	clothing	n	n	n	s	s	t	t	t	t	t	s	n	–
	stress				o	o	f	ff	ff	ff	f	o		–
	p	13	7	11	1	3	0	0	0	0	12	32	27	103 mm
	dp	2	2	2	0	1	0	0	0	0	1	1	4	13 days

Port Said

latitude: 31°17′n longitude: 32°14′e height: 7 m

BMhl	081-5	Jan	Feb	Mar	Apr	May	Jun	Jul	Aug	Sep	Oct	Nov	Dec	Year
	tx	23	26	29	32	34	34	34	34	33	32	30	25	38 °C
	t	15	16	17	19	23	26	27	28	27	25	21	17	22 °C
	tn	7	8	9	12	15	18	21	22	21	18	13	8	6 °C
	clothing	n	n	s	s	t	t	t	t	t	t	s	n	–
	stress				o	f	ff	ff	ff	ff	ff	o		–
	p	12	11	7	2	4	0	0	0	0	3	9	15	63 mm
	dp	3	3	2	1	0	0	0	0	0	0	2	2	13 days
	sunshine	226	215	257	279	316	351	357	350	312	295	240	201	3399 hours
	%	71	69	69	72	74	83	82	85	84	83	76	64	76 %

Cairo

latitude: 30°08'n longitude: 31°34'e height: 74 m

BWhl	081-6	Jan	Feb	Mar	Apr	May	Jun	Jul	Aug	Sep	Oct	Nov	Dec	Year
	tx	23	27	33	38	40	41	39	38	36	36	32	26	42 °C
	t	14	15	18	21	25	28	29	29	26	24	20	16	22 °C
	tn	4	5	7	9	13	17	20	20	18	15	10	6	4 °C
	clothing	n	n	s	s	t	t	t	t	t	t	s	n	–
	stress			o	o	f	ff	ff	ff	ff	f	o		–
	p	4	4	3	1	1	0	0	0	0	1	1	7	22 mm
	dp	1	1	0	0	0	0	0	0	0	0	0	1	5 days
	sunshine	233	229	276	291	329	354	360	347	315	288	252	220	3504 hours
	%	72	73	74	75	78	84	83	84	85	81	79	69	79 %

Siwa

latitude: 29°12'n longitude: 25°29'e height: -14 m

BWhl	081-7	Jan	Feb	Mar	Apr	May	Jun	Jul	Aug	Sep	Oct	Nov	Dec	Year
	tx	25	29	33	39	43	44	42	42	41	38	33	26	46 °C
	t	12	14	17	21	26	28	29	29	27	23	18	14	22 °C
	tn	-1	1	3	7	11	15	17	17	14	10	5	1	-1 °C
	clothing	n	n	s	s	t	t	t	t	t	t	s	n	–
	stress			o	o	f	ff	ff	ff	f	f	o		–
	p	1	3	0	0	2	0	0	0	0	0	0	2	9 mm

Baharia

latitude: 28°20'n longitude: 28°54'e height: 130 m

BWhl	081-8	Jan	Feb	Mar	Apr	May	Jun	Jul	Aug	Sep	Oct	Nov	Dec	Year
	tx	26	30	34	39	43	42	42	41	39	36	32	27	45 °C
	t	12	14	17	21	26	28	29	29	26	23	19	14	22 °C
	tn	1	3	4	8	12	16	17	18	16	12	7	2	0 °C
	clothing	n	n	s	s	t	t	t	t	t	t	s	n	–
	stress			o	o	f	f	ff	ff	f	f	o		–
	p	0	1	0	1	0	0	0	0	0	0	1	1	4 mm

Minya

latitude: 28°05'n longitude: 30°44'e height: 40 m

BWhl	081-9	Jan	Feb	Mar	Apr	May	Jun	Jul	Aug	Sep	Oct	Nov	Dec	Year
	t	12	14	17	21	26	28	29	28	26	23	19	15	22 °C
	clothing	n	n	s	s	t	t	t	t	t	t	s	n	–
	stress			o	o	f	f	ff	ff	f	f	o		–
	p	1	1	0	1	1	0	0	0	0	1	0	1	6 mm

Hurghada

latitude: 27°17'n longitude: 33°46'e height: 3 m

BMhl	081-10	Jan	Feb	Mar	Apr	May	Jun	Jul	Aug	Sep	Oct	Nov	Dec	Year
	t	15	16	21	22	26	28	28	29	27	24	21	17	23 °C
	clothing	n	n	s	s	t	t	t	t	t	t	s	s	–
	stress			o	o	f	ff	ff	ff	ff	f	o	o	–
	p	0	0	0	0	0	0	0	0	0	0	2	2	4 mm

Manqabad

latitude: 27°11'n longitude: 31°06'e height: 70 m

BWhl	081-11	Jan	Feb	Mar	Apr	May	Jun	Jul	Aug	Sep	Oct	Nov	Dec	Year
	tx	26	28	34	39	43	43	42	41	38	33	31	26	44 °C
	t	13	15	18	23	28	29	30	30	28	25	20	15	23 °C
	tn	3	4	6	10	14	18	21	21	18	14	8	4	2 °C
	clothing	n	n	s	t	t	t	t	t	t	t	s	n	–
	stress			o	o	f	ff	ff	ff	ff	f	o		–
	p	1	1	0	0	1	0	0	0	0	0	1	1	5 mm

Luxor

latitude: 25°40'n longitude: 32°42'e height: 89 m

BWhl	081-12	Jan	Feb	Mar	Apr	May	Jun	Jul	Aug	Sep	Oct	Nov	Dec	Year
	tx	29	33	38	43	47	46	46	46	43	42	37	31	47 °C
	t	14	16	20	25	30	32	32	32	30	27	21	16	25 °C
	tn	2	3	6	9	15	18	20	21	19	14	8	3	1 °C
	clothing	n	n	s	t	t	t	t	t	t	t	s	n	–
	stress			o	f	f	ff	ff	ff	ff	f	o		–
	p	0	0	0	0	0	0	0	0	0	0	0	0	0 mm

Dakhla

latitude: 25°29'n longitude: 29°00'e height: 112 m

BWhl	081-13	Jan	Feb	Mar	Apr	May	Jun	Jul	Aug	Sep	Oct	Nov	Dec	Year
	tx	29	33	37	43	45	45	44	43	41	39	35	29	47 °C
	t	13	15	19	24	29	30	31	31	28	25	20	15	23 °C
	tn	1	2	4	8	13	18	19	19	17	13	7	2	-1 °C
	clothing	n	n	s	t	t	t	t	t	t	t	s	n	–
	stress			o	f	f	ff	ff	ff	ff	f	o		–
	p	0	0	0	0	0	0	0	0	0	0	0	0	0 mm

Kharga

latitude: 25°26'n longitude: 30°34'e height: 72 m

BWhl	081-14	Jan	Feb	Mar	Apr	May	Jun	Jul	Aug	Sep	Oct	Nov	Dec	Year
	t	14	16	20	24	29	31	31	31	29	26	21	16	24 °C
	clothing	n	n	s	t	t	t	t	t	t	t	s	n	–
	stress			o	f	ff	ff	ff	ff	ff	f	o		–
	p	0	0	0	0	0	0	0	0	0	0	0	0	0 mm

Aswan

latitude: 23°58'n longitude: 32°47'e height: 194 m

BWhl	081-15	Jan	Feb	Mar	Apr	May	Jun	Jul	Aug	Sep	Oct	Nov	Dec	Year
	tx	32	34	40	44	46	47	46	40	39	37	38	33	48 °C
	t	17	19	22	27	32	34	34	34	32	30	24	20	27 °C
	tn	6	6	9	13	17	22	23	23	21	17	12	6	5 °C
	clothing	s	s	s	t	t	t	t	t	t	t	t	s	–
	stress	o	o	o	f	ff	ff	ff	ff	ff	f	f	o	–
	p	0	0	0	0	1	0	0	0	0	0	0	0	1 mm

4.3.81.2. Predominant general weather situations are described in 4.3.0.1. under North Africa. A hygrothermal diagram of Cairo is to be found in 4.3.0.2.
Wind names: Khamsin is a hot dry wind blowing from the south and carrying large quantity of dust and sand, lasting usually 2–4 days in Egypt.
Shamal on the Red Sea is a northerly or northwesterly wind which ends a Khamsin situation by a marked squall belonging to a cold front.

4.3.82. Sudan

4.3.82.1. Climatic descriptions and tables

1. Climatic area: Northern Province, Kassala, Khartoum, northern parts of Dafur, Kordofar, Blue Nile
1.1. Desert Climate BW and Desert Mountain Climate GBW, coastal area: Marine Desert Climate BM
1.2. Summer severely or very hot; winter mild to warm in the north, hot in the south.
1.3. Generally dry and sunny; high sun period: now and then scattered showers in the afternoon, except northernmost region, Red Sea region.
1.4. Wind: occasional gales and dust-storms.

Wadi Halfa

latitude: 21°50'n longitude: 31°18'e height: 155 m

BWhl	082-1	Jan	Feb	Mar	Apr	May	Jun	Jul	Aug	Sep	Oct	Nov	Dec	Year
	tx	32	36	39	43	46	46	45	44	43	42	37	33	47 °C
	t	13	16	21	26	31	32	32	32	30	28	22	17	25 °C
	tn	3	4	7	11	16	18	20	21	18	14	10	4	2 °C
	clothing	n	n	s	t	t	t	t	t	t	t	s	s	–
	stress			o	f	ff	ff	ff	ff	ff	f	o		–
	p	0	0	0	0	1	0	1	0	0	1	0	0	3 mm
	sunshine	313	291	319	321	350	354	363	344	321	329	312	307	3924 hours
	%	92	90	86	85	86	88	88	86	87	91	94	92	88 %

Port Sudan

latitude: 19°35'n longitude: 37°13'e height: 2 m

BMhb	082-2	Jan	Feb	Mar	Apr	May	Jun	Jul	Aug	Sep	Oct	Nov	Dec	Year
	tx	30	30	32	35	40	44	46	45	43	37	33	31	46 °C
	t	24	22	24	26	29	32	34	34	32	29	27	25	28 °C
	tn	15	15	15	17	20	23	24	25	22	22	20	17	14 °C
	clothing	t	s	t	t	t	t	t	t	t	t	t	t	–
	stress	f	f	f	f	ff	ff	ff	ff	ff	ff	f	f	–
	p	4	1	1	1	2	0	2	3	0	12	52	25	110 mm
0.1 mm	dp	2	1	0	0	1	0	1	1	0	1	6	4	17 days
	sunshine	217	223	264	306	335	306	301	301	300	307	249	208	3316 hours
	%	63	69	71	81	83	77	73	76	81	84	74	61	75 %

Abu Hamed

latitude: 19°32'n longitude: 33°20'e height: 312 m

BWib	082-3	Jan	Feb	Mar	Apr	May	Jun	Jul	Aug	Sep	Oct	Nov	Dec	Year
	t	20	21	25	29	33	35	34	34	34	31	26	21	28 °C
	clothing	s	s	t	t	t	t	t	t	t	t	t	s	–
	stress	o	o	f	ff	ff	ff	ff	ff	ff	ff	f	o	–
	p	0	0	0	0	1	0	5	10	1	0	0	0	17 mm

Gebeit

latitude: 18°57′n longitude: 36°50′e height: 796 m

GBWhb	082-4	Jan	Feb	Mar	Apr	May	Jun	Jul	Aug	Sep	Oct	Nov	Dec	Year
	t	20	20	22	24	28	31	31	30	30	26	23	21	25 °C
	clothing	s	s	s	t	t	t	t	t	t	t	s	s	–
	stress	o	o	o	f	ff	ff	ff	ff	ff	f	o	o	–
	p	1	0	1	2	9	5	32	55	13	6	6	1	131 mm

Karima

latitude: 18°33′n longitude: 31°51′e height: 249 m

BWhb	082-5	Jan	Feb	Mar	Apr	May	Jun	Jul	Aug	Sep	Oct	Nov	Dec	Year
	t	20	21	25	29	33	34	34	33	33	32	26	21	28 °C
	clothing	s	s	t	t	t	t	t	t	t	t	t	s	–
	stress	o	o	f	ff	ff	ff	ff	ff	ff	ff	f	o	–
	p	0	0	0	0	1	0	11	20	5	1	0	0	38 mm

Tokar

latitude: 18°26′n longitude: 37°44′e height: 19 m

BWia	082-6	Jan	Feb	Mar	Apr	May	Jun	Jul	Aug	Sep	Oct	Nov	Dec	Year
	t	24	24	26	28	31	34	36	36	35	31	28	26	30 °C
	clothing	t	t	t	t	t	t	t	t	t	t	t	t	–
	stress	f	f	f	ff	ff	ff	ff	ff	ff	ff	ff	f	–
	p	16	5	1	2	4	0	7	3	1	4	31	14	88 mm

Atbara

latitude: 17°42′n longitude: 33°58′e height: 345 m

BWib	082-7	Jan	Feb	Mar	Apr	May	Jun	Jul	Aug	Sep	Oct	Nov	Dec	Year
	tx	36	38	41	44	45	46	44	43	43	42	38	36	46 °C
	t	22	23	26	30	34	35	33	32	33	32	27	23	29 °C
	tn	9	10	13	16	20	23	23	22	23	21	16	12	9 °C
	clothing	s	t	t	t	t	t	t	t	t	t	t	t	–
	stress	o	f	f	ff	ff	ff	ff	ff	ff	ff	f	f	–
	p	0	0	0	1	4	1	20	38	7	1	0	0	72 mm
0.1 mm	dp					1	1	2	3	1				8 days
	sunshine	332	297	319	345	333	294	273	282	303	332	321	326	3756 hours
	%	96	91	86	92	83	74	67	71	82	91	95	95	85 %

Shendi

latitude: 16°42′n longitude: 33°26′e height: 360 m

BWia	082-8	Jan	Feb	Mar	Apr	May	Jun	Jul	Aug	Sep	Oct	Nov	Dec	Year
	t	23	24	28	31	34	35	33	32	33	32	28	24	30 °C
	clothing	t	t	t	t	t	t	t	t	t	t	t	t	–
	stress	o	f	ff	ff	ff	ff	ff	ff	ff	ff	ff	f	–
	p	0	0	0	1	3	2	43	64	2	3	0	0	118 mm

Khartoum

BWhb	082-9	Jan	Feb	Mar	Apr	May	Jun	Jul	Aug	Sep	Oct	Nov	Dec	Year
		latitude: 15° 36′n				longitude: 32° 33′e				height: 380 m				
	tx	36	39	42	44	45	45	43	41	42	42	39	36	46 °C
	t	22	24	27	31	33	33	31	29	31	31	28	24	29 °C
	tn	9	11	13	16	20	22	21	21	21	21	16	11	8 °C
	clothing	s	t	t	t	t	t	t	t	t	t	t	t	–
	stress	o	f	f	ff	ff	ff	ff	ff	ff	ff	ff	f	–
	p	0	0	0	1	5	7	48	72	27	4	0	0	164 mm
0.1 mm	dp	0	0	0	0	1	2	5	7	3	1	0	0	19 days
	sunshine	329	300	322	321	310	294	260	267	279	319	324	326	3650 hours
	%	94	91	86	86	78	75	65	68	76	87	95	94	82 %

Kassala

GBWha	082-10	Jan	Feb	Mar	Apr	May	Jun	Jul	Aug	Sep	Oct	Nov	Dec	Year
		latitude: 15° 28′n				longitude: 36° 24′e				height: 500 m				
	tx	39	40	43	44	44	43	39	37	39	41	40	38	44 °C
	t	24	25	28	31	32	31	28	27	28	30	28	25	28 °C
	tn	12	11	14	17	20	21	20	19	20	21	16	13	10 °C
	clothing	t	t	t	t	t	t	t	t	t	t	t	t	–
	stress	f	f	f	ff	ff	ff	ff	ff	ff	ff	ff	f	–
	p	0	1	1	5	14	27	100	124	60	7	2	0	341 mm
0.1 mm	dp	0	0	0	1	2	3	12	12	7	1	0	0	38 days
	sunshine	310	288	310	321	322	297	245	239	288	319	309	313	3562 hours
	%	88	88	83	86	81	76	61	61	78	87	90	90	80 %

Wad Medani

BWhb	082-11	Jan	Feb	Mar	Apr	May	Jun	Jul	Aug	Sep	Oct	Nov	Dec	Year
		latitude: 14° 24′n				longitude: 33° 29′e				height: 408 m				
	t	22	24	27	30	32	31	28	26	28	29	26	23	27 °C
	clothing	s	t	t	t	t	t	t	t	t	t	t	t	–
	stress	o	f	f	ff	ff	ff	ff	ff	ff	ff	f	f	–
	p	0	0	0	3	16	31	122	129	55	16	1	0	373 mm

Ed Dueim

BWha	082-12	Jan	Feb	Mar	Apr	May	Jun	Jul	Aug	Sep	Oct	Nov	Dec	Year
		latitude: 14° 00′n				longitude: 32° 20′e				height: 378 m				
	t	25	26	28	31	33	32	30	28	29	31	28	25	29 °C
	clothing	t	t	t	t	t	t	t	t	t	t	t	t	–
	stress	f	f	ff	ff	ff	ff	ff	ff	ff	ff	ff	f	–
	p	1	0	1	2	9	5	32	55	13	6	3	1	128 mm

El Fasher

latitude: 13° 37′n longitude: 25° 20′e height: 730 m

GBWhb	082-13	Jan	Feb	Mar	Apr	May	Jun	Jul	Aug	Sep	Oct	Nov	Dec	Year
	tx	36	37	41	42	43	42	40	39	39	39	38	36	44 °C
	t	19	20	24	27	29	30	27	26	27	26	22	19	25 °C
	tn	5	6	8	13	16	19	19	18	17	13	9	6	4 °C
	clothing	s	s	t	t	t	t	t	t	t	t	s	s	–
	stress	o	o	f	f	ff	ff	ff	ff	f	f	o	o	–
	p	0	0	0	1	10	13	89	133	36	5	0	0	287 mm
0.1 mm	dp	0	0	0	1	1	3	10	13	5	1	0	0	34 days
	sunshine	319	305	310	306	316	273	232	236	264	316	327	304	3509 hours
	%	90	92	83	82	80	70	58	60	72	86	95	86	79 %

El Obeid

latitude: 13° 10′n longitude: 30° 14′e height: 574 m

GBWhb	082-14	Jan	Feb	Mar	Apr	May	Jun	Jul	Aug	Sep	Oct	Nov	Dec	Year
	tx	36	38	41	42	42	41	38	36	38	39	37	34	43 °C
	t	21	22	26	29	30	30	27	26	26	28	25	22	26 °C
	tn	8	8	11	15	18	19	19	19	19	18	13	9	7 °C
	clothing	s	s	t	t	t	t	t	t	t	t	t	s	–
	stress	o	o	f	ff	ff	ff	ff	ff	ff	ff	f	o	–
	p	0	0	0	3	20	33	121	145	77	19	0	0	418 mm
0.1 mm	dp	0	0	0	1	2	4	10	10	7	2	0	0	37 days
	sunshine	322	297	304	303	313	246	211	211	243	298	321	329	3397 hours
	%	91	90	81	81	79	64	53	54	66	81	93	93	77 %

Kosti

latitude: 13° 10′n longitude: 32° 40′e height: 381 m

BWha	082-15	Jan	Feb	Mar	Apr	May	Jun	Jul	Aug	Sep	Oct	Nov	Dec	Year
	t	23	24	28	31	31	30	28	26	27	29	28	24	28 °C
	clothing	t	t	t	t	t	t	t	t	t	t	t	t	–
	stress	f	f	ff	ff	ff	ff	ff	ff	ff	ff	ff	f	–
	p	0	0	1	4	18	47	111	143	60	22	1	0	407 mm

En Nahud

latitude: 12° 42′n longitude: 28° 26′e height: 564 m

GBWhb	082-16	Jan	Feb	Mar	Apr	May	Jun	Jul	Aug	Sep	Oct	Nov	Dec	Year
	t	21	23	26	29	30	29	27	26	27	28	25	22	26 °C
	clothing	s	t	t	t	t	t	t	t	t	t	t	s	–
	stress	o	f	f	ff	ff	ff	ff	ff	ff	ff	f	o	–
	p	0	0	0	2	24	44	113	134	85	21	0	0	423 mm

2. *Climatic area:* Southern parts of the provinces of Dafur, Kordofan, Blue Nile and northern part of Upper Nile

2.1. Steppe Climate BS and Steppe Mountain Climate GBS

2.2. Summer very hot, winter hot.

2.3. High sun period: generally dry and sunny in the morning, often in the afternoon showers; the odd rainy day. Low sun period: dry and sunny.

2.4. Wind: occasional gales and dust-storms.

Geneina

GBShb	082-17	Jan	Feb	Mar	Apr	May	Jun	Jul	Aug	Sep	Oct	Nov	Dec	Year
		latitude: 13°29′n					longitude: 22°27′e				height: 805 m			
	t	20	22	25	28	28	28	25	23	24	24	23	20	24 °C
	clothing	s	s	t	t	t	t	t	t	t	t	t	s	–
	stress	o	o	f	ff	ff	ff	ff	ff	ff	ff	f	o	–
	p	0	0	1	4	27	36	162	230	79	10	0	0	549 mm

Malakal

BSha	082-18	Jan	Feb	Mar	Apr	May	Jun	Jul	Aug	Sep	Oct	Nov	Dec	Year
		latitude: 9°33′n					longitude: 31°39′e				height: 388 m			
	tx	39	41	42	42	40	37	35	34	35	37	38	38	42 °C
	t	26	27	30	30	28	26	25	25	26	26	26	26	27 °C
	tn	14	14	16	18	19	21	19	18	19	19	16	15	13 °C
	clothing	t	t	t	t	t	t	t	t	t	t	t	t	–
	stress	f	f	ff	ff	ff	ff	ff	ff	ff	ff	f	f	–
	p	0	0	3	24	95	115	153	167	144	75	6	1	783 mm
0.1 mm	dp	0	0	1	4	9	10	16	17	13	8	1	0	79 days
	sunshine	307	260	276	243	242	156	155	167	177	226	285	304	2798 hours
	%	85	78	74	66	62	41	40	43	48	61	81	85	63 %

3. Climatic area: Bahr el Ghazal, southern part of Upper Nile, Equatoria

3.1. Tropical Summer Rain Climate Aw and
Tropical Summer Rain Mountain Climate GAw

3.2. Summer very hot, winter hot.

3.3. High sun period: often dry and sunny in the morning, in the afternoon showers, some rainy days. Low sun period: mostly dry and sunny.

3.4. Wind: occasional gales.

Wau

Awha	082-19	Jan	Feb	Mar	Apr	May	Jun	Jul	Aug	Sep	Oct	Nov	Dec	Year
		latitude: 7°42′n					longitude: 28°01′e				height: 438 m			
	tx	39	41	42	41	39	36	34	34	35	37	38	38	42 °C
	t	26	27	29	29	28	26	25	25	25	26	26	26	26 °C
	tn	14	16	17	20	19	19	19	19	19	19	16	15	14 °C
	clothing	t	t	t	t	t	t	t	t	t	t	t	t	–
	stress	f	f	ff	ff	ff	ff	ff	ff	ff	ff	f	f	–
	p	0	4	20	69	132	170	199	234	179	130	8	0	1145 mm
0.1 mm	dp	0	1	4	7	12	13	16	18	15	13	1	0	100 days
	sunshine	298	257	257	219	254	213	174	195	204	220	270	298	2859 hours
	%	82	77	69	59	66	56	45	51	56	59	76	82	65 %

Juba

Awha	082-20	Jan	Feb	Mar	Apr	May	Jun	Jul	Aug	Sep	Oct	Nov	Dec	Year
		latitude: 4°52′n					longitude: 31°36′e				height: 457 m			
	tx	40	41	41	39	37	36	34	34	36	37	38	39	41 °C
	t	27	28	28	28	26	25	24	24	25	26	26	27	26 °C
	tn	17	18	19	20	19	19	19	18	19	19	18	17	17 °C
	clothing	t	t	t	t	t	t	t	t	t	t	t	t	–
	stress	ff	ff	ff	ff	ff	ff	ff	ff	ff	ff	ff	f	–
	p	5	10	43	107	157	116	136	154	105	101	35	13	982 mm
0.1 mm	dp	1	2	8	12	13	12	12	13	10	10	6	3	102 days
	sunshine	285	229	226	201	242	222	189	208	234	242	240	276	2794 hours
	%	77	68	60	55	63	60	49	55	64	65	67	75	63 %

4.3.82.2. Predominant general weather situations are described in 4.3.0.1. under Sudan. A hygrothermal diagram of Khartoum is to be found in 4.3.0.2.
Wind names: Hababai are dry and hot winds blowing from east to northeast in October and November on the west coast of the Red Sea near Port Sudan.
Haboob are hot dust-storms and sandstorms blowing predominantly from the north.
Kharif is a wind blowing from south to southwest during the high sun period.

4.3.83. Ethiopia

4.3.83.1. Climatic descriptions and tables

1. Climatic area: roughly the western half of the country
1.1. Tropical Summer Rain Climate Aw and Tropical Summer Rain Mountain Climate GAw and Tropical Summer Rain High Mountain Climate H Aw
1.2. In the lowlands permanently very hot.
1.3. High sun period: in the morning often dry and sunny or cloudy, in the afternoon showers; at times rainy days. Low sun period: mostly dry and sunny.
1.4. Wind: occasional gales, squalls in thunderstorms.

Gondar

latitude: 12°35′n longitude: 37°26′e height: 1966 m

GAwab	083-1	Jan	Feb	Mar	Apr	May	Jun	Jul	Aug	Sep	Oct	Nov	Dec	Year
	t	21	22	23	23	23	20	18	18	19	20	21	20	21 °C
	clothing	s	s	s	s	s	s	s	s	s	s	s	s	–
	stress	o	o	f	f	f	f	f	f	f	o	o	o	–
	p	4	13	23	63	73	192	370	350	127	42	25	20	1302 mm

Combolcha

latitude: 11°05′n longitude: 39°43′e height: 1916 m

GAwal	083-2	Jan	Feb	Mar	Apr	May	Jun	Jul	Aug	Sep	Oct	Nov	Dec	Year
	t	18	20	23	22	23	24	21	21	20	19	18	17	21 °C
	clothing	s	s	s	s	s	s	s	s	s	s	s	s	–
	stress	o	o	f	f	f	f	o	o	o				–
	p	55	46	84	87	43	71	299	271	139	32	21	26	1178 mm

Addis Ababa

latitude: 9°00′n longitude: 38°44′e height: 2360 m

GAwbl	083-3	Jan	Feb	Mar	Apr	May	Jun	Jul	Aug	Sep	Oct	Nov	Dec	Year
	tx	26	27	27	28	28	27	25	24	24	26	25	25	29 °C
	t	16	18	19	19	20	18	15	15	15	15	15	15	17 °C
	tn	3	5	7	7	7	8	8	8	7	5	3	2	2 °C
	clothing		s	s	s	s	s	n	n	n	n	n		–
	stress		o	o	o	o	o							–
	p	24	25	68	93	50	105	228	263	174	41	3	15	1089 mm
0.1 mm	dp	2	5	8	10	10	20	28	27	21	3	2	2	138 days
	sunshine	275	253	245	216	238	179	86	83	143	258	259	275	2510 hours
	%	76	76	65	58	61	47	22	22	39	70	74	76	57 %

Jimma

latitude: 7°39′n longitude: 36°50′e height: 1676 m

GAwbl	083-4	Jan	Feb	Mar	Apr	May	Jun	Jul	Aug	Sep	Oct	Nov	Dec	Year
	t	18	19	20	20	20	19	18	19	19	19	18	17	19 °C
	clothing	s	s	s	s	s	s	s	s	s	s	s	s	–
	stress	o	o	o	o	o	o	o	o	o	o	o	o	–
	p	29	56	82	176	146	240	230	208	198	80	13	26	1484 mm
0.1 mm	dp	6	14	16	19	21	25	28	27	24	16	7	4	207 days

2. Climatic area: Eastern half of the country, except the coastal regions of Eritrea, the Danakil Desert and parts of the southeast frontier zone
2.1. Steppe Climate BS and Steppe Mountain Climate GBS
2.2. In the low lands permanently very hot.
2.3. High sun period: in the morning often dry and sunny or cloudy, in the afternoon showers; at times rainy days. Low sun period: mostly dry and sunny.
2.4. Wind: occasional gales, squalls in thunderstorms, dust-storms.

Asmara – Eritrea

latitude: 15°17′n longitude: 38°55′e height: 2321 m

GBSbl	083-5	Jan	Feb	Mar	Apr	May	Jun	Jul	Aug	Sep	Oct	Nov	Dec	Year
	tx	26	27	28	28	28	28	26	25	26	25	24	25	29 °C
	t	14	16	17	18	19	18	16	16	16	15	15	14	16 °C
	tn	3	4	7	8	9	9	9	9	8	6	5	3	2 °C
	clothing	n	n	n	s	s	s	n	n	n	n	n	n	–
	stress				o	o	o							–
	p	1	0	10	29	55	44	197	146	36	9	23	8	558 mm
	dp	1	0	3	5	5	5	17	14	5	2	2	1	60 days

Dire Dawa

latitude: 9°45′n longitude: 41°50′e height: 1162 m

GBShb	083-6	Jan	Feb	Mar	Apr	May	Jun	Jul	Aug	Sep	Oct	Nov	Dec	Year
	t	22	23	25	26	27	28	25	24	25	25	23	22	25 °C
	clothing	s	t	t	t	t	t	t	t	t	t	t	s	–
	stress	o	f	f	f	f	f	f	f	f	f	f	o	–
	p	22	47	54	82	19	34	87	168	77	11	6	10	617 mm
0.1 mm	dp	1	2	7	9	9	6	12	14	10	1	3	2	76 days

Awash

latitude: 8°56′n longitude: 40°05′e height: 1052 m

GBSha	083-7	Jan	Feb	Mar	Apr	May	Jun	Jul	Aug	Sep	Oct	Nov	Dec	Year
	t	24	25	26	28	30	29	26	26	28	28	26	25	27 °C
	clothing	t	t	t	t	t	t	t	t	t	t	t	t	–
	stress	o	f	f	f	f	f	f	f	f	f	f	f	–
	p	44	?	60	33	29	32	126	165	39	15	13	?	? mm

3. *Climatic area:* coastal regions of Eritrea, the Danakil Desert and parts of the southeast frontier zone
3.1. Desert Climate BW and Desert Mountain Climate GBW, coastal area: Marine Desert Climate BM
3.2. At low levels permanently very hot.
3.3. Generally dry and sunny; high sun period: now and then scattered showers or rain, cloudy.
3.4. Wind: occasional gales and dust-storms.

4.3.83.2. Predominant general weather situations are described in 4.3.0.1. under Ethiopia. A hygrothermal diagram of Addis Ababa is to be found in 4.3.0.2.

4.3.84. Djibouti (formerly Terr. Afar and Issas)

4.3.84.1. Climatic description and table

1. Desert Climate BW and Desert Mountain Climate GBW, coastal area: Marine Desert Climate BM
2. Permanently very hot.
3. Generally dry and sunny; now and then scattered showers or rain.
4. Wind: occasional gales and dust-storms.

Djibouti

		latitude: 11°36'n					longitude: 43°09'e				height: 7 m			
BMha	084-1	Jan	Feb	Mar	Apr	May	Jun	Jul	Aug	Sep	Oct	Nov	Dec	Year
	tx	31	31	33	34	38	44	44	44	43	36	33	32	45 °C
	t	25	25	27	28	31	33	34	33	32	30	27	26	29 °C
	tn	21	22	23	24	26	28	28	27	27	24	23	21	20 °C
	clothing	t	t	t	t	t	t	t	t	t	t	t	t	–
	stress	ff	ff	ff	ff	ff	ff	ff	ff	ff	ff	ff	ff	–
	p	13	7	24	11	6	0	7	8	3	13	30	13	133 mm
	dp	3	1	1	1	0	0	1	3	1	1	3	2	17 days
	sunshine	241	242	253	280	310	250	231	262	271	299	299	277	3215 hours
	%	67	73	68	75	79	65	58	68	74	81	86	78	73 %

4.3.84.2. Predominant general weather situations are described in 4.3.0.1. under Sudan, Ethiopia, Somalia.

4.3.85. Somalia

4.3.85.1. Climatic descriptions and tables

1. *Climatic area:* the greater part of the country, except parts of the southern lowlands and the northeastern mountains

1.1. Desert Climate BW and Desert Mountain Climate GBW, coastal areas: Marine Desert Climate BM

1.2. Very hot throughout the year.

1.3. Generally dry and sunny; now and then scattered showers, particularly in the afternoon.

1.4. Wind: occasional gales and dust-storms.

Bosaco

		latitude: 11°17'n					longitude: 49°11'e				height: 2 m			
BMia	085-1	Jan	Feb	Mar	Apr	May	Jun	Jul	Aug	Sep	Oct	Nov	Dec	Year
	tx	32	33	36	38	41	42	41	42	42	36	33	33	43 °C
	t	25	25	26	28	30	35	36	35	33	28	25	24	29 °C
	tn	18	18	18	23	23	28	29	27	25	19	17	18	17 °C
	clothing	t	t	t	t	t	t	t	t	t	t	t	t	–
	stress	f	f	ff	ff	ff	ff	ff	ff	ff	ff	ff	f	–
	p	0	0	0	0	0	0	0	0	0	6	5	0	12 mm

Galcayo

BWha	085-2	Jan	Feb	Mar	Apr	May	Jun	Jul	Aug	Sep	Oct	Nov	Dec	Year
	latitude: 6°51'n							longitude: 47°16'e			height: 302 m			
	tx	34	35	36	36	37	36	35	34	36	36	35	34	37 °C
	t	26	26	28	29	29	28	27	27	28	27	27	26	27 °C
	tn	15	16	18	19	19	19	21	21	21	19	15	13	13 °C
	clothing	t	t	t	t	t	t	t	t	t	t	t	t	–
	stress	f	f	f	ff	ff	ff	ff	ff	ff	ff	f	f	–
	p	0	0	0	34	20	1	0	0	2	25	16	1	98 mm
	dp	0	0	0	2	5	0	0	0	0	3	1	0	11 days

Lugh Farrandi

BWhh	085-3	Jan	Feb	Mar	Apr	May	Jun	Jul	Aug	Sep	Oct	Nov	Dec	Year
	latitude: 3°45'n							longitude: 42°35'e			height: 193 m			
	t	32	32	33	32	30	30	28	28	30	30	31	31	30 °C
	clothing	t	t	t	t	t	t	t	t	t	t	t	t	–
	stress	ff	ff	ff	ff	ff	ff	ff	ff	ff	ff	ff	ff	–
	p	2	4	28	113	40	1	3	0	1	47	56	15	310 mm
0.1 mm	dp	1	1	3	9	3	0	1	0	0	4	5	2	29 days

2. *Climatic area:* parts of the southern lowlands and the northeastern mountains
2.1. Steppe Climate BS and Steppe Mountain Climate GBS
2.2. Very hot throughout the year.
2.3. Generally dry and sunny; now and then showers, particularly in the afternoon.
2.4. Wind: occasional gales and dust-storms.

Erigavo

GBSbl	085-4	Jan	Feb	Mar	Apr	May	Jun	Jul	Aug	Sep	Oct	Nov	Dec	Year
	latitude: 10°40'n							longitude: 47°25'e			height: 1737 m			
	t	15	16	17	18	19	20	20	20	18	16	16	15	17 °C
	clothing	n	n	s	s	s	s	s	s	s	n	n	n	–
	stress	o	o	o	o	o	o	o	o	o	o	o	o	–
	p	18	13	33	38	81	64	10	41	114	8	13	2	435 mm
0.1 mm	dp	1	3	6	5	8	9	1	5	15	1	2	0	56 days

Hargeisa

GBSab	085-5	Jan	Feb	Mar	Apr	May	Jun	Jul	Aug	Sep	Oct	Nov	Dec	Year
	latitude: 9°29'n							longitude: 44°05'e			height: 1334 m			
	t	18	20	22	23	24	24	23	23	24	22	20	18	22 °C
	clothing	s	s	s	t	t	t	t	t	t	s	s	s	–
	stress	o	o	o	o	f	f	f	f	f	o	o	o	–
	p	0	0	25	53	53	66	33	79	56	20	2	0	387 mm
	dp	0	0	2	4	6	8	5	9	8	3	1	0	46 days

Mogadiscio

latitude: 2°02′n longitude: 45°21′e height: 10 m

BSha	085-6	Jan	Feb	Mar	Apr	May	Jun	Jul	Aug	Sep	Oct	Nov	Dec	Year
	tx	32	31	32	33	33	31	30	29	30	31	31	32	33 °C
	t	26	27	28	29	29	27	26	26	27	27	28	27	27 °C
	tn	22	22	22	23	23	22	21	21	21	22	22	22	19 °C
	clothing	t	t	t	t	t	t	t	t	t	t	t	t	–
	stress	ff	ff	ff	ff	ff	ff	ff	ff	ff	ff	ff	ff	–
	p	1	0	11	103	63	72	60	26	10	18	46	12	422 mm
0.1 mm	dp	0	0	1	5	7	13	13	10	5	4	4	2	64 days
	sunshine	267	254	282	261	273	219	226	254	264	267	261	257	3085 hours
	%	72	75	75	72	72	60	60	67	73	71	72	69	70 %

Chisimaio

latitude: 0°22′s longitude: 42°26′e height: 10 m

BSha	085-7	Jan	Feb	Mar	Apr	May	Jun	Jul	Aug	Sep	Oct	Nov	Dec	Year
	t	27	28	28	29	28	26	26	26	26	27	28	28	27 °C
	clothing	t	t	t	t	t	t	t	t	t	t	t	t	–
	stress	ff	ff	ff	ff	ff	ff	ff	ff	ff	ff	ff	ff	–
	p	0	0	0	22	102	63	59	12	8	8	9	4	288 mm
0.1 mm	dp	0	0	0	4	7	11	9	5	3	2	2	1	44 days

4.3.85.2. Predominant general weather situations are described in 4.3.0.1. under Somalia. A hygrothermal diagram of Mogadiscio is to be found in 4.3.0.2.
Wind name: Kharif is a wind blowing from the south or the southwest during the high sun period.

4.3.86. Kenya

4.3.86.1. Climatic descriptions and tables

1. Climatic area: Northernmost part of Rift Valley Province, Northern Province
1.1. Desert Climate BW and Desert Mountain Climate GBW
1.2. Permanently very hot.
1.3. Mostly dry and sunny; now and then scattered showers, particularly in the afternoon.
1.4. Wind: occasional gales and dust-storms.

Mandera

latitude: 3°57′n longitude: 41°52′e height: 331 m

BWhh	086-1	Jan	Feb	Mar	Apr	May	Jun	Jul	Aug	Sep	Oct	Nov	Dec	Year
	tx	37	38	39	37	36	36	36	35	36	36	36	36	39 °C
	t	29	30	31	30	29	29	28	28	29	28	29	29	29 °C
	tn	17	19	21	21	22	22	21	20	22	21	21	18	17 °C
	clothing	t	t	t	t	t	t	t	t	t	t	t	t	–
	stress	ff	ff	ff	ff	ff	ff	ff	ff	ff	ff	ff	ff	–
	p	1	4	24	83	24	0	0	1	2	43	26	9	218 mm
	dp	0	0	2	5	2	0	0	0	0	3	4	1	17 days

Lodwar

GBWhh	086-2	Jan	Feb	Mar	Apr	May	Jun	Jul	Aug	Sep	Oct	Nov	Dec	Year
				latitude: 3°07′n				longitude: 35°37′e				height: 506 m		
	tx	37	38	38	37	37	36	35	35	36	37	37	36	39 °C
	t	29	30	30	30	30	29	28	29	29	30	29	29	29 °C
	tn	19	20	21	21	22	22	21	21	22	21	21	19	18 °C
	clothing	t	t	t	t	t	t	t	t	t	t	t	t	–
	stress	ff	ff	ff	ff	ff	ff	ff	ff	ff	ff	ff	ff	–
	p	8	5	22	41	25	10	15	9	4	7	7	10	162 mm
	dp	1	1	2	4	3	1	2	1	1	1	1	1	19 days
	sunshine	322	287	285	266	314	301	285	310	316	313	287	306	3592 hours
	%	87	84	76	73	83	82	75	82	87	84	80	83	81 %

Wajir

BWha	086-3	Jan	Feb	Mar	Apr	May	Jun	Jul	Aug	Sep	Oct	Nov	Dec	Year
				latitude: 1°45′n				longitude: 40°04′e				height: 244 m		
	tx	37	37	37	36	34	34	33	33	34	35	35	35	38 °C
	t	29	30	30	29	28	27	26	26	27	28	28	28	28 °C
	tn	19	20	22	21	21	19	18	18	19	20	20	20	18 °C
	clothing	t	t	t	t	t	t	t	t	t	t	t	t	–
	stress	ff	ff	ff	ff	ff	ff	ff	ff	ff	ff	ff	ff	–
	p	6	13	19	68	35	1	4	2	4	25	41	22	239 mm
	dp	1	1	3	6	3	0	0	0	0	2	5	3	24 days

Garissa

BWha	086-4	Jan	Feb	Mar	Apr	May	Jun	Jul	Aug	Sep	Oct	Nov	Dec	Year
				latitude: 0°29′s				longitude: 39°38′e				height: 128 m		
	tx	37	38	38	38	37	36	34	36	36	36	36	36	39 °C
	t	29	29	30	30	29	27	26	27	28	29	29	28	28 °C
	tn	19	20	22	22	22	19	19	18	19	21	21	20	18 °C
	clothing	t	t	t	t	t	t	t	t	t	t	t	t	–
	stress	ff	ff	ff	ff	ff	ff	ff	ff	ff	ff	ff	ff	–
	p	10	5	26	55	17	5	2	6	5	20	64	65	280 mm
	dp	1	1	2	4	2	1	1	1	1	2	6	6	28 days

2. *Climatic area:* Northern parts of Rift Valley Province, and Central Province, and Coast Province

2.1. Steppe Climate BS and Steppe Mountain Climate GBS

2.2. Permanently very hot.

2.3. Mostly dry and sunny; now and then scattered showers or rain.

2.4. Wind: occasional gales and dust-storms. Squalls in thunderstorms.

3. *Climatic area:* Southern part of Rift Valley Province, Nyanza Province, Central Province, Masai Province, Coast Province

3.1. Tropical Summer Rain Climate Aw and Tropical Summer Rain Mountain Climate GAw (In some parts Tropical Winter Rain Climate etc. and in other parts in the west Tropical Rain Climate Ar etc.)

3.2 In the lowlands very hot throughout the year.

3.3. In the morning often dry and sunny or cloudy, in the afternoon frequent showers, on the coast from April to June, and inland from March to May and November to December; at times rainy days.

3.4. Wind: occasional gales, squalls in thunderstorms.

Kitale

GAwbl	086-5	Jan	Feb	Mar	Apr	May	Jun	Jul	Aug	Sep	Oct	Nov	Dec	Year
				latitude: 1°00′n				longitude: 35°00′e			height: 1896 m			
	t	19	19	19	19	18	17	17	17	17	18	18	18	18 °C
	clothing	s	s	s	s	s	s	s	s	s	s	s	s	–
	stress	o	o	o	o	o	o	o	o	o	o	o	o	–
	p	32	29	74	147	155	126	164	163	113	106	82	61	1251 mm
0.2 mm	dp	5	7	14	21	20	16	21	22	14	16	16	12	184 days

Nanyuki

GAwll	086-6	Jan	Feb	Mar	Apr	May	Jun	Jul	Aug	Sep	Oct	Nov	Dec	Year
				latitude: 0°01′n				longitude: 37°04′e			height: 1946 m			
	tx	27	28	27	26	26	26	25	25	27	27	25	26	28 °C
	t	16	17	17	17	17	16	15	15	16	16	16	16	16 °C
	tn	4	4	7	7	7	7	6	6	6	6	7	6	3 °C
	clothing	n	s	s	s	s	n	n	n	n	n	n	n	–
	stress		o	o	o	o								–
	p	19	22	53	119	83	44	65	70	49	78	92	43	737 mm
0.2 mm	dp	4	4	9	17	14	8	8	10	10	14	13	9	120 days
	sunshine	232	209	195	153	183	198	177	186	189	164	132	174	2192 hours
	%	62	61	52	42	49	54	47	49	52	44	36	46	49 %

Equator Station

HAsll	086-7	Jan	Feb	Mar	Apr	May	Jun	Jul	Aug	Sep	Oct	Nov	Dec	Year
				latitude: 0°01′s				longitude: 35°33′e			height: 2762 m			
	tx	22	23	23	22	21	19	18	19	20	21	20	21	24 °C
	t	14	14	14	14	13	12	12	12	13	13	13	13	13 °C
	tn	5	6	6	7	7	6	6	7	6	6	6	6	4 °C
	clothing	n	n	n	n	n	n	n	n	n	n	n	n	–
	stress													–
	p	31	35	72	170	138	126	162	206	110	53	57	55	1215 mm
0.2 mm	dp	7	7	11	17	18	16	19	23	15	11	11	9	164 days
	sunshine	273	257	257	216	236	201	161	161	222	251	228	254	2717 hours
	%	73	75	68	59	63	55	43	43	61	67	63	68	61 %

Nairobi/Dagoretti

GAwbl	086-8	Jan	Feb	Mar	Apr	May	Jun	Jul	Aug	Sep	Oct	Nov	Dec	Year
				latitude: 1°18′s				longitude: 36°46′e			height: 1798 m			
	tx	27	28	27	26	24	24	24	25	27	27	26	26	28 °C
	t	18	18	19	19	18	16	15	16	17	19	18	18	18 °C
	tn	9	10	11	13	11	9	8	8	8	9	10	10	7 °C
	clothing	s	s	s	s	s	s	s	s	s	s	s	s	–
	stress	o	o	o	o	o					o	o	o	–
	p	45	51	101	206	160	46	19	26	26	54	109	82	926 mm
	dp	4	4	9	15	14	6	3	4	4	6	12	8	89 days
	sunshine	282	267	270	211	186	172	133	136	180	220	212	256	2525 hours
	%	75	78	72	58	50	48	36	36	50	58	58	68	57 %

Makindu

GAwab	086-9	latitude: 2°17's Jan	Feb	Mar	Apr	May	Jun	longitude: 37°50'e Jul	Aug	Sep	Oct	height: 998 m Nov	Dec	Year
	tx	32	34	34	33	31	30	29	30	31	32	31	30	34 °C
	t	23	24	25	24	23	21	20	21	22	23	23	23	23 °C
	tn	14	15	16	16	14	12	11	11	11	13	15	15	10 °C
	clothing	t	t	t	t	t	t	t	t	t	t	t	t	–
	stress	f	f	f	f	f	f	f	f	f	f	f	f	–
	p	40	28	78	112	31	2	0	0	1	29	168	120	609 mm
0.2 mm	dp	3	2	5	8	4	1	0	1	1	3	10	11	49 days

Mombasa

Asha	086-10	latitude: 4°02's Jan	Feb	Mar	Apr	May	Jun	longitude: 39°37'e Jul	Aug	Sep	Oct	height: 58 m Nov	Dec	Year
	tx	32	32	33	32	30	29	28	28	29	30	31	32	33 °C
	t	28	28	28	28	26	25	24	24	25	26	27	28	26 °C
	tn	23	23	23	23	22	21	20	20	21	22	22	23	19 °C
	clothing	t	t	t	t	t	t	t	t	t	t	t	t	–
	stress	ff	ff	ff	ff	ff	ff	ff	ff	ff	ff	ff	ff	–
	p	30	14	59	192	319	100	72	69	71	86	74	76	1163 mm
	dp	3	2	5	13	17	10	10	11	9	8	7	6	101 days
	sunshine	267	259	282	222	205	225	217	248	252	267	273	273	2990 hours
	%	70	75	75	62	55	63	59	67	69	71	74	71	68 %

4.3.86.2. Predominant general weather situations are described in 4.3.0.1. under Equatorial South Africa. A hygrothermal diagram of Nairobi is to be found in 4.3.0.2.

4.3.87. Tanzania

4.3.87.1. Climatic descriptions and tables

1. Climatic area: Most of the country except some central parts

1.1. Tropical Summer Rain Climate Aw and Tropical Summer Rain Mountain Climate GAw (in the west, partly Tropical Monsoonal Rain Climate Am etc.)

1.2. Permanently very hot, in the lowlands.

1.3. High sun period: in the morning often dry and sunny, showers in the afternoon and at night; at times rainy days. Low sun period: more often dry and sunny days.

1.4. Wind: occasional gales and squalls in thunderstorms.

Bukoba

GAmbb	087-1	latitude: 1°20's Jan	Feb	Mar	Apr	May	Jun	longitude: 31°49'e Jul	Aug	Sep	Oct	height: 1137 m Nov	Dec	Year
	tx	29	29	29	28	28	28	28	28	28	28	29	29	30 °C
	t	21	21	21	21	21	21	20	20	21	21	21	21	21 °C
	tn	14	14	14	15	14	14	13	13	13	13	14	14	12 °C
	clothing	s	s	s	s	s	s	s	s	s	s	s	s	–
	stress	o	o	o	o	o	o	o	o	o	o	o	o	–
	p	151	159	247	361	317	85	49	80	107	134	163	190	2043 mm
0.2 mm	dp	14	13	17	22	19	7	5	10	11	13	17	18	166 days

Mwanza

latitude: 2°31′s longitude: 32°54′e height: 1140 m

GAwab	087-2	Jan	Feb	Mar	Apr	May	Jun	Jul	Aug	Sep	Oct	Nov	Dec	Year
	tx	32	32	32	31	31	31	31	31	31	32	32	31	33 °C
	t	24	23	24	23	23	23	22	23	24	24	24	23	23 °C
	tn	15	16	16	16	16	14	13	14	16	17	17	16	13 °C
	clothing	t	t	t	t	t	t	t	t	t	t	t	t	–
	stress	f	f	f	f	f	f	f	f	f	f	f	f	–
	p	93	102	152	149	90	17	13	18	41	61	121	143	1000 mm
0.2 mm	dp	12	10	15	17	13	3	1	3	4	6	15	12	111 days
	sunshine	239	218	223	258	270	294	291	291	258	251	249	229	3071 hours
	%	63	63	59	71	73	82	78	78	71	67	68	60	69 %

Moshi

latitude: 3°21′s longitude: 37°20′e height: 831 m

GAwab	087-3	Jan	Feb	Mar	Apr	May	Jun	Jul	Aug	Sep	Oct	Nov	Dec	Year
	tx	36	36	35	32	29	29	28	29	32	33	34	34	37 °C
	t	25	26	25	24	22	21	21	21	22	24	25	25	23 °C
	tn	14	15	15	17	16	13	12	12	12	13	14	15	11 °C
	clothing	t	t	t	t	s	s	s	s	s	t	t	t	–
	stress	f	f	f	f	f	o	o	o	o	f	f	f	–
	p	40	42	119	297	167	31	13	13	13	35	52	48	870 mm
0.2 mm	dp	3	3	9	18	16	4	3	3	3	3	6	6	77 days
	sunshine	276	257	229	198	167	155	124	177	225	245	267	276	2596 hours
	%	72	74	61	55	45	43	33	48	62	65	73	72	59 %

Kigoma

latitude: 4°53′s longitude: 29°38′e height: 885 m

GAwaa	087-4	Jan	Feb	Mar	Apr	May	Jun	Jul	Aug	Sep	Oct	Nov	Dec	Year
	tx	29	30	30	30	31	30	31	32	32	33	30	29	33 °C
	t	23	23	24	24	24	23	23	24	25	25	23	23	24 °C
	tn	17	17	17	18	17	15	14	16	17	18	17	17	14 °C
	clothing	t	t	t	t	t	t	t	t	t	t	t	t	–
	stress	f	f	f	f	f	f	f	f	f	f	f	f	–
	p	127	118	158	144	50	5	3	2	13	60	129	152	961 mm
0.2 mm	dp	14	12	17	17	8	1	1	1	3	8	17	19	118 days

Tabora

latitude: 5°05′s longitude: 32°50′e height: 1190 m

GAwab	087-5	Jan	Feb	Mar	Apr	May	Jun	Jul	Aug	Sep	Oct	Nov	Dec	Year
	tx	31	31	31	30	30	29	30	32	33	34	34	32	34 °C
	t	23	23	23	23	22	22	21	23	25	26	25	23	23 °C
	tn	15	16	16	16	14	12	13	14	16	17	17	16	12 °C
	clothing	s	s	s	s	s	s	s	s	t	t	t	s	–
	stress	o	o	o	o	o	o	o	o	f	f	f	f	–
	p	135	126	170	133	34	3	0	1	5	14	81	181	882 mm
	dp	12	11	13	10	3	0	0	0	1	2	9	15	76 days
	sunshine	220	194	217	240	273	306	322	316	279	267	213	195	3042 hours
	%	57	56	58	67	74	86	88	85	77	70	57	51	69 %

Kisaumi (Airport in Zanzibar)

latitude: 6°13′s longitude: 39°13′e height: 18 m

Awha	087-6	Jan	Feb	Mar	Apr	May	Jun	Jul	Aug	Sep	Oct	Nov	Dec	Year
	tx	34	36	36	33	31	30	29	30	32	33	34	34	37 °C
	t	27	27	28	27	26	25	24	24	25	25	26	27	26 °C
	tn	23	23	23	23	22	21	21	21	21	21	22	23	20 °C
	clothing	t	t	t	t	t	t	t	t	t	t	t	t	–
	stress	ff	ff	ff	ff	ff	f	f	f	f	f	ff	ff	–
	p	93	92	130	356	267	53	51	33	32	106	190	136	1539 mm
0.2 mm	dp	11	8	13	21	15	8	8	7	7	9	15	16	138 days
	sunshine	248	232	223	177	195	216	242	254	252	267	252	254	2812 hours
	%	64	67	59	49	53	61	66	69	70	70	68	66	64 %

Morogoro

latitude: 6°51′s longitude: 37°40′e height: 579 m

GAwab	087-7	Jan	Feb	Mar	Apr	May	Jun	Jul	Aug	Sep	Oct	Nov	Dec	Year
	tx	34	34	34	33	31	29	30	31	32	33	34	34	36 °C
	t	26	26	26	25	24	22	21	22	23	24	26	26	24 °C
	tn	19	18	18	18	15	12	11	12	13	15	17	18	11 °C
	clothing	t	t	t	t	t	t	t	t	t	t	t	t	–
	stress	ff	ff	ff	f	f	f	f	f	f	f	ff	ff	–
	p	97	116	128	206	100	21	17	13	16	27	41	75	857 mm
0.2 mm	dp	10	10	16	22	16	8	5	5	5	8	8	11	124 days
	sunshine	183	170	174	129	115	123	124	124	144	180	186	186	1836 hours
	%	47	49	46	36	31	35	34	34	40	47	50	48	41 %

Dar-Es-Salaam

latitude: 6°53′s longitude: 39°12′e height: 58 m

Awha	087-8	Jan	Feb	Mar	Apr	May	Jun	Jul	Aug	Sep	Oct	Nov	Dec	Year
	tx	32	32	33	32	32	31	30	31	31	31	31	31	33 °C
	t	27	27	28	27	26	24	23	24	24	25	26	27	26 °C
	tn	22	22	22	22	19	18	17	17	18	19	21	21	17 °C
	clothing	t	t	t	t	t	t	t	t	t	t	t	t	–
	stress	ff	ff	ff	ff	ff	ff	f	f	f	ff	ff	ff	–
	p	58	68	128	264	219	33	21	24	28	48	68	83	1043 mm
	dp	6	6	11	18	13	5	4	4	5	5	7	8	92 days
	sunshine	242	229	220	158	222	227	233	268	255	275	253	254	2836 hours
	%	63	66	58	44	61	64	64	73	70	72	68	65	64 %

Mbeya

latitude: 8°56′s longitude: 33°28′e height: 1707 m

GAwbl	087-9	Jan	Feb	Mar	Apr	May	Jun	Jul	Aug	Sep	Oct	Nov	Dec	Year
	tx	26	26	26	25	25	24	25	26	28	29	29	28	30 °C
	t	19	18	18	18	16	15	14	16	18	20	20	19	18 °C
	tn	12	12	12	11	8	5	4	7	8	10	11	12	4 °C
	clothing	s	s	s	s	n	n	n	n	s	s	s	s	–
	stress	o	o	o	o					o	o	o	o	–
	p	194	165	160	118	18	1	0	1	3	15	53	152	880 mm
0.2 mm	dp	23	19	20	17	5	0	0	1	0	3	10	19	117 days
	sunshine	127	102	118	165	257	273	304	294	270	264	204	143	2521 hours
	%	33	29	31	46	71	78	84	80	75	69	54	36	57 %

Lindi

Awha	087-10	Jan	Feb	Mar	Apr	May	Jun	Jul	Aug	Sep	Oct	Nov	Dec	Year
		latitude: 10°00's						longitude: 39°42'e				height: 41 m		
	tx	33	33	33	33	32	32	31	33	32	32	33	33	34 °C
	t	28	27	27	27	26	25	24	24	25	26	27	28	26 °C
	tn	22	22	22	22	18	16	16	17	18	19	21	22	15 °C
	clothing	t	t	t	t	t	t	t	t	t	t	t	t	–
	stress	ff	ff	ff	ff	f	f	f	f	f	f	ff	ff	–
	p	145	117	170	173	38	10	8	5	12	15	53	150	897 mm
	dp	12	10	13	14	6	2	1	1	2	2	6	11	80 days

Songea

GAwal	087-11	Jan	Feb	Mar	Apr	May	Jun	Jul	Aug	Sep	Oct	Nov	Dec	Year
		latitude: 10°41's						longitude: 35°35'e				height: 1067 m		
	tx	29	29	29	27	27	26	26	26	28	31	32	31	32 °C
	t	22	22	22	21	20	18	17	18	20	22	24	23	21 °C
	tn	16	17	16	15	12	9	9	11	12	15	16	16	8 °C
	clothing	s	s	s	s	s	s	s	s	s	s	s	s	–
	stress	f	f	f	o	o				o	o	f	f	–
	p	290	241	238	106	14	1	0	2	2	7	46	177	1126 mm
	dp	19	17	18	12	3	0	0	0	1	1	4	12	86 days

2. Climatic area: approximately the Central Province

2.1. Steppe Mountain Climate GBS

2.2. Summer hot, at higher levels warm; winter warm, at higher levels mild.

2.3. High sun period: in the morning often dry and sunny, in the afternoon showers; at times rainy days. Low sun period: more frequent dry and sunny days.

2.4. Wind: occasional gales and squalls in thunderstorms.

Dodoma

GBSab	087-12	Jan	Feb	Mar	Apr	May	Jun	Jul	Aug	Sep	Oct	Nov	Dec	Year
		latitude: 6°10's						longitude: 35°46'e				height: 1120 m		
	tx	33	33	32	31	31	30	29	30	32	33	34	34	34 °C
	t	24	23	23	23	22	21	20	21	22	24	25	25	23 °C
	tn	17	16	16	16	13	11	11	11	13	14	16	17	10 °C
	clothing	t	t	t	t	s	s	s	s	s	t	t	t	–
	stress	f	f	f	f	o	o	o	o	o	f	f	f	–
	p	148	116	124	51	5	1	0	0	1	5	20	105	576 mm
0.2 mm	dp	12	9	11	7	2	0	0	0	0	1	4	9	55 days

4.3.87.2. Predominant general weather situations are described in 4.3.0.1. under Equatorial South Africa. A hygrothermal diagram of Dar-Es-Salaam is to be found in 4.3.0.2.

4.3.88. Uganda

4.3.88.1. Climatic description and tables

1. Tropical Summer Rain Mountain Climate GAw; in some parts Tropical Rain Mountain Climate GAr
2. High sun period: hot; low sun period: warm at about 1100 m.
3. In the morning often dry and sunny; during the low sun period often in the afternoon too, during the high sun period often showers, at times rainy days.
4. Wind: occasional gales, squalls in thunderstorms.

Gulu

latitude: 2°45′n longitude: 32°20′e height: 1109 m

GAwab	088-1	Jan	Feb	Mar	Apr	May	Jun	Jul	Aug	Sep	Oct	Nov	Dec	Year
	tx	34	36	36	33	30	29	28	29	29	31	32	33	36 °C
	t	24	25	25	24	23	22	22	22	22	23	23	23	23 °C
	tn	13	14	15	17	16	16	15	15	15	15	14	13	12 °C
	clothing	t	t	t	t	s	s	s	s	s	s	s	s	–
	stress	f	f	f	f	o	o	o	o	o	o	o	o	–
	p	16	48	91	176	185	137	164	253	178	157	88	47	1540 mm
	dp	3	4	9	14	16	13	14	17	14	14	9	5	132 days
	sunshine	285	244	244	212	251	235	195	215	241	251	246	277	2896 hours
	%	77	72	65	58	66	64	51	57	66	67	68	75	65 %

Moroto

latitude: 2°33′n longitude: 34°36′e height: 1347 m

GAwab	088-2	Jan	Feb	Mar	Apr	May	Jun	Jul	Aug	Sep	Oct	Nov	Dec	Year
	t	24	24	23	23	22	21	21	21	22	22	23	23	23 °C
	clothing	t	t	s	s	s	s	s	s	s	s	s	s	–
	stress	f	f	f	o	o	o	o	o	o	o	o	o	–
	p	9	31	74	124	150	90	136	107	51	48	49	25	894 mm
0.2 mm	dp	1	4	6	10	7	9	12	10	5	5	4	2	74 days

Masindi

latitude: 1°41′n longitude: 31°43′e height: 1146 m

GAwab	088-3	Jan	Feb	Mar	Apr	May	Jun	Jul	Aug	Sep	Oct	Nov	Dec	Year
	tx	34	35	35	32	31	30	29	29	30	31	32	32	36 °C
	t	24	24	24	23	23	22	22	22	22	23	23	23	23 °C
	tn	15	16	16	16	16	15	14	14	14	15	15	14	13 °C
	clothing	t	t	t	s	s	s	s	s	s	s	s	s	–
	stress	f	f	f	f	o	o	o	o	o	o	o	o	–
	p	29	56	103	157	145	99	112	138	143	144	118	44	1288 mm
0.2 mm	dp	5	6	9	14	16	11	10	15	13	14	13	6	132 days
	sunshine	195	186	183	153	158	141	130	149	171	192	195	202	2055 hours
	%	52	55	49	42	42	38	34	39	47	51	54	54	46 %

Entebbe

GArbb	088-4	latitude: 0°03′n						longitude: 32°27′e				height: 1146 m		
		Jan	Feb	Mar	Apr	May	Jun	Jul	Aug	Sep	Oct	Nov	Dec	Year
	tx	29	28	29	28	27	27	27	27	28	28	29	28	31 °C
	t	22	22	22	22	22	21	21	21	21	22	22	22	22 °C
	tn	15	15	16	16	16	16	15	14	15	15	15	15	14 °C
	clothing	s	s	s	s	s	s	s	s	s	s	s	s	–
	stress	f	f	f	f	f	f	f	f	f	f	f	f	–
	p	79	85	170	278	279	113	73	84	77	84	137	115	1574 mm
	dp	7	8	13	18	17	10	7	8	7	9	12	11	127 days
	sunshine	234	204	205	181	191	187	197	194	194	205	202	214	2408 hours
	%	62	60	55	50	51	51	52	52	53	55	56	57	54 %

Kabale

GAwll	088-5	latitude: 1°15′s						longitude: 29°59′e				height: 1871 m		
		Jan	Feb	Mar	Apr	May	Jun	Jul	Aug	Sep	Oct	Nov	Dec	Year
	tx	27	27	27	25	24	24	25	26	26	26	25	25	28 °C
	t	17	17	17	17	17	16	16	17	17	17	17	17	17 °C
	tn	7	7	8	8	8	6	6	6	7	8	7	7	5 °C
	clothing	s	s	s	s	s	n	n	s	s	s	s	s	–
	stress	o	o	o	o	o			o	o	o	o	o	–
	p	61	91	116	136	93	26	20	54	94	97	107	92	987 mm
0.2 mm	dp	11	13	16	20	16	5	3	8	15	18	19	15	159 days
	sunshine	149	158	149	132	140	177	152	158	177	161	153	146	1851 hours
	%	40	46	40	36	37	48	40	42	49	43	42	39	42 %

4.3.88.2. Predominant general weather situations are described in 4.3.0.1. under Equatorial South Africa.

4.3.89. Rwanda

4.3.89.1. Climatic description and tables

1. Tropical Summer Rain Mountain Climate GAw
2. Warm at lower levels throughout the year.
3. In the morning often dry and sunny; in the afternoon also during low sun period, during high sun period often showers; at times rainy days.
4. Wind: occasional gales and squalls in thunderstorms.

Kigali

GAwbb	089-1	latitude: 1°58′s						longitude: 30°08′e				height: 1465 m		
		Jan	Feb	Mar	Apr	May	Jun	Jul	Aug	Sep	Oct	Nov	Dec	Year
	t	21	21	21	20	20	20	20	21	21	21	20	20	21 °C
	clothing	s	s	s	s	s	s	s	s	s	s	s	s	–
	stress	o	o	o	o	o	o	o	o	o	o	o	o	–
	p	90	92	103	165	128	26	7	22	61	102	101	89	986 mm
0.1 mm	dp	11	11	14	19	14	3	2	5	9	15	16	13	132 days

Rubona – Collina

latitude: 2°29′s longitude: 29°46′e height: 1706 m

GAwbb	089-2	Jan	Feb	Mar	Apr	May	Jun	Jul	Aug	Sep	Oct	Nov	Dec	Year
	t	19	19	19	19	19	19	19	20	20	20	19	19	19 °C
	clothing	s	s	s	s	s	s	s	s	s	s	s	s	–
	stress	o	o	o	o	o	o	o	o	o	o	o	o	–
	p	106	114	140	179	63	25	8	24	61	112	111	84	1027 mm
0.1 mm	dp	15	15	18	21	18	3	2	5	12	16	19	18	162 days
	sunshine	177	164	183	162	164	222	236	226	204	186	174	177	2275 hours
	%	47	48	49	45	44	62	63	61	56	49	47	47	51 %

4.3.89.2. Predominant general weather situations are described in 4.3.0.1. under Equatorial South Africa.

4.3.90. Burundi

4.3.90.1. Climatic description and tables

1. Tropical Summer Rain Mountain Climate GAw
2. Warm in the lower levels throughout the year.
3. In the morning often dry and sunny; in the afternoon also during low sun period, during high sun period often showers; at times rainy days.
4. Wind: occasional gales, squalls in thunderstorms.

Bujumbura

latitude: 3°19′s longitude: 29°19′e height: 783 m

GAwab	090-1	Jan	Feb	Mar	Apr	May	Jun	Jul	Aug	Sep	Oct	Nov	Dec	Year
	t	23	23	23	23	23	23	22	23	24	24	23	23	23 °C
	clothing	s	s	s	s	s	s	s	s	t	t	s	s	–
	stress	f	f	f	f	f	f	f	f	f	f	f	f	–
	p	90	110	120	125	54	11	7	10	37	62	97	115	838 mm
0.1 mm	dp	17	16	18	21	13	3	1	2	8	13	20	20	152 days
	sunshine	158	141	170	153	195	243	270	248	201	174	141	149	2243 hours
	%	41	41	45	42	53	68	73	67	55	46	38	39	51 %

Kisozi – Collina

latitude: 3°33′s longitude: 29°41′e height: 2155 m

GAwll	090-2	Jan	Feb	Mar	Apr	May	Jun	Jul	Aug	Sep	Oct	Nov	Dec	Year
	t	16	17	17	17	16	15	15	16	17	17	17	17	16 °C
	clothing	n	s	s	s	n	n	n	n	s	s	s	s	–
	stress		o	o	o					o	o	o	o	–
	p	164	156	199	232	120	15	7	14	64	117	181	181	1450 mm
0.1 mm	dp	21	19	20	23	16	3	1	3	9	15	21	21	172 days
	sunshine	149	127	167	153	149	216	236	229	204	183	153	136	2102 hours
	%	39	37	44	42	40	60	64	62	56	48	42	36	47 %

4.3.90.2. Predominant general weather situations are described in 4.3.0.1. under Equatorial South Africa.

4.3.91. Zaire

4.3.91.1. Climatic descriptions and tables

1. Climatic area: Most of Equatorial Province, East Province and Manjema
1.1. Tropical Rain Climate Ar and Tropical Rain Mountain Climate GAr
1.2. In the lowlands very hot all the year.
1.3. In the morning often dry and sunny, in the afternoon showers; during the high sun period, rainy days more often than during the low sun period.
1.4. Wind: occasional gales and squalls in thunderstorms.

Bambesa

		latitude: 3°27′n						longitude: 25°43′e				height: 621 m	
GAraa	091-1	Jan	Feb	Mar	Apr	May	Jun	Jul	Aug	Sep	Oct	Nov	Dec	Year
	tx	36	38	36	35	33	32	32	32	32	33	33	34	38 °C
	t	24	25	25	25	25	24	24	24	24	24	25	24	24 °C
	tn	13	12	15	15	17	17	16	17	17	17	14	13	11 °C
	clothing	t	t	t	t	t	t	t	t	t	t	t	t	–
	stress	ff	ff	ff	ff	ff	ff	ff	ff	ff	ff	ff	ff	–
	p	34	76	132	197	206	154	181	209	212	212	129	36	1778 mm
0.1 mm	dp	4	9	11	15	16	15	14	17	18	21	13	6	159 days
	sunshine	236	220	195	185	205	179	161	143	175	181	201	223	2304 hours
	%	64	65	52	51	54	48	42	38	48	48	56	60	52 %

Lisala

		latitude: 2°19′n						longitude: 21°34′e				height: 460 m	
Araa	091-2	Jan	Feb	Mar	Apr	May	Jun	Jul	Aug	Sep	Oct	Nov	Dec	Year
	t	25	26	26	26	25	25	24	24	24	25	25	25	25 °C
	clothing	t	t	t	t	t	t	t	t	t	t	t	t	–
	stress	ff	ff	ff	ff	ff	ff	ff	ff	ff	ff	ff	ff	–
	p	63	50	127	154	119	155	190	181	177	195	136	79	1626 mm
0.1 mm	dp	4	4	10	10	12	12	13	14	13	16	13	6	127 days
	sunshine	211	195	177	180	186	159	155	155	150	161	156	189	2069 hours
	%	57	57	47	49	49	43	41	41	41	43	43	51	47 %

Kisangani

		latitude: 0°31′n						longitude: 25°11′e				height: 415 m	
Araa	091-3	Jan	Feb	Mar	Apr	May	Jun	Jul	Aug	Sep	Oct	Nov	Dec	Year
	t	25	25	25	25	25	24	24	24	24	24	24	24	24 °C
	clothing	t	t	t	t	t	t	t	t	t	t	t	t	–
	stress	ff	ff	ff	ff	ff	ff	ff	ff	ff	ff	ff	ff	–
	p	83	104	146	172	181	112	116	179	178	215	161	114	1761 mm
0.1 mm	dp	7	9	14	13	14	11	11	14	15	17	15	12	152 days
	sunshine	189	178	177	177	183	150	140	127	156	167	162	167	1973 hours
	%	50	52	47	49	49	41	37	34	43	44	45	45	45 %

Mbandaka (Coquilhatville)

latitude: 0°03'n longitude: 18°17'e height: 317 m

Araa	091-4	Jan	Feb	Mar	Apr	May	Jun	Jul	Aug	Sep	Oct	Nov	Dec	Year
	t	25	25	25	25	25	24	24	24	24	24	24	25	24 °C
	clothing	t	t	t	t	t	t	t	t	t	t	t	t	–
	stress	ff	ff	ff	ff	ff	ff	ff	ff	ff	ff	ff	ff	–
	p	82	102	156	141	131	120	100	110	206	213	195	122	1678 mm
0.1 mm	dp	9	9	14	14	15	15	6	12	16	16	18	14	158 days
	sunshine	186	178	183	177	192	168	167	158	162	164	153	164	2053 hours
	%	49	52	49	49	51	46	44	42	45	44	42	44	46 %

Mukumari

latitude: 2°50's longitude: 23°11'e height: 535 m

GAraa	091-5	Jan	Feb	Mar	Apr	May	Jun	Jul	Aug	Sep	Oct	Nov	Dec	Year
	t	24	24	25	25	25	24	23	24	24	24	24	24	24 °C
	clothing	t	t	t	t	t	t	t	t	t	t	t	t	–
	stress	ff	ff	ff	ff	ff	ff	ff	ff	ff	ff	ff	ff	–
	p	175	158	209	174	137	52	78	93	162	169	231	201	1839 mm
0.1 mm	dp	13	13	14	15	12	8	11	12	15	17	20	18	168 days
	sunshine	149	155	161	156	180	180	158	174	180	177	153	133	1956 hours
	%	39	45	43	43	48	50	43	47	50	47	42	35	44 %

Kindu

latitude: 2°57's longitude: 23°55'e height: 497 m

Araa	091-6	Jan	Feb	Mar	Apr	May	Jun	Jul	Aug	Sep	Oct	Nov	Dec	Year
	t	24	25	25	25	25	25	24	24	24	24	24	24	24 °C
	clothing	t	t	t	t	t	t	t	t	t	t	t	t	–
	stress	ff	ff	ff	ff	ff	ff	ff	ff	ff	ff	ff	ff	–
	p	178	151	193	157	107	30	32	72	110	158	197	209	1594 mm
0.1 mm	dp	17	14	17	17	11	4	5	7	11	16	19	19	157 days
	sunshine	155	153	174	171	186	177	152	155	162	167	156	146	1953 hours
	%	41	44	46	47	50	49	41	42	45	44	42	38	44 %

2. *Climatic area:* greater part of the country except Equatorial Province, East Province and Manjema

2.1. Tropical Summer Rain Climate Aw, in part Tropical Monsoonal Rain Climate Am and Tropical Summer Rain Mountain Climate GAw

2.2. At lower levels very hot in summer, hot in winter.

2.3. In the morning often dry and sunny, in the afternoon showers – very frequent during the high sun period, less so during low sun period.

2.4. Wind: occasional gales, squalls in thunderstorms.

Libenge

latitude: 3°38'n longitude: 18°38'e height: 360 m

Awaa	091-7	Jan	Feb	Mar	Apr	May	Jun	Jul	Aug	Sep	Oct	Nov	Dec	Year
	t	24	25	25	25	25	24	24	24	24	24	24	24	24 °C
	clothing	t	t	t	t	t	t	t	t	t	t	t	t	–
	stress	ff	ff	ff	ff	ff	ff	ff	ff	ff	ff	ff	ff	–
	p	28	51	113	138	160	157	156	204	176	190	122	44	1539 mm
	dp	3	4	8	10	12	10	12	14	14	14	10	4	115 days
	sunshine	202	201	205	177	189	159	152	136	156	164	165	195	2101 hours
	%	55	59	55	48	50	43	40	36	43	44	46	53	47 %

Watsa

		latitude: 3°04′n				longitude: 29°30′e				height: 1030 m			
GAmab	091-8	Jan	Feb	Mar	Apr	May	Jun	Jul	Aug	Sep	Oct	Nov	Dec	Year
	t	24	24	24	24	23	22	22	22	22	23	24	24	23 °C
	clothing	t	t	t	t	t	s	s	s	s	t	t	t	–
	stress	f	f	f	f	f	f	f	f	f	f	f	f	–
	p	47	59	129	206	206	190	190	249	214	243	137	54	1924 mm
0.1 mm	dp	5	7	14	17	19	20	18	20	20	23	15	9	187 days
	sunshine	242	206	198	189	189	162	140	140	171	155	213	236	2240 hours
	%	65	61	53	52	50	44	37	37	47	41	59	64	51 %

Inogo

		latitude: 1°58′s				longitude: 18°16′e				height: 300 m			
Awaa	091-9	Jan	Feb	Mar	Apr	May	Jun	Jul	Aug	Sep	Oct	Nov	Dec	Year
	t	26	26	26	26	26	25	25	25	25	25	25	25	25 °C
	clothing	t	t	t	t	t	t	t	t	t	t	t	t	–
	stress	ff	ff	ff	ff	ff	ff	ff	ff	ff	ff	ff	ff	–
	p	188	128	153	179	121	41	19	82	205	219	229	158	1722 mm

Bukavu

		latitude: 2°28′s				longitude: 28°55′e				height: 1582 m			
GAwbb	091-10	Jan	Feb	Mar	Apr	May	Jun	Jul	Aug	Sep	Oct	Nov	Dec	Year
	t	19	19	19	19	19	19	19	20	20	19	19	19	19 °C
	clothing	s	s	s	s	s	s	s	s	s	s	s	s	–
	stress	o	o	o	o	o	o	o	o	o	o	o	o	–
	p	138	156	162	141	86	17	14	36	100	146	168	150	1314 mm
0.1 mm	dp	20	19	24	24	14	5	3	4	14	19	24	23	193 days
	sunshine	152	136	149	147	161	216	242	223	180	155	135	130	2026 hours
	%	40	39	40	41	43	60	65	60	50	41	37	34	46 %

Kinshasa (Leopoldville)

		latitude: 4°23′s				longitude: 15°26′e				height: 309 m			
Awab	091-11	Jan	Feb	Mar	Apr	May	Jun	Jul	Aug	Sep	Oct	Nov	Dec	Year
	tx	34	35	34	35	33	33	32	33	34	34	33	33	36 °C
	t	25	26	26	26	25	23	22	23	25	25	25	25	25 °C
	tn	20	19	19	20	19	17	15	15	17	19	19	19	15 °C
	clothing	t	t	t	t	t	t	t	t	t	t	t	t	–
	stress	ff	ff	ff	ff	ff	f	f	f	f	ff	ff	ff	–
	p	126	141	169	212	137	5	1	3	33	139	235	170	1371 mm
0.1 mm	dp	9	9	12	14	10	1	0	1	1	10	15	12	93 days
	sunshine	133	136	149	165	140	138	124	146	129	133	141	136	1670 hours
	%	35	39	40	46	38	39	34	39	36	35	38	35	38 %

Kikwit

		latitude: 5°02′s				longitude: 18°48′e				height: 518 m			
GAwaa	091-12	Jan	Feb	Mar	Apr	May	Jun	Jul	Aug	Sep	Oct	Nov	Dec	Year
	t	24	24	24	24	24	24	24	24	24	24	24	24	24 °C
	clothing	t	t	t	t	t	t	t	t	t	t	t	t	–
	stress	ff	ff	ff	ff	ff	f	f	f	ff	ff	ff	ff	–
	p	196	110	189	248	103	6	10	22	159	211	217	175	1646 mm
0.1 mm	dp	12	11	16	18	12	1	2	3	10	14	18	15	132 days
	sunshine	143	138	149	156	192	252	248	223	174	180	150	143	2148 hours
	%	37	40	40	43	52	71	67	60	48	48	40	37	49 %

Kananga (Luluabourg)

GAwaa	091-13	latitude: 5°45′s					longitude: 22°25′e				height: 657 m			
		Jan	Feb	Mar	Apr	May	Jun	Jul	Aug	Sep	Oct	Nov	Dec	Year
	tx	33	33	33	33	33	33	32	33	33	33	32	32	34 °C
	t	24	24	24	24	25	24	23	23	24	24	23	23	24 °C
	tn	18	18	18	18	18	15	14	16	17	18	18	18	14 °C
	clothing	t	t	t	t	t	t	t	t	t	t	t	t	–
	stress	ff	ff	ff	ff	ff	f	f	f	f	ff	ff	ff	–
	p	120	115	186	155	82	13	17	51	116	146	233	206	1440 mm
	dp	12	12	14	15	8	2	2	5	10	13	15	15	122 days
	sunshine	149	133	152	162	217	257	226	192	183	192	159	140	2163 hours
	%	39	38	40	45	59	73	62	52	51	51	43	36	49 %

Kalemie (Albertville)

GAwab	091-14	latitude: 5°53′s					longitude: 29°11′e				height: 790 m			
		Jan	Feb	Mar	Apr	May	Jun	Jul	Aug	Sep	Oct	Nov	Dec	Year
	tx	32	31	31	29	30	29	29	31	32	32	31	30	33 °C
	t	23	23	23	23	23	22	21	22	24	24	23	23	23 °C
	tn	17	17	17	18	16	13	12	13	14	14	17	17	12 °C
	clothing	t	t	t	t	t	s	s	s	t	t	t	t	–
	stress	f	f	f	f	f	o	o	f	f	f	f	f	–
	p	120	78	130	210	88	9	1	7	48	49	140	183	1062 mm
0.1 mm	dp	13	11	12	15	8	1	0	1	3	6	16	17	103 days
	sunshine	155	158	183	195	248	285	310	288	246	214	162	146	2590 hours
	%	40	45	49	54	67	81	85	78	68	56	44	38	59 %

Moanda (Banana)

Awab	091-15	latitude: 6°00′s					longitude: 12°25′e				height: 27 m			
		Jan	Feb	Mar	Apr	May	Jun	Jul	Aug	Sep	Oct	Nov	Dec	Year
	tx	32	32	33	33	32	29	27	27	29	31	31	32	33 °C
	t	27	27	27	27	26	24	22	22	24	26	27	26	25 °C
	tn	22	22	22	22	21	18	17	17	19	22	22	22	17 °C
	clothing	t	t	t	t	t	t	t	t	t	t	t	t	–
	stress	ff	ff	ff	ff	ff	ff	f	f	ff	ff	ff	ff	–
	p	69	125	130	168	60	0	0	1	6	28	121	81	788 mm
0.1 mm	dp	6	9	10	12	5	1	1	2	5	9	12	7	79 days
	sunshine	170	172	180	174	149	168	133	121	90	105	138	155	1755 hours
	%	44	49	48	48	41	48	36	33	25	28	37	40	40 %

Kamina – Baka

GAwab	091-16	latitude: 8°38′s					longitude: 25°15′e				height: 1120 m			
		Jan	Feb	Mar	Apr	May	Jun	Jul	Aug	Sep	Oct	Nov	Dec	Year
	t	22	22	23	23	23	21	22	23	24	23	23	22	23 °C
	clothing	s	s	t	t	t	s	s	t	t	t	t	s	–
	stress	f	f	f	f	f	f	f	f	f	f	f	f	–
	p	245	220	198	141	5	0	0	13	51	92	162	240	1367 mm
0.1 mm	dp	21	20	19	13	2	0	0	1	6	12	18	22	134 days
	sunshine	121	110	149	207	291	294	319	294	222	195	156	127	2486 hours
	%	31	31	40	58	80	84	88	80	61	51	42	32	56 %

Lulumbashi (Elisabethville)

GAwbl	091-17	Jan	Feb	Mar	Apr	May	Jun	Jul	Aug	Sep	Oct	Nov	Dec	Year
		latitude: 11°35's				longitude: 27°31'e				height: 1300 m				
	tx	30	30	31	31	30	29	29	32	35	36	34	32	36 °C
	t	20	20	21	22	19	17	17	19	22	23	22	20	20 °C
	tn	14	14	13	9	7	3	3	4	7	11	14	14	3 °C
	clothing	s	s	s	s	s	s	s	s	s	s	s	s	–
	stress	f	f	f	o	o	o	o	o	f	f	f	f	–
	p	248	263	204	57	4	0	0	1	3	27	164	258	1229 mm
	dp	24	23	21	9	2	0	0	0	0	5	17	23	124 days
	sunshine	130	124	177	234	291	294	316	319	294	282	195	130	2787 hours
	%	33	35	47	66	81	86	89	88	81	74	51	33	63 %

4.3.91.2. Predominant general weather situations are described in 4.3.0.1. under Equatorial South Africa. A hygrothermal diagram of Kinshasa is to be found in 4.3.0.2.

4.3.92. Congo

4.3.92.1. Climatic descriptions and tables

1. Climatic area: northern part including equatorial region
1.1. Tropical Rain Climate Ar and Tropical Rain Mountain Climate GAr
1.2. At low levels permanently very hot.
1.3. In the morning often dry and sunny, in the afternoon showers, more frequent about the equinoxes than at other times, now and then rainy days.
1.4. Wind: occasional gales, squalls in thunderstorms.

Souanke

GAraa	092-1	Jan	Feb	Mar	Apr	May	Jun	Jul	Aug	Sep	Oct	Nov	Dec	Year
		latitude: 2°04'n				longitude: 14°02'e				height: 547 m				
	tx	32	32	33	33	33	31	31	31	31	32	32	32	24 °C
	t	23	24	24	24	24	23	23	23	23	24	24	24	23 °C
	tn	15	15	16	16	14	16	15	14	16	16	16	15	13 °C
	clothing	t	t	t	t	t	t	t	t	t	t	t	t	–
	stress	f	f	f	f	f	f	f	f	f	f	f	f	–
	p	58	89	166	149	211	114	64	74	209	232	173	58	1597 mm
0.1 mm	dp	4	7	13	13	14	10	6	7	15	16	13	6	124 days

Ouesso

Araa	092-2	Jan	Feb	Mar	Apr	May	Jun	Jul	Aug	Sep	Oct	Nov	Dec	Year
		latitude: 1°37'n				longitude: 16°04'e				height: 351 m				
	tx	34	36	37	36	34	33	32	33	33	34	34	32	37 °C
	t	26	26	26	26	26	25	25	25	25	25	25	25	25 °C
	tn	18	18	19	19	19	19	18	18	18	18	19	18	17 °C
	clothing	t	t	t	t	t	t	t	t	t	t	t	t	–
	stress	ff	ff	ff	ff	ff	ff	ff	ff	ff	ff	ff	ff	–
	p	56	79	150	127	152	117	79	155	220	226	152	83	1596 mm
0.1 mm	dp	4	6	9	9	11	8	6	9	14	14	12	6	108 days
	sunshine	165	157	157	166	189	153	137	135	122	139	151	160	1831 hours
	%	44	46	42	46	50	42	36	36	34	37	42	43	41 %

Impfondo

Araa	092-3	Jan	Feb	Mar	Apr	May	Jun	Jul	Aug	Sep	Oct	Nov	Dec	Year
		latitude: 1°37′n				longitude: 18°04′e				height: 326 m				
	t	25	26	26	26	26	25	25	25	25	25	25	25	25 °C
	clothing	t	t	t	t	t	t	t	t	t	t	t	t	–
	stress	ff	ff	ff	ff	ff	ff	ff	ff	ff	ff	ff	ff	–
	p	73	78	159	171	188	151	132	169	169	220	176	107	1793 mm
0.1 mm	dp	5	5	9	10	12	10	8	10	12	14	12	6	114 days
	sunshine	191	169	168	156	175	145	146	143	133	152	160	184	1922 hours
	%	51	50	45	43	46	40	39	38	37	41	44	49	43 %

2. *Climatic area:* southern part south of the equator

2.1. Tropical Summer Rain Climate Aw and Tropical Summer Rain Mountain Climate GAw

2.2. At low levels summer very hot, winter hot.

2.3. In the morning often dry and sunny, in the afternoon showers – very frequent during the high sun period, less so during low sun period.

2.4. Wind: occasional gales, squalls in thunderstorms.

Gamboma

Awaa	092-4	Jan	Feb	Mar	Apr	May	Jun	Jul	Aug	Sep	Oct	Nov	Dec	Year
		latitude: 1°52′s				longitude: 15°51′e				height: 376 m				
	t	26	26	26	26	25	25	25	26	25	25	25	25	25 °C
	clothing	t	t	t	t	t	t	t	t	t	t	t	t	–
	stress	ff	ff	ff	ff	ff	ff	ff	ff	ff	ff	ff	ff	–
	p	141	168	171	152	152	41	21	57	160	260	205	175	1703 mm
0.1 mm	dp	9	10	11	9	10	3	2	4	12	15	13	11	109 days

Djambala

GAwab	092-5	Jan	Feb	Mar	Apr	May	Jun	Jul	Aug	Sep	Oct	Nov	Dec	Year
		latitude: 2°32′s				longitude: 14°46′e				height: 790 m				
	tx	31	31	31	31	30	29	29	31	30	29	29	29	32 °C
	t	23	23	24	24	23	22	21	22	23	23	23	23	23 °C
	tn	17	17	17	17	18	16	14	14	17	17	17	17	14 °C
	clothing	t	t	t	t	t	s	s	s	t	t	t	t	–
	stress	f	f	f	f	f	f	f	f	f	f	f	f	–
	p	174	191	242	244	183	24	10	38	139	150	246	248	1889 mm
0.1 mm	dp	13	12	15	14	11	2	1	3	9	16	18	17	131 days

Mpouya

Awaa	092-6	Jan	Feb	Mar	Apr	May	Jun	Jul	Aug	Sep	Oct	Nov	Dec	Year
		latitude: 2°37′s				longitude: 16°13′e				height: 312 m				
	t	25	26	26	26	26	25	24	26	26	25	25	25	25 °C
	clothing	t	t	t	t	t	t	t	t	t	t	t	t	–
	stress	ff	ff	ff	ff	ff	ff	ff	ff	ff	ff	ff	ff	–
	p	138	140	184	160	154	19	4	23	93	208	197	134	1454 mm
0.1 mm	dp	10	11	12	12	11	2	1	3	8	15	14	12	111 days

Sibiti

latitude: 3°41's longitude: 13°21'e height: 531 m

GAwab	092-7	Jan	Feb	Mar	Apr	May	Jun	Jul	Aug	Sep	Oct	Nov	Dec	Year
	t	24	24	24	24	23	21	20	20	22	24	24	24	23 °C
	clothing	t	t	t	t	t	s	s	s	s	t	t	t	–
	stress	ff	ff	ff	ff	ff	f	f	f	f	ff	ff	ff	–
	p	135	140	230	245	138	1	1	2	13	132	296	231	1564 mm
0.1 mm	dp	10	11	14	15	11	1	3	3	5	13	22	17	125 days

Mouyondzi

latitude: 4°00's longitude: 13°57'e height: 512 m

GAwab	092-8	Jan	Feb	Mar	Apr	May	Jun	Jul	Aug	Sep	Oct	Nov	Dec	Year
	t	24	24	25	25	24	22	21	22	23	24	24	24	23 °C
	clothing	t	t	t	t	t	s	s	s	t	t	t	t	–
	stress	ff	ff	ff	ff	ff	f	f	f	ff	ff	ff	ff	–
	p	136	85	158	175	133	1	0	1	17	102	257	194	1259 mm
0.1 mm	dp	11	8	13	14	11	1	0	0	3	10	17	13	101 days

Dolisie

latitude: 4°12's longitude: 12°40'e height: 330 m

Awab	092-9	Jan	Feb	Mar	Apr	May	Jun	Jul	Aug	Sep	Oct	Nov	Dec	Year
	t	25	26	26	26	25	23	21	22	23	25	25	25	24 °C
	clothing	t	t	t	t	t	t	s	s	t	t	t	t	–
	stress	ff	ff	ff	ff	ff	ff	f	f	ff	ff	ff	ff	–
	p	124	167	205	203	93	0	0	0	7	71	214	168	1252 mm
0.1 mm	dp	11	11	15	20	8	0	0	0	1	7	15	13	101 days

Brazzaville

latitude: 4°15's longitude: 15°14'e height: 314 m

Awab	092-10	Jan	Feb	Mar	Apr	May	Jun	Jul	Aug	Sep	Oct	Nov	Dec	Year
	tx	34	34	35	34	34	32	31	33	33	34	33	34	36 °C
	t	26	26	26	26	25	25	22	23	25	26	26	26	25 °C
	tn	19	19	19	20	19	15	13	14	17	18	19	19	13 °C
	clothing	t	t	t	t	t	t	s	t	t	t	t	t	–
	stress	ff	ff	ff	ff	ff	f	f	f	f	ff	ff	ff	–
	p	120	123	185	210	133	2	1	2	33	141	227	194	1371 mm
	dp	9	8	10	13	8	1	0	0	3	9	13	11	87 days
	sunshine	139	136	156	166	136	129	123	146	123	130	144	139	1667 hours
	%	36	39	41	46	37	36	33	39	34	34	39	36	38 %

Pointe – Noire

latitude: 4°49's longitude: 11°54'e height: 17 m

Awab	092-11	Jan	Feb	Mar	Apr	May	Jun	Jul	Aug	Sep	Oct	Nov	Dec	Year
	t	26	27	27	27	26	23	21	22	23	25	26	26	25 °C
	clothing	t	t	t	t	t	t	s	s	t	t	t	t	–
	stress	ff	ff	ff	ff	ff	f	f	f	f	ff	ff	ff	–
	p	152	210	227	179	89	1	0	2	13	71	167	143	1254 mm
	dp	9	10	12	11	5	0	0	1	4	10	12	9	83 days

4.3.92.2. Predominant general weather situations are described in 4.3.0.1. under Equatorial South Africa.

4.3.93. Gabon

4.3.93.1. Climatic description and tables

1. Tropical Summer or Winter Rain Climate Aw or As and Tropical Summer or Winter Rain Mountain Climate GAw or GAs (Maxima of rainfall mostly October/November.)
2. Permanently very hot at low levels.
3. In the morning often dry and sunny or cloudy, in the afternoon showers, repeatedly rainy days.
4. Wind: occasional gales, squalls in thunderstorms.

Bitam

latitude: 2°05′n longitude: 11°29′e height: 599 m

GAsaa	093-1	Jan	Feb	Mar	Apr	May	Jun	Jul	Aug	Sep	Oct	Nov	Dec	Year
	t	25	25	25	25	25	24	23	23	24	24	24	25	24 °C
	clothing	t	t	t	t	t	t	t	t	t	t	t	t	–
	stress	ff	ff	ff	ff	ff	f	f	f	f	ff	ff	ff	–
	p	53	68	201	200	247	130	23	43	267	329	226	70	1857 mm
0.1 mm	dp	4	5	10	10	12	8	2	4	12	17	12	5	101 days
	sunshine	118	121	124	123	140	102	93	71	93	112	120	124	1341 hours
	%	32	36	33	34	37	28	25	19	26	30	33	33	30 %

Mitzic

latitude: 0°47′n longitude: 11°32′e height: 583 m

GAsab	093-2	Jan	Feb	Mar	Apr	May	Jun	Jul	Aug	Sep	Oct	Nov	Dec	Year
	tx	32	32	33	33	32	31	30	30	31	32	31	31	33 °C
	t	24	25	25	25	24	23	22	22	24	24	24	24	24 °C
	tn	17	17	17	18	18	17	16	15	17	18	17	18	14 °C
	clothing	t	t	t	t	t	t	s	s	t	t	t	t	–
	stress	ff	ff	ff	ff	ff	ff	f	f	f	ff	ff	ff	–
	p	118	110	226	207	222	46	10	14	150	346	247	146	1842 mm

Makokou

latitude: 0°34′n longitude: 12°52′e height: 516 m

GAsab	093-3	Jan	Feb	Mar	Apr	May	Jun	Jul	Aug	Sep	Oct	Nov	Dec	Year
	t	24	25	25	25	25	23	22	22	24	24	24	24	24 °C
	clothing	t	t	t	t	t	t	s	s	t	t	t	t	–
	stress	ff	ff	ff	ff	ff	f	f	f	ff	ff	ff	ff	–
	p	62	137	245	236	174	48	7	17	141	321	224	144	1756 mm

Libreville

latitude: 0°27′n longitude: 9°25′e height: 10 m

Amaa	093-4	Jan	Feb	Mar	Apr	May	Jun	Jul	Aug	Sep	Oct	Nov	Dec	Year
	tx	32	33	33	33	33	32	31	31	31	32	32	32	34 °C
	t	27	27	27	27	27	25	24	24	26	26	26	26	26 °C
	tn	21	21	21	21	21	19	19	19	20	20	20	21	18 °C
	clothing	t	t	t	t	t	t	t	t	t	t	t	t	–
	stress	ff	ff	ff	ff	ff	ff	ff	ff	ff	ff	ff	ff	–
	p	331	305	410	363	291	18	1	9	113	384	506	389	3120 mm
0.1 mm	dp	15	16	18	19	14	3	1	4	13	23	22	17	165 days
	sunshine	164	175	167	171	149	126	127	118	90	105	132	167	1692 hours
	%	44	51	44	47	40	35	34	31	25	28	36	45	38 %

Port Gentil

latitude: 0°42′s longitude: 8°45′e height: 6 m

Awaa	093-5	Jan	Feb	Mar	Apr	May	Jun	Jul	Aug	Sep	Oct	Nov	Dec	Year
	t	27	27	27	27	26	24	23	24	25	26	26	26	26 °C
	clothing	t	t	t	t	t	t	t	t	t	t	t	t	–
	stress	ff	ff	ff	ff	ff	ff	ff	ff	ff	ff	ff	ff	–
	p	255	236	371	266	212	0	1	3	32	192	315	297	2180 mm
0.1 mm	dp	13	14	14	15	11	1	1	3	9	17	18	13	129 days
	sunshine	149	141	130	129	127	144	153	146	111	99	99	140	1568 hours
	%	40	41	35	36	34	40	41	39	31	26	27	37	35 %

Lambarene

latitude: 0°45′s longitude: 10°14′e height: 27 m

Awaa	093-6	Jan	Feb	Mar	Apr	May	Jun	Jul	Aug	Sep	Oct	Nov	Dec	Year
	t	27	27	27	27	26	24	23	24	25	26	26	26	26 °C
	clothing	t	t	t	t	t	t	t	t	t	t	t	t	–
	stress	ff	ff	ff	ff	ff	ff	f	ff	ff	ff	ff	ff	–
	p	156	134	261	246	231	8	4	4	67	348	372	244	2074 mm
0.1 mm	dp	10	10	16	13	12	1	1	2	8	21	21	14	129 days
	sunshine	136	133	140	141	133	60	62	65	54	99	117	105	1246 hours
	%	36	39	37	39	35	17	17	17	15	26	32	28	28 %

Lastoursville

latitude: 0°50′s longitude: 12°43′e height: 483 m

Awab	093-7	Jan	Feb	Mar	Apr	May	Jun	Jul	Aug	Sep	Oct	Nov	Dec	Year
	t	24	25	25	25	25	23	22	23	24	25	25	25	24 °C
	clothing	t	t	t	t	t	t	s	t	t	t	t	t	–
	stress	ff	ff	ff	ff	ff	f	f	f	ff	ff	ff	ff	–
	p	127	152	237	150	166	30	4	13	96	226	225	180	1606 mm

Franceville

latitude: 1°38′s longitude: 13°34′e height: 426 m

Awaa	093-8	Jan	Feb	Mar	Apr	May	Jun	Jul	Aug	Sep	Oct	Nov	Dec	Year
	t	25	25	25	26	25	23	23	23	25	25	25	25	24 °C
	clothing	t	t	t	t	t	t	t	t	t	t	t	t	–
	stress	ff	ff	ff	ff	ff	ff	ff	ff	ff	ff	ff	ff	–
	p	160	191	217	210	205	32	7	21	107	265	255	193	1863 mm
0.1 mm	dp	10	12	13	13	14	3	1	2	8	16	16	15	123 days
	sunshine	161	147	143	150	124	114	93	112	123	127	138	136	1568 hours
	%	43	43	38	41	33	32	25	30	34	34	38	36	35 %

Mouila

latitude: 1°53′s longitude: 11°01′e height: 89 m

Awaa	093-9	Jan	Feb	Mar	Apr	May	Jun	Jul	Aug	Sep	Oct	Nov	Dec	Year
	tx	34	35	36	35	34	32	31	32	34	34	36	34	36 °C
	t	27	27	28	28	27	24	23	24	25	27	27	27	26 °C
	tn	21	21	21	21	21	18	16	17	18	20	21	21	15 °C
	clothing	t	t	t	t	t	t	t	t	t	t	t	t	–
	stress	ff	ff	ff	ff	ff	ff	f	f	ff	ff	ff	ff	–
	p	225	224	253	232	172	23	4	6	44	395	396	278	2252 mm
0.1 mm	dp	15	15	15	14	15	4	2	4	7	21	20	16	148 days

Mayumba

Awab	093-10	Jan	Feb	Mar	Apr	May	Jun	Jul	Aug	Sep	Oct	Nov	Dec	Year
				latitude: 3°25′s				longitude: 10°39′e			height: 35 m			
	tx	31	31	32	31	31	27	27	28	27	29	29	29	32 °C
	t	26	26	27	27	26	23	22	22	24	25	25	26	25 °C
	tn	21	21	21	21	21	19	18	19	19	21	22	21	17 °C
	clothing	t	t	t	t	t	t	s	s	t	t	t	t	–
	stress	ff	ff	ff	ff	ff	ff	ff	ff	ff	ff	ff	ff	–
	p	250	326	310	188	107	0	0	3	38	259	411	206	2098 mm
0.1 mm	dp	12	13	15	9	7	0	0	3	11	21	20	13	124 days

4.3.93.2. Predominant general weather situations are described in 4.3.0.1. under Equatorial West Africa. A hygrothermal diagram of Libreville is to be found in 4.3.0.2.

4.3.94. Central Africa

4.3.94.1. Climatic descriptions and tables

1. Climatic area: small parts of the northern frontier zone
1.1. Steppe Climate BS and Steppe Mountain Climate GBS
1.2. Very hot throughout the year.
1.3. High sun period: generally dry and sunny in the morning, often showers in the afternoon; at times a rainy day. Low sun period: dry and sunny.
1.4. Wind occasional gales and dust-storms.

Birao

BSha	094-1	Jan	Feb	Mar	Apr	May	Jun	Jul	Aug	Sep	Oct	Nov	Dec	Year
				latitude: 10°17′n				longitude: 22°47′e			height: 465 m			
	tx	39	41	43	42	41	37	34	33	34	36	36	37	43 °C
	t	25	25	29	30	30	28	26	26	26	27	25	24	27 °C
	tn	8	11	14	13	19	17	19	19	19	16	10	7	6 °C
	clothing	t	t	t	t	t	t	t	t	t	t	t	t	–
	stress	f	f	f	f	ff	ff	ff	ff	ff	ff	f	f	–
	p	0	0	2	19	97	112	217	204	171	37	1	0	860 mm
0.1 mm	dp	0	0	1	2	7	9	12	15	10	4	0	0	60 days

2. Climatic area: most of the country (except parts of the northern frontier zone)
2.1. Tropical Summer Rain Climate Aw and Tropical Summer Rain Mountain Climate GAw
2.2. At low levels permanently very hot.
2.3. In the morning often dry and sunny, in the afternoon showers – very frequent during the high sun period, less during low sun period.
2.4. Wind: occasional gales, squalls in thunderstorms.

Ndele

GAwha	094-2	latitude: 8°24′n				longitude: 20°39′e				height: 510 m				
		Jan	Feb	Mar	Apr	May	Jun	Jul	Aug	Sep	Oct	Nov	Dec	Year
	tx	39	39	41	40	36	34	33	32	34	35	37	38	41 °C
	t	27	28	29	29	27	26	25	25	25	26	26	26	26 °C
	tn	18	18	19	20	19	19	19	19	18	18	17	16	15 °C
	clothing	t	t	t	t	t	t	t	t	t	t	t	t	–
	stress	ff	ff	ff	ff	ff	ff	ff	ff	ff	ff	f	f	–
	p	1	6	34	42	140	140	194	255	258	171	15	2	1258 mm
0.1 mm	dp	0	1	2	4	11	12	13	18	17	13	2	0	93 days

Bria

GAwaa	094-3	latitude: 6°32′n				longitude: 21°59′e				height: 584 m				
		Jan	Feb	Mar	Apr	May	Jun	Jul	Aug	Sep	Oct	Nov	Dec	Year
	tx	38	39	39	38	35	33	32	32	33	33	34	36	40 °C
	t	24	26	27	26	26	24	24	24	25	25	25	24	25 °C
	tn	11	12	14	16	18	18	18	18	17	17	13	11	9 °C
	clothing	t	t	t	t	t	t	t	t	t	t	t	t	–
	stress	f	f	f	f	ff	ff	ff	ff	ff	ff	f	f	–
	p	4	16	86	107	188	182	222	301	240	183	51	6	1586 mm
0.1 mm	dp	1	2	6	8	12	13	13	16	16	13	3	0	103 days

Bouca

Awha	094-4	latitude: 6°30′n				longitude: 18°16′e				height: 458 m				
		Jan	Feb	Mar	Apr	May	Jun	Jul	Aug	Sep	Oct	Nov	Dec	Year
	tx	38	41	39	39	37	34	33	33	33	34	35	36	41 °C
	t	24	27	28	28	27	26	25	25	25	26	26	24	26 °C
	tn	10	13	17	19	19	18	19	18	18	18	15	12	9 °C
	clothing	t	t	t	t	t	t	t	t	t	t	t	t	–
	stress	f	f	ff	ff	ff	ff	ff	ff	ff	ff	f	f	–
	p	2	7	48	94	136	165	242	241	232	195	34	5	1401 mm
0.1 mm	dp	0	1	4	6	9	10	13	14	14	13	3	1	88 days

Yalinda

GAwaa	094-5	latitude: 6°30′n				longitude: 23°16′e				height: 601 m				
		Jan	Feb	Mar	Apr	May	Jun	Jul	Aug	Sep	Oct	Nov	Dec	Year
	t	24	25	26	26	26	25	24	24	24	25	25	24	25 °C
	clothing	t	t	t	t	t	t	t	t	t	t	t	t	–
	stress	f	f	f	f	ff	ff	ff	ff	ff	ff	f	f	–
	p	3	21	73	97	151	205	214	252	273	226	39	7	1561 mm

Bossangoa

Awha	094-6	latitude: 6°29′n				longitude: 17°26′e				height: 465 m				
		Jan	Feb	Mar	Apr	May	Jun	Jul	Aug	Sep	Oct	Nov	Dec	Year
	t	25	27	29	28	27	26	26	25	26	26	26	26	26 °C
	clothing	t	t	t	t	t	t	t	t	t	t	t	t	–
	stress	f	f	ff	ff	ff	ff	ff	ff	ff	ff	ff	f	–
	p	3	10	44	77	138	158	217	305	231	199	20	2	1404 mm

Bouar

latitude: 5°56′n longitude: 15°35′e height: 936 m

GAwaa	094-7	Jan	Feb	Mar	Apr	May	Jun	Jul	Aug	Sep	Oct	Nov	Dec	Year
	tx	35	36	35	35	34	33	33	31	31	31	33	34	37 °C
	t	25	26	26	25	24	24	23	23	23	23	24	25	24 °C
	tn	16	17	17	17	17	16	16	16	16	16	16	15	15 °C
	clothing	t	t	t	t	t	t	t	t	t	t	t	t	–
	stress	f	f	f	f	f	f	f	f	f	f	f	f	–
	p	5	28	77	123	139	182	199	320	274	201	35	5	1588 mm
0.1 mm	dp	1	2	6	8	12	13	12	17	18	14	2	0	105 days

Bombari

latitude: 5°46′n longitude: 20°40′e height: 448 m

Awha	094-8	Jan	Feb	Mar	Apr	May	Jun	Jul	Aug	Sep	Oct	Nov	Dec	Year
	t	26	27	28	28	27	26	25	25	26	26	26	26	26 °C
	clothing	t	t	t	t	t	t	t	t	t	t	t	t	–
	stress	f	ff	ff	ff	ff	ff	ff	ff	ff	ff	ff	f	–
	p	7	40	98	103	161	169	214	264	162	227	57	20	1522 mm

Obo

latitude: 5°24′n longitude: 26°30′e height: 651 m

GAwaa	094-9	Jan	Feb	Mar	Apr	May	Jun	Jul	Aug	Sep	Oct	Nov	Dec	Year
	t	26	26	26	26	26	25	24	24	25	25	26	26	25 °C
	clothing	t	t	t	t	t	t	t	t	t	t	t	t	–
	stress	f	f	f	f	ff	ff	ff	ff	ff	ff	ff	f	–
	p	11	17	83	121	177	193	176	193	185	176	58	8	1398 mm

Bossembelé

latitude: 5°16′n longitude: 17°38′e height: 674 m

GAwaa	094-10	Jan	Feb	Mar	Apr	May	Jun	Jul	Aug	Sep	Oct	Nov	Dec	Year
	t	25	26	26	26	25	24	23	24	24	24	24	24	25 °C
	clothing	t	t	t	t	t	t	t	t	t	t	t	t	–
	stress	f	f	f	f	ff	ff	ff	ff	ff	ff	ff	f	–
	p	10	33	76	147	168	182	231	307	284	230	66	21	1755 mm

Bangassou

latitude: 4°44′n longitude: 22°50′e height: 500 m

GAwaa	094-11	Jan	Feb	Mar	Apr	May	Jun	Jul	Aug	Sep	Oct	Nov	Dec	Year
	tx	37	38	38	37	36	33	33	33	35	34	34	35	39 °C
	t	26	27	27	26	26	25	25	25	25	25	26	25	26 °C
	tn	14	15	17	18	19	18	18	18	17	17	16	13	13 °C
	clothing	t	t	t	t	t	t	t	t	t	t	t	t	–
	stress	f	f	ff	ff	ff	ff	ff	ff	ff	ff	f	f	–
	p	22	43	119	138	229	184	183	206	194	263	97	37	1715 mm
0.1 mm	dp	2	3	7	9	13	11	12	14	12	16	7	3	109 days
	sunshine	242	246	239	198	214	180	133	170	168	183	195	211	2379 hours
	%	66	73	64	54	56	48	35	45	46	49	55	57	54 %

Bangui

Awaa	094-12	latitude: 4°23'n Jan	Feb	Mar	Apr	longitude: 18°34'e May	Jun	Jul	Aug	Sep	height: 381 m Oct	Nov	Dec	Year
	tx	35	36	37	35	35	34	33	33	33	34	34	35	38 °C
	t	26	27	27	27	27	26	25	25	25	26	26	26	26 °C
	tn	16	18	19	19	19	19	18	18	19	18	18	16	14 °C
	clothing	t	t	t	t	t	t	t	t	t	t	t	t	–
	stress	f	ff	ff	ff	ff	ff	ff	ff	ff	ff	ff	f	–
	p	21	47	124	128	175	135	185	225	185	202	101	34	1560 mm
0.1 mm	dp	3	5	9	10	14	12	15	17	16	17	10	4	132 days
	sunshine	170	178	183	186	174	147	118	124	147	152	168	183	1930 hours
	%	46	53	49	51	46	40	31	33	40	41	47	50	44 %

Berbérati

GAwaa	094-13	latitude: 4°13'n Jan	Feb	Mar	Apr	longitude: 15°47'e May	Jun	Jul	Aug	Sep	height: 583 m Oct	Nov	Dec	Year
	tx	34	34	35	34	34	32	32	32	32	32	33	34	36 °C
	t	25	25	26	25	24	24	24	24	24	24	25	24	25 °C
	tn	14	15	18	18	18	18	17	17	17	18	16	14	13 °C
	clothing	t	t	t	t	t	t	t	t	t	t	t	t	–
	stress	f	f	ff	ff	ff	ff	ff	ff	ff	ff	f	f	–
	p	21	53	90	134	163	153	137	192	215	268	90	23	1539 mm
0.1 mm	dp	2	4	6	10	12	11	9	12	15	16	8	2	107 days
	sunshine	202	203	170	192	211	174	158	130	153	164	195	214	2167 hours
	%	55	60	45	52	55	47	41	34	42	44	54	58	49 %

4.3.94.2. Predominant general weather situations are described in 4.3.0.1. under Equatorial West Africa. A hygrothermal diagram of Bangui is to be found in 4.3.0.2.
Wind name: Harmattan is a dry and hot wind blowing from the northeast and carrying sand and laterite dust.

4.3.95. Chad

4.3.95.1. Climatic descriptions and tables

1. Climatic area: Northern half of the country
1.1. Desert Climate BW and Desert Mountain Climate GBW
1.2. Very hot throughout the year.
1.3. Generally dry and sunny; high sun period: frequent showers in the afternoon, less frequent in the north.
1.4. Wind: occasional gales and dust-storms.

Faya-Largeau

BWhb	095-1	latitude: 18°00'n Jan	Feb	Mar	Apr	longitude: 19°10'e May	Jun	Jul	Aug	Sep	height: 234 m Oct	Nov	Dec	Year
	tx	37	38	41	47	47	47	45	44	44	43	38	36	47 °C
	t	20	23	27	31	34	34	34	33	33	30	26	21	29 °C
	tn	7	8	13	16	19	19	19	20	21	17	12	7	6 °C
	clothing	s	t	t	t	t	t	t	t	t	t	t	s	–
	stress	o	o	f	ff	ff	ff	ff	ff	ff	ff	f	o	–
	p	0	0	0	0	0	2	1	11	2	0	0	0	16 mm
	dp	0	0	0	0	0	0	1	2	0	0	0	0	3 days

Mao

BWha	095-2	latitude: 14°07'n					longitude: 15°19'e				height: 356 m			
		Jan	Feb	Mar	Apr	May	Jun	Jul	Aug	Sep	Oct	Nov	Dec	Year
	t	23	25	30	33	34	33	30	28	30	30	28	26	29 °C
	clothing	t	t	t	t	t	t	t	t	t	t	t	t	–
	stress	o	o	f	f	ff	ff	ff	ff	f	f	f	f	–
	p	0	0	0	0	10	18	107	148	46	6	0	0	335 mm
	dp	0	0	0	0	1	3	8	12	5	1	0	0	30 days

Ati

BWha	095-3	latitude: 13°13'n					longitude: 18°19'e				height: 334 m			
		Jan	Feb	Mar	Apr	May	Jun	Jul	Aug	Sep	Oct	Nov	Dec	Year
	tx	40	43	46	46	46	44	39	36	37	41	41	40	47 °C
	t	24	26	31	33	34	33	29	28	29	29	28	26	29 °C
	tn	9	11	13	17	19	19	19	19	19	17	13	10	8 °C
	clothing	t	t	t	t	t	t	t	t	t	t	t	t	–
	stress	f	f	f	ff	ff	ff	ff	ff	ff	ff	f	f	–
	p	0	0	0	1	23	25	121	206	73	6	0	0	457 mm
	dp	0	0	0	0	2	4	10	15	6	1	0	0	38 days

2. *Climatic area:* central part of the country
2.1. Steppe Climate BS and Steppe Mountain Climate GBS
2.2. Very hot all the time.
2.3. Generally dry and sunny; high sun period: frequent showers in the afternoon.
2.4. Wind: occasional gales and dust-storms.

Abéché

GBSha	095-4	latitude: 13°51'n					longitude: 20°51'e				height: 549 m			
		Jan	Feb	Mar	Apr	May	Jun	Jul	Aug	Sep	Oct	Nov	Dec	Year
	tx	43	43	45	46	45	44	41	37	39	42	41	40	46 °C
	t	25	28	31	33	33	32	29	26	27	29	28	26	29 °C
	tn	11	13	15	17	19	21	18	18	17	16	14	12	11 °C
	clothing	t	t	t	t	t	t	t	t	t	t	t	t	–
	stress	f	f	f	f	ff	ff	ff	ff	f	f	f	f	–
	p	0	0	0	1	24	26	141	232	67	14	0	0	505 mm
	dp	0	0	0	1	3	6	12	17	7	1	0	0	47 days
	sunshine	313	294	295	297	307	258	239	202	261	310	306	316	3398 hours
	%	89	89	79	80	78	66	60	52	71	84	89	90	77 %

Mongo

BSha	095-5	latitude: 12°11'n					longitude: 18°41'e				height: 428 m			
		Jan	Feb	Mar	Apr	May	Jun	Jul	Aug	Sep	Oct	Nov	Dec	Year
	t	27	28	32	34	33	30	28	26	27	29	29	28	29 °C
	clothing	t	t	t	t	t	t	t	t	t	t	t	t	–
	stress	f	f	f	f	f	ff	ff	ff	f	f	f	f	–
	p	0	0	2	12	57	64	207	325	155	35	1	0	856 mm
	dp	0	0	1	2	5	9	15	19	12	4	0	0	67 days

Ndjemena (Fort Lamy)

BSha	095-6	Jan	Feb	Mar	Apr	May	Jun	Jul	Aug	Sep	Oct	Nov	Dec	Year
		latitude: 12°08'n				longitude: 15°02'e				height: 295 m				
	tx	39	41	43	44	43	41	39	34	36	39	39	37	45 °C
	t	24	26	30	33	32	30	28	26	27	29	27	24	28 °C
	tn	10	12	16	18	21	21	19	20	20	17	13	10	10 °C
	clothing	t	t	t	t	t	t	t	t	t	t	t	t	–
	stress	f	f	f	ff	ff	ff	ff	ff	ff	ff	f	f	–
	p	0	0	0	5	36	66	156	257	104	23	1	0	646 mm
	dp	0	0	0	1	5	8	13	18	10	3	0	0	58 days
	sunshine	304	288	282	270	288	252	214	177	228	291	303	298	3195 hours
	%	85	87	75	73	73	65	54	46	62	79	87	84	72 %

3. Climatic area: South of the country

3.1. Tropical Summer Rain Climate Aw and Tropical Summer Rain Mountain Climate GAw

3.2. Summer very hot, winter hot.

3.3. High sun period: often dry and sunny in the morning, showers in the afternoon; at times rainy days. Low sun period: generally dry and sunny.

3.4. Wind: occasional gales and dust-storms, squalls in thunderstorms.

Am-Timan

Awha	095-7	Jan	Feb	Mar	Apr	May	Jun	Jul	Aug	Sep	Oct	Nov	Dec	Year
		latitude: 11°02'n				longitude: 20°17'e				height: 436 m				
	tx	41	42	43	43	41	38	36	34	34	38	38	39	43 °C
	t	25	25	29	30	30	28	26	25	26	26	26	25	27 °C
	tn	9	11	15	14	18	19	18	18	18	16	11	8	7 °C
	clothing	t	t	t	t	t	t	t	t	t	t	t	t	–
	stress	f	f	f	f	ff	ff	ff	ff	ff	f	f	f	–
	p	0	4	8	19	80	123	182	311	154	38	0	0	918 mm
	dp	0	0	1	2	7	11	15	18	11	4	0	0	69 days

Bousso

Awha	095-8	Jan	Feb	Mar	Apr	May	Jun	Jul	Aug	Sep	Oct	Nov	Dec	Year
		latitude: 10°29'n				longitude: 16°43'e				height: 336 m				
	t	25	27	31	33	32	29	27	26	26	28	27	26	28 °C
	clothing	t	t	t	t	t	t	t	t	t	t	t	t	–
	stress	f	f	f	ff	ff	ff	ff	ff	ff	ff	f	f	–
	p	0	0	3	21	64	105	280	326	189	47	6	0	1040 mm
	dp	0	0	1	3	6	9	12	15	12	3	0	0	61 days
	sunshine	307	280	270	246	264	222	198	174	210	251	278	299	2999 hours
	%	86	84	72	66	68	58	50	45	57	68	79	84	68 %

Pala

Awha	095-9	Jan	Feb	Mar	Apr	May	Jun	Jul	Aug	Sep	Oct	Nov	Dec	Year
		latitude: 9°22'n				longitude: 14°55'e				height: 455 m				
	t	26	27	31	31	30	27	26	25	25	27	28	26	27 °C
	clothing	t	t	t	t	t	t	t	t	t	t	t	t	–
	stress	f	f	f	f	ff	ff	ff	ff	ff	f	f	f	–
	p	0	0	10	32	94	156	230	224	210	84	4	0	1043 mm
	dp	0	0	1	5	10	12	15	18	17	7	1	0	86 days
	sunshine	319	285	285	243	267	213	195	158	201	248	291	316	3021 hours
	%	88	85	76	66	69	56	50	41	55	67	83	88	68 %

Sarh (Fort Archambault)

Awha	095-10	Jan	Feb	Mar	Apr	May	Jun	Jul	Aug	Sep	Oct	Nov	Dec	Year
				latitude: 9°08′n				longitude: 18°23′e				height: 365 m		
	t	27	28	31	32	30	27	26	26	26	27	27	26	28 °C
	clothing	t	t	t	t	t	t	t	t	t	t	t	t	–
	stress	f	f	f	ff	ff	ff	ff	ff	ff	ff	f	f	–
	p	0	1	8	39	105	143	226	311	225	80	3	0	1140 mm
	dp	0	0	1	5	9	12	16	20	17	8	1	0	89 days
	sunshine	276	260	239	222	236	195	164	158	180	226	270	276	2702 hours
	%	76	78	64	60	61	51	42	41	49	61	77	77	61 %

Moundou

Awha	095-11	Jan	Feb	Mar	Apr	May	Jun	Jul	Aug	Sep	Oct	Nov	Dec	Year
				latitude: 8°37′n				longitude: 16°04′e				height: 422 m		
	t	25	27	30	31	29	27	26	25	25	27	27	25	27 °C
	clothing	t	t	t	t	t	t	t	t	t	t	t	t	–
	stress	f	f	f	f	ff	ff	ff	ff	ff	ff	f	f	–
	p	0	0	2	40	118	171	244	303	250	96	4	0	1228 mm
	dp	0	0	1	5	8	11	14	17	16	8	1	0	81 days
	sunshine	288	260	251	213	239	198	183	158	177	226	285	298	2776 hours
	%	80	78	67	58	62	52	47	41	48	61	81	83	63 %

4.3.95.2. Predominant general weather situations are described in 4.3.0.1. under Equatorial West Africa. A hygrothermal diagram of Ndjemena is to be found in 4.3.0.2.
Wind name: Harmattan is a hot dry wind blowing from the northeast and carrying sand and laterite dust.

4.3.96. Equatorial Guinea

4.3.96.1. Climatic description and tables

1. Tropical Rain Climate Ar and Tropical Rain Mountain Climate GAr and Tropical Monsoonal Rain Climate Am and Tropical Monsoonal Rain Mountain Climate GAm
2. Very hot throughout the year.
3. Now and then rainy days; often showers in the afternoon.
4. Wind: squalls in thunderstorms, gales rare.

Santa Isabel / Fernando Po

Amaa	096-1	Jan	Feb	Mar	Apr	May	Jun	Jul	Aug	Sep	Oct	Nov	Dec	Year
				latitude: 3°41′n				longitude: 8°46′e				height: 50 m		
	t	25	26	26	26	25	25	24	24	24	24	25	25	25 °C
	clothing	t	t	t	t	t	t	t	t	t	t	t	t	–
	stress	ff	ff	ff	ff	ff	ff	ff	ff	ff	ff	ff	ff	–
	p	42	33	110	187	179	224	284	188	277	238	92	36	1890 mm
0.1 mm	dp	4	4	11	14	18	21	21	18	23	20	13	5	172 days
	sunshine	149	153	108	120	118	69	46	59	48	68	99	140	1177 hours
	%	40	45	29	33	31	19	12	16	13	18	28	38	27 %

Bata

Amaa	096-2	Jan	Feb	Mar	Apr	May	Jun	Jul	Aug	Sep	Oct	Nov	Dec	Year
		latitude: 1°54′n				longitude: 9°48′e				height: 2 m				
	t	26	26	26	26	26	25	24	24	25	25	25	25	25 °C
	clothing	t	t	t	t	t	t	t	t	t	t	t	t	–
	stress	ff	ff	ff	ff	ff	ff	ff	ff	ff	ff	ff	ff	–
	p	128	94	223	288	247	65	25	28	240	436	307	128	2209 mm
0.1 mm	dp	11	9	14	17	17	8	3	6	18	25	19	11	159 days
	sunshine	202	192	174	177	189	147	143	143	114	115	141	186	1921 hours
	%	54	56	46	49	50	40	38	38	31	31	39	50	43 %

4.3.96.2. Predominant general weather situations are described in 4.3.0.1. under Equatorial West Africa.

4.3.97. Cameroun

4.3.97.1. Climatic descriptions and tables

1. Climatic area: Greater part of the country except the south-west

1.1. Tropical Summer Rain Climate Aw and
Tropical Summer Rain Mountain Climate GAw

1.2. Permanently very hot.

1.3. Generally in the morning dry and sunny, in the afternoon during the high sun period very often showers, fewer during low sun period.

1.4. Wind: occasional gales, squalls in thunderstorms.

Garoua

Awha	097-1	Jan	Feb	Mar	Apr	May	Jun	Jul	Aug	Sep	Oct	Nov	Dec	Year
		latitude: 9°20′n				longitude: 13°23′e				height: 247 m				
	t	26	28	32	33	30	28	26	26	26	28	28	26	28 °C
	clothing	t	t	t	t	t	t	t	t	t	t	t	t	–
	stress	f	ff	ff	ff	ff	ff	ff	ff	ff	ff	f	f	–
	p	0	1	4	35	135	157	159	199	220	76	2	2	990 mm
	dp	0	0	1	3	9	10	10	13	14	7	0	0	67 days
	sunshine	281	269	260	236	256	219	194	170	195	266	287	299	2952 hours
	%	78	80	69	64	66	58	50	44	53	72	82	83	67 %

Ngaundéré

GAwab	097-2	Jan	Feb	Mar	Apr	May	Jun	Jul	Aug	Sep	Oct	Nov	Dec	Year
		latitude: 7°17′n				longitude: 13°21′e				height: 1119 m				
	tx	33	34	34	34	31	29	28	28	29	31	32	31	36 °C
	t	22	23	24	24	23	22	21	21	22	22	22	21	22 °C
	tn	10	11	12	15	16	15	15	15	14	14	11	10	9 °C
	clothing	s	s	s	s	s	s	s	s	s	s	s	s	–
	stress	o	o	o	o	o	o	o	o	o	o	o	o	–
	p	2	3	43	137	205	207	256	270	237	133	16	2	1511 mm
	dp	0	1	4	11	18	19	19	20	20	14	2	0	128 days
	sunshine	274	254	218	178	183	149	117	118	128	173	251	286	2329 hours
	%	75	76	58	48	47	40	30	31	35	47	71	79	53 %

Koundja

GAwab	097-3	Jan	Feb	Mar	Apr	May	Jun	Jul	Aug	Sep	Oct	Nov	Dec	Year
		latitude: 5°37′n						longitude: 10°45′e				height: 1217 m		
	tx	32	32	32	31	28	27	27	27	27	28	29	30	33 °C
	t	22	23	23	22	22	20	20	20	20	21	21	22	21 °C
	tn	13	14	14	15	14	14	14	13	14	14	14	13	13 °C
	clothing	s	s	s	s	s	s	s	s	s	s	s	s	–
	stress	o	o	o	o	o	o	o	o	o	o	o	o	–
	p	7	48	118	155	208	218	302	324	385	248	65	18	2096 mm
	dp	0	4	10	14	17	17	19	22	23	22	8	2	156 days
	sunshine	256	239	214	198	213	171	129	127	132	178	233	268	2358 hours
	%	70	71	57	54	55	46	33	33	36	48	65	73	53 %

Batouri

GAwaa	097-4	Jan	Feb	Mar	Apr	May	Jun	Jul	Aug	Sep	Oct	Nov	Dec	Year
		latitude: 4°25′n						longitude: 14°24′e				height: 655 m		
	tx	33	34	35	34	34	32	31	31	32	32	33	33	35 °C
	t	23	23	25	25	25	24	23	23	24	24	24	23	24 °C
	tn	13	14	16	17	17	17	16	17	17	16	16	14	13 °C
	clothing	t	t	t	t	t	t	t	t	t	t	t	t	–
	stress	f	f	f	f	f	f	f	f	f	f	f	f	–
	p	34	55	114	150	202	170	102	157	221	281	127	35	1648 mm
	dp	2	4	8	10	13	13	9	11	16	19	9	3	117 days
	sunshine	161	165	156	174	192	139	105	94	112	138	178	183	1797 hours
	%	44	49	42	47	50	37	27	25	31	37	50	50	41 %

2. Climatic area: the south-west of the country

2.1. Tropical Rain Climate Ar and Tropical Rain Mountain Climate GAr

2.2. Permanently very hot.

2.3. At times rainy days during the high sun period; otherwise in the morning often dry and sunny or cloudy, in the afternoon frequent showers. Low sun period: showers now and then.

2.4. Wind: occasional gales, squalls in thunderstorms.

Douala

Araa	097-5	Jan	Feb	Mar	Apr	May	Jun	Jul	Aug	Sep	Oct	Nov	Dec	Year
		latitude: 4°07′n						longitude: 9°43′e				height: 13 m		
	tx	31	32	32	32	32	31	29	29	30	31	31	31	33 °C
	t	27	27	27	27	27	26	25	25	25	26	26	27	26 °C
	tn	21	21	21	21	21	21	21	21	21	21	21	21	20 °C
	clothing	t	t	t	t	t	t	t	t	t	t	t	t	–
	stress	ff	ff	ff	ff	ff	ff	ff	ff	ff	ff	ff	ff	–
	p	61	88	226	240	353	472	710	726	628	399	146	60	4109 mm
	dp	5	7	13	16	19	22	28	29	26	24	12	5	206 days
	sunshine	124	140	134	149	132	88	41	39	68	110	122	127	1274 hours
	%	34	41	36	41	35	24	11	10	19	29	34	34	29 %

Yaounde

GArab	097-6	Jan	Feb	Mar	Apr	May	Jun	Jul	Aug	Sep	Oct	Nov	Dec	Year
		latitude: 3°52'n			longitude: 11°32'e				height: 760 m					
	tx	32	32	32	32	31	30	29	29	29	31	30	31	33 °C
	t	24	25	24	24	24	23	22	22	23	23	23	24	24 °C
	tn	16	16	17	17	17	17	17	17	17	16	17	17	15 °C
	clothing	t	t	t	t	t	t	t	t	t	t	t	t	–
	stress	f	f	f	f	f	f	f	f	f	f	f	f	–
	p	22	63	146	182	204	151	56	74	202	300	127	20	1547 mm
	dp	2	5	12	14	17	13	6	6	17	22	13	3	130 days
	sunshine	169	160	151	152	149	118	89	88	103	128	153	177	1637 hours
	%	46	47	40	41	39	32	23	23	28	34	43	48	37 %

4.3.97.2. Predominant general weather situations are described in 4.3.0.1. under Equatorial West Africa. A hygrothermal diagram of Yaounde is to be found in 4.3.0.2.

4.3.98. Nigeria

4.3.98.1. Climatic descriptions and tables

1. Climatic area: Northernmost part of the country

1.1. Steppe Climate BS and Steppe Mountain Climate GBS

1.2. Summer very hot, winter hot, in lowlands.

1.3. Generally dry and sunny in the morning; high sun period: frequent showers in the afternoon; at times rainy days. Low sun period: dry and sunny.

1.4. Wind: occasional gales and dust-storms, squalls in thunderstorms.

Kano

BShb	098-1	Jan	Feb	Mar	Apr	May	Jun	Jul	Aug	Sep	Oct	Nov	Dec	Year
		latitude: 12°03'n			longitude: 8°32'e				height: 481 m					
	tx	36	38	41	42	41	38	34	33	34	35	37	35	41 °C
	t	21	24	28	31	30	28	26	25	26	27	25	22	26 °C
	tn	9	11	14	19	21	19	19	18	19	16	13	11	9 °C
	clothing	s	t	t	t	t	t	t	t	t	t	t	s	–
	stress	o	f	f	ff	ff	ff	ff	ff	ff	f	f	o	–
	p	0	0	2	8	71	119	209	311	137	14	0	0	872 mm
	dp	0	0	0	1	5	9	13	18	11	1	0	0	58 days
	sunshine	273	249	264	246	276	267	232	180	240	298	297	291	3112 hours
	%	77	75	71	66	70	69	59	46	66	81	85	82	70 %

Maiduguri

BShb	098-2	Jan	Feb	Mar	Apr	May	Jun	Jul	Aug	Sep	Oct	Nov	Dec	Year
		latitude: 11°51'n			longitude: 11°05'e				height: 354 m					
	tx	38	39	41	43	42	39	37	34	36	38	38	36	43 °C
	t	22	24	29	31	32	30	27	26	27	28	25	22	27 °C
	tn	9	10	12	18	21	21	19	19	19	16	11	9	8 °C
	clothing	s	t	t	t	t	t	t	t	t	t	t	s	–
	stress	f	f	f	ff	ff	ff	ff	ff	ff	f	f	f	–
	p	0	0	1	4	34	78	180	227	112	23	0	0	659 mm
	dp	0	0	0	1	5	8	13	17	10	3	0	0	57 days
	sunshine	291	277	276	264	273	258	214	180	225	294	303	307	3162 hours
	%	82	83	74	71	69	67	54	46	61	80	87	87	71 %

2. *Climatic area:* greater part of the country except northernmost part
2.1. Tropical Summer Rain Climate Aw and Tropical Summer Rain Mountain Climate GAw (coastal area partly Tropical Rain Climate Ar or Tropical Monsoonal Rain Climate Am)
2.2. Permanently very hot.
2.3. In the morning generally dry and sunny, in the afternoon during the high sun period often showers; at times rainy the whole day. Low sun period: often dry and sunny the whole day.
2.4. Wind: occasional gales, squalls in thunderstorms.

Jos

latitude: 9°52′n longitude: 8°54′e height: 1285 m

GAwab	098-3	Jan	Feb	Mar	Apr	May	Jun	Jul	Aug	Sep	Oct	Nov	Dec	Year
	tx	32	33	34	34	33	30	28	28	29	31	31	31	35 °C
	t	21	22	24	24	23	22	20	20	21	22	22	21	22 °C
	tn	11	12	15	17	16	16	16	16	16	16	13	11	10 °C
	clothing	s	s	t	t	t	s	s	s	s	s	s	s	–
	stress	o	o	f	f	f	f	f	f	f	f	f	o	–
	p	2	4	24	93	205	229	318	274	219	39	5	2	1414 mm
	dp	0	0	2	7	15	17	23	22	19	6	0	0	111 days
	sunshine	301	274	264	207	208	195	143	124	165	248	297	313	2738 hours
	%	84	82	71	56	53	51	36	32	45	67	85	87	62 %

Minna

latitude: 9°37′n longitude: 6°32′e height: 260 m

Awha	098-4	Jan	Feb	Mar	Apr	May	Jun	Jul	Aug	Sep	Oct	Nov	Dec	Year
	tx	38	39	41	39	37	35	33	32	33	35	37	38	41 °C
	t	28	29	30	30	28	26	25	25	25	26	27	27	27 °C
	tn	16	17	19	20	21	19	19	19	19	19	16	15	14 °C
	clothing	t	t	t	t	t	t	t	t	t	t	t	t	–
	stress	f	f	ff	ff	ff	ff	ff	ff	ff	ff	f	f	–
	p	1	5	22	58	138	177	210	241	276	140	10	4	1281 mm
	dp	0	0	1	5	11	14	16	19	20	13	1	0	100 days
	sunshine	264	246	254	234	239	195	124	112	159	254	288	285	2653 hours
	%	73	74	68	63	61	51	32	29	43	69	82	79	60 %

Yola

latitude: 9°14′n longitude: 12°28′e height: 174 m

Awha	098-5	Jan	Feb	Mar	Apr	May	Jun	Jul	Aug	Sep	Oct	Nov	Dec	Year
	tx	39	41	42	42	39	36	34	33	33	33	38	38	42 °C
	t	26	28	32	32	30	27	26	26	26	28	27	26	28 °C
	tn	14	16	19	22	21	20	20	20	20	19	15	14	13 °C
	clothing	t	t	t	t	t	t	t	t	t	t	t	t	–
	stress	f	f	ff	ff	ff	ff	ff	ff	ff	ff	f	f	–
	p	0	0	8	39	137	154	154	180	203	64	4	1	944 mm
	dp	0	0	0	4	9	10	11	16	13	7	0	0	70 days
	sunshine	288	268	254	234	257	231	192	167	186	267	297	304	2946 hours
	%	80	80	68	63	66	61	49	43	51	72	85	85	66 %

Ilorin

Awha	098-6	Jan	Feb	Mar	Apr	May	Jun	Jul	Aug	Sep	Oct	Nov	Dec	Year
latitude: 8°29′n	longitude: 4°35′e	height: 305 m												
	tx	37	38	38	37	35	33	32	32	32	33	35	36	38 °C
	t	26	28	29	28	27	26	25	25	25	26	27	26	27 °C
	tn	14	14	19	20	19	19	19	19	19	19	15	13	12 °C
	clothing	t	t	t	t	t	t	t	t	t	t	t	t	–
	stress	f	f	ff	ff	ff	ff	ff	ff	ff	ff	f	f	–
	p	12	15	68	93	167	193	131	138	251	171	28	11	1277 mm
	dp	1	1	5	7	11	14	10	9	16	10	3	1	88 days
	sunshine	229	226	229	204	214	186	127	112	123	192	228	245	2314 hours
	%	63	67	61	55	55	49	33	29	34	52	65	68	52 %

Makurdi

Awha	098-7	Jan	Feb	Mar	Apr	May	Jun	Jul	Aug	Sep	Oct	Nov	Dec	Year
latitude: 7°41′n	longitude: 8°37′e	height: 97 m												
	tx	37	38	39	38	36	33	32	32	32	33	35	36	39 °C
	t	26	28	30	29	28	26	26	26	26	26	27	25	27 °C
	tn	14	16	19	21	20	19	20	20	20	20	16	13	12 °C
	clothing	t	t	t	t	t	t	t	t	t	t	t	t	–
	stress	f	f	ff	ff	ff	ff	ff	ff	ff	ff	f	f	–
	p	6	6	33	107	186	195	156	160	279	155	15	5	1299 mm
	dp	0	0	2	5	10	11	11	14	16	11	1	0	81 days
	sunshine	236	226	223	216	223	177	143	118	144	198	228	248	2380 hours
	%	65	67	60	59	58	47	37	31	39	53	64	68	54 %

Lagos / Ikeja

Amha	098-8	Jan	Feb	Mar	Apr	May	Jun	Jul	Aug	Sep	Oct	Nov	Dec	Year
latitude: 6°35′n	longitude: 3°20′e	height: 38 m												
	tx	33	33	34	33	33	31	30	30	31	31	33	33	35 °C
	t	27	28	28	27	27	26	24	24	25	26	27	27	27 °C
	tn	20	22	22	22	22	22	22	21	22	21	22	21	19 °C
	clothing	t	t	t	t	t	t	t	t	t	t	t	t	–
	stress	ff	ff	ff	ff	ff	ff	ff	ff	ff	ff	ff	ff	–
	p	40	57	100	115	215	336	150	59	214	222	77	41	1625 mm
	dp	1	3	7	8	14	18	13	9	16	15	7	2	113 days
	sunshine	183	192	195	183	174	114	87	102	90	158	198	202	1878 hours
	%	50	57	52	50	45	30	22	27	25	43	56	55	42 %

Enugu

Awha	098-9	Jan	Feb	Mar	Apr	May	Jun	Jul	Aug	Sep	Oct	Nov	Dec	Year
latitude: 6°28′n	longitude: 7°33′e	height: 137 m												
	tx	34	36	36	36	34	32	31	31	31	33	33	34	36 °C
	t	26	28	29	28	27	26	26	26	26	26	27	26	27 °C
	tn	18	19	21	21	20	19	20	20	20	19	19	17	17 °C
	clothing	t	t	t	t	t	t	t	t	t	t	t	t	–
	stress	ff	ff	ff	ff	ff	ff	ff	ff	ff	ff	ff	f	–
	p	19	15	81	209	195	166	182	190	282	246	53	23	1661 mm
	dp	1	2	5	9	14	16	16	15	19	16	3	1	117 days
	sunshine	208	212	195	186	198	159	121	108	117	167	216	208	2096 hours
	%	57	63	52	51	51	42	31	28	32	45	61	57	47 %

Warri

latitude: 5°31′n longitude: 5°44′e height: 6 m

Arha	098-10	Jan	Feb	Mar	Apr	May	Jun	Jul	Aug	Sep	Oct	Nov	Dec	Year
	t	27	28	28	28	28	26	25	25	26	26	28	27	27 °C
	clothing	t	t	t	t	t	t	t	t	t	t	t	t	–
	stress	ff	ff	ff	ff	ff	ff	ff	ff	ff	ff	ff	ff	–
	p	36	63	151	228	240	334	492	298	573	312	110	58	2897 mm
0.2 mm	dp	3	4	10	13	19	23	24	19	25	23	10	3	176 days
	sunshine	133	141	133	150	155	105	81	105	72	127	180	186	1569 hours
	%	36	42	35	41	40	28	21	28	20	34	51	51	35 %

Port Harcourt

latitude: 4°51′n longitude: 7°01′e height: 18 m

Araa	098-11	Jan	Feb	Mar	Apr	May	Jun	Jul	Aug	Sep	Oct	Nov	Dec	Year
	tx	33	34	34	34	33	32	32	31	32	32	33	33	36 °C
	t	26	27	27	27	27	26	25	25	25	26	26	26	26 °C
	tn	18	20	21	20	21	20	21	20	21	21	22	19	17 °C
	clothing	t	t	t	t	t	t	t	t	t	t	t	t	–
	stress	ff	ff	ff	ff	ff	ff	ff	ff	ff	ff	ff	ff	–
	p	29	68	187	175	249	269	332	278	442	266	93	32	2421 mm
	dp	3	6	12	14	18	21	22	21	25	21	12	4	179 days
	sunshine	161	161	140	141	149	99	65	78	63	108	138	180	1482 hours
	%	44	48	37	38	39	27	17	20	17	29	39	49	33 %

4.3.98.2. Predominant general weather situations are described in 4.3.0.1. under Equatorial West Africa. A hygrothermal diagram of Lagos is to be found in 4.3.0.2.
Wind name: Harmattan is a hot dry wind blowing from the northeast and carrying sand and laterite dust.

4.3.99. Benin

4.3.99.1. Climatic description and tables

1. Tropical Summer Rain Climate Aw
2. Permanently very hot.
3. Apart from some rainy days in the high sun period, often dry and sunny in the morning and frequent showers in the afternoon, which are mostly lacking in the low sun period.
4. Wind: occasional gales, squalls in thunderstorms.

Kandi

latitude: 11°08′n longitude: 2°56′e height: 292 m

Awha	099-1	Jan	Feb	Mar	Apr	May	Jun	Jul	Aug	Sep	Oct	Nov	Dec	Year
	t	25	28	31	32	30	28	26	25	26	28	27	25	28 °C
	clothing	t	t	t	t	t	t	t	t	t	t	t	t	–
	stress	f	f	f	f	ff	ff	ff	ff	ff	ff	f	f	–
	p	0	1	8	29	92	150	191	282	219	52	2	0	1026 mm
	dp	0	0	1	2	7	10	13	18	16	5	0	0	74 days
	sunshine	282	260	267	246	257	234	195	158	195	276	282	282	2934 hours
	%	79	78	71	66	66	61	49	41	53	75	81	79	66 %

Natitingou

Awha	099-2	Jan	Feb	Mar	Apr	May	Jun	Jul	Aug	Sep	Oct	Nov	Dec	Year
				latitude: 10°19′n				longitude: 1°23′e			height: 461 m			
	tx	37	38	39	38	36	33	33	30	31	33	33	34	40 °C
	t	27	28	30	29	28	26	25	24	25	26	26	26	27 °C
	tn	15	17	18	19	19	19	18	19	18	18	18	17	14 °C
	clothing	t	t	t	t	t	t	t	t	t	t	t	t	–
	stress	f	f	ff	ff	ff	ff	ff	ff	ff	ff	ff	ff	–
	p	3	8	26	75	126	162	221	254	311	118	33	5	1340 mm
	dp	0	1	2	6	10	12	14	16	19	10	3	0	94 days
	sunshine	294	260	254	234	239	195	152	108	144	229	273	273	2656 hours
	%	82	78	68	63	61	51	39	28	39	62	78	76	60 %

Parakou

Awha	099-3	Jan	Feb	Mar	Apr	May	Jun	Jul	Aug	Sep	Oct	Nov	Dec	Year
				latitude: 9°21′n				longitude: 2°37′e			height: 393 m			
	tx	37	36	43	43	42	39	37	35	36	39	39	37	43 °C
	t	27	28	29	28	27	26	25	24	25	26	26	26	27 °C
	tn	14	16	19	19	19	19	20	19	19	18	17	15	13 °C
	clothing	t	t	t	t	t	t	t	t	t	t	t	t	–
	stress	f	f	ff	ff	ff	ff	ff	ff	ff	ff	ff	f	–
	p	3	12	39	80	131	162	180	198	247	111	10	4	1177 mm
	dp	0	1	3	5	8	10	10	11	13	8	1	0	71 days

Tchaourou

Awha	099-4	Jan	Feb	Mar	Apr	May	Jun	Jul	Aug	Sep	Oct	Nov	Dec	Year
				latitude: 8°52′n				longitude: 2°36′e			height: 327 m			
	tx	37	38	38	37	36	34	31	31	32	33	34	36	38 °C
	t	27	28	29	28	27	26	25	24	25	26	27	26	27 °C
	tn	15	17	21	20	20	19	19	19	19	19	15	13	13 °C
	clothing	t	t	t	t	t	t	t	t	t	t	t	t	–
	stress	f	f	ff	ff	ff	ff	ff	ff	ff	ff	f	f	–
	p	7	15	58	104	141	161	165	162	214	161	16	7	1211 mm
0.1 mm	dp	1	1	4	8	12	16	16	15	18	15	2	1	109 days
	sunshine	242	237	239	219	214	171	108	87	120	186	225	242	2290 hours
	%	67	71	64	59	55	45	28	23	33	50	64	67	52 %

Save

Awha	099-5	Jan	Feb	Mar	Apr	May	Jun	Jul	Aug	Sep	Oct	Nov	Dec	Year
				latitude: 7°59′n				longitude: 2°26′e			height: 200 m			
	t	28	29	29	28	27	26	25	25	26	26	28	28	27 °C
	clothing	t	t	t	t	t	t	t	t	t	t	t	t	–
	stress	ff	ff	ff	ff	ff	ff	ff	ff	ff	ff	ff	ff	–
	p	7	19	79	108	145	141	139	111	174	133	33	8	1097 mm
	dp	1	2	5	8	10	11	10	8	13	11	2	1	82 days
	sunshine	242	240	239	228	232	183	108	93	129	195	234	241	2366 hours
	%	67	72	64	62	60	48	28	24	35	53	66	67	53 %

Bohicon

Awha	099-6	Jan	Feb	Mar	Apr	May	Jun	Jul	Aug	Sep	Oct	Nov	Dec	Year
				latitude: 7°10'n				longitude: 2°04'e				height: 167 m		
	t	28	29	29	28	28	26	25	25	26	27	28	27	27 °C
	clothing	t	t	t	t	t	t	t	t	t	t	t	t	–
	stress	ff	ff	ff	ff	ff	ff	ff	ff	ff	ff	ff	ff	–
	p	11	25	98	134	147	148	116	68	140	149	40	14	1090 mm
	dp	1	2	7	8	11	10	8	5	11	12	4	1	79 days

Cotonou

Awha	099-7	Jan	Feb	Mar	Apr	May	Jun	Jul	Aug	Sep	Oct	Nov	Dec	Year
				latitude: 6°21'n				longitude: 2°23'e				height: 13 m		
	tx	31	32	32	32	32	30	29	28	29	31	32	32	34 °C
	t	27	28	28	28	27	26	25	25	26	26	27	27	27 °C
	tn	21	22	22	23	22	21	22	22	22	22	23	21	19 °C
	clothing	t	t	t	t	t	t	t	t	t	t	t	t	–
	stress	ff	ff	ff	ff	ff	ff	ff	ff	ff	ff	ff	ff	–
	p	36	51	104	134	201	338	120	22	82	164	68	19	1339 mm
	dp	1	2	5	8	12	21	12	5	6	8	4	1	83 days
	sunshine	217	212	220	210	211	138	121	155	159	195	237	229	2304 hours
	%	59	63	59	57	55	37	31	41	44	52	67	63	52 %

4.3.99.2. Predominant general weather situations are described in 4.3.0.1. under Equatorial West Africa. A hygrothermal diagram of Cotonou is to be found in 4.3.0.2.

4.3.100. Togo

4.3.100.1. Climatic description and tables

1. Tropical Summer Rain Climate Aw and Tropical Summer Rain Mountain Climate GAw
2. Permanently very hot.
3. Often dry and sunny in the morning; frequent showers in the afternoon in the high sun period which has some rainy days too. Low sun period: mostly dry and sunny, scattered showers.
4. Wind: occasional gales, squalls in thunderstorms.

Mango

Awha	100-1	Jan	Feb	Mar	Apr	May	Jun	Jul	Aug	Sep	Oct	Nov	Dec	Year
				latitude: 10°22'n				longitude: 0°28'e				height: 146 m		
	t	27	30	32	32	30	27	26	26	27	28	28	27	28 °C
	clothing	t	t	t	t	t	t	t	t	t	t	t	t	–
	stress	ff	ff	ff	ff	ff	ff	ff	ff	ff	ff	ff	ff	–
	p	2	5	19	51	109	145	172	238	247	88	12	1	1088 mm
0.1 mm	dp	0	0	2	5	8	10	12	15	15	8	1	0	76 days
	sunshine	288	266	282	246	251	213	170	130	162	279	294	288	2870 hours
	%	80	80	75	66	64	56	43	34	44	76	84	81	65 %

Sokodé

Awha	100-2	Jan	Feb	Mar	Apr	May	Jun	Jul	Aug	Sep	Oct	Nov	Dec	Year
				latitude: 8°59'n				longitude: 1°08'e				height: 403 m		
	t	26	27	28	28	27	25	25	24	25	26	26	25	26 °C
	clothing	t	t	t	t	t	t	t	t	t	t	t	t	–
	stress	f	ff	ff	ff	ff	ff	ff	ff	ff	ff	ff	f	–
	p	14	13	44	105	155	188	222	245	258	125	28	17	1413 mm
0.1 mm	dp	1	1	5	8	11	13	16	17	17	10	2	1	102 days
	sunshine	264	254	244	222	229	174	121	105	135	208	246	267	2470 hours
	%	73	76	65	60	59	46	31	27	37	56	70	74	56 %

Atakpamé

Awha	100-3	Jan	Feb	Mar	Apr	May	Jun	Jul	Aug	Sep	Oct	Nov	Dec	Year
				latitude: 7°35'n				longitude: 1°07'e				height: 402 m		
	t	27	28	28	27	26	25	24	24	25	25	26	26	26 °C
	clothing	t	t	t	t	t	t	t	t	t	t	t	t	–
	stress	ff	ff	ff	ff	ff	ff	ff	ff	ff	ff	ff	ff	–
	p	19	50	114	133	164	188	202	158	200	149	47	28	1451 mm
0.1 mm	dp	1	4	10	11	13	17	15	15	19	15	4	2	124 days
	sunshine	248	234	236	216	217	165	105	87	105	195	255	254	2318 hours
	%	68	70	63	59	56	44	27	23	29	53	72	70	52 %

Tabligho

Awha	100-4	Jan	Feb	Mar	Apr	May	Jun	Jul	Aug	Sep	Oct	Nov	Dec	Year
				latitude: 6°36'n				longitude: 1°30'e				height: 52 m		
	t	27	29	28	28	27	26	25	25	26	26	27	27	27 °C
	clothing	t	t	t	t	t	t	t	t	t	t	t	t	–
	stress	ff	ff	ff	ff	ff	ff	ff	ff	ff	ff	ff	ff	–
	p	15	40	111	126	170	155	62	39	106	182	84	19	1109 mm

Lomé

Awha	100-5	Jan	Feb	Mar	Apr	May	Jun	Jul	Aug	Sep	Oct	Nov	Dec	Year
				latitude: 6°10'n				longitude: 1°15'e				height: 22 m		
	t	27	28	28	28	27	26	25	25	26	26	27	27	26 °C
	clothing	t	t	t	t	t	t	t	t	t	t	t	t	–
	stress	ff	ff	ff	ff	ff	ff	ff	ff	ff	ff	ff	ff	–
	p	17	37	80	104	153	202	55	16	48	153	21	6	893 mm
0.1 mm	dp	2	3	7	7	12	15	8	5	9	13	4	2	87 days
	sunshine	205	209	217	207	198	147	140	149	159	205	240	229	2304 hours
	%	56	62	58	56	51	39	36	39	44	55	68	63	52 %

4.3.100.2. Predominant general weather situations are described in 4.3.0.1. under Equatorial West Africa.

4.3.101. Ghana

4.3.101.1. Climatic description and tables

1. Tropical Summer Rain Climate Aw and Tropical Summer Rain Mountain Climate GAw (partly Tropical Rain Climate Ar and Tropical Rain Mountain Climate GAr)
2. Permanently very hot.
3. Generally dry and sunny in the morning, frequent showers in the afternoon, less frequent in low sun period. In high sun period rainy days at times.
4. Wind: occasional gales, squalls in thunderstorms.

Tamale

latitude: 9°25′n longitude: 0°53′w height: 188 m

Awha	101-1	Jan	Feb	Mar	Apr	May	Jun	Jul	Aug	Sep	Oct	Nov	Dec	Year
	tx	38	39	40	39	37	34	33	32	33	34	36	37	40 °C
	t	28	30	31	30	28	27	26	25	26	27	28	28	28 °C
	tn	17	19	21	21	21	20	20	20	20	20	19	17	17 °C
	clothing	t	t	t	t	t	t	t	t	t	t	t	t	–
	stress	f	ff	ff	ff	ff	ff	ff	ff	ff	ff	ff	f	–
	p	2	9	51	88	121	132	128	189	217	98	13	5	1053 mm
	dp	0	1	3	6	8	10	10	12	16	9	1	1	77 days
	sunshine	263	238	251	243	248	210	164	136	159	254	288	270	2724 hours
	%	73	71	67	66	64	55	42	35	43	69	82	75	61 %

Kumasi

latitude: 6°43′n longitude: 1°36′w height: 293 m

Araa	101-2	Jan	Feb	Mar	Apr	May	Jun	Jul	Aug	Sep	Oct	Nov	Dec	Year
	tx	33	35	34	34	33	32	30	29	31	32	32	32	36 °C
	t	25	27	27	27	27	25	24	24	25	25	26	25	26 °C
	tn	13	16	19	20	20	19	19	18	19	19	18	15	12 °C
	clothing	t	t	t	t	t	t	t	t	t	t	t	t	–
	stress	f	f	ff	ff	ff	ff	ff	ff	ff	ff	ff	f	–
	p	26	65	136	139	191	223	113	74	170	201	95	32	1465 mm
	dp	2	5	10	10	12	15	11	10	14	17	11	3	120 days
	sunshine	180	182	208	207	201	144	99	71	102	158	201	192	1945 hours
	%	49	45	56	56	52	38	26	19	28	43	57	53	44 %

Accra

latitude: 5°36′n longitude: 0°10′w height: 65 m

Awha	101-3	Jan	Feb	Mar	Apr	May	Jun	Jul	Aug	Sep	Oct	Nov	Dec	Year
	tx	33	33	33	33	33	32	29	28	29	31	32	32	33 °C
	t	27	28	28	28	27	26	25	24	25	26	27	27	26 °C
	tn	19	21	22	22	22	21	21	20	21	21	22	21	18 °C
	clothing	t	t	t	t	t	t	t	t	t	t	t	t	–
	stress	ff	ff	ff	ff	ff	ff	ff	ff	ff	ff	ff	ff	–
	p	16	37	73	82	145	193	49	16	40	80	38	18	787 mm
	dp	1	2	5	5	9	11	5	3	5	7	3	2	58 days
	sunshine	208	210	226	219	214	147	149	149	168	214	246	233	2383 hours
	%	57	62	60	60	56	39	39	39	46	57	69	64	54 %

Takoradi

latitude: 4°53′n longitude: 1°46′w height: 9 m

Awab	101-4	Jan	Feb	Mar	Apr	May	Jun	Jul	Aug	Sep	Oct	Nov	Dec	Year
	tx	32	33	33	33	32	31	28	28	29	30	31	32	33 °C
	t	26	27	27	27	26	26	24	24	25	23	22	24	25 °C
	tn	18	21	21	21	21	21	19	18	20	21	21	19	17 °C
	clothing	t	t	t	t	t	t	t	t	t	t	t	t	–
	stress	ff	ff	ff	ff	ff	ff	ff	ff	ff	ff	ff	ff	–
	p	31	38	80	102	250	288	87	35	48	131	77	38	1205 mm
	dp	2	3	5	7	13	15	8	6	8	11	7	3	88 days
	sunshine	205	207	229	216	189	120	158	130	126	192	243	226	2241 hours
	%	56	61	61	59	49	32	41	34	35	52	68	62	51 %

4.3.101.2. Predominant general weather situations are described in 4.3.0.1. under Equatorial West Africa. A hygrothermal diagram of Accra is to be found in 4.3.0.2.

4.3.102. Upper Volta

4.3.102.1. Climatic descriptions and tables

1. Climatic area: greater northern part of the country
1.1. Steppe Climate BS and Steppe Mountain Climate GBS
1.2. In lowlands summer very hot, winter hot.
1.3. Generally dry and sunny; in high sun period frequent showers in the afternoon.
1.4. Wind: occasional gales and dust-storms.

Dori

latitude: 14°02′n longitude: 0°03′e height: 277 m

BSha	102-1	Jan	Feb	Mar	Apr	May	Jun	Jul	Aug	Sep	Oct	Nov	Dec	Year
	tx	37	39	41	43	43	42	38	36	38	39	39	36	44 °C
	t	24	26	30	33	34	31	28	27	28	30	28	24	28 °C
	tn	9	12	17	18	21	20	19	19	19	19	14	10	9 °C
	clothing	t	t	t	t	t	t	t	t	t	t	t	t	–
	stress	f	f	f	ff	ff	ff	ff	ff	ff	ff	f	f	–
	p	0	1	1	4	26	59	151	189	96	15	0	0	542 mm
0.1 mm	dp	0	0	0	1	4	7	11	13	8	2	0	0	46 days
	sunshine	307	291	310	276	301	285.	273	251	270	310	306	307	3486 hours
	%	87	88	83	74	76	73	68	64	74	84	89	88	79 %

Ouagadougou

latitude: 12°21′n longitude: 1°31′w height: 314 m

BSha	102-2	Jan	Feb	Mar	Apr	May	Jun	Jul	Aug	Sep	Oct	Nov	Dec	Year
	tx	38	42	42	43	42	39	36	33	35	38	38	38	45 °C
	t	25	28	31	33	31	29	27	26	27	29	28	26	28 °C
	tn	11	15	19	21	21	20	20	19	21	21	18	13	10 °C
	clothing	t	t	t	t	t	t	t	t	t	t	t	t	–
	stress	f	f	ff	ff	ff	ff	ff	ff	ff	ff	ff	f	–
	p	0	3	8	19	84	118	193	265	153	37	2	0	879 mm
0.1 mm	dp	0	0	1	3	8	10	14	17	14	5	0	0	72 days
	sunshine	279	266	279	243	254	246	226	155	213	282	282	285	3010 hours
	%	78	80	75	65	65	64	57	40	58	77	81	81	68 %

Fada N'Gourma

BSha	102-3	Jan	Feb	Mar	Apr	May	Jun	Jul	Aug	Sep	Oct	Nov	Dec	Year
		latitude: 12°04'n				longitude: 0°21'e				height: 309 m				
	tx	37	42	43	44	43	39	37	34	36	39	40	39	46 °C
	t	26	28	31	32	31	29	27	26	26	27	27	26	28 °C
	tn	12	13	18	21	21	19	19	19	19	19	14	13	11 °C
	clothing	t	t	t	t	t	t	t	t	t	t	t	t	–
	stress	f	f	ff	ff	ff	ff	ff	ff	ff	ff	f	f	–
	p	0	2	6	22	87	127	171	276	171	25	2	1	890 mm
0.1 mm	dp	0	0	1	2	8	10	11	15	14	4	1	0	66 days
	sunshine	298	277	291	255	279	258	236	195	219	298	297	298	3200 hours
	%	84	83	78	69	71	67	60	50	60	81	85	84	72 %

2. *Climatic area:* South of the country

2.1. Tropical Summer Rain Climate Aw and
Tropical Summer Rain Mountain Climate GAw

2.2. Permanently very hot.

2.3. Generally dry and sunny in the morning, frequent showers in the afternoon, less frequent in low sun period. In high sun period rainy days at times.

2.4. Wind: Occasional gales, squalls in thunderstorms.

Boromo

Awha	102-4	Jan	Feb	Mar	Apr	May	Jun	Jul	Aug	Sep	Oct	Nov	Dec	Year
		latitude: 11°40'n				longitude: 2°55'w				height: 271 m				
	t	26	28	31	32	30	28	27	26	26	28	28	27	28 °C
	clothing	t	t	t	t	t	t	t	t	t	t	t	t	–
	stress	f	f	f	f	ff	ff	ff	ff	ff	ff	f	f	–
	p	0	3	9	32	91	120	228	291	196	48	9	2	1029 mm
0.1 mm	dp	0	0	2	4	7	10	12	17	13	4	1	0	70 days

Bobo Dioulasso

Awha	102-5	Jan	Feb	Mar	Apr	May	Jun	Jul	Aug	Sep	Oct	Nov	Dec	Year
		latitude: 11°10'n				longitude: 4°18'w				height: 467 m				
	tx	36	39	44	41	38	35	33	34	34	36	36	37	44 °C
	t	25	27	30	30	29	27	26	25	25	27	27	25	27 °C
	tn	11	13	17	18	19	18	18	19	18	19	16	13	11 °C
	clothing	t	t	t	t	t	t	t	t	t	t	t	t	–
	stress	f	f	f	ff	ff	ff	ff	ff	ff	ff	f	f	–
	p	1	3	20	47	116	132	229	336	211	75	13	2	1185 mm
0.1 mm	dp	0	1	2	4	8	11	13	18	14	9	2	0	82 days
	sunshine	304	263	270	222	242	222	198	152	192	254	267	276	2861 hours
	%	85	79	72	60	62	58	50	39	52	69	77	78	65 %

Gaoua

Awha	102-6	Jan	Feb	Mar	Apr	May	Jun	Jul	Aug	Sep	Oct	Nov	Dec	Year
		latitude: 10°20'n				longitude: 3°35'w				height: 335 m				
	t	27	29	30	30	29	27	26	25	26	27	28	27	28 °C
	clothing	t	t	t	t	t	t	t	t	t	t	t	t	–
	stress	f	f	f	ff	ff	ff	ff	ff	ff	ff	f	f	–
	p	3	7	37	72	139	141	172	250	222	85	28	5	1161 mm
0.1 mm	dp	1	1	3	6	9	10	11	15	16	8	2	0	82 days

4.3.102.2. Predominant general weather situations are described in 4.3.0.1. under Equatorial West Africa. A hygrothermal diagram of Ouagadougou is to be found in 4.3.0.2. *Wind name:* Harmattan is a hot dry wind blowing from the northeast and carrying sand and laterite dust.

4.3.103. Ivory Coast

4.3.103.1. Climatic descriptions and tables

1. Climatic area: north and centre of the country

1.1. Tropical Summer Rain Climate Aw and Tropical Summer Rain Mountain Climate GAw

1.2. Permanently very hot.

1.3. Generally dry and sunny in the morning; frequent showers in the afternoon, less frequent in the low sun period. In high sun period rainy days at times.

1.4. Wind: occasional gales, squalls in thunderstorms and dust-storms in dry period.

Ferkessedougou

latitude: 9°35'n longitude: 5°12'w height: 323 m

Awha	103-1	Jan	Feb	Mar	Apr	May	Jun	Jul	Aug	Sep	Oct	Nov	Dec	Year
	tx	35	38	37	37	36	34	32	31	33	33	34	34	38 °C
	t	26	28	29	29	28	26	25	25	26	27	27	25	27 °C
	tn	12	14	17	20	20	19	19	19	19	18	16	12	11 °C
	clothing	t	t	t	t	t	t	t	t	t	t	t	t	–
	stress	f	f	ff	ff	ff	ff	ff	ff	ff	ff	f	f	–
	p	3	38	72	90	161	141	204	292	240	138	42	18	1441 mm
0.1 mm	dp	1	2	5	7	11	12	14	18	19	13	4	1	107 days
	sunshine	276	254	254	231	242	213	167	146	171	242	264	267	2727 hours
	%	77	76	68	62	62	56	43	38	47	65	75	74	63 %

Odienné

latitude: 9°30'n longitude: 7°34'w height: 434 m

Awha	103-2	Jan	Feb	Mar	Apr	May	Jun	Jul	Aug	Sep	Oct	Nov	Dec	Year
	t	26	28	29	29	28	26	25	25	25	26	26	25	26 °C
	clothing	t	t	t	t	t	t	t	t	t	t	t	t	–
	stress	f	f	ff	ff	ff	ff	ff	ff	ff	ff	ff	f	–
	p	0	14	42	77	126	173	282	381	281	152	58	10	1596 mm
0.1 mm	dp	2	3	5	7	15	16	8	6	7	10	10	6	95 days
	sunshine	223	215	223	216	189	123	118	140	162	217	216	226	2268 hours
	%	62	64	60	58	49	32	30	36	44	59	62	63	51 %

Bouake

latitude: 7°44'n longitude: 5°04'w height: 376 m

Awha	103-3	Jan	Feb	Mar	Apr	May	Jun	Jul	Aug	Sep	Oct	Nov	Dec	Year
	tx	36	37	37	36	36	34	34	33	33	34	35	36	39 °C
	t	27	28	28	28	27	26	24	24	25	26	26	26	26 °C
	tn	17	18	19	19	19	18	18	18	18	18	19	17	16 °C
	clothing	t	t	t	t	t	t	t	t	t	t	t	t	–
	stress	f	ff	ff	ff	ff	ff	ff	ff	ff	ff	ff	f	–
	p	10	57	88	127	145	123	121	78	169	173	32	13	1137 mm
0.1 mm	dp	1	4	6	8	11	12	10	13	16	13	4	2	100 days
	sunshine	208	209	211	195	195	129	81	74	120	170	183	189	1964 hours
	%	57	62	56	53	50	34	21	19	33	46	52	52	44 %

Dimbokro

		latitude: 6°39′n					longitude: 4°42′w				height: 91 m		
Awha	103-4	Jan	Feb	Mar	Apr	May	Jun	Jul	Aug	Sep	Oct	Nov	Dec	Year
	t	28	29	29	29	28	27	26	26	27	28	28	28	28 °C
	clothing	t	t	t	t	t	t	t	t	t	t	t	t	–
	stress	ff	ff	ff	ff	ff	ff	ff	ff	ff	ff	ff	ff	–
	p	12	59	76	153	172	162	123	43	123	136	64	23	1146 mm
0.1 mm	dp	1	4	7	11	11	14	7	7	10	13	6	2	93 days

2. *Climatic area:* Coastal area
2.1. Tropical Rain Climate Ar
2.2. Permanently very hot.
2.3. Generally dry and sunny in the morning; frequent showers in the afternoon. At times – particularly after the equinoxes – rainy days.
2.4. Wind: occasional gales, squalls in thunderstorms.

Abidjan

		latitude: 5°15′n					longitude: 3°56′w				height: 7 m		
Arha	103-5	Jan	Feb	Mar	Apr	May	Jun	Jul	Aug	Sep	Oct	Nov	Dec	Year
	tx	32	33	33	33	33	32	31	29	30	31	32	33	34 °C
	t	27	27	28	28	27	26	25	24	25	26	27	27	26 °C
	tn	19	20	21	22	22	21	20	19	21	21	22	21	18 °C
	clothing	t	t	t	t	t	t	t	t	t	t	t	t	–
	stress	ff	ff	ff	ff	ff	ff	ff	ff	ff	ff	ff	ff	–
	p	26	42	120	169	366	608	200	34	55	225	188	111	2144 mm
0.1 mm	dp	4	5	10	12	19	21	13	10	14	20	17	9	154 days
	sunshine	183	212	226	210	192	117	115	121	141	202	225	208	2151 hours
	%	50	63	60	57	50	31	30	32	39	54	63	57	59 %

Tabou

		latitude: 4°25′n					longitude: 7°22′w				height: 6 m		
Araa	103-6	Jan	Feb	Mar	Apr	May	Jun	Jul	Aug	Sep	Oct	Nov	Dec	Year
	tx	31	31	32	32	31	29	29	28	28	29	30	31	32 °C
	t	26	27	27	27	27	26	25	24	24	26	26	26	26 °C
	tn	21	21	21	22	22	21	21	21	21	22	22	21	18 °C
	clothing	t	t	t	t	t	t	t	t	t	t	t	t	–
	stress	ff	ff	ff	ff	ff	ff	ff	ff	ff	ff	ff	ff	–
	p	64	50	98	153	438	545	109	90	257	229	180	170	2383 mm
0.1 mm	dp	5	5	9	9	19	17	9	10	19	19	14	9	144 days
	sunshine	226	229	236	216	174	93	90	108	105	174	210	220	2080 hours
	%	61	68	63	59	46	25	23	28	29	47	59	60	47 %

4.3.103.2. Predominant general weather situations are described in 4.3.0.1. under Equatorial West Africa.
Wind name: Harmattan is a hot dry wind blowing from the northeast and carrying sand and laterite dust.

4.3.104. Liberia

4.3.104.1. Climatic description and table

1. Tropical Rain Climate Ar and Tropical Rain Mountain Climate GAr
2. Permanently very hot.
3. Generally dry and sunny in the morning; frequent showers in the afternoon. Low sun period: less shower activity.
4. Wind: occasional gales, squalls in thunderstorms.

Monrovia

				latitude: 6°18′n				longitude: 10°48′w			height: 7 m		
Araa	104-1	Jan	Feb	Mar	Apr	May	Jun	Jul	Aug	Sep	Oct	Nov	Dec	Year
	tx	32	32	32	32	32	30	29	29	29	30	32	32	33 °C
	t	27	26	27	27	27	26	25	25	25	26	26	25	26 °C
	tn	19	21	21	20	19	21	19	21	20	21	20	19	17 °C
	clothing	t	t	t	t	t	t	t	t	t	t	t	t	–
	stress	ff	ff	ff	ff	ff	ff	ff	ff	ff	ff	ff	ff	–
	p	43	62	100	156	422	928	920	474	719	619	221	129	4793 mm
0.2 mm	dp	4	4	7	9	17	24	20	17	20	21	15	10	168 days
	sunshine	158	167	198	195	155	105	84	81	96	121	147	155	1661 hours
	%	43	50	53	53	40	28	22	21	26	33	41	42	37 %

4.3.104.2. Predominant general weather situations are described in 4.3.0.1. under Equatorial West Africa.

4.3.105. Angola

4.3.105.1. Climatic descriptions and tables

1. Climatic area: greater part of the country, except the coastal area and bordering the western slopes of the mountains, as well as the southernmost part of the country

1.1. Tropical Summer Rain Climate Aw and Tropical Summer Rain Mountain Climate GAw

1.2. In low levels summer hot, winter warm or hot.

1.3. Generally dry and sunny in the morning; high sun period: showers frequent in the afternoon; low sun period: often dry and cloudy.

1.4. Wind: occasional gales, squalls in thunderstorms.

Cabinda

				latitude: 5°33′s				longitude: 12°11′e			height: 20 m		
Awab	105-1	Jan	Feb	Mar	Apr	May	Jun	Jul	Aug	Sep	Oct	Nov	Dec	Year
	tx	32	32	33	32	31	29	27	27	29	30	31	32	33 °C
	t	26	26	27	27	25	23	21	22	23	25	26	26	26 °C
	tn	21	21	21	21	20	17	16	16	18	21	22	22	14 °C
	clothing	t	t	t	t	t	t	s	s	t	t	t	t	–
	stress	ff	ff	ff	ff	ff	f	o	o	f	ff	ff	ff	–
	p	80	116	176	157	56	0	0	2	6	52	148	100	892 mm
0.1 mm	dp	7	9	10	12	4	0	0	1	5	11	14	10	83 days
	sunshine	127	136	152	150	105	99	99	81	60	68	99	118	1294 hours
	%	33	39	40	42	29	28	27	22	17	18	27	31	29 %

Malange

GAwbb	105-2	Jan	Feb	Mar	Apr	May	Jun	Jul	Aug	Sep	Oct	Nov	Dec	Year
				latitude: 9°33′s				longitude: 16°22′e			height: 1142 m			
	tx	31	31	31	30	31	31	31	32	32	31	29	30	33 °C
	t	22	22	22	22	21	19	19	20	21	22	21	21	21 °C
	tn	15	14	14	14	8	6	5	9	13	14	15	15	4 °C
	clothing	s	s	s	s	s	s	s	s	s	s	s	s	–
	stress	o	o	o	o					o	o	o	o	–
	p	101	128	208	187	43	0	0	2	45	122	228	130	1194 mm
	dp	8	8	13	14	4	0	0	0	5	11	16	12	91 days
	sunshine	152	146	152	144	228	272	253	227	185	157	130	149	2195 hours
	%	39	42	40	40	63	78	70	62	51	41	35	38	50 %

Henrique de Carvalho

GAwab	105-3	Jan	Feb	Mar	Apr	May	Jun	Jul	Aug	Sep	Oct	Nov	Dec	Year
				latitude: 9°39′s				longitude: 20°24′e			height: 1080 m			
	t	22	22	22	22	22	21	21	23	23	23	22	22	22 °C
	clothing	s	s	s	s	s	s	s	s	s	s	s	s	–
	stress	o	o	o	o	o	o	o	o	o	o	o	o	–
	p	237	175	245	124	12	0	0	20	94	127	240	183	1457 mm
	dp	15	13	16	14	1	0	0	2	7	10	17	15	110 days
	sunshine	161	158	152	194	288	297	310	274	219	211	169	157	2590 hours
	%	41	45	40	54	80	86	86	75	61	55	45	40	59 %

Luso

GAwbl	105-4	Jan	Feb	Mar	Apr	May	Jun	Jul	Aug	Sep	Oct	Nov	Dec	Year
				latitude: 11°47′s				longitude: 19°55′e			height: 1326 m			
	tx	31	31	31	30	30	28	29	32	33	33	32	31	34 °C
	t	21	21	21	21	19	17	17	20	22	22	21	21	20 °C
	tn	14	15	14	12	8	6	6	8	11	14	14	14	5 °C
	clothing	s	s	s	s	s	s	s	s	s	s	s	s	–
	stress	f	f	f	f	o	o	o	o	f	f	f	f	–
	p	204	160	209	122	12	0	0	0	25	107	144	210	1193 mm
	dp	1	14	18	10	1	0	0	0	3	9	14	18	104 days
	sunshine	135	139	144	194	275	288	302	293	234	195	148	138	2485 hours
	%	34	39	38	55	77	84	85	81	65	51	39	35	56 %

Nova Lisboa

GAwbb	105-5	Jan	Feb	Mar	Apr	May	Jun	Jul	Aug	Sep	Oct	Nov	Dec	Year
				latitude: 12°48′s				longitude: 15°45′e			height: 1701 m			
	tx	29	28	28	28	28	27	27	30	31	31	29	29	32 °C
	t	19	19	19	19	18	16	16	19	21	20	19	19	19 °C
	tn	11	12	12	11	7	5	6	8	9	12	12	12	4 °C
	clothing	s	s	s	s	s	n	n	s	s	s	s	s	–
	stress	o	o	o	o	o			o	o	o	o	o	–
	p	199	158	218	178	27	0	0	0	17	109	220	217	1343 mm
	dp	15	12	18	13	2	0	0	0	3	12	18	16	109 days
	sunshine	138	156	144	164	249	239	269	265	214	169	137	141	2285 hours
	%	35	44	38	46	70	70	76	73	59	44	36	35	52 %

Serpa Pinto

latitude: 14°39's longitude: 17°41'e height: 1343 m

GAwbl	105-6	Jan	Feb	Mar	Apr	May	Jun	Jul	Aug	Sep	Oct	Nov	Dec	Year
	tx	31	32	31	31	31	29	29	32	34	36	37	32	37 °C
	t	22	21	22	20	17	14	14	17	22	23	22	22	20 °C
	tn	13	14	16	8	3	1	1	2	7	12	12	13	1 °C
	clothing	s	s	s	s	s	n	n	s	s	s	s	s	–
	stress	o	o	o	o	o			o	o	f	f	o	–
	p	190	165	181	58	7	0	0	0	8	27	83	149	868 mm
	dp	18	13	16	6	2	0	0	0	1	3	10	14	83 days

Sa Da Bandeira

latitude: 14°56's longitude: 13°34'e height: 1758 m

GAwbl	105-7	Jan	Feb	Mar	Apr	May	Jun	Jul	Aug	Sep	Oct	Nov	Dec	Year
	tx	28	28	28	27	27	26	27	28	30	31	31	28	32 °C
	t	19	18	19	19	17	15	16	18	20	20	20	19	18 °C
	tn	9	10	9	9	7	4	4	7	9	9	10	10	3 °C
	clothing	s	s	s	s	n	n	n	s	s	s	s	s	–
	stress	o	o	o					o	o	o	o	o	–
	p	130	164	182	83	6	0	0	0	6	70	117	126	884 mm
	dp	10	9	16	8	1	0	0	0	1	8	12	11	76 days
	sunshine	172	186	173	206	269	293	282	290	252	214	200	222	2758 hours
	%	43	52	46	58	76	87	80	81	70	56	52	55	62 %

2. Climatic area: Coastal area between 7 and 10 degrees southern latitude and western slopes of the mountains bordering the coastal area as well as the southernmost part of the country

2.1. Steppe Climate BS and Steppe Mountain Climate GBS

2.2. Summer hot, winter warm.

2.3. Predominantly dry and sunny or cloudy, in high sun period frequent showers in the afternoon, the occasional rainy day.

2.4. Wind: occasional gales, squalls in thunderstorms.

Luanda

latitude: 8°51's longitude: 13°14'e height: 70 m

BSab	105-8	Jan	Feb	Mar	Apr	May	Jun	Jul	Aug	Sep	Oct	Nov	Dec	Year
	tx	31	32	32	32	31	27	26	26	27	28	30	30	33 °C
	t	26	27	27	26	25	22	20	20	22	24	25	26	24 °C
	tn	22	22	22	22	20	17	16	16	18	20	22	22	16 °C
	clothing	t	t	t	t	t	s	s	s	s	t	t	t	–
	stress	ff	ff	ff	ff	ff	f	f	f	ff	ff	ff	ff	–
	p	26	35	97	124	19	0	0	1	2	6	34	26	367 mm
	dp	2	3	6	9	2	0	0	0	1	2	4	3	32 days
	sunshine	215	209	212	195	227	217	167	146	146	167	192	208	2301 hours
	%	54	60	56	55	62	62	46	40	40	44	51	53	52 %

3. Climatic area: coastal area south of 10 degrees southern latitude

3.1. Marine Desert Climate BM

3.2. Summer hot, winter warm, in the south mild.

3.3. Predominantly dry and cloudy, in low sun period repeatedly overcast days, in high sun period often dry and sunny in the morning, at times showers in the afternoon, less frequent in the south.

3.4. Wind: occasional gales and dust-storms. Fog: frequent haze and low stratus during the night and the morning, particularly in the low sun period.

Lobito

BMab	105-9	Jan	Feb	Mar	Apr	May	Jun	Jul	Aug	Sep	Oct	Nov	Dec	Year
				latitude: 12°22's				longitude: 13°32'e				height: 3 m		
	tx	31	32	32	32	32	29	26	26	27	28	30	30	33 °C
	t	25	26	27	26	24	21	20	19	21	23	25	26	24 °C
	tn	19	20	21	22	19	16	15	15	16	18	20	20	14 °C
	clothing	t	t	t	t	t	s	s	s	s	t	t	t	–
	stress	ff	ff	ff	ff	ff	f	o	o	o	ff	ff	ff	–
	p	8	31	82	49	5	0	0	0	2	8	16	20	221 mm
	dp	2	2	7	5	0	0	0	0	1	3	3	4	27 days
	sunshine	185	198	203	201	197	163	121	130	157	154	171	191	2071 hours
	%	47	56	54	57	55	48	34	36	43	40	45	48	47 %

Mocamedes

BMal	105-10	Jan	Feb	Mar	Apr	May	Jun	Jul	Aug	Sep	Oct	Nov	Dec	Year
				latitude: 15°12's				longitude: 12°09'e				height: 45 m		
	tx	29	31	32	33	32	29	23	24	25	27	28	29	34 °C
	t	22	23	24	23	19	17	16	17	18	20	22	22	20 °C
	tn	16	17	17	15	12	11	9	11	12	13	14	15	9 °C
	clothing	s	s	t	t	s	s	s	s	s	s	s	s	–
	stress	f	f	f	f	o	o	o	o	o	o	o	f	–
	p	8	15	14	7	0	0	0	0	0	1	2	2	49 mm
	dp	1	1	2	1	0	0	0	0	0	0	1	1	7 days
	sunshine	210	207	232	232	217	142	106	113	130	155	214	223	2181 hours
	%	52	58	61	66	61	42	30	31	36	40	56	55	49 %

4.3.105.2. Predominant general weather situations are described in 4.3.0.1. under Equatorial South Africa. A hygrothermal diagram of Luanda is to be found in 4.3.0.2.

4.3.106. Mauritius

4.3.106.1. Climatic description and tables

1. Tropical Rain Climate Ar and Tropical Rain Mountain Climate GAr
   partly Tropical Summer Rain Climate Aw
2. Permanently very hot.
3. Generally dry and sunny in the morning, often showers in the afternoon. At times rainy days.
4. Wind: occasional gales and tropical cyclones, which occur particularly from January to March.

Diego Gracia

Arha	106-1	Jan	Feb	Mar	Apr	May	Jun	Jul	Aug	Sep	Oct	Nov	Dec	Year
				latitude: 7°14's				longitude: 72°26'e				height: 2 m		
	t	28	28	28	28	28	26	26	26	26	26	27	28	27 °C
	clothing	t	t	t	t	t	t	t	t	t	t	t	t	–
	stress	ff	ff	ff	ff	ff	ff	ff	ff	ff	ff	ff	ff	–
	p	322	233	228	212	141	183	140	161	180	233	218	211	2462 mm
	dp	18	18	15	15	13	13	13	11	12	16	14	14	172 days

Agalega

Arha	106-2	Jan	Feb	Mar	Apr	May	Jun	Jul	Aug	Sep	Oct	Nov	Dec	Year
				latitude: 10°35′s				longitude: 56°45′e				height: 3 m		
	tx	32	32	32	32	31	30	29	29	29	30	32	32	33 °C
	t	27	27	28	28	27	26	25	25	25	26	26	27	26 °C
	tn	23	23	23	23	23	22	22	21	22	22	23	23	21 °C
	clothing	t	t	t	t	t	t	t	t	t	t	t	t	–
	stress	ff	ff	ff	ff	ff	ff	ff	ff	ff	ff	ff	ff	–
	p	272	198	195	168	175	83	90	87	91	143	113	225	1840 mm
	dp	17	18	15	13	13	13	13	13	10	10	9	15	159 days

St. Brandon

Awha	106-3	Jan	Feb	Mar	Apr	May	Jun	Jul	Aug	Sep	Oct	Nov	Dec	Year
				latitude: 16°27′s				longitude: 59°33′e				height: 4 m		
	tx	34	34	34	33	32	29	30	29	28	31	33	34	34 °C
	t	28	28	28	27	26	24	23	23	24	24	26	28	26 °C
	tn	23	23	23	22	22	20	19	19	19	21	22	22	18 °C
	clothing	t	t	t	t	t	t	t	t	t	t	t	t	–
	stress	ff	ff	ff	ff	ff	ff	ff	ff	ff	ff	ff	ff	–
	p	196	191	213	110	59	58	49	40	27	21	39	47	1050 mm
	dp	12	13	14	11	10	12	11	11	8	6	7	7	122 days

Rodrigues

Awab	106-4	Jan	Feb	Mar	Apr	May	Jun	Jul	Aug	Sep	Oct	Nov	Dec	Year
				latitude: 19°41′s				longitude: 63°25′e				height: 59 m		
	tx	31	31	32	31	29	28	27	27	27	28	29	31	32 °C
	t	27	27	27	26	24	23	22	22	22	23	24	26	24 °C
	tn	23	22	23	22	19	18	18	17	17	18	20	21	17 °C
	clothing	t	t	t	t	t	t	s	s	s	t	t	t	–
	stress	ff	ff	ff	ff	ff	ff	ff	f	f	ff	ff	ff	–
	p	170	126	117	164	147	78	79	74	49	52	52	97	1205 mm
	dp	15	13	15	14	15	14	15	15	11	10	8	9	154 days

Plaisance / Mauritius

Arab	106-5	Jan	Feb	Mar	Apr	May	Jun	Jul	Aug	Sep	Oct	Nov	Dec	Year
				latitude: 20°26′s				longitude: 57°40′e				height: 57 m		
	tx	32	31	31	29	28	26	25	26	27	29	31	32	33 °C
	t	26	26	26	24	23	21	21	21	21	22	24	25	23 °C
	tn	20	20	19	18	15	13	13	13	14	15	16	19	12 °C
	clothing	t	t	t	t	t	s	s	s	s	s	t	t	–
	stress	ff	ff	ff	ff	f	f	f	f	f	f	ff	ff	–
	p	237	241	383	206	175	117	130	85	84	56	80	172	1966 mm
	dp	18	18	21	16	19	16	16	16	13	9	10	13	185 days

4.3.106.2. Predominant general weather situations are described in 4.3.0.1. under Indian Ocean.

4.3.107. Islands in the Indian Ocean south of the Equator

4.3.107.1. Climatic descriptions and tables

1. Climatic area: Seychelles, Amirantes Islands and Comoro Islands
1.1. Tropical Rain Climate Ar partly Tropical Summer Rain Climate Aw
1.2. Permanently very hot.
1.3. Generally dry and sunny in the morning, frequent showers in the afternoon. Rainy days at times.
1.4. Wind: Squalls in thunderstorms, occasionally storm.

Mahe – Seychelles Islands

		latitude: 4°37′s			longitude: 55°27′e				height: 3 m					
Arha	107-1	Jan	Feb	Mar	Apr	May	Jun	Jul	Aug	Sep	Oct	Nov	Dec	Year
	tx	30	30	31	31	31	29	28	28	29	29	30	30	32 °C
	t	27	27	28	28	28	27	26	26	26	26	27	27	27 °C
	tn	23	23	23	24	23	23	22	22	22	22	22	23	21 °C
	clothing	t	t	t	t	t	t	t	t	t	t	t	t	–
	stress	ff	ff	ff	ff	ff	ff	ff	ff	ff	ff	ff	ff	–
	p	358	242	204	170	161	85	83	78	116	167	206	332	2203 mm
	dp	15	10	11	10	9	9	8	7	9	9	12	15	123 days

Moroni – Comoro Islands

		latitude: 11°42′s			longitude: 43°14′e				height: 17 m					
Arab	107-2	Jan	Feb	Mar	Apr	May	Jun	Jul	Aug	Sep	Oct	Nov	Dec	Year
	tx	32	32	32	32	31	29	29	29	29	31	32	32	33 °C
	t	26	26	26	26	25	24	23	23	22	24	26	26	25 °C
	tn	22	22	22	22	20	18	17	18	18	19	21	22	17 °C
	clothing	t	t	t	t	t	t	t	t	s	t	t	t	–
	stress	ff	ff	ff	ff	ff	ff	ff	ff	ff	ff	ff	ff	–
	p	357	291	284	309	211	217	271	141	131	89	105	233	2639 mm
	dp	16	14	15	14	9	10	10	8	9	9	9	13	136 days

Dzaoudzi/Pamanzi – Mayotte – Comoro Islands

		latitude: 12°48′s			longitude: 45°17′e				height: 7 m					
Awaa	107-3	Jan	Feb	Mar	Apr	May	Jun	Jul	Aug	Sep	Oct	Nov	Dec	Year
	tx	33	33	33	32	32	30	28	29	31	32	33	33	33 °C
	t	27	27	27	27	26	25	24	24	24	25	26	27	26 °C
	tn	22	23	22	23	22	21	20	19	21	21	22	23	19 °C
	clothing	t	t	t	t	t	t	t	t	t	t	t	t	–
	stress	ff	ff	ff	ff	ff	ff	ff	ff	ff	ff	ff	ff	–
	p	264	254	163	94	13	12	13	10	29	50	93	177	1172 mm
	dp	13	15	12	7	3	2	3	3	4	6	7	12	87 days

2. Climatic area: Tromelin Island and Reunion Island
2.1. Tropical Summer Rain Climate Aw
2.2. Summer very hot, winter hot or warm.
2.3. Generally dry and sunny in the morning, frequent showers in the afternoon, less frequent during low sun period. Rainy days at times.
2.4. Wind: Squalls in thunderstorms, occasionally tropical cyclones, which occur particularly from January to March.

Serge Frolow – Tromelin Island

Awaa	107-4	Jan	Feb	Mar	Apr	May	Jun	Jul	Aug	Sep	Oct	Nov	Dec	Year
		latitude: 15°53's				longitude: 54°31'e				height: 8 m				
	t	28	28	28	27	26	24	24	23	24	25	26	27	26 °C
	clothing	t	t	t	t	t	t	t	t	t	t	t	t	–
	stress	ff	ff	ff	ff	ff	ff	ff	ff	ff	ff	ff	ff	–
	p	166	138	187	108	70	56	64	44	34	30	52	82	1031 mm

Saint-Denis/Gillot – Reunion Island

Awab	107-5	Jan	Feb	Mar	Apr	May	Jun	Jul	Aug	Sep	Oct	Nov	Dec	Year
		latitude: 20°52's				longitude: 55°31'e				height: 12 m				
	t	26	26	26	24	23	22	21	21	21	22	24	25	23 °C
	clothing	t	t	t	t	t	s	s	s	s	s	t	t	–
	stress	ff	ff	ff	ff	f	f	f	f	f	f	ff	ff	–
	p	252	217	292	162	71	80	72	47	44	50	99	140	1526 mm

Hellburg – Reunion Island

GArbl	107-6	Jan	Feb	Mar	Apr	May	Jun	Jul	Aug	Sep	Oct	Nov	Dec	Year
		latitude: 21°04's				longitude: 55°22'e				height: 936 m				
	tx	26	27	27	26	24	22	21	22	22	24	24	26	28 °C
	t	19	20	20	18	16	15	14	14	14	16	17	18	17 °C
	tn	12	13	13	10	8	7	6	6	7	8	9	11	5 °C
	clothing	s	s	s	s	n	n	n	n	n	n	n	s	–
	stress	o	o	o	o								o	–
	p	569	203	417	183	135	112	79	76	51	58	89	328	2299 mm
0.1 mm	dp	16	15	18	13	11	9	13	11	10	10	10	15	151 days

3. Climatic area: Ile Europa

3.1. Steppe Climate BS

3.2. Summer hot, winter warm.

3.3. Generally dry and sunny in the morning; in the low sun period often in the afternoon too, in the high sun period often showers in the afternoon, and at times rainy days.

3.4. Wind: squalls in thunderstorms, occasional gales and tropical cyclones, which occur particularly from January to March.

Ile Europa

BShb	107-7	Jan	Feb	Mar	Apr	May	Jun	Jul	Aug	Sep	Oct	Nov	Dec	Year
		latitude: 22°21's				longitude: 40°21'e				height: 7 m				
	t	27	28	27	25	24	22	21	22	23	24	25	26	24 °C
	clothing	t	t	t	t	t	s	s	s	t	t	t	t	–
	stress	ff	ff	ff	ff	f	f	f	f	f	f	ff	ff	–
	p	136	111	73	15	23	19	13	8	14	7	53	140	612 mm

4. Climatic area: New Amsterdam

4.1. Subtropical Rain Climate Cr

4.2. Mild throughout the year.

4.3. Variable weather prevails.

4.4. Wind: frequent gales. Fog: frequent.

New Amsterdam

Crll	107-8	Jan	Feb	Mar	Apr	May	Jun	Jul	Aug	Sep	Oct	Nov	Dec	Year
		latitude: 37°50′s					longitude: 77°34′e				height: 27 m			
	t	17	17	16	15	13	12	11	11	12	12	13	16	14 °C
	clothing stress	n	n	n	n	n	n	n	n	n	n	n	n	–
	p	109	93	115	90	110	126	116	81	65	73	93	86	1157 mm

5. Climatic area: Prince Edward Islands, Crozet Islands, Kerguelen Land
5.1. Tundra Climate FT
5.2. Cool throughout the year.
5.3. Variable weather prevails.
5.4. Wind: frequent gales. Fog: frequent. Snow cover: at times.

Marion Island – Prince Edward Islands

FTkk	107-9	Jan	Feb	Mar	Apr	May	Jun	Jul	Aug	Sep	Oct	Nov	Dec	Year
		latitude: 46°53′s					longitude: 37°52′e				height: 0 m			
	t	7	7	7	6	5	4	4	3	3	4	5	6	5 °C
	clothing stress	n	n	n	n	n	w	w	w	w	w	w	w	–
	p	194	185	216	221	235	221	228	178	190	167	190	226	2451 mm
	sunshine	153	134	114	79	68	53	61	75	89	136	150	157	1269 hours
	%	32	34	29	24	23	21	22	24	25	33	33	32	29 %

Port aux Francais – Kerguelen Island

FTkk	107-10	Jan	Feb	Mar	Apr	May	Jun	Jul	Aúg	Sep	Oct	Nov	Dec	Year
		latitude: 49°20′s					longitude: 70°13′e				height: 14 m			
	t	7	8	7	6	4	2	2	2	2	4	5	6	5 °C
	clothing stress	n	n	n	n	w	w	w	w	w	w	n	n	–
	p	75	50	66	90	129	117	113	100	107	85	94	93	1121 mm
0.2 mm	dp	19	21	21	18	27	28	26	26	22	29	25	21	283 days

Heard Island

FTko	107-11	Jan	Feb	Mar	Apr	May	Jun	Jul	Aug	Sep	Oct	Nov	Dec	Year
		latitude: 53°01′s					longitude: 73°23′e				height: 5 m			
	tx	9	9	9	9	7	6	6	7	4	4	5	8	12 °C
	t	3	3	4	3	1	0	-1	-1	-1	0	1	2	1 °C
	tn	0	-1	-1	-3	-4	-6	-6	-8	-7	-7	-3	-1	-8 °C
	clothing stress	w	w	w	w	w	w	w	w	w	w	w	w	–
	p	147	147	145	155	147	99	91	56	64	94	102	130	1379 mm
0.2 mm	dp	26	25	27	27	25	23	21	21	20	20	21	24	280 days

4.3.107.2. Predominant general weather situations are described in 4.3.0.1. under Indian Ocean.

4.3.108. Madagascar

4.3.108.1. Climatic descriptions and tables

1. Climatic area: Eastern Coast and adjacent slopes of the mountains
1.1. Tropical Rain Climate Ar and Tropical Rain Mountain Climate GAr
1.2. Summer hot, winter mild.
1.3. In the morning often dry and sunny, in the afternoon often showers; at times rainy days.
1.4. Wind: Squalls in thunderstorms, occasional gales and tropical cyclones, which occur particularly from December to May. Fog: occasionally.

Antalaha

latitude: 15°00′s longitude: 50°20′e height: 24 m

Arab	108-1	Jan	Feb	Mar	Apr	May	Jun	Jul	Aug	Sep	Oct	Nov	Dec	Year
	tx	33	33	33	32	31	29	28	28	28	29	31	32	33 °C
	t	26	26	26	25	23	22	21	21	22	23	24	25	24 °C
	tn	21	21	21	20	19	18	17	16	17	18	19	21	16 °C
	clothing	t	t	t	t	t	s	s	s	s	t	t	t	–
	stress	ff	ff	ff	ff	ff	ff	f	f	f	ff	ff	ff	–
	p	260	246	267	281	150	162	148	132	104	76	114	211	2151 mm
	dp	15	15	14	15	13	16	18	17	14	11	11	14	173 days
	sunshine	215	213	201	206	211	189	184	197	225	247	233	246	2564 hours
	%	54	60	53	58	60	56	52	55	62	64	61	61	58 %

Tamatave

latitude: 18°07′s longitude: 49°24′e height: 5 m

Arab	108-2	Jan	Feb	Mar	Apr	May	Jun	Jul	Aug	Sep	Oct	Nov	Dec	Year
	tx	33	32	32	31	29	27	26	26	27	29	31	32	34 °C
	t	26	26	26	25	23	21	21	21	22	23	25	26	24 °C
	tn	22	22	22	20	18	17	16	16	16	17	19	21	15 °C
	clothing	t	t	t	t	t	s	s	s	s	t	t	t	–
	stress	ff	ff	ff	ff	ff	f	f	f	f	ff	ff	ff	–
	p	420	442	528	403	303	300	257	208	135	91	184	259	3530 mm
	dp	19	18	19	19	17	20	20	20	14	11	12	16	205 days
	sunshine	216	194	183	196	203	167	173	187	215	245	247	247	2473 hours
	%	53	54	48	56	58	50	50	52	60	63	63	60	56 %

Mananjary

latitude: 21°12′s longitude: 48°22′e height: 6 m

Arab	108-3	Jan	Feb	Mar	Apr	May	Jun	Jul	Aug	Sep	Oct	Nov	Dec	Year
	t	26	26	25	24	22	20	20	20	22	23	24	25	23 °C
	clothing	t	t	t	t	s	s	s	s	s	t	t	t	–
	stress	ff	ff	ff	ff	f	f	f	f	f	ff	ff	ff	–
	p	389	361	452	240	192	227	168	144	106	75	172	206	2732 mm
	dp	17	17	19	15	15	15	14	13	10	8	11	13	167 days
	sunshine	216	194	174	213	210	181	196	236	246	245	233	229	2573 hours
	%	52	54	46	61	61	56	58	67	69	63	59	55	58 %

Fort Dauphin

Arab	108-4	Jan	Feb	Mar	Apr	May	Jun	Jul	Aug	Sep	Oct	Nov	Dec	Year
latitude: 25°02's	longitude: 46°57'e	height: 8 m												
	tx	32	32	31	30	28	27	26	26	28	29	31	31	33 °C
	t	25	26	25	24	22	20	20	20	21	22	24	25	23 °C
	tn	19	19	19	18	16	14	13	13	14	16	17	18	12 °C
	clothing	t	t	t	t	s	s	s	s	s	s	t	t	–
	stress	ff	ff	ff	ff	f	o	o	o	o	f	f	ff	–
	p	199	184	236	113	117	135	109	94	61	73	91	124	1534 mm
	dp	13	13	14	9	9	11	10	8	6	6	9	10	118 days
	sunshine	221	230	199	228	243	205	210	237	231	259	242	240	2745 hours
	%	53	63	52	66	72	54	63	68	65	66	60	57	62 %

2. *Climatic area:* Greater part of the country, except the eastern coast and southwest
2.1. Tropical Summer Rain Climate Aw and Tropical Summer Rain Mountain Climate GAw
2.2. At low levels very hot in summer, hot in winter.
2.3. Generally dry and sunny in the morning; in the low sun period often in the afternoon too; in the high sun period, often showers in the afternoon, and at times rainy days.
2.4. Wind: squalls in thunderstorms, occasional gales and tropical cyclones, which occur particularly from December to May. Fog: occasionally.

Diégo-Suarez

Awaa	108-5	Jan	Feb	Mar	Apr	May	Jun	Jul	Aug	Sep	Oct	Nov	Dec	Year
latitude: 12°21's	longitude: 49°18'e	height: 105 m												
	tx	34	34	34	33	33	32	31	31	31	32	34	34	35 °C
	t	27	27	27	27	26	25	24	24	25	26	27	27	26 °C
	tn	22	23	22	22	21	19	19	19	19	21	22	22	18 °C
	clothing	t	t	t	t	t	t	t	t	t	t	t	t	–
	stress	ff	ff	ff	ff	ff	ff	ff	ff	ff	ff	ff	ff	–
	p	277	211	187	56	8	8	7	7	5	11	28	111	915 mm
	dp	16	15	11	5	1	2	2	2	1	2	3	9	69 days
	sunshine	185	171	209	227	273	260	250	279	293	303	288	249	2987 hours
	%	47	48	55	64	77	76	70	77	81	79	76	62	68 %

Analalava

Awha	108-6	Jan	Feb	Mar	Apr	May	Jun	Jul	Aug	Sep	Oct	Nov	Dec	Year
latitude: 14°38's	longitude: 47°46'e	height: 57 m												
	tx	33	33	34	35	34	33	33	33	35	36	35	34	36 °C
	t	27	27	27	28	27	26	25	26	26	27	27	27	26 °C
	tn	21	22	22	21	19	18	17	18	18	19	21	21	17 °C
	clothing	t	t	t	t	t	t	t	t	t	t	t	t	–
	stress	ff	ff	ff	ff	ff	ff	ff	ff	ff	ff	ff	ff	–
	p	521	455	256	77	10	6	5	3	12	57	202	308	1912 mm
	dp	20	16	13	5	1	1	1	1	1	4	10	14	87 days
	sunshine	220	205	228	282	313	293	304	321	317	334	291	230	3338 hours
	%	55	58	80	80	88	87	86	89	88	87	76	57	75 %

Majunga

latitude: 15°40′s longitude: 46°21′e height: 22 m

Awha 108-7	Jan	Feb	Mar	Apr	May	Jun	Jul	Aug	Sep	Oct	Nov	Dec	Year
tx	34	34	34	34	34	32	32	33	34	36	36	35	36 °C
t	27	27	28	28	26	25	24	25	26	27	28	28	27 °C
tn	21	22	22	22	20	18	18	18	19	20	21	21	17 °C
clothing	t	t	t	t	t	t	t	t	t	t	t	t	–
stress	ff	ff	ff	ff	ff	ff	ff	ff	ff	ff	ff	ff	–
p	472	370	287	57	9	3	1	2	3	25	112	246	1587 mm
dp	18	16	12	4	1	1	1	1	1	2	6	12	75 days
sunshine	183	173	208	297	325	297	308	319	320	334	289	231	3284 hours
%	46	49	55	84	92	88	88	89	89	87	75	57	74 %

Tananarive

latitude: 18°54′s longitude: 47°32′e height: 1310 m

GAwbl 108-8	Jan	Feb	Mar	Apr	May	Jun	Jul	Aug	Sep	Oct	Nov	Dec	Year
tx	29	29	29	28	26	23	23	25	28	31	31	30	32 °C
t	19	19	19	18	16	14	13	14	15	17	19	19	17 °C
tn	14	14	13	11	8	6	6	6	7	8	11	13	3 °C
clothing	s	s	s	s	n	n	n	n	n	s	s	s	–
stress	o	o	o	o						o	o	o	–
p	286	218	231	36	13	9	10	10	15	43	143	257	1270 mm
dp	18	15	14	5	3	2	2	2	2	4	10	17	94 days
sunshine	195	194	189	234	229	206	219	239	247	269	224	219	2664 hours
%	48	54	50	67	66	62	64	67	69	69	57	53	60 %

Maintirano

latitude: 18°03′s longitude: 44°02′e height: 25 m

Awab 108-9	Jan	Feb	Mar	Apr	May	Jun	Jul	Aug	Sep	Oct	Nov	Dec	Year
tx	33	33	33	33	32	30	29	31	32	33	34	34	35 °C
t	27	27	27	27	25	23	22	23	25	26	27	27	26 °C
tn	22	22	22	20	17	16	16	16	18	19	20	21	14 °C
clothing	t	t	t	t	t	t	s	t	t	t	t	t	–
stress	ff	ff	ff	ff	f	f	f	f	ff	ff	ff	ff	–
p	302	220	158	32	9	4	4	4	9	17	63	177	999 mm
dp	14	11	10	3	1	1	1	1	1	2	4	10	59 days
sunshine	259	239	267	301	306	291	302	319	312	327	309	286	3518 hours
%	64	67	70	86	87	88	87	89	87	84	79	70	80 %

3. Climatic area: Menabe and hinterland of Fiherenana and Mahafaly

3.1. Steppe Climate BS

3.2. Summer very hot, winter warm.

3.3. Generally dry and sunny; in the high sun period, frequent showers in the afternoon and occasionally rainy days.

3.4. Wind: occasional gales and tropical cyclones, which occur particularly from December to May. Fog: occasionally.

Morondava

BShb	108-10	Jan	Feb	Mar	Apr	May	Jun	Jul	Aug	Sep	Oct	Nov	Dec	Year
		latitude: 20°17′s						longitude: 44°17′e				height: 10 m		
	tx	36	35	34	34	33	32	32	32	33	33	34	34	37 °C
	t	28	28	27	26	23	22	21	22	23	25	27	27	25 °C
	tn	22	22	20	17	13	12	11	12	14	17	19	21	11 °C
	clothing	t	t	t	t	t	s	s	s	t	t	t	t	–
	stress	ff	ff	ff	ff	f	f	f	f	f	ff	ff	ff	–
	p	257	209	114	15	6	5	1	1	5	8	24	132	777 mm
0.1 mm	dp	14	12	8	2	1	1	1	0	1	1	3	8	52 days

4. Climatic area: Fiherenana and Mahafaly
4.1. Desert Climate BW
4.2. Summer hot, winter warm.
4.3. Predominantly dry and sunny; high sun period: showers now and then, particularly in the afternoon.
4.4. Wind: occasional gales and tropical cyclones, which occur particularly from December to May. Fog: occasionally.

Tulear

BWab	108-11	Jan	Feb	Mar	Apr	May	Jun	Jul	Aug	Sep	Oct	Nov	Dec	Year
		latitude: 23°23′s						longitude: 43°44′e				height: 9 m		
	tx	37	37	35	34	33	31	29	32	33	33	34	34	37 °C
	t	27	27	26	25	22	20	20	20	22	24	25	26	24 °C
	tn	18	18	17	13	11	9	8	9	11	13	15	17	7 °C
	clothing	t	t	t	t	s	s	s	s	s	t	t	t	–
	stress	ff	ff	f	f	f	o	o	o	o	f	f	f	–
	p	71	71	42	7	18	11	4	3	10	14	34	57	342 mm
	dp	5	5	3	1	2	2	1	1	1	1	2	4	28 days
	sunshine	300	292	286	306	300	273	300	310	315	336	316	315	3649 hours
	%	64	67	70	86	87	88	87	89	87	84	79	70	80 %

4.3.108.2. Predominant general weather situations are described in 4.3.0.1. under Equatorial South Africa. A hygrothermal diagram of Tananarive is to be found in 4.3.0.2.

4.3.109. Mozambique

4.3.109.1. Climatic descriptions and tables

1. Climatic area: Greater part of the country except north of the Province of Manica e Sofala
1.1. Tropical Summer Rain Climate Aw and
Tropical Summer Rain Mountain Climate GAw
1.2. Summer very hot, winter warm, in the north hot.
1.3. Generally dry and sunny in the morning, in the afternoon often showers, less frequent in low sun period. At times rainy days.
1.4. Wind: occasional gales, squalls in thunderstorms. Coastal areas can be influenced by tropical cyclones, which occur particularly from December to May.

Porto Amélia

latitude: 12°58′s longitude: 40°30′e height: 50 m

Awha	109-1	Jan	Feb	Mar	Apr	May	Jun	Jul	Aug	Sep	Oct	Nov	Dec	Year
	tx	34	34	34	33	32	31	31	31	32	33	34	34	34 °C
	t	27	27	27	27	26	24	24	24	25	26	27	28	26 °C
	tn	22	22	22	22	19	18	18	18	19	21	22	22	17 °C
	clothing	t	t	t	t	t	t	t	t	t	t	t	t	–
	stress	ff	ff	ff	ff	ff	ff	ff	ff	ff	ff	ff	ff	–
	p	155	170	200	127	22	25	18	4	7	6	18	113	865 mm
0.1 mm	dp	10	12	12	9	3	3	2	0	2	1	2	8	64 days
	sunshine	194	182	207	214	260	221	230	254	272	298	287	237	2856 hours
	%	49	51	55	60	73	65	79	70	75	78	75	59	65 %

Villa Cabral

latitude: 13°18′s longitude: 35°14′e height: 1357 m

GAwbl	109-2	Jan	Feb	Mar	Apr	May	Jun	Jul	Aug	Sep	Oct	Nov	Dec	Year
	tx	28	28	28	28	28	26	26	28	29	32	32	30	33 °C
	t	20	20	19	19	17	15	14	16	18	21	22	20	18 °C
	tn	14	14	13	12	9	8	7	8	9	12	13	13	7 °C
	clothing	s	s	s	s	n	n	n	n	s	s	s	s	–
	stress	f	f	f	o	o			o	o	o	f	f	–
	p	262	259	173	103	26	6	1	3	3	22	37	224	1119 mm
	dp	23	19	14	10	2	2	0	0	1	2	5	19	97 days
	sunshine	125	112	159	157	229	205	232	225	220	269	227	140	2300 hours
	%	31	32	42	44	64	60	66	62	61	70	59	35	52 %

Quelimane

latitude: 17°53′s longitude: 36°53′e height: 7 m

Awhb	109-3	Jan	Feb	Mar	Apr	May	Jun	Jul	Aug	Sep	Oct	Nov	Dec	Year
	tx	37	36	35	34	33	31	31	33	37	38	39	38	40 °C
	t	28	27	27	26	24	21	21	22	24	26	27	27	25 °C
	tn	21	21	21	18	15	13	13	13	14	16	18	19	12 °C
	clothing	t	t	t	t	t	s	s	s	t	t	t	t	–
	stress	ff	ff	ff	ff	f	f	f	f	f	f	ff	ff	–
	p	248	254	261	99	76	67	56	34	20	16	81	169	1381 mm
	dp	12	14	14	9	8	9	7	4	3	2	5	11	98 days
	sunshine	193	195	201	215	221	178	196	252	235	291	236	220	2633 hours
	%	48	54	53	61	63	54	57	71	65	75	61	54	60 %

Beira

latitude: 19°48′s longitude: 34°54′e height: 8 m

Awhb	109-4	Jan	Feb	Mar	Apr	May	Jun	Jul	Aug	Sep	Oct	Nov	Dec	Year
	tx	34	34	33	34	32	31	31	30	32	33	36	35	38 °C
	t	27	28	27	25	23	21	20	21	23	25	26	27	24 °C
	tn	21	21	20	18	15	13	13	13	14	16	18	19	12 °C
	clothing	t	t	t	t	t	s	s	s	s	t	t	t	–
	stress	ff	ff	ff	ff	f	f	f	f	f	f	ff	ff	–
	p	265	225	244	105	58	42	37	30	27	29	133	234	1429 mm
	dp	12	11	12	7	6	5	4	3	3	3	7	10	83 days
	sunshine	245	224	243	246	252	222	231	253	243	258	229	236	2883 hours
	%	60	62	64	70	73	68	67	71	68	66	58	57	65 %

Inhambane

					latitude: 23°52′s			longitude: 35°23′e			height: 15 m			
Awab	109-5		Jan	Feb	Mar	Apr	May	Jun	Jul	Aug	Sep	Oct	Nov	Dec	Year
		tx	34	33	33	33	31	29	28	29	30	32	32	33	35 °C
		t	27	27	26	24	22	20	20	21	22	24	25	26	24 °C
		tn	21	21	20	18	16	14	13	14	15	17	18	19	13 °C
		clothing	t	t	t	t	s	s	s	s	s	t	t	t	–
		stress	ff	ff	ff	ff	f	f	f	f	f	f	ff	ff	–
		p	135	146	107	71	58	63	45	30	31	29	80	124	919 mm
		dp	9	9	10	7	6	6	6	4	3	3	6	8	77 days
		sunshine	223	209	219	225	247	217	236	252	225	246	217	229	2745 hours
		%	53	57	58	65	73	68	70	72	63	63	55	54	62 %

Lourenço Marques

					latitude: 25°58′s			longitude: 32°36′e			height: 59 m			
Awab	109-6		Jan	Feb	Mar	Apr	May	Jun	Jul	Aug	Sep	Oct	Nov	Dec	Year
		tx	37	36	35	34	33	31	32	34	37	38	38	38	41 °C
		t	25	26	25	23	21	19	18	20	21	22	24	25	22 °C
		tn	18	19	17	15	12	10	10	11	12	14	16	17	9 °C
		clothing	t	t	t	t	s	s	s	s	s	s	t	t	–
		stress	ff	ff	ff	f	f	o	f	f	f	f	f	ff	–
		p	130	124	97	64	28	27	12	13	38	46	86	103	768 mm
		dp	9	8	9	6	3	2	2	2	4	6	8	8	67 days
		sunshine	222	210	225	229	253	245	255	251	227	210	198	220	2745 hours
		%	53	57	59	66	75	78	77	72	63	53	49	52	62 %

2. *Climatic area:* North of the Province of Manica e Sofala
2.1. Steppe Climate BS and Steppe Mountain Climate GBS
2.2. In low level very hot in summer, warm in winter.
2.3. Generally dry and sunny; high sun period: often showers in the afternoon; the occasional rainy day.
2.4. Wind: occasional gales.

Tete

					latitude: 16°11′s			longitude: 33°35′e			height: 150 m			
BShb	109-7		Jan	Feb	Mar	Apr	May	Jun	Jul	Aug	Sep	Oct	Nov	Dec	Year
		tx	40	39	37	38	38	36	34	37	41	43	43	42	44 °C
		t	28	28	27	27	25	22	22	24	27	30	30	28	26 °C
		tn	17	18	18	16	13	11	11	12	14	17	17	17	9 °C
		clothing	t	t	t	t	t	s	s	t	t	t	t	t	–
		stress	ff	ff	ff	f	f	f	o	f	f	ff	ff	ff	–
		p	157	188	55	18	1	6	2	4	4	5	69	171	679 mm
0.1 mm		dp	14	15	10	4	1	3	2	1	1	1	8	14	74 days
		sunshine	215	187	238	239	261	226	250	270	255	275	247	209	2872 hours
		%	53	52	63	68	74	67	72	75	71	71	64	51	65 %

4.3.109.2. Predominant general weather situations are described in 4.3.0.1. under Equatorial South Africa. A hygrothermal diagram of Lourenco Marques is to be found in 4.3.0.2.

4.3.110. Zambia

4.3.110.1. Climatic description and tables

1. Tropical Summer Rain Climate Aw and Tropical Summer Rain Mountain Climate GAw except the southernmost part of the country with Steppe Climate BS and Steppe Mountain Climate GBS
2. About 1000 m level warm throughout the year, partly mild in winter.
3. Generally dry and sunny in the morning; in the high sun period often showers in the afternoon, at times rainy days; low sun period: mostly dry and sunny or cloudy.
4. Wind: occasional gales.

Abercorn

latitude: 8°49′s longitude: 31°23′e height: 1659 m

GAwbl	110-1	Jan	Feb	Mar	Apr	May	Jun	Jul	Aug	Sep	Oct	Nov	Dec	Year
	tx	27	28	27	27	27	27	27	28	31	31	30	28	31 °C
	t	19	20	20	20	19	18	18	19	21	22	21	20	20 °C
	tn	13	13	14	13	11	8	7	9	11	13	13	13	7 °C
	clothing	s	s	s	s	s	s	s	s	s	s	s	s	–
	stress	f	f	f	f	f	o	o	o	f	f	f	f	–
	p	201	206	239	102	28	2	0	1	3	25	135	213	1154 mm
0.25 mm	dp	21	20	22	13	4	0	0	0	1	4	15	20	120 days

Kasama

latitude: 10°13′s longitude: 31°08′e height: 1384 m

GAwbl	110-2	Jan	Feb	Mar	Apr	May	Jun	Jul	Aug	Sep	Oct	Nov	Dec	Year
	tx	28	29	28	28	28	27	27	29	32	33	32	29	33 °C
	t	20	20	20	20	19	17	17	19	22	22	22	20	20 °C
	tn	15	15	15	13	9	7	7	8	11	14	14	14	7 °C
	clothing	s	s	s	s	s	s	s	s	s	s	s	s	–
	stress	f	f	f	f	o			o	o	f	f	f	–
	p	266	250	258	71	8	0	0	1	1	17	137	236	1244 mm
0.2 mm	dp	24	19	21	7	1	0	0	0	0	3	16	22	113 days
	sunshine	112	110	167	237	279	273	316	307	285	270	213	161	2730 hours
	%	29	31	44	66	77	79	88	84	79	71	56	41	62 %

Mpika

latitude: 11°51′s longitude: 31°27′e height: 1402 m

GAwal	110-3	Jan	Feb	Mar	Apr	May	Jun	Jul	Aug	Sep	Oct	Nov	Dec	Year
	tx	29	28	28	27	27	26	26	27	31	33	32	29	33 °C
	t	21	21	21	20	18	16	15	17	21	23	22	21	19 °C
	tn	14	14	14	12	9	5	3	6	8	12	13	14	3 °C
	clothing	s	s	s	s	s	n	n	n	s	s	s	s	–
	stress	f	f	f	o	o				o	f	f	f	–
	p	307	244	254	53	3	1	1	1	5	58	201	292	1421 mm
0.25 mm	dp	17	15	16	5	1	0	0	0	1	4	12	16	87 days

Ndola

latitude: 13°00's longitude: 28°39'e height: 1270 m

GAwal	110-4	Jan	Feb	Mar	Apr	May	Jun	Jul	Aug	Sep	Oct	Nov	Dec	Year
	tx	29	29	29	29	29	27	28	30	33	33	33	30	34 °C
	t	21	22	21	20	18	15	15	17	21	23	23	22	20 °C
	tn	14	14	13	9	4	2	0	3	7	12	13	14	0 °C
	clothing	s	s	s	s	s	n	n	n	s	t	t	s	–
	stress	f	f	f	f	o			o	o	f	f	f	–
	p	288	252	182	39	5	0	0	1	1	19	130	249	1166 mm
0.2 mm	dp	22	19	17	4	1	0	0	0	0	3	13	19	98 days
	sunshine	136	138	167	243	273	282	294	301	282	267	192	164	2740 hours
	%	34	39	44	69	76	83	83	83	78	70	50	41	62 %

Balovale

latitude: 13°34's longitude: 23°06'e height: 1091 m

GAwab	110-5	Jan	Feb	Mar	Apr	May	Jun	Jul	Aug	Sep	Oct	Nov	Dec	Year
	tx	34	33	33	33	33	31	33	36	39	39	37	34	39 °C
	t	23	23	23	22	20	18	18	21	24	25	24	23	22 °C
	tn	16	16	15	12	7	6	6	7	9	14	15	16	4 °C
	clothing	s	s	s	s	s	s	s	s	s	s	s	s	–
	stress	f	f	f	f	f	f	f	f	f	f	f	f	–
	p	216	175	147	30	1	0	0	1	8	58	112	226	975 mm
0.25 mm	dp	15	13	13	4	0	0	0	0	1	6	11	17	80 days

Broken Hill

latitude: 14°25's longitude: 28°29'e height: 1208 m

GAwal	110-6	Jan	Feb	Mar	Apr	May	Jun	Jul	Aug	Sep	Oct	Nov	Dec	Year
	tx	30	29	29	30	29	27	28	30	33	36	34	31	36 °C
	t	21	21	21	20	18	16	15	18	22	25	23	21	20 °C
	tn	16	15	13	11	7	4	4	6	10	14	14	15	4 °C
	clothing	s	s	s	s	s	n	n	s	s	t	t	s	–
	stress	f	f	f	f	o			o	f	f	f	f	–
	p	226	207	121	20	1	0	0	2	2	18	96	232	925 mm
0.2 mm	dp	22	18	14	3	0	0	0	0	0	3	12	19	92 days

Mongu

latitude: 15°15's longitude: 23°10'e height: 1053 m

GAwal	110-7	Jan	Feb	Mar	Apr	May	Jun	Jul	Aug	Sep	Oct	Nov	Dec	Year
	tx	32	32	31	32	31	29	30	33	36	37	36	33	38 °C
	t	22	22	23	22	20	17	18	20	24	26	24	22	22 °C
	tn	17	17	16	14	8	5	5	9	11	16	15	16	4 °C
	clothing	s	s	s	s	s	s	s	s	t	t	t	s	–
	stress	f	f	f	f	o	o	o	f	f	f	f	f	–
	p	217	211	145	37	1	1	0	0	2	35	102	219	969 mm
0.2 mm	dp	21	18	15	4	0	0	0	0	1	8	13	19	99 days
	sunshine	152	147	202	255	285	276	304	313	282	260	204	170	2850 hours
	%	38	41	53	72	81	82	87	87	78	67	53	42	64 %

Lusaka

latitude: 15°25′s longitude: 28°19′e height: 1279 m

GAwal	110-8	Jan	Feb	Mar	Apr	May	Jun	Jul	Aug	Sep	Oct	Nov	Dec	Year
	tx	29	28	28	29	28	26	27	29	33	35	34	31	35 °C
	t	22	22	21	21	19	16	16	19	22	25	23	22	21 °C
	tn	16	15	14	12	9	7	7	8	11	14	14	15	6 °C
	clothing	s	s	s	s	s	n	n	s	s	t	t	s	–
	stress	f	f	f	f	o	o	o	o	f	f	f	f	–
	p	224	180	114	17	2	0	0	0	0	14	87	191	829 mm
0.2 mm	dp	16	14	10	2	0	0	0	0	0	2	9	15	68 days
	sunshine	152	147	198	255	282	267	288	301	285	276	207	174	2832 hours
	%	38	41	52	72	80	79	82	84	79	72	54	43	64 %

Livingstone

latitude: 17°49′s longitude: 25°49′e height: 987 m

GBSal	110-9	Jan	Feb	Mar	Apr	May	Jun	Jul	Aug	Sep	Oct	Nov	Dec	Year
	tx	34	33	32	33	32	29	31	32	36	38	38	35	39 °C
	t	23	23	23	22	19	16	16	19	24	27	25	24	22 °C
	tn	16	16	13	11	6	3	3	6	9	15	16	16	1 °C
	clothing	t	t	t	s	s	n	n	s	t	t	t	t	–
	stress	f	f	f	f	o	o	o	o	f	f	f	f	–
	p	186	175	101	28	4	1	0	0	2	26	92	164	781 mm
0.2 mm	dp	18	15	11	3	1	0	0	0	0	4	10	15	77 days
	sunshine	178	175	239	255	294	273	298	313	297	276	198	183	2979 hours
	%	44	49	63	73	84	82	86	88	83	71	51	45	67 %

4.3.110.2. Predominant general weather situations are described in 4.3.0.1. under Equatorial South Africa. A hygrothermal diagram of Lusaka is to be found in 4.3.0.2.

4.3.111. Malawi

4.3.111.1. Climatic description and tables

1. Tropical Summer Rain Climate Aw and Tropical Summer Rain Mountain Climate GAw
2. According to height summer hot or warm, winter warm or mild.
3. High sun period: rainy days at times, often showers in the afternoon, in the morning frequently dry and sunny; low sun period: generally dry and sunny or cloudy.
4. Wind: occasional gales.

Karonga

latitude: 9°57′s longitude: 33°54′e height: 529 m

GAwab	111-1	Jan	Feb	Mar	Apr	May	Jun	Jul	Aug	Sep	Oct	Nov	Dec	Year
	tx	33	32	32	31	31	29	29	31	33	36	36	34	36 °C
	t	26	25	25	25	23	22	21	22	23	25	26	26	24 °C
	tn	20	20	20	19	15	13	12	13	14	16	19	20	12 °C
	clothing	t	t	t	t	t	s	s	s	t	t	t	t	–
	stress	ff	ff	ff	ff	f	f	o	o	f	f	f	ff	–
	p	190	178	326	201	35	6	0	3	1	4	32	159	1135 mm
0.2 mm	dp	16	17	23	15	5	2	0	1	0	1	6	17	103 days
	sunshine	177	164	220	210	257	258	264	319	306	322	267	167	2932 hours
	%	45	47	58	59	71	74	73	87	85	84	71	42	66 %

Lilongwe

GAwal	111-2	Jan	Feb	Mar	Apr	May	Jun	Jul	Aug	Sep	Oct	Nov	Dec	Year
				latitude: 13°58′s				longitude: 33°42′e				height: 1136 m		
	tx	30	29	29	29	28	26	27	29	31	32	33	32	33 °C
	t	21	21	21	20	18	16	15	17	20	23	23	22	20 °C
	tn	14	15	13	11	6	4	3	4	8	11	14	15	2 °C
	clothing	s	s	s	s	s	n	n	n	s	s	t	s	–
	stress	f	f	f	f	o	o	o	o	o	f	f	f	–
	p	204	210	135	37	5	2	0	2	4	5	75	171	849 mm
	dp	15	14	11	5	1	0	0	0	0	1	5	13	66 days
	sunshine	140	138	192	234	254	231	239	260	261	298	216	152	2615 hours
	%	35	39	51	66	71	68	68	72	72	77	56	38	59 %

Zomba

GAwal	111-3	Jan	Feb	Mar	Apr	May	Jun	Jul	Aug	Sep	Oct	Nov	Dec	Year
				latitude: 15°23′s				longitude: 35°19′e				height: 957 m		
	tx	30	30	29	29	28	27	26	28	31	33	34	32	34 °C
	t	23	23	22	22	20	17	17	19	21	24	24	23	21 °C
	tn	16	16	15	13	11	9	8	9	11	13	14	16	8 °C
	clothing	t	t	s	s	s	n	n	s	s	t	t	t	–
	stress	f	f	f	f	o	o		o	f	f	f	f	–
	p	276	290	200	77	27	13	6	8	18	18	135	279	1346 mm
0.2 mm	dp	21	18	19	10	4	4	3	2	1	3	10	19	114 days

Chileka

GAwab	111-4	Jan	Feb	Mar	Apr	May	Jun	Jul	Aug	Sep	Oct	Nov	Dec	Year
				latitude: 15°41′s				longitude: 34°58′e				height: 767 m		
	t	23	23	23	22	20	18	18	20	23	25	25	24	22 °C
	clothing	t	t	t	s	s	s	s	s	t	t	t	t	–
	stress	f	f	f	f	o	o	o	o	o	f	f	f	–
	p	210	169	122	39	8	4	3	1	4	22	80	166	828 mm
	dp	15	13	11	5	1	1	1	0	0	3	8	12	70 days
	sunshine	177	172	198	228	245	216	220	242	243	267	210	183	2601 hours
	%	44	48	52	65	69	64	63	67	67	69	54	45	59 %

4.3.111.2. Predominant general weather situations are described in 4.3.0.1. under Equatorial South Africa.

4.3.112. Zimbabwe (Rhodesia)

4.3.112.1. Climatic descriptions and tables

1. Climatic area: East of the country, except southernmost part

1.1. Tropical Summer Rain Climate Aw and Tropical Summer Rain Mountain Climate GAw

1.2. According to height, summer hot to warm, winter warm to mild.

1.3. Predominantly dry and sunny; high sun period: often showers in the afternoon, rainy days at times.

1.4. Wind: occasional gales.

Salisbury

latitude: 17°56′s longitude: 31°06′e height: 1479 m

GAwbl	112-1	Jan	Feb	Mar	Apr	May	Jun	Jul	Aug	Sep	Oct	Nov	Dec	Year
	tx	29	29	28	28	27	24	26	27	31	32	32	30	33 °C
	t	20	20	19	19	16	14	14	16	19	21	21	20	18 °C
	tn	13	12	11	9	5	3	2	3	6	10	12	12	2 °C
	clothing	s	s	s	s	n	n	n	n	s	s	s	s	–
	stress	o	o	o	o					o	o	o	o	–
	p	216	172	99	36	11	4	1	3	5	30	100	186	863 mm
	dp	15	13	9	4	1	1	0	0	1	3	10	13	71 days
	sunshine	195	178	214	246	267	252	273	288	285	282	207	189	2876 hours
	%	48	50	56	70	76	76	79	81	79	73	53	46	65 %

Umtali

latitude: 18°58′s longitude: 32°40′e height: 1119 m

GAwal	112-2	Jan	Feb	Mar	Apr	May	Jun	Jul	Aug	Sep	Oct	Nov	Dec	Year
	tx	33	32	31	31	29	26	27	29	33	36	36	33	37 °C
	t	23	22	21	20	18	15	15	16	19	22	22	22	20 °C
	tn	14	14	13	10	7	4	4	6	7	11	13	13	3 °C
	clothing	s	s	s	s	s	n	n	n	s	s	s	s	–
	stress	f	f	f	f	o				o	o	f	f	–
	p	198	158	127	31	13	8	8	8	8	25	91	145	820 mm
0.2 mm	dp	16	15	14	7	4	3	3	3	2	5	10	14	96 days
	sunshine	229	189	223	246	257	225	232	257	261	270	222	217	2830 hours
	%	56	53	59	70	74	68	67	72	73	70	57	53	64 %

Chipinga

latitude: 20°12′s longitude: 32°37′e height: 1132 m

GAwbl	112-3	Jan	Feb	Mar	Apr	May	Jun	Jul	Aug	Sep	Oct	Nov	Dec	Year
	t	20	20	20	19	17	14	14	16	18	20	20	20	18 °C
	clothing	s	s	s	s	s	n	n	n	s	s	s	s	–
	stress	o	o	o	o						o	o	o	–
	p	230	199	145	50	21	29	20	24	20	46	130	203	1116 mm
	dp	13	13	11	6	3	4	3	3	3	5	9	12	87 days
	sunshine	211	223	226	240	260	231	254	260	249	260	232	214	2860 hours
	%	51	62	60	69	75	70	74	73	69	67	59	52	65 %

2. Climatic area: West of the country, except southernmost part
2.1. Steppe Climate BS and Steppe Mountain Climate GBS
2.2. Mostly summer very hot or hot, winter warm to mild.
2.3. Generally dry and sunny, in the high sun period frequent showers in the afternoon.
2.4. Wind: occasional gales.

Wankie

latitude: 18°22′s longitude: 26°29′e height: 783 m

GBSab	112-4	Jan	Feb	Mar	Apr	May	Jun	Jul	Aug	Sep	Oct	Nov	Dec	Year
	tx	38	36	36	35	34	31	32	33	38	41	41	38	41 °C
	t	26	26	26	26	22	19	19	22	26	29	29	27	25 °C
	tn	18	18	17	14	9	7	7	9	12	18	18	18	6 °C
	clothing	t	t	t	t	s	s	s	s	t	t	t	t	–
	stress	ff	ff	ff	f	f	o	o	o	f	ff	ff	ff	–
	p	147	147	79	18	5	0	0	0	0	18	58	119	592 mm
0.2 mm	dp	16	13	9	3	1	0	0	0	0	3	9	13	68 days

Bulawayo

latitude: 20°09′s longitude: 28°37′e height: 1345 m

GBSbl	112-5	Jan	Feb	Mar	Apr	May	Jun	Jul	Aug	Sep	Oct	Nov	Dec	Year
	tx	32	31	30	30	28	26	26	28	32	35	34	33	36 °C
	t	21	21	20	19	16	13	14	16	19	22	22	21	19 °C
	tn	13	13	11	9	6	3	3	5	7	11	12	13	2 °C
	clothing	s	s	s	s	n	n	n	n	s	s	s	s	–
	stress	o	o	o						o	o	o	o	–
	p	134	111	65	21	9	3	0	1	5	55	89	125	589 mm
	dp	10	8	5	3	1	0	0	0	1	3	8	10	50 days
	sunshine	217	209	239	255	279	267	285	301	273	267	232	217	3041 hours
	%	53	58	63	73	81	81	83	85	76	69	59	52	69 %

3. *Climatic area:* Southernmost part of the country
3.1. Desert Climate BW and Desert Mountain Climate GBW
3.2. Summer very hot, winter warm to mild, in lowlands.
3.3. Generally dry and sunny; in the high sun period showers now and then in the afternoon; at times a rainy day.
3.4. Wind: occasional gales and dust-storms.

Beit Bridge

latitude: 22°13′s longitude: 29°59′e height: 459 m

BWal	112-6	Jan	Feb	Mar	Apr	May	Jun	Jul	Aug	Sep	Oct	Nov	Dec	Year
	tx	39	38	37	36	34	30	31	33	37	41	41	41	42 °C
	t	27	27	26	24	20	17	17	19	22	26	27	27	22 °C
	tn	17	18	14	12	7	3	3	4	8	13	16	16	2 °C
	clothing	t	t	t	t	s	s	s	s	s	t	t	t	–
	stress	ff	ff	f	f	f	o	o	f	f	ff	ff	ff	–
	p	69	51	41	13	10	3	2	1	8	25	51	53	325 mm
0.2 mm	dp	6	6	5	2	1	1	0	0	1	2	5	6	35 days

4.3.112.2. Predominant general weather situations are described in 4.3.0.1. under Equatorial South Africa. A hygrothermal diagram of Salisbury is to be found in 4.3.0.2.

4.3.113. South Africa

4.3.113.1. Climatic descriptions and tables

1. *Climatic area:* Central and southeast Transvaal, north of Orange State and Natal
1.1. Tropical Summer Rain Climate Aw and
Tropical Summer Rain Mountain Climate GAw
1.2. Mostly summer warm, winter mild.
1.3. Predominantly dry and sunny; high sun period: frequent showers in the afternoon, at times rainy days.
1.4. Wind: occasional gales. Fog: occasionally.

Pretoria

latitude: 25°45′s longitude: 28°14′e height: 1460 m

GAwbl	113-1	Jan	Feb	Mar	Apr	May	Jun	Jul	Aug	Sep	Oct	Nov	Dec	Year
	tx	32	31	29	28	25	23	23	26	29	32	32	32	33 °C
	t	21	21	20	17	13	10	10	13	16	20	20	21	17 °C
	tn	12	12	9	4	1	-2	-2	1	4	8	9	11	-3 °C
	clothing	s	s	s	s	n	n	n	n	n	s	s	s	–
	stress	o	o	o	o						o	o	o	–
	p	125	108	90	41	25	10	10	5	21	56	126	129	746 mm
0.2 mm	dp	14	11	10	7	3	2	2	1	4	8	13	13	88 days
	sunshine	267	236	251	257	283	272	288	307	285	286	260	270	3262 hours
	%	63	64	66	74	84	86	87	88	80	73	65	63	74 %

Johannesburg

latitude: 26°08′s longitude: 28°14′e height: 1700 m

GAwbl	113-2	Jan	Feb	Mar	Apr	May	Jun	Jul	Aug	Sep	Oct	Nov	Dec	Year
	tx	31	29	28	26	24	21	22	25	28	31	31	31	32 °C
	t	19	19	18	16	13	10	10	13	15	18	18	18	16 °C
	tn	11	10	8	3	0	-2	-3	-1	2	5	7	8	-4 °C
	clothing	s	s	s	n	n	n	n	n	n	s	s	s	–
	stress	o	o	o							o	o	o	–
	p	123	106	89	51	22	8	9	7	25	61	116	135	752 mm
0.2 mm	dp	14	11	11	6	4	2	2	1	3	9	12	14	89 days
	sunshine	254	237	249	250	279	266	283	304	283	283	241	261	3190 hours
	%	60	65	65	73	83	84	86	88	79	72	60	61	72 %

Newcastle

latitude: 27°45′s longitude: 29°56′e height: 1200 m

GAwbl	113-3	Jan	Feb	Mar	Apr	May	Jun	Jul	Aug	Sep	Oct	Nov	Dec	Year
	tx	36	34	33	31	28	26	26	29	33	36	35	36	37 °C
	t	22	22	20	17	14	11	11	14	17	19	21	22	17 °C
	tn	10	11	8	3	-2	-5	-5	-2	1	6	7	9	-6 °C
	clothing	s	s	s	n	n	n	n	n	n	s	s	s	–
	stress	o	o	o							o	o	o	–
	p	168	140	130	43	25	13	15	13	33	76	130	135	921 mm
	dp	11	10	9	5	2	1	2	1	4	8	10	11	74 days

2. *Climatic area:* Transvaal (except centre and south-east), north-east and centre of Cape Province as far as bordering mountains, south of Orange

2.1. Steppe Climate BS and Steppe Mountain Climate GBS

2.2. Summer warm, partly hot, winter mild.

2.3. Predominantly dry and sunny; in summer frequent showers or rain.

2.4. Wind: occasional gales and dust-storms.

Pietersburg

				latitude: 23°52's				longitude: 29°27'e				height: 1250 m		
GBSbl	113-4	Jan	Feb	Mar	Apr	May	Jun	Jul	Aug	Sep	Oct	Nov	Dec	Year
	tx	32	31	31	29	27	24	24	27	31	33	33	32	34 °C
	t	21	21	20	18	15	12	12	14	17	20	20	21	17 °C
	tn	12	11	9	5	1	-2	-2	0	2	7	9	11	-3 °C
	clothing	s	s	s	s	n	n	n	n	n	s	s	s	–
	stress	o	o	o							o	o	o	–
	p	82	71	56	31	11	6	3	1	12	31	76	100	480 mm
0.2 mm	dp	10	8	8	5	2	1	1	1	2	5	9	10	62 days
	sunshine	248	213	232	243	269	266	271	286	269	270	251	247	3066 hours
	%	59	58	61	70	79	83	81	82	75	69	63	59	69 %

Mafeking

				latitude: 25°52's				longitude: 25°38'e				height: 1300 m		
GBSal	113-5	Jan	Feb	Mar	Apr	May	Jun	Jul	Aug	Sep	Oct	Nov	Dec	Year
	tx	35	33	32	30	27	24	24	28	32	34	34	35	36 °C
	t	23	22	21	17	14	11	11	14	18	22	22	23	18 °C
	tn	13	12	9	4	-1	-4	-4	-1	2	7	8	11	-5 °C
	clothing	s	s	s	n	n	n	n	n	s	s	s	s	–
	stress	f	o	o						o	o	o	f	–
	p	86	70	73	39	16	8	3	3	9	37	56	98	498 mm
	dp	9	8	8	5	2	1	1	0	1	5	7	7	54 days

Kimberley

				latitude: 28°48's				longitude: 24°46'e				height: 1250 m		
GBSal	113-6	Jan	Feb	Mar	Apr	May	Jun	Jul	Aug	Sep	Oct	Nov	Dec	Year
	tx	37	36	33	31	27	24	24	28	32	33	36	37	38 °C
	t	24	23	21	17	13	10	10	12	16	20	22	24	18 °C
	tn	12	12	9	3	-1	-3	-4	-3	0	4	7	9	-5 °C
	clothing	t	s	s	n	n	n	n	n	n	s	s	t	–
	stress	f	o	o							o	o	f	–
	p	53	61	71	43	19	9	6	11	11	30	45	58	417 mm
	dp	6	7	8	5	3	1	1	1	1	4	5	5	47 days
	sunshine	310	265	271	267	269	264	281	301	295	309	311	323	3468 hours
	%	73	72	71	78	81	85	87	88	83	78	76	75	78 %

Bloemfontein

				latitude: 29°06's				longitude: 26°18'e				height: 1400 m		
GBSbk	113-7	Jan	Feb	Mar	Apr	May	Jun	Jul	Aug	Sep	Oct	Nov	Dec	Year
	tx	35	32	31	28	24	21	21	25	29	33	33	34	36 °C
	t	22	21	19	15	11	8	8	10	14	18	20	21	16 °C
	tn	10	9	7	2	-2	-4	-5	-4	-2	3	6	7	-6 °C
	clothing	s	s	s	n	n	n	n	n	n	s	s	s	–
	stress	o	o	o							o	o	o	–
	p	83	81	76	51	22	7	10	14	19	48	68	68	547 mm
0.2 mm	dp	11	11	10	8	5	2	2	3	3	7	7	8	77 days
	sunshine	306	263	269	262	267	256	273	293	289	299	301	320	3398 hours
	%	71	71	71	77	81	83	84	85	81	76	74	74	77 %

Okiep

latitude: 29° 36′s longitude: 17° 52′e height: 918 m

GBSal	113-8	Jan	Feb	Mar	Apr	May	Jun	Jul	Aug	Sep	Oct	Nov	Dec	Year
	t	23	23	22	18	14	12	12	12	15	17	20	21	18 °C
	clothing	s	s	s	s	n	n	n	n	n	n	s	s	–
	stress	o	o	o	o							o	o	–
	p	2	13	12	15	40	26	32	24	3	8	7	8	190 mm

Calvinia

latitude: 31° 28′s longitude: 19° 46′e height: 982 m

GBSak	113-9	Jan	Feb	Mar	Apr	May	Jun	Jul	Aug	Sep	Oct	Nov	Dec	Year
	tx	38	38	37	33	28	23	24	28	32	33	36	38	39 °C
	t	23	22	21	16	13	11	9	11	13	17	18	21	16 °C
	tn	7	7	5	1	-1	-3	-4	-3	-1	1	3	5	-5 °C
	clothing	s	s	s	n	n	n	n	n	n	n	s	s	–
	stress	o	o	o								o	o	–
	p	8	13	18	20	28	28	23	23	18	10	13	8	210 mm
	dp	1	2	2	3	4	4	5	5	4	2	2	1	35 days

3\. *Climatic area:* north-west of Cape Province as far as slopes of Mountains in the east and west

3.1. Desert Climate BW and Desert Mountain Climate GBW, coastal areas: Marine Desert Climate BM

3.2. Summer hot or very hot, at the coast warm; winter generally mild.

3.3. Generally dry and sunny, partly cloudy, in coastal areas also overcast; at times scattered showers or rain.

3.4. Wind: occasional gales and dust-storms. Fog: frequent on the coast.

Upington

latitude: 28° 26′s longitude: 21° 16′e height: 800 m

GBWhl	113-10	Jan	Feb	Mar	Apr	May	Jun	Jul	Aug	Sep	Oct	Nov	Dec	Year
	tx	40	39	38	34	31	27	27	31	34	37	39	40	41 °C
	t	28	26	25	20	14	12	12	14	18	21	26	27	20 °C
	tn	13	14	12	6	2	-2	-2	-1	2	7	9	12	-3 °C
	clothing	t	t	t	s	n	n	n	n	s	s	t	t	–
	stress	f	f	f	o					o	o	f	f	–
	p	24	37	36	27	11	3	5	3	3	11	16	21	197 mm
	dp	3	4	4	2	1	1	1	1	1	2	2	2	23 days
	sunshine	355	302	300	290	286	270	285	309	311	332	340	370	3754 hours
	%	83	82	79	85	86	87	88	90	87	84	84	86	85 %

Alexander Bay

latitude: 28° 34′s longitude: 16° 32′e height: 0 m

BMbl	113-11	Jan	Feb	Mar	Apr	May	Jun	Jul	Aug	Sep	Oct	Nov	Dec	Year
	t	19	19	18	16	14	14	13	14	14	15	17	18	16 °C
	clothing	s	s	s	n	n	n	n	n	n	n	n	s	–
	stress	o	o	o									o	–
	p	1	1	6	4	7	4	8	4	3	2	2	2	44 m
	sunshine	321	283	287	263	252	246	242	259	263	295	310	325	3346 hours
	%	75	77	75	77	76	79	74	75	74	75	76	75	76 %

Beaufort West

latitude: 32°18′s longitude: 22°40′e height: 900 m

GBWal	113-12	Jan	Feb	Mar	Apr	May	Jun	Jul	Aug	Sep	Oct	Nov	Dec	Year
	t	23	23	21	17	14	11	11	12	15	17	20	22	17 °C
	clothing	s	s	s	n	n	n	n	n	n	n	s	s	–
	stress	o	o	o								o	o	–
	p	14	32	30	16	13	6	10	12	11	15	27	25	211 mm

4. *Climatic area:* Coastal area in south-west and south of Cape Province as far as bordering mountains
4.1. Subtropical Winter Rain Climate Cs
4.2. Summer warm, winter mild.
4.3. Summer generally dry and sunny; winter frequently variable weather.
4.4. Wind: occasional gales. Fog: occasionally.

Cape Town

latitude: 33°56′s longitude: 18°29′e height: 12 m

Csbl	113-13	Jan	Feb	Mar	Apr	May	Jun	Jul	Aug	Sep	Oct	Nov	Dec	Year
	tx	33	34	34	32	29	27	25	27	28	29	31	32	36 °C
	t	22	22	20	17	15	14	13	13	15	16	19	20	17 °C
	tn	9	10	8	6	3	1	1	2	3	4	7	8	0 °C
	clothing	s	s	s	s	n	n	n	n	n	n	s	s	–
	stress	o	o	o	o							o	o	–
	p	9	23	19	76	123	92	96	96	41	40	21	14	652 mm
0.2 mm	dp	4	3	5	8	11	12	13	13	9	7	5	5	95 days
	sunshine	349	293	287	220	180	181	185	208	227	281	301	344	3059 hours
	%	79	78	75	65	56	61	59	62	64	70	72	77	69 %

5. *Climatic area:* Coastal area and mountains in the east of Cape Province, south of Natal
5.1. Subtropical Rain Climate Cr and Subtropical Rain Mountain Climate GCr
5.2. Summer hot or warm, winter mild.
5.3. Partly dry and sunny, partly variable weather.
5.4. Wind: occasional gales. Fog: occasionally.

Durban

latitude: 29°58′s longitude: 30°57′e height: 8 m

Cral	113-14	Jan	Feb	Mar	Apr	May	Jun	Jul	Aug	Sep	Oct	Nov	Dec	Year
	tx	31	31	30	30	28	27	28	27	28	30	31	31	35 °C
	t	23	24	23	21	19	16	17	17	18	20	21	22	20 °C
	tn	17	17	16	13	9	7	7	7	9	12	13	15	6 °C
	clothing	s	t	s	s	s	n	n	n	s	s	s	s	–
	stress	o	o	o	o	o				o	o	o	o	–
	p	119	127	126	94	59	36	29	43	63	94	125	129	1044 mm
0.2 mm	dp	13	12	12	9	5	4	4	7	9	13	15	14	117 days
	sunshine	197	190	200	209	226	221	230	220	185	167	165	183	2393 hours
	%	46	51	52	61	69	72	71	64	52	42	40	42	54 %

East London

latitude: 33°02's longitude: 27°50'e height: 0 m

Crbl	113-15	Jan	Feb	Mar	Apr	May	Jun	Jul	Aug	Sep	Oct	Nov	Dec	Year
	tx	31	32	31	32	32	28	31	33	33	31	29	30	36 °C
	t	21	21	20	19	17	15	15	15	16	17	19	20	18 °C
	tn	14	14	13	10	7	6	5	6	8	9	11	12	4 °C
	clothing	s	s	s	s	s	n	n	n	n	n	s	s	–
	stress	o	o	o	o	o						o	o	–
	p	75	87	109	74	64	37	47	43	90	115	118	88	947 mm
	dp	8	9	9	6	4	2	2	4	6	9	8	9	76 days
	sunshine	221	204	211	215	225	226	237	228	218	210	225	240	2660 hours
	%	51	54	55	63	69	75	75	68	61	53	54	54	60 %

George

latitude: 33°58's longitude: 22°25'e height: 0 m

Crbl	113-16	Jan	Feb	Mar	Apr	May	Jun	Jul	Aug	Sep	Oct	Nov	Dec	Year
	t	19	19	18	16	14	13	12	12	13	14	16	18	15 °C
	clothing	s	s	s	n	n	n	n	n	n	n	n	s	–
	stress	o	o	o									o	–
	p	75	65	80	57	56	34	50	56	76	83	78	67	777 mm

Port Elizabeth

latitude: 33°59's longitude: 25°36'e height: 0 m

Crbl	113-17	Jan	Feb	Mar	Apr	May	Jun	Jul	Aug	Sep	Oct	Nov	Dec	Year
	tx	32	32	34	33	31	28	28	32	32	28	31	31	37 °C
	t	20	21	20	17	15	14	13	14	15	16	18	20	17 °C
	tn	11	11	9	7	4	2	2	2	4	5	8	9	1 °C
	clothing	s	s	s	n	n	n	n	n	n	n	s	s	–
	stress	o	o	o								o	o	–
	p	38	33	48	44	66	59	59	60	70	61	61	42	641 mm
0.2 mm	dp	8	8	8	8	8	8	8	9	10	11	9	9	104 days
	sunshine	268	228	234	225	218	211	225	237	226	234	255	285	2846 hours
	%	61	61	61	67	68	71	72	70	63	58	61	64	64 %

4.3.113.2. Predominant general weather situations are described in 4.3.0.1. under South Africa. Hygrothermal diagrams of Pretoria and Cape Town are to be found in 4.3.0.2. *Wind name:* Berg wind is a hot dry foehn-like easterly wind blowing in winter from the high plateau down to the coastal lowlands, especially in Southwest Africa.

4.3.114. Namibia

4.3.114.1. Climatic descriptions and tables

1. Climatic area: Amboland, Dmaraland

1.1. Steppe Climate BS and Steppe Mountain Climate GBS

1.2. Summer hot, winter mild, at higher levels.

1.3. Predominantly dry and sunny; high sun period; repeatedly showers or rain.

1.4. Wind: occasional gales and dust-storms.

Tsumeb

		latitude: 19°14′s				longitude: 17°43′e				height: 1312 m				
GBSal	114-1	Jan	Feb	Mar	Apr	May	Jun	Jul	Aug	Sep	Oct	Nov	Dec	Year
	tx	36	34	33	32	30	27	28	31	35	37	36	36	37 °C
	t	25	24	23	22	19	17	17	19	23	26	26	25	22 °C
	tn	15	15	13	11	6	4	4	5	9	14	14	15	2 °C
	clothing	t	t	s	s	s	n	n	s	s	t	t	t	–
	stress	f	f	o	o	o			o	o	f	f	f	–
	p	117	119	81	38	5	1	1	1	1	18	56	84	522 mm
	dp	12	11	9	4	1	0	0	0	0	3	7	10	57 days

Windhoek

		latitude: 22°34′s				longitude: 17°06′e				height: 1720 m				
GBSal	114-2	Jan	Feb	Mar	Apr	May	Jun	Jul	Aug	Sep	Oct	Nov	Dec	Year
	tx	33	32	30	28	26	23	23	27	30	32	33	33	35 °C
	t	23	21	20	19	16	13	13	16	20	22	23	23	19 °C
	tn	13	12	11	8	4	2	1	2	4	8	8	11	1 °C
	clothing	t	s	s	s	n	n	n	n	s	s	t	t	–
	stress	o	o	o	o					o	o	o	o	–
	p	77	73	81	38	6	1	1	0	1	12	38	47	370 mm
0.2 mm	dp	10	10	10	5	2	1	1	1	1	3	5	8	55 days
	sunshine	276	240	259	277	310	300	321	330	312	317	294	300	3536 hours
	%	67	66	68	80	90	93	95	94	87	81	74	72	80 %

2. *Climatic area:* Namaland and coastal area

2.1. Desert Climate BW and Desert Mountain Climate GBW, coastal area: Marine Desert Climate BM

2.2. Summer very hot, winter warm or mild, on the coast: summer warm, winter mild.

2.3. Generally dry and sunny, partly cloudy, in coastal areas partly overcast, at times scattered showers or rain.

2.4. Wind: occasional gales and dust storms. Fog: frequent on the coast.

Walvis Bay

		latitude: 22°57′s				longitude: 14°30′e				height: 3 m				
BMbl	114-3	Jan	Feb	Mar	Apr	May	Jun	Jul	Aug	Sep	Oct	Nov	Dec	Year
	tx	29	29	30	33	36	34	32	31	29	26	29	28	37 °C
	t	19	20	19	18	17	16	15	14	14	15	17	18	17 °C
	tn	11	12	12	9	6	3	2	3	4	6	8	11	2 °C
	clothing	s	s	s	s	n	n	n	n	n	n	n	s	–
	stress	o	o	o	o	o	o	o	o	o		o	o	–
	p	0	5	8	2	2	1	1	3	1	0	0	0	23 mm
0.2 mm	dp	1	1	2	1	1	1	1	3	2	1	1	1	16 days
	sunshine	232	189	211	237	251	231	236	220	189	226	210	214	2647 hours
	%	56	52	56	68	73	72	70	63	53	58	53	51	60 %

Keetmanshoop

latitude: 26°34′s longitude: 18°07′e height: 1100 m

GBWal	114-4	Jan	Feb	Mar	Apr	May	Jun	Jul	Aug	Sep	Oct	Nov	Dec	Year
	tx	39	39	37	34	32	27	27	32	34	37	38	39	41 °C
	t	27	26	24	21	16	14	13	15	19	21	24	26	20 °C
	tn	11	13	11	7	3	1	-1	1	2	6	8	10	-1 °C
	clothing	t	t	t	s	n	n	n	n	s	s	t	t	–
	stress	f	f	f	o					o	o	f	f	–
	p	22	31	35	13	5	2	1	1	2	4	14	17	147 mm
0.2 mm	dp	4	4	5	2	1	0	1	0	1	1	2	3	25 days
	sunshine	353	300	312	306	304	287	305	323	319	343	348	370	3870 hours
	%	84	82	82	89	91	91	93	93	89	87	86	86	87 %

4.3.114.2. Predominant general weather situations are described in 4.3.0.1. under South Africa.

4.3.115. Botswana

4.3.115.1. Climatic descriptions and tables

1. Climatic area: almost the whole country, except the southernmost part

1.1. Steppe Climate BS and Steppe Mountain Climate GBS

1.2. Summer hot, winter mild in about 1000 m height.

1.3. Predominantly dry and sunny; high sun period: often showers in the afternoon, the odd rainy day.

1.4. Wind: occasional gales and dust-storms.

Maun

latitude: 19°59′s longitude: 23°25′ e height: 945 m

GBSal	115-1	Jan	Feb	Mar	Apr	May	Jun	Jul	Aug	Sep	Oct	Nov	Dec	Year
	tx	37	36	35	34	32	28	29	33	37	39	38	37	39 °C
	t	24	24	23	22	18	16	16	19	23	27	26	25	22 °C
	tn	15	16	13	8	3	0	0	3	7	12	14	15	-1 °C
	clothing	t	t	t	s	s	n	n	s	t	t	t	t	–
	stress	f	f	f	f	o			o	f	f	f	f	–
	p	110	104	83	25	5	1	0	0	1	16	46	80	471 mm
0.1 mm	dp	12	11	9	4	1	1	0	0	1	3	7	10	59 days
	sunshine	245	209	257	282	310	300	313	332	315	291	264	214	3332 hours
	%	60	58	68	81	89	91	91	94	88	75	67	52	75 %

Ghanzi

latitude: 21°30′s longitude: 21°45′e height: 1132 m

GBSab	115-2	Jan	Feb	Mar	Apr	May	Jun	Jul	Aug	Sep	Oct	Nov	Dec	Year
	tx	37	36	34	33	31	27	28	32	35	37	38	37	39 °C
	t	24	24	24	23	20	18	18	18	24	27	27	25	23 °C
	tn	14	13	11	6	1	-2	-2	0	3	8	10	12	-4 °C
	clothing	t	t	t	s	s	s	s	s	t	t	t	t	–
	stress	f	f	f	o	o	o	o	o	f	f	f	f	–
	p	104	91	91	38	8	1	1	1	3	23	43	66	470 mm
	dp	9	7	6	4	1	0	0	0	1	3	5	7	43 days

Mahalapye

GBSal	115-3	latitude: 23°04's					longitude: 26°48'e				height: 1005 m			
		Jan	Feb	Mar	Apr	May	Jun	Jul	Aug	Sep	Oct	Nov	Dec	Year
	tx	38	36	34	33	31	27	28	31	35	38	38	37	39 °C
	t	25	25	23	20	16	13	13	16	20	24	25	25	20 °C
	tn	15	14	12	7	2	-1	-2	1	4	9	13	14	-3 °C
	clothing	t	t	t	s	n	n	n	n	s	t	t	t	–
	stress	f	f	f	o					o	f	f	f	–
	p	99	69	75	20	10	4	3	3	6	29	63	76	457 mm
0.2 mm	dp	10	9	7	4	2	1	1	0	2	4	7	9	56 days

2. *Climatic area:* Southernmost part of the country
2.1. Desert Climate BW and Desert Mountain Climate GBW
2.2. Summer very hot, winter warm at lower levels.
2.3. Most of the year dry and sunny. Now and then, particularly in the high sun period, scattered showers or rain.
2.4. Wind: occasional gales and dust-storms.

Tsabong

GBWal	115-4	latitude: 26°03's					longitude: 22°27'e				height: 962 m			
		Jan	Feb	Mar	Apr	May	Jun	Jul	Aug	Sep	Oct	Nov	Dec	Year
	tx	39	37	36	33	30	27	27	31	34	37	38	39	39 °C
	t	26	26	23	20	15	12	11	14	18	22	24	26	20 °C
	tn	13	13	10	1	-2	-7	-4	-4	-2	4	8	9	-7 °
	clothing	t	t	t	s	n	n	n	n	s	s	t	t	–
	stress	f	f	f	o					o	o	f	f	–
	p	51	48	49	34	10	11	2	1	5	17	29	38	295 mm
0.1 mm	dp	5	5	6	4	2	1	1	1	1	2	3	5	36 days
	sunshine	332	288	313	288	294	291	301	332	321	344	333	332	3769 hours
	%	79	78	82	84	87	92	91	96	90	87	83	78	85 %

4.3.115.2. Predominant general weather situations are described in 4.3.0.1. under South Africa.

4.3.116 Ngwane (Swaziland)

4.3.116.1. Climatic description and table

1. Tropical Summer Rain Mountain Climate AwG and Tropical Summer Rain Climate Aw
2. Summer warm, winter mild. In the lowlands, summer very hot, winter warm or hot.
3. High sun period: often dry and sunny in the morning, in the afternoon frequent showers; the odd rainy day. Low sun period: generally dry and sunny, occasionally scattered showers.
4. Wind: occasional gales. Fog: occasionally.

Mbabane

GAwbl	116-1	Jan	Feb	Mar	Apr	May	Jun	Jul	Aug	Sep	Oct	Nov	Dec	Year
latitude: 26°19's	longitude: 31°08'e	height: 1163 m												
	tx	31	31	29	29	27	24	26	34	32	33	32	32	34 °C
	t	20	20	19	17	15	12	12	14	16	18	19	20	17 °C
	tn	11	11	9	6	3	0	0	1	3	7	8	10	-1 °C
	clothing	s	s	s	n	n	n	n	n	n	s	s	s	–
	stress	o	o	o					o	o	o	o	o	–
	p	243	212	187	71	34	19	23	28	62	126	169	213	1387 mm
0.1 mm	dp	17	15	15	10	6	3	3	5	9	14	17	18	132 days

4.3.116.2. Predominant general weather situations are described in 4.3.0.1. under South Africa.

4.3.117. Lesotho

4.3.117.1. Climatic description and table

1. Subtropical Rain Mountain Climate GCr
2. Summer warm, winter mild.
3. Partly dry and sunny, partly variable weather. High sun period: occasionally some rainy days.
4. Wind: occasional gales. Fog: occasionally.

Maseru

GCrbk	117-1	Jan	Feb	Mar	Apr	May	Jun	Jul	Aug	Sep	Oct	Nov	Dec	Year
latitude: 29°19's	longitude: 27°29'e	height: 1528 m												
	t	21	20	19	15	11	8	8	11	14	17	19	21	15 °C
	clothing	s	s	s	n	n	n	n	n	n	n	s	s	–
	stress	o	o	o								o	o	–
	p	100	95	94	55	26	12	13	15	20	57	79	81	647 mm
0.1 mm	dp	10	10	10	7	5	2	2	2	3	7	8	9	75 days

4.3.117.2. Predominant general weather situations are described in 4.3.0.1. under South Africa.

4.4. North and Central America

4.4.0.1. Predominant general weather situations

Alaska, Canada and Greenland

In summer, a trough of low pressure extends between the weakly developed centres of action of the atmospheric circulation in this region, which are the Aleutian depression in the Pacific and the Icelandic depression in the Atlantic. In this trough, mostly small disturbances migrate from west to east, causing in their passage rain and heavy cloud for a short time. In winter, both centres of action are highly developed. The pressure field is restless. Almost always eastward-tracking depressions and anticyclones alternate. However, occasionally an anticyclone becomes stationary, its sinking air governing the weather over an extensive area for several days.

United States of America

In summer, the strong North Atlantic anticyclone extends a wedge beyond the east coast into the country. A trough of low pressure along the eastern slopes of the Rocky Mountains joins a depression over Mexico with a weakly developed trough to the north of the continent. Especially in spring and early summer, when this situation is building up, there are advances of warm humid air from the Gulf of Mexico, which are followed by invasions of polar air in the rear of secondary depressions migrating from west to east. The withdrawal of this position in late summer and autumn is characterised by sinking air more often than rising air. In winter, there often exists a trough between the strongly developed centres of action which are the Aleutian and Icelandic depressions. In this trough, depressions and secondary depressions and anticyclones alternate, migrating from west to east. These migrations are frequently blocked by the building-up of a strong continental anticyclone, the sinking air of which often shapes the weather over an extensive area for some time. The constant alternations of the pressure field are the characteristic feature determining variable weather.

Mexico

In summer, the Equatorial Trough often has a fairly stationary extension as far as Mexico. The anticyclone near the Bermudas extends a wedge into the Gulf of Mexico. Equatorial disturbances frequently invade the country, coming both from the Gulf and the Pacific Ocean. In winter, an anticyclone over the United States extends a wedge southwards, the sinking air of which often determines the weather. However, secondary depressions often traverse the country on their way south, and now and then bring polar air a short way into the tropics.

Central America (Guatemala, El Salvador, Honduras, Nicaragua, Costa Rica, Panama) *and West Indies*

In summer, these countries are mostly influenced by the Equatorial Trough, which generally extends near the Pacific coast and often over the Isthmus of Panama into the Caribbean Sea. Along the southern slope of a weak wedge of the Bermuda anticyclone, which often extends into the Gulf of Mexico, tropical disturbances migrate from east to west. They have a decisive influence on the shower activity, especially on the Atlantic side of the region. In front of them, showers are widespread and heavy; in the rear, they are scattered. In winter, the influence of sinking air predominates, belonging to a strong wedge extending from an anticyclone over the continent into the Gulf of Mexico. The wedge moves away at times, when secondary depressions coming from the continent advance south and south-eastwards, often causing strong invasions of polar air.

4.4.0.2. Hygrothermal diagrams

In the following figures, hygrothermal diagrams are given for the stations

Ottawa, Vancouver, Anchorage, New York, Washington, St. Louis, New Orleans, San Francisco, Phoenix, Honolulu, Mexico City, Acapulco, San Salvador, Balboa Heights (Panama Canal Zone), Havana, Kingston.

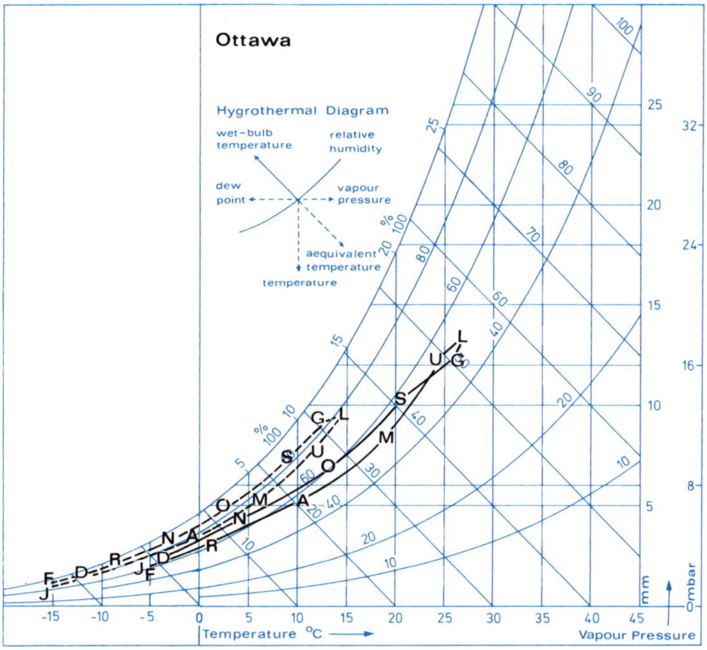
Ottawa
Hygrothermal Diagram
wet-bulb temperature
relative humidity
dew point
vapour pressure
aequivalent temperature
temperature
Temperature °C
Vapour Pressure
mm
mbar

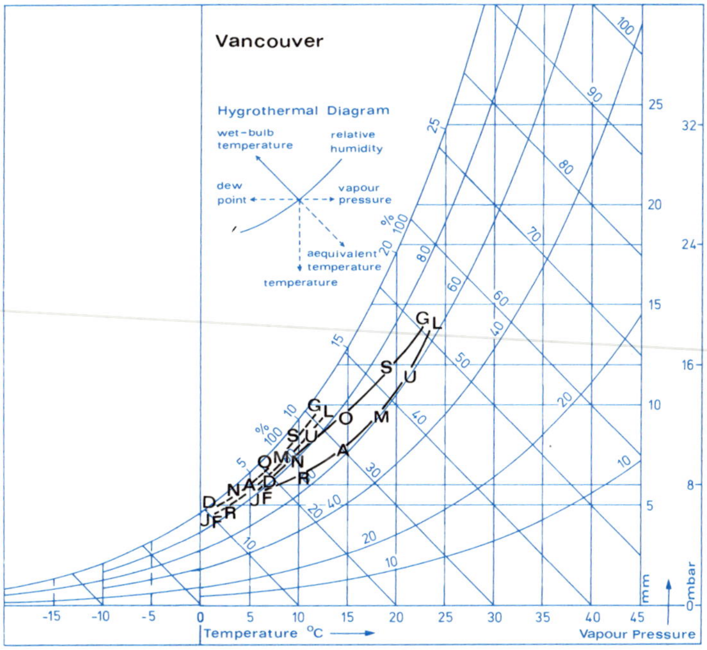
Vancouver
Hygrothermal Diagram
wet-bulb temperature
relative humidity
dew point
vapour pressure
aequivalent temperature
temperature
Temperature °C
Vapour Pressure
mm
mbar

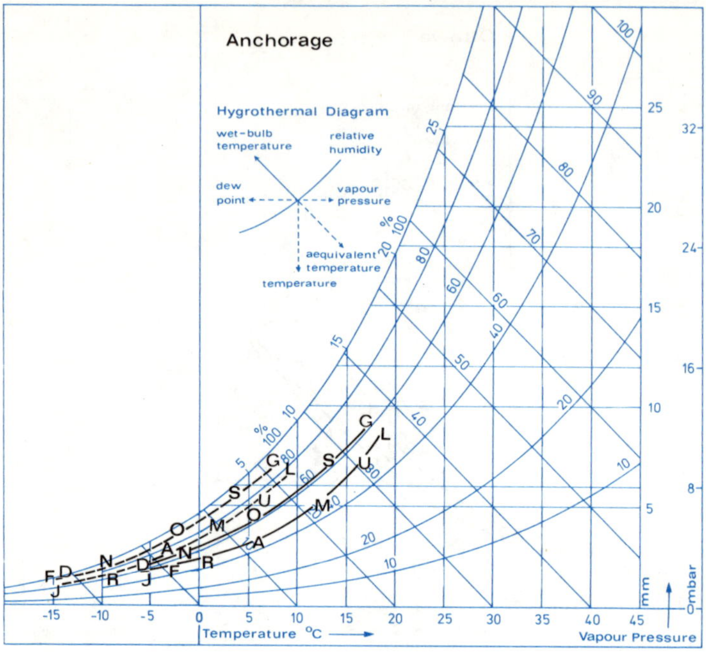
Anchorage
Hygrothermal Diagram
wet-bulb temperature
relative humidity
dew point
vapour pressure
aequivalent temperature
temperature
Temperature °C
Vapour Pressure
mm
mbar

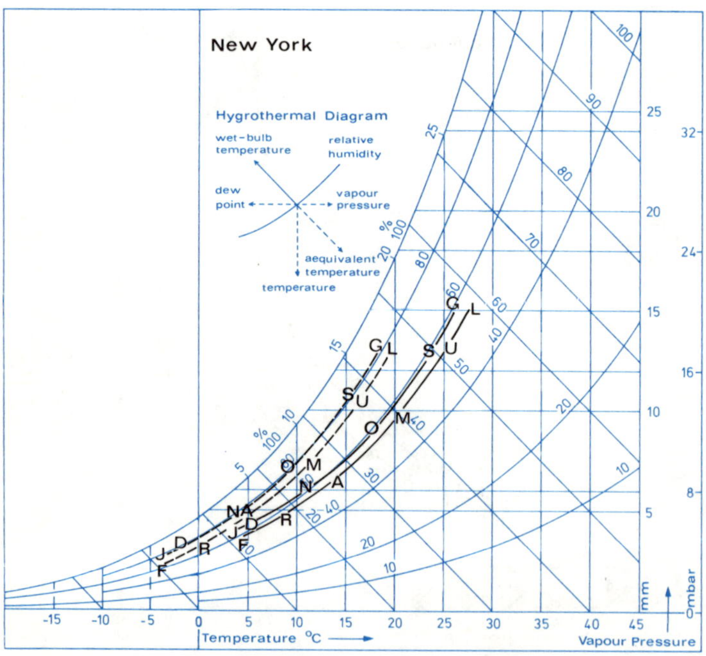
New York
Hygrothermal Diagram
wet-bulb temperature
relative humidity
dew point
vapour pressure
aequivalent temperature
temperature
Temperature °C
Vapour Pressure
mm
mbar

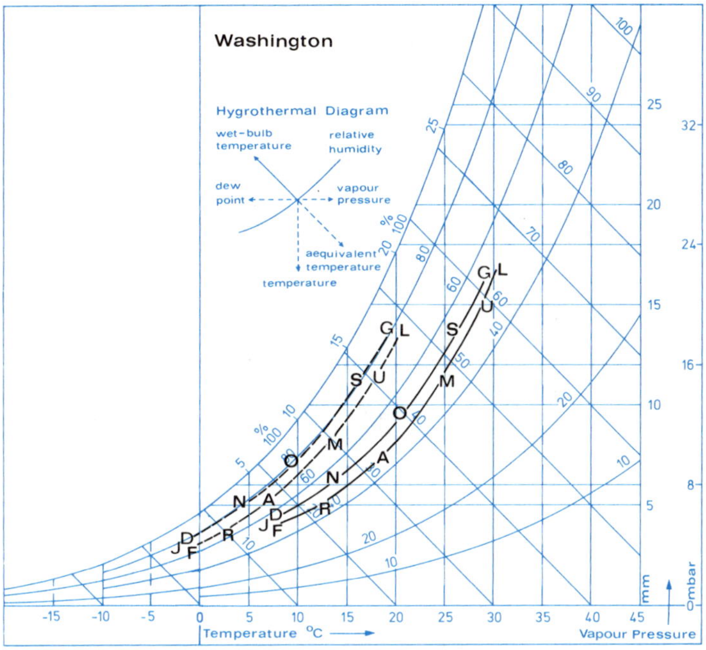
Washington
Hygrothermal Diagram
wet-bulb temperature
relative humidity
dew point
vapour pressure
aequivalent temperature
temperature
Temperature °C
Vapour Pressure
mm
mbar

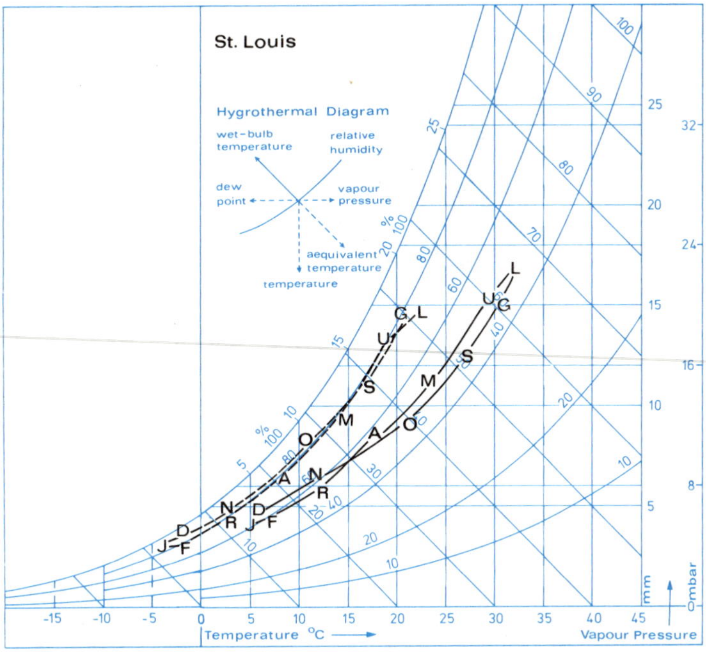
St. Louis
Hygrothermal Diagram
wet-bulb temperature
relative humidity
dew point
vapour pressure
aequivalent temperature
temperature
Temperature °C
Vapour Pressure
mm
mbar

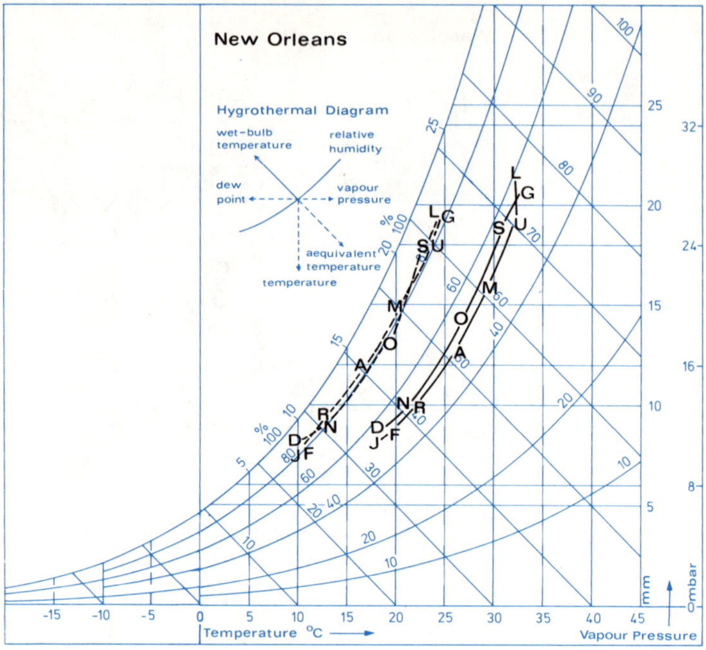
New Orleans
Hygrothermal Diagram
wet-bulb temperature
relative humidity
dew point
vapour pressure
aequivalent temperature
temperature
Temperature °C
Vapour Pressure
mm
mbar

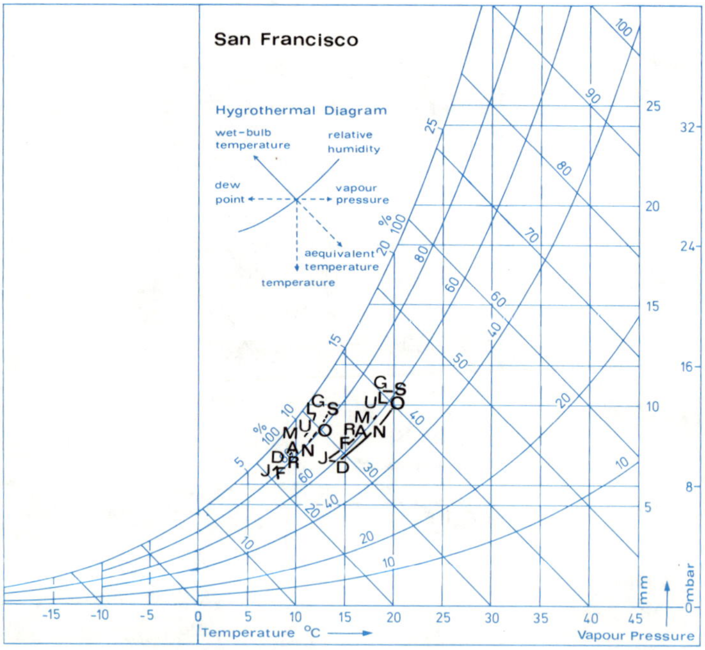
San Francisco
Hygrothermal Diagram
wet-bulb temperature
relative humidity
dew point
vapour pressure
aequivalent temperature
temperature
Temperature °C
Vapour Pressure
mm
mbar

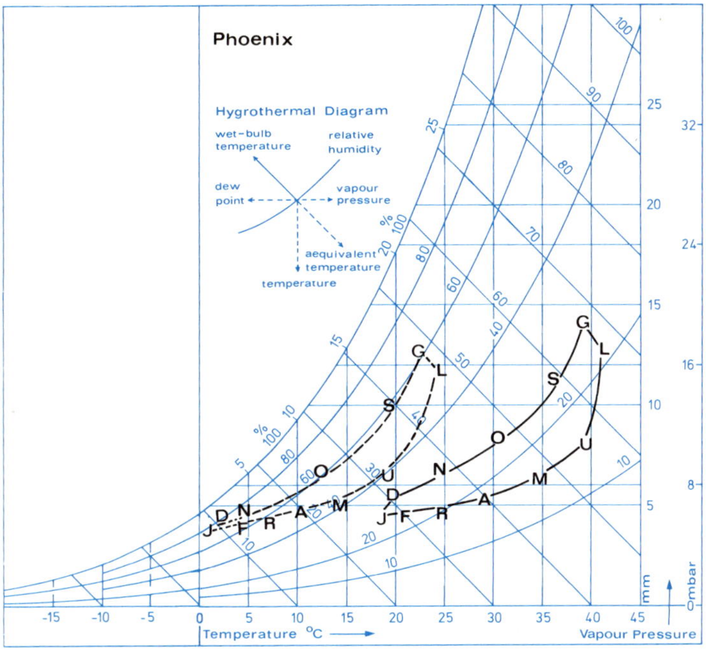
Phoenix
Hygrothermal Diagram
wet-bulb temperature
relative humidity
dew point
vapour pressure
aequivalent temperature
temperature
Temperature °C
Vapour Pressure
mm
mbar

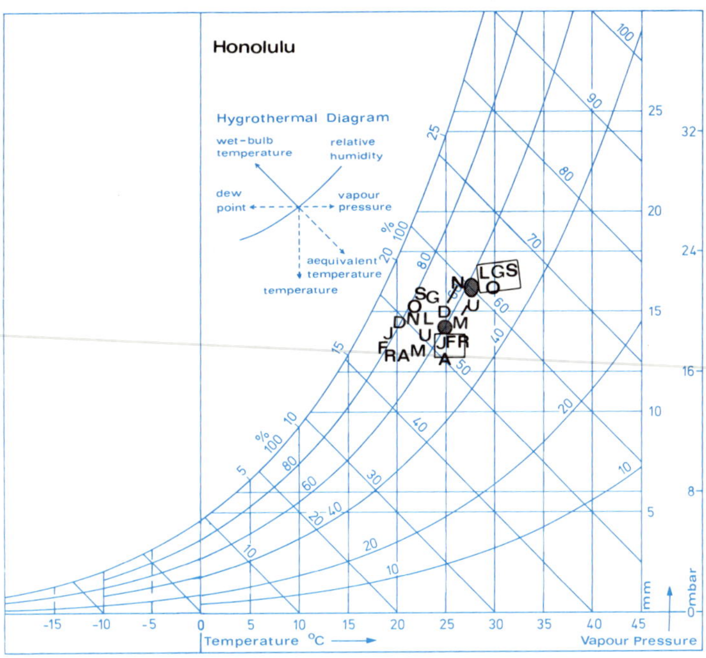
Honolulu
Hygrothermal Diagram
wet-bulb temperature
relative humidity
dew point
vapour pressure
aequivalent temperature
temperature
Temperature °C
Vapour Pressure
mm
mbar

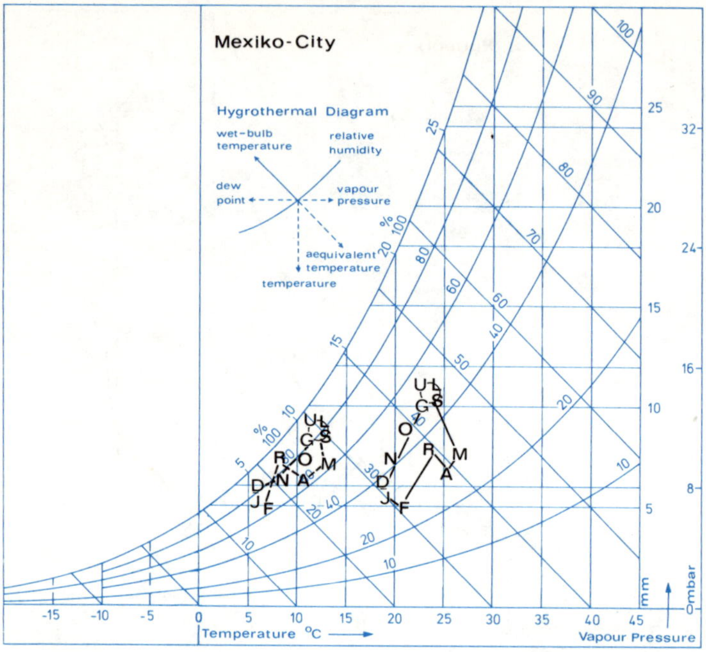
Mexiko-City
Hygrothermal Diagram
wet-bulb temperature
relative humidity
dew point
vapour pressure
aequivalent temperature
temperature
Temperature °C
Vapour Pressure
mm
mbar

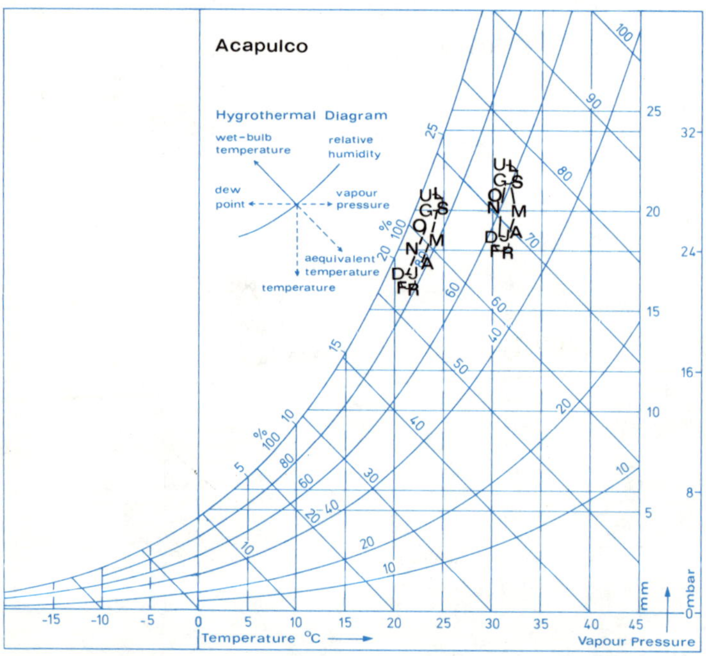
Acapulco
Hygrothermal Diagram
wet-bulb temperature
relative humidity
dew point
vapour pressure
aequivalent temperature
temperature
Temperature °C
Vapour Pressure
mm
mbar

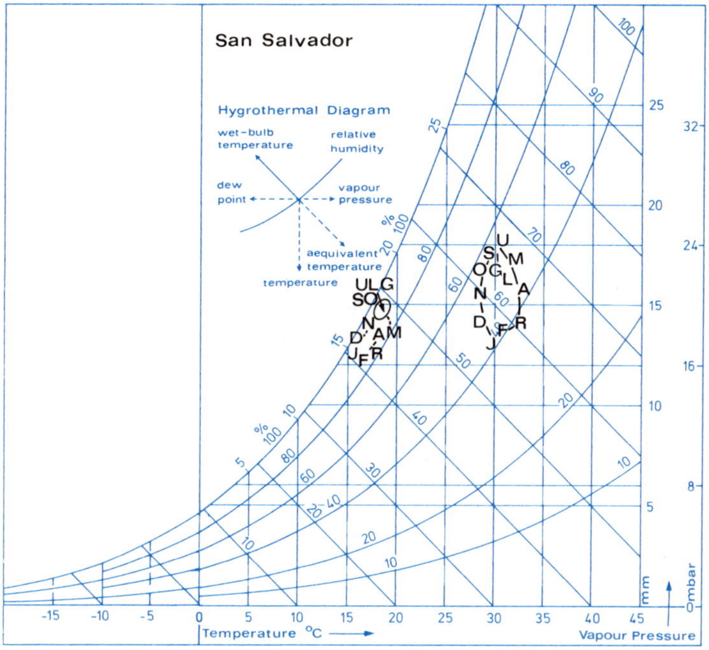
San Salvador
Hygrothermal Diagram
wet-bulb temperature
relative humidity
dew point
vapour pressure
aequivalent temperature
temperature
Temperature °C
Vapour Pressure
mm
mbar

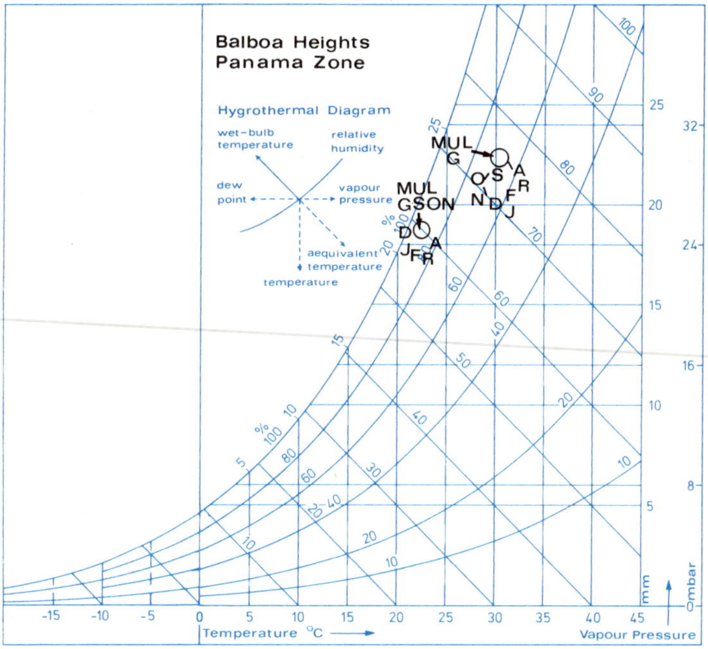
Balboa Heights
Panama Zone
Hygrothermal Diagram
wet-bulb temperature
relative humidity
dew point
vapour pressure
aequivalent temperature
temperature
Temperature °C
Vapour Pressure
mm
mbar

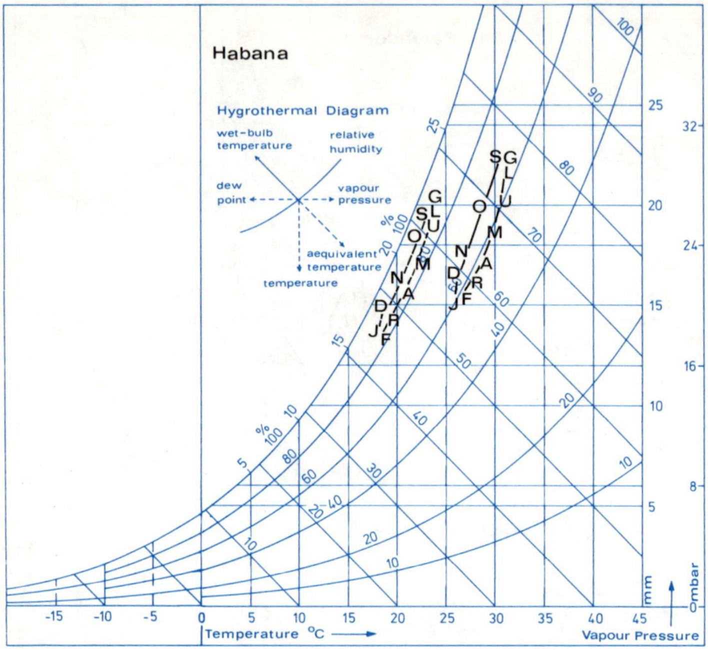
Habana
Hygrothermal Diagram
wet-bulb temperature
relative humidity
dew point
vapour pressure
aequivalent temperature
temperature
Temperature °C
Vapour Pressure
mm
mbar

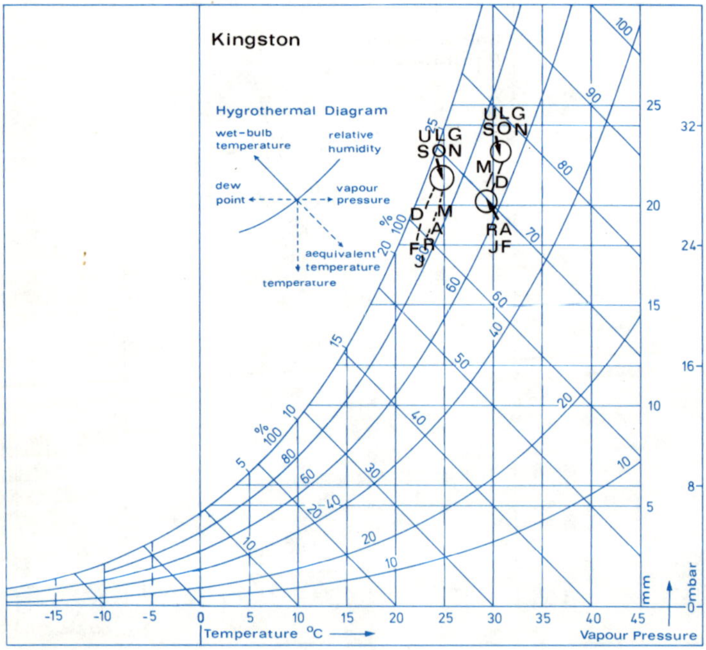
Kingston
Hygrothermal Diagram
wet-bulb temperature
relative humidity
dew point
vapour pressure
aequivalent temperature
temperature
Temperature °C
Vapour Pressure
mm
mbar

4.4.118. Canada

4.4.118.1. Climatic descriptions and tables

1. *Climatic area:* Arctic Archipelago and along the continental coast of the Northwest Territories and northwest of Quebec
1.1. Tundra Climate FT
1.2. Summer short and cold, winter mostly severely cold.
1.3. Variable weather prevails.
1.4. Wind: frequent gales. Fog: prevailing in summer. Snow cover: most of the year.

Alert – Northwest Territories (NWT)

latitude: 82° 30′n longitude: 62° 20′w height: 62 m

FTkd	118-1	Jan	Feb	Mar	Apr	May	Jun	Jul	Aug	Sep	Oct	Nov	Dec	Year
	tx	-17	-11	-18	-2	5	10	14	11	2	-9	-12	-13	15 °C
	t	-32	-33	-33	-24	-11	0	4	1	-10	-20	-26	-30	-18 °C
	tn	-44	-46	-42	-38	-21	-9	-3	-11	-22	-33	-38	-42	-47 °C
	clothing	p	p	p	p	p	w	w	w	p	p	p	p	–
	stress													–
	p	6	6	6	6	9	12	15	29	30	16	6	7	148 mm
0.2 mm	dp	10	8	9	7	9	7	6	10	10	10	10	12	108 days

Eureka – NWT

latitude: 80° 00′n longitude: 85° 56′w height: 11 m

FTkd	118-2	Jan	Feb	Mar	Apr	May	Jun	Jul	Aug	Sep	Oct	Nov	Dec	Year
	tx	-19	-22	-16	-9	3	12	17	12	3	-8	-13	-20	-17 °C
	t	-36	-37	-38	-27	-10	3	6	4	-7	-22	-31	-35	-19 °C
	tn	-49	-49	-48	-41	-23	-5	1	-4	-21	-35	-42	-46	-51 °C
	clothing	p	p	p	p	p	w	n	w	w	p	p	p	–
	stress													–
	p	3	2	2	2	3	3	16	14	11	9	2	2	69 mm
0.2 mm	dp	3	3	2	3	4	2	4	5	7	5	5	3	46 days

Isachsen – NWT

latitude: 78° 47′n longitude: 103° 32′w height: 25 m

FTkd	118-3	Jan	Feb	Mar	Apr	May	Jun	Jul	Aug	Sep	Oct	Nov	Dec	Year
	tx	-18	-24	-19	-8	0	8	14	9	0	-6	-12	-19	15 °C
	t	-35	-37	-35	-24	-12	0	4	1	-8	-19	-28	-32	-19 °C
	tn	-49	-48	-46	-37	-23	-9	-2	-8	-21	-32	-41	-43	-50 °C
	clothing	p	p	p	p	p	w	w	w	w	p	p	p	–
	stress													–
	p	2	2	1	4	8	3	22	23	18	10	4	2	99 mm
0.2 mm	dp	3	3	3	5	8	6	6	8	12	7	4	3	68 days

Mould Bay – NWT

FTkd	118-4	latitude: 76°14'n						longitude: 119°20'w				height: 15 m		
		Jan	Feb	Mar	Apr	May	Jun	Jul	Aug	Sep	Oct	Nov	Dec	Year
	tx	-16	-22	-16	-5	0	8	13	11	2	-5	-10	-21	14 °C
	t	-33	-35	-33	-23	-11	0	4	2	-6	-17	-27	-31	-17 °C
	tn	-46	-46	-44	-36	-20	-8	-2	-7	-20	-30	-39	-43	-48 °C
	clothing stress	p	p	p	p	p	w	w	w	w	p	p	p	–
	p	4	2	3	3	7	4	17	21	11	8	3	3	86 mm
0.2 mm	dp	4	3	4	3	11	6	6	8	10	6	4	6	71 days

Resolute – NWT

FTkd	118-5	latitude: 74°43'n						longitude: 94°59'w				height: 64 m		
		Jan	Feb	Mar	Apr	May	Jun	Jul	Aug	Sep	Oct	Nov	Dec	Year
	tx	-16	-22	-13	-7	2	8	14	12	3	-3	-9	-13	14 °C
	t	-32	-34	-31	-22	-10	1	5	3	-4	-14	-25	-29	-16 °C
	tn	-44	-46	-45	-37	-23	-7	-1	-5	-16	-28	-36	-42	-47 °C
	clothing stress	p	p	p	p	p	w	n	w	w	p	p	p	–
	p	3	3	3	6	7	11	21	34	16	16	6	5	131 mm
0.2 mm	dp	3	4	5	6	10	8	11	10	12	12	5	4	90 days
	sunshine	0	24	153	272	258	258	279	158	68	23	1	0	1494 hours
	%	–	16	43	51	35	36	37	23	16	9	2	–	32 %

Sachs Harbour – NWT

FTkd	118-6	latitude: 71°57'n						longitude: 124°44'w				height: 84 m		
		Jan	Feb	Mar	Apr	May	Jun	Jul	Aug	Sep	Oct	Nov	Dec	Year
	t	-30	-32	-28	-20	-9	2	6	5	-2	-12	-24	-28	-14 °C
	clothing stress	p	p	p	p	w	w	n	n	w	p	p	p	–
	p	2	2	4	3	6	5	25	17	17	11	5	4	101 mm
	sunshine	0	59	187	260	263	344	291	202	71	46	3	0	1726 hours
	%	–	31	52	53	35	48	39	34	17	17	5	–	38 %

Clyde – NWT

FTkd	118-7	latitude: 70°27'n						longitude: 68°33'w				height: 0 m		
		Jan	Feb	Mar	Apr	May	Jun	Jul	Aug	Sep	Oct	Nov	Dec	Year
	t	-27	-29	-26	-17	-6	1	5	4	0	-7	-18	-26	-12 °C
	clothing stress	p	p	p	p	w	w	n	w	w	w	p	p	–
	p	11	7	5	13	14	10	22	33	42	32	14	4	207 mm

Hall Beach – NWT

FTkd	118-8	latitude: 68°47'n						longitude: 81°15'w				height: 0 m		
		Jan	Feb	Mar	Apr	May	Jun	Jul	Aug	Sep	Oct	Nov	Dec	Year
	t	-28	-30	-28	-21	-9	0	6	5	0	-10	-22	-27	-14 °C
	clothing stress	p	p	p	p	w	w	n	n	w	p	p	p	–
	p	6	7	3	7	16	13	48	35	23	10	11	5	184 mm

Coppermine – NWT

latitude: 67°49′n longitude: 115°05′w height: 0 m

FTkd	118-9	Jan	Feb	Mar	Apr	May	Jun	Jul	Aug	Sep	Oct	Nov	Dec	Year
	tx	-12	-12	-7	1	8	22	26	22	14	6	-4	-10	26 °C
	t	-29	-30	-26	-17	-6	3	9	8	3	-7	-20	-26	-11 °C
	tn	-43	-42	-40	-36	-23	-6	1	-1	-8	-28	-34	-39	-44 °C
	clothing stress	p	p	p	p	w	w	n	n	w	w	p	p	–
	p	13	8	15	14	12	20	34	44	29	27	17	13	246 mm
0.2 mm	dp	8	6	7	6	7	6	10	15	10	10	10	8	103 days
	sunshine	5	78	149	219	210	275	283	187	72	43	12	0	1533 hours
	%	5	35	41	47	34	38	38	34	18	15	8	–	33 %

Coral Harbour – NWT

latitude: 64°12′n longitude: 83°22′w height: 59 m

FTkd	118-10	Jan	Feb	Mar	Apr	May	Jun	Jul	Aug	Sep	Oct	Nov	Dec	Year
	t	-30	-29	-23	-15	-5	3	9	8	1	-8	-18	-26	-11 °C
	clothing stress	p	p	p	p	w	w	n	n	w	w	p	p	–
	p	9	10	10	13	22	26	34	40	38	33	17	12	264 mm

Frobisher Bay – NWT

latitude: 63°45′n longitude: 68°33′w height: 21 m

FTkd	118-11	Jan	Feb	Mar	Apr	May	Jun	Jul	Aug	Sep	Oct	Nov	Dec	Year
	t	-26	-25	-21	-13	-2	4	8	7	3	-5	-14	-22	-9 °C
	clothing stress	p	p	p	p	w	w	n	n	w	w	p	p	–
	p	32	35	24	26	21	42	70	58	38	43	40	28	457 mm
	sunshine	26	75	184	245	174	142	190	130	80	58	38	11	1353 hours
	%	15	31	51	54	31	23	31	26	20	19	19	8	30 %

Fort Harrison – Quebec

latitude: 58°27′n longitude: 78°08′w height: 20 m

FTkd	118-12	Jan	Feb	Mar	Apr	May	Jun	Jul	Aug	Sep	Oct	Nov	Dec	Year
	tx	-7	-8	-4	2	7	17	20	19	14	7	2	-4	22 °C
	t	-25	-25	-20	-11	-2	4	9	9	5	0	-8	-18	-7 °C
	tn	-39	-38	-36	-28	-16	-5	-1	1	-3	-12	-24	-35	-41 °C
	clothing stress	p	p	p	p	w	w	n	n	n	w	w	p	–
	p	14	9	16	17	23	30	51	54	62	49	47	23	395 mm
0.2 mm	dp	11	6	9	12	11	7	12	9	16	17	16	11	137 days
	sunshine	60	110	155	162	135	186	198	156	76	50	27	27	1342 hours
	%	27	42	43	38	26	34	36	33	20	16	11	14	30 %

2. *Climatic area:* Yukon Territory except southern part, continental part of the Northwest Territories, north of Alberta, Saskatchewan, Manitoba

2.1. Subarctic Continental Climate EC

2.2. Summer cool or mild, winter severely cold.

2.3. Variable weather prevails.

2.4. Wind: frequent gales. Fog: occasionally in summer. Snow cover: in winter.

Inuvik – NWT

latitude: 68°18′n longitude: 133°29′w height: 61 m

ECld	118-13	Jan	Feb	Mar	Apr	May	Jun	Jul	Aug	Sep	Oct	Nov	Dec	Year
	tx	-9	-8	-4	6	17	25	25	24	17	6	-3	-6	28 °C
	t	-29	-28	-23	-13	-1	9	14	10	3	-7	-20	-27	-9 °C
	tn	-44	-43	-39	-32	-17	-2	2	1	-6	-14	-34	-42	-46 °C
	clothing stress	p	p	p	p	w	n	n	n	w	w	p	p	–
	p	14	11	10	12	9	19	32	33	27	27	19	13	226 mm
0.2 mm	dp	4	5	4	4	4	5	9	10	10	8	8	5	76 days

Norman Wells – NWT

latitude: 65°17′n longitude: 126°48′w height: 64 m

ECld	118-14	Jan	Feb	Mar	Apr	May	Jun	Jul	Aug	Sep	Oct	Nov	Dec	Year
	tx	-9	-9	4	11	21	27	31	28	21	10	-2	-11	31 °C
	t	-30	-26	-20	-8	5	13	16	13	6	-5	-18	-26	-7 °C
	tn	-46	-44	-37	-26	-9	1	2	0	-8	-18	-35	-42	-48 °C
	clothing stress	p	p	p	w	n	n	n	n	n	w	p	p	–
	p	18	18	13	15	13	35	57	61	37	27	30	24	348 mm
0.2 mm	dp	14	11	8	8	6	10	10	13	13	11	14	12	130 days

Baker Lake – NWT

latitude: 64°18′n longitude: 96°00′w height: 0 m

ECld	118-15	Jan	Feb	Mar	Apr	May	Jun	Jul	Aug	Sep	Oct	Nov	Dec	Year
	tx	-18	-17	-10	-1	5	16	24	25	15	5	-1	-9	27 °C
	t	-33	-33	-26	-16	-6	4	11	10	3	-8	-20	-28	-11 °C
	tn	-47	-44	-42	-31	-18	-7	0	-1	-8	-19	-34	-42	-47 °C
	clothing stress	p	p	p	p	w	w	n	n	w	w	p	p	–
	p	5	4	6	9	8	21	40	45	34	20	9	7	208 mm
0.2 mm	dp	6	6	8	8	7	9	8	9	7	11	9	9	97 days

Fort Smith – NWT

latitude: 60°01′n longitude: 111°58′w height: 203 m

ECld	118-16	Jan	Feb	Mar	Apr	May	Jun	Jul	Aug	Sep	Oct	Nov	Dec	Year
	tx	-4	-1	6	17	26	28	31	28	24	17	4	1	32 °C
	t	-25	-22	-14	-3	8	13	16	14	8	0	-12	-21	-3 °C
	tn	-46	-43	-40	-28	-9	-2	1	-2	-7	-18	-32	-43	-47 °C
	clothing stress	p	p	p	w	n	n	n	n	n	w	p	p	–
	p	16	17	19	17	26	31	53	35	42	30	26	27	339 mm
0.2 mm	dp	7	6	7	4	8	6	11	10	11	10	9	9	98 days
	sunshine	61	117	167	240	282	317	296	268	133	86	46	27	2040 hours
	%	29	45	46	55	53	56	53	55	34	27	20	15	45 %

Churchill – Manitoba

latitude: 58°45′n longitude: 94°04′w height: 35 m

ECld	118-17	Jan	Feb	Mar	Apr	May	Jun	Jul	Aug	Sep	Oct	Nov	Dec	Year
	tx	-7	-9	-1	6	16	23	28	26	21	11	0	-6	29 °C
	t	-28	-26	-20	-10	-2	6	12	12	6	-1	-12	-22	-7 °C
	tn	-40	-38	-36	-28	-16	-3	2	3	-3	-14	-29	-37	-41 °C
	clothing	p	p	p	p	w	n	n	n	n	w	p	p	–
	stress													–
	p	17	17	21	36	40	48	39	60	52	42	46	25	443 mm
0.2 mm	dp	5	6	6	6	7	9	10	12	11	12	9	8	101 days
	sunshine	79	127	180	186	166	212	285	232	100	67	44	54	1732 hours
	%	36	48	49	43	32	38	52	48	26	21	19	28	38 %

3. *Climatic area:* South of Yukon Territory, Central part of Alberta, Saskatchewan, Manitoba, north and central Ontario, central Quebec

3.1. Subarctic Continental Climate EC

3.2. Summer mild, winter very cold.

3.3. Variable weather prevails.

3.4. Wind: frequent gales. Fog: occasionally in summer. Snow cover: in winter.

White Horse – Yukon Terr.

latitude: 60°43′n longitude: 135°04′w height: 698 m

GEClc	118-18	Jan	Feb	Mar	Apr	May	Jun	Jul	Aug	Sep	Oct	Nov	Dec	Year
	tx	3	4	7	11	21	27	28	27	22	13	7	2	29 °C
	t	-18	-14	-8	0	8	13	14	12	8	1	-8	-15	-1 °C
	tn	-40	-39	-28	-18	-6	0	2	0	-6	-13	-27	-36	-46 °C
	clothing	p	p	w	w	n	n	n	n	n	w	w	p	–
	stress													–
	p	18	14	15	11	13	27	35	37	25	19	23	20	257 mm
0.2 mm	dp	8	7	8	5	6	9	13	10	10	10	11	9	106 days
	sunshine	48	74	164	246	265	295	241	219	148	114	56	28	1898 hours
	%	24	29	45	56	49	51	42	45	38	36	25	16	42 %

The Pas – Manitoba

latitude: 53°58′n longitude: 101°06′w height: 272 m

ECbc	118-19	Jan	Feb	Mar	Apr	May	Jun	Jul	Aug	Sep	Oct	Nov	Dec	Year
	tx	-4	3	8	19	27	29	31	29	26	19	7	1	33 °C
	t	-22	-17	-12	0	8	14	18	16	10	3	-8	-16	0 °C
	tn	-41	-38	-33	-19	-6	1	5	2	-3	-12	-26	-36	-42 °C
	clothing	p	p	p	w	n	n	s	n	n	w	w	p	–
	stress							o						–
	p	21	16	28	27	48	59	67	46	58	29	33	24	456 mm
0.2 mm	dp	6	7	8	6	8	11	10	10	10	9	9	8	102 days
	sunshine	94	123	165	211	258	244	300	259	158	125	64	66	2067 hours
	%	38	44	45	50	52	48	58	56	41	38	25	29	46 %

Trout Lake – Ontario

latitude: 53°50′n longitude: 89°52′w height: 217 m

EClc	118-20	Jan	Feb	Mar	Apr	May	Jun	Jul	Aug	Sep	Oct	Nov	Dec	Year
	tx	-2	-4	6	14	22	26	28	27	25	16	3	-1	28 °C
	t	-24	-21	-14	-4	4	11	16	15	9	2	-9	-20	-3 °C
	tn	-40	-41	-38	-26	-12	-3	3	2	-4	-12	-26	-38	-42 °C
	clothing stress	p	p	p	w	w	n	n	n	n	w	w	p	–
	p	23	21	16	26	49	72	99	95	78	52	50	30	611 mm
0.2 mm	dp	13	12	8	6	10	11	14	12	13	12	14	12	137 days

Nitchequon – Quebec

latitude: 53°12′n longitude: 70°54′w height: 537 m

GEClc	118-21	Jan	Feb	Mar	Apr	May	Jun	Jul	Aug	Sep	Oct	Nov	Dec	Year
	tx	-3	-3	3	8	18	24	27	25	21	14	4	1	28 °C
	t	-22	-19	-13	-5	2	10	14	12	7	0	-8	-18	-3 °C
	tn	-42	-40	-38	-23	-12	-3	3	3	-2	-12	-25	-37	-44 °C
	clothing stress	p	p	p	w	w	n	n	n	n	w	w	p	–
	p	37	30	21	33	48	87	95	94	93	81	62	42	723 mm
0.2 mm	dp	14	12	13	12	15	17	17	18	17	17	18	18	188 days

Moonsonee – Ontario

latitude: 51°16′n longitude: 80°39′w height: 0 m

EClc	118-22	Jan	Feb	Mar	Apr	May	Jun	Jul	Aug	Sep	Oct	Nov	Dec	Year
	t	-20	-17	-11	-2	5	12	16	15	10	5	-4	-15	-1 °C
	clothing stress	p	p	p	w	n	n	n	n	n	n	w	p	–
	p	48	52	45	45	78	95	86	79	85	70	78	56	817 mm
	sunshine	78	109	135	160	188	179	241	195	119	80	43	45	1572 hours
	%	30	39	37	39	39	36	48	43	31	24	16	18	35 %

Armstrong – Ontario

latitude: 50°17′n longitude: 88°54′w height: 325 m

EClc	118-23	Jan	Feb	Mar	Apr	May	Jun	Jul	Aug	Sep	Oct	Nov	Dec	Year
	t	-20	-17	-11	0	7	13	17	15	9	4	-6	-16	0 °C
	clothing stress	p	p	p	w	n	n	n	n	n	w	w	p	–
	p	43	36	34	46	58	89	87	80	91	59	69	40	732 mm
	sunshine	97	127	164	180	214	219	263	232	133	109	48	76	1862 hours
	%	37	45	45	44	45	45	53	51	35	33	18	30	42 %

4. *Climatic area:* British Columbia east of the Rocky Mountains, central and south Alberta, southern Saskatchewan, Manitoba, Ontario, Quebec

4.1. Temperate Continental Climate DC

4.2. Summer mild, in the south warm; winter very cold.

4.3. Variable weather prevails.

4.4. Wind: frequent gales. Fog: occasionally in summer. Snow cover: in winter.

Fort Nelson – British Columbia

latitude: 58°50′n longitude: 122°35′w height: 375 m

DClc	118-24	Jan	Feb	Mar	Apr	May	Jun	Jul	Aug	Sep	Oct	Nov	Dec	Year
	tx	-2	7	12	19	27	28	31	30	26	19	3	-2	32 °C
	t	-22	-17	-9	1	10	14	17	15	9	1	-12	-21	-1 °C
	tn	-39	-38	-31	-18	-3	2	4	2	-4	-12	-29	-37	-43 °C
	clothing	p	p	w	w	n	n	n	n	n	w	p	p	–
	stress													–
	p	25	27	28	20	38	68	70	55	36	24	30	30	451 mm
0.2 mm	dp	6	8	7	4	8	11	9	8	7	8	10	9	95 days

Prince George – British Columbia

latitude: 53°53′n longitude: 122°41′w height: 676 m

GDClc	118-25	Jan	Feb	Mar	Apr	May	Jun	Jul	Aug	Sep	Oct	Nov	Dec	Year
	tx	7	9	13	22	27	29	32	31	27	21	12	7	33 °C
	t	-10	-7	-2	4	10	13	15	14	10	5	-2	-7	4 °C
	tn	-36	-33	-22	-12	-6	-1	1	0	-5	-10	-19	-31	-40 °C
	clothing	p	w	w	w	n	n	n	n	n	n	w	w	–
	stress													–
	p	56	44	36	28	43	62	64	65	56	59	57	56	626 mm
0.2 mm	dp	14	12	13	12	11	15	15	14	13	16	14	13	162 days
	sunshine	54	87	130	183	248	240	267	243	166	101	56	41	1816 hours
	%	22	31	36	44	50	47	52	53	43	31	22	18	40 %

Edmonton – Alberta

latitude: 53°34′n longitude: 113°31′w height: 676 m

GDClc	118-26	Jan	Feb	Mar	Apr	May	Jun	Jul	Aug	Sep	Oct	Nov	Dec	Year
	tx	6	9	12	22	27	29	31	29	27	23	13	7	32 °C
	t	-14	-12	-6	4	11	14	17	16	11	5	-4	-10	3 °C
	tn	-38	-36	-27	-12	-4	1	4	2	-4	-11	-24	-27	-41 °C
	clothing	p	p	w	w	n	n	n	n	n	n	w	p	–
	stress													–
	p	24	20	21	28	47	80	85	65	34	23	22	25	474 mm
0.2 mm	dp	12	9	10	8	12	15	14	12	9	9	11	12	133 days
	sunshine	86	119	163	221	258	251	315	269	186	157	100	78	2203 hours
	%	35	43	45	53	52	49	61	58	49	48	39	33	49 %

Goose – Newfoundland

latitude: 53°19′n longitude: 60°25′w height: 44 m

DClc	118-27	Jan	Feb	Mar	Apr	May	Jun	Jul	Aug	Sep	Oct	Nov	Dec	Year
	t	-16	-14	-8	-2	5	11	16	15	10	3	-4	-13	0 °C
	clothing	p	p	w	w	n	n	n	n	n	w	w	p	–
	stress													–
	p	72	63	68	62	56	72	84	91	76	63	67	63	837 mm
	sunshine	90	111	143	136	176	198	194	187	124	91	69	66	1585 hours
	%	36	40	39	32	36	39	38	41	32	28	26	28	35 %

Regina – Saskatchewan

latitude: 50°26′n longitude: 104°40′w height: 577 m

GDCbc	118-28	Jan	Feb	Mar	Apr	May	Jun	Jul	Aug	Sep	Oct	Nov	Dec	Year
	tx	2	3	8	23	29	32	34	33	29	24	13	5	36 °C
	t	-17	-15	-8	3	11	15	19	18	12	5	-5	-12	2 °C
	tn	-39	-37	-30	-15	-7	0	3	1	-6	-13	-26	-34	-41 °C
	clothing	p	p	w	w	n	n	s	s	n	n	w	p	–
	stress							o	o					–
	p	19	17	21	21	40	83	55	49	34	18	20	17	394 mm
0.2 mm	dp	9	13	9	7	8	12	11	9	6	5	10	10	109 days
	sunshine	98	118	152	215	266	249	334	286	197	170	96	85	2266 hours
	%	37	41	41	52	55	51	67	63	52	51	35	34	51 %

Sept Iles – Quebec

latitude: 50°13′n longitude: 66°16′w height: 58 m

DClc	118-29	Jan	Feb	Mar	Apr	May	Jun	Jul	Aug	Sep	Oct	Nov	Dec	Year
	t	-13	-11	-6	0	6	12	15	14	10	4	-2	-11	2 °C
	clothing	p	p	w	w	n	n	n	n	n	w	w	p	–
	stress													–
	p	100	96	75	66	61	94	102	93	105	85	111	91	1079 mm

Winnipeg – Manitoba

latitude: 49°54′n longitude: 97°14′w height: 248 m

DCbc	118-30	Jan	Feb	Mar	Apr	May	Jun	Jul	Aug	Sep	Oct	Nov	Dec	Year
	tx	-1	2	8	22	29	32	33	33	29	22	11	3	34 °C
	t	-18	-16	-8	3	11	16	20	19	13	6	-5	-13	2 °C
	tn	-38	-36	-29	-14	-5	2	6	3	-3	-9	-24	-34	-39 °C
	clothing	p	p	w	w	n	n	s	s	n	n	w	p	–
	stress							o	o					–
	p	26	21	27	30	50	81	69	70	55	37	29	22	517 mm
0.2 mm	dp	12	11	9	9	10	12	10	10	9	6	9	11	118 days
	sunshine	101	133	167	207	244	248	310	270	181	153	82	81	2177 hours
	%	38	47	45	50	51	51	63	60	48	46	30	32	49 %

Kapuskasing – Ontario

latitude: 49°25′n longitude: 82°28′w height: 29 m

DClc	118-31	Jan	Feb	Mar	Apr	May	Jun	Jul	Aug	Sep	Oct	Nov	Dec	Year
	tx	2	1	10	20	28	32	32	30	28	22	12	4	33 °C
	t	-17	-15	-9	1	8	14	17	16	10	5	-4	-14	1 °C
	tn	-37	-37	-32	-19	-6	-1	3	2	-4	-9	-25	-34	-39 °C
	clothing	p	p	w	w	n	n	n	n	n	n	w	p	–
	stress							o						–
	p	53	47	55	52	79	87	82	83	91	76	87	55	847 mm
0.2 mm	dp	19	16	13	10	12	13	14	16	16	16	18	19	182 days
	sunshine	72	104	135	167	195	208	238	202	126	90	44	54	1635 hours
	%	27	36	37	41	41	43	48	45	33	27	16	21	36 %

North Bay – Ontario

latitude: 46°22′n longitude: 79°25′w height: 369 m

DCbc	118-32	Jan	Feb	Mar	Apr	May	Jun	Jul	Aug	Sep	Oct	Nov	Dec	Year
	tx	3	4	11	19	26	29	32	31	27	19	13	5	33 °C
	t	-13	-11	-5	3	10	16	18	17	12	7	-1	-9	4 °C
	tn	-32	-32	-24	-13	-3	2	6	4	-1	-7	-16	-29	-35 °C
	clothing	p	p	w	w	n	n	s	n	n	n	w	w	–
	stress							o						–
	p	53	47	55	52	79	87	82	83	91	76	87	55	847 mm
0.2 mm	dp	18	19	16	12	13	10	12	12	14	15	17	19	177 days

Montreal – Quebec

latitude: 45°28′n longitude: 73°45′w height: 30 m

DCbc	118-33	Jan	Feb	Mar	Apr	May	Jun	Jul	Aug	Sep	Oct	Nov	Dec	Year
	tx	6	5	9	21	27	30	32	31	28	22	14	7	32 °C
	t	-10	-8	-2	6	13	19	21	20	15	9	2	-6	7 °C
	tn	-27	-24	-19	-7	1	7	11	9	3	-3	-13	-23	-28 °C
	clothing	p	w	w	n	n	s	s	s	n	n	w	w	–
	stress						o	o	o					–
	p	83	81	78	72	72	85	89	77	82	78	85	89	971 mm
0.2 mm	dp	15	14	14	12	12	13	12	11	12	13	14	15	157 days
	sunshine	81	105	146	168	205	222	245	223	171	127	69	62	1823 hours
	%	29	36	40	41	44	47	51	51	45	37	24	28	41 %

5. *Climatic area:* Newfoundland, Prince Edward Island, New Brunswick, Nova Scotia
5.1. Temperate Continental Climate DC
5.2. Summer warm, winter cold.
5.3. Variable weather prevails.
5.4. Wind: frequent gales. Fog: occasionally. Snow cover: in winter.

Lethbridge – Alberta

latitude: 49°38′n longitude: 112°48′w height: 929 m

GDCbo	118-34	Jan	Feb	Mar	Apr	May	Jun	Jul	Aug	Sep	Oct	Nov	Dec	Year
	tx	11	12	18	24	28	31	33	32	28	24	15	12	34 °C
	t	-8	-7	-2	5	11	15	19	17	13	7	0	-5	5 °C
	tn	-33	-32	-23	-12	-4	1	4	2	-4	-8	-22	-32	-37 °C
	clothing	w	w	w	n	n	n	s	n	n	n	w	w	–
	stress							o	o					–
	p	21	23	29	35	52	74	39	40	35	29	26	21	424 mm
0.2 mm	dp	6	11	10	8	10	12	9	6	7	4	7	6	96 days
	sunshine	103	126	162	210	265	271	345	300	216	177	113	96	2384 hours
	%	39	44	44	51	56	55	70	67	57	53	41	38	58 %

Gander – Newfoundland

latitude: 48°57′n longitude: 54°34′w height: 147 m

DClo	118-35	Jan	Feb	Mar	Apr	May	Jun	Jul	Aug	Sep	Oct	Nov	Dec	Year
	t	-6	-6	-4	1	6	12	17	16	12	6	2	-4	4 °C
	clothing	w	w	w	w	n	n	n	n	n	n	w	w	–
	stress													–
	p	83	88	81	69	59	76	84	101	86	97	107	87	1018 mm
0.2 mm	dp	19	22	19	19	17	14	13	15	15	20	21	21	215 days
	sunshine	69	82	107	115	150	160	200	178	143	112	63	58	1437 hours
	%	26	29	29	28	32	33	41	40	38	33	23	23	32 %

Sydney – Nova Scotia

latitude: 46°10′n longitude: 60°03′w height: 60 m

DCbo	118-36	Jan	Feb	Mar	Apr	May	Jun	Jul	Aug	Sep	Oct	Nov	Dec	Year
	t	-4	-5	-2	3	8	13	18	18	14	9	4	-2	6 °C
	clothing	w	w	w	w	n	n	s	s	n	n	w	w	–
	stress							o	o					–
	p	130	119	112	98	94	89	71	103	104	115	143	128	1306 mm
	sunshine	72	100	119	158	201	210	258	221	173	142	73	69	1796 hours
	%	26	34	32	39	43	44	54	50	46	42	26	26	40 %

Halifax – Nova Scotia

latitude: 44°39′n longitude: 63°36′w height: 30 m

DCbo	118-37	Jan	Feb	Mar	Apr	May	Jun	Jul	Aug	Sep	Oct	Nov	Dec	Year
	tx	9	7	11	13	24	28	30	29	27	22	16	11	32 °C
	t	-5	-5	-1	4	10	14	18	18	15	9	4	-2	7 °C
	tn	-21	-19	-14	-6	-1	3	8	8	3	-2	-8	-16	-22 °C
	clothing	w	w	w	w	n	n	s	s	n	n	w	w	–
	stress							o	o					–
	p	137	109	124	114	104	102	96	112	104	137	135	137	1412 mm
0.2 mm	dp	17	14	15	14	14	14	13	12	12	13	14	15	167 days
	sunshine	96	116	149	153	202	213	239	220	168	152	90	90	1887 hours
	%	34	39	40	38	44	46	51	50	45	44	31	33	42 %

Sable Island – Nova Scotia

latitude: 43°56′n longitude: 60°21′w height: 4 m

DCbo	118-38	Jan	Feb	Mar	Apr	May	Jun	Jul	Aug	Sep	Oct	Nov	Dec	Year
	tx	9	8	8	10	13	18	22	23	22	18	15	11	23 °C
	t	0	-1	0	3	7	11	16	18	16	12	7	2	8 °C
	tn	-12	-12	-8	-3	1	4	8	12	8	3	-2	-8	-14 °C
	clothing	w	w	w	w	n	n	n	s	n	n	n	w	–
	stress													–
	p	125	114	106	97	87	81	74	101	97	110	126	133	1251 mm
	dp	14	12	12	11	11	10	11	9	10	13	14	15	142 days

Toronto – Ontario

latitude: 43°41′n longitude: 79°38′w height: 176 m

DCbo	118-39	Jan	Feb	Mar	Apr	May	Jun	Jul	Aug	Sep	Oct	Nov	Dec	Year
	tx	7	7	12	21	27	31	32	31	29	22	15	9	34 °C
	t	-6	-5	-1	7	12	18	21	20	16	10	3	-3	8 °C
	tn	-21	-21	-17	-6	0	6	9	8	2	-3	-12	-18	-24 °C
	clothing	w	w	w	n	n	s	s	s	n	n	w	w	–
	stress						o	o	o					–
	p	58	52	65	63	81	61	74	67	66	62	56	55	760 mm
0.2 mm	dp	16	13	14	12	12	10	10	9	11	10	13	14	144 days
	sunshine	78	105	139	170	220	257	287	259	198	154	84	75	2026 hours
	%	27	35	38	42	48	56	61	60	53	45	29	27	45 %

6. *Climatic area:* British Columbia, except the part east of the Rocky Mountains
6.1. Temperate Oceanic Climate DO and Temperate Oceanic Mountain Climate GDO
6.2. Summer mild, winter cool.
6.3. Variable weather prevails.
6.4. Wind: frequent gales, particularly in winter. Fog: frequent. Snow cover: variable.

Sandspit – British Columbia

latitude: 53°15'n longitude: 131°49'w height: 0 m

DOlk	118-40	Jan	Feb	Mar	Apr	May	Jun	Jul	Aug	Sep	Oct	Nov	Dec	Year
	t	2	3	3	6	9	12	14	15	13	9	6	4	8 °C
	clothing	w	w	w	n	n	n	n	n	n	n	n	w	–
	stress													–
	p	153	103	105	86	44	47	46	48	80	181	188	187	1268 mm
	sunshine	56	81	116	154	224	166	214	147	139	85	59	38	1479 hours
	%	22	29	32	37	46	33	42	32	36	26	23	16	33 %

Port Hardy – British Columbia

latitude: 50°41'n longitude: 127°22'w height: 23 m

DOlk	118-41	Jan	Feb	Mar	Apr	May	Jun	Jul	Aug	Sep	Oct	Nov	Dec	Year
	t	2	4	4	7	10	12	14	14	12	9	5	4	8 °C
	clothing	w	w	w	n	n	n	n	n	n	n	n	w	–
	stress													–
	p	181	159	133	95	61	75	41	67	122	205	229	272	1640 mm

Vancouver – British Columbia

latitude: 49°11'n longitude: 123°10'w height: 0 m

DObk	118-42	Jan	Feb	Mar	Apr	May	Jun	Jul	Aug	Sep	Oct	Nov	Dec	Year
	tx	11	12	16	20	24	27	29	28	24	19	14	11	29 °C
	t	2	4	6	9	13	15	18	17	14	10	6	4	10 °C
	tn	-8	-6	-2	0	2	6	8	8	4	1	-3	-6	-8 °C
	clothing	w	w	n	n	n	n	s	n	n	n	n	w	–
	stress							o						–
	p	139	121	96	60	48	51	26	36	56	117	142	156	1048 mm
0.2 mm	dp	20	17	17	14	12	11	7	8	9	16	19	22	172 days
	sunshine	43	75	111	168	225	209	277	249	173	109	51	33	1723 hours
	%	16	26	30	41	47	43	56	56	46	32	19	13	38 %

4.4.118.2. Predominant general weather situations are described in 4.4.0.1. under Alaska, Canada and Greenland. Hygrothermal diagrams of Ottawa and Vancouver are to be found in 4.4.0.2.

Wind names: Barber is a snowstorm blowing from northwesterly directions in eastern Canada.

A special phenomenon is the "débacle", the breaking up in the spring of the ice in the rivers. The term is chiefly applied to the great rivers of the North American Continent and of Russia and Siberia. The débacle lasts from two to six weeks; during this period the rivers often overflow their banks, inundating the surrounding country. In Ontario, the débacle takes place in March and the water is free by April, in the St. Lawrence a little later, the river being free of ice in May.

4.4.119. United States of America

4.4.119.1. Climatic descriptions and tables

1. Climatic area: Alaska: north coast, part of west coast and islands in the Bering Sea
1.1. Tundra Climate FT
1.2. Summer cool, winter severely cold on the north coast, very cold to the north of the Bering Sea and cold to the south of the Bering Sea.
1.3. Variable weather prevails.
1.4. Wind: frequent gales. Fog: predominantly in summer. Snow cover: in the north most of the year.

Barrow – Alaska

latitude: 71°18'n longitude: 156°47'w height: 4 m

FTkd	119-1		Jan	Feb	Mar	Apr	May	Jun	Jul	Aug	Sep	Oct	Nov	Dec	Year
		tx	-8	-12	-11	-4	2	12	17	17	8	1	-2	-8	19 °C
		t	-27	-28	-26	-18	-8	1	4	3	-1	-9	-18	-24	-12 °C
		tn	-40	-41	-39	-32	-21	-7	-2	-3	-8	-20	-29	-36	-43 °C
		clothing stress	p	p	p	p	w	w	w	w	w	w	p	p	–
		p	5	4	3	3	3	9	20	23	16	13	6	4	110 mm
		dp	2	2	1	1	1	2	6	7	5	6	3	2	38 days

Barter Island – Alaska

latitude: 70°08'n longitude: 143°38'w height: 15 m

FTkd	119-2		Jan	Feb	Mar	Apr	May	Jun	Jul	Aug	Sep	Oct	Nov	Dec	Year
		t	-27	-29	-26	-17	-6	1	5	4	0	-8	-18	-24	-12 °C
		clothing stress	p	p	p	p	w	w	n	w	w	w	p	p	–
		p	10	9	5	4	6	13	22	27	24	21	10	7	160 mm
		dp	3	3	2	2	3	4	6	7	6	8	3	2	49 days

Gambell – Alaska

latitude: 63°51'n longitude: 171°36'w height: 8 m

FTkc	119-3		Jan	Feb	Mar	Apr	May	Jun	Jul	Aug	Sep	Oct	Nov	Dec	Year
		tx	-1	-2	-1	2	6	12	14	14	9	6	2	-1	15 °C
		t	-15	-16	-13	-8	-2	3	7	7	4	0	-4	-12	-4 °C
		tn	-27	-28	-27	-19	-11	-2	2	3	-1	-7	-14	-23	-30 °C
		clothing stress	p	p	p	w	w	w	n	n	w	w	w	p	–
		p	25	30	30	30	18	15	46	66	36	36	33	36	401 mm
0.2 mm		dp	10	11	11	12	11	9	14	16	12	15	13	15	149 days

St. Paul Island – Alaska

latitude: 57°09'n longitude: 170°13'w height: 9 m

FTko	119-4		Jan	Feb	Mar	Apr	May	Jun	Jul	Aug	Sep	Oct	Nov	Dec	Year
		tx	4	3	3	4	8	12	13	13	12	9	7	4	14 °C
		t	-4	-5	-4	-2	2	5	8	9	7	3	0	-3	1 °C
		tn	-16	-16	-18	-8	-4	-1	4	4	1	-3	-7	-12	-19 °C
		clothing stress	w	w	w	w	w	n	n	n	n	w	w	w	–
		p	46	31	27	24	33	29	57	84	78	80	64	46	599 mm
		dp	10	9	8	6	9	8	10	14	13	15	14	10	126 days

2. *Climatic area:* Alaska: northern polar part, except coast and central and southwestern part
2.1. Subarctic Continental Climate EC
2.2. Summer mild, winter very cold, beyond the arctic circle severely cold.
2.3. Variable weather prevails.
2.4. Wind: frequent gales. Fog: particularly in summer. Snow cover: in the north most of the year.

Kotzebue – Alaska

latitude: 66°52'n longitude: 162°38'w height: 5 m

EClc	119-5	Jan	Feb	Mar	Apr	May	Jun	Jul	Aug	Sep	Oct	Nov	Dec	Year
	t	-21	-20	-19	-10	-1	7	12	10	5	-4	-14	-20	-6 °C
	clothing	p	p	p	p	w	n	n	n	n	w	p	p	–
	stress													–
	p	10	8	7	8	8	12	37	55	31	15	9	7	208 mm
	dp	3	3	3	2	3	3	8	11	9	5	4	3	57 days
	sunshine	62	105	177	270	285	261	192	115	108	90	54	25	1743 hours
	%	50	46	49	58	47	36	28	22	27	30	32	28	37 %

Fairbanks – Alaska

latitude: 64°49'n longitude: 147°52'w height: 138 m

EClc	119-6	Jan	Feb	Mar	Apr	May	Jun	Jul	Aug	Sep	Oct	Nov	Dec	Year
	tx	-3	-3	-2	3	12	21	24	20	13	7	1	-3	24 °C
	t	-24	-19	-13	-1	8	15	15	12	6	-3	-16	-22	-3 °C
	tn	-37	-38	-37	-27	-15	-2	3	3	-3	-16	-28	-36	-41 °C
	clothing	p	p	p	w	n	n	n	n	n	w	p	p	–
	stress													–
	p	23	13	10	6	18	35	47	56	28	22	15	14	287 mm
	dp	5	4	2	1	3	6	8	8	5	4	4	5	55 days
	sunshine	105	119	220	300	301	318	267	174	114	84	75	40	2116 hours
	%	66	50	61	66	52	49	43	34	29	28	39	33	46 %

Nome – Alaska

latitude: 64°30'n longitude: 165°26'w height: 7 m

EClc	119-7	Jan	Feb	Mar	Apr	May	Jun	Jul	Aug	Sep	Oct	Nov	Dec	Year
	tx	1	1	1	6	14	22	21	18	14	10	3	0	23 °C
	t	-15	-15	-13	-6	2	8	10	9	6	-1	-9	-14	-3 °C
	tn	-32	-33	-32	-22	-12	-1	2	1	-4	-13	-23	-31	-36 °C
	clothing	p	p	p	w	w	n	n	n	n	w	w	p	–
	stress													–
	p	26	24	22	20	18	24	58	97	68	43	29	25	454 mm
	dp	5	5	6	5	5	5	9	12	11	6	6	4	79 days
	sunshine	62	112	176	219	254	254	190	132	113	99	52	44	1707 hours
	%	39	47	49	48	44	39	30	26	29	32	27	37	37 %

McGrath – Alaska

latitude: 62°58'n longitude: 155°37'w height: 103 m

EClc	119-8	Jan	Feb	Mar	Apr	May	Jun	Jul	Aug	Sep	Oct	Nov	Dec	Year
	tx	-1	1	7	10	21	26	29	24	18	14	5	-2	28 °C
	t	-23	-18	-13	-2	7	14	15	12	7	-3	-15	-22	-3 °C
	tn	-46	-43	-36	-23	-9	2	4	1	-6	-17	-19	-42	-46 °C
	clothing	p	p	p	w	n	n	n	n	n	w	p	p	–
	stress													–
	p	32	29	24	12	22	42	62	96	66	34	27	26	472 mm
	dp	6	6	5	3	5	7	10	15	12	6	7	6	88 days

Anchorage – Alaska

latitude: 61° 10′n longitude: 147° 59′w height: 40 m

EClc	119-9	Jan	Feb	Mar	Apr	May	Jun	Jul	Aug	Sep	Oct	Nov	Dec	Year
	tx	4	4	8	11	19	24	26	22	18	14	9	2	27 °C
	t	-11	-8	-5	2	8	12	14	13	9	2	-5	-10	2 °C
	tn	-29	-33	-21	-14	-3	2	5	3	-1	-8	-20	-25	-36 °C
	clothing	p	w	w	w	n	n	n	n	n	w	w	p	–
	stress													–
	p	20	18	13	11	13	25	47	65	64	47	26	24	374 mm
	dp	5	6	3	4	3	5	10	11	11	7	5	6	76 days
	sunshine	79	104	205	248	282	265	244	186	151	124	79	63	2030 hours
	%	40	41	56	56	52	46	43	38	39	39	36	36	45 %

Bethel – Alaska

latitude: 60° 47′n longitude: 161° 48′w height: 46 m

EClc	119-10	Jan	Feb	Mar	Apr	May	Jun	Jul	Aug	Sep	Oct	Nov	Dec	Year
	tx	3	2	3	8	18	23	26	21	18	12	5	3	27 °C
	t	-16	-13	-11	-3	4	11	13	11	7	0	-8	-15	-2 °C
	tn	-36	-37	-30	-22	-8	2	6	2	-2	-10	-22	-34	-41 °C
	clothing	p	p	p	w	w	n	n	n	n	w	w	p	–
	stress													–
	p	28	28	26	15	24	30	52	107	66	39	27	26	468 mm
	dp	6	8	8	5	6	8	11	17	13	6	7	6	101 days

King Salmon – Alaska

latitude: 58° 41′n longitude: 156° 39′w height: 15 m

EClc	119-11	Jan	Feb	Mar	Apr	May	Jun	Jul	Aug	Sep	Oct	Nov	Dec	Year
	t	-10	-8	-7	0	6	11	13	13	9	2	-6	-10	1 °C
	clothing	p	w	w	w	n	n	n	n	n	w	w	p	–
	stress													–
	p	27	24	24	16	24	36	52	87	78	55	37	26	486 mm
	dp	6	7	6	6	7	7	10	14	12	8	8	8	99 days

3. Climatic area: Alaska: southeastern part and Aleutian Islands, except Alexander Archipelago

3.1. Subarctic Oceanic Climate EO

3.2. Summer mild or cool, winter cold.

3.3. Variable weather prevails.

3.4. Wind: frequent gales. Fog: particularly in summer. Snow cover: in winter; in transitional seasons variable.

Cordova / Mile 13 – Alaska

latitude: 60° 30′n longitude: 145° 30′w height: 15 m

EOlo	119-12	Jan	Feb	Mar	Apr	May	Jun	Jul	Aug	Sep	Oct	Nov	Dec	Year
	tx	6	6	7	10	18	21	23	22	18	14	9	6	25 °C
	t	-5	-3	-2	2	6	10	12	11	9	4	0	-4	3 °C
	tn	-24	-23	-14	-11	-3	2	3	1	-2	-6	-15	-21	-28 °C
	clothing	w	w	w	w	n	n	n	n	n	w	w	w	–
	stress													–
	p	155	118	98	110	129	88	159	205	318	302	204	172	2057 mm
0.2 mm	dp	17	17	17	16	18	15	17	18	19	22	18	18	212 days

Yakutat – Alaska

latitude: 59°31′n longitude: 139°40′w height: 9 m

EOlo	119-13	Jan	Feb	Mar	Apr	May	Jun	Jul	Aug	Sep	Oct	Nov	Dec	Year
	t	-3	-2	0	3	7	10	12	12	10	6	1	-2	4 °C
	clothing	w	w	w	w	n	n	n	n	n	n	w	w	–
	stress													–
	p	276	208	221	184	203	129	214	277	420	498	407	312	3348 mm
	dp	15	17	15	14	15	11	13	17	19	21	20	23	200 days
	sunshine	59	51	84	81	71	84	90	71	54	50	36	28	758 hours
	%	28	20	23	19	13	15	16	15	14	16	16	15	17 %

Juneau – Alaska

latitude: 58°22′n longitude: 134°35′w height: 7 m

EOlo	119-14	Jan	Feb	Mar	Apr	May	Jun	Jul	Aug	Sep	Oct	Nov	Dec	Year
	tx	8	8	10	15	21	25	24	24	20	14	11	9	27 °C
	t	-4	-5	-1	3	8	11	13	12	9	5	1	-2	4 °C
	tn	-11	-12	-8	-3	1	5	7	7	3	-2	-6	-11	-19 °C
	clothing	w	w	w	w	n	n	n	n	n	n	w	w	–
	stress													–
	p	102	78	83	73	82	86	114	128	168	212	154	107	1387 mm
	dp	12	14	14	12	17	11	13	16	17	20	19	18	183 days
	sunshine	70	81	140	169	198	185	163	147	92	57	57	38	1397 hours
	%	32	31	38	39	38	34	30	31	34	18	24	19	31 %

Kodiak – Alaska

latitude: 57°48′n longitude: 152°21′w height: 46 m

EOlo	119-15	Jan	Feb	Mar	Apr	May	Jun	Jul	Aug	Sep	Oct	Nov	Dec	Year
	tx	7	8	8	11	16	20	22	22	18	13	9	6	24 °C
	t	-1	-1	1	3	7	10	12	13	10	7	2	1	6 °C
	tn	-11	-10	-11	-6	-1	3	6	6	2	-2	-7	-10	-15 °C
	clothing	w	w	w	w	n	n	n	n	n	n	w	w	–
	stress													–
	p	119	116	96	96	150	119	86	130	135	183	147	152	1532 mm
0.2 mm	dp	14	14	12	13	16	13	12	15	13	16	15	15	168 days

Cold Bay – Alaska

latitude: 55°12′n longitude: 162°45′w height: 31 m

EOlo	119-16	Jan	Feb	Mar	Apr	May	Jun	Jul	Aug	Sep	Oct	Nov	Dec	Year
	t	-2	-2	-2	1	4	7	10	11	9	4	1	-2	3 °C
	clothing	w	w	w	w	w	n	n	n	n	w	w	w	–
	stress													–
	p	59	81	44	37	58	50	46	108	110	117	96	66	871 mm
	dp	12	12	12	10	12	10	12	16	14	18	17	13	158 days

Shemya Island – Alaska

latitude: 52°43′n longitude: 174°06′w height: 31 m

EOlo	119-17	Jan	Feb	Mar	Apr	May	Jun	Jul	Aug	Sep	Oct	Nov	Dec	Year
	t	0	-1	0	2	3	5	8	10	9	4	2	0	4 °C
	clothing	w	w	w	w	w	n	n	n	n	w	w	w	–
	stress													–
	p	64	58	65	52	61	34	55	54	57	70	69	54	693 mm
	dp	15	12	14	11	12	8	9	9	10	17	16	14	147 days

4. Climatic area: Alexander Archipelago (Alaska), Washington, Oregon and west Montana
4.1. Temperate Oceanic Climate DO and Temperate Oceanic Mountain Climate GDO
4.2. Summer warm, in the islands mild, winter cool.
4.3. Variable weather prevails.
4.4. Wind: frequent gales. Fog: particularly in summer. Snow cover: variable in winter.

Annette Island – Alaska

latitude: 55°02'n longitude: 131°34'w height: 34 m

DOlk	119-18	Jan	Feb	Mar	Apr	May	Jun	Jul	Aug	Sep	Oct	Nov	Dec	Year
	tx	10	11	12	17	23	24	25	26	22	17	13	11	28 °C
	t	1	2	3	6	9	12	14	14	12	8	5	2	7 °C
	tn	-8	-8	-5	-2	1	5	7	8	9	-1	-3	-7	-12 °C
	clothing	w	w	w	n	n	n	n	n	n	n	n	w	–
	stress													–
	p	289	217	243	232	180	144	153	191	251	429	372	308	3006 mm
	dp	18	16	18	15	14	12	11	13	15	23	21	22	198 days

Spokane – Washington

latitude: 47°37'n longitude: 117°31'w height: 721 m

GDObo	119-19	Jan	Feb	Mar	Apr	May	Jun	Jul	Aug	Sep	Oct	Nov	Dec	Year
	tx	9	11	18	24	29	33	37	36	31	24	14	10	37 °C
	t	-4	-1	3	8	13	16	21	20	16	9	2	-1	8 °C
	tn	-18	-15	-8	-2	1	5	8	7	1	-3	-9	-13	-22 °C
	clothing	w	w	w	n	n	n	s	s	n	n	w	w	–
	stress							o	o					–
	p	62	47	38	23	31	38	10	10	19	40	57	62	437 mm
	dp	10	8	8	6	6	6	2	2	4	7	9	10	78 days
	sunshine	72	118	199	250	291	315	395	335	271	172	81	53	2552 hours
	%	26	41	54	61	62	66	82	75	72	51	29	20	57 %

Seattle – Washington

latitude: 47°36'n longitude: 122°00'w height: 4 m

DObk	119-20	Jan	Feb	Mar	Apr	May	Jun	Jul	Aug	Sep	Oct	Nov	Dec	Year
	tx	13	14	18	24	27	29	31	30	27	22	17	14	32 °C
	t	5	6	8	11	14	17	19	18	16	12	8	6	12 °C
	tn	-4	-3	-1	2	5	8	10	10	7	3	1	-2	-6 °C
	clothing	n	n	n	n	n	n	s	s	n	n	n	n	–
	stress							o	o					–
	p	114	95	78	49	41	32	13	22	40	78	113	136	815 mm
0.2 mm	dp	18	15	16	13	11	10	5	6	7	14	17	19	151 days

Great Falls – Montana

latitude: 47°29'n longitude: 111°21'w height: 1115 m

GDObo	119-21	Jan	Feb	Mar	Apr	May	Jun	Jul	Aug	Sep	Oct	Nov	Dec	Year
	t	-5	-4	0	7	12	16	21	20	14	9	2	-2	7 °C
	clothing	w	w	w	n	n	n	s	s	n	n	w	w	–
	stress							o	o					–
	p	15	19	23	25	53	74	33	32	30	19	19	15	357 mm
	dp	4	4	6	5	8	10	5	4	5	5	5	4	65 days
	sunshine	144	170	251	258	300	310	390	339	249	209	129	127	2875 hours
	%	52	59	68	63	64	65	81	76	66	62	46	48	64 %

Portland – Oregon

latitude: 45°36′n longitude: 122°36′w height: 12 m

DObk	119-22	Jan	Feb	Mar	Apr	May	Jun	Jul	Aug	Sep	Oct	Nov	Dec	Year
	tx	13	16	20	26	29	32	34	32	30	23	18	14	35 °C
	t	4	6	8	10	14	17	20	19	17	12	7	5	12 °C
	tn	-6	-4	-1	1	3	7	9	9	6	3	-1	-3	-9 °C
	clothing	w	n	n	n	n	n	s	s	n	n	n	n	–
	stress							o	o					–
	p	136	107	97	53	51	42	10	17	41	92	135	162	944 mm
	dp	15	14	15	11	8	7	2	3	5	11	15	17	123 days
	sunshine	68	99	141	191	240	225	322	266	218	129	83	57	2039 hours
	%	24	34	38	47	52	48	68	61	58	38	29	21	46 %

Medford – Oregon

latitude: 42°22′n longitude: 122°52′w height: 405 m

DObk	119-23	Jan	Feb	Mar	Apr	May	Jun	Jul	Aug	Sep	Oct	Nov	Dec	Year
	t	3	5	8	11	14	18	22	22	18	12	6	4	12 °C
	clothing	w	n	n	n	n	s	s	s	s	n	n	w	–
	stress						o	o	o	o				–
	dp	80	61	45	27	37	26	5	5	15	49	66	86	502 mm
	dp	10	8	8	7	7	4	1	0	3	6	10	10	74 days

5. *Climatic area:* Parts of Idaho, Wyoming, Nevada, Utah, Colorado, eastern North Dakota, South Dakota, Nebraska, Kansas and Minnesota, Wisconsin, Michigan, Iowa, Missouri, Illinois, Indiana, Ohio, Kentucky, Maine, Vermont, New Hampshire, Massachusetts, Rhode Island, Connecticut, New York, New Jersey, Pennsylvania, West Virginia, Virginia, Maryland, Delaware, Oklahoma, Tennessee, western North Carolina, northwestern South Carolina, a small northern part of Georgia

5.1. Temperate Continental Climate DC and Temperate Continental Mountain Climate GDC

5.2. Summer warm, in most of the southern part hot, winter very cold in the north, otherwise cold, in the southern parts cool.

5.3. Variable weather prevails, in summer often sunny and dry days.

5.4. Wind: gales at times, tornadoes particularly in the southern and eastern parts. Snow cover: in southern parts variable in winter.

Caribou – Maine

latitude: 46°52′n longitude: 68°01′w height: 191 m

DCbc	119-24	Jan	Feb	Mar	Apr	May	Jun	Jul	Aug	Sep	Oct	Nov	Dec	Year
	t	-12	-11	-5	2	10	15	18	17	12	6	-1	-9	4 °C
	clothing	p	p	w	w	n	n	s	n	n	n	w	w	–
	stress							o						–
	p	54	51	60	67	77	103	103	93	90	85	77	62	922 mm
	dp	9	9	8	10	10	11	11	10	8	9	10	9	114 days

Duluth – Minnesota

latitude: 46°50′n longitude: 92°11′w height: 432 m

DCbc	119-25	Jan	Feb	Mar	Apr	May	Jun	Jul	Aug	Sep	Oct	Nov	Dec	Year
	tx	4	6	11	20	26	30	32	31	28	23	13	6	33 °C
	t	-13	-12	-6	3	10	15	19	18	13	7	-3	-10	3 °C
	tn	-32	-29	-22	-9	-2	4	8	8	1	-6	-18	-27	-33 °C
	clothing	p	p	w	w	n	n	s	s	n	n	w	p	–
	stress							o	o					–
	p	29	24	41	60	84	108	90	97	73	55	45	29	736 mm
	dp	7	5	7	8	10	10	8	8	7	6	7	5	88 days
	sunshine	143	162	214	220	261	285	321	268	196	172	99	112	2453 hours
	%	52	56	58	54	56	60	67	61	52	51	35	42	55 %

Bismarck – North Dakota

latitude: 46°46′n longitude: 100°45′w height: 506 m

GDCbc	119-26	Jan	Feb	Mar	Apr	May	Jun	Jul	Aug	Sep	Oct	Nov	Dec	Year
	tx	7	9	16	24	30	33	36	36	33	27	17	9	38 °C
	t	-13	-11	-4	6	13	18	22	21	15	8	-2	-8	5 °C
	tn	-33	-31	-24	-10	-3	4	7	5	-2	-8	-16	-29	-36 °C
	clothing	p	p	w	n	n	s	s	s	n	n	w	w	–
	stress						o	o	o					–
	p	11	11	20	31	50	86	56	44	30	22	15	9	385 mm
	dp	3	3	4	6	6	8	7	5	5	4	4	3	58 days
	sunshine	154	168	218	237	289	294	363	321	245	209	130	128	2755 hours
	%	55	58	59	58	62	62	75	73	65	61	46	48	62 %

Sault Ste. Marie – Michigan

latitude: 46°28′n longitude: 84°22′w height: 221 m

DCbo	119-27	Jan	Feb	Mar	Apr	May	Jun	Jul	Aug	Sep	Oct	Nov	Dec	Year
	t	-9	-9	-5	3	10	15	18	18	13	7	0	-6	4 °C
	clothing	w	w	w	w	n	n	s	s	n	n	w	w	–
	stress							o	o					–
	p	53	38	46	55	70	84	63	73	97	72	85	58	793 mm
	dp	13	11	9	8	10	9	7	7	10	10	12	13	119 days
	sunshine	96	127	196	216	256	269	301	254	169	145	62	72	2163 hours
	%	34	43	53	53	55	57	63	58	45	43	22	27	48 %

Burlington-Vermont

latitude: 44°28′n longitude: 73°09′w height: 104 m

DCbo	119-28	Jan	Feb	Mar	Apr	May	Jun	Jul	Aug	Sep	Oct	Nov	Dec	Year
	t	-8	-7	-2	6	13	19	21	20	16	9	3	-5	7 °C
	clothing	w	w	w	n	n	s	s	s	n	n	w	w	–
	stress						o	o	o					–
	p	50	45	54	67	76	89	98	86	84	75	67	54	845 mm
	dp	10	8	10	10	10	9	10	8	9	8	10	8	110 days
	sunshine	115	137	192	198	257	279	305	269	210	170	89	91	2312 hours
	%	41	47	52	49	56	59	64	61	56	50	31	34	52 %

Sheridan – Wyoming

latitude: 44°46'n longitude: 106°58'w height: 1209 m

GDCbo	119-29	Jan	Feb	Mar	Apr	May	Jun	Jul	Aug	Sep	Oct	Nov	Dec	Year
	tx	12	14	19	26	31	34	38	37	34	28	19	16	38 °C
	t	-6	-5	0	6	12	17	22	21	15	9	1	-3	7 °C
	tn	-27	-26	-19	-9	-2	2	7	6	-1	-6	-18	-22	-31 °C
	clothing	w	w	w	n	n	n	s	s	n	n	w	w	–
	stress							o	o					–
	p	16	19	36	55	65	65	30	23	30	29	20	16	404 mm
	dp	5	5	8	8	9	8	5	3	5	5	5	4	70 days
	sunshine	163	169	226	235	262	284	358	322	248	221	153	146	2787 hours
	%	57	57	61	58	57	61	76	74	66	65	53	53	62 %

Boise – Idaho

latitude: 43°34'n longitude: 116°13'w height: 871 m

GDCao	119-30	Jan	Feb	Mar	Apr	May	Jun	Jul	Aug	Sep	Oct	Nov	Dec	Year
	tx	12	14	20	26	31	36	39	38	33	27	18	13	40 °C
	t	-2	1	5	10	14	18	24	22	17	11	4	0	10 °C
	tn	-16	-11	-6	-2	1	4	8	8	2	-3	-7	-13	-19 °C
	clothing	w	w	n	n	n	s	t	s	n	n	w	w	–
	stress						o	f	o					–
	p	34	34	34	29	33	23	5	4	10	21	30	34	290 mm
	dp	8	8	8	5	5	4	1	1	2	5	7	8	62 days
	sunshine	122	150	226	270	319	351	415	367	308	232	134	112	3008 hours
	%	42	51	61	67	70	76	88	84	82	68	46	40	67 %

Buffalo – New York

latitude: 42°56'n longitude: 78°44'w height: 215 m

DCbo	119-31	Jan	Feb	Mar	Apr	May	Jun	Jul	Aug	Sep	Oct	Nov	Dec	Year
	t	-5	-5	-1	6	12	18	21	20	16	10	4	-3	8 °C
	clothing	w	w	w	n	n	s	s	s	n	n	w	w	–
	stress						o	o	o					–
	p	72	69	82	76	75	65	65	77	80	76	91	76	906 mm
	dp	13	12	12	10	10	8	7	7	8	8	11	12	118 days
	sunshine	97	118	174	205	268	312	325	284	232	188	88	82	2372 hours
	%	33	40	47	51	59	68	70	66	62	55	30	29	53 %

Boston – Massachusetts

latitude: 42°22'n longitude: 71°01'w height: 9 m

DCao	119-32	Jan	Feb	Mar	Apr	May	Jun	Jul	Aug	Sep	Oct	Nov	Dec	Year
	tx	14	12	18	27	31	34	35	34	32	28	22	13	35 °C
	t	-1	-1	3	9	15	20	23	22	18	13	7	1	11 °C
	tn	-15	-16	-10	-2	2	9	13	12	7	1	-7	-15	-16 °C
	clothing	w	w	w	n	n	s	t	s	s	n	n	w	–
	stress						o	f	o	o				–
	p	100	84	107	96	85	88	73	93	88	80	100	92	1086 mm
	dp	9	8	9	9	9	9	7	7	7	6	8	8	96 days
	sunshine	156	169	211	225	263	288	301	279	240	209	153	153	2647 hours
	%	53	57	57	56	58	63	65	65	64	61	52	54	59 %

Blue Hill – Massachusetts

DCbo	119-33	Jan	Feb	Mar	Apr	May	Jun	Jul	Aug	Sep	Oct	Nov	Dec	Year
		latitude: 42°13′n				longitude: 71°07′w				height: 195 m				
	t	-3	-3	2	8	14	18	22	21	17	12	6	-1	9 °C
	clothing	w	w	w	n	n	s	s	s	n	n	n	w	–
	stress						o	o	o					–
	p	114	95	115	102	88	95	83	103	100	95	115	101	1206 mm
	dp	10	8	10	9	9	9	9	7	8	7	8	9	103 days
	sunshine	129	140	172	194	227	242	253	240	201	182	133	126	2239 hours
	%	44	47	47	48	50	53	54	56	53	53	45	45	50 %

Chicago – Illinois

DCao	119-34	Jan	Feb	Mar	Apr	May	Jun	Jul	Aug	Sep	Oct	Nov	Dec	Year
		latitude: 41°47′n				longitude: 87°45′w				height: 190 m				
	tx	11	12	20	26	29	33	34	33	32	26	20	12	35 °C
	t	-3	-2	2	10	16	22	24	24	19	13	4	-2	10 °C
	tn	-21	-19	-12	-3	4	8	13	13	6	-1	-9	-18	-23 °C
	clothing	w	w	w	n	n	s	t	t	s	n	w	w	–
	stress						o	o	o	o				–
	p	47	41	70	77	95	103	86	80	69	71	56	48	843 mm
	dp	7	6	9	8	11	8	6	7	7	6	7	7	89 days
	sunshine	130	139	189	208	275	305	323	291	240	213	124	117	2554 hours
	%	44	46	51	52	61	67	70	68	64	62	42	41	57 %

Des Moines – Iowa

DCao	119-35	Jan	Feb	Mar	Apr	May	Jun	Jul	Aug	Sep	Oct	Nov	Dec	Year
		latitude: 41°32′n				longitude: 93°39′w				height: 294 m				
	tx	11	14	23	29	32	38	37	36	33	29	21	13	38 °C
	t	-6	-4	2	10	17	22	25	24	19	13	3	-3	10 °C
	tn	-24	-20	-14	-3	3	9	13	11	3	-4	-11	-21	-26 °C
	clothing	w	w	w	n	n	s	t	t	s	n	w	w	–
	stress						o	f	f	o				–
	p	33	28	53	64	103	120	78	93	73	52	45	29	771 mm
	dp	6	4	7	7	9	9	7	8	7	6	5	5	80 days
	sunshine	151	163	204	216	270	300	331	304	247	227	154	132	2699 hours
	%	51	54	55	54	60	66	72	71	66	66	52	46	60 %

Nantucket – Massachusetts

DObk	119-36	Jan	Feb	Mar	Apr	May	Jun	Jul	Aug	Sep	Oct	Nov	Dec	Year
		latitude: 41°15′n				longitude: 70°04′w				height: 4 m				
	t	0	0	2	7	11	16	20	20	17	12	7	2	10 °C
	clothing	w	w	w	n	n	n	s	s	n	n	n	w	–
	stress							o	o					–
	p	107	96	115	96	73	74	69	93	89	94	103	100	1109 mm
	dp	11	10	10	9	8	7	6	7	6	7	9	10	100 days
	sunshine	125	146	207	224	265	281	280	257	225	199	125	121	2544 hours
	%	42	49	56	56	59	62	61	60	60	58	42	42	55 %

Winnemucca – Nevada

latitude: 40°54′n longitude: 117°48′w height: 1322 m

GDCbo	119-37	Jan	Feb	Mar	Apr	May	Jun	Jul	Aug	Sep	Oct	Nov	Dec	Year
	tx	11	13	19	26	32	36	38	37	34	28	19	16	39 °C
	t	-3	0	3	8	12	16	22	20	15	9	2	-1	9 °C
	tn	-17	-13	-11	-6	-1	2	8	4	-1	-7	-11	-15	-22 °C
	clothing	w	w	w	n	n	n	s	s	n	n	w	w	–
	stress						o	f	o					–
	p	27	24	22	21	24	19	7	4	9	21	20	24	219 mm
	dp	7	6	6	5	4	3	1	1	2	4	5	6	50 days
	sunshine	134	155	200	248	292	330	389	358	303	255	161	141	2966 hours
	%	47	52	54	62	65	73	85	83	81	74	54	49	66 %

Salt Lake City – Utah

latitude: 40°46′n longitude: 111°58′w height: 1288 m

GDCao	119-38	Jan	Feb	Mar	Apr	May	Jun	Jul	Aug	Sep	Oct	Nov	Dec	Year
	tx	10	13	20	27	31	36	39	37	33	27	19	14	39 °C
	t	-2	1	5	10	15	19	25	24	18	12	3	0	11 °C
	tn	-17	-8	-8	-3	1	5	11	9	2	-1	-8	-11	-18 °C
	clothing	w	w	n	n	n	s	t	t	s	n	w	w	–
	stress						o	f	f	o				–
	p	34	30	40	45	36	25	15	22	13	29	33	31	353 mm
	dp	7	7	8	7	5	4	3	3	3	5	6	7	65 days
	sunshine	143	161	233	263	323	352	380	350	310	255	161	127	3058 hours
	%	48	53	63	66	72	78	83	82	83	74	54	44	69 %

La Guardia Airport (New York)

latitude: 40°46′n longitude: 74°01′w height: 16 m

DOak	119-39	Jan	Feb	Mar	Apr	May	Jun	Jul	Aug	Sep	Oct	Nov	Dec	Year
	tx	14	13	21	27	31	35	36	34	32	27	21	16	37 °C
	t	1	1	5	11	17	22	25	24	20	15	8	2	13 °C
	tn	-14	-14	-8	-1	6	11	15	14	8	2	-4	-11	-17 °C
	clothing	w	w	n	n	n	s	t	t	s	n	n	w	–
	stress						o	f	f	o				–
	p	84	78	107	91	91	86	94	129	100	86	91	86	1123 mm
	dp	9	8	10	9	10	8	8	8	6	6	7	8	97 days
	sunshine	140	167	211	237	257	285	285	260	225	202	162	133	2564 hours
	%	47	55	57	59	57	63	62	61	60	58	54	46	57 %

Columbus – Ohio

latitude: 40°00′n longitude: 83°53′w height: 254 m

DCao	119-40	Jan	Feb	Mar	Apr	May	Jun	Jul	Aug	Sep	Oct	Nov	Dec	Year
	tx	14	16	22	28	31	33	35	34	32	28	21	15	36 °C
	t	-1	0	4	11	17	22	24	23	19	13	6	0	12 °C
	tn	-18	-17	-10	-4	3	9	12	11	5	-1	-8	-13	-21 °C
	clothing	w	w	w	n	n	s	t	t	s	n	n	w	–
	stress						o	f	o	o				–
	p	80	59	80	89	102	106	100	73	67	54	64	59	931 mm
	dp	9	8	10	10	10	9	8	7	7	6	8	7	99 days
	sunshine	120	129	170	207	268	296	310	285	250	211	126	99	2471 hours
	%	40	43	46	52	60	66	68	67	67	61	42	34	55 %

Dayton – Ohio

Drao	119-41	Jan	Feb	Mar	Apr	May	Jun	Jul	Aug	Sep	Oct	Nov	Dec	Year
		latitude: 39°54′n						longitude: 84°13′w				height: 306 m		
	t	-1	-1	4	10	16	22	24	23	19	13	5	0	11 °C
	clothing	w	w	w	n	n	s	t	t	s	n	n	w	–
	stress						o	f	f	o				–
	p	81	59	79	84	95	104	90	73	66	57	68	60	915 mm
	dp	9	7	10	10	10	9	8	7	6	6	8	8	98 days
	sunshine	123	135	182	207	268	305	314	306	258	221	132	114	2565 hours
	%	41	45	49	50	60	68	69	72	69	64	44	39	57 %

Denver – Colorado

GDCao	119-42	Jan	Feb	Mar	Apr	May	Jun	Jul	Aug	Sep	Oct	Nov	Dec	Year
		latitude: 39°46′n						longitude: 104°52′w				height: 1625 m		
	t	-1	0	3	9	14	19	23	22	18	11	4	1	10 °C
	clothing	w	w	w	n	n	s	t	s	s	n	w	w	–
	stress						o	o	o	o				–
	p	14	18	31	54	69	37	39	33	29	26	174	118	640 mm
	dp	3	4	6	7	8	5	5	5	4	4	3	3	57 days
	sunshine	220	213	263	263	285	314	323	306	284	259	198	198	3126 hours
	%	73	70	71	66	64	70	71	72	76	75	66	68	70 %

Colombia – Missouri

DCao	119-43	Jan	Feb	Mar	Apr	May	Jun	Jul	Aug	Sep	Oct	Nov	Dec	Year
		latitude: 38°58′n						longitude: 92°22′w				height: 257 m		
	t	-1	1	6	13	18	23	26	25	21	15	6	1	13 °C
	clothing	w	w	n	n	s	t	t	t	s	n	n	w	–
	stress					o	o	f	f	o				–
	p	43	46	67	84	119	110	87	97	99	79	58	50	939 mm
	dp	6	6	8	9	10	9	7	8	8	8	6	5	88 days
	sunshine	158	165	207	230	288	303	335	322	258	229	169	147	2811 hours
	%	52	54	56	58	65	68	74	76	69	66	56	50	63 %

Washington – District of Columbia

DOak	119-44	Jan	Feb	Mar	Apr	May	Jun	Jul	Aug	Sep	Oct	Nov	Dec	Year
		latitude: 38°51′n						longitude: 77°02′w				height: 20 m		
	tx	17	18	23	29	32	34	36	36	33	28	23	18	37 °C
	t	3	3	7	13	19	23	26	25	21	15	9	3	14 °C
	tn	-13	-12	-7	-1	5	10	14	13	7	1	-5	-11	-16 °C
	clothing	w	w	n	n	s	t	t	t	s	n	n	w	–
	stress					o	o	f	f	o				–
	p	77	63	82	80	105	82	105	124	97	78	72	71	1036 mm
	dp	8	7	10	8	10	8	10	8	7	6	6	8	96 days
	sunshine	152	157	211	218	253	294	285	262	235	211	179	144	2601 hours
	%	50	52	57	55	57	66	63	62	63	61	59	49	58 %

St. Louis – Missouri

DCak	119-45	Jan	Feb	Mar	Apr	May	Jun	Jul	Aug	Sep	Oct	Nov	Dec	Year
				latitude: 38°45′n				longitude: 90°23′w				height: 172 m		
	tx	17	19	25	29	32	34	36	36	34	29	23	18	37 °C
	t	0	2	6	13	19	24	26	25	21	15	7	2	13 °C
	tn	-18	-14	-8	-1	6	12	16	14	8	1	-7	-14	-21 °C
	clothing	w	w	n	n	s	t	t	t	s	n	n	w	–
	stress					o	f	f	f	o				–
	p	50	52	78	94	96	109	84	77	70	73	65	50	897 mm
	dp	7	6	9	9	9	9	5	6	7	6	7	7	87 days
	sunshine	158	154	197	214	279	307	321	292	239	225	151	133	2670 hours
	%	52	51	53	54	63	69	71	69	64	65	50	45	60 %

Wichita – Kansas

DCak	119-46	Jan	Feb	Mar	Apr	May	Jun	Jul	Aug	Sep	Oct	Nov	Dec	Year
				latitude: 37°39′n				longitude: 97°25′w				height: 408 m		
	t	0	2	7	14	19	25	27	27	22	16	7	2	14 °C
	clothing	w	w	n	n	s	t	t	t	s	n	n	w	–
	stress					o	f	f	f	o				–
	p	21	23	42	58	101	107	92	73	82	61	38	24	722 mm
	dp	4	5	5	6	8	8	5	6	6	5	4	4	66 days
	sunshine	181	179	230	241	273	305	328	311	257	243	186	176	2910 hours
	%	59	59	62	61	62	69	73	74	69	70	61	59	65 %

Asheville – North Carolina

GDCak	119-47	Jan	Feb	Mar	Apr	May	Jun	Jul	Aug	Sep	Oct	Nov	Dec	Year
				latitude: 35°36′n				longitude: 82°32′w				height: 687 m		
	tx	19	21	26	29	32	34	36	36	33	29	24	19	37 °C
	t	4	5	8	13	18	22	24	23	20	14	8	4	14 °C
	tn	-13	-11	-6	1	7	13	16	15	8	1	-6	-10	-16 °C
	clothing	w	n	n	n	s	s	t	t	s	n	n	w	–
	stress					o	o	f	f	o				–
	p	81	77	95	81	73	89	109	92	71	63	56	74	962 mm
0.2 mm	dp	11	11	12	10	12	13	15	13	9	7	8	10	131 days
	sunshine	175	180	267	255	266	279	257	238	216	220	170	176	2699 hours
	%	56	58	72	65	61	64	58	57	58	63	55	58	61 %

Oklahoma – Oklahoma

DCak	119-48	Jan	Feb	Mar	Apr	May	Jun	Jul	Aug	Sep	Oct	Nov	Dec	Year
				latitude: 35°24′n				longitude: 97°36′w				height: 397 m		
	tx	20	24	28	31	33	36	39	39	36	32	27	19	40 °C
	t	2	5	9	15	20	25	27	27	23	17	9	4	15 °C
	tn	-13	-11	-5	-1	7	13	18	17	8	1	-6	-9	-15 °C
	clothing	w	n	n	n	s	t	t	t	t	n	n	w	–
	stress					o	f	ff	ff	f	o			–
	p	33	35	50	79	132	114	60	64	77	64	40	36	783 mm
	dp	4	5	5	6	8	8	5	5	6	5	4	4	65 days
	sunshine	185	186	241	247	278	318	332	325	272	249	192	183	3007 hours
	%	59	60	65	63	64	73	75	78	73	71	62	60	67 %

6. *Climatic area:* eastern Texas; Arkansas, Louisiana, Mississippi, Alabama, eastern North Carolina, southeastern South Carolina, greater southern part of Georgia, northern Florida
6.1. Subtropical Rain Climate Cr and Subtropical Mountain Rain Climate GCr
6.2. Summer hot, in the south very hot; winter cool, in the south mild.
6.3. Variable weather prevails; frequent dry and sunny days.
6.4. Wind: at times gales and tornadoes, in coastal areas tropical storms or hurricanes.

Nashville – Tennessee

				latitude: 36°07′n				longitude: 86°41′w				height: 184 m		
Crak	119-49	Jan	Feb	Mar	Apr	May	Jun	Jul	Aug	Sep	Oct	Nov	Dec	Year
	tx	19	21	26	29	32	34	36	36	33	29	24	19	37 °C
	t	4	6	10	15	20	25	27	26	23	16	9	5	16 °C
	tn	-13	-11	-6	1	7	13	16	15	8	1	-6	-10	-16 °C
	clothing	w	n	n	n	s	t	t	t	t	n	n	n	–
	stress					o	f	ff	f	f				–
	p	139	115	132	95	94	83	94	73	73	59	83	106	1147 mm
	dp	9	9	10	8	9	7	9	7	6	5	7	9	95 days
	sunshine	127	143	193	228	271	297	285	276	238	227	160	127	2572 hours
	%	41	47	52	58	62	68	64	66	64	65	52	42	58 %

Hatteras – North Carolina

				latitude: 35°16′n				longitude: 75°33′w				height: 3 m		
Crak	119-50	Jan	Feb	Mar	Apr	May	Jun	Jul	Aug	Sep	Oct	Nov	Dec	Year
	t	8	8	11	15	20	24	26	25	23	19	13	9	17 °C
	clothing	n	n	n	n	s	t	t	t	t	s	n	n	–
	stress					o	f	f	f	o	o			–
	p	99	100	106	58	101	105	156	163	150	108	104	116	1366 mm
	dp	9	8	9	7	8	6	10	10	8	7	7	9	98 days
	sunshine	176	168	249	255	287	283	301	284	253	253	235	195	2939 hours
	%	57	54	67	65	66	65	68	68	68	72	76	64	66 %

Little Rock – Arkansas

				latitude: 34°44′n				longitude: 92°14′w				height: 81 m		
Crhk	119-51	Jan	Feb	Mar	Apr	May	Jun	Jul	Aug	Sep	Oct	Nov	Dec	Year
	tx	21	23	27	30	32	35	37	36	34	31	25	21	38 °C
	t	5	7	11	17	21	26	28	27	24	17	10	6	16 °C
	tn	-10	-8	-3	3	9	15	18	17	11	3	-3	-7	-12 °C
	clothing	n	n	n	n	s	t	t	t	t	n	n	n	–
	stress					o	f	ff	ff	f				–
	p	133	110	122	125	134	92	85	72	82	73	105	104	1236 mm
	dp	8	8	8	8	9	7	7	7	5	5	7	8	87 days
	sunshine	141	162	208	231	282	308	304	296	245	245	174	147	2743 hours
	%	45	52	56	59	65	71	69	71	66	70	56	48	62 %

Atlanta – Georgia

latitude: 33°39′n longitude: 84°25′w height: 315 m

Crak	119-52	Jan	Feb	Mar	Apr	May	Jun	Jul	Aug	Sep	Oct	Nov	Dec	Year
	tx	20	21	25	29	32	34	35	34	33	28	23	19	36 °C
	t	7	8	11	16	21	25	26	26	23	17	11	7	16 °C
	tn	-9	-8	-3	2	8	13	18	17	11	3	-4	-7	-9 °C
	clothing	n	n	n	n	s	t	t	t	t	n	n	n	–
	stress					o	f	ff	ff	f				–
	p	113	115	136	114	80	97	120	91	83	62	75	111	1197 mm
	dp	9	9	10	8	8	9	10	9	6	5	6	9	98 days
	sunshine	152	160	212	254	297	288	272	274	238	239	185	154	2725 hours
	%	48	52	57	65	69	67	62	66	64	68	59	50	61 %

Charleston – South Carolina

latitude: 32°54′n longitude: 80°02′w height: 15 m

Cral	119-53	Jan	Feb	Mar	Apr	May	Jun	Jul	Aug	Sep	Oct	Nov	Dec	Year
	tx	14	11	12	16	18	19	21	20	19	17	14	12	23 °C
	t	10	11	14	18	22	26	27	26	24	19	13	10	18 °C
	tn	-2	-2	0	3	5	8	9	8	7	6	2	0	-5 °C
	clothing	n	n	n	s	s	t	t	t	t	s	n	n	–
	stress				o	o	f	ff	ff	f	o			–
	p	65	84	100	73	92	126	196	168	148	72	53	72	1249 mm
	dp	7	7	7	6	6	9	11	10	8	5	6	7	89 days
	sunshine	181	176	245	261	296	266	271	269	219	233	198	193	2808 hours
	%	57	57	66	67	69	62	62	65	59	66	63	62	63 %

Shreveport – Louisiana

latitude: 32°28′n longitude: 93°49′w height: 79 m

Crhk	119-54	Jan	Feb	Mar	Apr	May	Jun	Jul	Aug	Sep	Oct	Nov	Dec	Year
	t	9	10	14	18	23	27	28	28	25	20	13	10	19 °C
	clothing	n	n	n	s	t	t	t	t	t	s	n	n	–
	stress				o	f	ff	ff	ff	f	o			–
	p	122	104	105	116	122	85	95	65	58	71	107	125	1175 mm
	dp	9	8	8	8	7	6	7	6	5	4	7	8	83 days
	sunshine	163	173	219	218	279	303	322	297	260	265	192	168	2860 hours
	%	51	55	59	56	65	71	74	72	70	75	61	54	64 %

Abilene – Texas

latitude: 32°26′n longitude: 99°41′w height: 534 m

GCwhk	119-55	Jan	Feb	Mar	Apr	May	Jun	Jul	Aug	Sep	Oct	Nov	Dec	Year
	tx	24	26	29	31	34	37	38	39	36	32	27	26	39 °C
	t	7	9	13	18	22	27	28	28	24	19	12	8	18 °C
	tn	-8	-7	-2	4	10	14	18	17	12	5	-1	-8	-11 °C
	clothing	n	n	n	s	s	t	t	t	t	s	n	n	–
	stress				o	o	f	ff	ff	f	o			–
	p	22	28	26	58	100	68	58	37	53	72	28	32	582 mm
	dp	4	4	4	5	7	6	4	4	4	4	3	4	53 days
	sunshine	198	198	256	269	291	329	339	322	256	264	230	203	3155 hours
	%	62	63	69	69	68	77	78	78	69	75	73	65	71 %

Mobile – Alabama

latitude: 30°42'n longitude: 88°02'w height: 3 m

Crhl	119-56	Jan	Feb	Mar	Apr	May	Jun	Jul	Aug	Sep	Oct	Nov	Dec	Year
	tx	23	24	27	29	33	36	36	36	35	31	27	24	37 °C
	t	11	12	16	19	23	27	28	28	26	21	15	12	20 °C
	tn	-3	-2	2	7	12	18	20	21	15	7	1	-2	-5 °C
	clothing	n	n	n	s	t	t	t	t	t	s	n	n	–
	stress				o	f	ff	ff	ff	f	o			–
	p	122	135	165	119	109	140	170	173	132	89	99	125	1578 mm
	dp	10	10	10	8	8	12	15	14	10	6	7	10	120 days

Jacksonville – Florida

latitude: 30°25'n longitude: 81°31'w height: 9 m

Crhl	119-57	Jan	Feb	Mar	Apr	May	Jun	Jul	Aug	Sep	Oct	Nov	Dec	Year
	tx	25	26	29	31	34	36	36	35	34	31	28	25	36 °C
	t	13	14	17	20	24	27	28	28	26	22	16	13	21 °C
	tn	-2	0	3	8	13	18	21	21	17	9	3	-1	-4 °C
	clothing	n	n	n	s	t	t	t	t	t	s	n	n	–
	stress				o	f	ff	ff	ff	f	o			–
	p	62	74	89	90	88	161	195	174	192	131	43	56	1355 mm
	dp	6	6	7	6	7	11	14	12	11	8	5	6	99 days
	sunshine	188	190	242	275	293	249	254	238	181	198	191	174	2673 hours
	%	58	60	65	71	69	59	59	58	49	56	60	55	60 %

New Orleans – Louisiana

latitude: 29°59'n longitude: 90°15'w height: 9 m

Cral	119-58	Jan	Feb	Mar	Apr	May	Jun	Jul	Aug	Sep	Oct	Nov	Dec	Year
	tx	25	26	28	29	32	34	35	34	34	31	27	25	36 °C
	t	12	13	16	19	23	26	27	27	25	21	15	13	20 °C
	tn	-1	1	4	9	14	20	21	21	17	11	4	0	-3 °C
	clothing	n	n	n	s	t	t	t	t	t	s	n	n	–
	stress				o	f	ff	ff	ff	f	o			–
	p	98	101	136	116	111	113	171	136	128	72	85	104	1369 mm
	dp	8	7	7	7	7	10	13	13	9	5	6	8	100 days
	sunshine	155	164	211	240	279	294	245	239	234	239	192	140	2630 hours
	%	48	52	57	62	66	70	57	58	63	67	60	44	59 %

Houston – Texas

latitude: 29°46'n longitude: 95°22'w height: 12 m

Crhl	119-59	Jan	Feb	Mar	Apr	May	Jun	Jul	Aug	Sep	Oct	Nov	Dec	Year
	tx	25	27	29	30	33	36	37	37	36	33	28	26	38 °C
	t	12	13	17	21	24	27	28	29	26	22	19	12	21 °C
	tn	-4	-1	2	7	13	19	21	21	16	8	3	-2	-6 °C
	clothing	n	n	n	s	t	t	t	t	t	s	s	n	–
	stress				o	f	ff	ff	ff	f	o	o		–
	p	89	76	84	91	119	117	99	99	104	94	89	109	1171 mm
0.2 mm	dp	9	8	8	7	7	8	10	10	8	6	8	10	99 days
	sunshine	140	144	183	207	251	306	285	279	234	239	180	155	2602 hours
	%	43	46	49	54	59	73	66	68	63	67	56	49	58 %

San Antonio – Texas

Crhl	119-60	Jan	Feb	Mar	Apr	May	Jun	Jul	Aug	Sep	Oct	Nov	Dec	Year
				latitude: 29°32′n				longitude: 98°28′w			height: 242 m			
	t	11	13	16	20	24	28	29	29	26	21	15	12	20 °C
	clothing	n	n	n	s	t	t	t	t	t	s	n	n	–
	stress				o	f	ff	ff	ff	f	o			–
	p	44	42	42	72	88	75	53	60	89	64	35	44	707 mm
	dp	6	6	5	6	6	5	4	4	6	4	4	5	61 days
	sunshine	159	165	212	216	241	286	330	311	259	241	183	169	2772 hours
	%	49	52	57	56	57	68	77	76	70	68	57	53	62 %

Tampa – Florida

Crhl	119-61	Jan	Feb	Mar	Apr	May	Jun	Jul	Aug	Sep	Oct	Nov	Dec	Year
				latitude: 27°57′n				longitude: 82°27′w			height: 11 m			
	tx	27	27	29	31	33	34	34	34	34	32	29	27	35 °C
	t	16	17	19	22	25	27	28	28	27	24	19	17	22 °C
	tn	2	2	6	10	15	18	21	21	19	12	6	2	-1 °C
	clothing	n	n	s	s	t	t	t	t	t	t	s	n	–
	stress			o	o	f	ff	ff	ff	ff	f	o		–
	p	66	69	69	51	76	183	221	218	160	71	43	58	1285 mm
0.2 mm	dp	7	7	6	5	7	15	17	17	15	8	5	7	116 days
	sunshine	223	220	260	273	316	279	257	239	225	232	222	202	2949 hours
	%	68	69	70	71	75	67	60	59	61	65	69	63	66 %

Brownsville – Texas

Crhl	119-62	Jan	Feb	Mar	Apr	May	Jun	Jul	Aug	Sep	Oct	Nov	Dec	Year
				latitude: 25°54′n				longitude: 97°26′w			height: 6 m			
	tx	28	29	31	34	35	35	36	37	36	33	30	28	37 °C
	t	16	18	20	23	26	28	29	29	27	24	20	17	23 °C
	tn	-1	2	5	9	10	14	22	22	17	11	6	1	-2 °C
	clothing	n	s	s	t	t	t	t	t	t	t	s	n	–
	stress		o	o	f	f	f	ff	ff	ff	f	o		–
	p	34	38	26	39	60	75	43	70	127	90	34	44	670 mm
	dp	5	4	4	3	5	4	5	6	8	5	5	6	60 days
	sunshine	153	155	186	211	274	297	341	307	251	240	173	151	2739 hours
	%	46	49	50	55	66	72	81	76	68	67	53	46	62 %

7. *Climatic area:* West of California

7.1. Subtropical Winter Rain Climate Cs and
Subtropical Winter Rain Mountain Climate GCs

7.2. Summer warm, inland hot; winter cool in the north, otherwise mild.

7.3. Summer generally dry and sunny; winter often dry and sunny, now and then variable weather.

7.4. Wind: occasional gales. Fog: frequent in coastal areas.

Sacramento – California

Csak	119-63	Jan	Feb	Mar	Apr	May	Jun	Jul	Aug	Sep	Oct	Nov	Dec	Year
latitude: 38°31′n	longitude: 121°30′w	height: 8 m												
	tx	17	21	24	27	33	37	39	38	36	31	24	17	40 °C
	t	8	10	12	16	19	22	25	24	23	18	12	9	17 °C
	tn	-1	1	3	4	7	9	11	11	9	6	1	-1	-2 °C
	clothing	n	n	n	n	s	s	t	t	t	s	n	n	–
	stress					o	o	f	f	f	o			–
	p	81	76	60	36	15	3	0	1	5	20	37	82	414 mm
	dp	8	8	7	4	3	1	0	0	0	3	4	8	46 days
	sunshine	137	184	256	313	376	405	438	401	347	298	194	133	3482 hours
	%	45	60	69	79	85	91	97	94	93	85	64	45	78 %

Oakland – California

Csbk	119-64	Jan	Feb	Mar	Apr	May	Jun	Jul	Aug	Sep	Oct	Nov	Dec	Year
latitude: 37°44′n	longitude: 122°12′w	height: 3 m												
	t	9	10	12	14	15	17	18	18	18	16	13	10	14 °C
	clothing	n	n	n	n	n	n	s	s	s	n	n	n	–
	stress							o	o	o				–
	p	97	82	61	35	17	3	0	1	5	20	44	91	456 mm
	dp	8	9	7	5	3	1	0	0	1	3	5	8	50 days

San Francisco – California

Csbk	119-65	Jan	Feb	Mar	Apr	May	Jun	Jul	Aug	Sep	Oct	Nov	Dec	Year
latitude: 37°37′n	longitude: 122°23′w	height: 5 m												
	tx	17	19	22	24	27	28	26	26	29	28	23	18	32 °C
	t	9	10	12	13	15	16	17	17	18	16	13	10	14 °C
	tn	4	5	6	7	8	9	10	10	11	9	7	4	3 °C
	clothing	n	n	n	n	n	n	s	s	s	n	n	n	–
	stress							o	o	o				–
	p	102	88	68	33	12	3	0	1	5	19	40	104	475 mm
	dp	9	9	7	5	3	1	0	0	1	3	5	8	51 days
	sunshine	172	188	252	285	317	323	296	274	268	247	192	161	2975 hours
	%	56	62	68	72	72	73	66	65	72	71	63	54	67 %

Santa Maria – California

Csll	119-66	Jan	Feb	Mar	Apr	May	Jun	Jul	Aug	Sep	Oct	Nov	Dec	Year
latitude: 34°54′n	longitude: 120°27′w	height: 73 m												
	t	10	11	12	13	14	15	17	17	17	16	13	11	14 °C
	clothing	n	n	n	n	n	n	s	s	s	n	n	n	–
	stress							o	o	o				–
	p	72	64	52	30	6	4	1	1	4	15	26	66	340 mm
	dp	7	7	7	3	1	0	0	0	0	2	3	7	37 days

Los Angeles – California

		latitude: 34°03′n					longitude: 118°14′w				height: 95 m			
Csal	119-67	Jan	Feb	Mar	Apr	May	Jun	Jul	Aug	Sep	Oct	Nov	Dec	Year
	tx	26	27	28	30	31	32	33	34	36	33	31	27	38 °C
	t	13	14	15	16	18	20	22	23	22	19	17	14	18 °C
	tn	3	3	4	6	8	11	12	12	11	8	6	3	2 °C
	clothing	n	n	n	n	s	s	s	t	s	s	n	n	–
	stress						o	o	o	o				–
	p	60	85	60	30	6	2	0	0	7	13	26	79	368 mm
0.2 mm	dp	7	5	6	4	2	1	0	1	1	2	4	6	39 days
	sunshine	217	232	270	249	279	282	335	332	279	226	240	245	3185 hours
	%	69	75	73	64	65	65	76	80	75	64	77	80	72 %

San Diego – California

		latitude: 32°44′n					longitude: 117°10′w				height: 9 m			
Csbl	119-68	Jan	Feb	Mar	Apr	May	Jun	Jul	Aug	Sep	Oct	Nov	Dec	Year
	tx	23	24	24	26	24	26	27	28	29	29	28	24	33 °C
	t	13	14	15	16	18	19	21	22	21	19	16	14	17 °C
	tn	3	4	6	7	10	12	14	15	13	9	7	4	2 °C
	clothing	n	n	n	n	s	s	s	s	s	s	n	n	–
	stress							o	o	o	o			–
	p	51	55	40	20	4	1	0	2	4	12	23	52	264 mm
	dp	5	7	5	3	1	0	0	0	1	2	2	6	32 days
	sunshine	226	222	260	246	253	247	297	285	256	236	233	221	2982 hours
	%	71	71	70	63	59	58	68	69	69	67	74	71	67 %

8. *Climatic area:* northeastern California, eastern Montana, most of Idaho, Wyoming, Utah, Colorado, northern Nevada and New Mexico, western North Dakota, South Dakota, Nebraska, Kansas, western Texas

8.1. Steppe Mountain Climate GBS and Steppe High Mountain Climate HBS

8.2. At about 1000 m, summer hot, winter cold.

8.3. Generally dry and sunny, at times variable weather.

8.4. Wind: occasional gales. Snow cover: according to height.

Rapid City – South Dakota

		latitude: 44°03′n					longitude: 103°04′w				height: 966 m			
GBSao	119-69	Jan	Feb	Mar	Apr	May	Jun	Jul	Aug	Sep	Oct	Nov	Dec	Year
	t	-6	-4	0	7	13	18	23	22	16	10	2	-3	8 °C
	clothing	w	w	w	n	n	s	t	s	n	n	w	w	–
	stress						o	o	o					–
	p	9	12	26	42	68	78	45	31	24	20	10	8	374 mm
	dp	3	3	5	7	9	9	6	5	5	4	2	3	61 days
	sunshine	159	173	226	242	256	273	333	307	244	225	162	144	2744 hours
	%	55	58	61	60	56	59	71	70	65	66	56	52	61 %

North Platte – Nebraska

latitude: 41°08'n longitude: 100°41'w height: 849 m

GBSao 119-70	Jan	Feb	Mar	Apr	May	Jun	Jul	Aug	Sep	Oct	Nov	Dec	Year
t	-4	-2	2	9	15	21	24	24	18	11	2	-3	10 °C
clothing	w	w	w	n	n	s	t	t	s	n	w	w	–
stress							o	o					–
p	11	13	25	51	75	83	64	54	42	23	13	10	464 mm
dp	3	2	4	6	8	8	7	6	5	2	2	2	55 days
sunshine	181	170	222	248	269	308	340	308	258	251	184	170	2909 hours
%	61	57	60	62	60	68	74	72	69	73	62	59	65 %

Ely – Nevada

latitude: 39°17'n longitude: 114°51'w height: 1909 m

GBSbo 119-71	Jan	Feb	Mar	Apr	May	Jun	Jul	Aug	Sep	Oct	Nov	Dec	Year
t	-5	-3	1	6	10	15	20	19	14	8	1	-3	7 °C
clothing	w	w	w	n	n	n	s	s	n	n	w	w	–
stress							o	o					–
p	20	18	22	24	22	13	17	12	14	19	15	17	213 mm
dp	4	4	5	5	5	3	3	3	2	3	3	4	44 days
sunshine	197	198	259	266	311	353	367	343	306	263	202	188	3253 hours
%	65	65	70	67	70	79	81	81	82	76	67	64	73 %

Grand Junction – Colorado

latitude: 39°07'n longitude: 108°32'w height: 1475 m

GBSao 119-72	Jan	Feb	Mar	Apr	May	Jun	Jul	Aug	Sep	Oct	Nov	Dec	Year
tx	10	14	21	27	31	36	37	36	32	27	18	11	38 °C
t	-3	0	5	10	17	22	26	24	20	13	4	-2	11 °C
tn	-19	-14	-7	-3	2	7	13	12	4	-2	-8	-17	-22 °C
clothing	w	w	n	n	n	s	t	t	s	n	w	w	–
stress						o	f	f	o				–
p	16	18	19	19	15	11	14	27	23	19	15	14	211 mm
dp	4	5	5	4	3	3	3	5	4	3	3	4	46 days
sunshine	182	189	237	262	311	348	349	313	295	263	193	176	3118 hours
%	60	62	64	66	70	78	77	74	79	76	64	60	70 %

9. *Climatic area:* eastern California, southern Nevada and New Mexico, Arizona, westernmost Texas

9.1. Desert Climate BW, Desert Mountain Climate GBW, Desert High Mountain Climate HBW

9.2. Summer very hot at lower levels, winter cool at lower levels.

9.3. Predominantly dry and sunny; now and then scattered showers, the odd rainy day.

9.4. Wind: occasional gales and dust-storms.

Las Vegas – Nevada

latitude: 36°05′n longitude: 115°10′w height: 661 m

GBWhk	119-73	Jan	Feb	Mar	Apr	May	Jun	Jul	Aug	Sep	Oct	Nov	Dec	Year
	tx	22	26	29	36	39	42	43	42	41	36	27	23	43 °C
	t	6	9	13	18	23	29	32	31	27	20	12	7	19 °C
	tn	-7	-5	-2	3	7	10	16	13	11	3	-3	-6	-8 °C
	clothing	n	n	n	s	t	t	t	t	t	s	n	n	–
	stress				o	f	f	f	f	f	o			–
	p	13	11	9	6	2	1	13	12	9	5	8	10	99 mm
	dp	2	2	2	1	1	0	1	1	2	1	1	2	16 days
	sunshine	246	250	308	335	384	407	382	368	339	304	250	239	3810 hours
	%	80	81	83	85	88	93	86	88	91	87	81	79	86 %

Albuquerque – New Mexico

latitude: 35°03′n longitude: 106°37′w height: 1620 m

GBWak	119-74	Jan	Feb	Mar	Apr	May	Jun	Jul	Aug	Sep	Oct	Nov	Dec	Year
	t	2	4	8	13	18	24	26	25	21	15	7	3	14 °C
	clothing	w	w	n	n	s	t	t	t	s	n	n	w	–
	stress					o	f	f	f	o				–
	p	10	10	12	12	19	14	30	34	24	19	10	12	207 mm
	dp	2	3	2	3	3	3	5	6	5	3	2	3	40 days
	sunshine	226	223	271	298	347	365	336	317	301	280	239	217	3420 hours
	%	72	72	73	76	80	84	76	76	81	80	77	71	77 %

Phoenix – Arizona

latitude: 33°26′n longitude: 112°01′w height: 337 m

BWhl	119-75	Jan	Feb	Mar	Apr	May	Jun	Jul	Aug	Sep	Oct	Nov	Dec	Year
	tx	24	27	31	36	40	44	45	43	41	37	31	26	46 °C
	t	10	12	16	20	25	30	33	32	29	22	15	11	21 °C
	tn	-1	1	3	6	11	16	20	21	17	8	2	0	-2 °C
	clothing	n	n	n	s	t	t	t	t	t	s	n	n	–
	stress				o	f	f	ff	ff	f	o			–
	p	19	22	17	8	3	2	20	28	19	12	12	22	183 mm
	dp	2	4	2	1	1	0	3	4	3	1	2	3	26 days
	sunshine	244	246	308	344	401	404	368	352	331	310	263	238	3809 hours
	%	77	79	83	88	93	94	84	85	89	88	84	77	86 %

El Paso – Texas

latitude: 31°48′n longitude: 106°24′w height: 1194 m

GBWak	119-76	Jan	Feb	Mar	Apr	May	Jun	Jul	Aug	Sep	Oct	Nov	Dec	Year
	tx	22	23	28	31	35	38	38	36	35	32	26	21	39 °C
	t	7	10	13	17	22	27	27	27	24	19	11	7	18 °C
	tn	-8	-6	-2	3	8	13	17	16	11	3	-3	-7	-10 °C
	clothing	n	n	n	n	s	t	t	t	t	s	n	n	–
	stress					o	f	f	f	f	o			–
	p	12	10	9	7	10	18	33	30	29	22	8	12	200 mm
	dp	3	2	2	1	1	3	5	5	4	3	2	3	34 days
	sunshine	247	251	309	335	380	375	343	329	308	300	259	241	3677 hours
	%	77	80	83	86	89	88	79	80	83	85	82	77	83 %

10. Climatic area: South of Florida
10.1. Tropical Summer Rain Climate Aw
10.2. Summer very hot, winter warm.
10.3. Often dry and sunny. Summer: frequent showers in the afternoon, now and then rainy days. Winter: variable weather at times.
10.4. Wind: occasional gales, hurricane season June – November.

Miami – Florida

latitude: 25°48′n longitude: 80°16′w height: 3 m

Awhb	119-77	Jan	Feb	Mar	Apr	May	Jun	Jul	Aug	Sep	Oct	Nov	Dec	Year
	tx	27	28	29	30	32	32	33	33	32	31	29	28	33 °C
	t	19	20	21	23	25	27	28	28	27	26	22	20	24 °C
	tn	7	6	9	12	17	20	21	22	21	17	9	6	4 °C
	clothing	s	s	s	t	t	t	t	t	t	t	s	s	–
	stress	o	o	o	o	f	ff	ff	ff	ff	f	o	o	–
	p	52	47	58	99	164	187	171	177	241	209	72	42	1518 mm
	dp	5	5	5	6	8	10	12	12	15	12	7	6	103 days
	sunshine	232	240	270	279	282	255	260	257	219	211	225	214	2945 hours
	%	70	75	73	73	68	62	62	63	59	59	69	65	66 %

11. Climatic area: Hawaii
11.1. Tropical Rain Climate Ar, partly Tropical Winter Rain Climate As and corresponding Mountain or High Mountain Climate GAr, GAs, HAr, HAs
NB: in some restricted areas of the islands in the lee of the mountains where rain is lacking, there is Steppe Climate BS or even Desert Climate BW
11.2. Summer hot, winter warm.
11.3. Often dry and sunny, frequent showers in the afternoon, occasionally a rainy day.
11.4. Wind: occasional gales, season of tropical cyclones May – November.

Lihue/Kauai – Hawaii

latitude: 21°59′n longitude: 159°21′w height: 45 m

Arab	119-78	Jan	Feb	Mar	Apr	May	Jun	Jul	Aug	Sep	Oct	Nov	Dec	Year
	t	21	22	22	23	24	25	25	26	26	25	24	22	24 °C
	clothing	s	s	s	t	t	t	t	t	t	t	t	s	–
	stress	o	o	o	f	f	ff	ff	ff	ff	ff	f	o	–
	p	140	135	116	85	66	37	49	62	53	102	115	132	1092 mm
	dp	11	11	15	13	13	13	15	15	12	14	14	14	160 days
	sunshine	177	170	183	190	225	250	260	251	250	209	166	161	2492 hours
	%	52	53	49	50	55	62	63	63	68	58	50	48	56 %

Honolulu/Oahu – Hawaii

latitude: 21°21′n longitude: 157°56′w height: 5 m

Asab	119-79	Jan	Feb	Mar	Apr	May	Jun	Jul	Aug	Sep	Oct	Nov	Dec	Year
	tx	27	27	27	27	28	29	29	29	29	29	28	27	30 °C
	t	22	22	23	23	24	26	26	26	26	26	24	23	24 °C
	tn	16	16	16	17	19	20	21	21	21	20	18	17	15 °C
	clothing	s	s	t	t	t	t	t	t	t	t	t	t	–
	stress	o	o	f	f	ff	ff	ff	ff	ff	ff	ff	f	–
	p	96	84	73	33	25	8	11	23	25	47	55	76	556 mm
	dp	8	8	8	6	6	4	6	6	5	7	8	9	81 days
	sunshine	222	205	254	262	285	298	313	307	276	246	200	199	3067 hours
	%	65	63	68	69	70	74	76	77	75	68	60	59	69 %

Lanai City/Lanai – Hawaii

Asbb	119-80	latitude: 20°50'n				longitude: 156°55'w					height: 495 m			
		Jan	Feb	Mar	Apr	May	Jun	Jul	Aug	Sep	Oct	Nov	Dec	Year
	t	19	19	19	19	20	21	22	22	22	22	20	19	20 °C
	clothing	s	s	s	s	s	s	s	s	s	s	s	s	–
	stress	o	o	o	o	o	o	o	o	o	o	o	o	–
	p	117	95	110	74	68	36	44	45	58	75	73	95	890 mm

Hilo/Hawaii – Hawaii

Arab	119-81	latitude: 19°43'n				longitude: 155°04'w					height: 11 m			
		Jan	Feb	Mar	Apr	May	Jun	Jul	Aug	Sep	Oct	Nov	Dec	Year
	t	22	21	21	22	23	24	24	24	24	24	23	22	23 °C
	clothing	s	s	s	s	t	t	t	t	t	t	t	s	–
	stress	f	f	f	f	f	ff	ff	ff	ff	ff	f	f	–
	p	300	329	373	303	237	172	249	291	216	274	340	386	3470 mm
	dp	17	15	20	22	21	22	25	25	22	21	20	19	249 days
	sunshine	155	139	138	129	122	160	168	151	143	141	111	119	1676 hours
	%	45	43	37	34	30	40	41	38	39	39	33	35	38 %

4.4.119.2. Predominant general weather situations are described in 4.4.0.1. under United States of America and in 4.6.0.1. under Oceania. Hygrothermal diagrams of Anchorage, Washington, New York, St. Louis, New Orleans, San Francisco, Phoenix, and Honolulu are to be found in 4.4.0.2.
Wind names: Black Blizzard is a stormy cold dry wind blowing from the northwest and drifting the bare upper soil in the prairies.
Blizzard is a snowstorm causing a remarkable fall of temperature.
Chinook is 1) a warm dry wind falling off the eastern slopes of the Rocky Mountains. Even in winter it can rapidly melt the snow; 2) a warm humid wind blowing from the southwest in the coastal areas of Washington and Oregon.
Kona is a rainy storm in the Hawaiian region blowing from the southsouth-west.
Sonora storm is the name of thunderstorm squalls in summer occurring in the mountains and deserts of California.
A special phenomenon is the "débacle": the breaking up in the spring of the ice in the rivers. The term is chiefly applied to the great rivers of the North American Continent and of Russia and Siberia. The débacle lasts from two to six weeks; during this period the rivers often overflow their banks, inundating the surrounding country.

4.4.120 Mexico

4.4.120.1. Climatic descriptions and tables

1. Climatic area: Parts of Baja California, Sonora, Sinaloa, Chihuahua, Coahuila

1.1. Desert Climate BW and Desert Mountain Climate GBW;
coastal areas Marine Desert Climate BM
1.2. In lowlands summer very hot, winter warm.
1.3. Almost the whole year dry and sunny; scattered showers here and there.
1.4. Wind: occasional gales and dust-storms. Fog or low stratus frequent, particularly in winter on the coast.

Guayamas

BMhb	120-1	latitude: 27°55′n						longitude: 110°53′w				height: 4 m		
		Jan	Feb	Mar	Apr	May	Jun	Jul	Aug	Sep	Oct	Nov	Dec	Year
	t	18	19	21	23	26	29	31	30	30	27	23	19	24 °C
	clothing	s	s	s	t	t	t	t	t	t	t	t	s	–
	stress	o	o	o	f	f	ff	ff	ff	ff	f	f	o	–
	p	8	6	5	3	3	0	47	76	54	10	11	29	252 mm
0.1 mm	dp	1	2	0	0	0	0	5	7	4	1	1	3	28 days
	sunshine	192	206	239	279	276	294	236	232	249	264	210	195	2872 hours
	%	59	65	64	72	66	70	55	57	67	74	65	61	65 %

La Paz

BMhb	120-2	latitude: 24°10′n						longitude: 110°18′w				height: 18 m		
		Jan	Feb	Mar	Apr	May	Jun	Jul	Aug	Sep	Oct	Nov	Dec	Year
	t	18	20	21	23	26	27	31	31	29	27	22	19	24 °C
	clothing	s	s	s	t	t	t	t	t	t	t	s	s	–
	stress	o	o	o	f	f	f	ff	ff	ff	f	o	o	–
	p	3	11	0	0	0	0	6	42	52	10	13	34	171 mm
0.1 mm	dp	0	1	0	0	0	0	0	4	5	0	1	3	15 days

2. Climatic area: other parts of the states named in climatic area 1 and Durango, Zacatecas, Jalisco (eastern part), St. Luis Potosi (western part), Guanajuato, Queretano

2.1. Steppe Climate BS and Steppe Mountain Climate GBS

2.2. At low levels summer very hot, winter warm.

2.3. Generally dry and sunny; summer: scattered showers in the afternoon and at times a rainy day; winter: now and then a rainy day.

2.4. Wind: occasional gales; in winter at times "nortes".

Zacatecas

HBSll	120-3	latitude: 22°47′n						longitude: 102°35′w				height: 2612 m		
		Jan	Feb	Mar	Apr	May	Jun	Jul	Aug	Sep	Oct	Nov	Dec	Year
	t	10	11	13	15	17	16	14	15	14	13	11	10	13 °C
	clothing	n	n	n	n	n	n	n	n	n	n	n	n	–
	stress													–
	p	7	3	2	3	14	50	69	62	60	23	13	7	313 mm

3. Climatic area: rest of the country (except parts mentioned in climatic areas 1 and 2)

3.1. Tropical Summer Rain Climate Aw and Tropical Summer Rain Mountain Climate GAw

3.2. In lowlands summer very hot, winter warm or hot.

3.3. High sun period: variable weather prevails, at times periods of rainy days; otherwise generally dry and sunny or cloudy in the morning, frequent showers in the afternoon. Low sun period: generally dry and sunny, now and then showers.

3.4. Wind: occasional gales, hurricanes when tropical cyclones pass over the coasts of the Gulf of Mexico and the Pacific, particularly from May to November.

Monterrey

latitude: 25°40′n longitude: 100°18′w height: 538 m

GAwhl	120-4	Jan	Feb	Mar	Apr	May	Jun	Jul	Aug	Sep	Oct	Nov	Dec	Year
	t	14	17	20	23	26	27	27	28	25	22	18	14	22 °C
	clothing	n	n	s	t	t	t	t	t	t	s	s	n	–
	stress			o	f	ff	ff	ff	ff	f	o	o		–
	p	22	23	16	29	42	85	73	63	205	110	26	24	718 mm
0.1 mm	dp	4	5	4	6	7	6	7	5	11	7	6	6	74 days
	sunshine	140	138	143	153	170	204	242	214	162	155	120	118	1962 hours
	%	42	43	38	40	41	50	57	53	44	43	37	36	44 %

Mazatlan

latitude: 23°12′n longitude: 106°25′w height: 78 m

Awhb	120-5	Jan	Feb	Mar	Apr	May	Jun	Jul	Aug	Sep	Oct	Nov	Dec	Year
	tx	24	24	26	27	29	31	32	32	31	31	29	26	32 °C
	t	20	20	20	22	24	27	28	28	28	27	24	21	24 °C
	tn	14	13	14	16	18	22	22	22	22	22	19	14	13 °C
	clothing	s	s	s	s	t	t	t	t	t	t	t	s	–
	stress	o	o	o	f	ff	ff	ff	ff	ff	ff	ff	o	–
	p	12	8	3	0	1	34	174	215	250	63	17	27	805 mm
0.1 mm	dp	1	1	0	0	0	4	14	15	14	4	2	2	57 days
	sunshine	195	219	239	255	288	252	220	214	204	242	216	180	2724 hours
	%	58	68	64	67	70	62	53	53	55	67	65	54	61 %

Tampico

latitude: 22°13′n longitude: 97°51′w height: 13 m

Awhb	120-6	Jan	Feb	Mar	Apr	May	Jun	Jul	Aug	Sep	Oct	Nov	Dec	Year
	t	19	20	22	25	27	28	28	28	26	26	22	20	24 °C
	clothing	s	s	s	t	t	t	t	t	t	t	s	s	–
	stress	o	o	o	f	ff	ff	ff	ff	ff	ff	f	o	–
	p	39	19	13	19	49	143	151	130	297	146	48	30	1083 mm

Guanajuato

latitude: 21°01′n longitude: 101°15′w height: 2037 m

GAwbl	120-7	Jan	Feb	Mar	Apr	May	Jun	Jul	Aug	Sep	Oct	Nov	Dec	Year
	t	14	16	18	20	21	20	19	19	18	18	16	15	18 °C
	clothing	n	n	s	s	s	s	s	s	s	s	n	n	–
	stress						o	o	o	o				–
	p	12	5	6	14	31	130	144	132	130	46	18	13	683 mm

Mérida

latitude: 20°59′n longitude: 89°39′w height: 22 m

Awha	120-8	Jan	Feb	Mar	Apr	May	Jun	Jul	Aug	Sep	Oct	Nov	Dec	Year
	tx	31	33	35	36	37	35	34	34	33	32	31	31	37 °C
	t	23	24	26	27	28	28	27	27	27	26	24	23	26 °C
	tn	14	15	16	16	20	22	21	22	22	20	16	15	13 °C
	clothing	t	t	t	t	t	t	t	t	t	t	t	t	–
	stress	o	o	f	f	ff	ff	ff	ff	ff	ff	f	o	–
	p	31	23	17	21	82	241	133	143	173	97	34	32	930 mm
0.1 mm	dp	5	2	2	3	7	13	15	13	14	10	4	4	92 days
	sunshine	174	164	217	213	242	210	236	226	183	180	168	167	2379 hours
	%	51	51	58	56	59	52	57	57	50	50	50	50	54 %

Morelia

latitude: 19°42′n longitude: 101°11′w height: 1923 m

GAwbl	120-9	Jan	Feb	Mar	Apr	May	Jun	Jul	Aug	Sep	Oct	Nov	Dec	Year
	t	14	16	18	20	21	20	18	18	18	17	16	14	18 °C
	clothing	n	n	s	s	s	s	s	s	s	n	n	n	–
	stress						o	o	o	o				–
	p	13	6	7	15	42	135	171	153	134	59	19	9	763 mm

Mexico-Tacubaya

latitude: 19°24′n longitude: 99°12′w height: 2306 m

GAwll	120-10	Jan	Feb	Mar	Apr	May	Jun	Jul	Aug	Sep	Oct	Nov	Dec	Year
	tx	23	24	27	28	29	28	26	26	24	24	23	22	29 °C
	t	12	14	16	17	17	17	16	16	16	15	13	12	15 °C
	tn	2	3	5	8	8	11	10	10	9	6	4	2	1 °C
	clothing	n	n	n	n	n	n	s	s	n	n	n	n	–
	stress							o	o					–
	p	8	5	10	23	55	118	160	145	129	49	17	6	726 mm
0.1 mm	dp	2	3	3	8	12	18	24	24	22	10	5	2	133 days
	sunshine	208	218	220	201	217	171	189	211	165	189	192	186	2366 hours
	%	60	67	59	53	54	43	46	53	45	52	57	55	53 %

Veracruz

latitude: 19°12′n longitude: 96°08′w height: 16 m

Awhb	120-11	Jan	Feb	Mar	Apr	May	Jun	Jul	Aug	Sep	Oct	Nov	Dec	Year
	tx	28	30	31	33	33	33	33	33	33	32	31	29	34 °C
	t	21	22	23	25	27	28	28	28	27	26	24	22	25 °C
	tn	14	16	16	18	21	22	21	21	21	20	17	16	13 °C
	clothing	s	s	t	t	t	t	t	t	t	t	t	s	–
	stress	o	o	f	ff	ff	ff	ff	ff	ff	ff	f	o	–
	p	22	16	14	19	65	263	358	283	353	175	76	26	1672 mm
0.1 mm	dp	5	3	2	2	6	14	19	18	19	13	10	5	116 days

Manzanillo

latitude: 19°04′n longitude: 104°20′w height: 8 m

Awha	120-12	Jan	Feb	Mar	Apr	May	Jun	Jul	Aug	Sep	Oct	Nov	Dec	Year
	tx	33	32	32	33	34	36	36	37	35	35	34	31	37 °C
	t	25	24	24	25	27	28	29	29	28	28	27	26	27 °C
	tn	17	17	16	17	19	22	22	22	22	21	20	17	15 °C
	clothing	t	t	t	t	t	t	t	t	t	t	t	t	–
	stress	f	f	f	f	ff	ff	ff	ff	ff	ff	ff	f	–
	p	3	5	1	0	3	119	145	163	368	129	23	46	1003 mm
	dp	1	0	0	0	0	8	10	9	11	5	1	1	46 days

Acapulco

latitude: 16°50′n longitude: 99°55′w height: 3 m

Awha	120-13	Jan	Feb	Mar	Apr	May	Jun	Jul	Aug	Sep	Oct	Nov	Dec	Year
	t	26	26	27	27	28	29	29	29	29	28	28	27	28 °C
	clothing	t	t	t	t	t	t	t	t	t	t	t	t	–
	stress	f	f	f	f	f	ff	ff	ff	ff	ff	f	f	–
	p	8	1	0	1	36	325	230	236	353	170	30	9	1401 mm
0.1 mm	dp	1	1	0	1	3	13	12	12	15	9	2	1	70 days
	sunshine	239	254	273	261	232	195	161	161	150	186	204	217	2534 hours
	%	69	77	73	69	58	49	40	41	41	51	60	63	57 %

Salina Cruz

Awha	120-14	Jan	Feb	Mar	Apr	May	Jun	Jul	Aug	Sep	Oct	Nov	Dec	Year
				latitude: 16° 10′n				longitude: 95° 12′w				height: 4 m		
	tx	32	32	33	35	36	35	34	34	33	33	33	32	36 °C
	t	25	26	27	28	29	28	28	28	27	27	27	27	27 °C
	tn	19	19	20	21	23	22	22	23	22	22	21	19	18 °C
	clothing	t	t	t	t	t	t	t	t	t	t	t	t	–
	stress	ff	ff	ff	ff	ff	ff	ff	ff	ff	ff	ff	ff	–
	p	4	3	1	3	56	266	166	153	246	94	8	2	1003 mm
	dp	0	1	1	1	3	12	11	9	12	4	1	1	55 days

4.4.120.2. Predominant general weather situations are described in 4.4.0.1. under Mexico. Hygrothermal diagrams of Mexico City, Acapulco are to be found in 4.4.0.2.
Wind names: Cordonazo (meaning scourge) is a local name by which the tropical cyclones experienced on the Pacific coast of Mexico are known. Tropical cyclones can occur from May to November on the coasts of the Gulf of Mexico and the Pacific.
Norte is a strong northerly wind blowing from the United States, often causing an abrupt fall of temperature accompanied by heavy cloud and rain on the Atlantic side of Mexico, in January and February even snow at times. When descending the Pacific slopes the air becomes hot and dry; the strong northerly wind arising in the Gulf of Tehuantepec in winter is called Tehuantepecer. A similar northerly wind on the Pacific coast of Central America is called "Norte" or "Papagayo".

4.4.121. Bermuda Islands

4.4.121.1. Climatic description and table

1. Subtropical Rain Climate Cr
2. Summer hot, winter warm.
3. Variable weather prevails; often dry and sunny, at times rainy days.
4. Wind: occasional gales and passing hurricanes, particularly from June to November.

Kindley Air Force Base – Bermuda

Cral	121-1	Jan	Feb	Mar	Apr	May	Jun	Jul	Aug	Sep	Oct	Nov	Dec	Year
				latitude: 33° 22′n				longitude: 64° 41′w				height: 6 m		
	tx	23	23	23	25	27	30	32	32	32	29	27	24	33 °C
	t	18	17	18	19	22	24	27	27	25	24	22	19	22 °C
	tn	9	9	9	11	14	17	21	21	19	17	13	11	8 °C
	clothing	s	s	s	s	s	t	t	t	t	t	s	s	–
	stress				o	f	f	ff	ff	ff	f	o		–
	p	126	86	107	73	74	116	117	125	115	183	97	131	1350 mm
0.2 mm	dp	17	16	15	11	11	11	13	15	13	15	15	16	168 mm
	sunshine	150	153	195	225	254	258	307	279	237	198	174	152	2587 hours
	%	50	49	53	58	59	60	70	67	64	56	55	49	58 %

4.4.121.2. Predominant general weather situations are described in 4.4.0.1. under West Indies.

4.4.122. Central America

4.4.122. Climatic descriptions and tables

1. Climatic area: Most of the region between the Atlantic slopes of the mountains and the coast of the Caribbean Sea

1.1. Tropical Rain Climate Ar and Tropical Rain Mountain Climate GAr

1.2. In the lowlands, summer very hot, winter hot.

1.3. Generally dry and sunny in the morning, often showers in the afternoon; now and then rainy days.

1.4. Wind: occasional gales, squalls in thunderstorms. Tropical cyclones or hurricanes can occur, particularly from May to November.

Belize – Belize

latitude: 17°31'n longitude: 88°12'w height: 1 m

Arha	122-1	Jan	Feb	Mar	Apr	May	Jun	Jul	Aug	Sep	Oct	Nov	Dec	Year
	tx	29	31	31	32	32	32	32	32	32	32	31	30	33 °C
	t	24	25	26	27	28	29	28	29	29	27	26	24	27 °C
	tn	14	16	17	19	20	21	21	20	20	19	16	15	13 °C
	clothing	t	t	t	t	t	t	t	t	t	t	t	t	–
	stress	ff	ff	ff	ff	ff	ff	ff	ff	ff	ff	ff	ff	–
	p	96	56	35	38	101	214	208	148	229	292	196	149	1761 mm
0.2 mm	dp	11	6	4	4	7	14	17	13	15	16	14	14	135 days

Tela – Honduras

latitude: 15°43'n longitude: 87°29'w height: 3 m

Araa	122-2	Jan	Feb	Mar	Apr	May	Jun	Jul	Aug	Sep	Oct	Nov	Dec	Year
	t	23	24	25	26	27	27	27	27	27	26	25	24	26 °C
	clothing	t	t	t	t	t	t	t	t	t	t	t	t	–
	stress	ff	ff	ff	ff	ff	ff	ff	ff	ff	ff	ff	ff	–
	p	250	152	75	87	99	128	179	234	204	335	409	363	2515 mm

Cobán – Guatemala

latitude: 15°29'n longitude: 90°16'w height: 1307 m

GArbl	122-3	Jan	Feb	Mar	Apr	May	Jun	Jul	Aug	Sep	Oct	Nov	Dec	Year
	tx	27	28	30	29	29	28	27	27	27	27	26	27	31 °C
	t	17	18	18	20	21	21	20	20	20	19	18	18	19 °C
	tn	6	6	7	8	11	13	12	12	13	11	9	7	4 °C
	clothing	s	s	s	s	s	s	s	s	s	s	s	s	–
	stress	o	o	o	o	o	o	o	o	o	o	o	o	–
	p	147	114	112	124	201	312	252	264	267	335	224	165	2517 mm
0.2 mm	dp	14	10	9	10	16	22	25	22	23	23	21	17	212 days

San Pedro Sula – Honduras

latitude: 15°28'n longitude: 88°01'w height: 76 m

Arha	122-4	Jan	Feb	Mar	Apr	May	Jun	Jul	Aug	Sep	Oct	Nov	Dec	Year
	t	23	24	26	27	27	28	27	27	27	26	25	24	26 °C
	clothing	t	t	t	t	t	t	t	t	t	t	t	t	–
	stress	f	f	f	f	ff	ff	ff	ff	ff	ff	f	f	–
	p	74	49	47	37	93	148	154	122	206	166	132	104	1332 mm

Puerto Cabezas – Nicaragua

latitude: 14°00′n longitude: 83°24′w height: 13 m

Araa	122-5	Jan	Feb	Mar	Apr	May	Jun	Jul	Aug	Sep	Oct	Nov	Dec	Year
	t	25	25	26	27	27	27	26	27	27	27	26	25	26 °C
	clothing	t	t	t	t	t	t	t	t	t	t	t	t	–
	stress	f	f	f	f	ff	ff	ff	ff	ff	ff	ff	ff	–
	p	166	83	38	51	265	431	478	366	278	335	354	280	3125 mm

El Recreo – Nicaragua

latitude: 12°09′n longitude: 84°26′w height: – m

Arha	122-6	Jan	Feb	Mar	Apr	May	Jun	Jul	Aug	Sep	Oct	Nov	Dec	Year
	t	27	28	29	30	30	29	29	28	30	29	28	27	28 °C
	clothing	t	t	t	t	t	t	t	t	t	t	t	t	–
	stress	ff	ff	ff	ff	ff	ff	ff	ff	ff	ff	ff	ff	–
	p	220	136	72	76	269	406	499	453	229	260	263	222	3103 mm

Bluefields – Nicaragua

latitude: 12°00′n longitude: 83°43′w height: 12 m

Araa	122-7	Jan	Feb	Mar	Apr	May	Jun	Jul	Aug	Sep	Oct	Nov	Dec	Year
	t	25	25	26	27	27	27	26	27	27	26	26	25	26 °C
	clothing	t	t	t	t	t	t	t	t	t	t	t	t	–
	stress	ff	ff	ff	ff	ff	ff	ff	ff	ff	ff	ff	ff	–
	p	264	111	80	102	399	466	746	609	298	389	481	423	4370 mm

La Cola – Costa Rica

latitude: 10°06′n longitude: 83°25′w height: 39 m

Araa	122-8	Jan	Feb	Mar	Apr	May	Jun	Jul	Aug	Sep	Oct	Nov	Dec	Year
	t	24	24	25	25	26	26	26	26	26	26	25	25	25 °C
	clothing	t	t	t	t	t	t	t	t	t	t	t	t	–
	stress	ff	ff	ff	ff	ff	ff	ff	ff	ff	ff	ff	ff	–
	p	315	182	165	183	281	258	378	280	127	254	337	482	3242 mm
0.1 mm	dp	16	15	15	15	21	21	21	19	18	21	17	22	221 days
	sunshine	149	147	158	153	140	117	118	143	153	133	132	130	1672 hours
	%	41	44	42	41	36	31	30	37	42	36	38	36	38 %

Turialba – Costa Rica

latitude: 9°53′n longitude: 83°38′w height: 602 m

GArab	122-9	Jan	Feb	Mar	Apr	May	Jun	Jul	Aug	Sep	Oct	Nov	Dec	Year
	t	20	20	21	24	23	23	22	22	22	22	22	21	22 °C
	clothing	s	s	s	t	t	t	s	s	s	s	s	s	–
	stress	f	f	f	f	ff	ff	ff	ff	f	f	f	f	–
	p	171	113	79	99	217	247	259	207	207	241	272	300	2412 mm
0.1 mm	dp	16	13	11	13	20	22	22	26	20	21	19	19	216 days
	sunshine	146	158	167	153	133	117	118	130	144	146	135	136	1684 hours
	%	41	47	45	41	34	31	30	34	39	39	39	38	37 %

Cristobal – Canal Zone

Araa	122-10	Jan	Feb	Mar	Apr	May	Jun	Jul	Aug	Sep	Oct	Nov	Dec	Year
latitude: 9°21'n longitude: 79°54'w height: 7 m														
	tx	29	29	29	31	32	32	31	31	32	31	31	30	33 °C
	t	26	26	27	27	27	27	27	27	27	27	26	26	27 °C
	tn	22	22	23	23	23	23	23	23	23	23	23	22	22 °C
	clothing	t	t	t	t	t	t	t	t	t	t	t	t	–
	stress	ff	ff	ff	ff	ff	ff	ff	ff	ff	ff	ff	ff	–
	p	120	45	51	113	330	285	406	378	291	412	571	349	3352 mm
0.2 mm	dp	16	14	12	15	21	24	25	25	23	24	25	22	246 days
	sunshine	270	243	273	258	198	171	170	167	168	152	147	208	2425 hours
	%	75	73	73	70	51	45	43	43	46	41	42	58	55 %

2. *Climatic area:* Most of the region between the Pacific slopes of the mountains and the Pacific coast

2.1. Tropical Summer Rain Climate Aw and Tropical Summer Rain Mountain Climate GAw

2.2. At low levels, very hot all the year round.

2.3. High sun period: generally dry and sunny in the morning, in the afternoon often showers; at times rainy days. Low sun period: most of the time dry and sunny; occasionally showers or some rain.

2.4. Wind: occasional gales, squalls in thunderstorms. Tropical cyclones can occur, particularly from May to November.

Guatemala – Guatemala

GAwbl	122-11	Jan	Feb	Mar	Apr	May	Jun	Jul	Aug	Sep	Oct	Nov	Dec	Year
latitude: 14°35'n longitude: 90°32'w height: 1502 m														
	tx	27	28	30	29	29	28	27	27	27	27	26	27	31 °C
	t	16	17	18	20	20	19	18	19	18	18	17	16	18 °C
	tn	7	7	7	8	11	13	12	12	13	11	9	7	4 °C
	clothing	s	s	s	s	s	s	s	s	s	s	s	s	–
	stress	o	o	o	o	o	f	f	f	f	f	o	o	–
	p	3	2	7	19	141	265	211	187	257	159	23	7	1281 mm
0.2 mm	dp	4	2	3	5	15	23	22	21	22	18	7	4	145 days

San Salvador – El Salvador

GAwab	122-12	Jan	Feb	Mar	Apr	May	Jun	Jul	Aug	Sep	Oct	Nov	Dec	Year
latitude: 13°43'n longitude: 89°12'w height: 689 m														
	tx	30	31	32	32	31	30	30	30	29	29	29	29	33 °C
	t	22	22	24	24	24	23	23	23	22	22	22	22	23 °C
	tn	16	16	17	18	19	19	18	18	19	18	17	16	15 °C
	clothing	s	s	t	t	t	t	t	t	s	s	s	s	–
	stress	o	o	f	f	f	f	f	f	f	f	f	o	–
	p	5	5	9	54	187	323	317	297	318	228	36	8	1793 mm
	dp	1	0	1	3	12	18	18	19	19	14	4	1	110 days
	sunshine	304	280	299	248	235	182	252	261	188	218	267	295	3029 hours
	%	86	85	80	66	59	47	63	67	51	59	77	84	68 %

Santa Tecla – El Salvador

					latitude: 13°39′n			longitude: 89°13′w			height: 955 m		
GAwbb	122-13	Jan	Feb	Mar	Apr	May	Jun	Jul	Aug	Sep	Oct	Nov	Dec	Year
	tx	28	30	30	30	29	28	28	29	27	27	28	28	31 °C
	t	20	20	21	22	22	21	21	22	21	21	20	20	21 °C
	tn	14	14	15	16	17	17	17	17	17	17	16	15	13 °C
	clothing	s	s	s	s	s	s	s	s	s	s	s	s	–
	stress	o	o	o	f	f	f	f	f	f	f	o	o	–
	p	4	4	7	54	158	329	334	311	388	236	44	8	1877 mm
0.1 mm	dp	1	1	1	6	15	22	24	22	22	14	5	1	134 days
	sunshine	316	280	298	255	229	177	248	251	183	217	267	310	3030 hours
	%	89	85	80	68	58	46	62	64	50	59	77	88	68 %

Acajutla – El Salvador

					latitude: 13°36′n			longitude: 89°50′w			height: 5 m		
Awha	122-14	Jan	Feb	Mar	Apr	May	Jun	Jul	Aug	Sep	Oct	Nov	Dec	Year
	tx	32	32	32	32	32	31	32	32	31	31	32	32	33 °C
	t	26	26	27	28	28	27	27	27	26	26	26	26	27 °C
	tn	21	22	23	24	24	23	23	23	23	23	22	22	20 °C
	clothing	t	t	t	t	t	t	t	t	t	t	t	t	–
	stress	ff	ff	ff	ff	ff	ff	ff	ff	ff	ff	ff	ff	–
	p	1	1	4	50	172	302	277	263	321	273	42	4	1710 mm
0.1 mm	dp	0	0	0	4	10	17	18	17	17	12	2	0	97 days
	sunshine	303	283	301	261	247	204	259	263	207	235	273	303	3139 hours
	%	86	86	81	70	62	53	65	67	56	64	79	86	71 %

Santa Cruz Porillo – El Salvador

					latitude: 13°40′n			longitude: 88°56′w			height: 30 m		
Awha	122-15	Jan	Feb	Mar	Apr	May	Jun	Jul	Aug	Sep	Oct	Nov	Dec	Year
	tx	36	36	36	36	35	33	34	34	33	34	35	35	– °C
	t	26	26	27	28	28	27	27	27	26	26	26	26	27 °C
	tn	19	20	21	23	23	22	22	22	22	22	20	19	– °C
	clothing	t	t	t	t	t	t	t	t	t	t	t	t	–
	stress	ff	ff	ff	ff	ff	ff	ff	ff	ff	ff	ff	ff	–
	p	1	0	6	31	178	297	284	263	353	284	53	3	1753 mm
0.1 mm	dp	0	1	1	4	12	19	18	19	21	17	7	1	111 days
	sunshine	294	282	298	276	270	225	273	285	222	242	252	288	3207 hours
	%	83	85	80	74	68	58	68	73	61	66	73	82	72 %

Santa Rosa de Copán – Honduras

					latitude: 14°47′n			longitude: 88°48′w			height: 1076 m		
GAwbl	122-16	Jan	Feb	Mar	Apr	May	Jun	Jul	Aug	Sep	Oct	Nov	Dec	Year
	t	17	19	20	22	22	22	21	21	21	20	19	17	20 °C
	clothing	s	s	s	s	s	s	s	s	s	s	s	s	–
	stress	o	o	o	o	o	f	f	f	f	f	o	o	–
	p	37	16	25	30	132	334	246	180	332	198	89	69	1686 mm

Catacamas – Honduras

Awab	122-17	Jan	Feb	Mar	Apr	May	Jun	Jul	Aug	Sep	Oct	Nov	Dec	Year
latitude: 14°45′n longitude: 85°56′w height: 412 m														
	t	22	23	24	26	27	26	24	25	25	25	24	23	24 °C
	clothing	s	t	t	t	t	t	t	t	t	t	t	t	–
	stress	f	f	f	f	ff	ff	ff	ff	ff	ff	f	f	–
	p	39	17	15	24	120	205	228	157	188	165	64	33	1254 mm

Tegucigalpa – Honduras

GAwab	122-18	Jan	Feb	Mar	Apr	May	Jun	Jul	Aug	Sep	Oct	Nov	Dec	Year
latitude: 14°04′n longitude: 87°13′w height: 1007 m														
	t	19	20	22	23	24	23	23	23	23	22	21	20	22 °C
	clothing	s	s	s	t	t	t	t	t	t	s	s	s	–
	stress	o	o	o	o	f	f	f	f	f	f	o	o	–
	p	14	5	11	28	156	166	92	111	176	141	40	15	954 mm

Managua – Nicaragua

Awha	122-19	Jan	Feb	Mar	Apr	May	Jun	Jul	Aug	Sep	Oct	Nov	Dec	Year
latitude: 12°07′n longitude: 86°11′w height: 53 m														
	t	26	26	27	28	28	27	27	27	27	26	26	26	27 °C
	clothing	t	t	t	t	t	t	t	t	t	t	t	t	–
	stress	f	f	f	f	ff	ff	ff	ff	ff	ff	f	f	–
	p	2	4	4	3	149	234	127	113	212	306	42	10	1207 mm

San José – Costa Rica

GAwbb	122-20	Jan	Feb	Mar	Apr	May	Jun	Jul	Aug	Sep	Oct	Nov	Dec	Year
latitude: 9°56′n longitude: 84°05′w height: 1158 m														
	tx	27	28	29	29	29	28	27	28	28	27	28	27	31 °C
	t	19	19	20	21	21	21	21	21	21	21	20	19	20 °C
	tn	11	12	12	13	14	14	14	14	14	14	13	12	11 °C
	clothing	s	s	s	s	s	s	s	s	s	s	s	s	–
	stress	o	o	o	o	f	f	f	f	f	f	f	o	–
	p	8	5	10	37	244	284	230	233	342	333	172	46	1944 mm
0.1 mm	dp	3	2	2	6	20	23	21	22	24	26	16	9	174 days
	sunshine	217	220	248	210	161	120	124	136	150	136	135	183	2040 hours
	%	60	66	66	57	41	31	32	35	41	37	39	51	46 %

Balboa Heights / Panama – Canal Zone

Awha	122-21	Jan	Feb	Mar	Apr	May	Jun	Jul	Aug	Sep	Oct	Nov	Dec	Year
latitude: 8°57′n longitude: 79°33′w height: 36 m														
	tx	32	33	34	34	34	33	33	33	33	32	32	33	34 °C
	t	27	27	28	28	27	27	27	27	26	26	26	27	27 °C
	tn	20	20	21	21	22	22	22	22	22	22	21	21	19 °C
	clothing	t	t	t	t	t	t	t	t	t	t	t	t	–
	stress	ff	ff	ff	ff	ff	ff	ff	ff	ff	ff	ff	ff	–
	p	48	25	10	63	239	180	199	216	200	305	268	151	1903 mm
0.2 mm	dp	4	2	2	7	17	20	17	19	20	21	20	15	164 days
	sunshine	279	254	260	195	164	132	149	149	144	149	147	217	2239 hours
	%	77	76	69	53	42	35	38	39	39	40	42	60	51 %

4.4.122.2. Predominant general weather situations are described in 4.4.0.1. under Central America. Hygrothermal diagrams of San Salvador and Balboa Heights are to be found in 4.4.0.2.
Wind names: Norte or Norder or Norther is a strong northerly wind blowing in Central America from the United States, often causing an abrupt fall of temperature on the Atlantic side of Central America, accompanied by heavy cloud and rain. When descending the Pacific slopes of the Central American mountains, the air becomes hot and dry. A strong or stormy northerly wind on the Pacific coast is in some parts called "Papagayo". Temporal is a period of some rainy days, sometimes with strong southerly winds, on the Pacific side of Central America, occurring once or more a year, particularly in June and September/October.

4.4.123. West Indies

4.4.123.1. Climatic descriptions and tables

1. Climatic area: North of Puerto Rico and Hispaniola (Haiti and Dominican Republic), Windward Islands
1.1. Tropical Rain Climate Ar and Tropical Rain Mountain Climate GAr
1.2. Summer very hot, winter hot.
1.3. High sun period: generally dry and sunny in the morning, often showers in the afternoon; at times rainy days. Low sun period: fewer showers.
1.4. Wind: occasional gales, squalls in thunderstorms; tropical cyclones or hurricanes can occur, particularly from May to November.

Puerto Plata – Dominican Republic

latitude: 19°47'n longitude: 70°40'w height: 16 m

Arab	123-1	Jan	Feb	Mar	Apr	May	Jun	Jul	Aug	Sep	Oct	Nov	Dec	Year
	t	22	22	23	24	25	26	26	27	27	26	25	22	25 °C
	clothing	s	s	t	t	t	t	t	t	t	t	t	s	–
	stress	f	f	f	ff	ff	ff	ff	ff	ff	ff	ff	f	–
	p	209	162	134	160	168	82	78	88	97	158	283	307	1924 mm

San Juan – Puerto Rico

latitude: 18°26'n longitude: 66°00'w height: 19 m

Araa	123-2	Jan	Feb	Mar	Apr	May	Jun	Jul	Aug	Sep	Oct	Nov	Dec	Year
	tx	29	29	30	31	32	32	31	32	32	32	31	29	33 °C
	t	24	24	24	25	26	27	27	27	27	27	26	25	26 °C
	tn	19	19	19	20	21	22	22	22	22	22	21	20	18 °C
	clothing	t	t	t	t	t	t	t	t	t	t	t	t	–
	stress	f	f	ff	ff	ff	ff	ff	ff	ff	ff	ff	ff	–
	p	119	71	56	94	181	144	159	181	172	148	165	138	1631 mm
	dp	15	9	9	10	13	12	15	16	15	15	16	16	161 days
	sunshine	229	217	272	256	246	226	260	265	213	233	210	202	2829 hours
	%	66	67	73	68	61	57	64	67	58	64	62	59	64 %

Pointe-à-Pitre / Raizet – Guadeloupe

latitude: 16°16′n longitude: 61°31′w height: 8 m

Araa	123-3	Jan	Feb	Mar	Apr	May	Jun	Jul	Aug	Sep	Oct	Nov	Dec	Year
	t	23	24	24	26	26	27	27	27	27	26	25	24	26 °C
	clothing	t	t	t	t	t	t	t	t	t	t	t	t	–
	stress	f	f	f	ff	ff	ff	ff	ff	ff	ff	ff	ff	–
	p	91	66	66	86	135	148	179	244	196	229	231	143	1814 mm
	dp	15	11	10	11	14	14	18	17	18	18	17	17	180 days
	sunshine	223	219	257	239	231	228	231	232	210	220	210	220	2720 hours
	%	64	67	69	64	58	58	57	59	57	60	61	64	61 %

Fort-de-France / Desaix – Martinique

latitude: 14°37′n longitude: 61°04′w height: 146 m

Araa	123-4	Jan	Feb	Mar	Apr	May	Jun	Jul	Aug	Sep	Oct	Nov	Dec	Year
	tx	29	30	31	32	32	32	32	32	33	32	32	31	33 °C
	t	24	24	24	25	25	26	26	26	26	26	25	24	25 °C
	tn	18	18	18	19	21	22	22	22	22	21	21	18	17 °C
	clothing	t	t	t	t	t	t	t	t	t	t	t	t	–
	stress	f	f	f	ff	ff	ff	ff	ff	ff	ff	ff	f	–
	p	96	68	58	82	126	160	214	227	232	221	230	126	1840 mm
	dp	17	12	13	13	15	18	22	21	19	18	18	17	203 days
	sunshine	258	236	274	251	246	222	239	262	226	227	227	238	2906 hours
	%	73	72	73	67	62	57	60	67	62	62	66	68	66 %

Seawell Airport – Barbados

latitude: 13°04′n longitude: 59°29′w height: 56 m

Araa	123-5	Jan	Feb	Mar	Apr	May	Jun	Jul	Aug	Sep	Oct	Nov	Dec	Year
	tx	29	30	31	31	31	31	31	32	32	32	31	29	32 °C
	t	25	25	26	27	27	27	27	27	27	27	26	26	26 °C
	tn	18	18	19	20	21	22	22	22	22	21	21	19	18 °C
	clothing	t	t	t	t	t	t	t	t	t	t	t	t	–
	stress	f	f	f	ff	ff	ff	ff	ff	ff	ff	ff	f	–
	p	68	62	37	51	70	103	141	144	168	176	160	93	1273 mm
	dp	12	10	7	7	7	14	19	15	15	15	11	13	145 days
	sunshine	271	249	275	268	267	246	264	277	242	235	234	249	3077 hours
	%	76	75	74	72	68	64	66	71	66	64	68	71	69 %

Piarco Airport – Trinidad

latitude: 10°37′n longitude: 61°21′w height: 14 m

Araa	123-6	Jan	Feb	Mar	Apr	May	Jun	Jul	Aug	Sep	Oct	Nov	Dec	Year
	tx	32	32	33	33	34	33	33	32	33	33	33	32	35 °C
	t	24	25	25	26	27	26	26	26	26	26	25	25	26 °C
	tn	17	17	17	18	18	19	19	19	19	19	19	18	16 °C
	clothing	t	t	t	t	t	t	t	t	t	t	t	t	–
	stress	f	f	f	f	ff	ff	ff	ff	ff	ff	ff	ff	–
	p	79	51	26	55	131	291	252	236	159	142	198	152	1772 mm
	dp	11	9	6	7	13	20	20	19	15	15	18	15	168 days
	sunshine	230	219	251	241	233	190	217	220	202	207	196	202	2608 hours
	%	64	66	67	65	60	50	55	57	55	56	56	57	59 %

2. *Climatic area:* Bahama Islands, South of Puerto Rico and Hispaniola (Haiti and Dominican Republic), Jamaica, Cuba (exccept southeasterly coastal area), Leeward Islands in parts.
2.1. Tropical Summer Rain Climate Aw and Tropical Summer Rain Mountain Climate GAw, in parts Tropical Monsoonal Rain Climate Am
2.2. Summer very hot, winter hot.
2.3. In the morning generally dry and sunny; high sun period: often showers in the afternoon; at times rainy days; low sun period: now and then showers, in southern regions more frequent.
2.4. Wind: occasional gales, squalls in thunderstorms; tropical cyclones or hurricanes can occur, particularly from May to November.

Nassau – Bahama Islands

latitude: 25°03'n longitude: 77°28'w height: 10 m

Awhb	123-7	Jan	Feb	Mar	Apr	May	Jun	Jul	Aug	Sep	Oct	Nov	Dec	Year
	tx	28	28	29	30	31	32	33	33	32	32	29	28	33 °C
	t	20	21	22	24	25	27	27	28	27	26	24	22	24 °C
	tn	13	13	14	16	17	21	21	22	21	19	17	14	12 °C
	clothing	s	s	s	t	t	t	t	t	t	t	t	s	–
	stress	o	f	f	f	f	ff	ff	ff	ff	ff	f	f	–
	p	36	43	45	78	117	159	150	135	165	164	85	39	1216 mm
0.2 mm	dp	6	7	6	8	12	16	16	17	18	17	10	9	142 days
	sunshine	232	234	264	255	264	243	276	264	216	208	222	214	2892 hours
	%	70	73	71	67	64	59	66	65	58	58	68	65	65 %

Havana – Cuba

latitude: 23°10'n longitude: 82°21'w height: 49 m

Awab	123-8	Jan	Feb	Mar	Apr	May	Jun	Jul	Aug	Sep	Oct	Nov	Dec	Year
	tx	30	31	31	32	33	33	33	33	33	33	31	31	34 °C
	t	21	22	23	24	25	26	27	27	27	25	24	22	24 °C
	tn	14	13	15	17	19	21	22	22	22	21	16	14	12 °C
	clothing	s	s	t	t	t	t	t	t	t	t	t	s	–
	stress	f	f	f	f	ff	ff	ff	ff	ff	ff	f	f	–
	p	51	38	54	53	155	130	93	108	167	183	73	64	1167 mm
0.2 mm	dp	6	5	5	5	8	11	10	11	14	13	9	6	103 days

Cienfuegos – Cuba

latitude: 22°04'n longitude: 80°27'w height: 30 m

Awab	123-9	Jan	Feb	Mar	Apr	May	Jun	Jul	Aug	Sep	Oct	Nov	Dec	Year
	tx	30	31	31	32	33	34	34	34	34	33	31	31	34 °C
	t	22	22	23	24	26	27	27	27	27	26	24	23	25 °C
	tn	12	11	13	16	17	20	21	21	21	19	14	13	10 °C
	clothing	s	s	t	t	t	t	t	t	t	t	t	t	–
	stress	f	f	f	f	f	ff	ff	ff	ff	ff	f	f	–
	p	18	25	33	46	119	152	122	160	173	160	41	23	1072 mm
	dp	2	3	3	4	9	12	11	13	13	10	4	2	86 days

Camaguey – Cuba

latitude: 21°24'n longitude: 77°55'w height: 121 m

Awab	123-10	Jan	Feb	Mar	Apr	May	Jun	Jul	Aug	Sep	Oct	Nov	Dec	Year
	t	22	23	24	25	26	27	27	27	27	26	24	22	25 °C
	clothing	s	t	t	t	t	t	t	t	t	t	t	s	–
	stress	f	f	f	f	ff	ff	ff	ff	ff	ff	f	f	–
	p	26	43	47	85	156	210	133	188	185	182	65	23	1342 mm

Port-au-Prince – Haiti

latitude: 18°33'n longitude: 72°20'w height: 41 m

Awha	123-11	Jan	Feb	Mar	Apr	May	Jun	Jul	Aug	Sep	Oct	Nov	Dec	Year
	tx	33	33	34	34	34	36	37	36	35	34	33	33	37 °C
	t	25	26	26	27	28	28	28	29	28	27	27	26	27 °C
	tn	18	18	19	20	21	21	21	21	21	22	20	18	17 °C
	clothing	t	t	t	t	t	t	t	t	t	t	t	t	–
	stress	f	f	f	ff	ff	ff	ff	ff	ff	ff	ff	f	–
	p	29	34	76	156	195	110	67	135	166	161	82	53	1263 mm
0.1 mm	dp	4	5	7	13	14	9	8	11	15	13	9	4	112 days
	sunshine	276	257	282	267	251	243	270	270	234	236	243	260	3089 hours
	%	80	79	76	71	62	61	66	68	64	65	72	76	70 %

Santo Domingo – Dominican Republic

latitude: 18°28'n longitude: 69°53'w height: 14 m

Amaa	123-12	Jan	Feb	Mar	Apr	May	Jun	Jul	Aug	Sep	Oct	Nov	Dec	Year
	tx	31	31	32	32	32	33	33	34	34	33	32	32	35 °C
	t	24	24	24	25	26	26	27	27	27	26	26	25	26 °C
	tn	17	17	17	18	20	21	21	21	21	21	19	18	16 °C
	clothing	t	t	t	t	t	t	t	t	t	t	t	t	–
	stress	f	f	f	f	ff	ff	ff	ff	ff	ff	ff	ff	–
	p	47	45	45	65	190	175	158	147	168	165	113	67	1386 mm
0.2 mm	dp	9	6	5	9	13	13	14	13	13	13	11	11	130 days
	sunshine	178	178	229	203	185	185	188	213	198	200	185	176	2317 hours
	%	52	55	61	54	46	47	46	54	54	55	55	51	52 %

Kingston – Jamaica

latitude: 17°56'n longitude: 76°47'w height: 12 m

Awha	123-13	Jan	Feb	Mar	Apr	May	Jun	Jul	Aug	Sep	Oct	Nov	Dec	Year
	tx	32	32	32	32	33	33	34	34	34	33	33	32	34 °C
	t	25	25	26	27	28	28	28	28	28	28	27	26	27 °C
	tn	17	17	18	19	21	22	21	22	22	21	19	18	17 °C
	clothing	t	t	t	t	t	t	t	t	t	t	t	t	–
	stress	f	f	ff	ff	ff	ff	ff	ff	ff	ff	ff	ff	–
	p	20	18	10	37	138	114	51	92	86	168	52	25	811 mm
0.2 mm	dp	4	3	3	5	7	7	5	6	11	11	5	4	69 days
	sunshine	257	249	270	261	257	234	264	263	228	226	255	239	3003 hours
	%	74	76	72	69	64	59	65	67	62	62	75	70	68 %

Swan Island

latitude: 17°24'n longitude: 83°56'w height: 11 m

Awha	123-14	Jan	Feb	Mar	Apr	May	Jun	Jul	Aug	Sep	Oct	Nov	Dec	Year
	t	26	26	27	28	28	28	28	28	28	28	27	26	27 °C
	clothing	t	t	t	t	t	t	t	t	t	t	t	t	–
	stress	f	f	f	ff	ff	ff	ff	ff	ff	ff	ff	f	–
	p	87	29	16	21	84	164	107	93	132	246	188	143	1310 mm
	dp	10	5	3	4	8	12	12	10	11	15	15	11	116 days

Dr. Plesman Airport – Curaçao (tx, tn Willemstad)

Asha	123-15	Jan	Feb	Mar	Apr	May	Jun	Jul	Aug	Sep	Oct	Nov	Dec	Year
latitude: 12°11'n longitude: 68°59'w height: 0 m														
	tx	30	31	31	32	32	32	32	32	34	33	32	31	34 °C
	t	26	26	26	27	28	28	28	28	29	28	28	27	28 °C
	tn	21	22	22	23	23	23	23	23	23	23	22	22	21 °C
	clothing	t	t	t	t	t	t	t	t	t	t	t	t	–
	stress	ff	ff	ff	ff	ff	ff	ff	ff	ff	ff	ff	ff	–
	p	63	29	16	15	18	27	29	40	33	78	128	106	580 mm
	dp	9	6	4	3	3	4	6	5	3	7	10	12	72 days
	sunshine	251	240	269	239	218	245	281	289	262	259	226	227	3006 hours
	%	70	72	72	64	55	64	71	74	72	70	65	64	68 %

3. Climatic area: Southeasterly coastal area of Cuba, greater part of the Leeward Islands

3.1. Steppe Climate BS and Desert Marine Climate BM

3.2. Summer very hot, winter hot.

3.3. Generally dry and sunny in the morning; in the afternoon often scattered showers, at times cloudy and rainy.

3.4. Wind: occasional gales. Tropical cyclones or hurricanes can occur, particularly from May to November.

Guantanamo – Cuba

BSha	123-16	Jan	Feb	Mar	Apr	May	Jun	Jul	Aug	Sep	Oct	Nov	Dec	Year
latitude: 19°54'n longitude: 75°09'w height: 23 m														
	t	24	24	25	26	27	28	29	29	28	27	26	25	27 °C
	clothing	t	t	t	t	t	t	t	t	t	t	t	t	–
	stress	f	f	f	ff	ff	ff	ff	ff	ff	ff	ff	f	–
	p	12	18	27	40	62	53	23	40	96	158	24	25	578 mm

La Orchila Island – Venezuela

BMha	123-17	Jan	Feb	Mar	Apr	May	Jun	Jul	Aug	Sep	Oct	Nov	Dec	Year
latitude: 11°49'n longitude: 66°11'w height: 3 m														
	t	26	26	27	27	28	28	28	28	29	28	28	27	28 °C
	clothing	t	t	t	t	t	t	t	t	t	t	t	t	–
	stress	ff	ff	ff	ff	ff	ff	ff	ff	ff	ff	ff	ff	–
	p	23	4	1	0	26	4	25	14	3	13	21	16	150 mm
0.1 mm	dp	13	3	1	0	1	3	9	7	1	2	7	7	54 days
	sunshine	285	285	307	270	276	291	298	301	315	304	285	251	3468 hours
	%	80	85	82	73	70	76	75	78	86	83	82	71	78 %

4.4.123.2. Predominant general weather situations are describedin 4.4.0.1. under West Indies. Hygrothermal diagrams of Havana and Kingston are to be found in 4.4.0.2.
Wind name: Bayamo is the name of a heavy tropical thunderstorm accompanied by torrential rain on the southern coast of Cuba.

4.5. South America

4.5.0.1. Predominant general weather situations

Colombia

The south of the country is nearly always influenced by the equatorial trough and its disturbances approaching from the east. The trough goes north and south once a year.

In summer, it often advances as far as the Caribbean. Then the south of the country is influenced by the sinking air of a wedge of the anticyclone over the South Pacific, extending to Peru. In winter, most of the time the north of the country is influenced by the sinking air of a wedge of the Bermuda anticyclone, extending to the Caribbean Sea. Only occasionally are there weak invasions of cold air from the North American continent into the north of the country.

Venezuela

In summer, the equatorial trough moves away far to the north of the continent. Its disturbances, migrating from east to west, sometimes influence the coastal region. There are times with intensified cyclonic activity over the Caribbean, which demonstrate different frequencies in different years. Some tropical cyclones – most of them are tracking from the Atlantic towards Central America – influence the weather of the north of the country. In winter, the north is influenced by the sinking air of the wedge of the Bermuda anticyclone extending to the Caribbean Sea. This influence extends in weaker form as far as the south of the country. But more frequently, westward-tracking disturbances of the equatorial trough, which lies farther to the south, determine the weather with their rising air. Occasionally invasions of polar air from the north American Continent invade the country; then sudden torrential rainfall can happen.

Guyana, Surinam, Guiana

The equatorial trough determines the weather. It is in its southernmost position in March. Then sinking air predominates, under the influence of the anticyclonic belt of high pressure over the north Atlantic. On its southern slope, weak cold front troughs invade the countries at times. In summer the equatorial trough is over the region or somewhat the north of it, so that disturbances advancing from the east intensify general shower activity at their front and weaken it at their rear.

Brazil

In summer (November to April), the equatorial trough is mostly south of the Amazon. Its east-west tracking disturbances intensify the normal shower activity in front of them and weaken it at the rear of them. A weak wedge of the south Atlantic anticyclone extends to the southern coast and an extension of the belt of low pressure in the temperate latitudes extends to the Chaca. In winter (May to October), the equatorial trough is to be found far to the north of the country. A strong wedge of the south Atlantic anticyclone extends to the south of the country. Occasionally it moves away when secondary depressions of the belt of low pressure in temperate latitudes traverse the region and cold polar air makes a brief incursion.

Ecuador

The equatorial trough determines the weather of the north of the country all the year with its rising air, which in detail is formed by disturbances approaching from the east. In summer (October/November to April/May), the equatorial trough traverses the country east of the Andes, while on the Pacific side it stays more or less in the same position. In winter (May/June to November/December), the south of the country is influenced by the sinking air of a wedge of the anticyclone over the South Pacific, extending to Peru. Occasionally disturbances from the equatorial trough expand southward to the north of the country.

Peru

In summer (November to April), the equatorial trough lies south of the Equator east of the Andes, so that its west-tracking disturbances directly encounter the eastern slopes of the mountains and cause abundant rain. The anticyclone over the South Pacific is weakened and has a position south of this in winter. In winter (May to October), the strong anticyclone over the South Pacific Ocean has a position nearer to the equator. The influence of its sinking air predominates in the greater part of the country and causes dry weather, except that the northeast of the country is influenced by the proximity of the equatorial trough and its disturbances, which normally intensify shower activity when they approach from the east and lessen it when departing. Then it can happen for some days that there are no showers at all.

Bolivia

In summer (October to March), the equatorial trough lies in the region of the Amazon south of the equator, so that the general rising air influences the weather of the country, and more or less regularly there are showers in the afternoon and evening, lasting at times far into the night. In the case of stronger disturbances, which are not frequent, there are some rainy days. In winter (April to September), the equatorial trough is far north of the equator. Sinking air from a strong anticyclone over the South Pacific extends its influence into the country. Now and then there are invasions of cold air from higher latitudes along the eastern slopes of the Andes.

Chile

The south Pacific anticyclone determines the weather of a great part of the country. On its eastern slope, sinking air predominates, especially in summer (November to April), when secondary depressions of the belt of low pressure in the temperate latitudes rarely move so far north. The south of the country is influenced all the year round by the rising air of the disturbances of the middle latitude westerlies, which cause rapid changes of weather with their frontes, tracking east with great speed, and accompanying rain and showers. In winter (May to October), there are often invasions of cold air on the eastern slope of the south Pacific anticyclone, going far to the north, where in coastal areas they cause a complete cloud cover of low stratus and some drizzle at times.

Paraguay, Uruguay, Argentina

In summer (October to March), a trough of the middle-latitude belt of low pressure extends to the Chaco and sometimes connects with the equatorial trough. Its east-west tracking disturbances sometimes extend beyond the tropic. A wedge of the South Atlantic anticyclone extends to the Rio de la Plata, but frequently weakens and moves away when secondary depressions or depressions of the westerlies approach from the west or southwest. In winter (April to September), there is an anticyclonic bridge south of 30° southern latitude. This is the link between the strong anticyclone over the South Atlantic and the anticyclone over the South Pacific west of Chile. Its sinking air determines the weather most of the time. Invasions of cold air can occasionally advance far to the north.

South Atlantic

On the southern slope of the subtropical anticyclonic belt, depressions and anticyclones migrate in the middle-latitude westerlies all the year round. In summer, this drift is generally somewhat weaker than in winter. At times the westerlies are interrupted when the depressions and anticyclones become stationary for a while; this is an exception to the rule.

4.5.0.2. Hygrothermal diagrams

In the following figures hygrothermal diagrams are given for the stations

Bogota, Caracas, Cayenne, Rio de Janeiro, Quito, Lima, La Paz, Santiago de Chile, Asuncion, Montevideo, Buenos Aires, Ushuaia

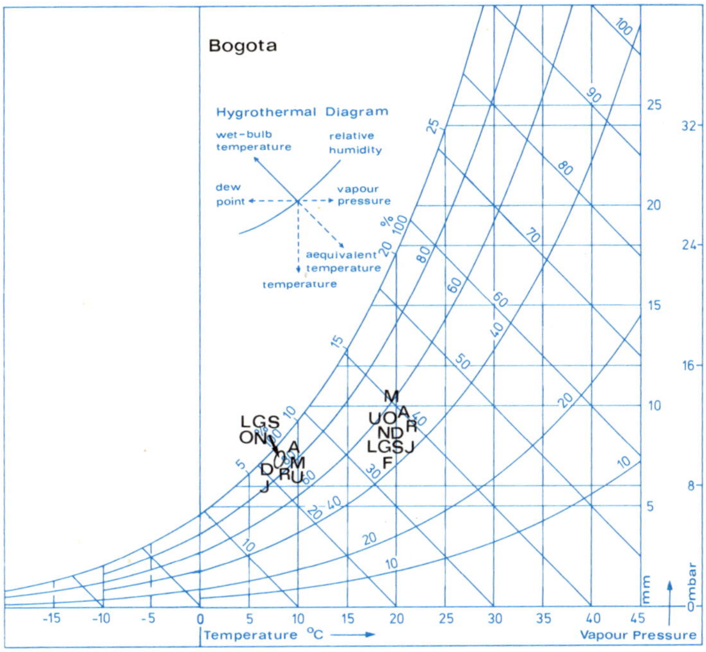
Bogota
Hygrothermal Diagram
wet-bulb temperature
relative humidity
dew point
vapour pressure
aequivalent temperature
temperature
Temperature °C
Vapour Pressure
mm
mbar

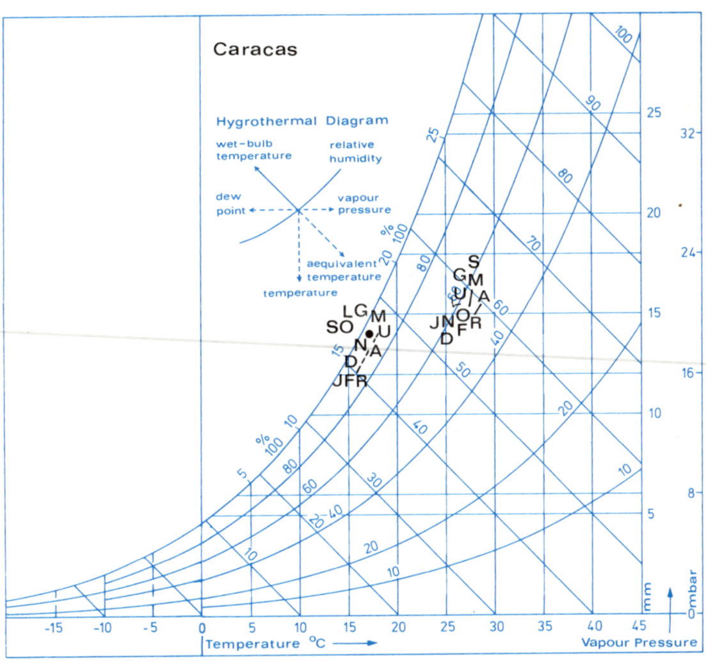
Caracas
Hygrothermal Diagram
wet-bulb temperature
relative humidity
dew point
vapour pressure
aequivalent temperature
temperature
Temperature °C
Vapour Pressure
mm
mbar

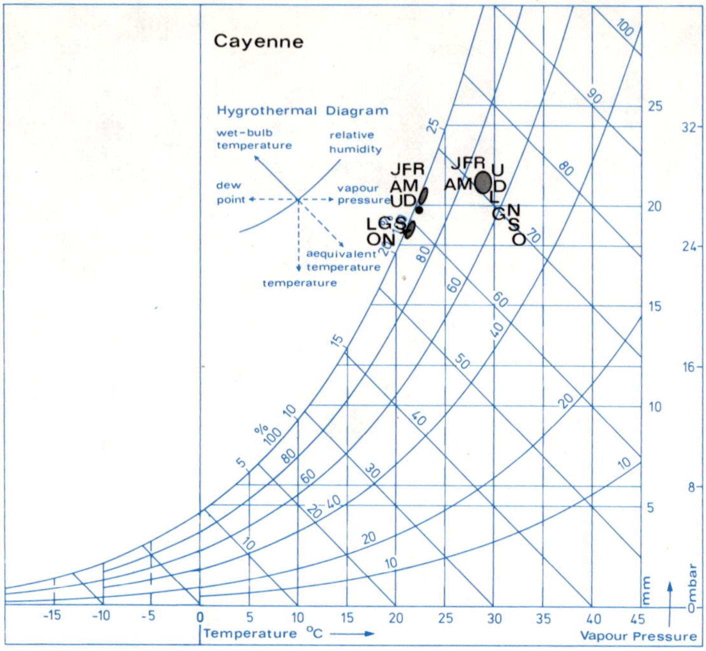
Cayenne
Hygrothermal Diagram
wet-bulb temperature
relative humidity
dew point
vapour pressure
aequivalent temperature
temperature
Temperature °C
Vapour Pressure
mm
mbar

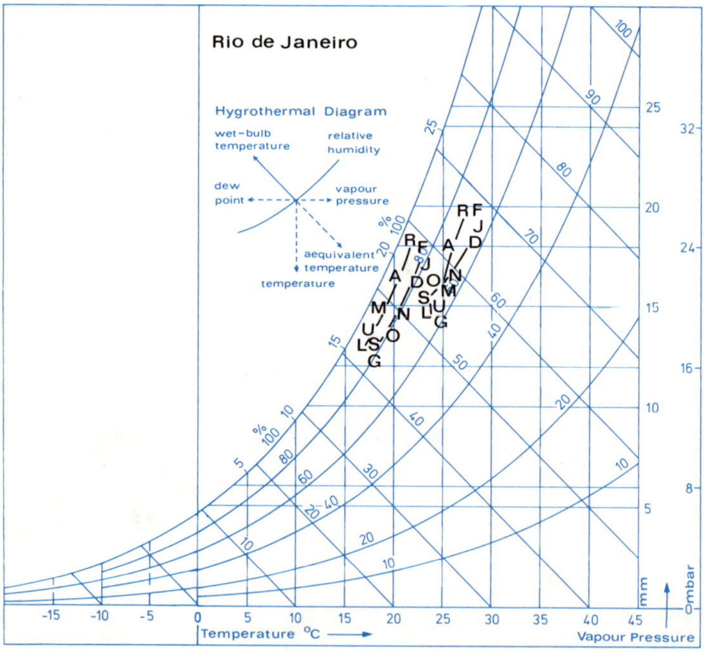
Rio de Janeiro
Hygrothermal Diagram
wet-bulb temperature
relative humidity
dew point
vapour pressure
aequivalent temperature
temperature
Temperature °C
Vapour Pressure
mm
mbar

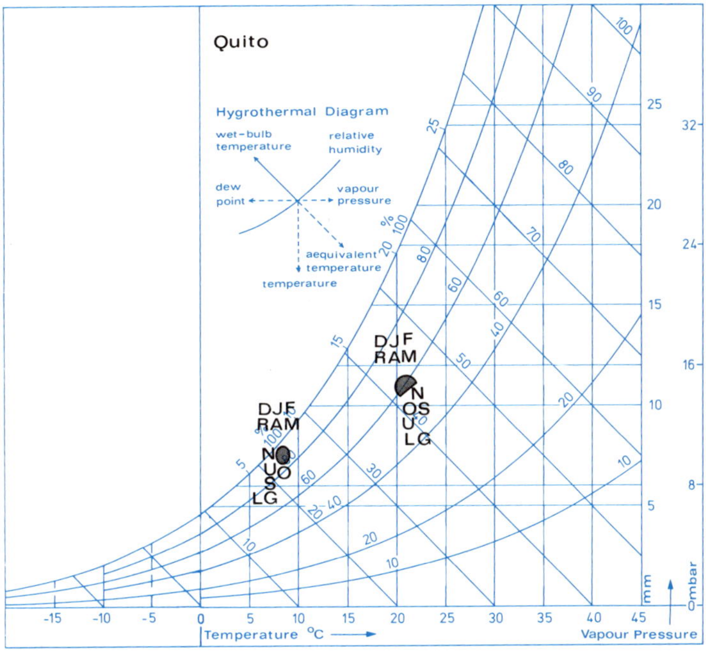
Quito
Hygrothermal Diagram
wet-bulb temperature
relative humidity
dew point
vapour pressure
aequivalent temperature
temperature
Temperature °C
Vapour Pressure
mm
mbar

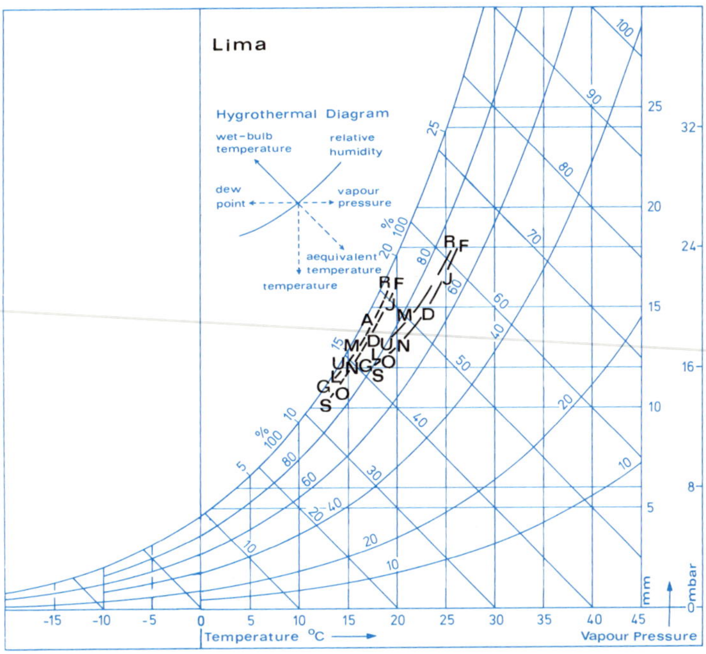
Lima
Hygrothermal Diagram
wet-bulb temperature
relative humidity
dew point
vapour pressure
aequivalent temperature
temperature
Temperature °C
Vapour Pressure
mm
mbar

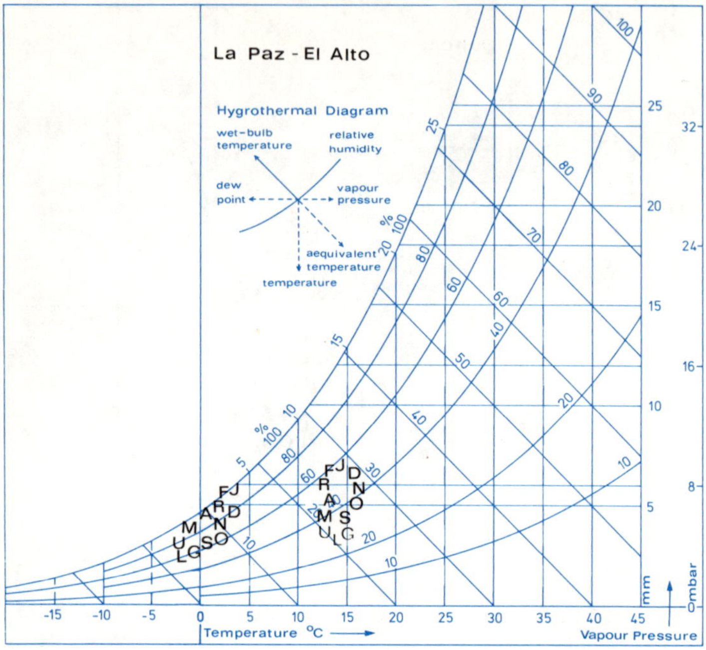
La Paz - El Alto
Hygrothermal Diagram
wet-bulb temperature
relative humidity
dew point
vapour pressure
aequivalent temperature
temperature
Temperature °C
Vapour Pressure
mm
mbar

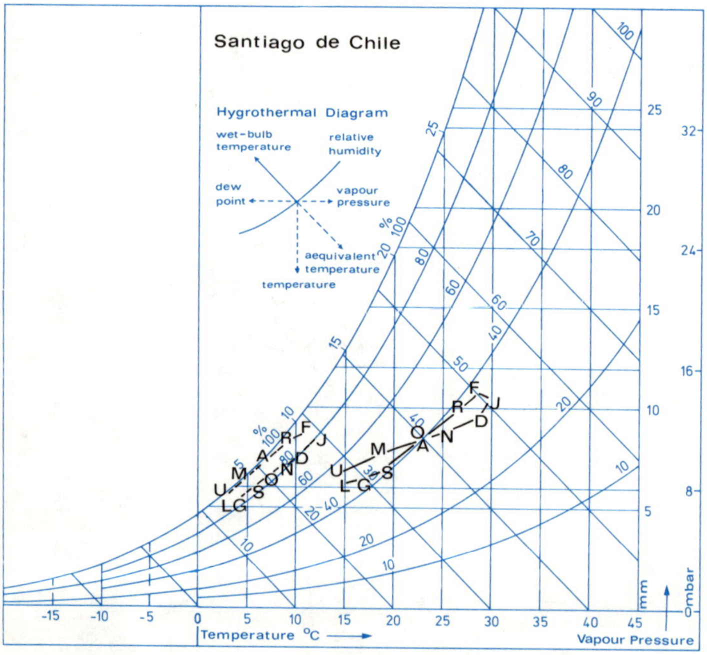
Santiago de Chile
Hygrothermal Diagram
wet-bulb temperature
relative humidity
dew point
vapour pressure
aequivalent temperature
temperature
Temperature °C
Vapour Pressure
mm
mbar

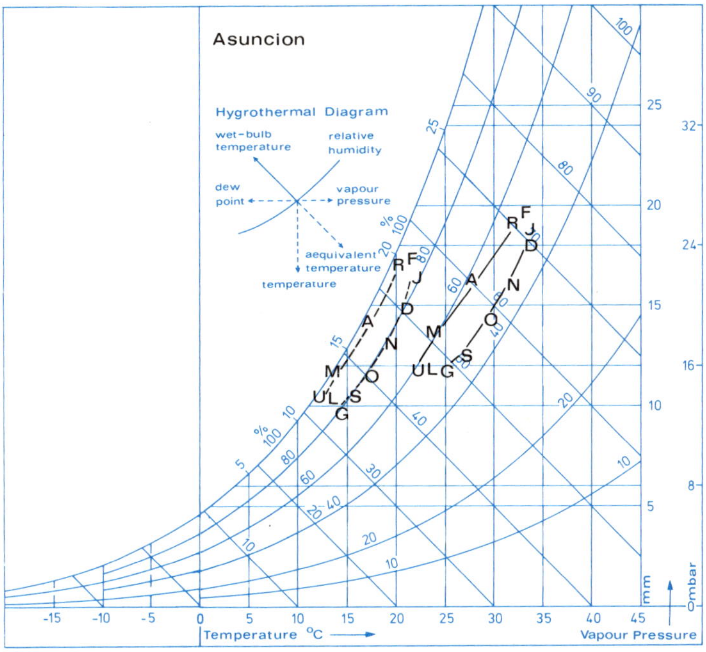
Asuncion
Hygrothermal Diagram
wet-bulb temperature
relative humidity
dew point
vapour pressure
aequivalent temperature
temperature
Temperature °C
Vapour Pressure
mm
mbar

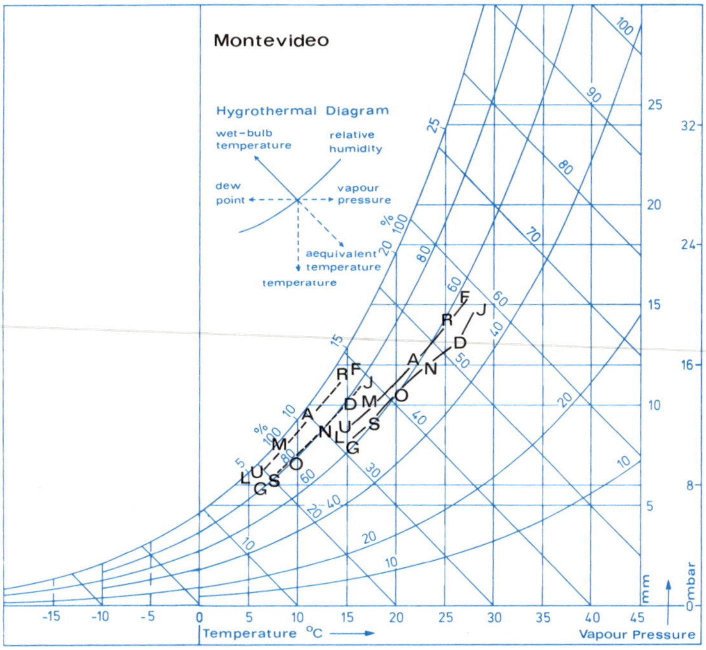
Montevideo
Hygrothermal Diagram
wet-bulb temperature
relative humidity
dew point
vapour pressure
aequivalent temperature
temperature
Temperature °C
Vapour Pressure
mm
mbar

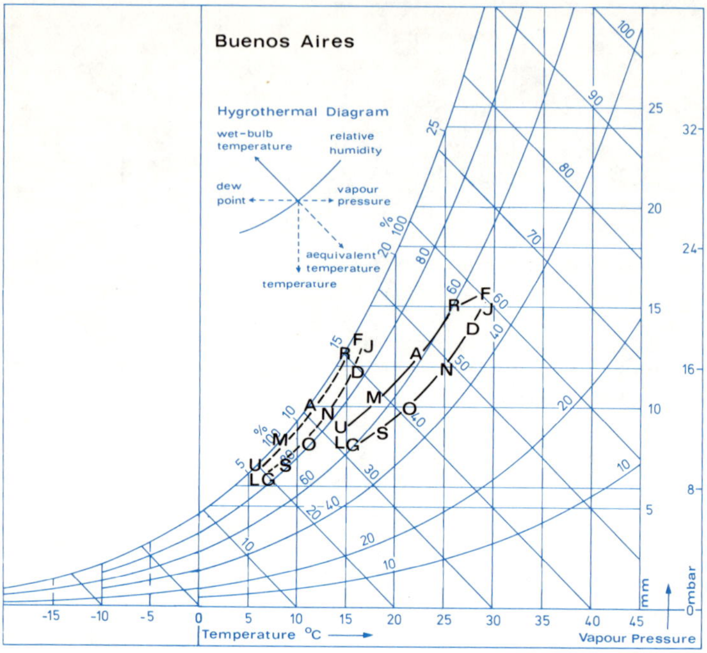
Buenos Aires
Hygrothermal Diagram
wet-bulb temperature
relative humidity
dew point
vapour pressure
aequivalent temperature
temperature
Temperature °C
Vapour Pressure
mm
mbar

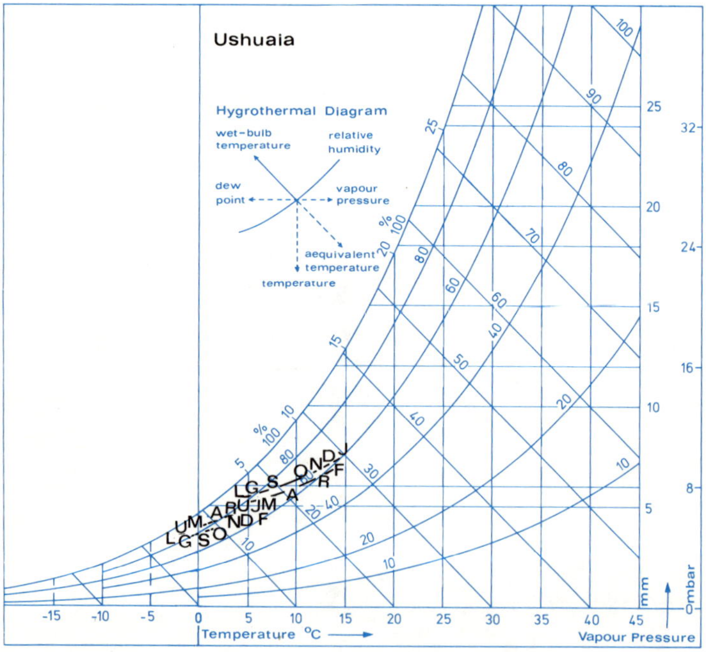
Ushuaia
Hygrothermal Diagram
wet-bulb temperature
relative humidity
dew point
vapour pressure
aequivalent temperature
temperature
Temperature °C
Vapour Pressure
mm
mbar

4.5.124. Colombia

4.5.124.1. Climatic descriptions and tables

1. Climatic area: Guajira Peninsula
1.1. Steppe Climate BS
1.2. Hot throughout the year.
1.3. Generally dry and sunny. High sun period: repeatedly showers, particularly in the afternoon; the odd rainy day.
1.4. Wind: occasional gales.

2. Climatic area: lowlands in the north (except the Guajira Peninsula), Sierra Nevada de Sta. Marta, and llanos or broad plains in the east, north of Rio Guaviare
2.1. Tropical Summer Rain Climate Aw, in parts Tropical Monsoonal Rain Climate Am
2.2. Very hot or hot throughout the year.
2.3. High sun period: often dry and sunny in the morning, in the afternoon showers, frequent rainy days. Low sun period: fewer showers.
2.4. Wind: occasional gales, squalls in thunderstorms.

Cartagena

latitude: 10°27′n longitude: 75°31′w height: 2 m

Awha	124-1	Jan	Feb	Mar	Apr	May	Jun	Jul	Aug	Sep	Oct	Nov	Dec	Year
	t	26	26	27	27	28	27	27	27	27	28	28	27	27 °C
	clothing	t	t	t	t	t	t	t	t	t	t	t	t	–
	stress	ff	ff	ff	ff	ff	ff	ff	ff	ff	ff	ff	ff	–
	p	0	0	0	9	103	132	128	38	157	215	97	16	895 mm
0.1 mm	dp	0	0	0	1	6	10	10	6	13	14	9	1	70 days
	sunshine	260	246	249	225	267	189	260	242	219	223	228	282	2886 hours
	%	71	74	67	61	68	49	66	63	60	60	65	79	65 %

Pueblo Bello

latitude: 10°26′n longitude: 73°35′w height: 1175 m

GAmbb	124-2	Jan	Feb	Mar	Apr	May	Jun	Jul	Aug	Sep	Oct	Nov	Dec	Year
	t	20	20	21	21	21	21	21	21	22	20	20	20	21 °C
	clothing	s	s	s	s	s	s	s	s	s	s	s	s	–
	stress	f	f	f	f	f	f	f	f	f	f	f	f	–
	p	29	31	56	122	299	248	192	220	272	401	243	33	2146 mm
0.1 mm	dp	1	4	7	10	20	16	14	16	20	25	17	5	157 days
	sunshine	285	254	232	192	170	174	211	205	180	180	198	248	2529 hours
	%	79	76	62	52	43	45	54	53	49	48	57	69	57 %

3. *Climatic area:* Smaller parts of the ranges of the Andes in the Cordillera Oriental and Central
3.1. Tropical Summer or Winter Rain Mountain Climate GAw/GAs or High Mountain Climate HAw/HAs
3.2. At about 1000 m hot, at 1500 m warm, at 2000–2500 m mild.
3.3. High sun period: often rain and cloudy or overcast, occasionally dry and sunny or cloudy. Low sun period: predominantly dry and cloudy, now and then rain.
3.4. Wind: occasional gales, squalls in thunderstorms. Fog: frequently at high levels. Snow cover: permanent above 4800 m.

Bogotá

latitude: 4°42′n longitude: 74°08′w height: 2547 m

HAsll	124-3	Jan	Feb	Mar	Apr	May	Jun	Jul	Aug	Sep	Oct	Nov	Dec	Year
	tx	23	23	23	22	22	21	21	21	22	22	22	22	23 °C
	t	13	14	14	14	14	14	13	13	13	14	14	14	14 °C
	tn	6	7	7	8	8	8	8	8	7	8	8	6	6 °C
	clothing	n	n	n	n	n	n	n	n	n	n	n	n	–
	stress													–
	p	38	48	66	96	103	60	48	38	53	158	140	93	942 mm
0.1 mm	dp	9	10	13	20	21	18	18	15	15	21	19	14	193 days
	sunshine	177	144	133	87	96	108	130	130	132	102	117	146	1503 hours
	%	48	43	35	24	25	29	34	34	36	27	33	40	34 %

Palmira

latitude: 3°32′n longitude: 76°17′w height: 1006 m

GAsaa	124-4	Jan	Feb	Mar	Apr	May	Jun	Jul	Aug	Sep	Oct	Nov	Dec	Year
	t	24	24	24	24	24	24	24	24	24	24	24	24	24 °C
	clothing	t	t	t	t	t	t	t	t	t	t	t	t	–
	stress	f	f	f	f	f	f	f	f	f	f	f	f	–
	p	77	67	86	133	117	67	28	34	55	146	112	87	1009 mm
0.1 mm	dp	7	8	9	12	11	8	5	6	7	14	13	11	111 days
	sunshine	211	189	198	165	170	195	211	208	183	170	168	161	2229 hours
	%	57	56	53	45	45	53	55	55	50	46	47	44	50 %

4. *Climatic area:* Pacific coastal region, Valley of Rio Atrato and llanos or broad plains in the east, south of the Rio Gaviare
4.1. Tropical Rain Climate Ar
4.2. Very hot throughout the year.
4.3. In the morning at times dry and sunny or cloudy, in the afternoon often showers; frequent rainy days.
4.4. Wind: occasional gales, squalls in thunderstorms.

Andagoya

latitude: 5°06′n longitude: 76°40′w height: 60 m

Arha	124-5	Jan	Feb	Mar	Apr	May	Jun	Jul	Aug	Sep	Oct	Nov	Dec	Year
	tx	34	34	34	34	34	34	34	34	34	34	34	34	35 °C
	t	28	28	28	28	28	27	27	27	28	28	27	27	28 °C
	tn	22	22	22	22	22	21	21	22	22	21	21	22	20 °C
	clothing	t	t	t	t	t	t	t	t	t	t	t	t	–
	stress	ff	ff	ff	ff	ff	ff	ff	ff	ff	ff	ff	ff	–
	p	635	544	495	663	648	655	592	643	625	577	569	495	7140 mm
0.2 mm	dp	26	21	23	25	26	25	27	27	27	25	27	27	306 days

5. Climatic area: most of the ranges of the Andes
5.1. Tropical Rain Mountain Climate GAr and Tropical Rain High Mountain Climate HAr
5.2. At about 1000 m hot, at 1500 m warm, at 2000–2500 m mild.
5.3. Occasionally dry and sunny, often showers in the afternoon, at times rainy days.
5.4. Wind: occasional gales. Fog: frequent at high levels. Snow cover: permanent above 4800 m.

Chinácota

latitude: 7°37′n longitude: 72°37′w height: 1235 m

GArbb	124-6	Jan	Feb	Mar	Apr	May	Jun	Jul	Aug	Sep	Oct	Nov	Dec	Year
	t	19	19	20	20	20	20	20	20	20	20	20	19	20 °C
	clothing	s	s	s	s	s	s	s	s	s	s	s	s	–
	stress	f	f	f	f	f	f	f	f	f	f	f	f	–
	p	155	148	132	95	137	137	156	166	157	142	137	141	1703 mm
0.1 mm	dp	10	9	11	16	17	18	18	19	18	20	19	14	189 days
	sunshine	155	150	133	96	136	138	155	167	156	143	138	143	1710 hours
	%	43	45	36	26	35	37	40	44	43	39	39	39	39 %

Yolombo

latitude: 6°36′n longitude: 75°03′w height: 1680 m

GArbb	124-7	Jan	Feb	Mar	Apr	May	Jun	Jul	Aug	Sep	Oct	Nov	Dec	Year
	t	20	20	20	20	20	20	20	20	20	19	19	19	20 °C
	clothing	s	s	s	s	s	s	s	s	s	s	s	s	–
	stress	f	f	f	f	f	f	f	f	f	f	f	f	–
	p	50	83	105	279	367	204	192	270	349	398	292	124	2713 mm
0.1 mm	dp	8	10	14	21	24	19	18	21	21	24	20	15	215 days
	sunshine	186	172	152	138	180	207	245	232	210	161	135	149	2167 hours
	%	51	51	41	37	47	55	63	61	58	43	38	41	49 %

Manzanares

latitude: 5°13′n longitude: 75°09′w height: 1450 m

GArbb	124-8	Jan	Feb	Mar	Apr	May	Jun	Jul	Aug	Sep	Oct	Nov	Dec	Year
	t	19	20	20	20	20	20	20	20	20	19	19	19	20 °C
	clothing	s	s	s	s	s	s	s	s	s	s	s	s	–
	stress	f	f	f	f	f	f	f	f	f	f	f	f	–
	p	194	230	335	340	345	174	130	152	219	379	367	274	3139 mm
0.1 mm	dp	17	16	21	22	22	16	13	14	18	22	23	21	225 days
	sunshine	149	150	140	120	140	150	174	180	165	136	120	127	1749 hours
	%	41	44	37	33	37	40	45	47	45	37	34	35	39 %

Chinchina

latitude: 4°53′n longitude: 75°36′w height: 1360 m

GArbb	124-9	Jan	Feb	Mar	Apr	May	Jun	Jul	Aug	Sep	Oct	Nov	Dec	Year
	t	21	21	21	21	21	20	21	21	20	20	20	20	20 °C
	clothing	s	s	s	s	s	s	s	s	s	s	s	s	–
	stress	f	f	f	f	f	f	f	f	f	f	f	f	–
	p	134	144	204	279	285	201	163	159	172	313	265	205	2524 mm
0.1 mm	dp	16	15	19	22	24	21	19	17	18	25	22	19	237 days
	sunshine	177	161	217	228	226	216	211	205	201	236	225	208	2510 hours
	%	48	48	58	62	59	58	55	54	55	63	63	57	57 %

Bajo Calima

latitude: 3°49′n longitude: 76°52′w height: 40 m

Araa	124-10	Jan	Feb	Mar	Apr	May	Jun	Jul	Aug	Sep	Oct	Nov	Dec	Year
	t	26	27	27	27	27	27	27	27	27	26	26	26	26 °C
	clothing	t	t	t	t	t	t	t	t	t	t	t	t	–
	stress	ff	ff	ff	ff	ff	ff	ff	ff	ff	ff	ff	ff	–
	p	344	284	379	508	621	491	475	462	642	665	616	434	5921 mm
0.1 mm	dp	23	22	25	26	29	27	26	28	27	28	26	27	314 days
	sunshine	114	110	87	75	99	111	127	96	105	102	93	118	1238 hours
	%	31	32	23	20	26	30	33	25	29	27	26	32	28 %

Popayan

latitude: 2°27′n longitude: 76°34′w height: 1850 m

GArbl	124-11	Jan	Feb	Mar	Apr	May	Jun	Jul	Aug	Sep	Oct	Nov	Dec	Year
	t	17	17	17	17	17	17	17	18	18	17	17	17	17 °C
	clothing	s	s	s	s	s	s	s	s	s	s	s	s	–
	stress	o	o	o	o	o	o	o	o	o	o	o	o	–
	p	158	149	169	177	139	93	40	31	89	262	296	287	1890 mm
0.1 mm	dp	19	16	18	21	19	16	12	9	13	23	25	24	215 days
	sunshine	183	170	152	117	130	156	180	180	165	167	132	152	1883 hours
	%	49	50	40	32	34	42	47	48	45	45	37	41	43 %

Consaca

latitude: 1°13′n longitude: 77°29′w height: 1700 m

GArbb	124-12	Jan	Feb	Mar	Apr	May	Jun	Jul	Aug	Sep	Oct	Nov	Dec	Year
	t	18	19	19	19	19	19	19	19	19	19	18	18	19 °C
	clothing	s	s	s	s	s	s	s	s	s	s	s	s	–
	stress	o	o	o	o	o	o	o	o	o	o	o	o	–
	p	119	90	113	173	158	89	41	42	62	69	187	150	1293 mm
0.1 mm	dp	18	15	18	21	20	16	10	9	11	21	24	23	206 days
	sunshine	164	144	130	114	152	171	186	211	168	130	135	149	1854 hours
	%	44	42	35	31	40	47	49	56	46	35	37	40	42 %

4.5.124.2. Predominant general weather situations are described in 4.5.0.1. under Colombia. A hygrothermal diagram of Bogota is to be found in 4.5.0.2.
Wind name: Paramito is a cold easterly wind accompanied by drizzling fog in the Andes at Bogota.

4.5.125. Venezuela

4.5.125.1. Climatic descriptions and tables

1. Climatic area: most of the lowlands and mountains near the Caribbean Sea
1.1. Steppe Climate BS and Steppe Mountain Climate GBS
1.2. Summer very hot, winter hot.
1.3. Generally dry and sunny; high sun period: frequent showers in the afternoon, at times rainy days.
1.4. Wind: occasional gales.

Maracaibo

BSha	125-1	latitude: 10°39′n							longitude: 71°36′w			height: 48 m		
		Jan	Feb	Mar	Apr	May	Jun	Jul	Aug	Sep	Oct	Nov	Dec	Year
	tx	34	34	35	36	36	36	36	37	37	36	35	34	37 °C
	t	26	27	27	28	28	29	29	29	29	28	28	27	28 °C
	tn	21	22	22	23	22	23	23	22	22	21	23	22	21 °C
	clothing	t	t	t	t	t	t	t	t	t	t	t	t	–
	stress	ff	ff	ff	ff	ff	ff	ff	ff	ff	ff	ff	ff	–
	p	3	2	3	27	73	43	28	42	40	99	22	5	387 mm
0.1 mm	dp	1	1	1	3	9	8	7	7	8	11	7	2	65 days
	sunshine	264	240	251	180	192	207	254	254	216	198	219	242	2717 hours
	%	74	72	67	49	49	54	65	66	59	54	63	68	61 %

Maiquetia

BSha	125-2	latitude: 10°36′n							longitude: 66°59′w			height: 43 m		
		Jan	Feb	Mar	Apr	May	Jun	Jul	Aug	Sep	Oct	Nov	Dec	Year
	tx	29	30	31	32	33	33	33	33	34	33	32	31	34 °C
	t	25	25	26	27	27	27	27	28	28	28	28	26	26 °C
	tn	21	21	22	23	23	23	24	24	24	23	23	22	20 °C
	clothing	t	t	t	t	t	t	t	t	t	t	t	t	–
	stress	ff	ff	ff	ff	ff	ff	ff	ff	ff	ff	ff	ff	–
	p	63	57	18	14	36	43	45	57	51	46	47	81	558 mm
0.1 mm	dp	7	4	3	3	8	10	10	10	10	10	10	8	93 days
	sunshine	214	220	239	198	223	225	257	264	243	211	198	195	2687 hours
	%	60	66	64	53	57	59	65	68	66	57	57	55	61 %

Barcelona

BSha	125-3	latitude: 10°07′n							longitude: 64°41′w			height: 14 m		
		Jan	Feb	Mar	Apr	May	Jun	Jul	Aug	Sep	Oct	Nov	Dec	Year
	t	25	26	26	27	28	27	26	26	27	27	27	26	26 °C
	clothing	t	t	t	t	t	t	t	t	t	t	t	t	–
	stress	ff	ff	ff	ff	ff	ff	ff	ff	ff	ff	ff	ff	–
	p	8	3	6	6	36	108	141	117	63	49	52	22	611 mm
	dp													days
	sunshine	291	280	301	261	260	210	245	239	252	257	261	273	3130 hours
	%	81	84	80	70	67	55	62	62	69	70	75	76	71 %

Barquisimeto

GBSaa	125-4	latitude: 10°04′n							longitude: 69°19′w			height: 590 m		
		Jan	Feb	Mar	Apr	May	Jun	Jul	Aug	Sep	Oct	Nov	Dec	Year
	t	23	23	24	25	24	24	23	24	24	24	24	23	24 °C
	clothing	t	t	t	t	t	t	t	t	t	t	t	t	–
	stress	f	f	f	f	f	f	f	f	f	f	f	f	–
	p	15	9	7	51	82	86	71	54	44	48	57	25	549 mm
0.1 mm	dp	3	2	1	5	13	14	18	11	8	9	8	7	99 days
	sunshine	242	218	239	183	183	198	220	229	219	205	219	223	2577 hours
	%	67	65	64	49	47	52	56	59	60	55	63	62	58 %

2. *Climatic area:* Northern half of the country, except regions near the Caribbean
2.1. Tropical Summer Rain Climate Aw and Tropical Summer Rain Mountain Climate GAw and Tropical Summer Rain High Mountain Climate H Aw
2.2. In the lowlands very hot or hot all the year round.
2.3. Generally dry and sunny in the morning. High sun period: frequent showers in the afternoon; at times rainy days. Low sun period: showers now and then.
2.4. Wind: occasional gales, squalls in thunderstorms.

Guiria

latitude: 10°35′n longitude: 62°18′w height: 8 m

Awha	125-5	Jan	Feb	Mar	Apr	May	Jun	Jul	Aug	Sep	Oct	Nov	Dec	Year
	t	26	26	27	28	28	27	27	27	28	28	27	26	27 °C
	clothing	t	t	t	t	t	t	t	t	t	t	t	t	–
	stress	ff	ff	ff	ff	ff	ff	ff	ff	ff	ff	ff	ff	–
	p	44	27	17	26	62	122	108	136	96	99	95	72	904 mm
0.1 mm	dp	10	8	6	6	11	17	15	16	13	13	13	11	139 days
	sunshine	236	226	251	237	239	210	242	232	225	229	219	223	2769 hours
	%	64	68	67	64	61	55	61	60	61	62	63	62	62 %

Caracas

latitude: 10°30′n longitude: 66°55′w height: 865 m

GAwab	125-6	Jan	Feb	Mar	Apr	May	Jun	Jul	Aug	Sep	Oct	Nov	Dec	Year
	tx	27	27	29	30	29	28	27	28	29	29	28	27	31 °C
	t	20	20	21	22	23	22	22	22	22	22	21	20	22 °C
	tn	10	10	11	13	14	14	14	14	14	14	13	11	9 °C
	clothing	s	s	s	s	t	s	s	s	s	s	s	s	–
	stress	o	o	o	f	f	f	f	f	f	f	o	o	–
	p	23	3	13	27	75	151	109	120	115	104	74	40	854 mm
0.1 mm	dp	6	4	3	6	12	18	17	15	15	14	12	8	130 days
	sunshine	238	246	257	219	198	201	239	236	213	214	210	217	2689 hours
	%	66	74	69	59	51	53	61	61	58	58	60	61	61 %

Maracay

latitude: 10°15′n longitude: 67°39′w height: 443 m

Awaa	125-7	Jan	Feb	Mar	Apr	May	Jun	Jul	Aug	Sep	Oct	Nov	Dec	Year
	t	23	24	25	26	26	25	24	25	25	25	24	24	25 °C
	clothing	t	t	t	t	t	t	t	t	t	t	t	t	–
	stress	f	f	f	f	f	f	f	f	f	f	f	f	–
	p	8	9	1	26	104	150	131	201	135	84	40	25	914 mm
0.1 mm	dp	3	2	1	6	14	17	18	20	15	13	8	5	122 days
	sunshine	276	263	285	225	214	207	236	226	228	229	246	257	2892 hours
	%	77	79	76	61	55	54	60	58	62	62	70	72	65 %

Maturin

latitude: 9°45′n longitude: 63°11′w height: 70 m

Awaa	125-8	Jan	Feb	Mar	Apr	May	Jun	Jul	Aug	Sep	Oct	Nov	Dec	Year
	t	25	25	26	27	27	26	26	26	26	26	26	25	26 °C
	clothing	t	t	t	t	t	t	t	t	t	t	t	t	–
	stress	f	f	f	ff	ff	ff	ff	ff	ff	ff	ff	ff	–
	p	67	41	36	47	117	206	196	149	122	110	109	120	1320 mm
0.1 mm	dp	14	10	8	7	15	22	23	19	14	16	18	18	184 days
	sunshine	238	229	251	219	208	162	192	205	222	223	219	220	2588 hours
	%	66	69	67	59	53	43	49	53	61	58	62	61	58 %

Merida

GAwbb	125-9	Jan	Feb	Mar	Apr	May	Jun	Jul	Aug	Sep	Oct	Nov	Dec	Year
				latitude: 8°36′n				longitude: 71°10′w			height: 1498 m			
	tx	26	27	27	27	27	27	27	27	28	27	26	25	28 °C
	t	18	18	19	19	19	19	19	19	19	19	18	18	19 °C
	tn	11	12	13	13	14	13	13	13	13	13	13	12	11 °C
	clothing	s	s	s	s	s	s	s	s	s	s	s	s	–
	stress			o	o	o	o	o	o	o	o	o		–
	p	55	42	52	163	262	164	125	106	148	224	192	100	1633 mm
0.1 mm	dp	11	10	11	17	23	21	20	20	20	23	19	15	210 days
	sunshine	245	220	229	186	186	174	205	208	198	189	204	229	2472 hours
	%	68	66	61	50	48	46	53	54	54	51	58	63	56 %

Bolivar

Awha	125-10	Jan	Feb	Mar	Apr	May	Jun	Jul	Aug	Sep	Oct	Nov	Dec	Year
				latitude: 8°09′n				longitude: 63°33′w			height: 55 m			
	t	26	27	28	28	28	27	27	28	28	28	28	27	28 °C
	clothing	t	t	t	t	t	t	t	t	t	t	t	t	–
	stress	ff	ff	ff	ff	ff	ff	ff	ff	ff	ff	ff	ff	–
	p	36	19	17	25	97	139	160	180	92	101	72	34	972 mm
0.1 mm	dp	10	5	3	4	12	20	19	16	11	8	8	8	124 days
	sunshine	239	234	264	231	239	204	242	254	258	264	252	245	2926 hours
	%	66	70	71	63	62	54	62	66	71	71	71	68	66 %

San Fernando

Awha	125-11	Jan	Feb	Mar	Apr	May	Jun	Jul	Aug	Sep	Oct	Nov	Dec	Year
				latitude: 7°54′n				longitude: 67°25′w			height: 74 m			
	t	27	28	29	29	27	26	26	26	27	27	27	27	27 °C
	clothing	t	t	t	t	t	t	t	t	t	t	t	t	–
	stress	ff	ff	ff	ff	ff	ff	ff	ff	ff	ff	ff	ff	–
	p	1	1	5	87	174	274	329	297	175	120	50	20	1533 mm
0.1 mm	dp	1	1	1	6	14	20	25	23	16	11	6	3	127 days
	sunshine	276	257	279	213	195	159	183	180	207	236	258	273	2716 hours
	%	76	77	75	58	50	42	47	47	57	64	73	75	61 %

San Antonio

Awaa	125-12	Jan	Feb	Mar	Apr	May	Jun	Jul	Aug	Sep	Oct	Nov	Dec	Year
				latitude: 7°51′n				longitude: 72°27′w			height: 405 m			
	t	24	25	26	26	27	27	27	27	27	26	25	24	26 °C
	clothing	t	t	t	t	t	t	t	t	t	t	t	t	–
	stress	f	f	f	ff	ff	ff	ff	ff	ff	ff	ff	f	–
	p	56	35	29	40	73	49	18	30	40	111	94	80	655 mm
0.1 mm	dp	7	7	8	10	13	13	11	12	10	13	13	11	128 days
	sunshine	195	181	183	144	177	153	180	205	207	195	192	192	2201 hours
	%	54	54	49	39	46	41	46	53	57	53	54	53	50 %

3. Climatic area: Southern half of the country, south of the Orinoco
3.1. Tropical Rain Climate Ar or Tropical Rain Mountain Climate GAr
3.2. In the lowlands hot.
3.3. Often showers in the afternoon, at times rainy days. Now and then dry and sunny or cloudy.
3.4. Wind: occasional gales, squalls in thunderstorms.

Timeremos

latitude: 7°18'n longitude: 61°27'w height: 180 m

Araa	125-13	Jan	Feb	Mar	Apr	May	Jun	Jul	Aug	Sep	Oct	Nov	Dec	Year
	t	24	24	25	25	25	25	25	26	26	26	25	24	25 °C
	clothing	t	t	t	t	t	t	t	t	t	t	t	t	–
	stress	f	f	f	f	ff	ff	ff	ff	ff	ff	f	f	–
	p	100	78	61	77	155	183	152	124	76	55	64	111	1236 mm
0.1 mm	dp	19	14	12	11	20	23	22	17	11	9	9	17	184 days
	sunshine	174	170	192	186	180	159	208	232	243	232	210	180	2366 hours
	%	48	51	51	50	47	42	54	61	67	62	59	50	53 %

Santa Elena

latitude: 4°36'n longitude: 61°07'w height: 907 m

GArbb	125-14	Jan	Feb	Mar	Apr	May	Jun	Jul	Aug	Sep	Oct	Nov	Dec	Year
	tx	32	31	33	32	32	29	29	31	32	32	32	31	33 °C
	t	21	22	22	22	22	21	21	21	21	22	22	22	22 °C
	tn	13	14	14	15	15	16	15	15	14	14	14	14	13 °C
	clothing	s	s	s	s	s	s	s	s	s	s	s	s	–
	stress	f	f	f	f	f	f	f	f	f	f	f	f	–
	p	81	81	80	145	245	242	231	192	134	125	125	115	1796 mm
0.1 mm	dp	14	12	12	14	23	26	26	23	15	14	13	16	208 days
	sunshine	189	184	202	183	164	147	174	189	237	229	210	195	2303 hours
	%	51	54	54	50	43	40	45	50	65	61	59	53	52 %

4.5.125.2. Predominant general weather situations are described in 4.5.0.1. under Venezuela. A hygrothermal diagram of Caracas is to be found in 4.5.0.2.

4.5.126. Guyana, Surinam and French Guiana

4.5.126.1. Climatic description and tables

1. Tropical Rain Climate Ar and Tropical Rain Mountain Climate GAr
2. Very hot all the year round.
3. Occasionally dry and sunny days, otherwise often showers in the afternoon; at times rainy days.
4. Wind: occasional gales, squalls in thunderstorms.

Georgetown – Guyana

Arha	126-1	Jan	Feb	Mar	Apr	May	Jun	Jul	Aug	Sep	Oct	Nov	Dec	Year
				latitude: 6°49'n				longitude: 58°11'w				height: 1 m		
	tx	30	30	30	31	31	31	31	31	32	32	32	31	32 °C
	t	26	26	27	27	27	27	27	28	28	28	28	27	27 °C
	tn	22	22	22	23	23	23	22	23	23	23	23	22	21 °C
	clothing	t	t	t	t	t	t	t	t	t	t	t	t	–
	stress	ff	ff	ff	ff	ff	ff	ff	ff	ff	ff	ff	ff	–
	p	251	122	113	178	296	346	281	185	88	98	147	313	2418 mm
	dp	18	13	11	14	20	24	22	15	7	8	11	19	182 days
	sunshine	188	186	209	196	182	165	215	245	239	226	212	181	2444 hours
	%	52	55	56	53	47	44	56	64	65	61	60	50	55 %

Paramaribo – Surinam

Arha	126-2	Jan	Feb	Mar	Apr	May	Jun	Jul	Aug	Sep	Oct	Nov	Dec	Year
				latitude: 5°50'n				longitude: 55°10'w				height: 2 m		
	tx	32	32	33	33	33	33	33	34	34	35	34	33	36 °C
	t	26	27	27	27	27	27	27	28	28	28	28	27	27 °C
	tn	20	20	20	21	22	22	22	22	22	22	22	21	19 °C
	clothing	t	t	t	t	t	t	t	t	t	t	t	t	–
	stress	ff	ff	ff	ff	ff	ff	ff	ff	ff	ff	ff	ff	–
	p	193	150	162	232	321	303	226	167	86	87	109	174	2208 mm
0.1 mm	dp	22	18	19	20	20	27	25	19	12	12	15	22	235 days
	sunshine	174	170	195	177	167	189	239	270	279	282	228	180	2550 hours
	%	47	50	52	48	43	51	62	71	76	76	64	49	58 %

Cayenne – French Guiana

Araa	126-3	Jan	Feb	Mar	Apr	May	Jun	Jul	Aug	Sep	Oct	Nov	Dec	Year
				latitude: 4°50'n				longitude: 52°22'w				height: 9 m		
	tx	31	31	32	32	32	32	33	33	34	34	33	32	34 °C
	t	25	25	26	26	25	25	25	26	26	26	26	25	26 °C
	tn	22	22	22	22	22	22	21	22	22	22	22	22	21 °C
	clothing	t	t	t	t	t	t	t	t	t	t	t	t	–
	stress	ff	ff	ff	ff	ff	ff	ff	ff	ff	ff	ff	ff	–
	p	431	423	432	480	590	457	274	144	32	42	122	317	3744 mm
	dp	23	20	21	19	25	26	21	14	6	5	11	21	212 days
	sunshine	143	113	137	130	137	170	208	233	261	266	234	192	2224 hours
	%	39	33	37	35	36	46	54	61	72	76	68	52	50 %

4.5.126.2. Predominant general weather situations are described in 4.5.0.1. under Guyana, Surinam, French Guiana. A hygrothermal diagram of Cayenne is to be found in 4.5.0.2.

4.5.127. Brazil

4.5.127.1. Climatic descriptions and tables

1. *Climatic area:* Rio Branco, Amazonas, Territorio do Acre, Guapore, Para, Amapa, coastal region of Bahia, small part of the southernmost Mato Grosso
1.1. Tropical Rain Climate Ar and Tropical Rain Mountain Climate GAr
1.2. Very hot or hot throughout the year.
1.3. In the morning often dry and sunny, showers in the afternoon; rainy days at times.
1.4. Wind: occasional gales, squalls in thunderstorms.

Boa Vista – Rio Branco

latitude: 2°49′n longitude: 60°40′w height: 74 m

Arha	127-1	Jan	Feb	Mar	Apr	May	Jun	Jul	Aug	Sep	Oct	Nov	Dec	Year
	t	28	28	28	28	27	26	26	26	28	28	28	28	27 °C
	clothing	t	t	t	t	t	t	t	t	t	t	t	t	–
	stress	ff	ff	ff	ff	ff	ff	ff	ff	ff	ff	ff	ff	–
	p	75	75	80	151	304	365	346	226	110	74	66	69	1941 mm

Uaupés – Amazonas

latitude: 0°08′s longitude: 67°05′w height: 86 m

Araa	127-2	Jan	Feb	Mar	Apr	May	Jun	Jul	Aug	Sep	Oct	Nov	Dec	Year
	t	25	26	26	25	25	24	24	25	25	26	26	26	25 °C
	clothing	t	t	t	t	t	t	t	t	t	t	t	t	–
	stress	ff	ff	ff	ff	ff	ff	ff	ff	ff	ff	ff	ff	–
	p	275	250	285	267	317	250	246	195	148	173	202	309	2917 mm
	dp	12	12	12	16	16	16	16	11	11	11	14	12	159 days
	sunshine	163	154	163	144	152	148	164	187	198	190	180	174	2018 hours
	%	43	45	43	40	40	41	44	50	54	51	49	46	46 %

Belém – Pará

latitude: 1°28′s longitude: 48°29′w height: 24 m

Araa	127-3	Jan	Feb	Mar	Apr	May	Jun	Jul	Aug	Sep	Oct	Nov	Dec	Year
	tx	33	32	33	32	33	33	32	33	33	33	34	33	34 °C
	t	26	25	25	26	26	26	26	26	26	26	26	26	26 °C
	tn	21	21	21	22	22	21	21	21	21	21	21	21	19 °C
	clothing	t	t	t	t	t	t	t	t	t	t	t	t	–
	stress	ff	ff	ff	ff	ff	ff	ff	ff	ff	ff	ff	ff	–
	p	318	407	436	382	265	165	161	116	116	105	94	197	2762 mm
0.2 mm	dp	27	26	28	27	24	22	19	16	16	15	12	19	251 days
	sunshine	157	112	102	131	196	239	268	267	235	247	213	213	2380 hours
	%	41	33	27	36	52	66	72	71	65	66	58	56	54 %

Manaus – Amazonas

latitude: 3°08′s longitude: 60°01′w height: 83 m

Arha	127-4	Jan	Feb	Mar	Apr	May	Jun	Jul	Aug	Sep	Oct	Nov	Dec	Year
	tx	33	33	33	33	33	33	33	35	36	36	35	34	36 °C
	t	26	26	26	26	26	27	27	28	28	28	27	27	27 °C
	tn	21	21	21	21	21	21	20	21	21	21	21	21	19 °C
	clothing	t	t	t	t	t	t	t	t	t	t	t	t	–
	stress	ff	ff	ff	ff	ff	ff	ff	ff	ff	ff	ff	ff	–
	p	276	277	301	287	193	99	61	41	62	112	165	228	2102 mm
	dp	20	19	20	19	18	11	8	6	7	11	12	16	167 days
	sunshine	119	112	111	118	168	208	245	256	225	206	177	153	2098 hours
	%	31	32	30	33	45	58	66	69	62	55	48	40	47 %

Coari – Amazonas

latitude: 4°05′s longitude: 63°08′w height: 48 m

Araa	127-5	Jan	Feb	Mar	Apr	May	Jun	Jul	Aug	Sep	Oct	Nov	Dec	Year
	t	25	25	25	25	25	25	25	26	26	26	26	26	26 °C
	clothing	t	t	t	t	t	t	t	t	t	t	t	t	–
	stress	ff	ff	ff	ff	ff	ff	ff	ff	ff	ff	ff	ff	–
	p	361	274	281	288	226	134	88	75	99	158	188	222	2349 mm

Cruzeiro do Sul – Territorio do Acre

latitude: 7°38′s longitude: 72°40′w height: 170 m

Araa	127-6	Jan	Feb	Mar	Apr	May	Jun	Jul	Aug	Sep	Oct	Nov	Dec	Year
	t	24	25	24	24	24	23	23	24	24	25	25	25	24 °C
	clothing	t	t	t	t	t	t	t	t	t	t	t	t	–
	stress	ff	ff	ff	ff	f	f	f	f	f	ff	ff	ff	–
	p	246	244	269	241	138	105	47	85	147	251	216	241	2230 mm

Seno Madureira – Territorio do Acre

latitude: 9°08′s longitude: 68°40′w height: 135 m

Araa	127-7	Jan	Feb	Mar	Apr	May	Jun	Jul	Aug	Sep	Oct	Nov	Dec	Year
	tx	36	36	36	36	36	36	36	36	36	36	36	37	37 °C
	t	25	25	25	25	24	23	23	24	25	25	26	26	25 °C
	tn	18	18	18	18	14	12	13	17	16	17	18	18	11 °C
	clothing	t	t	t	t	t	t	t	t	t	t	t	t	–
	stress	ff	ff	ff	ff	f	f	f	f	f	ff	ff	ff	–
	p	301	259	264	217	104	69	31	32	157	185	207	257	2083 mm
0.2 mm	dp	18	16	17	15	8	7	5	5	8	11	12	16	138 days

Salvador (Ondina) – Bahia

latitude: 13°00′s longitude: 38°31′w height: 45 m

Araa	127-8	Jan	Feb	Mar	Apr	May	Jun	Jul	Aug	Sep	Oct	Nov	Dec	Year
	tx	32	32	32	31	30	29	28	28	29	31	31	31	33 °C
	t	26	26	26	26	25	24	23	23	24	24	25	26	25 °C
	tn	22	22	22	22	21	20	19	19	19	21	21	21	18 °C
	clothing	t	t	t	t	t	t	t	t	t	t	t	t	–
	stress	ff	ff	ff	ff	ff	ff	ff	ff	ff	ff	ff	ff	–
	p	77	85	153	291	293	247	207	97	83	94	135	104	1863 mm
0.2 mm	dp	14	17	19	22	24	23	25	20	17	14	15	15	225 days
	sunshine	254	234	245	228	208	192	208	223	219	229	228	226	2695 hours
	%	64	66	65	64	58	56	59	62	61	60	60	57	61 %

Caravelas – Bahia

latitude: 17°38′s longitude: 39°15′w height: 4 m

Arab	127-9	Jan	Feb	Mar	Apr	May	Jun	Jul	Aug	Sep	Oct	Nov	Dec	Year
	t	26	26	25	24	23	22	21	21	22	24	24	25	24 °C
	clothing	t	t	t	t	t	s	s	s	s	t	t	t	–
	stress	ff	ff	ff	ff	ff	f	f	f	f	ff	ff	ff	–
	p	125	97	183	218	141	113	119	81	98	144	239	173	1731 mm

Ponta Pora – Mato Grosso

latitude: 22°33′s longitude: 55°42′w height: 660 m

GAral	127-10	Jan	Feb	Mar	Apr	May	Jun	Jul	Aug	Sep	Oct	Nov	Dec	Year
	t	23	23	22	19	17	16	16	18	20	21	22	23	20 °C
	clothing	t	t	s	s	n	n	n	s	s	s	s	t	–
	stress	f	f	f	o	o			o	o	o	f	f	–
	p	189	180	158	138	119	103	62	51	120	207	161	170	1658 mm

Santos – São Paulo

latitude: 23°56′s longitude: 46°20′w height: 2 m

Arab	127-11	Jan	Feb	Mar	Apr	May	Jun	Jul	Aug	Sep	Oct	Nov	Dec	Year
	t	25	25	25	23	21	19	18	19	20	21	22	24	22 °C
	clothing	t	t	t	t	s	s	s	s	s	s	s	t	–
	stress	ff	ff	ff	f	o	o	o	o	o	o	f	f	–
	p	248	321	248	203	158	120	91	113	145	176	158	204	2185 mm
	dp	15	13	13	12	9	10	6	8	10	13	13	12	134 days

2. *Climatic area:* Maranhao, Goias, Mato Grosso (except southernmost part), Piau (except central Part), West of Ceara, Bahia (northwest part and coastal region), greater east Rio Grande do Norte, Fernando do Noronha, Paraiba, central and east Pernambuco, Alagoas, Sergipe, Minas Gerais, Espiritu Santo, Rio de Janeiro, Sao Paulo (except southernmost part)

2.1. Tropical Summer Rain Climate Aw and Tropical Summer Rain Mountain Climate GAw, in parts Tropical Monsoonal Rain Climate Am and Tropical Monsoonal Rain Mountain Climate GAm

2.2. At low levels summer very hot or hot, winter hot or warm.

2.3. In the morning often dry and sunny; high sun period: often showers in the afternoon, at times rainy days. Low sun period: fewer showers.

2.4. Wind: occasional gales, squalls in thunderstorms.

Turiaçu – Maranhao

latitude: 1°41′s longitude: 45°22′w height: 8 m

Asha	127-12	Jan	Feb	Mar	Apr	May	Jun	Jul	Aug	Sep	Oct	Nov	Dec	Year
	t	27	26	26	26	26	26	26	27	27	27	28	28	27 °C
	clothing	t	t	t	t	t	t	t	t	t	t	t	t	–
	stress	ff	ff	ff	ff	ff	ff	ff	ff	ff	ff	ff	ff	–
	p	141	260	446	425	328	221	181	69	7	10	20	59	2177 mm
	sunshine	197	135	127	124	162	203	237	268	260	263	244	239	2459 hours
	%	52	39	34	34	43	56	63	72	72	70	67	63	56 %

Santarém – Pará

latitude: 2°25′s longitude: 54°43′w height: 72 m

Amaa	127-13	Jan	Feb	Mar	Apr	May	Jun	Jul	Aug	Sep	Oct	Nov	Dec	Year
	tx	32	32	31	31	32	32	33	33	34	35	34	33	35 °C
	t	26	26	26	26	26	25	26	26	27	27	27	26	26 °C
	tn	21	21	22	22	22	21	20	21	21	22	21	21	20 °C
	clothing	t	t	t	t	t	t	t	t	t	t	t	t	–
	stress	ff	ff	ff	ff	ff	ff	ff	ff	ff	ff	ff	ff	–
	p	179	275	258	262	293	174	112	50	39	46	85	123	1896 mm
0.2 mm	dp	21	23	26	26	26	22	15	10	8	7	8	13	205 days
	sunshine	142	106	108	118	147	177	214	244	223	230	195	187	2091 hours
	%	37	31	29	33	29	49	58	65	61	61	53	49	47 %

Fernando Noronha – Fernando Noronha

latitude: 3°51′s longitude: 32°25′w height: 45 m

Asaa	127-14	Jan	Feb	Mar	Apr	May	Jun	Jul	Aug	Sep	Oct	Nov	Dec	Year
	tx	29	29	29	29	29	28	27	28	28	29	29	29	30 °C
	t	26	27	27	26	26	26	25	25	26	26	26	26	26 °C
	tn	22	22	22	22	21	22	21	22	22	22	22	23	21 °C
	clothing	t	t	t	t	t	t	t	t	t	t	t	t	–
	stress	ff	ff	ff	ff	ff	ff	ff	ff	ff	ff	ff	ff	–
	p	35	124	158	254	244	170	174	43	20	10	12	15	1259 mm

Quixeramobim – Ceará

Asha	127-15	Jan	Feb	Mar	Apr	May	Jun	Jul	Aug	Sep	Oct	Nov	Dec	Year
				latitude: 5°12's				longitude: 39°18'w				height: 198 m		
	t	29	28	28	27	27	26	27	28	28	29	29	29	28 °C
	clothing	t	t	t	t	t	t	t	t	t	t	t	t	–
	stress	ff	ff	ff	ff	ff	ff	ff	ff	ff	ff	ff	ff	–
	p	43	88	171	160	102	39	20	9	4	3	6	16	661 mm
0.2 mm	dp	11	10	14	12	10	8	5	3	1	0	2	4	80 days
	sunshine	232	198	202	206	230	233	266	296	283	294	262	257	2959 hours
	%	60	57	54	57	62	66	72	80	78	78	71	67	67 %

Barra do Corda – Maranhao

Awha	127-16	Jan	Feb	Mar	Apr	May	Jun	Jul	Aug	Sep	Oct	Nov	Dec	Year
				latitude: 5°30's				longitude: 45°16'w				height: 82 m		
	tx	34	33	33	34	34	33	34	36	37	37	37	36	37 °C
	t	26	26	26	26	25	25	24	26	28	28	27	26	26 °C
	tn	20	20	20	20	18	16	14	15	17	19	19	20	14 °C
	clothing	t	t	t	t	t	t	t	t	t	t	t	t	–
	stress	ff	ff	ff	ff	f	f	f	f	f	ff	ff	ff	–
	p	185	175	228	156	53	14	5	7	16	42	75	118	1074 mm
0.2 mm	dp	16	18	18	14	11	5	2	3	4	7	8	12	116 days
	sunshine	179	154	172	194	235	268	297	293	241	214	194	193	2634 hours
	%	47	44	46	54	64	76	81	79	66	56	52	50	60 %

Terezina – Maranhao

Awha	127-17	Jan	Feb	Mar	Apr	May	Jun	Jul	Aug	Sep	Oct	Nov	Dec	Year
				latitude: 5°03's				longitude: 42°49'w				height: 50 m		
	t	27	26	26	26	26	26	26	28	29	29	29	28	27 °C
	clothing	t	t	t	t	t	t	t	t	t	t	t	t	–
	stress	ff	ff	ff	ff	f	f	f	ff	ff	ff	ff	ff	–
	p	171	235	311	254	91	15	8	6	9	28	60	105	1293 mm
	sunshine	238	186	166	169	222	234	266	288	278	292	275	267	2883 hours
	%	62	54	44	47	60	66	72	78	77	77	74	69	65 %

Natal – Rio Grande do Norte

Asaa	127-18	Jan	Feb	Mar	Apr	May	Jun	Jul	Aug	Sep	Oct	Nov	Dec	Year
				latitude: 5°55's				longitude: 35°15'w				height: 49 m		
	tx	31	31	31	31	31	29	29	29	29	30	30	30	32 °C
	t	27	27	27	27	26	25	24	24	26	26	27	27	26 °C
	tn	22	21	21	22	21	19	18	18	19	21	22	22	18 °C
	clothing	t	t	t	t	t	t	t	t	t	t	t	t	–
	stress	ff	ff	ff	ff	ff	ff	ff	ff	ff	ff	ff	ff	–
	p	43	86	189	255	263	284	208	124	46	15	14	20	1547 mm
0.2 mm	dp	10	12	15	17	17	18	18	14	8	5	6	7	147 days
	sunshine	273	233	231	205	211	195	217	246	262	297	291	296	2957 hours
	%	71	67	61	57	57	55	59	67	72	78	78	77	67 %

Alto Tapajos – Pará

Amaa	127-19	Jan	Feb	Mar	Apr	May	Jun	Jul	Aug	Sep	Oct	Nov	Dec	Year
		latitude: 7°21's				longitude: 57°31'w				height: 125 m				
	t	25	25	26	26	26	25	24	26	26	26	26	25	25 °C
	clothing	t	t	t	t	t	t	t	t	t	t	t	t	–
	stress	ff	ff	ff	ff	ff	f	f	f	ff	ff	ff	ff	–
	p	408	375	435	285	128	26	11	33	138	238	215	329	2621 mm
	sunshine	100	97	106	129	188	237	264	258	173	157	127	108	1944 hours
	%	26	28	28	36	51	67	73	70	48	41	34	28	44 %

Olinda/Recife – Pernambuco

Asaa	127-20	Jan	Feb	Mar	Apr	May	Jun	Jul	Aug	Sep	Oct	Nov	Dec	Year
		latitude: 8°01's				longitude: 34°55'w				height: 10 m				
	tx	32	32	32	32	31	29	28	28	29	31	31	31	32 °C
	t	27	27	27	26	26	25	24	24	25	26	26	27	26 °C
	tn	22	22	22	22	21	21	20	19	21	22	22	22	19 °C
	clothing	t	t	t	t	t	t	t	t	t	t	t	t	–
	stress	ff	ff	ff	ff	ff	ff	ff	ff	ff	ff	ff	ff	–
	p	41	90	197	241	335	318	224	147	62	37	25	40	1757 mm
0.2 mm	dp	10	12	14	17	21	21	22	19	11	8	7	6	167 days
	sunshine	220	201	245	237	214	141	192	205	201	285	267	248	2655 hours
	%	57	58	65	66	59	40	53	56	56	75	71	64	60 %

Conceiçao do Araguaia – Pará

Awaa	127-21	Jan	Feb	Mar	Apr	May	Jun	Jul	Aug	Sep	Oct	Nov	Dec	Year
		latitude: 8°16's				longitude: 49°17'w				height: 157 m				
	tx	34	33	34	34	34	35	36	37	37	36	36	35	38 °C
	t	25	25	25	25	26	25	24	26	26	26	25	25	25 °C
	tn	19	20	19	19	18	16	14	15	18	19	19	19	14 °C
	clothing	t	t	t	t	t	t	t	t	t	t	t	t	–
	stress	ff	ff	ff	ff	ff	f	f	f	f	ff	ff	ff	–
	p	252	225	256	193	52	10	5	6	46	133	215	246	1639 mm
0.2 mm	dp	17	18	18	13	5	1	1	2	6	10	14	13	118 days

Porto Velho – Guaporé

Amaa	127-22	Jan	Feb	Mar	Apr	May	Jun	Jul	Aug	Sep	Oct	Nov	Dec	Year
		latitude: 8°46's				longitude: 63°55'w				height: 105 m				
	t	25	25	25	25	25	25	24	26	26	26	26	25	25 °C
	clothing	t	t	t	t	t	t	t	t	t	t	t	t	–
	stress	ff	ff	ff	ff	ff	f	f	f	ff	ff	ff	ff	–
	p	337	305	337	225	108	32	11	26	99	207	241	317	2245 mm

Porto Nacional – Goias

Awaa	127-23	Jan	Feb	Mar	Apr	May	Jun	Jul	Aug	Sep	Oct	Nov	Dec	Year
		latitude: 10°42's				longitude: 48°25'w				height: 238 m				
	t	25	25	25	26	26	25	25	26	28	27	26	25	26 °C
	clothing	t	t	t	t	t	t	t	t	t	t	t	t	–
	stress	ff	ff	ff	ff	f	f	f	f	ff	ff	ff	ff	–
	p	274	229	273	150	37	1	2	3	35	142	233	284	1663 mm
	sunshine	164	137	150	202	277	292	315	314	227	196	156	149	2579 hours
	%	42	39	40	57	77	85	88	86	63	51	41	38	58 %

Aracaju – Sergipe

latitude: 10°55′s longitude: 37°03′w height: 6 m

Asaa	127-24	Jan	Feb	Mar	Apr	May	Jun	Jul	Aug	Sep	Oct	Nov	Dec	Year
	t	26	27	27	26	25	24	24	24	24	25	26	26	25 °C
	clothing	t	t	t	t	t	t	t	t	t	t	t	t	–
	stress	ff	ff	ff	ff	ff	ff	ff	ff	ff	ff	ff	ff	–
	p	43	46	107	162	263	177	161	104	54	50	49	38	1254 mm
	sunshine	253	226	224	214	190	186	179	201	214	242	246	268	2648 hours
	%	64	64	59	60	53	54	50	55	59	63	65	68	60 %

Utiariti – Mato Grosso

latitude: 13°02′s longitude: 58°17′w height: 385 m

Awab	127-25	Jan	Feb	Mar	Apr	May	Jun	Jul	Aug	Sep	Oct	Nov	Dec	Year
	t	23	24	24	24	23	21	21	23	25	25	24	24	24 °C
	clothing	t	t	t	t	t	s	s	t	t	t	t	t	–
	stress	ff	ff	ff	ff	f	f	f	f	f	ff	ff	ff	–
	p	335	338	289	203	46	22	8	17	81	194	278	323	2134 mm

Caetité – Bahia

latitutde: 14°04′s longitude: 42°29′w height: 878 m

GAwab	127-26	Jan	Feb	Mar	Apr	May	Jun	Jul	Aug	Sep	Oct	Nov	Dec	Year
	tx	32	31	31	30	29	28	30	31	33	33	32	32	34 °C
	t	22	22	22	22	20	19	19	20	22	23	22	22	21 °C
	tn	16	16	16	16	13	12	12	12	13	14	15	16	11 °C
	clothing	s	s	s	s	s	s	s	s	s	t	s	s	–
	stress	f	f	f	f	o	o	o	o	o	f	f	f	–
	p	86	93	127	70	19	11	12	10	13	59	183	161	846 mm
0.2 mm	dp	13	8	10	9	7	7	5	5	4	7	15	14	103 days

Formosa – Goias

latitude: 15°32′s longitude: 47°20′w height: 906 m

GAwab	127-27	Jan	Feb	Mar	Apr	May	Jun	Jul	Aug	Sep	Oct	Nov	Dec	Year
	t	22	22	22	22	20	19	19	21	23	23	22	22	21 °C
	clothing	s	s	s	s	s	s	s	s	s	s	s	s	–
	stress	f	f	f	f	o				o	f	f	f	–
	p	252	204	227	93	17	3	5	3	30	127	255	343	1559 mm
	sunshine	180	159	187	222	270	280	298	303	236	201	143	145	2624 hours
	%	45	45	49	63	76	83	85	84	66	52	37	36	59 %

Cuiabá – Mato Grosso

latitude: 15°36′s longitude: 56°06′w height: 165 m

Awaa	127-28	Jan	Feb	Mar	Apr	May	Jun	Jul	Aug	Sep	Oct	Nov	Dec	Year
	tx	36	33	35	35	34	33	34	37	39	38	37	36	39 °C
	t	26	26	26	26	24	23	23	25	27	27	27	27	26 °C
	tn	21	19	19	17	12	8	7	9	11	16	17	18	5 °C
	clothing	t	t	t	t	t	t	t	t	t	t	t	t	–
	stress	ff	ff	ff	ff	f	f	f	f	f	ff	ff	ff	–
	p	216	198	232	115	52	14	6	12	39	130	165	194	1373 mm
0.2 mm	dp	26	23	25	21	15	10	5	2	8	17	21	25	198 days
	sunshine	167	151	175	201	232	236	261	249	199	195	187	171	2424 hours
	%	42	42	46	57	66	70	75	69	55	51	49	42	55 %

Santa Cruz (Meruri) – Mato Grosso

latitude: 15°43′s longitude: 52°43′w height: 416 m

Awab	127-29	Jan	Feb	Mar	Apr	May	Jun	Jul	Aug	Sep	Oct	Nov	Dec	Year
	t	24	24	24	24	22	20	20	22	24	25	24	24	23 °C
	clothing	t	t	t	t	s	s	s	s	t	t	t	t	–
	stress	ff	ff	ff	f	f	o	o	o	f	ff	ff	ff	–
	p	286	251	214	95	32	12	4	5	53	130	212	273	1567 mm

Goiania – Goias

latitude: 16°38′s longitude: 49°13′w height: 747 m

GAwab	127-30	Jan	Feb	Mar	Apr	May	Jun	Jul	Aug	Sep	Oct	Nov	Dec	Year
	t	23	23	23	22	20	18	18	21	23	23	23	23	22 °C
	clothing	t	t	t	s	s	s	s	s	t	t	t	t	–
	stress	f	f	f	o	o	o	o	o	f	f	f	f	–
	p	243	220	192	91	34	7	8	6	37	162	219	274	1493 mm
	sunshine	176	163	194	231	264	273	291	296	210	203	172	156	2529 hours
	%	44	46	51	66	75	82	84	83	58	53	44	38	59 %

Corumba – Mato Grosso

latitude: 19°00′s longitude: 57°39′w height: 145 m

Awab	127-31	Jan	Feb	Mar	Apr	May	Jun	Jul	Aug	Sep	Oct	Nov	Dec	Year
	tx	37	36	36	35	34	33	33	36	38	38	38	38	39 °C
	t	27	27	26	25	23	22	21	23	25	26	27	27	25 °C
	tn	20	19	18	17	12	7	9	10	11	15	17	18	5 °C
	clothing	t	t	t	t	t	s	s	t	t	t	t	t	–
	stress	ff	ff	ff	f	f	f	f	f	f	f	ff	ff	–
	p	170	158	119	71	62	35	28	20	52	85	118	145	1063 mm
0.2 mm	dp	12	10	10	7	5	4	2	3	5	7	8	11	84 days
	sunshine	212	188	216	218	216	194	227	241	210	227	229	236	2614 hours
	%	52	52	57	62	62	59	66	68	58	58	59	57	59 %

Araxá – Minas Gerais

latitude: 19°34′s longitude: 46°56′w height: 961 m

GAwbl	127-32	Jan	Feb	Mar	Apr	May	Jun	Jul	Aug	Sep	Oct	Nov	Dec	Year
	t	22	22	21	20	18	18	17	19	21	21	21	21	20 °C
	clothing	s	s	s	s	s	s	n	s	s	s	s	s	–
	stress	f	f	f	o	o	o		o	o	f	f	f	–
	p	318	261	234	96	43	21	10	14	46	153	226	370	1792 mm
	sunshine	169	161	183	219	253	251	269	274	212	182	171	132	2474 hours
	%	41	45	48	63	73	76	78	77	59	47	44	32	56 %

Belo Horizonte – Minas Gerais

latitude: 19°51′s longitude: 43°57′w height: 785 m

GAwab	127-33	Jan	Feb	Mar	Apr	May	Jun	Jul	Aug	Sep	Oct	Nov	Dec	Year
	t	23	23	22	21	19	18	18	19	21	22	22	22	21 °C
	clothing	t	t	s	s	s	s	s	s	s	s	s	s	–
	stress	f	f	f	f	o	o	o	o	o	o	o	o	–
	p	268	194	165	77	22	10	7	9	38	113	215	354	1472 mm
0.2 mm	dp	21	13	12	12	4	4	2	4	4	12	16	19	123 days
	sunshine	192	174	205	219	240	249	257	263	199	182	171	142	2493 hours
	%	47	48	54	63	69	76	75	74	55	47	44	34	56 %

Campo Grande – Mato Grosso

latitude: 20°28′s longitude: 54°40′w height: 552 m

GAwab	127-34	Jan	Feb	Mar	Apr	May	Jun	Jul	Aug	Sep	Oct	Nov	Dec	Year
	t	24	24	24	22	20	19	19	21	23	24	24	24	22 °C
	clothing	t	t	t	s	s	s	s	s	t	t	t	t	–
	stress	ff	ff	ff	f	f	f	f	f	f	ff	ff	ff	–
	p	229	199	140	101	81	50	36	29	62	162	157	191	1437 mm

Tres Lagos – Mato Grosso

latitude: 20°47′s longitude: 51°42′w height: 314 m

Awab	127-35	Jan	Feb	Mar	Apr	May	Jun	Jul	Aug	Sep	Oct	Nov	Dec	Year
	t	26	26	25	23	20	19	19	21	23	25	25	26	23 °C
	clothing	t	t	t	t	s	s	s	s	t	t	t	t	–
	stress	f	f	f	f	o	o	o	o	f	f	f	f	–
	p	238	202	139	92	63	41	24	19	47	109	128	205	1308 mm
0.2 mm	dp	15	13	10	6	4	4	3	4	7	8	9	15	98 days
	sunshine	195	186	211	239	252	212	256	260	200	196	224	233	2664 hours
	%	47	51	56	68	73	65	75	74	56	50	57	56	60 %

Rio de Janeiro – Rio de Janeiro

latitude: 22°54′s longitude: 43°10′w height: 27 m

Awab	127-36	Jan	Feb	Mar	Apr	May	Jun	Jul	Aug	Sep	Oct	Nov	Dec	Year
	tx	34	34	33	32	31	29	29	31	32	32	33	34	36 °C
	t	26	26	26	24	22	22	21	21	22	22	23	24	23 °C
	tn	19	21	20	18	16	15	14	14	16	16	17	18	13 °C
	clothing	t	t	t	t	s	s	s	s	s	s	t	t	–
	stress	ff	ff	ff	ff	f	o	o	o	o	f	f	ff	–
	p	137	137	143	116	73	43	43	43	53	74	97	127	1086 mm
0.2 mm	dp	13	11	12	10	10	7	7	7	11	13	13	14	128 days
	sunshine	215	195	210	190	192	189	201	204	153	157	170	174	2250 hours
	%	52	54	55	55	56	59	60	58	43	40	43	41	51 %

São Paulo – São Paulo

latitude: 23°37′s longitude: 46°39′w height: 807 m

GAwbl	127-37	Jan	Feb	Mar	Apr	May	Jun	Jul	Aug	Sep	Oct	Nov	Dec	Year
	t	22	21	21	20	17	16	15	16	17	19	19	21	19 °C
	clothing	s	s	s	s	s	n	n	n	s	s	s	s	–
	stress	f	f	f	o						o	o	o	–
	p	248	289	152	62	48	41	36	29	48	113	134	186	1386 mm
	sunshine	155	142	153	179	170	181	185	197	141	122	147	176	1947 hours
	%	37	39	40	52	50	56	55	56	39	31	37	42	44 %

3. *Climatic area:* eastern Ceara, parts of Rio Grande do Norte, central Piau, western Pernambuco, northwestern Bahia

3.1. Steppe Climate BS and Steppe Mountain Climate GBS

3.2. At low levels permanently very hot.

3.3. In the morning often dry and sunny. High sun period: frequent showers in the afternoon, now and then rainy days. Low sun period: fewer showers.

3.4. Wind: occasional gales. At times years of drought.

Remanso – Bahia

latitude: 9°41's longitude: 42°04'w height: 411 m

BSha	127-38	Jan	Feb	Mar	Apr	May	Jun	Jul	Aug	Sep	Oct	Nov	Dec	Year
	t	27	28	27	27	26	26	26	26	27	28	28	28	27 °C
	clothing	t	t	t	t	t	t	t	t	t	t	t	t	–
	stress	ff	ff	ff	ff	f	f	f	f	f	ff	ff	ff	–
	p	92	65	108	35	10	2	1	0	4	10	78	93	498 mm

4. *Climatic area:* Southernmost part of Sao Paulo, Parana, Santa Catarina, Rio Grande do Sul

4.1. Subtropical Rain Climate Cr and Subtropical Rain Mountain Climate GCr

4.2. Summer hot, in the south warm, winter warm, in the south mild.

4.3. Often dry and sunny in the morning, frequent showers in the afternoon; now and then rainy days.

4.4. Wind: occasional gales. From May to August in the south, invasions of cold air common.

Iguape – São Paulo

latitude: 24°43's longitude: 47°33'w height: 4 m

Cral	127-39	Jan	Feb	Mar	Apr	May	Jun	Jul	Aug	Sep	Oct	Nov	Dec	Year
	t	25	25	24	22	19	18	17	18	19	21	22	24	21 °C
	clothing	t	t	t	s	s	s	n	s	s	s	s	t	–
	stress	f	f	f	f	o	o		o	o	f	f	f	–
	p	246	242	233	150	140	98	99	75	103	137	98	138	1757 mm

Curitiba – Parana

latitude: 25°31's longitude: 49°10'w height: 908 m

GCrbl	127-40	Jan	Feb	Mar	Apr	May	Jun	Jul	Aug	Sep	Oct	Nov	Dec	Year
	t	20	20	19	17	14	13	13	14	15	16	17	19	16 °C
	clothing	s	s	s	n	n	n	n	n	n	n	n	s	–
	stress	f	f	o	o							o	o	–
	p	191	163	124	78	85	89	81	83	119	130	105	147	1395 mm
0.2 mm	dp	20	17	18	15	14	13	11	11	15	14	15	16	179 days
	sunshine	177	156	175	164	174	163	186	185	141	148	169	187	2025 hours
	%	42	43	46	47	52	51	56	53	39	38	42	44	46 %

Foz do Iguaçu – Santa Catarina

latitude: 25°31's longitude: 54°35'w height: 180 m

Cral	127-41	Jan	Feb	Mar	Apr	May	Jun	Jul	Aug	Sep	Oct	Nov	Dec	Year
	t	25	24	23	20	17	16	14	16	18	20	22	24	20 °C
	clothing	t	t	t	s	n	n	n	n	s	s	s	t	–
	stress	f	f	f	o						o	f	f	–
	p	152	136	166	142	154	157	130	81	132	146	124	140	1660 mm

Brusque – Santa Catarina

latitude: 27°06's longitude: 48°56'w height: 24 m

Cral	127-42	Jan	Feb	Mar	Apr	May	Jun	Jul	Aug	Sep	Oct	Nov	Dec	Year
	t	25	25	24	20	18	16	15	17	18	20	22	24	20 °C
	clothing	t	t	t	s	s	n	n	n	s	s	s	t	–
	stress	f	f	f	o	o				o	o	f	f	–
	p	165	190	154	128	102	82	97	104	151	166	124	152	1615 mm

Alegrete – Rio Grande do Sul

latitude: 29°47′s longitude: 55°47′w height: 103 m

Cral	127-43	Jan	Feb	Mar	Apr	May	Jun	Jul	Aug	Sep	Oct	Nov	Dec	Year
	t	25	24	22	18	15	13	13	14	16	18	21	23	19 °C
	clothing	t	t	s	s	n	n	n	n	n	s	s	t	–
	stress	f	f	f	o						o	f	f	–
	p	145	109	122	165	149	129	109	108	128	191	133	123	1610 mm
	sunshine	283	252	246	214	189	147	181	204	197	230	275	299	2717 hours
	%	66	68	64	63	57	48	56	60	55	58	67	69	61 %

Porto Alegre – Rio Grande do Sul

latitude: 30°00′s longitude: 51°11′w height: 3 m

Cral	127-44	Jan	Feb	Mar	Apr	May	Jun	Jul	Aug	Sep	Oct	Nov	Dec	Year
	tx	37	36	35	32	29	28	28	29	31	32	34	37	38 °C
	t	25	24	23	20	17	15	14	15	16	19	21	23	20 °C
	tn	14	15	13	9	4	2	2	3	5	7	10	12	0 °C
	clothing	t	t	t	s	n	n	n	n	n	s	s	t	–
	stress	f	f	f	o						o	o	f	–
	p	119	104	89	103	115	140	128	114	124	115	75	88	1314 mm
0.2 mm	dp	8	9	8	9	9	10	9	11	10	10	8	7	105 days
	sunshine	261	214	228	192	174	143	165	180	150	208	246	282	2443 hours
	%	61	58	60	56	53	47	51	53	42	52	60	65	55 %

Santa Victoria do Palmar – Rio Grande do Sul

latitude: 33°31′s longitude: 53°22′w height: 6 m

Crbl	127-45	Jan	Feb	Mar	Apr	May	Jun	Jul	Aug	Sep	Oct	Nov	Dec	Year
	t	22	22	21	17	14	12	11	12	13	16	18	20	17 °C
	clothing	s	s	s	n	n	n	n	n	n	n	s	s	–
	stress	f	f	o								o	o	–
	p	92	91	128	124	106	119	97	109	119	101	70	80	1236 mm
	sunshine	268	233	221	197	167	137	147	172	175	213	249	274	2452 hours
	%	61	62	58	58	52	46	47	51	49	53	60	62	55 %

4.5.127.2. Predominant general weather situations are described in 4.5.0.1. under Brazil. A hygrothermal diagram of Rio de Janeiro is to be found in 4.5.0.2.

Wind names: Abrolhos squalls are rainy squalls in eastern Brazil in winter (May to August). Marajos are strong squalls from the northeast in the Amazon basin during the rainy season.

Minuano is a cold and often strong wind on the eastern coast of Brazil, particularly of Rio Grande do Sul, blowing from the west and southwest.

Reboyo is a storm from the southwest on the eastern coast of Brazil, blowing during the rainy season from December to June.

Vento de cima is a cool, often rainy, westerly wind in the upper Amazon basin.

4.5.128. Ecuador

4.5.128.1. Climatic descriptions and tables

1. Climatic area: North and east of the country

1.1. Tropical Rain Climate Ar and Tropical Rain Mountain Climate GAr and Tropical Rain High Mountain Climate HAr

1.2. Hot throughout the year.

1.3. In the morning at times dry and sunny or cloudy; in the afternoon very often showers. Frequent rainy days too.

1.4. Wind: occasional gales, squalls in thunderstorms. Fog; often at higher levels. Snow cover: at about 4800 m, sometimes lower.

San Lorenzo

latitude: 1°16′n longitude: 78°50′w height: 6 m

Araa	128-1	Jan	Feb	Mar	Apr	May	Jun	Jul	Aug	Sep	Oct	Nov	Dec	Year
	t	26	26	26	26	26	26	26	26	26	26	26	26	26 °C
	clothing	t	t	t	t	t	t	t	t	t	t	t	t	–
	stress	ff	ff	ff	ff	ff	ff	ff	ff	ff	ff	ff	ff	–
	p	240	247	269	301	272	245	179	86	89	102	92	136	2257 mm
0.1 mm	dp	20	17	20	20	23	21	21	18	16	14	13	12	215 days
	sunshine	96	116	108	126	93	78	65	93	84	115	90	81	1145 hours
	%	26	34	29	35	25	21	17	25	23	31	25	22	26 %

2. Climatic area: West of the country, except small coastal part

2.1. Tropical Summer Rain Climate Aw and Tropical Summer Rain Mountain Climate GAw and Tropical Summer Rain High Mountain Climate HAw

2.2. In the lowlands, hot throughout the year.

2.3. Generally dry and sunny or cloudy in the morning. High sun period: often showers in the afternoon, at times rainy days. Low sun period: fewer showers.

2.4. Wind: occasional gales, squalls in thunderstorms. Fog: often at higher levels. Snow cover: at about 4800 m.

Esmeraldas

latitude: 0°59′n longitude: 79°39′w height: 6 m

Asaa	128-2	Jan	Feb	Mar	Apr	May	Jun	Jul	Aug	Sep	Oct	Nov	Dec	Year
	t	25	26	26	26	26	25	25	25	25	25	25	25	25 °C
	clothing	t	t	t	t	t	t	t	t	t	t	t	t	–
	stress	ff	ff	ff	ff	ff	ff	ff	ff	ff	ff	ff	ff	–
	p	110	144	134	124	80	73	58	34	29	17	13	44	861 mm
	dp	16	15	14	14	13	12	11	8	7	5	4	7	126 days

Quito

latitude: 0°08′s longitude: 78°29′w height: 2812 m

HAwll	128-3	Jan	Feb	Mar	Apr	May	Jun	Jul	Aug	Sep	Oct	Nov	Dec	Year
	tx	25	24	25	24	24	24	24	26	27	26	25	25	27 °C
	t	13	13	13	13	13	13	13	13	13	13	13	13	13 °C
	tn	4	5	6	6	5	4	3	4	3	4	3	4	2 °C
	clothing	n	n	n	n	n	n	n	n	n	n	n	n	–
	stress													–
	p	119	131	154	185	130	54	20	25	81	134	96	104	1233 mm
	dp	13	14	17	19	16	9	4	4	9	16	12	12	145 days
	sunshine	166	139	137	140	167	188	220	222	256	168	171	175	2149 hours
	%	44	41	36	39	44	52	59	59	70	45	47	47	49 %

Guayaquil

						latitude: 2°12′s		longitude: 79°53′w			height: 6 m			
Awaa	128-4	Jan	Feb	Mar	Apr	May	Jun	Jul	Aug	Sep	Oct	Nov	Dec	Year
	tx	34	33	33	33	34	33	32	32	33	33	34	35	36 °C
	t	26	26	26	26	26	24	24	23	24	24	25	25	25 °C
	tn	20	20	19	21	18	18	17	17	17	18	18	19	16 °C
	clothing	t	t	t	t	t	t	t	t	t	t	t	t	–
	stress	ff	ff	ff	ff	ff	ff	f	f	f	ff	ff	ff	–
	p	212	289	292	207	54	11	4	0	0	1	2	28	1100 mm
	dp	15	17	17	12	4	1	1	0	0	0	1	4	72 days
	sunshine	101	102	141	150	168	123	128	132	145	135	119	136	1580 hours
	%	27	30	37	41	45	34	34	35	40	36	32	36	36 %

3. *Climatic area:* Small coastal parts of the Gulf of Guayaquil and Manta Bay, Galapagos Archipelago
3.1. Steppe Climate BS and Desert Climate BW, coastal areas: Marine Desert Climate BM
3.2. Permanently hot; islands hot in summer, warm in winter.
3.3. After December solstice, frequent showers or rain on the coast, particularly in the afternoon; otherwise mostly dry and cloudy or sunny.
3.4. Wind: occasional gales, squalls in thunderstorms.

Manta

						latitude: 0°57′s		longitude: 80°42′w			height: 6 m			
BMaa	128-5	Jan	Feb	Mar	Apr	May	Jun	Jul	Aug	Sep	Oct	Nov	Dec	Year
	t	26	26	26	26	25	24	23	23	23	24	24	25	25 °C
	clothing	t	t	t	t	t	t	t	t	t	t	t	t	–
	stress	ff	ff	ff	ff	ff	f	f	f	f	f	ff	ff	–
	p	36	48	71	27	6	3	3	0	2	0	0	4	200 mm
0.1 mm	dp	9	9	11	7	2	2	2	1	1	1	1	2	47 days

Puerto Baquerizo/San Cristobal – Galapagos Islands

						latitude: 0°54′s		longitude: 89°37′w			height: 6 m			
BSab	128-6	Jan	Feb	Mar	Apr	May	Jun	Jul	Aug	Sep	Oct	Nov	Dec	Year
	t	25	26	26	26	25	24	23	22	22	22	23	24	24 °C
	clothing	t	t	t	t	t	t	t	s	s	s	t	t	–
	stress	ff	ff	ff	ff	ff	f	f	f	f	f	f	f	–
	p	38	155	124	106	28	4	12	7	5	7	6	14	507 mm
0.1 mm	dp	11	11	10	9	6	5	12	14	14	12	9	9	122 days
	sunshine	177	234	226	237	239	231	208	189	159	161	159	177	2397 hours
	%	47	68	60	65	64	64	56	50	44	43	44	47	54 %

4.5.128.2. Predominant general weather situations are described in 4.5.0.1. under Ecuador. A hygrothermal diagram of Quito is to be found in 4.5.0.2.
Wind name: Nevado is a cold fall-wind occurring in the Andes, blowing from the snow and ice-covered slopes to the elevated plateau of Quito.

4.5.129. Peru

4.5.129.1. Climatic descriptions and tables

1. Climatic area: Coastal district and adjacent slopes of the Andes

1.1. Desert Climate BW and Desert Mountain Climate GBW, on the coast Marine Desert Climate BM

1.2. At low levels: in the north summer hot, winter warm, in the south summer warm, winter mild.

1.3. November – April: often dry and cloudy, in the afternoon sunny, at times scattered showers. May – October: especially in the southern part drizzle at night, in the afternoon at times cloudy or sunny.

1.4. Wind: occasional gales and dust-storms. Fog: or low stratus during the night and morning often May – October.

El Alto

latitude: 4°15′s longitude: 81°14′w height: 270 m

BMab	129-1	Jan	Feb	Mar	Apr	May	Jun	Jul	Aug	Sep	Oct	Nov	Dec	Year
	t	23	24	25	24	22	20	19	19	19	19	20	22	21 °C
	clothing	t	t	t	t	s	s	s	s	s	s	s	s	–
	stress	f	f	f	f	f	o	o	o	o	o	o	f	–
	p	5	9	12	11	2	0	0	0	0	0	0	0	41 mm
0.2 mm	dp	3	5	3	3	1	0	0	0	0	0	0	0	18 days
	sunshine	170	153	164	144	155	129	133	130	126	130	132	164	1731 hours
	%	44	44	44	40	42	36	36	35	35	34	36	43	39 %

Lambayeque

latitude: 6°42′s longitude: 79°54′w height: 18 m

BMab	129-2	Jan	Feb	Mar	Apr	May	Jun	Jul	Aug	Sep	Oct	Nov	Dec	Year
	t	23	24	24	22	21	19	18	18	18	19	20	21	20 °C
	clothing	t	t	t	s	s	s	s	s	s	s	s	s	–
	stress	f	f	f	f	f	o	o	o	o	o	o	f	–
	p	2	3	1	4	1	0	0	0	0	1	2	1	15 mm
0.2 mm	dp	2	1	1	2	2	1	2	0	0	2	1	1	16 days
	sunshine	226	198	205	213	214	159	149	167	186	195	204	223	2339 hours
	%	58	57	54	59	58	45	41	45	51	51	55	57	53 %

Lima

latitude: 12°06′s longitude: 77°02′w height: 155 m

BMbl	129-3	Jan	Feb	Mar	Apr	May	Jun	Jul	Aug	Sep	Oct	Nov	Dec	Year
	tx	31	31	31	29	27	24	23	22	23	24	26	28	32 °C
	t	21	22	22	20	18	16	15	15	15	16	17	19	18 °C
	tn	17	17	17	16	13	12	11	12	12	13	13	15	11 °C
	clothing	s	s	s	s	s	n	n	n	n	n	n	s	–
	stress	f	f	f	o	o							o	–
	p	1	0	1	0	6	4	6	7	6	2	1	1	36 mm
0.2 mm	dp	4	4	8	2	12	14	19	22	22	16	8	6	137 days
	sunshine	195	192	214	201	124	42	34	31	33	78	123	155	1422 hours
	%	49	54	57	57	35	12	10	9	9	20	32	39	32 %

Arequipa

GBWl1	129-4	latitude: 16°24′s				longitude: 71°32′w					height: 2332 m			
			Jan	Feb	Mar	Apr	May	Jun	Jul	Aug	Sep	Oct	Nov	Dec	Year
		t	15	15	15	15	14	14	13	14	16	16	16	16	15 °C
		clothing stress	n	n	n	n	n	n	n	n	n	n	n	n	–
		p	22	45	49	1	1	0	0	0	0	0	0	1	119 mm
0.2 mm		dp	5	25	4	1	0	0	0	0	0	0	0	2	39 days
		sunshine	223	189	245	294	288	291	291	310	297	304	309	291	3333 hours
		%	55	53	65	83	82	87	83	86	82	79	80	82	75 %

2. Climatic area: Andes, except some central and southern parts

2.1. Tropical Summer Rain Mountain Climate GAw and Tropical Summer Rain High Mountain Climate HAw

2.2. Mild at about 3000 m.

2.3. High sun period: often dry and sunny in the morning, in the afternoon frequent showers, at times rainy days. Low sun period: generally dry and sunny, now and then showers.

2.4. Wind: frequent gales. Fog: often at higher levels during rainy season. Snow cover: at about 4800 m.

Cajamarca

latitude: 7°09′s longitude: 78°30′w height: 2642 m

HAwl1	129-5	Jan	Feb	Mar	Apr	May	Jun	Jul	Aug	Sep	Oct	Nov	Dec	Year
	tx	24	24	24	23	24	23	23	24	24	25	24	25	26 °C
	t	15	15	15	15	14	13	13	14	14	15	15	15	14 °C
	tn	5	4	5	4	2	2	1	2	4	4	1	2	0 °C
	clothing stress	n	n	n	n	n	n	n	n	n	n	n	n	–
	p	91	107	117	86	43	13	5	8	58	58	48	81	716 mm
	dp	13	17	17	14	9	4	2	2	9	9	8	11	115 days

Jauja

latitude: 11°43′s longitude: 75°28′w height: 3389 m

HAwl1	129-6	Jan	Feb	Mar	Apr	May	Jun	Jul	Aug	Sep	Oct	Nov	Dec	Year
	tx	18	19	19	19	19	20	18	19	21	21	22	21	23 °C
	t	13	13	13	12	12	12	11	12	13	13	14	13	12 °C
	tn	7	8	8	6	4	3	3	4	5	6	6	6	2 °C
	clothing stress	n	n	n	n	n	n	n	n	n	n	n	n	–
	p	114	112	81	38	15	3	2	3	33	43	53	86	584 mm
	dp	18	16	17	9	5	1	0	1	6	9	10	15	107 days

Huancayo

latitude: 12°02′s longitude: 75°20′w height: 3350 m

HAwl1	129-7	Jan	Feb	Mar	Apr	May	Jun	Jul	Aug	Sep	Oct	Nov	Dec	Year
	t	12	12	12	12	11	10	10	11	12	12	12	12	12 °C
	clothing stress	n	n	n	n	n	n	n	n	n	n	n	n	–
	p	134	155	102	51	22	13	5	9	48	67	80	90	776 mm
0.2 mm	dp	22	24	20	15	9	2	2	4	12	14	14	20	158 days
	sunshine	167	136	174	189	220	240	257	245	192	214	201	198	2433 hours
	%	42	39	46	53	61	70	72	67	53	56	53	50	55 %

Cuzco

HAwll	129-8	Jan	Feb	Mar	Apr	May	Jun	Jul	Aug	Sep	Oct	Nov	Dec	Year
		latitude: 13°11′s						longitude: 71°57′w				height: 3365 m		
	tx	24	24	24	25	24	23	23	24	25	26	26	25	28 °C
	t	14	14	14	13	11	11	10	11	13	14	14	14	13 °C
	tn	4	4	3	1	-2	-2	-5	-2	1	2	2	3	-5 °C
	clothing stress	n	n	n	n	n	n	n	n	n	n	n	n	–
	p	163	150	109	51	15	5	5	10	25	66	76	137	812 mm
0.2 mm	dp	25	21	17	16	7	1	2	3	9	12	18	23	154 days
	sunshine	143	121	170	210	239	228	257	236	195	198	195	158	2347 hours
	%	36	34	45	59	67	67	73	65	54	52	51	39	53 %

Imata

HAwkk	129-9	Jan	Feb	Mar	Apr	May	Jun	Jul	Aug	Sep	Oct	Nov	Dec	Year
		latitude: 15°49′s						longitude: 71°04′w				height: 4405 m		
	t	5	6	5	4	2	0	0	0	2	3	4	5	3 °C
	clothing stress	n	n	n	w	w	w	w	w	w	w	w	n	–
	p	116	139	93	20	5	3	1	3	13	9	29	83	515 mm

Puno

HAwlk	129-10	Jan	Feb	Mar	Apr	May	Jun	Jul	Aug	Sep	Oct	Nov	Dec	Year
		latitude: 15°50′s						longitude: 70°01′w				height: 3852 m		
	t	9	9	9	7	7	6	6	7	8	9	10	10	8 °C
	clothing stress	n	n	n	n	n	n	n	n	n	n	n	n	–
	p	139	132	105	36	10	3	3	6	24	36	28	83	605 mm

3. Climatic area: some central and southern parts of the Andes

3.1. Steppe Mountain Climate GBS and Steppe High Mountain Climate HBS
NB In some restricted areas there may be Desert Mountain Climate GBW and Desert High Mountain Climate HBW.

3.2. At 3500–4000 m, summer mild, winter cool.

3.3. Generally dry and sunny or cloudy. High sun period: frequent showers in the afternoon, the odd a rainy day. Low sun period: scattered showers.

3.4. Wind: gales frequent, especially coupled with invasions of cold air. Fog: frequent above 3000 m.

Paucarany

HBSko	129-11	Jan	Feb	Mar	Apr	May	Jun	Jul	Aug	Sep	Oct	Nov	Dec	Year
		latitude: 17°30′s						longitude: 69°46′w				height: 4451 m		
	t	4	4	4	2	0	-2	-1	0	1	2	2	3	2 °C
	clothing stress	w	w	w	w	w	w	w	w	w	w	w	w	–
	p	60	79	60	4	0	1	0	0	2	2	19	58	286 mm

4. Climatic area: extensive plains in northeast, together with eastern slopes of Andes
4.1. Tropical Rain Climate Ar and Tropical Rain Mountain Climate GAr
4.2. At low levels permanently hot.
4.3. Occasionally dry and sunny in the morning, often showers in the afternoon, at times rainy days.
4.4. Wind: occasional gales, squalls in thunderstorms.

Iquitos

		latitude: 3°45′s				longitude: 73°12′w				height: 117 m				
Araa	129-12	Jan	Feb	Mar	Apr	May	Jun	Jul	Aug	Sep	Oct	Nov	Dec	Year
	t	26	26	26	25	26	25	25	25	26	26	27	26	26 °C
	clothing	t	t	t	t	t	t	t	t	t	t	t	t	–
	stress	ff	ff	ff	ff	ff	ff	ff	ff	ff	ff	ff	ff	–
	p	251	383	455	294	271	103	141	200	177	214	220	190	2959 mm
0.1 mm	dp	19	19	22	20	20	19	17	16	16	16	18	20	222 days

Tingo Maria

		latitude: 9°08′s				longitude: 75°57′w				height: 641 m				
GAraa	129-13	Jan	Feb	Mar	Apr	May	Jun	Jul	Aug	Sep	Oct	Nov	Dec	Year
	t	23	23	24	24	23	23	23	23	23	24	24	24	23 °C
	clothing	t	t	t	t	t	t	t	t	t	t	t	t	–
	stress	f	f	f	f	f	f	f	f	f	f	f	f	–
	p	417	352	414	278	241	126	139	113	168	293	303	348	3193 mm
0.2 mm	dp	20	19	17	15	15	10	12	9	12	16	15	18	178 days
	sunshine	112	102	115	129	161	159	180	208	168	164	138	121	1756 hours
	%	29	29	30	36	44	46	50	57	46	43	37	31	40 %

4.5.129.2. Predominant general weather situations are described in 4.5.0.1. under Peru. A hygrothermal diagram of Lima is to be found in 4.5.0.2.
Wind name: Yalka is a gusty fall-wind blowing in the passes of the Andes in northern Peru.

4.5.130. Bolivia

4.5.130.1. Climatic descriptions and tables

1. Climatic area: Departments of Oruro, Potosi, Tarija, West of Chuquisaca
1.1. Steppe Mountain Climate GBS and Steppe High Mountain Climate HBS
NB In some restricted areas there may be Desert Mountain Climate GBW and Desert High Mountain Climate HBW.
1.2. At 3500–4000 m, summer mild, winter cool.
1.3. Generally dry and sunny or cloudy. High sun period: frequent showers in the afternoon, the odd a rainy day. Low sun period: scattered showers.
1.4. Wind: gales frequent, especially coupled with invasions of cold air ("suaroz"). Fog: frequent above 3000 m.

Cochabamba

latitude: 17°23′s longitude: 66°10′w height: 2570 m

HBSbl	130-1	Jan	Feb	Mar	Apr	May	Jun	Jul	Aug	Sep	Oct	Nov	Dec	Year
	t	18	18	18	16	14	12	12	14	16	18	19	18	16 °C
	clothing	s	s	s	n	n	n	n	n	n	s	s	s	–
	stress	o	o	o							o	o	o	–
	p	124	98	53	16	4	1	4	9	10	26	50	89	484 mm
0.1 mm	dp	14	15	10	4	1	1	1	1	4	4	8	14	77 days

Charana

latitude: 17°36′s longitude: 69°28′w height: 4059 m

HBSlk	130-2	Jan	Feb	Mar	Apr	May	Jun	Jul	Aug	Sep	Oct	Nov	Dec	Year
	t	10	10	10	8	6	3	3	5	7	9	10	10	8 °C
	clothing	n	n	n	n	n	w	w	n	n	n	n	n	–
	stress													–
	p	73	71	53	5	1	4	1	8	5	3	13	47	284 mm

Oruro

latitude: 17°58′s longitude: 67°07′w height: 3706 m

HBSlk	130-3	Jan	Feb	Mar	Apr	May	Jun	Jul	Aug	Sep	Oct	Nov	Dec	Year
	t	12	12	12	10	7	4	4	6	9	11	12	13	10 °C
	clothing	n	n	n	n	n	w	w	n	n	n	n	n	–
	stress													–
	p	86	69	34	9	3	1	1	7	13	16	20	36	295 mm

2. *Climatic area:* Departments of Pando, Beni (except southern part), La Paz (except southwestern part), Cochabamba (except Yunga region), Santa Cruz (except Llanos de Chiquitos)

2.1. Tropical Summer Rain Climate Aw and Tropical Summer Rain Mountain Climate GAw and Tropical Summer Rain High Mountain Climate HAw

2.2. In the lowlands: in the north, hot throughout the year; in the south, summer hot, winter warm.

2.3. High sun period: often showers in the afternoon, at times rainy days; low sun period: generally dry and sunny or cloudy, now and then showers.

2.4. Wind: frequent gales, especially coupled with invasions of cold air ("suaroz"). Fog: frequent above 3000 m.

Riberalta

latitude: 11°00′s longitude: 66°05′w height: 172 m

Awha	130-4	Jan	Feb	Mar	Apr	May	Jun	Jul	Aug	Sep	Oct	Nov	Dec	Year
	t	26	26	27	26	26	25	25	27	28	27	27	27	26 °C
	clothing	t	t	t	t	t	t	t	t	t	t	t	t	–
	stress	ff	ff	ff	ff	ff	f	f	f	f	f	ff	ff	–
	p	299	245	282	145	67	28	8	21	60	136	161	272	1724 mm
0.1 mm	dp	17	17	18	13	7	3	2	2	5	10	12	16	122 days

Cobija

Awab	130-5	Jan	Feb	Mar	Apr	May	Jun	Jul	Aug	Sep	Oct	Nov	Dec	Year
		latitude: 11°04's				longitude: 68°44'w				height: 260 m				
	t	26	25	25	24	24	23	22	24	26	26	26	26	25 °C
	clothing	t	t	t	t	t	t	s	t	t	t	t	t	–
	stress	ff	ff	ff	ff	f	f	f	f	f	ff	ff	ff	–
	p	236	259	226	179	85	18	18	41	87	162	194	232	1737 mm
0.1 mm	dp	17	13	13	11	6	3	2	3	6	9	11	11	105 days

San Joaquín

Awha	130-6	Jan	Feb	Mar	Apr	May	Jun	Jul	Aug	Sep	Oct	Nov	Dec	Year
		latitude: 13°04's				longitude: 64°48'w				height: 200 m				
	t	26	24	27	26	25	25	25	26	28	28	27	27	26 °C
	clothing	t	t	t	t	t	t	t	t	t	t	t	t	–
	stress	ff	ff	ff	ff	f	f	f	f	ff	ff	ff	ff	–
	p	300	289	180	101	51	20	16	36	60	136	167	242	1598 mm

Concepción

Awab	130-7	Jan	Feb	Mar	Apr	May	Jun	Jul	Aug	Sep	Oct	Nov	Dec	Year
		latitude: 16°15's				longitude: 62°03'w				height: 490 m				
	tx	33	33	33	33	32	31	32	33	37	36	37	33	37 °C
	t	25	26	25	23	22	21	21	23	26	26	26	26	24 °C
	tn	16	15	13	12	7	4	6	7	9	12	15	13	4 °C
	clothing	t	t	t	t	s	s	s	t	t	t	t	t	–
	stress	ff	ff	ff	f	f	f	f	f	f	f	ff	ff	–
	p	213	159	145	66	63	41	38	37	49	93	153	155	1212 mm

San Ignacio de Velasco

Awab	130-8	Jan	Feb	Mar	Apr	May	Jun	Jul	Aug	Sep	Oct	Nov	Dec	Year
		latitude: 16°22's				longitude: 60°59'w				height: 370 m				
	t	25	25	25	23	21	20	20	22	25	26	26	26	24 °C
	clothing	t	t	t	t	s	s	s	s	t	t	t	t	–
	stress	ff	ff	ff	f	f	f	f	f	f	ff	ff	ff	–
	p	204	172	178	73	57	41	34	30	51	95	140	156	1231 mm

El Alto – La Paz Airport (tx and tn La Paz)

HAwlk	130-9	Jan	Feb	Mar	Apr	May	Jun	Jul	Aug	Sep	Oct	Nov	Dec	Year
		latitude: 16°30's				longitude: 68°10'w				height: 4103 m				
	tx	22	22	22	22	21	19	19	21	22	23	23	23	24 °C
	t	9	9	9	9	8	7	8	8	9	10	10	10	9 °C
	tn	4	4	4	3	0	-1	-2	-1	1	2	3	4	-2 °C
	clothing	n	n	n	n	n	n	n	n	n	n	n	n	–
	stress													–
	p	139	108	56	22	9	5	4	17	34	37	48	76	555 mm
0.1 mm	dp	19	16	14	7	3	1	2	3	9	9	9	15	107 days
	sunshine	183	153	149	165	223	240	236	217	189	180	171	186	2292 hours
	%	45	43	39	47	63	72	68	61	52	47	44	46	52 %

San José

latitude: 17°51′s longitude: 60°47′w height: 297 m

Awab	130-10	Jan	Feb	Mar	Apr	May	Jun	Jul	Aug	Sep	Oct	Nov	Dec	Year
	t	27	27	27	25	22	21	22	24	26	27	27	27	25 °C
	clothing	t	t	t	t	s	s	s	t	t	t	t	t	–
	stress	ff	ff	ff	f	f	o	o	f	f	ff	ff	ff	–
	p	176	138	107	37	37	51	23	10	29	43	119	128	898 mm

Puerto Suárez

latitude: 18°57′s longitude: 57°52′w height: 154 m

Awhb	130-11	Jan	Feb	Mar	Apr	May	Jun	Jul	Aug	Sep	Oct	Nov	Dec	Year
	t	28	28	28	25	23	22	21	23	26	28	28	29	26 °C
	clothing	t	t	t	t	t	s	s	t	t	t	t	t	–
	stress	ff	ff	ff	f	f	f	o	f	f	ff	ff	ff	–
	p	170	192	97	80	55	44	31	28	58	86	122	137	1100 mm

Camiri

latitude: 20°06′s longitude: 63°33′w height: 914 m

GAwal	130-12	Jan	Feb	Mar	Apr	May	Jun	Jul	Aug	Sep	Oct	Nov	Dec	Year
	t	26	25	24	21	19	17	18	19	24	26	26	26	23 °C
	clothing	t	t	t	s	s	n	s	s	t	t	t	t	–
	stress	ff	ff	f	o	o		o	o	f	f	ff	ff	–
	p	190	151	94	40	18	14	9	9	11	46	93	150	825 mm

Yacuiba

latitude: 22°01′s longitude: 63°43′w height: 580 m

GAwal	130-13	Jan	Feb	Mar	Apr	May	Jun	Jul	Aug	Sep	Oct	Nov	Dec	Year
	t	27	25	24	20	18	16	16	19	23	25	26	27	22 °C
	clothing	t	t	t	s	s	n	n	s	t	t	t	t	–
	stress	ff	ff	f	o	o			o	f	f	ff	ff	–
	p	224	205	187	61	24	13	6	11	7	75	129	184	1126 mm

3. Climatic area: South of Beni, Yunga area of Cochabamba, Llanos of Santa Cruz

3.1. Tropical Rain Climate Ar and Tropical Rain Mountain Climate GAr

3.2. Summer very hot or hot, winter hot or warm.

3.3. High sun period: dry and cloudy or sunny in the morning, showers in the afternoon and night; at times rainy days. Low sun period: less rain activity.

3.4. Wind: occasional gales, squalls in thunderstorms. Invasions of cold air can cause a drop in temperature to 5 °C.

Rurrenabaque

latitude: 14°28′s longitude: 67°35′w height: 200 m

Arab	130-14	Jan	Feb	Mar	Apr	May	Jun	Jul	Aug	Sep	Oct	Nov	Dec	Year
	t	26	27	27	26	24	23	22	24	26	27	27	27	26 °C
	clothing	t	t	t	t	t	t	s	t	t	t	t	t	–
	stress	ff	ff	ff	ff	ff	f	f	f	ff	ff	ff	ff	–
	p	318	217	214	108	105	115	103	77	88	114	151	189	1799 mm

Trinidad

Arha	130-15	Jan	Feb	Mar	Apr	May	Jun	Jul	Aug	Sep	Oct	Nov	Dec	Year
		latitude: 14°45's						longitude: 64°48'w				height: 236 m		
	t	27	27	27	26	25	24	24	25	27	28	28	28	26 °C
	clothing	t	t	t	t	t	t	t	t	t	t	t	t	–
	stress	ff	ff	ff	ff	f	f	f	f	ff	ff	ff	ff	–
	p	336	296	188	147	70	60	49	40	97	175	185	259	1902 mm
0.1 mm	dp	16	14	12	7	7	4	3	2	6	8	9	13	101 days

Santa Cruz

Arab	130-16	Jan	Feb	Mar	Apr	May	Jun	Jul	Aug	Sep	Oct	Nov	Dec	Year
		latitude: 17°47's						longitude: 63°10'w				height: 439 m		
	t	25	26	25	23	21	19	20	21	24	24	25	26	23 °C
	clothing	t	t	t	t	s	s	s	s	t	t	t	t	–
	stress	f	f	f	f	o	o	o	o	f	f	f	f	–
	p	189	126	108	144	68	79	60	24	76	100	114	155	1243 mm
0.1 mm	dp	14	10	12	9	11	8	5	4	5	7	8	11	104 days

Roboré

Arhb	130-17	Jan	Feb	Mar	Apr	May	Jun	Jul	Aug	Sep	Oct	Nov	Dec	Year
		latitude: 18°20's						longitude: 59°45'w				height: 300 m		
	t	28	28	27	25	23	22	21	24	27	28	28	29	26 °C
	clothing	t	t	t	t	t	s	s	t	t	t	t	t	–
	stress	ff	ff	ff	f	f	f	o	f	ff	ff	ff	ff	–
	p	167	145	151	133	68	63	28	28	68	85	141	176	1253 mm

4.5.130.2. Predominant general weather situations are described in 4.5.0.1. under Bolivia. A hygrothermal diagram of El Alto (La Paz) is to be found in 4.5.0.2.
Wind names: Junta is the name of the stormy pass-wind in the Andes which blows up the rocky defile of the La Paz River in the daytime.
Suaroz is an invasion of cold air from Argentina, often accompanied by stormy winds and an abrupt fall of temperature by 20 °C to freezing point in the lowlands, and producing hoarfrost at higher levels.

4.5.131. Chile

4.5.131.1. Climatic descriptions and tables

1. Climatic area: Provinces of Tarapaca, Antofagasta, Atacama

1.1. Desert Climate BW and Desert Mountain Climate GBW and Desert High Mountain Climate HBW, near the coast: Marine Desert Climate BM

1.2. In the lowlands summer warm, winter mild.

1.3. Summer generally dry and sunny, in coastal regions at times dry and overcast with low stratus. Winter predominantly dry and sunny or cloudy, in coastal regions frequently overcast and drizzle.

1.4. Wind: occasional gales. Fog: frequent near the coast.

Arica

latitude: 18° 28′s longitude: 70° 20′w height: 29 m

BMbl	131-1	Jan	Feb	Mar	Apr	May	Jun	Jul	Aug	Sep	Oct	Nov	Dec	Year
	t	22	22	21	19	18	17	15	16	16	18	19	20	19 °C
	clothing	s	s	s	s	s	n	n	n	n	s	s	s	–
	stress	o	o	o	o	o					o	o	o	–
	p	0	0	0	0	0	0	0	0	0	0	0	0	0 mm
	dp	0	0	0	0	0	0	0	0	0	0	0	0	0 days
	sunshine	195	167	198	219	143	105	93	81	45	149	183	189	1766 hours
	%	48	47	52	62	41	32	27	23	13	38	47	46	40 %

Antofagasta

latitude: 23° 26′s longitude: 70° 28′w height: 122 m

BMbl	131-2	Jan	Feb	Mar	Apr	May	Jun	Jul	Aug	Sep	Oct	Nov	Dec	Year
	tx	27	27	26	24	22	21	20	19	19	21	22	24	27 °C
	t	20	20	19	16	15	14	13	14	14	15	17	18	16 °C
	tn	15	16	14	12	10	8	8	8	9	12	12	13	8 °C
	clothing	s	s	s	n	n	n	n	n	n	n	n	s	–
	stress	o	o	o									o	–
	p	0	0	0	0	0	0	0	0	0	0	0	0	0 mm
	dp	0	0	0	0	0	0	0	0	0	0	0	0	0 days
	sunshine	326	291	248	129	195	183	186	170	180	170	195	276	2640 hours
	%	78	80	65	63	57	57	55	49	50	43	49	66	60 %

2. *Climatic area:* Provinces of Coquimbo, Aconcagua, Valparaiso, Santiago, O'Higgins, Colchagua, Curico, Talca, Linares, Maule, Buble, Concepcion, Bio-Bio, Malleco, Aranco

2.1. Subtropical Winter Rain Climate Cs and Subtropical Winter Rain Mountain Climate GCs and Subtropical Winter Rain High Mountain Climate HCs

2.2. In the lowlands summer warm, winter mild.

2.3. Summer generally dry and sunny, now and then scattered showers and cloudy. Winter often dry and sunny or cloudy, frequent showers, at times rainy days or days with drizzle. Snow at higher levels.

2.4. Wind: occasional gales. Fog: occasionally, at higher levels frequently.

Valparaiso

latitude: 33°02′s longitude: 71° 38′w height: 41 m

Csbl	131-3	Jan	Feb	Mar	Apr	May	Jun	Jul	Aug	Sep	Oct	Nov	Dec	Year
	tx	27	27	26	24	23	20	21	23	22	25	26	27	29 °C
	t	18	18	17	14	13	12	12	12	13	14	15	17	14 °C
	tn	11	11	10	8	7	5	4	5	6	7	9	10	4 °C
	clothing	s	s	n	n	n	n	n	n	n	n	n	n	–
	stress	o	o											–
	p	2	0	3	18	82	96	72	50	19	4	0	4	349 mm
	dp	0	0	0	1	5	7	7	5	2	2	0	0	31 days
	sunshine	279	246	217	174	115	81	93	118	147	170	216	264	2119 hours
	%	64	66	57	51	36	27	29	35	41	43	52	59	48 %

Santiago

GCsbk	131-4	Jan	Feb	Mar	Apr	May	Jun	Jul	Aug	Sep	Oct	Nov	Dec	Year
latitude: 33° 27′s longitude: 70°42′w height: 520 m														
	tx	33	32	31	29	26	21	22	24	27	29	31	33	33 °C
	t	21	20	18	14	11	9	8	10	12	14	18	20	15 °C
	tn	9	9	7	3	2	-1	-2	-1	1	1	4	7	-3 °C
	clothing	s	s	s	n	n	n	n	n	n	n	s	s	–
	stress	o	o	o								o	o	–
	p	1	0	6	21	74	65	57	55	26	8	2	2	318 mm
	dp	0	0	1	1	5	6	6	5	3	3	1	0	31 days
	sunshine	331	251	264	165	112	99	102	112	144	189	261	313	2346 hours
	%	76	67	69	49	35	33	32	33	40	47	63	70	53 %

3. *Climatic area:* Provinces of Cautin, Valdivia, Osorno, Llanquihue, Chiloé, Territories of Aysen, North of Magallanes
3.1. Temperate Oceanic Climate DO and Temperate Oceanic Mountain Climate GDO and Temperate Oceanic High Mountain Climate HDO
3.2. Summer mild, winter cool in low levels.
3.3. Summer: in the north often dry and sunny, otherwise variable weather prevails. Winter: often rainy days.
3.4. Wind: gales frequent, about 30 days a year in the north, 50 days a year in the south in the mean. Fog: much fog on coast. Magallanes much cloud.

Valdivia

DOlk	131-5	Jan	Feb	Mar	Apr	May	Jun	Jul	Aug	Sep	Oct	Nov	Dec	Year
latitude: 39°48′s longitude: 73°14′w height: 13 m														
	tx	31	30	28	22	18	15	15	17	20	24	26	28	32 °C
	t	16	16	15	12	10	8	8	8	9	12	14	16	12 °C
	tn	6	6	4	2	0	-1	-1	-1	0	2	3	5	-2 °C
	clothing	s	n	n	n	n	n	n	n	n	n	n	n	–
	stress													–
	p	106	50	96	178	437	371	452	318	225	98	82	72	2486 mm
	dp	7	7	11	12	21	21	20	18	13	13	10	10	163 days
	sunshine	257	229	205	123	68	48	65	90	111	127	189	208	1720 hours
	%	56	60	53	37	22	17	22	28	31	31	44	45	39 %

Cabo Raper

DOlk	131-6	Jan	Feb	Mar	Apr	May	Jun	Jul	Aug	Sep	Oct	Nov	Dec	Year
latitude: 46°50′s longitude: 75°38′w height: 39 m														
	t	11	11	10	9	8	7	6	6	6	8	9	10	9 °C
	clothing	n	n	n	n	n	n	n	n	n	n	n	n	–
	stress													–
	p	198	147	180	195	190	201	241	190	142	178	170	178	2210 mm
	dp	23	19	22	21	22	22	25	22	21	21	21	22	261 days

4. *Climatic area:* South of Magallanes
4.1. Subarctic Oceanic Climate EO partly Tundra Climate FT
4.2. Summer mild or cool, winter cool.
4.3. Variable weather prevails, rainy and dry periods alternating rapidly. Dry and sunny often only for a few hours, especially in coastal areas.
4.4. Wind: frequent gales from the west. Fog: frequent. Snow cover: variable at low levels.

Punta Arenas

		latitude: 53°02′s				longitude: 70°51′w				height: 33 m				
EOlk	131-7	Jan	Feb	Mar	Apr	May	Jun	Jul	Aug	Sep	Oct	Nov	Dec	Year
	tx	21	21	18	14	12	9	8	9	13	16	17	19	22 °C
	t	11	10	9	6	4	2	2	3	4	7	9	10	7 °C
	tn	4	3	1	0	-2	-4	-4	-4	-3	-1	-1	2	-6 °C
	clothing	n	n	n	n	w	w	w	w	w	n	n	n	–
	stress													–
	p	37	30	46	43	44	26	37	52	25	23	39	41	442 mm
0.2 mm	dp	10	9	10	12	12	9	11	8	9	6	9	11	116 days

5. Climatic area: Pacific Islands

5.1. Easter Island: Tropical Rain Climate Ar; Isla de Mas a Tierra: Subtropical Rain Climate Cr

5.2. Summer hot or warm, winter warm or mild.

5.3. Summer often dry and sunny, frequent showers, at times rainy days. Winter: variable weather prevails.

5.4. Wind: occasional gales, more frequent in the south.

Easter Island

		latitude: 27°10′s				longitude: 109°26′w				height: 30 m				
Arab	131-8	Jan	Feb	Mar	Apr	May	Jun	Jul	Aug	Sep	Oct	Nov	Dec	Year
	tx	28	29	29	27	27	24	23	23	23	24	26	27	29 °C
	t	23	23	23	22	21	18	18	18	18	19	20	22	20 °C
	tn	17	17	18	14	13	12	11	12	12	12	14	16	11 °C
	clothing	t	t	t	s	s	s	s	s	s	s	s	s	–
	stress	f	ff	ff	f	f	o	o	o	o	o	o	f	–
	p	132	41	229	128	99	241	81	66	89	56	127	74	1364 mm
	dp	15	7	16	15	12	16	12	9	14	6	9	9	140 days

Juan Fernandez – Isla de Mas a Tierra

		latitude: 33°37′s				longitude: 78°52′w				height: 6 m				
Crbl	131-9	Jan	Feb	Mar	Apr	May	Jun	Jul	Aug	Sep	Oct	Nov	Dec	Year
	tx	26	26	25	23	22	19	19	18	19	19	22	24	27 °C
	t	19	19	17	17	16	14	13	12	13	13	15	17	16 °C
	tn	13	13	13	12	9	8	7	7	7	8	9	11	6 °C
	clothing	s	s	s	n	n	n	n	n	n	n	n	n	–
	stress	o	o	o										–
	p	20	30	41	86	150	162	147	112	73	48	41	25	935 mm
	dp	4	6	9	12	15	17	13	14	11	8	6	7	122 days
	sunshine	248	209	158	123	108	99	93	105	147	205	249	260	2005 hours
	%	57	56	41	36	33	33	30	31	41	51	60	58	45 %

4.5.131.2. Predominant general weather situations are described in 4.5.0.1. under Chile. A hygrothermal diagram of Santiago is to be found in 4.5.0.2.

Wind names: Puelche is a warm fall-wind in the Andes of southern Chile.

Rachas – also calles Williwas – are sudden strong gusts of cold air, occurring especially in the western part of the Strait of Magellan.

4.5.132. Paraguay

4.5.132.1. Climatic descriptions and tables

1. Climatic area: Chaco
1.1. Tropical Summer Rain Climate Aw
1.2. Summer very hot, winter warm.
1.3. Generally dry and sunny or cloudy in the morning, in summer often showers in the afternoon and night, at times rainy days; in winter scattered showers.
1.4. Wind: occasional gales, squalls in thunderstorms.

Mariscal Estigarribia

latitude: 22°01′s longitude: 60°37′w height: 181 m

Awhb	132-1	Jan	Feb	Mar	Apr	May	Jun	Jul	Aug	Sep	Oct	Nov	Dec	Year
	t	29	28	26	23	21	19	19	22	24	26	27	28	24 °C
	clothing	t	t	t	t	s	s	s	s	t	t	t	t	–
	stress	ff	ff	ff	f	f	o	o	f	f	ff	ff	ff	–
	p	112	112	92	67	42	25	17	4	26	94	84	99	774 mm

Puerto Casado

latitude: 22°17′s longitude: 57°52′w height: 87 m

Awhb	132-2	Jan	Feb	Mar	Apr	May	Jun	Jul	Aug	Sep	Oct	Nov	Dec	Year
	t	29	28	24	24	22	21	20	22	24	25	28	28	25 °C
	clothing	t	t	t	t	s	s	s	s	t	t	t	t	–
	stress	ff	ff	ff	f	f	f	f	f	f	ff	ff	ff	–
	p	143	127	145	106	79	57	47	28	71	125	148	126	1202 mm

2. Climatic area: East of country
2.1. Tropical Rain Climate Ar and Tropical Rain Mountain Climate GAr; southernmost part of the country: Subtropical Rain Climate Cr
2.2. In lowlands, summer very hot in the north, hot in the south; winter warm in the north, mild in the south.
2.3. Generally dry and sunny or cloudy in the morning, frequent showers in the afternoon and at night, some rainy days.
2.4. Wind: occasional gales, squalls in thunderstorms.

Asunción

latitude: 25°16′s longitude: 57°38′w height: 64 m

Arhb	132-3	Jan	Feb	Mar	Apr	May	Jun	Jul	Aug	Sep	Oct	Nov	Dec	Year
	tx	39	39	37	35	32	29	31	34	36	37	38	39	41 °C
	t	29	29	27	23	21	19	18	20	22	24	27	29	24 °C
	tn	16	17	15	11	7	3	3	4	6	9	13	14	2 °C
	clothing	t	t	t	t	s	s	s	s	s	t	t	t	–
	stress	ff	ff	f	f	o	o	o	o	f	f	f	ff	–
	p	145	137	154	138	124	84	53	35	82	133	124	131	1340 mm
0.1 mm	dp	8	6	6	7	6	6	5	4	7	8	8	7	78 days

Villarrica

latitude: 25°45′s longitude: 56°26′w height: 157 m

Crhl	132-4	Jan	Feb	Mar	Apr	May	Jun	Jul	Aug	Sep	Oct	Nov	Dec	Year
	t	28	27	26	22	19	18	17	19	21	23	25	27	22 °C
	clothing	t	t	t	s	s	s	n	s	s	t	t	t	–
	stress	ff	ff	f	o	o	o	o	o	o	f	f	ff	–
	p	158	130	136	162	141	94	84	80	123	225	129	123	1586 mm

Puerto Puente Franco

latitude: 25°36′s longitude: 54°33′w height: 125 m

Cral	132-5	Jan	Feb	Mar	Apr	May	Jun	Jul	Aug	Sep	Oct	Nov	Dec	Year
	t	26	26	24	21	18	16	16	18	20	21	23	25	21 °C
	clothing	t	t	t	s	s	n	n	s	s	s	t	t	–
	stress	ff	f	f	o	o			o	o	o	f	f	–
	p	147	120	162	138	148	125	100	72	156	161	143	139	1611 mm

Encarnación

latitude: 27°20′s longitude: 55°50′w height: 113 m

Cral	132-6	Jan	Feb	Mar	Apr	May	Jun	Jul	Aug	Sep	Oct	Nov	Dec	Year
	t	26	26	24	20	18	16	16	18	19	21	23	25	21 °C
	clothing	t	t	t	s	s	n	n	s	s	s	t	t	–
	stress	ff	ff	f	o	o			o	o	o	f	f	–
	p	132	185	151	171	162	153	111	76	142	206	136	123	1747 mm

4.5.132.2. Predominant general weather situations are described in 4.5.0.1. under Paraguay, Uruguay, Argentina. A hygrothermal diagram of Asuncion is to be found in 4.5.0.2.

4.5.133. Uruguay

4.5.133.1. Climatic description and tables

1. Subtropical Rain Climate Cr
2. Summer warm, hot in the north; winter mild.
3. Generally dry and sunny or cloudy in the morning, frequent showers in the afternoon and at night, some rainy days.
4. Wind: occasional gales.

Salto

latitude: 31°23′s longitude: 57°58′w height: 46 m

Cral	133-1	Jan	Feb	Mar	Apr	May	Jun	Jul	Aug	Sep	Oct	Nov	Dec	Year
	t	26	25	23	18	16	13	12	14	15	18	22	24	19 °C
	clothing	t	t	t	s	n	n	n	n	n	s	s	t	–
	stress	ff	f	f	o						o	f	f	–
	p	116	94	121	151	83	100	58	58	97	127	91	79	1175 mm

Rocha

latitude: 34°30′s longitude: 54°20′w height: 22 m

Crbl	133-2	Jan	Feb	Mar	Apr	May	Jun	Jul	Aug	Sep	Oct	Nov	Dec	Year
	t	22	21	20	16	13	11	11	11	12	15	18	20	16 °C
	clothing	s	s	s	n	n	n	n	n	n	n	s	s	–
	stress	f	o	o								o	o	–
	p	109	96	88	114	82	74	71	104	119	100	74	53	1085 mm

Montevideo

Crbl	133-3	Jan	Feb	Mar	Apr	May	Jun	Jul	Aug	Sep	Oct	Nov	Dec	Year
				latitude: 34°42′s				longitude: 56°12′w				height: 22 m		
	tx	33	31	29	27	23	19	21	23	24	26	29	31	33 °C
	t	22	22	20	17	14	11	10	11	13	15	18	21	16 °C
	tn	14	14	13	9	6	3	3	4	6	7	9	13	2 °C
	clothing	s	s	s	n	n	n	n	n	n	n	s	s	–
	stress	f	f	o								o	o	–
	p	83	74	104	102	91	88	73	87	84	73	79	77	1014 mm
	dp	6	5	5	6	6	5	6	7	6	6	6	7	71 days
	sunshine	329	285	282	225	195	162	158	192	213	242	300	322	2906 hours
	%	75	76	74	67	61	55	51	57	60	60	72	72	66 %

Punta del Este

Crbl	133-4	Jan	Feb	Mar	Apr	May	Jun	Jul	Aug	Sep	Oct	Nov	Dec	Year
				latitude: 34°58′s				longitude: 54°57′w				height: 16 m		
	t	22	22	20	18	15	12	11	12	13	15	17	20	16 °C
	clothing	s	s	s	s	n	n	n	n	n	n	n	s	–
	stress	f	f	o	o								o	–
	p	78	74	87	84	82	86	65	82	88	86	69	57	937 mm

4.5.133.2. Predominant general weather situations are described in 4.5.0.1. under Paraguay, Uruguay, Argentina. A hygrothermal diagram of Montevideo is to be found in 4.5.0.2.

4.5.134. Argentina

4.5.134.1. Climatic descriptions and tables

1. *Climatic area:* Provinces: Jujuy, Salta, eastern Catamarca and La Rioja, San Luis, western Eva Peron, eastern Rio Negro, western Chubut, Comodore Rivadavia, Santa Cruz
1.1. Steppe Climate BS and Steppe Mountain Climate GBS
1.2. Broad scale corresponding to level and latitude: hot in lowlands in the north and cold in winter in the south.
1.3. Generally dry and sunny, but in summer frequent showers, at times rainy; winter at times rainy and overcast.
1.4. Wind: especially in the mountains, but also in other parts, frequent gales. Fog: occasionally. Snow cover: at higher levels.

La Quiaca

HBSlk	134-1	Jan	Feb	Mar	Apr	May	Jun	Jul	Aug	Sep	Oct	Nov	Dec	Year
				latitude: 22°06′s				longitude: 65°36′w				height: 3459 m		
	tx	25	27	24	23	21	19	19	22	23	25	27	26	27 °C
	t	12	12	12	10	7	4	4	6	9	11	12	13	9 °C
	tn	2	2	1	-4	-10	-13	-13	-12	-8	-6	-2	2	-14 °C
	clothing	n	n	n	n	n	w	w	n	n	n	n	n	–
	stress													–
	p	89	77	43	5	1	2	1	0	2	9	31	63	323 mm
0.2 mm	dp	12	9	7	1	0	0	0	0	0	1	5	9	45 days
	sunshine	258	232	281	284	299	278	292	296	283	302	296	283	3384 hours
	%	62	64	74	82	87	86	86	84	79	77	75	68	76 %

Rivadavia

latitude: 24°10′s longitude: 62°54′w height: 204 m

BShl	134-2	Jan	Feb	Mar	Apr	May	Jun	Jul	Aug	Sep	Oct	Nov	Dec	Year
	t	28	27	25	22	19	17	16	19	22	24	26	28	23 °C
	clothing	t	t	t	s	s	n	n	s	s	t	t	t	–
	stress	ff	ff	f	f	o	o		o	f	f	ff	ff	–
	p	103	117	102	35	16	15	9	14	8	48	79	138	684 mm

Salta

latitude: 24°51′s longitude: 65°29′w height: 1182 m

GBSbl	134-3	Jan	Feb	Mar	Apr	May	Jun	Jul	Aug	Sep	Oct	Nov	Dec	Year
	t	21	20	19	16	14	11	11	13	16	18	20	22	17 °C
	clothing	s	s	s	n	n	n	n	n	n	s	s	s	–
	stress	o	o	o							o	o	o	–
	p	176	149	94	25	6	3	2	4	5	25	60	122	671 mm
	sunshine	196	150	138	138	158	143	188	216	189	187	188	198	2089 hours
	%	47	41	36	40	47	45	56	62	53	48	47	47	47 %

Tucumán

latitude: 26°50′s longitude: 65°12′w height: 420 m

BSal	134-4	Jan	Feb	Mar	Apr	May	Jun	Jul	Aug	Sep	Oct	Nov	Dec	Year
	t	25	24	22	18	15	12	12	14	17	20	22	25	19 °C
	clothing	t	t	s	s	n	n	n	n	n	s	s	t	–
	stress	f	f	o	o						o	o	f	–
	p	183	159	162	59	29	19	10	8	12	77	108	151	977 mm
0.2 mm	dp	10	10	11	6	4	4	2	2	3	7	8	10	77 days
	sunshine	226	185	180	177	160	145	193	190	187	200	207	219	2269 hours
	%	53	50	47	51	48	46	59	55	52	51	51	51	47 %

Santiago del Estero

latitude: 27°46′s longitude: 64°18′w height: 199 m

BSal	134-5	Jan	Feb	Mar	Apr	May	Jun	Jul	Aug	Sep	Oct	Nov	Dec	Year
	tx	43	41	39	35	31	27	31	35	38	39	42	42	44 °C
	t	27	26	24	20	16	14	13	15	19	22	24	27	20 °C
	tn	15	14	12	8	2	-1	-3	-1	3	7	11	13	-3 °C
	clothing	t	t	t	s	n	n	n	n	s	s	t	t	–
	stress	ff	f	f	o					o	o	f	f	–
	p	90	93	91	27	14	10	3	3	8	39	60	79	517 mm
0.2 mm	dp	6	6	6	4	2	2	1	1	2	4	5	5	44 days
	sunshine	252	204	190	172	162	126	180	210	205	231	236	263	2431 hours
	%	59	55	50	50	49	40	55	61	57	59	58	61	55 %

Catamarca

latitude: 28°26′s longitude: 65°46′w height: 547 m

GBShl	134-6	Jan	Feb	Mar	Apr	May	Jun	Jul	Aug	Sep	Oct	Nov	Dec	Year
	t	28	26	24	20	15	12	11	14	19	22	25	27	20 °C
	clothing	t	t	t	s	n	n	n	n	s	s	t	t	–
	stress	ff	f	f	o					o	o	f	ff	–
	p	66	82	49	20	11	6	4	4	7	29	41	51	370 mm

San Luis

latitude: 33°16′s longitude: 66°21′w height: 713 m

GBSak	134-7	Jan	Feb	Mar	Apr	May	Jun	Jul	Aug	Sep	Oct	Nov	Dec	Year
	t	24	23	20	16	12	9	9	11	14	17	20	23	17 °C
	clothing	t	t	s	n	n	n	n	n	n	n	s	t	–
	stress	f	f	o								o	f	–
	p	90	74	60	33	14	13	10	6	17	53	67	105	542 mm

San Antonio Oeste

latitude: 40°44′s longitude: 64°57′w height: 7 m

BSbk	134-8	Jan	Feb	Mar	Apr	May	Jun	Jul	Aug	Sep	Oct	Nov	Dec	Year
	t	22	22	19	15	11	8	8	9	12	15	19	21	15 °C
	clothing	s	s	s	n	n	n	n	n	n	n	s	s	–
	stress	o	o	o								o	o	–
	p	14	17	27	17	33	19	26	17	21	20	17	15	243 mm
	sunshine	342	284	258	202	151	130	146	180	193	263	308	323	2780 hours
	%	75	74	67	61	49	47	49	55	55	65	71	69	63 %

Sarmiento

latitude: 45°35′s longitude: 65°04′w height: 268 m

BSlk	134-9	Jan	Feb	Mar	Apr	May	Jun	Jul	Aug	Sep	Oct	Nov	Dec	Year
	tx	34	32	29	25	19	14	15	18	22	26	28	31	34 °C
	t	17	17	14	11	7	4	4	6	8	12	14	16	11 °C
	tn	6	5	1	-2	-8	-11	-9	-8	-4	-2	2	4	-12 °C
	clothing	n	n	n	n	n	w	w	n	n	n	n	n	–
	stress													–
	p	10	8	11	15	24	16	17	14	10	6	12	9	152 mm
	dp	1	1	2	3	4	4	3	3	3	2	2	2	30 days

Comodore Rivadavia

latitude: 45°47′s longitude: 67°30′w height: 61 m

BSbk	134-10	Jan	Feb	Mar	Apr	May	Jun	Jul	Aug	Sep	Oct	Nov	Dec	Year
	t	19	18	16	13	9	7	7	8	10	13	15	17	13 °C
	clothing	s	s	n	n	n	n	n	n	n	n	n	n	–
	stress	o	o											–
	p	16	11	21	22	35	20	21	18	15	10	16	13	218 mm
	sunshine	260	217	177	148	112	102	114	137	152	208	240	246	2113 hours
	%	55	55	46	46	38	39	41	43	43	50	54	51	48 %

Puerto Deseado

latitude: 47°44′s longitude: 65°55′w height: 79 m

BSlk	134-11	Jan	Feb	Mar	Apr	May	Jun	Jul	Aug	Sep	Oct	Nov	Dec	Year
	t	15	15	13	11	7	4	4	5	7	10	13	14	10 °C
	clothing	n	n	n	n	n	w	w	n	n	n	n	n	–
	stress													–
	p	16	13	19	14	27	20	18	17	17	5	11	15	192 mm

Santa Cruz

BSlk	134-12	Jan	Feb	Mar	Apr	May	Jun	Jul	Aug	Sep	Oct	Nov	Dec	Year
				latitude: 50°01's				longitude: 68°34'w			height: 111 m			
	tx	32	31	28	22	16	12	12	14	18	24	27	29	33 °C
	t	14	14	11	8	4	1	1	3	6	9	11	13	8 °C
	tn	4	3	0	-1	-3	-8	-9	-6	-4	-2	1	3	-11 °C
	clothing	n	n	n	n	w	w	w	w	n	n	n	n	–
	stress													–
	p	21	16	18	13	25	15	15	17	12	7	15	18	192 mm
0.2 mm	dp	4	3	2	4	4	5	5	6	2	3	4	7	49 days
	sunshine	236	206	180	144	118	96	81	143	159	205	210	223	2001 hours
	%	48	51	46	45	42	39	31	47	45	49	46	44	45 %

Rio Gallegos

BSlk	134-13	Jan	Feb	Mar	Apr	May	Jun	Jul	Aug	Sep	Oct	Nov	Dec	Year
				latitude: 51°37's				longitude: 69°17'w			height: 17 m			
	t	12	12	10	6	3	1	1	2	5	8	10	12	7 °C
	clothing	n	n	n	n	w	w	w	w	n	n	n	n	–
	stress													–
	p	32	23	23	25	26	17	20	14	11	11	22	25	249 mm

2. *Climatic area:* Provinces: western Catamarca and La Rioja, San Juan, Mendoza, Neuquen, western Rio Negro, eastern Chubut
2.1. Desert Climate BW and Desert Mountain Climate GBW; coastal areas: Marine Desert Climate BM
2.2. Broad scale corresponding to level and latitude: hot in lowlands in the north and cold in winter in the south, very cold at higher levels in winter.
2.3. Summer often dry and sunny, scattered showers or rain. Winter usually dry and sunny, in the mountains frequent rain, otherwise rainy now and then.
2.4. Wind: in the mountains frequent gales from the west. Fog: at higher levels often. Snow cover: at higher levels.

La Rioja

BWal	134-14	Jan	Feb	Mar	Apr	May	Jun	Jul	Aug	Sep	Oct	Nov	Dec	Year
				latitude: 29°23's				longitude: 66°49'w			height: 450 m			
	t	27	26	23	19	15	12	11	14	18	21	24	27	20 °C
	clothing	t	t	t	s	n	n	n	n	s	s	t	t	–
	stress	ff	ff	f	o					o	o	f	ff	–
	p	60	56	57	16	6	3	3	3	6	24	31	51	316 mm
	sunshine	280	231	228	313	200	181	217	241	244	259	275	289	2858 hours
	%	65	62	60	62	60	59	67	70	68	65	67	66	65 %

San Juan

GBWak	134-15	Jan	Feb	Mar	Apr	May	Jun	Jul	Aug	Sep	Oct	Nov	Dec	Year
				latitude: 31°36's				longitude: 68°33'w			height: 630 m			
	t	26	24	21	16	12	8	8	11	14	18	22	25	17 °C
	clothing	t	t	s	n	n	n	n	n	n	s	s	t	–
	stress	ff	f	o							o	o	f	–
	p	18	12	9	5	1	2	2	2	5	9	11	10	86 mm
	sunshine	336	290	282	259	219	194	220	245	259	295	322	338	3259 hours
	%	77	78	74	76	67	64	69	72	73	74	78	77	74 %

Mendoza

latitude: 32°53's longitude: 68°51'w height: 827 m

GBWak	134-16	Jan	Feb	Mar	Apr	May	Jun	Jul	Aug	Sep	Oct	Nov	Dec	Year
	tx	39	37	34	29	24	23	24	27	29	33	34	37	39 °C
	t	25	23	20	15	10	8	7	10	13	18	21	24	16 °C
	tn	11	8	5	2	-1	-4	-4	-3	-1	2	4	7	-5 °C
	clothing	t	t	s	n	n	n	n	n	n	s	s	t	–
	stress	f	f	o							o	o	f	–
	p	28	21	21	10	11	8	7	10	14	23	20	23	196 mm
0.2 mm	dp	5	5	4	3	2	2	2	2	2	4	4	5	40 days
	sunshine	309	265	243	224	198	173	194	216	231	263	296	312	2924 hours
	%	71	71	64	66	61	58	61	64	65	66	71	70	66 %

Neuquén

latitude: 38°57's longitude: 68°08'w height: 270 m

BWbk	134-17	Jan	Feb	Mar	Apr	May	Jun	Jul	Aug	Sep	Oct	Nov	Dec	Year
	t	22	21	18	12	8	5	5	7	11	14	19	22	14 °C
	clothing	s	s	s	n	n	n	n	n	n	n	s	s	–
	stress	o	o	o								o	o	–
	p	15	11	12	13	22	15	11	19	10	23	8	11	170 mm

Puerto Madryn

latitude: 42°46's longitude: 65°02'w height: 13 m

BMbk	134-18	Jan	Feb	Mar	Apr	May	Jun	Jul	Aug	Sep	Oct	Nov	Dec	Year
	t	20	19	17	14	10	8	7	8	10	13	16	18	14 °C
	clothing	s	s	n	n	n	n	n	n	n	n	n	s	–
	stress	o	o										o	–
	p	10	14	17	13	24	14	17	11	14	18	12	12	174 mm
0.2 mm	dp	2	2	2	4	4	4	3	2	2	3	2	2	32 days
	sunshine	316	266	257	195	158	123	149	164	267	239	270	298	2702 hours
	%	68	68	67	59	52	45	52	51	76	58	62	63	61 %

Trelew

latitude: 43°14's longitude: 65°18'w height: 39 m

BMbk	134-19	Jan	Feb	Mar	Apr	May	Jun	Jul	Aug	Sep	Oct	Nov	Dec	Year
	t	21	20	17	13	9	6	6	8	10	14	17	19	13 °C
	clothing	s	s	n	n	n	n	n	n	n	n	n	s	–
	stress	o	o										o	–
	p	8	14	18	14	22	11	17	14	13	12	16	18	177 mm

3. Climatic area: Provinces: Tucumán, Santiago, Cordoba, eastern Eva Peron

3.1. Subtropical Summer Rain Climate Cw and Subtropical Summer Rain Mountain Climate GCw

3.2. In lower levels summer hot or warm, winter mild.

3.3. Summer partly dry and sunny, partly showers, at times rainy. Winter often dry and sunny, at times showers or rain.

3.4. Wind: occasional gales (Pampero). Fog: frequent at higher levels. Snow cover: at higher levels.

Presidencia Roque Sáenz Peña

latitude: 26°49′s longitude: 60°27′w height: 92 m

Cwal	134-20	Jan	Feb	Mar	Apr	May	Jun	Jul	Aug	Sep	Oct	Nov	Dec	Year
	t	27	26	24	20	18	16	15	17	19	22	24	26	21 °C
	clothing	t	t	t	s	s	n	n	n	s	s	t	t	–
	stress	ff	ff	f	o	o				o	o	f	ff	–
	p	136	125	133	97	49	35	23	23	44	101	113	111	990 mm
	sunshine	270	229	234	202	187	142	189	224	203	241	269	288	2678 hours
	%	64	62	61	59	56	45	57	65	57	61	67	67	61 %

Córdoba

latitude: 31°24′s longitude: 64°11′w height: 425 m

Cwal	134-21	Jan	Feb	Mar	Apr	May	Jun	Jul	Aug	Sep	Oct	Nov	Dec	Year
	tx	40	38	37	32	29	28	29	32	34	36	37	38	41 °C
	t	24	23	21	17	14	11	11	13	15	18	21	22	18 °C
	tn	12	9	7	5	-2	-3	-5	-3	0	3	6	8	-6 °C
	clothing	t	t	s	n	n	n	n	n	n	s	s	s	–
	stress	f	f	f	o						o	o	o	–
	p	101	88	93	39	24	10	8	15	29	77	88	108	680 mm
0.2 mm	dp	10	9	9	5	3	2	2	2	4	8	10	7	71 days
	sunshine	279	240	221	205	175	154	190	213	207	239	264	272	2659 hours
	%	64	64	58	60	54	51	60	63	58	60	64	62	60 %

Pilar

latitude: 31°40′s longitude: 63°53′w height: 338 m

Cwal	134-22	Jan	Feb	Mar	Apr	May	Jun	Jul	Aug	Sep	Oct	Nov	Dec	Year
	t	24	23	20	16	13	10	10	11	14	17	20	23	17 °C
	clothing	t	t	s	n	n	n	n	n	n	n	s	t	–
	stress	f	f	o								o	f	–
	p	104	81	89	43	25	11	11	14	31	83	101	107	700 mm
	sunshine	292	246	234	215	195	168	198	219	223	247	281	293	2811 hours
	%	67	66	61	63	60	55	62	64	63	62	68	67	64 %

Laboulaye

latitude: 34°08′s longitude: 63°24′w height: 138 m

Cwak	134-23	Jan	Feb	Mar	Apr	May	Jun	Jul	Aug	Sep	Oct	Nov	Dec	Year
	t	24	23	20	16	12	9	9	10	13	19	20	23	16 °C
	clothing	t	t	s	n	n	n	n	n	n	s	s	t	–
	stress	f	f	o								o	f	–
	p	102	101	113	63	30	26	21	14	36	90	88	83	767 mm
	sunshine	304	263	235	213	170	131	160	194	202	243	283	300	2698 hours
	%	69	70	61	63	53	44	51	58	57	61	68	67	61 %

Santa Rosa

latitude: 36°34′s longitude: 64°16′w height: 189 m

Cwak	134-24	Jan	Feb	Mar	Apr	May	Jun	Jul	Aug	Sep	Oct	Nov	Dec	Year
	t	24	22	19	15	11	8	7	9	12	15	19	23	15 °C
	clothing	t	s	s	n	n	n	n	n	n	n	s	t	–
	stress	f	o	o								o	f	–
	sunshine	327	268	238	197	165	124	148	179	199	250	294	328	2717 hours
	%	73	71	62	59	52	43	48	54	56	62	70	72	61 %

4. Climatic area: Provinces: Formosa, Presidente Perón, Santa Fe, Misiones, Corrientes, Entrerios, Buenos Aires

4.1. Subtropical Rain Climate Cr

4.2. Summer hot, winter mild.

4.3. Summer partly dry and sunny, showers particularly in the afternoon, some rainy days. Winter partly dry and cloudy or sunny, partly rainy, for several days at a stretch.

4.4. Wind: occasional gales (Pampero). Fog: occasionally.

Formosa

latitude: 26°11's longitude: 58°13'w height: 60 m

Crhl	134-25	Jan	Feb	Mar	Apr	May	Jun	Jul	Aug	Sep	Oct	Nov	Dec	Year
	t	28	27	25	21	19	17	17	16	20	22	24	27	22 °C
	clothing	t	t	t	s	s	n	n	n	s	s	t	t	–
	stress	ff	ff	f	o	o				o	o	f	ff	–
	p	124	136	157	137	117	77	48	48	82	143	138	119	1326 mm
	sunshine	267	226	210	196	179	147	189	195	183	207	277	291	2567 hours
	%	63	62	55	57	53	47	57	56	51	53	69	68	58 %

Corrientes

latitude: 27°27's longitude: 58°46'w height: 67 m

Crhl	134-26	Jan	Feb	Mar	Apr	May	Jun	Jul	Aug	Sep	Oct	Nov	Dec	Year
	tx	39	38	37	33	31	28	29	32	33	36	37	39	39 °C
	t	28	27	25	21	19	16	16	17	19	22	24	26	22 °C
	tn	17	17	14	11	7	5	2	4	7	9	12	14	2 °C
	clothing	t	t	t	s	s	n	n	n	s	s	t	t	–
	stress	ff	ff	f	o	o				o	o	o	f	–
	p	149	127	151	135	86	60	47	42	74	139	139	119	1268 mm
0.2 mm	dp	5	5	5	5	4	4	3	3	4	6	6	6	56 days
	sunshine	282	241	239	209	197	147	185	217	196	235	274	294	2716 hours
	%	66	65	63	61	59	47	56	63	55	60	68	68	61 %

Resistencia

latitude: 27°28's longitude: 58°59'w height: 51 m

Cral	134-27	Jan	Feb	Mar	Apr	May	Jun	Jul	Aug	Sep	Oct	Nov	Dec	Year
	t	27	26	24	20	18	16	15	17	19	21	24	26	21 °C
	clothing	t	t	t	s	s	n	n	n	s	s	t	t	–
	stress	ff	ff	f	o	o				o	o	f	ff	–
	p	134	127	151	122	78	54	47	38	67	125	125	119	1187 mm
	sunshine	241	200	172	166	164	121	166	176	115	177	228	224	2150 hours
	%	57	54	45	48	49	39	51	51	32	45	56	52	49 %

Posadas

latitude: 27°22's longitude: 55°58'w height: 135 m

Cral	134-28	Jan	Feb	Mar	Apr	May	Jun	Jul	Aug	Sep	Oct	Nov	Dec	Year
	t	26	26	24	21	18	16	16	17	19	21	23	25	21 °C
	clothing	t	t	t	s	s	n	n	n	s	s	t	t	–
	stress	f	f	f	o	o				o	o	f	f	–
	p	137	163	134	181	129	156	114	83	164	240	130	120	1751 mm
	sunshine	248	215	203	194	175	138	172	187	151	198	234	257	2372 hours
	%	58	58	53	56	52	44	52	54	42	50	58	60	54 %

Ceres

latitude: 29°53′s longitude: 61°57′w height: 88 m

Cral	139-29	Jan	Feb	Mar	Apr	May	Jun	Jul	Aug	Sep	Oct	Nov	Dec	Year
	t	26	25	22	18	16	13	12	14	16	19	22	25	19 °C
	clothing	t	t	s	s	n	n	n	n	n	s	s	t	–
	stress	f	f	o	o						o	o	f	–
	p	113	101	124	58	37	27	22	20	36	84	103	110	835 mm
	sunshine	269	246	228	206	181	138	184	205	203	245	274	274	2653 hours
	%	63	66	60	60	55	45	57	60	57	62	57	63	60 %

Paso de los Libres

latitude: 29°41′s longitude: 57°09′w height: 70 m

Cral	134-30	Jan	Feb	Mar	Apr	May	Jun	Jul	Aug	Sep	Oct	Nov	Dec	Year
	t	26	26	24	19	17	15	14	15	17	19	22	25	20 °C
	clothing	t	t	t	s	n	n	n	n	n	s	s	t	–
	stress	ff	f	f	o						o	o	f	–
	p	130	107	129	169	114	98	70	67	115	148	107	107	1361 mm

Parana

latitude: 31°47′s longitude: 60°29′w height: 62 m

Cral	134-31	Jan	Feb	Mar	Apr	May	Jun	Jul	Aug	Sep	Oct	Nov	Dec	Year
	t	25	24	22	17	15	12	11	13	15	18	21	23	18 °C
	clothing	t	t	s	n	n	n	n	n	n	s	s	t	–
	stress	f	f	o							o	o	f	–
	p	136	101	154	94	52	46	31	37	55	101	108	101	1016 mm
	sunshine	272	256	238	210	187	137	173	205	203	247	280	271	2679 hours
	%	63	69	62	62	57	45	54	60	57	62	68	61	61 %

Rosario

latitude: 32°55′s longitude: 60°44′w height: 27 m

Cral	134-32	Jan	Feb	Mar	Apr	May	Jun	Jul	Aug	Sep	Oct	Nov	Dec	Year
	tx	38	37	35	28	28	24	24	28	31	31	33	36	39 °C
	t	24	23	20	16	13	11	10	11	14	17	20	22	17 °C
	tn	9	8	5	4	1	-3	-3	-2	0	1	6	8	-4 °C
	clothing	t	t	s	n	n	n	n	n	n	n	s	s	–
	stress	f	f	o								o	o	–
	p	141	102	138	92	60	42	37	34	45	104	114	84	993 mm
	dp	6	6	6	6	5	3	3	4	6	6	7	7	65 days
	sunshine	300	258	246	213	193	138	172	199	210	254	294	283	2760 hours
	%	69	69	64	63	60	46	54	59	59	64	71	64	62 %

Qualequaychu

latitude: 33°01′s longitude: 58°21′w height: 26 m

Cral	134-33	Jan	Feb	Mar	Apr	May	Jun	Jul	Aug	Sep	Oct	Nov	Dec	Year
	t	25	24	21	17	14	11	11	12	14	17	20	23	18 °C
	clothing	t	t	s	n	n	n	n	n	n	n	s	t	–
	stress	f	f	o								o	f	–
	p	112	72	144	98	67	71	53	70	75	95	89	70	1016 mm
	sunshine	257	231	197	194	178	99	146	166	152	196	255	250	2321 hours
	%	59	62	52	57	55	33	46	49	43	49	61	56	52 %

Buenos Aires

latitude: 34°35′s longitude: 58°29′w height: 25 m

Cral	134-34	Jan	Feb	Mar	Apr	May	Jun	Jul	Aug	Sep	Oct	Nov	Dec	Year
	tx	36	35	32	29	25	20	22	24	26	28	32	35	37 °C
	t	24	23	21	17	14	11	11	12	14	16	20	22	17 °C
	tn	10	9	7	4	2	-2	-2	-1	1	3	5	8	-3 °C
	clothing	t	t	s	n	n	n	n	n	n	n	s	s	–
	stress	f	f	o								o	o	–
	p	104	82	122	90	79	68	61	68	80	100	90	83	1027 mm
0.2 mm	dp	7	6	7	8	7	7	8	9	8	9	9	8	93 days
	sunshine	293	251	232	203	173	132	149	181	193	234	270	284	2595 hours
	%	66	67	61	60	54	45	48	54	54	58	65	63	59 %

Azul

latitude: 36°45′s longitude: 59°50′w height: 132 m

Crbk	134-35	Jan	Feb	Mar	Apr	May	Jun	Jul	Aug	Sep	Oct	Nov	Dec	Year
	t	22	21	18	13	10	8	7	8	10	13	17	20	14 °C
	clothing	s	s	s	n	n	n	n	n	n	n	n	s	–
	stress	o	o	o									o	–
	p	68	77	118	65	76	49	49	41	60	83	86	80	852 mm
	sunshine	303	254	226	196	146	114	135	166	182	244	257	272	2495 hours
	%	68	67	59	58	46	39	44	50	51	61	61	60	56 %

Mar del Plata

latitude: 37°56′s longitude: 57°35′w height: 24 m

Crbk	134-36	Jan	Feb	Mar	Apr	May	Jun	Jul	Aug	Sep	Oct	Nov	Dec	Year
	t	20	19	18	14	11	9	8	8	10	13	16	18	14 °C
	clothing	s	s	s	n	n	n	n	n	n	n	n	s	–
	stress	o	o											–
	p	86	77	94	65	91	38	67	56	61	60	65	77	837 mm
0.2 mm	dp	5	7	6	6	6	6	6	5	5	7	6	7	72 days
	sunshine	291	263	257	216	195	171	189	223	213	248	300	304	2871 hours
	%	65	69	67	65	62	60	62	68	60	61	71	66	65 %

Bahia Blanca

latitude: 38°44′s longitude: 62°11′w height: 75 m

Crak	134-37	Jan	Feb	Mar	Apr	May	Jun	Jul	Aug	Sep	Oct	Nov	Dec	Year
	tx	39	37	34	29	24	19	21	23	26	29	34	37	39 °C
	t	23	22	19	14	11	8	8	9	11	14	18	21	15 °C
	tn	9	8	6	2	-1	-3	-3	-3	-1	3	4	7	-5 °C
	clothing	t	s	s	n	n	n	n	n	n	n	s	s	–
	stress	o	o										o	–
	p	50	55	86	50	41	45	28	13	46	48	44	52	558 mm
0.2 mm	dp	5	5	6	5	4	3	4	4	5	6	6	5	58 days
	sunshine	282	260	229	159	105	93	96	152	186	205	237	245	2249 hours
	%	62	68	60	48	34	33	32	46	52	51	55	53	51 %

5. Climatic area: Westernmost parts of Neuquén and Rio Negro

5.1. Subtropical Winter Rain Mountain Climate Cs

5.2. At about 850 m summer mild, winter cool.

5.3. Summer generally dry and sunny, scattered showers. Winter partly dry and sunny, partly showers or rain and overcast.

5.4. Wind: in the mountains frequent gales. Fog: in winter often at higher levels. Snow cover: at higher levels.

Chos Malal

latitude: 37°23′s longitude: 70°17′w height: 858 m

GCsbk	134-38	Jan	Feb	Mar	Apr	May	Jun	Jul	Aug	Sep	Oct	Nov	Dec	Year
	t	21	20	17	12	9	6	6	7	10	14	17	20	13 °C
	clothing	s	s	n	n	n	n	n	n	n	n	n	s	–
	stress	o	o										o	–
	p	11	2	7	20	52	42	43	36	19	15	12	6	265 mm

Bariloche

latitude: 41°06′s longitude: 71°10′w height: 836 m

GCslk	134-39	Jan	Feb	Mar	Apr	May	Jun	Jul	Aug	Sep	Oct	Nov	Dec	Year
	t	14	14	12	8	6	4	3	3	5	8	11	13	8 °C
	clothing	n	n	n	n	n	w	w	w	n	n	n	n	–
	stress													–
	p	33	27	47	65	160	165	161	121	67	41	34	33	954 mm
0.2 mm	dp	3	3	5	8	10	10	9	9	5	5	3	4	74 days
	sunshine	338	270	251	185	119	101	118	148	180	246	294	324	2574 hours
	%	74	70	65	56	39	36	40	46	51	60	68	69	58 %

6. *Climatic area:* Province: Tierra de Fuego

6.1. Subarctic Oceanic Climate EO and Tundra Climate FT

6.2. Cool throughout the year. At times from June to August temperature below −10 °C.

6.3. Variable weather prevails. Snowfall in winter, often overcast days.

6.4. Wind: frequent gales from the west. Fog: frequent. Snow cover: now and then, at sealevel.

Ushuaia

latitude: 54°48′s longitude: 68°19′w height: 6 m

FTkk	134-40	Jan	Feb	Mar	Apr	May	Jun	Jul	Aug	Sep	Oct	Nov	Dec	Year
	tx	22	22	19	16	13	10	9	11	14	17	20	19	24 °C
	t	9	9	8	6	3	2	2	2	4	6	7	8	6 °C
	tn	1	0	-1	-3	-8	-11	-12	-9	-6	-2	-2	0	-13 °C
	clothing	n	n	n	n	w	w	w	w	w	n	n	n	–
	stress													–
	p	58	50	57	46	48	45	47	49	38	37	50	49	574 mm
0.2 mm	dp	13	12	11	12	13	20	9	8	7	11	11	13	140 days
	sunshine	195	170	167	150	112	102	112	136	144	161	171	180	1800 hours
	%	38	41	43	48	43	47	47	46	41	38	36	34	41 %

4.5.134.2. Predominant general weather situations are described in 4.5.0.1. under Paraguay, Uruguay, Argentina. Hygrothermal diagrams of Buenos Aires and Ushuaia are to be found in 4.5.0.2.

Wind names: Norte is a warm wet northerly wind. If it is strong it is sure that a Pampero will follow, in accordance with the popular saying "Norte duro, Pampero seguro".

Pampero is an abruptly-starting storm blowing from the south or southwest. Pampero seco is not accompanied by rain.

Pampero sucio is a dust-carrying storm. It is normally followed by the Pampero limpio, the clean cold air behind the cold front.

Sudestado – also called Suestado – is a storm from the southeast after a cold front passage on the Argentine coast. Sudestado prolongado often lasts for several days, accompanied by strong rain.

Travestas are westerly or northwesterly, at times stormy, winds in Patagonia blowing from the Pacific Ocean.

Turbonados are heavy thunderstorm squalls following a Pampero and blowing from the south or west in the La Plata area.
Zonda is a warm dry fall-wind (foehn) blowing from the north or northwest in the Andes of Argentina. It is called after the place Zonda. In the Puna – the Andean Highlands – these winterly northern winds are famous for their biting cold.

4.5.135. Islands in the Southern Atlantic Ocean and stations in the Antarctic

4.5.135.1. Climatic descriptions and tables

1. Climatic area: Tristan da Cunha and Gough Island
1.1. Subtropical Rain Climate Cr
1.2. Mild all the year round.
1.3. Variable weather prevails.
1.4. Wind: frequent gales. Fog: frequent.

Tristan da Cunha

latitude: 37°03's longitude: 12°19'w height: 0 m

Crbl	135-1	Jan	Feb	Mar	Apr	May	Jun	Jul	Aug	Sep	Oct	Nov	Dec	Year
	tx	22	22	22	21	19	17	17	16	17	17	19	21	23 °C
	t	17	18	17	16	14	13	12	12	12	13	14	16	15 °C
	tn	11	11	11	9	7	6	6	4	4	7	8	9	3 °C
	clothing stress	n	s	n	n	n	n	n	n	n	n	n	n	–
	p	103	88	88	153	144	187	147	193	129	139	117	167	1655 mm

Gough Island

latitude: 40°21's longitude: 9°55'w height: 0 m

Crll	135-2	Jan	Feb	Mar	Apr	May	Jun	Jul	Aug	Sep	Oct	Nov	Dec	Year
	t	14	15	14	13	12	11	10	10	10	10	12	14	12 °C
	clothing stress	n	n	n	n	n	n	n	n	n	n	n	n	–
	p	226	197	206	271	252	410	313	379	380	374	277	312	3397 mm
	sunshine	179	147	125	98	85	52	69	83	102	142	157	186	1425 hours
	%	36	36	32	31	31	32	27	27	29	34	34	36	32 %

2. Climatic area: Falkland Islands
2.1. Tundra Climate FT
2.2. Cool all the year round.
2.3. Variable weather prevails.
2.4. Wind: frequent gales. Fog: frequent. Snow cover: variable.

Stanley Falkland Islands

FTkk	135-3	Jan	Feb	Mar	Apr	May	Jun	Jul	Aug	Sep	Oct	Nov	Dec	Year
				latitude: 51°42′s				longitude: 57°52′w			height: 53 m			
	tx	19	18	17	14	11	8	8	9	12	14	17	18	21 °C
	t	9	9	8	6	4	2	2	3	3	6	7	8	6 °C
	tn	1	1	0	-2	-3	-6	-6	-5	-3	-2	-1	1	-8 °C
	clothing stress	n	n	n	n	w	w	w	w	w	n	n	n	–
	p	79	55	44	46	58	47	50	42	36	34	41	72	598 mm
	dp	15	12	10	11	13	11	12	9	9	7	10	14	135 days
	sunshine	198	161	169	115	77	57	69	90	128	189	200	198	1651 hours
	%	40	39	43	36	28	24	27	30	37	45	43	38	37 %

3. Climatic area: South Georgia
3.1. Tundra Climate FT
3.2. Summer cool, winter cold.
3.3. Variable weather prevails.
3.4. Wind: frequent gales. Fog: frequent. Snow cover: in winter, variable.

Grytviken – South Georgia

FTko	135-4	Jan	Feb	Mar	Apr	May	Jun	Jul	Aug	Sep	Oct	Nov	Dec	Year
				latitude: 54°14′s				longitude: 36°30′w			height: 2 m			
	t	4	5	5	3	1	-1	-2	-1	0	2	4	4	2 °C
	clothing stress	w	n	n	w	w	w	w	w	w	w	w	w	–
	p	118	115	139	126	174	123	113	121	102	89	92	119	1432 mm
	dp	14	13	14	15	17	11	13	11	10	11	12	13	154 days
	sunshine	149	145	131	67	34	1	14	75	123	166	173	155	1233 hours
	%	29	35	34	22	13	0	6	25	35	39	36	29	28 %

Cumberland Bay – South Georgia

FTko	135-5	Jan	Feb	Mar	Apr	May	Jun	Jul	Aug	Sep	Oct	Nov	Dec	Year
				latitude: 54°16′s				longitude: 36°30′w			height: 24 m			
	tx	16	18	18	13	10	8	8	8	11	13	14	15	21 °C
	t	5	6	5	2	0	-2	-2	-2	0	1	3	4	3 °C
	tn	-2	-2	-3	-6	-8	-10	-11	-11	-9	-7	-4	-3	-13 °C
	clothing stress	n	n	n	w	w	w	w	w	w	w	w	w	–
	p	84	109	135	137	132	124	140	135	89	66	86	76	1313 mm
	dp	12	13	14	14	12	15	15	14	11	12	11	11	154 days

4. Climatic area: Southern Orkney Islands
4.1. Tundra Climate FT
4.2. Summer cool, winter cold.
4.3. Variable weather prevails.
4.4. Wind; frequent gales. Fog: frequent. Snow cover: permanent in winter.

Signy Island – Southern Orkney Islands (by the United Kingdom)

latitude: 60°43′s longitude: 45°36′w height: 7 m

FTkc	135-6	Jan	Feb	Mar	Apr	May	Jun	Jul	Aug	Sep	Oct	Nov	Dec	Year
	t	0	1	0	-2	-6	-9	-10	-8	-5	-2	-1	0	-4 °C
	clothing stress	w	w	w	w	w	w	p	w	w	w	w	w	–
	sunshine	60	51	36	30	20	16	29	48	58	66	75	52	541 hours
	%	11	12	9	10	9	9	15	18	17	15	15	9	12 %

Dest. Naval Orcadas Laurie Island – Southern Orkney Islands (by Argentina)

latitude: 60°44′s longitude: 44°44′w height: 4 m

FTkc	135-7	Jan	Feb	Mar	Apr	May	Jun	Jul	Aug	Sep	Oct	Nov	Dec	Year
	t	0	0	0	-3	-7	-10	-10	-10	-6	-3	-2	0	-4 °C
	clothing stress	w	w	w	w	w	p	p	p	w	w	w	w	–
	p	35	41	55	42	35	27	28	27	32	31	32	27	412 mm
	dp	8	9	11	11	9	8	9	9	9	8	9	7	107 days
	sunshine	42	32	31	25	15	10	18	36	57	59	55	51	431 hours
	%	7	7	8	8	7	6	9	13	17	13	11	9	10 %

5. Climatic area: Southern Shetland Islands
5.1. Tundra Climate FT
5.2. Summer cold, winter very cold.
5.3. Variable weather prevails.
5.4. Wind: frequent gales. Fog: frequent. Snow cover: most of the year.

Deception Island (by Argentina)

latitude: 62°59′s longitude: 60°43′w height: 8 m

FTko	135-8	Jan	Feb	Mar	Apr	May	Jun	Jul	Aug	Sep	Oct	Nov	Dec	Year
	t	1	1	0	-2	-4	-6	-8	-7	-5	-2	-1	1	-3 °C
	clothing stress	w	w	w	w	w	w	w	w	w	w	w	w	–
	p	54	87	68	48	24	28	34	22	28	39	32	34	498 mm
	dp	11	10	11	10	1	2	3	5	7	9	8	9	85 days
	sunshine	109	68	49	25	3	0	0	18	41	75	99	129	616 hours
	%	18	15	12	9	1	0	0	7	12	17	19	21	14 %

Argentine Island (by United Kingdom)

latitude: 65°15′s longitude: 64°16′w height: 10 m

FTkc	135-9	Jan	Feb	Mar	Apr	May	Jun	Jul	Aug	Sep	Oct	Nov	Dec	Year
	t	0	0	-1	-5	-8	-10	-13	-11	-9	-5	-2	-1	-5 °C
	clothing stress	w	w	w	w	w	p	p	p	w	w	w	w	–
	sunshine	120	87	65	41	19	7	18	38	52	77	95	147	766 hours
	%	19	19	16	15	10	7	12	15	15	17	17	21	17 %

6. Climatic area: Antarctic Continent
6.1. Ice Climate FI, coastal regions often Tundra Climate FT
6.2. Coastal regions summer cold, winter very cold, otherwise colder.
6.3. Variable weather prevails.
6.4. Wind: in some regions very frequent gales.

Mean air temperature observed at the Stations in the Antarctic

Station	operated by	latitude	longitude	height		Climate	Jan	Feb	Mar	Apr	May	Jun	Jul	Aug	Sep	Oct	Nov	Dec	Year
Mawson	Australia	67°36's	62°53'e	14 m	135-10	FTkc	0	-5	-10	-14	-15	-16	-18	-18	-19	-13	-5	-1	-11 °C
Sanae	South Africa	70°19's	2°22'w	52 m	135-11	FIod	-4	-9	-14	-21	-24	-23	-26	-27	-27	-19	-10	-4	-17 °C
Scott Base	New Zealand	77°51's	166°45'e	16 m	135-12	FIod	-6	-12	-21	-24	-28	-27	-31	-32	-26	-24	-12	-6	-21 °C
Vostok	USSR	78°27's	106°52'e	3420 m	135-13	HFIde	-33	-43	-57	-64	-66	-68	-67	-71	-67	-58	-44	-33	-56 °C
Byrd Station	USA	80°00's	120°00'w	1500 m	135-14	GFIcd	-16	-20	-29	-30	-32	-34	-35	-37	-36	-31	-21	-15	-28 °C
Amundsen-Scott	USA	90°00's		2800 m	135-15	HFIde	-29	-40	-54	-59	-57	-56	-60	-59	-59	-51	-39	-28	-49 °C

4.5.135.2. Predominant general weather situations are described in 4.5.0.1. under Southern Atlantic Ocean and Antarctic.

4.6. Australia and Oceania

4.6.0.1. Predominant general weather situations

New Zealand
In summer (October to March), a wedge of the South Pacific anticyclone lying north of New Zealand governs the weather, with its sinking air, more often than eastward-tracking disturbances in the middle-latitude westerlies. These prevail in winter (April to September), when the wedge lies somewhat more to the north; it is also stronger than in summer. However, the weather mostly clears up without delay after the passage of the secondary depressions and the accompanying rain belts.

Australia
In summer (October to March), the weather in the north and north-east is dominated by the Equatorial Trough. Its disturbances, usually migrating from east to west, govern the weather, intensifying shower activity as they approach and weakening it as they recede. The anticyclone over the southern Indian Ocean – somewhat weakened and removed to the south in comparison with its position in winter – extends a wedge beyond the Great Australian Bight to Tasmania. It often moves away when depressions or secondary depressions approach from the west. At times these depressions extend further into the continent. In winter (April to September), the Equatorial Trough is far to the north, and does not usually influence the country. There is a bridge between the strong anticyclone over the southern Indian Ocean and an anticyclone over Australia, which dominates the weather with its sinking air. Especially in the south, this sinking air is frequently replaced by the rising air of the depressions and troughs in the strong winterly westerlies.

Oceania
The weather is governed by the Equatorial Trough. In southern summer (October to March) it is near the equator in the east and over the Coral Sea. In southern winter (April to September) it moves far to the north; in the east it is over the Line Is., in the west north of the Caroline Is. There are disturbances in the trough, migrating from east to west; they have intense general shower activity in front of them, and weak to the rear. Now and then they develop into tropical depressions, and sometimes further into tropical cyclones.

Temperate South Pacific marine zone
This is the zone of the middle-latitude westerlies in which, on the southern slope of the subtropical anticyclonic belt, depressions and anticyclones alternate in rapid succession. In winter, the drift is rather more strongly developed than in summer. At times, it happens that the drift is stopped, and wide areas of the zone are dominated by an extensive depression or anticyclone.

4.6.0.2. Hygrothermal diagrams

In the following figures, hygrothermal diagrams are given for the stations
Wellington, Darwin, Alice Springs, Perth, Adelaide, Sydney, Apia, and Papeete.

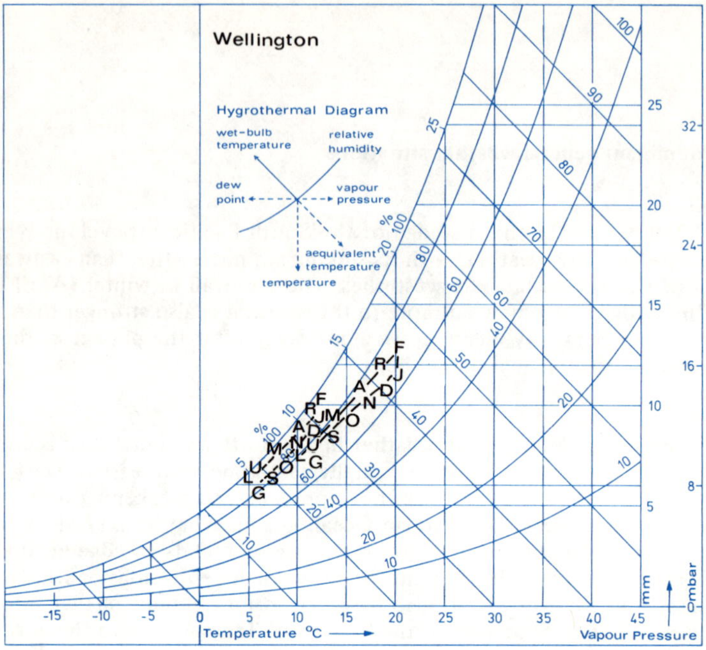
Wellington
Hygrothermal Diagram
wet-bulb temperature
relative humidity
dew point
vapour pressure
aequivalent temperature
temperature
Temperature °C
Vapour Pressure
mm
mbar

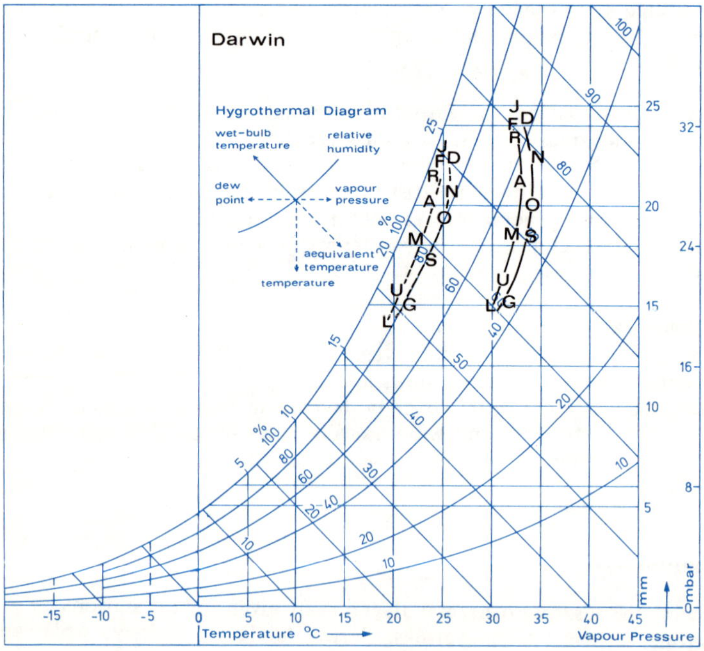
Darwin
Hygrothermal Diagram
wet-bulb temperature
relative humidity
dew point
vapour pressure
aequivalent temperature
temperature
Temperature °C
Vapour Pressure
mm
mbar

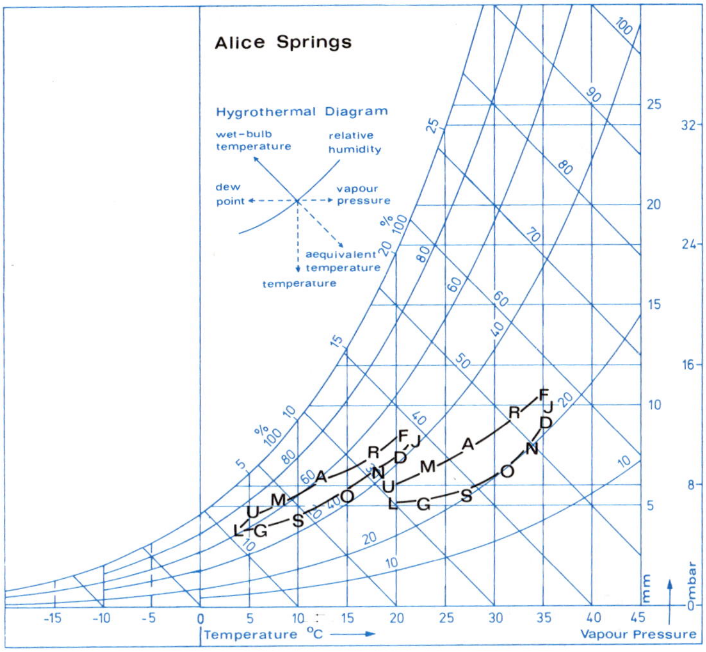
Alice Springs
Hygrothermal Diagram
wet-bulb temperature
relative humidity
dew point
vapour pressure
aequivalent temperature
temperature
Temperature °C
Vapour Pressure
mm
mbar

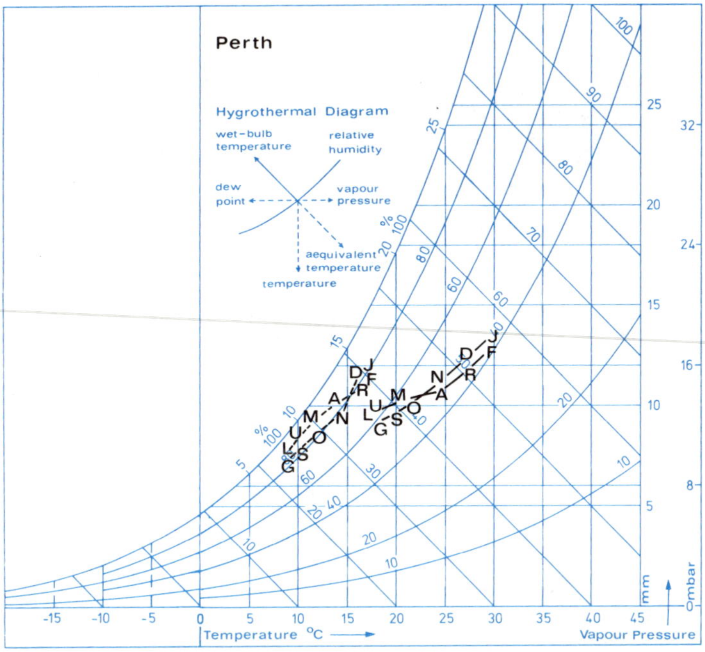
Perth
Hygrothermal Diagram
wet-bulb temperature
relative humidity
dew point
vapour pressure
aequivalent temperature
temperature
Temperature °C
Vapour Pressure
mm
mbar

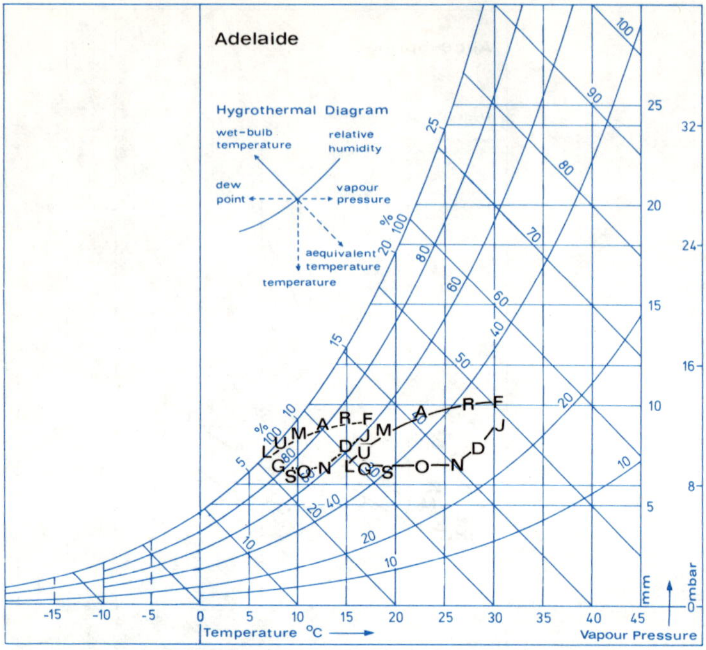
Adelaide
Hygrothermal Diagram
wet-bulb temperature
relative humidity
dew point
vapour pressure
aequivalent temperature
temperature
Temperature °C
Vapour Pressure
mm
mbar

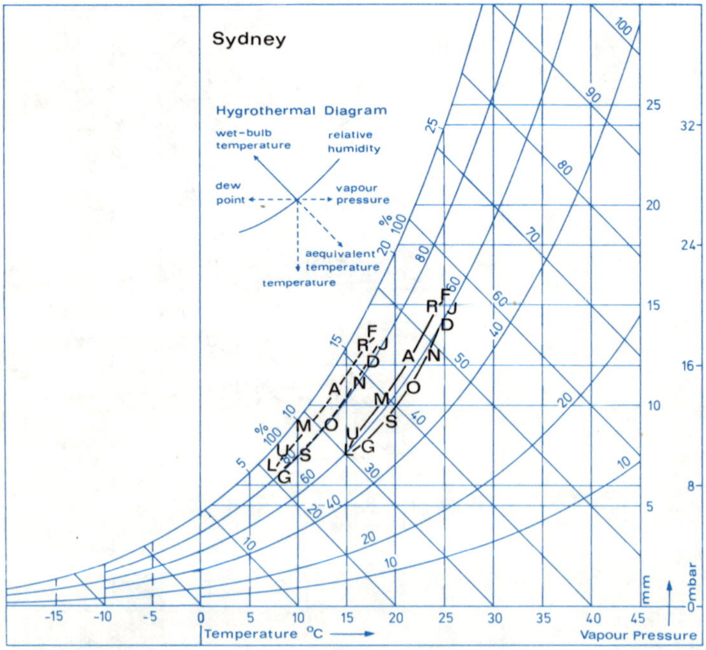
Sydney
Hygrothermal Diagram
wet-bulb temperature
relative humidity
dew point
vapour pressure
aequivalent temperature
temperature
Temperature °C
Vapour Pressure
mm
mbar

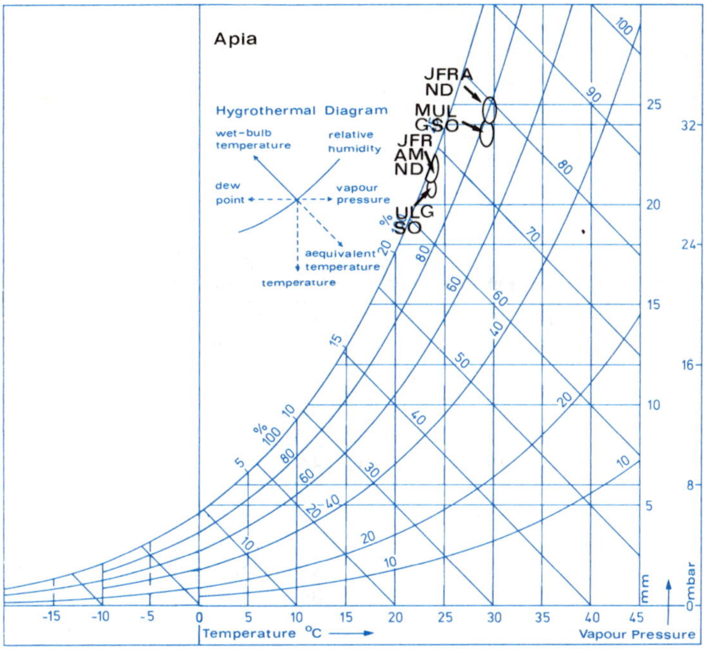
Apia
Hygrothermal Diagram
wet-bulb temperature
relative humidity
dew point
vapour pressure
aequivalent temperature
temperature
Temperature °C
Vapour Pressure
mm
mbar

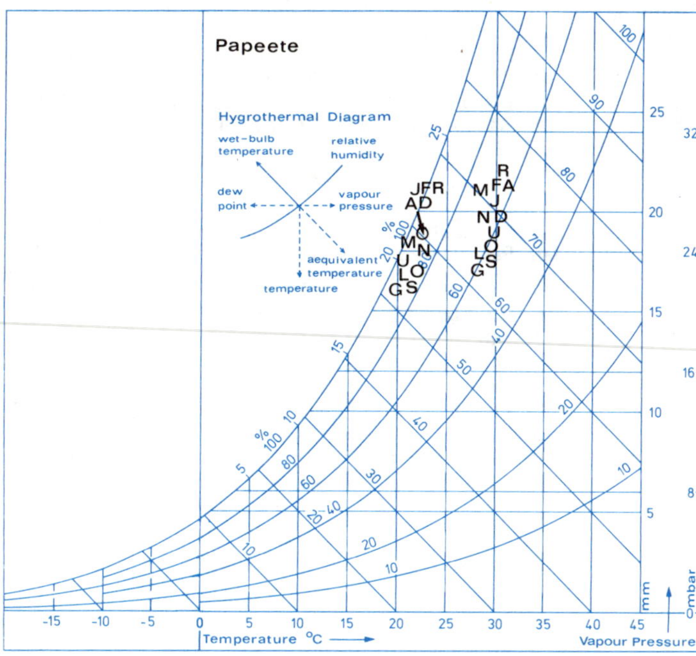
Papeete
Hygrothermal Diagram
wet-bulb temperature
relative humidity
dew point
vapour pressure
aequivalent temperature
temperature
Temperature °C
Vapour Pressure
mm
mbar

4.6.136. New Zealand

4.6.136.1. Climatic descriptions and tables

1. Climatic area: North Island
1.1. Subtropical Rain Climate Cr and Subtropical Rain Mountain Climate GCr
1.2. Summer warm, winter mild.
1.3. Variable weather prevails.
1.4. Wind: occasional gales. Fog: occasionally.

Kaitaia

latitude: 35°04′s longitude: 173°17′e height: 80 m

Crbl	136-1	Jan	Feb	Mar	Apr	May	Jun	Jul	Aug	Sep	Oct	Nov	Dec	Year
	t	19	20	18	16	14	12	11	12	13	14	16	17	15 °C
	clothing	s	s	s	n	n	n	n	n	n	n	n	n	–
	stress	o	o	o										–
	p	96	86	71	124	142	152	150	137	130	112	89	81	1370 mm
0.2 mm	dp	10	11	12	14	20	20	19	22	14	17	14	12	185 days
	sunshine	229	185	183	159	133	130	151	155	172	184	202	209	2092 hours
	%	52	49	48	47	42	44	49	46	48	46	48	46	47 %

Auckland

latitude: 36°51′s longitude: 174°46′e height: 38 m

Crbl	136-2	Jan	Feb	Mar	Apr	May	Jun	Jul	Aug	Sep	Oct	Nov	Dec	Year
	tx	26	26	25	23	19	17	17	17	18	20	23	24	27 °C
	t	20	20	19	17	14	12	11	12	13	15	17	18	16 °C
	tn	12	12	11	11	8	6	4	3	4	5	7	8	3 °C
	clothing	s	s	s	n	n	n	n	n	n	n	n	s	–
	stress	o	o	o									o	–
	p	46	123	87	136	138	126	139	142	93	110	84	97	1323 mm
0.2 mm	dp	10	10	12	15	19	20	20	19	16	17	15	12	185 days
	sunshine	233	194	190	155	138	123	129	151	168	181	207	224	2093 hours
	%	52	51	50	46	44	42	42	45	47	45	49	49	47 %

New Plymouth

latitude: 39°02′s longitude: 174°11′e height: 48 m

Crlk	136-3	Jan	Feb	Mar	Apr	May	Jun	Jul	Aug	Sep	Oct	Nov	Dec	Year
	tx	25	26	24	22	19	17	16	18	18	19	21	23	27 °C
	t	16	17	16	14	12	10	9	10	11	12	14	15	13 °C
	tn	7	8	7	5	2	1	0	1	2	3	5	7	-1 °C
	clothing	n	n	n	n	n	n	n	n	n	n	n	n	–
	stress	o	o											–
	p	112	114	86	130	152	165	152	147	107	140	114	127	1546 mm
0.2 mm	dp	13	11	12	14	17	18	19	18	18	18	16	14	188 days
	sunshine	245	194	196	165	145	116	135	154	168	175	201	216	2110 hours
	%	54	51	51	50	47	41	45	47	47	43	47	47	48 %

Gisborne

Crbl	136-4	Jan	Feb	Mar	Apr	May	Jun	Jul	Aug	Sep	Oct	Nov	Dec	Year
		latitude: 38°40′s				longitude: 177°59′e				height: 8 m				
	t	19	20	18	16	14	12	11	12	13	14	16	17	15 °C
	clothing	s	s	s	n	n	n	n	n	n	n	n	n	–
	stress	o	o	o										–
	p	66	78	86	99	114	91	130	109	81	71	61	53	1039 mm
0.2 mm	dp	10	10	12	14	14	15	16	16	12	14	11	10	154 days
	sunshine	252	207	196	158	139	136	130	150	179	200	220	245	2212 hours
	%	56	54	51	47	45	48	43	46	50	49	52	53	50 %

Wellington

Crlk	136-5	Jan	Feb	Mar	Apr	May	Jun	Jul	Aug	Sep	Oct	Nov	Dec	Year
		latitude: 41°17′s				longitude: 174°46′e				height: 119 m				
	tx	26	26	24	21	18	16	10	10	18	20	22	24	27 °C
	t	15	16	15	13	11	9	8	8	10	11	13	14	12 °C
	tn	8	8	7	5	3	3	1	1	2	3	5	7	0 °C
	clothing	n	n	n	n	n	n	n	n	n	n	n	n	–
	stress	o	o											–
	p	74	104	80	90	127	123	128	116	92	116	79	95	1224 mm
0.2 mm	dp	10	9	10	13	15	16	18	16	14	14	12	12	159 days
	sunshine	234	195	189	151	118	106	108	139	170	183	197	222	2012 hours
	%	51	50	49	46	39	38	37	43	48	45	45	47	45 %

2. Climatic area: South Island

2.1. Temperate Oceanic Climate DO and Temperate Oceanic Mountain Climate GDO

2.2. Summer mild, winter cool.

2.3. Variable weather prevails.

2.4. Wind: occasional gales. Fog: occasionally.

Hokitita South

DOlk	136-6	Jan	Feb	Mar	Apr	May	Jun	Jul	Aug	Sep	Oct	Nov	Dec	Year
		latitude: 42°43′s				longitude: 170°59′e				height: 40 m				
	tx	23	23	21	20	18	15	15	16	17	18	19	22	24 °C
	t	15	15	14	12	9	7	6	8	9	11	12	14	11 °C
	tn	6	7	5	4	0	-1	-2	-2	2	2	3	4	-2 °C
	clothing	n	n	n	n	n	n	n	n	n	n	n	n	–
	stress													–
	p	244	246	203	226	244	211	221	218	190	246	241	231	2721 mm
0.2 mm	dp	14	12	14	15	15	15	16	16	17	19	18	16	187 days
	sunshine	208	170	166	140	122	107	129	138	152	160	173	190	1855 hours
	%	45	44	43	43	40	39	45	43	43	39	40	40	42 %

Christchurch

DOlk	136-7	Jan	Feb	Mar	Apr	May	Jun	Jul	Aug	Sep	Oct	Nov	Dec	Year
		latitude: 43°29′s				longitude: 172°32′e				height: 34 m				
	tx	30	28	27	24	21	17	16	18	21	24	26	29	31 °C
	t	16	17	15	12	9	6	5	7	9	11	14	16	11 °C
	tn	5	5	3	1	-2	-3	-3	-3	-1	0	2	4	-4 °C
	clothing	n	n	n	n	n	n	n	n	n	n	n	n	–
	stress													–
	p	47	51	52	51	81	52	53	51	48	49	42	56	633 mm
	dp	10	8	9	10	12	13	14	11	10	11	10	11	129 days
	sunshine	215	182	174	143	133	116	126	145	169	187	206	194	1990 hours
	%	46	47	45	44	44	43	44	45	48	46	47	41	45 %

Dunedin

DOlk	136-8	Jan	Feb	Mar	Apr	May	Jun	Jul	Aug	Sep	Oct	Nov	Dec	Year
		latitude: 45°52′s						longitude: 170°32′e				height: 73 m		
	t	14	14	13	11	8	7	6	7	9	10	12	14	11 °C
	clothing stress	n	n	n	n	n	n	n	n	n	n	n	n	–
	p	86	71	76	71	83	83	79	76	69	76	81	89	940 mm
0.2 mm	dp	14	11	13	13	14	13	13	13	14	14	14	15	161 days
	sunshine	195	170	152	141	112	93	112	133	138	167	174	186	1773 hours
	%	41	43	39	43	38	36	40	42	39	40	39	38	40 %

Invercargill

DOlk	136-9	Jan	Feb	Mar	Apr	May	Jun	Jul	Aug	Sep	Oct	Nov	Dec	Year
		latitude: 46°25′s						longitude: 168°19′e				height: 1 m		
	tx	27	27	24	22	18	15	14	16	19	23	23	26	28 °C
	t	13	14	12	10	8	6	5	6	8	10	11	12	10 °C
	tn	3	3	1	0	-3	-3	-4	-3	-1	0	1	3	-4 °C
	clothing stress	n	n	n	n	n	n	n	n	n	n	n	n	–
	p	99	94	96	98	98	104	75	71	72	80	88	88	1063 mm
0.2 mm	dp	17	16	17	18	18	20	17	15	15	17	18	18	206 days
	sunshine	198	162	146	105	93	68	97	121	144	166	172	189	1661 hours
	%	41	41	38	32	32	26	35	38	41	39	38	39	37 %

4.6.136.2. Predominant general weather situations are described in 4.6.0.1. under New Zealand. A hygrothermal diagram of Wellington is to be found in 4.6.0.2.
Wind names: North-Easter or black North-Easter is the name of a stormy, gusty and rainy wind blowing from the northeast, especially on the east coast of the North Island. Northwester is a warm dry fall-wind blowing from the northwest on to the plain of Canterbury in the South Island.

4.6.137. Australia

4.6.137.1. Climatic descriptions and tables

1. *Climatic area:* Northern Coast of the Kimberley Division of Western Australia, Arnhem-Land and coastal lowlands on the Gulf of Carpentaria in the Northern Territory and Queensland, as well as the tropical east of Queensland east of the Great Dividing Range
1.1. Tropical Summer Rain Climate Aw and Tropical Summer Rain Mountain Climate GAw, in parts: Tropical Monsoonal Rain Climate Am
1.2. In the north summer very hot, winter hot; in the south summer hot, winter warm.
1.3. High sun period: in the morning often dry and sunny, showers in the afternoon, at times rainy days. Low sun period: generally dry and sunny, now and then showers, particularly on the east coast of Queensland.
1.4. Wind: occasional gales (January to March tropical cyclones possible), thunderstorm squalls.

Thursday Island – Queensland

Awha	137-1	latitude: 10° 35′s				longitude: 142° 13′e				height: 58 m				
		Jan	Feb	Mar	Apr	May	Jun	Jul	Aug	Sep	Oct	Nov	Dec	Year
	tx	33	32	32	32	31	30	29	30	31	32	33	34	34 °C
	t	28	28	27	28	27	26	26	26	26	28	28	28	27 °C
	tn	23	23	23	23	23	22	21	21	22	23	23	23	21 °C
	clothing	t	t	t	t	t	t	t	t	t	t	t	t	–
	stress	ff	ff	ff	ff	ff	ff	ff	ff	ff	ff	ff	ff	–
	p	385	335	350	201	64	22	10	11	3	11	39	175	1607 mm
0.2 mm	dp	21	20	20	14	10	8	7	5	4	3	4	13	129 days

Darwin – Northern Territory

Awha	137-2	latitude: 12° 26′s				longitude: 130° 52′e				height: 27 m				
		Jan	Feb	Mar	Apr	May	Jun	Jul	Aug	Sep	Oct	Nov	Dec	Year
	tx	36	35	36	36	35	34	33	34	36	37	37	36	37 °C
	t	28	28	28	28	27	25	25	26	28	29	29	29	28 °C
	tn	23	23	23	22	19	17	16	18	21	22	22	22	16 °C
	clothing	t	t	t	t	t	t	t	t	t	t	t	t	–
	stress	ff	ff	ff	ff	ff	ff	ff	ff	ff	ff	ff	ff	–
	p	341	338	274	121	9	1	2	5	17	66	156	233	1562 mm
0.2 mm	dp	20	18	17	6	1	1	0	0	2	5	10	15	95 days
	sunshine	176	169	214	254	292	295	304	321	297	295	254	218	3089 hours
	%	44	48	57	72	82	86	86	88	82	77	67	55	65 %

Cairns – Queensland

Amab	137-3	latitude: 16° 55′s				longitude: 145° 46′e				height: 7 m				
		Jan	Feb	Mar	Apr	May	Jun	Jul	Aug	Sep	Oct	Nov	Dec	Year
	tx	36	36	33	32	30	29	29	29	31	33	34	36	38 °C
	t	27	27	27	25	24	22	21	22	23	25	26	27	25 °C
	tn	21	21	20	18	14	12	11	11	14	16	18	19	9 °C
	clothing	t	t	t	t	t	s	s	s	t	t	t	t	–
	stress	ff	ff	ff	ff	f	f	o	f	f	f	ff	ff	–
	p	503	391	441	160	90	48	23	34	41	32	83	98	1944 mm
0.2 mm	dp	16	16	19	15	12	10	8	8	7	7	10	12	140 days

Townsville – Queensland

Awab	137-4	latitude: 19° 15′s				longitude: 146° 46′e				height: 4 m				
		Jan	Feb	Mar	Apr	May	Jun	Jul	Aug	Sep	Oct	Nov	Dec	Year
	tx	33	34	33	32	29	28	28	28	30	32	32	33	36 °C
	t	27	27	26	25	22	20	19	20	22	25	26	27	24 °C
	tn	22	22	21	18	13	11	10	12	15	18	21	22	9 °C
	clothing	t	t	t	t	s	s	s	s	s	t	t	t	–
	stress	ff	ff	ff	ff	f	f	o	f	f	ff	ff	ff	–
	p	332	364	275	83	34	26	22	10	10	21	55	102	1333 mm
0.2 mm	dp	14	16	13	7	6	5	3	2	2	4	6	8	86 days
	sunshine	223	201	214	255	226	240	273	285	282	294	276	260	3030 hours
	%	55	56	56	73	65	73	79	80	78	76	71	63	68 %

2. *Climatic area:* Western Australia: Kimberley Divison (except north coast), west of Northwest Division, northeast of Southwest Division, southernmost part of Central Division; north of Northern Territory (except Arnhem Land and the coastal plains); western Queensland (except the coastal plains and Gregory North and Gregory South); New South Wales west of the Great Dividing Range
2.1. Steppe Climate BS and Steppe Mountain Climate GBS
2.2. In the north: summer very hot, winter warm; in the south: summer hot, winter mild.
2.3. Generally dry and sunny; in northern regions during the high sun period frequent showers, in southern regions fewer showers; during the low sun period in the north scattered showers, in the south showers now and then.
2.4. Wind: occasional gales and dust-storms.

Daly Waters – Northern Territory

latitude: 16°16′s longitude: 133°23′e height: 211 m

BShb	137-5	Jan	Feb	Mar	Apr	May	Jun	Jul	Aug	Sep	Oct	Nov	Dec	Year
	t	30	30	29	27	24	21	21	23	26	30	31	31	27 °C
	clothing	t	t	t	t	t	s	s	t	t	t	t	t	–
	stress	ff	ff	ff	ff	f	f	f	f	f	ff	ff	ff	–
	p	165	152	122	23	5	8	2	2	5	20	51	102	667 mm
0.2 mm	dp	12	11	8	2	1	1	0	1	1	2	6	9	53 days
	sunshine	251	240	270	273	307	303	316	338	315	310	279	276	3478 hours
	%	62	67	71	77	87	90	91	94	87	80	72	68	79 %

Burketown – Queensland

latitude: 17°45′s longitude: 139°33′e height: 9 m

BShb	137-6	Jan	Feb	Mar	Apr	May	Jun	Jul	Aug	Sep	Oct	Nov	Dec	Year
	t	30	29	28	27	23	21	20	21	24	27	29	30	26 °C
	clothing	t	t	t	t	t	s	s	s	t	t	t	t	–
	stress	ff	ff	ff	ff	f	f	o	f	f	ff	ff	ff	–
	p	208	160	132	25	5	8	1	1	1	13	38	111	703 mm
0.2 mm	dp	10	10	7	2	1	1	0	0	0	1	3	6	41 days

Broome – Western Australia

latitude: 17°57′s longitude: 122°13′e height: 12 m

BShb	137-7	Jan	Feb	Mar	Apr	May	Jun	Jul	Aug	Sep	Oct	Nov	Dec	Year
	t	30	30	30	29	25	22	22	23	25	27	29	30	27 °C
	clothing	t	t	t	t	t	s	s	t	t	t	t	t	–
	stress	ff	ff	ff	ff	f	o	o	o	f	f	ff	ff	–
	p	121	148	75	32	38	22	8	2	2	1	10	34	492 mm
0.2 mm	dp	11	9	7	3	2	2	2	0	1	0	1	5	45 days

Cloncurry – Queensland

latitude: 20°40′s longitude: 140°30′e height: 188 m

BShb	137-8	Jan	Feb	Mar	Apr	May	Jun	Jul	Aug	Sep	Oct	Nov	Dec	Year
	tx	42	41	39	36	33	31	30	33	36	40	41	42	43 °C
	t	31	30	28	25	21	18	18	19	23	27	30	31	25 °C
	tn	21	20	18	14	9	7	6	7	11	14	18	19	6 °C
	clothing	t	t	t	t	s	s	s	s	t	t	t	t	–
	stress	ff	ff	ff	f	f	o	o	f	f	f	ff	ff	–
	p	112	130	78	16	26	9	9	6	8	18	26	68	505 mm
0.2 mm	dp	8	6	5	1	2	1	1	1	1	2	3	7	38 days

Carnarvon – Western Australia

latitude: 24°53′s longitude: 113°39′e height: 5 m

BSal	137-9	Jan	Feb	Mar	Apr	May	Jun	Jul	Aug	Sep	Oct	Nov	Dec	Year
	tx	41	41	40	37	32	28	27	28	31	34	37	40	43 °C
	t	24	25	24	22	18	16	15	15	18	19	21	22	20 °C
	tn	18	18	17	14	9	7	6	7	9	12	14	17	6 °C
	clothing	t	t	t	s	s	n	n	n	s	s	s	s	–
	stress	ff	ff	ff	f	o	o	o	o	o	f	f	ff	–
	p	5	22	23	7	49	50	52	16	5	4	1	1	233 mm
0.2 mm	dp	2	2	2	2	5	6	6	5	2	2	0	1	35 days

Bourke – New South Wales

latitude: 30°00′s longitude: 145°58′e height: 110 m

BShl	137-10	Jan	Feb	Mar	Apr	May	Jun	Jul	Aug	Sep	Oct	Nov	Dec	Year
	t	29	28	25	20	16	12	11	14	17	21	26	28	21 °C
	clothing	t	t	t	s	n	n	n	n	n	s	t	t	–
	stress	ff	ff	f	o					o	f	f	ff	–
	p	36	38	28	28	25	28	23	20	20	23	31	35	335 mm
0.2 mm	dp	3	3	3	3	4	4	5	3	3	4	4	5	44 days

Kalgoorlie – Western Australia

latitude: 30°46′s longitude: 121°27′e height: 361 m

BSal	137-11	Jan	Feb	Mar	Apr	May	Jun	Jul	Aug	Sep	Oct	Nov	Dec	Year
	tx	42	41	38	33	28	24	23	26	31	35	39	40	43 °C
	t	26	24	23	18	14	12	11	12	15	18	21	24	18 °C
	tn	12	12	11	7	3	3	1	1	3	4	8	11	1 °C
	clothing	t	t	t	s	n	n	n	n	n	s	s	t	–
	stress	f	f	f	o						o	f	f	–
	p	24	29	18	19	23	39	27	18	26	9	14	13	259 mm
0.2 mm	dp	3	2	4	4	5	6	7	6	3	4	3	2	49 days

Broken Hill – New South Wales

latitude: 31°57′s longitude: 141°28′e height: 300 m

BSal	137-12	Jan	Feb	Mar	Apr	May	Jun	Jul	Aug	Sep	Oct	Nov	Dec	Year
	t	25	25	22	18	14	11	10	12	15	18	22	24	18 °C
	clothing	t	t	s	s	n	n	n	n	n	s	s	t	–
	stress	f	f	o	o						o	o	f	–
	p	15	24	14	17	24	23	18	16	17	21	22	22	233 mm
0.2 mm	dp	3	3	2	3	4	5	6	5	4	4	4	3	46 days

Mildura – Victoria

latitude: 34°14′s longitude: 142°05′e height: 50 m

BSal	137-13	Jan	Feb	Mar	Apr	May	Jun	Jul	Aug	Sep	Oct	Nov	Dec	Year
	t	25	23	22	17	13	11	10	11	14	16	19	22	17 °C
	clothing	t	t	s	n	n	n	n	n	n	n	s	s	–
	stress	f	f	o								o	o	–
	p	21	19	16	17	35	28	30	33	28	38	31	13	309 mm
	sunshine	333	283	284	222	178	151	180	212	242	253	285	315	2938 hours
	%	76	75	74	66	55	51	58	63	68	63	68	70	66 %

Hay – New South Wales

latitude: 34°30′s longitude: 144°56′e height: 95 m

BSak	137-14	Jan	Feb	Mar	Apr	May	Jun	Jul	Aug	Sep	Oct	Nov	Dec	Year
	t	24	24	22	17	13	10	9	11	13	17	21	25	17 °C
	clothing	t	t	s	n	n	n	n	n	n	n	s	t	–
	stress	f	f	o								o	f	–
	p	23	22	25	31	36	40	30	33	31	30	23	25	349 mm
0.2 mm	dp	3	3	3	4	6	8	8	8	6	5	4	4	62 days

3. *Climatic area:* Western Australia: East of Northwest Division, Eastern Division with Great Sand Desert, Gibson Desert, Victoria Desert, most of Central Division; South of Northern Territory; Gregory North and Gregory South of Queensland; North and West of South Australia

3.1. Desert Climate BW and Desert Mountain Climate GBW, coastal areas: Marine Desert Climate BM

3.2. Summer very hot, warm near coast of the Great Australian Bight, winter warm in the north, mild in the south.

3.3. Generally dry and sunny; high sun period: now and then showers, in the south also in low sun period.

3.4. Wind: occasional gales and dust or sand storms.

Port Hedland – Western Australia

latitude: 20°23′s longitude: 118°57′e height: 8 m

BMhb	137-15	Jan	Feb	Mar	Apr	May	Jun	Jul	Aug	Sep	Oct	Nov	Dec	Year
	t	30	31	30	28	23	20	19	21	24	26	29	30	26 °C
	clothing	t	t	t	t	t	s	s	s	t	t	t	t	–
	stress	ff	ff	ff	ff	f	o	o	o	o	f	f	f	–
	p	56	86	36	15	45	21	15	4	1	1	1	19	299 mm

Onslow – Western Australia

latitude: 21°43′s longitude: 114°57′e height: 4 m

BMhb	137-16	Jan	Feb	Mar	Apr	May	Jun	Jul	Aug	Sep	Oct	Nov	Dec	Year
	t	30	30	29	26	22	19	18	19	22	24	27	28	24 °C
	clothing	t	t	t	t	s	s	s	s	s	t	t	t	–
	stress	ff	ff	ff	f	o	o	o	o	o	f	f	ff	–
	p	23	28	46	25	38	41	20	10	1	1	0	5	238 mm
0.2 mm	dp	3	3	4	2	3	3	2	2	0	0	0	1	23 days

Alice Springs – Northern Territory

latitude: 23°48′s longitude: 133°53′e height: 548 m

GBWha	137-17	Jan	Feb	Mar	Apr	May	Jun	Jul	Aug	Sep	Oct	Nov	Dec	Year
	tx	42	40	38	34	31	27	27	31	35	38	40	41	42 °C
	t	28	27	25	20	16	12	12	14	18	23	25	28	21 °C
	tn	13	13	10	4	1	-2	-3	-1	2	6	10	12	-3 °C
	clothing	t	t	t	s	n	n	n	n	s	t	t	t	–
	stress	f	f	f	o	o				o	f	f	f	–
	p	27	45	18	10	18	15	14	10	6	25	24	39	250 mm
0.2 mm	dp	4	3	3	2	2	2	1	2	1	3	4	4	31 days
	sunshine	317	300	302	300	257	260	276	306	307	322	331	314	3592 hours
	%	76	82	79	87	75	81	82	87	86	82	83	74	81 %

Mundivindi – West Australia

latitude: 23°52′s longitude: 120°10′e height: 561 m

GBWhl	137-18	Jan	Feb	Mar	Apr	May	Jun	Jul	Aug	Sep	Oct	Nov	Dec	Year
	t	28	30	28	23	18	14	13	16	20	23	28	30	23 °C
	clothing	t	t	t	t	s	n	n	n	s	t	t	t	–
	stress	f	f	f	o	o				o	o	f	f	–
	p	25	48	51	20	15	23	3	7	8	13	12	31	256 mm
0.2 mm	dp	6	5	6	3	3	3	2	2	0	1	3	4	38 days
	sunshine	291	257	260	297	251	213	273	285	297	301	324	319	3369 hours
	%	70	71	68	86	74	66	82	81	83	77	81	76	76 %

Windorah – Queensland

latitude: 25°26′s longitude: 142°36′e height: 119 m

BWhl	137-19	Jan	Feb	Mar	Apr	May	Jun	Jul	Aug	Sep	Oct	Nov	Dec	Year
	t	31	30	28	23	18	14	14	16	20	24	28	30	23 °C
	clothing	t	t	t	t	s	n	n	n	s	t	t	t	–
	stress	f	f	f	o	o				o	o	f	f	–
	p	35	41	40	23	20	20	13	10	12	15	23	36	288 mm
0.2 mm	dp	4	4	3	2	2	3	2	2	2	3	3	4	34 days

Oodnadatta – South Australia

latitude: 27°53′s longitude: 135°29′e height: 113 m

BWhl	137-20	Jan	Feb	Mar	Apr	May	Jun	Jul	Aug	Sep	Oct	Nov	Dec	Year
	clothing	t	t	t	t	s	n	n	n	n	s	t	t	–
	stress	f	f	f	o	o					o	f	f	–
	sunshine	340	303	310	269	236	218	244	275	298	308	320	333	3454 hours
	%	80	82	81	78	71	70	75	80	83	78	79	77	78 %

William Creek – South Australia

latitude: 28°55′s longitude: 136°21′e height: 75 m

BWhk	137-21	Jan	Feb	Mar	Apr	May	Jun	Jul	Aug	Sep	Oct	Nov	Dec	Year
	t	28	28	25	20	15	12	9	13	17	21	25	27	20 °C
	clothing	t	t	t	s	n	n	n	n	n	s	t	t	–
	stress	f	f	f	o					o	f	f	f	–
	p	13	15	8	7	8	13	5	7	8	12	13	18	127 mm
0.2 mm	dp	2	1	2	2	2	3	2	2	2	3	2	2	25 days

Woomera – South Australia

latitude: 31°09′s longitude: 136°48′e height: 166 m

BWhk	137-22	Jan	Feb	Mar	Apr	May	Jun	Jul	Aug	Sep	Oct	Nov	Dec	Year
	clothing	t	t	t	s	n	n	n	n	n	s	t	t	–
	stress	f	f	f	o						o	f	f	–
	sunshine	336	305	296	254	220	209	217	250	265	298	308	315	3273 hours
	%	78	82	78	75	67	69	68	73	74	75	75	72	74 %

Forrest – Western Australia

latitude: 30°50′s longitude: 128°06′e height: 157 m

BWal	137-23	Jan	Feb	Mar	Apr	May	Jun	Jul	Aug	Sep	Oct	Nov	Dec	Year
	t	23	23	21	17	14	12	11	11	14	17	20	21	17 °C
	clothing	t	t	s	n	n	n	n	n	n	n	s	s	–
	stress	f	f	o								o	o	–
	p	19	16	9	16	11	18	15	15	11	18	15	12	175 mm

4. *Climatic area:* West Australia: Southwest Division and western Eucla Division; southeastern South Australia; northwestern Victoria
4.1. Subtropical Winter Rain Climate Cs and Subtropical Winter Rain Mountain Climate GCs
4.2. Summer hot or warm, winter mild.
4.3. Summer generally dry and sunny, scattered showers; winter: variable weather prevails.
4.4. Wind: occasional gales. Fog: occasionally.

Perth – Western Australia

latitude: 31°57′s longitude: 115°51′e height: 60 m

Csal	137-24	Jan	Feb	Mar	Apr	May	Jun	Jul	Aug	Sep	Oct	Nov	Dec	Year
	tx	39	39	37	33	27	23	20	23	26	29	35	38	41 °C
	t	23	24	22	19	16	14	13	14	15	16	19	22	18 °C
	tn	12	12	11	8	7	5	4	4	6	7	9	11	3 °C
	clothing	t	t	s	s	n	n	n	n	n	n	s	s	–
	stress	f	f	f	o							o	f	–
	p	7	12	22	52	125	192	183	135	69	54	23	15	889 mm
0.2 mm	dp	3	3	5	8	15	17	19	19	15	12	7	5	128 days
	sunshine	322	277	270	222	177	144	167	186	210	245	288	326	2833 hours
	%	74	74	58	52	44	55	59	62	69	72	79	74	64 %

Esperance – Western Australia

latitude: 33°50′s longitude: 121°55′e height: 4 m

Csbl	137-25	Jan	Feb	Mar	Apr	May	Jun	Jul	Aug	Sep	Oct	Nov	Dec	Year
	t	20	21	19	17	15	13	12	12	14	15	17	19	16 °C
	clothing	s	s	s	n	n	n	n	n	n	n	n	s	–
	stress	o	o	o									o	–
	p	18	18	31	46	84	104	101	97	69	56	25	23	673 mm
0.2 mm	dp	5	4	7	9	15	15	16	15	13	12	7	6	124 days

Albany – Western Australia

latitude: 35°02′s longitude: 117°50′e height: 12 m

Csbl	137-26	Jan	Feb	Mar	Apr	May	Jun	Jul	Aug	Sep	Oct	Nov	Dec	Year
	t	19	19	19	17	15	13	12	12	13	15	16	18	16 °C
	clothing	s	s	s	n	n	n	n	n	n	n	n	s	–
	stress	o	o	o									o	–
	p	36	25	46	74	135	137	152	137	109	84	43	31	1009 mm
0.2 mm	dp	8	7	11	13	18	20	21	20	18	16	11	9	172 days
	sunshine	217	198	189	144	140	96	121	140	159	189	192	217	2001 hours
	%	49	52	49	43	44	33	39	42	45	47	46	48	45 %

Adelaide – South Australia

latitude: 34°56′s longitude: 138°35′e height: 44 m

Csal	137-27	Jan	Feb	Mar	Apr	May	Jun	Jul	Aug	Sep	Oct	Nov	Dec	Year
	tx	42	41	38	32	26	20	19	23	27	33	37	41	43 °C
	t	23	21	21	17	15	12	11	12	13	16	18	21	17 °C
	tn	11	11	9	8	6	3	3	3	4	5	7	9	2 °C
	clothing	t	s	s	n	n	n	n	n	n	n	s	s	–
	stress	f	f	f	o						o	f	f	–
	p	23	23	21	50	66	61	61	59	49	47	36	27	523 mm
0.2 mm	dp	5	5	5	10	13	15	16	16	13	10	8	6	122 days
	sunshine	304	264	245	175	143	123	125	158	182	210	242	279	2450 hours
	%	69	70	64	52	45	42	40	47	51	52	58	62	55 %

5. *Climatic area:* Subtropical Queensland and New South Wales east of the Great Dividing Range; Victoria (except northwest); Tasmania
5.1. Subtropical Rain Climate Cr and Subtropical Rain Mountain Climate GCr
5.2. Summer hot in the north, warm in the south; winter warm in the north, cool in the south.
5.3. Variable weather prevails.
5.4. Wind: occasional gales. Fog: occasionally.

Brisbane – Queensland

latitude: 27° 28′s longitude: 153° 02′e height: 41 m

Cral	137-28	Jan	Feb	Mar	Apr	May	Jun	Jul	Aug	Sep	Oct	Nov	Dec	Year
	tx	35	34	33	31	28	25	24	27	30	33	34	36	38 °C
	t	25	25	24	21	18	16	15	16	18	21	22	24	20 °C
	tn	17	17	16	12	8	6	4	5	8	11	14	16	4 °C
	clothing	t	t	t	s	s	n	n	n	s	s	s	t	–
	stress	f	f	f	o	o				o	o	o	f	–
	p	143	183	147	78	57	56	49	30	45	77	92	136	1092 mm
0.2 mm	dp	13	14	15	12	10	8	8	7	8	9	10	12	126 days
	sunshine	246	196	213	217	219	204	220	251	254	257	248	266	2792 hours
	%	58	53	56	63	66	65	67	73	71	65	61	62	63 %

Toowoomba

latitude: 27° 33′s longitude: 151° 57′e height: 586 m

GCrbl	137-29	Jan	Feb	Mar	Apr	May	Jun	Jul	Aug	Sep	Oct	Nov	Dec	Year
	tx	35	33	32	28	24	22	21	23	28	31	33	34	37 °C
	t	22	22	20	17	14	11	10	12	15	18	20	21	17 °C
	tn	12	12	9	4	2	-1	-2	-1	2	4	8	11	-4 °C
	clothing	s	s	s	n	n	n	n	n	n	s	s	s	–
	stress	f	f	o							o	o	f	–
	p	127	114	94	66	53	61	53	40	53	66	84	114	925 mm
0.2 mm	dp	12	10	10	8	8	8	8	6	7	8	9	11	105 days

Sydney – New South Wales

latitude: 33° 52′s longitude: 151° 02′e height: 42 m

Crbl	137-30	Jan	Feb	Mar	Apr	May	Jun	Jul	Aug	Sep	Oct	Nov	Dec	Year
	tx	35	33	32	28	23	21	21	23	28	32	33	35	38 °C
	t	22	22	21	18	16	13	12	13	15	18	19	21	18 °C
	tn	14	14	13	10	7	5	4	5	7	9	11	12	4 °C
	clothing	s	s	s	s	n	n	n	n	n	s	s	s	–
	stress	f	f	o	o						o	o	o	–
	p	104	125	129	101	115	141	94	83	72	80	77	86	1205 mm
0.2 mm	dp	14	13	14	14	13	12	12	11	12	12	12	13	152 days
	sunshine	223	189	196	184	182	159	189	213	214	228	230	233	2440 hours
	%	51	47	51	54	56	53	60	63	60	57	55	52	55 %

Canberra – Federal District

latitude: 35°18′s longitude: 149°12′e height: 560 m

GCrbk	137-31	Jan	Feb	Mar	Apr	May	Jun	Jul	Aug	Sep	Oct	Nov	Dec	Year
	tx	36	35	32	26	21	16	15	18	23	28	32	34	37 °C
	t	20	19	18	13	9	6	6	7	9	12	15	18	13 °C
	tn	7	7	4	0	-3	-5	-5	-4	-2	-1	2	5	-6 °C
	clothing	s	s	s	n	n	n	n	n	n	n	n	s	–
	stress	o	o										o	–
	p	63	59	60	47	59	47	43	40	43	72	63	55	651 mm
0.2 mm	dp	7	7	7	7	7	9	10	11	9	11	8	8	101 days
	sunshine	278	235	235	209	173	138	159	185	226	246	271	280	2635 hours
	%	63	62	61	62	54	47	51	55	64	61	65	62	60 %

Melbourne – Victoria

latitude: 37°49′s longitude: 144°58′e height: 28 m

Crbl	137-32	Jan	Feb	Mar	Apr	May	Jun	Jul	Aug	Sep	Oct	Nov	Dec	Year
	tx	39	38	35	29	23	18	17	21	24	29	33	37	41 °C
	t	20	20	18	15	12	10	10	10	12	14	16	18	15 °C
	tn	9	9	7	5	3	1	0	1	2	3	6	8	-1 °C
	clothing	s	s	s	n	n	n	n	n	n	n	n	s	–
	stress	o	o	o									o	–
	p	45	59	50	69	54	52	54	50	58	74	70	58	691 mm
0.2 mm	dp	9	8	9	13	14	16	17	17	15	14	13	11	156 days
	sunshine	252	211	200	144	118	95	109	139	161	179	186	229	2023 hours
	%	56	55	52	43	38	33	36	42	45	44	44	50	46 %

Mt. Gambier – Victoria

latitude: 37°49′s longitude: 140°46′e height: 65 m

Crlk	137-33	Jan	Feb	Mar	Apr	May	Jun	Jul	Aug	Sep	Oct	Nov	Dec	Year
	t	17	17	16	13	11	10	9	9	11	12	14	16	13 °C
	clothing	n	n	n	n	n	n	n	n	n	n	n	n	–
	stress													–
	p	28	38	37	64	82	77	76	94	73	69	47	36	721 mm

Stanley – Tasmania

latitude: 40°47′s longitude: 145°19′e height: 11 m

Crlk	137-34	Jan	Feb	Mar	Apr	May	Jun	Jul	Aug	Sep	Oct	Nov	Dec	Year
	t	16	17	15	13	12	10	9	10	11	12	13	15	13 °C
	clothing	n	n	n	n	n	n	n	n	n	n	n	n	–
	stress													–
	p	43	41	48	71	86	109	109	99	86	83	61	61	897 mm
2.5 mm	dp	5	4	5	8	8	10	11	14	10	8	8	7	98 days

Launceston – Tasmania

latitude: 41°27′s longitude: 147°10′e height: 77 m

Crbk	137-35	Jan	Feb	Mar	Apr	May	Jun	Jul	Aug	Sep	Oct	Nov	Dec	Year
	t	18	18	16	13	10	8	8	8	10	12	15	16	13 °C
	clothing	s	s	n	n	n	n	n	n	n	n	n	n	–
	stress		o											–
	p	46	31	43	56	69	81	76	76	74	69	46	56	723 mm
0.2 mm	dp	8	6	8	10	12	16	16	16	15	14	11	10	142 days

Hobart – Tasmania

latitude: 42° 53′s longitude: 147° 20′e height: 54 m

Crlk	137-36	Jan	Feb	Mar	Apr	May	Jun	Jul	Aug	Sep	Oct	Nov	Dec	Year
	tx	33	32	31	24	21	16	16	18	22	26	30	31	36 °C
	t	16	16	15	12	10	8	8	9	11	12	14	15	12 °C
	tn	7	8	6	4	2	1	0	1	2	2	4	6	-1 °C
	clothing	n	n	n	n	n	n	n	n	n	n	n	n	–
	stress													–
	p	42	47	52	63	51	66	47	53	53	72	58	64	668 mm
0.2 mm	dp	10	9	10	12	13	14	14	14	16	15	14	12	153 days
	sunshine	243	198	194	149	131	116	135	150	176	189	203	226	2110 hours
	%	52	51	50	45	43	43	47	47	50	46	46	48	48 %

4.6.137.2. Predominant general weather situations are described in 4.6.0.1. under Australia. Hygrothermal diagrams of Darwin, Alice Springs, Perth, and Sydney are to be found in 4.6.0.2.

Wind names: Brickfielder is the name of a gusty wind carrying dust and sand in the dry regions, particularly in the hot months.

Norther is a hot dry wind blowing from the desert on the southern coasts.

Queensland hurricanes are tropical cyclones approaching from the east. They are most frequent from January to March.

Southerly Burster is a name given in south and southeast Australia to a sudden change of wind from a northeasterly direction to a southerly or southeasterly direction, accompanied by a sudden fall in temperature. These storms may last only a few hours, but sometimes they continue for several days. They are frequently accompanied by thunder and lightning. They are most prevalent from October to March.

Willy-Willy is a name given in Western Australia to a severe cyclone.

4.6.138. Papua – Niugini

4.6.138.1. Climatic description and tables

1. Tropical Rain Climate Ar and Tropical Summer Rain Climate Aw and corresponding Mountain Climates GAr and GAw
2. Permanently very hot.
3. In the morning often dry and sunny; in the afternoon frequent showers; during low sun period: shower activity less frequent.
4. Wind: occasional gales, thunderstorm squalls.

Mandang

latitude: 5° 13′s longitude: 145° 48′e height: 12 m

Araa	138-1	Jan	Feb	Mar	Apr	May	Jun	Jul	Aug	Sep	Oct	Nov	Dec	Year
	tx	32	32	32	33	33	33	33	33	33	33	33	32	34 °C
	t	27	27	27	27	27	26	26	26	26	26	27	27	26 °C
	tn	22	22	22	22	22	22	22	22	22	22	22	22	21 °C
	clothing	t	t	t	t	t	t	t	t	t	t	t	t	–
	stress	ff	ff	ff	ff	ff	ff	ff	ff	ff	ff	ff	ff	–
	p	390	310	363	439	319	202	133	88	126	289	394	387	3439 mm
2.5 mm	dp	15	14	17	16	15	10	10	8	9	10	12	16	152 days
	sunshine	161	147	153	160	206	203	203	214	207	202	187	169	2212 hours
	%	42	42	41	44	56	57	55	58	57	53	50	44	50 %

Lae

Arha	138-2	Jan	Feb	Mar	Apr	May	Jun	Jul	Aug	Sep	Oct	Nov	Dec	Year
	latitude: 6°44′s							longitude: 147°00′e			height: 9 m			
	t	27	28	27	27	26	25	25	25	25	26	27	27	26 °C
	clothing	t	t	t	t	t	t	t	t	t	t	t	t	–
	stress	ff	ff	ff	ff	ff	ff	ff	ff	ff	ff	ff	ff	–
	p	252	243	330	420	387	414	538	542	415	320	326	351	4538 mm

Port Moresby

Awha	138-3	Jan	Feb	Mar	Apr	May	Jun	Jul	Aug	Sep	Oct	Nov	Dec	Year
	latitude: 9°26′s							longitude: 147°13′e			height: 48m			
	tx	34	34	33	33	32	31	30	30	31	33	33	34	35 °C
	t	28	27	27	27	27	26	26	26	27	27	28	28	27 °C
	tn	23	23	23	22	22	22	21	21	22	22	23	23	21 °C
	clothing	t	t	t	t	t	t	t	t	t	t	t	t	–
	stress	ff	ff	ff	ff	ff	ff	ff	ff	ff	ff	ff	ff	–
	p	150	194	170	173	49	31	12	37	53	20	97	164	1150 mm
2.5 mm	dp	8	7	9	5	2	3	2	2	2	2	3	6	51 days
	sunshine	174	150	186	222	223	216	211	229	201	214	249	214	2489 hours
	%	44	43	49	62	62	62	59	63	56	56	66	54	56 %

4.6.138.2. Predominant general weather situations are described in 4.6.0.1. under Oceania.

4.6.139. Islands in the Pacific Ocean north of the Equator

4.6.139.1. Climatic description and tables

Bonin Islands, Marianne Islands, Caroline Islands, Marshal Islands, Gilbert Islands, Line Islands

1. Tropical Rain Climate Ar; some islands: Tropical Summer Rain Climate Aw
2. Permanently very hot; except Bonin Islands: summer very hot, winter warm.
3. In the morning often dry and sunny, in the afternoon often showers, sometimes rainy days.
4. Wind: occasional gales, thunderstorm squalls. Tropical cyclones occur from May to January.

Titizima / Chichijima – Bonin Islands

Arab	139-1	Jan	Feb	Mar	Apr	May	Jun	Jul	Aug	Sep	Oct	Nov	Dec	Year
	latitude: 27°05′n							longitude: 142°11′e			height: 3 m			
	t	18	18	19	22	24	26	27	27	27	26	24	21	23 °C
	clothing	s	s	s	s	t	t	t	t	t	t	t	s	–
	stress	o	o	o	o	f	ff	ff	ff	ff	ff	f	o	–
	p	88	92	36	109	113	79	167	150	103	161	124	122	1344 mm

Wake Island

Awha	139-2	Jan	Feb	Mar	Apr	May	Jun	Jul	Aug	Sep	Oct	Nov	Dec	Year
	latitude: 19°17′n							longitude: 166°39′e			height: 4 m			
	t	25	25	25	26	26	27	28	28	28	28	27	26	27 °C
	clothing	t	t	t	t	t	t	t	t	t	t	t	t	–
	stress	ff	ff	ff	ff	ff	ff	ff	ff	ff	ff	ff	ff	–
	p	29	34	37	47	52	48	117	180	133	134	78	46	935 mm
	dp	5	4	7	9	9	11	13	14	13	14	9	6	116 days

Johnston Island

latitude: 16°44'n longitude: 169°31'w height: 5 m

Asha	139-3	Jan	Feb	Mar	Apr	May	Jun	Jul	Aug	Sep	Oct	Nov	Dec	Year
	t	25	25	25	26	26	27	27	28	28	27	26	26	26 °C
	clothing	t	t	t	t	t	t	t	t	t	t	t	t	–
	stress	ff	ff	ff	ff	ff	ff	ff	ff	ff	ff	ff	ff	–
	p	118	44	60	52	25	21	27	78	51	71	52	73	671 mm
	dp	7	7	7	8	6	6	6	7	8	10	9	9	90 days
	sunshine	241	231	243	229	252	291	299	263	239	211	180	200	2879 hours
	%	69	70	65	61	63	74	74	67	65	58	53	58	65 %

Guam – Mariana Islands

latitude: 13°34'n longitude: 144°55'e height: 162 m

Araa	139-4	Jan	Feb	Mar	Apr	May	Jun	Jul	Aug	Sep	Oct	Nov	Dec	Year
	tx	31	31	31	32	32	33	32	32	32	32	31	31	34 °C
	t	26	26	26	27	27	27	26	26	26	26	26	26	26 °C
	tn	22	22	22	23	23	23	23	23	23	23	23	23	22 °C
	clothing	t	t	t	t	t	t	t	t	t	t	t	t	–
	stress	ff	ff	ff	ff	ff	ff	ff	ff	ff	ff	ff	ff	–
	p	118	89	67	77	106	149	228	326	339	333	261	155	2249 mm
	dp	15	10	10	12	14	16	19	20	21	21	19	14	189 days
	sunshine	191	190	258	246	253	236	183	187	157	172	176	183	2428 hours
	%	54	57	69	66	64	61	46	48	43	47	51	52	55 %

Eniwetok – Marshal Islands

latitude: 11°21'n longitude: 162°21'e height: 6 m

Awha	139-5	Jan	Feb	Mar	Apr	May	Jun	Jul	Aug	Sep	Oct	Nov	Dec	Year
	t	27	27	27	28	28	28	28	28	28	28	28	28	28 °C
	clothing	t	t	t	t	t	t	t	t	t	t	t	t	–
	stress	ff	ff	ff	ff	ff	ff	ff	ff	ff	ff	ff	ff	–
	p	26	47	47	33	116	86	164	173	158	231	160	67	1308 mm
	dp	7	5	7	10	13	14	18	18	19	20	17	11	159 days

Yap – Caroline Islands

latitude: 9°31'n longitude: 138°08'e height: 17 m

Arha	139-6	Jan	Feb	Mar	Apr	May	Jun	Jul	Aug	Sep	Oct	Nov	Dec	Year
	t	27	27	27	28	28	28	27	27	27	28	28	27	27 °C
	clothing	t	t	t	t	t	t	t	t	t	t	t	t	–
	stress	ff	ff	ff	ff	ff	ff	ff	ff	ff	ff	ff	ff	–
	p	200	118	137	162	242	272	350	373	356	335	283	258	3086 mm
	dp	18	15	15	16	20	19	22	21	22	21	19	17	225 days
	sunshine	191	192	224	241	210	198	188	181	175	151	193	169	2313 hours
	%	53	57	60	65	54	52	48	47	48	41	55	47	52 %

Kwajalein – Marshall Islands

latitude: 8°43'n longitude: 167°44'e height: 8 m

Arha	139-7	Jan	Feb	Mar	Apr	May	Jun	Jul	Aug	Sep	Oct	Nov	Dec	Year
	t	27	27	27	27	27	27	28	28	28	28	27	27	27 °C
	clothing	t	t	t	t	t	t	t	t	t	t	t	t	–
	stress	ff	ff	ff	ff	ff	ff	ff	ff	ff	ff	ff	ff	–
	p	92	55	164	128	208	220	226	242	259	276	308	228	2405 mm
	dp	11	11	10	13	18	19	20	20	20	20	20	15	197 days

Truk – Caroline Islands

latitude: 7°28′n longitude: 151°51′e height: 2 m

Araa	139-8	Jan	Feb	Mar	Apr	May	Jun	Jul	Aug	Sep	Oct	Nov	Dec	Year
	t	27	27	27	27	27	27	27	27	27	27	27	27	27 °C
	clothing	t	t	t	t	t	t	t	t	t	t	t	t	–
	stress	ff	ff	ff	ff	ff	ff	ff	ff	ff	ff	ff	ff	–
	p	213	160	197	313	359	301	313	325	320	342	314	336	3492 mm
	dp	14	14	15	19	22	21	21	21	20	19	20	19	224 days
	sunshine	222	190	247	225	216	211	209	230	212	197	212	196	2567 hours
	%	61	57	66	61	56	56	54	60	58	53	60	54	58 %

Koror – Palau Islands – Caroline Islands

latitude: 7°22′n longitude: 134°29′e height: 33 m

Araa	139-9	Jan	Feb	Mar	Apr	May	Jun	Jul	Aug	Sep	Oct	Nov	Dec	Year
	tx	31	31	31	32	32	32	31	31	31	31	31	31	32 °C
	t	27	27	27	27	27	27	27	27	27	27	27	27	27 °C
	tn	22	23	23	23	23	23	22	22	23	22	23	23	22 °C
	clothing	t	t	t	t	t	t	t	t	t	t	t	t	–
	stress	ff	ff	ff	ff	ff	ff	ff	ff	ff	ff	ff	ff	–
	p	298	181	194	264	372	330	385	400	364	332	328	298	3746 mm
	dp	20	15	17	17	22	21	23	22	20	18	19	21	234 days
	sunshine	193	193	255	254	216	192	209	207	212	211	212	192	2546 hours
	%	53	57	68	69	56	51	54	54	58	57	60	53	57 %

Majuro – Marshall Islands

latitude: 7°05′n longitude: 171°23′e height: 3 m

Araa	139-10	Jan	Feb	Mar	Apr	May	Jun	Jul	Aug	Sep	Oct	Nov	Dec	Year
	t	27	27	27	27	27	27	27	27	27	27	27	27	27 °C
	clothing	t	t	t	t	t	t	t	t	t	t	t	t	–
	stress	ff	ff	ff	ff	ff	ff	ff	ff	ff	ff	ff	ff	–
	p	177	217	298	313	324	326	318	297	306	385	404	282	3647 mm
	dp	15	15	13	18	20	22	22	19	20	21	21	18	223 days
	sunshine	222	190	225	203	224	195	209	234	208	200	177	193	2480 hours
	%	61	56	60	55	58	52	54	61	57	54	50	53	56 %

Ponape – Caroline Islands

latitude: 6°58′n longitude: 158°13′e height: 46 m

Araa	139-11	Jan	Feb	Mar	Apr	May	Jun	Jul	Aug	Sep	Oct	Nov	Dec	Year
	t	27	27	27	27	27	27	26	26	26	26	27	27	27 °C
	clothing	t	t	t	t	t	t	t	t	t	t	t	t	–
	stress	ff	ff	ff	ff	ff	ff	ff	ff	ff	ff	ff	ff	–
	p	281	247	370	509	516	424	412	415	402	406	428	466	4876 mm
	dp	17	16	20	25	26	24	23	21	20	21	20	18	251 days
	sunshine	171	153	183	158	162	184	190	199	190	171	167	160	2088 hours
	%	47	45	49	43	42	49	49	52	52	46	17	44	47 %

Fanning Island – Line Islands

Araa	139-12	latitude: 3°45′n					longitude: 159°23′w				height: 3 m		
		Jan	Feb	Mar	Apr	May	Jun	Jul	Aug	Sep	Oct	Nov	Dec	Year
	tx	33	33	33	33	33	33	33	33	33	33	34	33	36 °C
	t	26	26	26	26	26	26	27	27	26	27	26	26	26 °C
	tn	23	23	23	23	23	23	23	23	23	22	22	22	22 °C
	clothing	t	t	t	t	t	t	t	t	t	t	t	t	–
	stress	ff	ff	ff	ff	ff	ff	ff	ff	ff	ff	ff	ff	–
	p	129	182	234	377	323	310	143	107	81	70	90	153	2198 mm
0.2 mm	dp	13	15	18	20	19	16	14	10	8	8	8	11	160 days

Tarawa – Gilbert Islands

Arhh	139-13	latitude: 1°21′n					longitude: 172°56′e				height: 4 m		
		Jan	Feb	Mar	Apr	May	Jun	Jul	Aug	Sep	Oct	Nov	Dec	Year
	tx	33	32	32	33	33	33	33	33	33	33	33	33	33 °C
	t	28	28	28	28	28	28	28	28	28	28	28	28	28 °C
	tn	23	24	24	24	23	23	23	23	23	23	23	24	22 °C
	clothing	t	t	t	t	t	t	t	t	t	t	t	t	–
	stress	ff	ff	ff	ff	ff	ff	ff	ff	ff	ff	ff	ff	–
	p	284	229	214	186	164	143	155	142	119	70	111	223	2040 mm

4.6.139.2. Predominant general weather situations are described in 4.6.0.1. under Oceania.

4.6.140. Islands in the Pacific Ocean south of the Equator

4.6.140.1. Climatic descriptions and tables

1. Climatic area: Melanesia and Polynesia
Gilbert Islands, Admirality Islands/Bismarck Archipelago, Solomon Islands, New Hebrides, New Caledonia, Willis Islets, Phoenix Islands, Ellice Islands, Samoa Islands, Fiji Islands, Tonga Islands, Cook Islands, Marquesas Islands, Tuamotu Islands, Gambier Islands, Society Islands, Tubuai Islands.

1.1. Tropical Rain Climate Ar, some islands Tropical Summer Rain Climate Aw
1.2. Generally summer very hot, winter hot, in the south warm.
1.3. Often dry and sunny in the morning, showers in the afternoon; some rainy days, more frequent in the high sun period.
1.4. Wind: occasional gales, thunderstorm squalls.

Nauru – Gilbert Islands

Arha	140-1	latitude: 0°32′s					longitude: 166°55′e				height: 37 m		
		Jan	Feb	Mar	Apr	May	Jun	Jul	Aug	Sep	Oct	Nov	Dec	Year
	tx	33	33	33	33	33	33	33	33	33	33	34	33	34 °C
	t	28	27	27	28	28	28	27	28	28	28	28	28	28 °C
	tn	21	22	22	22	22	21	22	21	21	21	21	22	19 °C
	clothing	t	t	t	t	t	t	t	t	t	t	t	t	–
	stress	ff	ff	ff	ff	ff	ff	ff	ff	ff	ff	ff	ff	–
	p	317	206	180	94	53	99	155	193	122	99	152	239	1909 mm
2.5 mm	dp	15	11	9	6	5	8	11	10	6	5	7	13	106 days

Ocean Island – Gilbert Islands

latitude: 0°54′s longitude: 169°32′e height: 24 m

Arha	140-2	Jan	Feb	Mar	Apr	May	Jun	Jul	Aug	Sep	Oct	Nov	Dec	Year
	tx	33	33	32	32	32	32	32	33	33	33	33	33	34 °C
	t	28	27	27	28	28	28	28	28	28	28	28	28	28 °C
	tn	23	23	23	23	23	23	23	23	23	23	23	23	22 °C
	clothing	t	t	t	t	t	t	t	t	t	t	t	t	–
	stress	ff	ff	ff	ff	ff	ff	ff	ff	ff	ff	ff	ff	–
	p	252	200	133	132	166	121	120	134	84	99	104	233	1778 mm

Canton Island – Phoenix Islands

latitude: 2°46′s longitude: 171°43′w height: 3 m

Ashh	140-3	Jan	Feb	Mar	Apr	May	Jun	Jul	Aug	Sep	Oct	Nov	Dec	Year
	t	28	28	28	29	29	29	29	29	29	29	29	29	29 °C
	clothing	t	t	t	t	t	t	t	t	t	t	t	t	–
	stress	ff	ff	ff	ff	ff	ff	ff	ff	ff	ff	ff	ff	–
	p	66	54	63	92	110	67	66	64	31	28	41	65	748 mm
0.2 mm	dp	6	4	14	16	13	10	10	8	4	4	4	4	97 days

Rabaul – New Britain

latitude: 4°13′s longitude: 152°11′e height: 6 m

Arha	140-4	Jan	Feb	Mar	Apr	May	Jun	Jul	Aug	Sep	Oct	Nov	Dec	Year
	tx	34	34	34	34	34	34	34	34	35	35	35	34	36 °C
	t	27	27	27	27	27	27	27	27	28	28	28	27	27 °C
	tn	22	22	22	22	22	22	21	21	21	22	21	22	21 °C
	clothing	t	t	t	t	t	t	t	t	t	t	t	t	–
	stress	ff	ff	ff	ff	ff	ff	ff	ff	ff	ff	ff	ff	–
	p	213	213	283	225	105	122	103	98	91	110	207	242	2012 mm
0.2 mm	dp	14	13	15	11	6	6	8	8	5	8	10	13	117 days

Funafuti – Ellice Islands

latitude: 8°31′s longitude: 179°12′e height: 2 m

Arha	140-5	Jan	Feb	Mar	Apr	May	Jun	Jul	Aug	Sep	Oct	Nov	Dec	Year
	tx	36	35	34	34	33	33	32	33	33	35	36	36	36 °C
	t	28	28	28	28	28	28	27	28	28	28	28	28	28 °C
	tn	23	23	23	22	23	23	23	22	22	23	23	23	21 °C
	clothing	t	t	t	t	t	t	t	t	t	t	t	t	–
	stress	ff	ff	ff	ff	ff	ff	ff	ff	ff	ff	ff	ff	–
	p	470	447	299	268	276	236	289	339	228	320	302	391	3864 mm
	dp	19	18	17	16	15	16	14	14	15	14	14	17	189 days

Honiara – Guadalcanal – Solomon Islands

latitude: 9°25′s longitude: 159°38′e height: 6 m

Araa	140-6	Jan	Feb	Mar	Apr	May	Jun	Jul	Aug	Sep	Oct	Nov	Dec	Year
	t	27	27	26	27	27	26	26	26	26	27	27	27	27 °C
	clothing	t	t	t	t	t	t	t	t	t	t	t	t	–
	stress	ff	ff	ff	ff	ff	ff	ff	ff	ff	ff	ff	ff	–
	p	245	276	404	219	118	95	80	74	93	139	131	222	2097 mm

Atuona – Hiva Oa Island – Marquesas Islands (French Polynesia)

latitude: 9°48′s longitude: 139°02′w height: 5 m

Araa	140-7	Jan	Feb	Mar	Apr	May	Jun	Jul	Aug	Sep	Oct	Nov	Dec	Year
	t	26	26	27	27	26	26	25	25	25	25	25	26	26 °C
	clothing	t	t	t	t	t	t	t	t	t	t	t	t	–
	stress	ff	ff	ff	ff	ff	ff	ff	ff	ff	ff	ff	ff	–
	p	88	92	106	97	98	183	118	95	102	68	53	68	1168 mm
	dp	10	10	13	11	11	14	13	12	11	10	8	9	132 days

Apia/Upolu – Samoa

latitude: 13°48′s longitude: 171°47′w height: 2 m

Araa	140-8	Jan	Feb	Mar	Apr	May	Jun	Jul	Aug	Sep	Oct	Nov	Dec	Year
	tx	32	32	32	32	31	31	31	31	31	32	32	32	32 °C
	t	27	27	27	27	27	26	26	26	26	26	27	27	26 °C
	tn	22	22	22	22	21	21	20	20	21	21	21	22	19 °C
	clothing	t	t	t	t	t	t	t	t	t	t	t	t	–
	stress	ff	ff	ff	ff	ff	ff	ff	ff	ff	ff	ff	ff	–
	p	424	364	352	214	186	130	115	111	147	221	279	385	2928 mm
	dp	22	19	19	14	12	7	9	9	11	14	16	19	171 days
	sunshine	187	165	199	216	226	226	244	247	242	225	208	190	2575 hours
	%	47	47	53	61	63	67	69	68	67	59	54	47	58 %

Takaroa – Tuamotu Islands and Gambier Islands (French Polynesia)

latitude: 14°29′s longitude: 145°02′w height: 2 m

Arha	140-9	Jan	Feb	Mar	Apr	May	Jun	Jul	Aug	Sep	Oct	Nov	Dec	Year
	t	27	28	28	28	27	27	26	26	26	26	27	27	27 °C
	clothing	t	t	t	t	t	t	t	t	t	t	t	t	–
	stress	ff	ff	ff	ff	ff	ff	ff	ff	ff	ff	ff	ff	–
	p	276	140	120	143	110	102	48	57	68	102	140	214	1520 mm
	dp	16	14	12	11	9	11	9	10	10	12	15	16	145 days

Lunganville – Espiritu Santo – New Hebrides

latitude: 15°31′s longitude: 167°13′e height: 41 m

Araa	140-10	Jan	Feb	Mar	Apr	May	Jun	Jul	Aug	Sep	Oct	Nov	Dec	Year
	t	26	26	25	25	24	23	23	23	24	24	25	25	24 °C
	clothing	t	t	t	t	t	t	t	t	t	t	t	t	–
	stress	ff	ff	ff	ff	ff	f	f	f	ff	ff	ff	ff	–
	p	283	273	397	276	214	160	246	116	205	183	223	338	2914 mm

Willis Island – Willis Islets

latitude: 16°18′s longitude: 149°59′e height: 6 m

Awha	140-11	Jan	Feb	Mar	Apr	May	Jun	Jul	Aug	Sep	Oct	Nov	Dec	Year
	tx	32	32	31	30	29	28	27	27	29	30	31	32	33 °C
	t	28	28	27	27	26	25	24	24	25	26	28	28	26 °C
	tn	23	23	23	22	21	20	19	19	20	22	22	23	19 °C
	clothing	t	t	t	t	t	t	t	t	t	t	t	t	–
	stress	ff	ff	ff	ff	ff	ff	f	f	ff	ff	ff	ff	–
	p	325	270	276	98	70	55	19	28	33	25	54	67	1320 mm
2.5 mm	dp	8	10	10	8	4	4	3	2	2	1	3	4	59 days

Bora-Bora – Society Islands (French Polynesia)

latitude: 16°29′s longitude: 151°45′w height: 3 m

Araa	140-12	Jan	Feb	Mar	Apr	May	Jun	Jul	Aug	Sep	Oct	Nov	Dec	Year
	t	27	27	27	27	26	26	25	25	25	26	26	27	26 °C
	clothing	t	t	t	t	t	t	t	t	t	t	t	t	–
	stress	ff	ff	ff	ff	ff	ff	ff	ff	ff	ff	ff	ff	–
	p	252	224	159	131	191	95	124	75	95	125	195	266	1935 mm
	dp	19	15	15	13	13	10	10	9	10	12	15	17	158 days

Mopelia – Society Islands (French Polynesia)

latitude: 16°46′s longitude: 153°57′w height: 3 m

Arha	140-13	Jan	Feb	Mar	Apr	May	Jun	Jul	Aug	Sep	Oct	Nov	Dec	Year
	t	28	27	28	28	27	26	26	26	26	26	27	27	27 °C
	clothing	t	t	t	t	t	t	t	t	t	t	t	t	–
	stress	ff	ff	ff	ff	ff	ff	ff	ff	ff	ff	ff	ff	–
	p	191	259	138	121	203	79	72	40	70	91	214	198	1676 mm
	dp	16	15	14	12	12	8	8	9	9	11	13	17	144 days

Tahiti-Faaa – Society Islands (French Polynesia)

latitude: 17°33′s longitude: 149°37′w height: 2 m

Araa	140-14	Jan	Feb	Mar	Apr	May	Jun	Jul	Aug	Sep	Oct	Nov	Dec	Year
	tx	33	33	33	33	32	31	31	31	32	32	33	33	34 °C
	t	26	26	26	26	26	25	24	24	24	25	26	26	25 °C
	tn	21	22	22	21	20	19	18	18	19	19	20	21	18 °C
	clothing	t	t	t	t	t	t	t	t	t	t	t	t	–
	stress	ff	ff	ff	ff	ff	ff	f	f	ff	ff	ff	ff	–
	p	423	240	122	92	144	60	70	47	103	81	166	252	1800 mm
	dp	15	12	10	8	8	5	6	4	5	6	10	15	104 days

Nandi – Vitu Levu – Fiji Islands

latitude: 17°45′s longitude: 177°27′e height: 16 m

Araa	140-15	Jan	Feb	Mar	Apr	May	Jun	Jul	Aug	Sep	Oct	Nov	Dec	Year
	t	26	26	26	25	24	24	23	23	24	24	25	26	25 °C
	clothing	t	t	t	t	t	t	t	t	t	t	t	t	–
	stress	ff	ff	ff	ff	ff	ff	f	f	ff	ff	ff	ff	–
	p	306	296	418	223	106	106	53	63	86	60	130	219	2066 mm
	sunshine	206	177	171	194	203	208	218	240	214	231	217	251	2530 hours
	%	51	49	45	55	58	63	63	67	59	60	56	61	57 %

Lauthala Bay – Suva – Fiji Islands

latitude: 18°09′s longitude: 178°27′e height: 9 m

Araa	140-16	Jan	Feb	Mar	Apr	May	Jun	Jul	Aug	Sep	Oct	Nov	Dec	Year
	tx	33	33	33	32	31	30	29	29	30	31	31	32	33 °C
	t	26	26	26	26	25	23	23	23	24	24	25	26	25 °C
	tn													°C
	clothing	t	t	t	t	t	t	t	t	t	t	t	t	–
	stress	ff	ff	ff	ff	ff	ff	ff	ff	ff	ff	ff	ff	–
	p	321	313	399	385	272	160	162	155	218	216	268	291	3160 mm
	dp	18	18	21	19	16	13	14	15	16	15	15	18	198 days
	sunshine	179	154	155	157	156	136	144	154	137	160	169	197	1898 hours
	%	44	43	41	45	45	41	42	43	38	41	43	48	43 %

Vavau – Tonga Islands

latitude: 18° 37′s longitude: 173° 59′w height: 9 m

Araa	140-17	Jan	Feb	Mar	Apr	May	Jun	Jul	Aug	Sep	Oct	Nov	Dec	Year
	tx	31	32	32	31	31	29	29	29	29	30	31	31	32 °C
	t	26	26	26	26	25	24	23	23	24	24	25	26	25 °C
	tn	21	22	21	21	19	18	17	18	17	19	19	21	16 °C
	clothing	t	t	t	t	t	t	t	t	t	t	t	t	–
	stress	ff	ff	ff	ff	ff	ff	f	f	ff	ff	ff	ff	–
	p	249	235	432	212	98	134	117	119	146	139	160	171	2203 mm
2.5 mm	dp	15	18	17	16	14	12	9	9	11	11	11	15	158 days

Koumac – New Caledonia

latitude: 20° 34′s longitude: 164° 17′e height: 18 m

Awab	140-18	Jan	Feb	Mar	Apr	May	Jun	Jul	Aug	Sep	Oct	Nov	Dec	Year
	t	26	26	25	24	22	21	20	20	21	23	24	26	23 °C
	clothing	t	t	t	t	s	s	s	s	s	t	t	t	–
	stress	ff	ff	ff	ff	f	f	f	f	f	f	ff	ff	–
	p	142	133	136	69	119	49	57	44	58	20	45	119	991 mm
	dp	10	10	10	6	6	5	4	2	4	2	4	7	71 days
	sunshine	227	199	201	202	207	201	221	226	204	262	242	237	2629 hours
	%	55	55	53	58	60	61	65	64	57	67	62	57	59 %

Roratonga Island – Cook Islands

latitude: 21° 12′s longitude: 159° 49′w height: 7 m

Arab	140-19	Jan	Feb	Mar	Apr	May	Jun	Jul	Aug	Sep	Oct	Nov	Dec	Year
	tx	31	31	31	31	29	27	27	28	28	28	29	30	32 °C
	t	26	26	26	25	23	22	22	22	22	23	25	25	24 °C
	tn	19	20	20	18	16	14	14	14	14	16	17	18	13 °C
	clothing	t	t	t	t	t	s	s	s	s	t	t	t	–
	stress	ff	ff	ff	ff	f	f	f	f	f	f	ff	ff	–
	p	300	205	266	197	164	124	98	97	107	124	176	261	2119 mm
	dp	15	15	16	12	13	11	9	10	8	10	11	12	142 days
	sunshine	167	167	179	174	162	165	178	191	182	183	175	173	2096 hours
	%	41	46	47	50	47	51	52	54	51	47	44	42	47 %

Nouméa – New Caledonia

latitude: 22° 16′s longitude: 166° 27′e height: 75 m

Arab	140-20	Jan	Feb	Mar	Apr	May	Jun	Jul	Aug	Sep	Oct	Nov	Dec	Year
	tx	34	33	32	32	29	29	29	28	30	31	32	33	34 °C
	t	26	26	25	23	22	20	19	20	21	22	23	25	23 °C
	tn	20	21	20	18	16	15	14	14	14	16	18	19	13 °C
	clothing	t	t	t	t	s	s	s	s	s	s	t	t	–
	stress	ff	ff	ff	f	f	o	o	o	o	f	f	f	–
	p	117	94	175	124	93	89	84	70	53	52	47	85	1083 mm
	dp	9	9	12	10	10	10	9	8	7	5	5	8	101 days
	sunshine	236	231	203	204	184	167	188	206	204	270	254	267	2614 hours
	%	57	64	53	59	54	52	56	59	59	69	64	64	59 %

Rikitea – Mangareva Island – Tuamotu Islands and Gambier Islands (French Polynesia)

latitude: 23°07′s longitude: 134°58′w height: 3 m

Arab	140-21	Jan	Feb	Mar	Apr	May	Jun	Jul	Aug	Sep	Oct	Nov	Dec	Year
	t	26	26	26	25	24	23	22	22	22	23	24	25	24 °C
	clothing	t	t	t	t	t	t	s	s	s	t	t	t	–
	stress	ff	ff	ff	ff	f	f	f	f	f	f	f	ff	–
	p	258	194	174	155	159	176	112	126	125	196	135	231	2041 mm
	dp	13	12	12	11	12	12	12	9	9	12	9	13	136 days

2. Climatic area: Austral Islands, Kermadec Islands, Lord Howe Islands, Norfolk Island, Chatham Islands

2.1. Subtropical Rain Climate Cr

2.2. Summer hot, winter warm or mild; Catham Islands summer mild, winter cool.

2.3. Variable weather prevails.

2.4. Wind: occasional gales. Fog: occasionally.

Rapa – Austral Islands (French Polynesia)

latitude: 27°37′s longitude: 144°20′w height: 7 m

Cral	140-22	Jan	Feb	Mar	Apr	May	Jun	Jul	Aug	Sep	Oct	Nov	Dec	Year
	t	23	24	23	22	20	19	18	17	18	19	20	22	20 °C
	clothing	t	t	t	s	s	s	s	n	s	s	s	s	–
	stress	f	ff	f	f	f	f	f	f	f	f	f	f	–
	p	311	219	257	254	203	149	279	225	193	202	171	193	2656 mm
	dp	14	12	15	15	14	13	15	14	14	11	12	12	161 days

Norfolk Island

latitude: 29°03′s longitude: 167°56′e height: 110 m

Crbl	140-23	Jan	Feb	Mar	Apr	May	Jun	Jul	Aug	Sep	Oct	Nov	Dec	Year
	t	22	22	22	20	18	16	16	16	16	17	19	21	19 °C
	clothing	s	s	s	s	s	n	n	n	n	n	s	s	–
	stress	f	f	f	o	o						o	o	–
	p	132	92	96	99	158	164	147	128	105	104	61	94	1381 mm

Raoul Island – Kermadec Islands

latitude: 29°15′s longitude: 177°55′w height: 49 m

Crbl	140-24	Jan	Feb	Mar	Apr	May	Jun	Jul	Aug	Sep	Oct	Nov	Dec	Year
	tx	27	27	27	26	24	22	22	21	21	23	24	26	27 °C
	t	21	22	21	20	19	17	16	16	16	17	19	21	19 °C
	tn	15	17	16	14	13	11	11	9	10	10	12	14	9 °C
	clothing	s	s	s	s	s	n	n	n	n	s	s	s	–
	stress	f	f	f	o	o	o				o	o	o	–
	p	82	166	137	126	162	155	170	143	112	82	77	91	1504 mm
2.5 mm	dp	7	6	7	7	10	10	13	10	6	6	4	5	91 days
	sunshine	205	170	168	165	154	131	155	170	166	192	203	215	2094 hours
	%	48	46	44	48	47	42	48	50	46	48	50	50	47 %

Lord Howe Island

latitude: 31°31's longitude: 159°04'e height: 11 m

Cral	140-25	Jan	Feb	Mar	Apr	May	Jun	Jul	Aug	Sep	Oct	Nov	Dec	Year
	tx	28	28	28	26	24	22	21	22	23	24	26	27	29 °C
	t	23	23	22	20	18	17	16	16	17	18	20	21	19 °C
	tn	14	15	14	12	11	9	8	8	8	9	11	13	7 °C
	clothing	t	t	s	s	s	n	n	n	n	s	s	s	–
	stress	f	f	o	o							o	o	–
	p	105	124	119	153	177	183	166	149	141	139	125	120	1701 mm
2.5 mm	dp	7	6	8	12	13	13	15	13	8	8	7	8	118 days

Chatham Islands

latitude: 43°57's longitude: 176°31'w height: 48 m

Crlk	140-26	Jan	Feb	Mar	Apr	May	Jun	Jul	Aug	Sep	Oct	Nov	Dec	Year
	tx	21	21	19	18	16	14	13	14	14	16	18	20	21 °C
	t	14	15	14	12	10	8	8	8	9	10	12	13	11 °C
	tn	6	6	5	3	2	1	1	0	1	2	3	4	-1 °C
	clothing	n	n	n	n	n	n	n	n	n	n	n	n	–
	stress													–
	p	66	65	66	65	102	95	86	79	51	51	53	59	838 mm
0.2 mm	dp	6	6	7	7	9	9	9	9	7	6	7	6	88 days
	sunshine	200	145	128	103	88	67	79	100	109	152	161	158	1470 hours
	%	43	37	33	31	29	25	28	31	31	32	37	33	33 %

3. *Climatic area:* Campbell Island, Macquerie Island
3.1. Tundra Climate FT
3.2. Cool throughout the year.
3.3. Variable weather prevails.
3.4. Wind: frequent gales. Fog: frequent. Snow cover: variable.

Campbell Island

latitude: 52°33's longitude: 169°07'e height: 19 m

FTkk	140-27	Jan	Feb	Mar	Apr	May	Jun	Jul	Aug	Sep	Oct	Nov	Dec	Year
	tx	19	18	17	13	12	9	9	10	11	12	16	18	19 °C
	t	9	9	9	7	6	5	5	6	6	6	7	8	7 °C
	tn	1	2	2	1	-3	-5	-2	-3	-1	-2	0	1	-6 °C
	clothing	n	n	n	n	n	n	n	n	n	n	n	n	–
	stress													–
	p	123	113	135	115	141	132	109	116	107	116	112	122	1441 mm
0.2 mm	dp	21	18	22	27	24	20	23	23	21	22	21	23	265 days
	sunshine	94	83	62	37	21	11	14	27	55	68	90	97	659 hours
	%	19	20	16	12	8	5	6	9	16	16	19	19	15 %

Hasselbough Bay – Macquerie Island

latitude: 54°30's longitude: 158°57'e height: 7 m

FTkk	140-28	Jan	Feb	Mar	Apr	May	Jun	Jul	Aug	Sep	Oct	Nov	Dec	Year
	tx	10	9	9	8	7	7	7	7	7	7	9	9	11 °C
	t	7	7	6	5	4	3	3	3	3	4	4	6	5 °C
	tn	2	1	1	-1	-2	-3	-3	-4	-4	-2	-1	3	-6 °C
	clothing	n	n	n	n	w	w	w	w	w	w	w	n	–
	stress													–
	p	99	81	103	94	81	81	68	68	68	72	64	72	952 mm
0.2 mm	dp	25	24	25	24	27	25	25	27	25	25	23	24	299 days

4.6.140.2. Predominant general weather situations are described in 4.6.0.1. under Oceania and Westwind Zone in the southern Pacific Ocean. Hygrothermal diagrams of Apia and Tahili (Papeete) are to be found in 4.6.0.2.

5. Appendix

5.1. Units of measurement in current use

The following basic dimensions are laid down, with the basic units given, in the International Unit System (SI):

basic dimension	basic unit	abbreviation
length	the metre	m
mass	the kilogram	kg
time	the second	s
electric current	the ampere	A
thermodynamic temperature	the kelvin	K
intensity of light	the candela	cd

When measuring temperature in centigrade, the particular name used for the kelvin is the degree Celsius, with the sign °C for the unit. 0°C corresponds to 273.16 K.

Decimal multiples and divisions of units are indicated by prefixes, as follows:

for the multiple of a billion (10^{12}) of the unit: tera (T)
for the multiple of a milliard (10^9) of the unit: giga (G)
for the multiple of a million (10^6) of the unit: mega (M)
for the multiple of a thousand (10^3) of the unit: kilo (k)
for the multiple of a hundred (10^2) of the unit: hecto (h)
for the multiple of ten (10) of the unit: deca (da)
for the tenth part (10^{-1}) of the unit: deci (d)
for the hundredth part (10^{-2}) of the unit: centi (c)
for the thousandth part (10^{-3}) of the unit: milli (m)
for the millionth part (10^{-6}) of the unit: micro (μ)
for the milliardth part (10^{-9}) of the unit: nano (n)
for the billionth part (10^{-12}) of the unit: pico (p)
for the billiardth part (10^{-15}) of the unit: femto (f)
for the trillionth part (10^{-18}) of the unit: atto (a)

Derived units of the International Unit System (SI) are:

of area	the square metre	m^2
of volume	the cubic metre	m^3
of the flat angle	the radiant (1 degree (1°) = (/180) rad)	rad
of mass	the gram	g
a particular name for the megagram (Mg) is	the tonne	t
of density	the kilogram per cubic metre	kg/m^3
of velocity	the metre per second	m/s
of acceleration	the metre per square second	m/s^2
of angular velocity	the radiant per second	rad/s
of angular acceleration	the radiant per square second	rad/s^2
of force	the newton	N

of pressure or mechanical tension	the pascal ($1 Pa = 1 N/m^2$)	Pa
the tenth of a megapascal is	the bar ($1 bar = 10^{-1} MPa$; $1 mbar = 100 Pa$)	bar
of energy, work and quantity of heat	the joule ($1J = 1 N \cdot m = 1 Ws$)	J
of energy, energy flow and heat flow	the watt ($1 W = 1 J/s$)	W

The calorie, a unit of heat based on the properties of water, is no longer in use. It was defined as the quantity of heat required to raise the temperature of 1 g of pure water from 14.5 °C to 15.5 °C at constant normal air pressure.
1 cal = 4.168 J = 1.163 mWh or 1 kcal = 1.163 Wh. 1 kWh corresponds to approx. 860 kcal.

5.2. Tables of sunrise and sunset

The seasons of the year are determined by the position of the earth and the inclination of its axis towards the sun. On March 21st and September 23rd, the sun is normally at its zenith over the equator at noon. On June 22nd, the summer solstice in the northern hemisphere, the sun is above places on the Tropic of Cancer at latitude 23.5 °N at its zenith at noon; on December 22nd, the winter solstice in the northern hemisphere, it is above places on the Tropic of Capricorn at latitude 23.5 °S.

Radiation conditions show the smallest seasonal variation at the equator, and the greatest at the poles.

Since it is sometimes important to know when the sun is vertically above a given parallel of latitude, the following table gives the days (day 1 and day 2) on which the sun is vertical at noon for a selection of parallels of latitude (P). Column D gives the number of days on which the sun is polewards of the parallel concerned; this is the time span between day 1 and day 2:

	Northern latitudes			Southern latitudes		
P	day 1	day 2	D	day 1	day 2	D
0	21st March	23rd Sept.	186	23rd Sept.	21st March	179
2	26th March	19th Sept.	176	29th Sept.	16th March	169
4	31st March	15th Sept.	166	4th Oct.	11th March	158
6	6th April	8th Sept.	155	9th Oct.	6th March	148
8	11th April	1st Sept.	145	14th Oct.	1st March	138
10	17th April	28th Aug.	133	20th Oct.	24th Feb.	127
12	22nd April	22nd Aug.	122	26th Oct.	18th Feb.	115
14	28th April	16th Aug.	110	1st Nov.	12th Feb.	104
16	5th May	9th Aug.	96	7th Nov.	6th Feb.	91
18	12th May	2nd Aug.	82	14th Nov.	30th Jan.	77
20	21st. May	24th July	64	22nd Nov.	22nd Jan.	60
21	26th May	19th July	54	27th Nov.	17th Jan.	51
22	1st June	13th July	41	3rd Dec.	11th Jan.	39
23	11th June	4th July	23	12th Dec.	2nd Jan.	21
23.5	22nd June	22nd June	0	22nd Dec.	22nd Dec.	0

For a number of purposes, the number of hours and minutes of the longest (L) and shortest (S) day of a given parallel of latitude (P) may be of interest; likewise the difference (L–S) between these two lengths of day. Half of this difference is the equivalent of the difference between the times of sunrise (and of sunset) on the longest and shortest days.

parallel	L	S	L–S
0	12 hrs 00 m	12 hrs 00 m	0 hrs 00 m
5	12 18	11 42	0 36
10	12 35	11 25	1 10
15	12 53	11 07	1 46
20	13 13	10 47	2 26
23.5	13 27	10 33	2 54
25	13 34	10 26	3 08
30	13 56	10 04	3 52
40	14 51	9 09	5 42
50	16 09	7 51	8 18
60	18 30	7 30	13 00
66.5	24 00	0 00	24 00

North of the Arctic Circle (66.5° of latitude), and south of the southern Polar Circle, is the zone of polar night, which lasts for six months at the poles themselves, whereas during the summer half-year the sun shines here continuously. Even on the Polar Circles, it does not set on the "longest day", so that this zone is referred to as the zone of the midnight sun.

The difference between the longest and the shortest day increases according to the distance from the equator. At 30° latitude, it amounts to almost 4 hours; here the sun rises at about 5 o'clock and sets at 19 in the summer, and rises at about 7 and sets at 17 in winter. The times of sunrise and sunset are of practical importance in the higher latitudes. In the pictorial figures to the following tables (Figs. 49 and 50), the local mean time of sunrise and sunset for all latitudes between 60 °N and 60 °S can be estimated to the nearest 5 minutes. (Figures are redrawn after Dr. E. Lange 1935/36, Ann. d. Hydr. Vol. 63/V, Vol. 64/VIII.)

Example 1: When does the sun rise on the 3rd May at Bonn (latitude approx. 50.5 °N)?

Follow the horizontal monthly scale for 3rd May vertically as far as latitude 50.5 °N, and from there horizontally to the left; on the vertical scale of sunrise time, 4.22 will show.

Example 2: When does the sun set on 23rd November in Buenos Aires (latitude approx. 35 °S)?

Follow 23rd November vertically up as far as 35 °S, then horizontally to the right; on the vertical sunset scale, 18.50 will show.

NB Local mean time (LMT) changes by one hour for every 15° longitude. Greenwich mean time (GMT) is used as a basis for calculating the time, and differs from a given LMT about $\lambda/15$ hours, λ being the degrees of longitude (counted positively to the east, negatively to the west of Greenwich): $LMT = GMT + \lambda/15$. Considering the time officially adopted for a country, see fig. 48.

5.3. Time Zones

Apart from the generally known influence of the seasons on man through climate, there is the influence of the time of day, which so far has been accorded little attention. This is a matter of the governing of the processes of life by daily rhythms, the most obvious of which is the alternation of waking by day and sleep by night. The body adjusts its functions to the daily rhythm obtaining at the place concerned. These functions may be disturbed when night becomes day and vice versa as, for instance, is the case when one flies from central Europe to New Zealand by jet without interrupting the journey en route. It takes even a healthy body several days to adapt completely to the new diurnal rhythm.

Even flying from Europe to America or vice versa is a strain in this respect, and it is said that a business-man or politician should avoid delicate negotiations on the day of arrival, as he is not on form physically or psychologically, and thus at a disadvantage compared with someone without such a journey behind him.

We recommend tourists by air covering long distances from east to west or vice versa to rest for about a day to give their bodies the chance to adapt to the change in rhythm. As was already mentioned, the influence concerned is at its greatest when night is turned to day, i.e. when the destination is 180° of longitude from the place of departure. In this case, the solar zeniths differ according to world time by 12 hours, as all 360 meridians of the earth turn beneath the sun once every 24 hours. In one hour, the sun traverses 15 degrees of longitude.

At the lower edge of the map in Fig. 48, a rapid calculation of the time difference between any two places on earth is possible; the map also contains a survey of the normal times legally applying in the various countries.

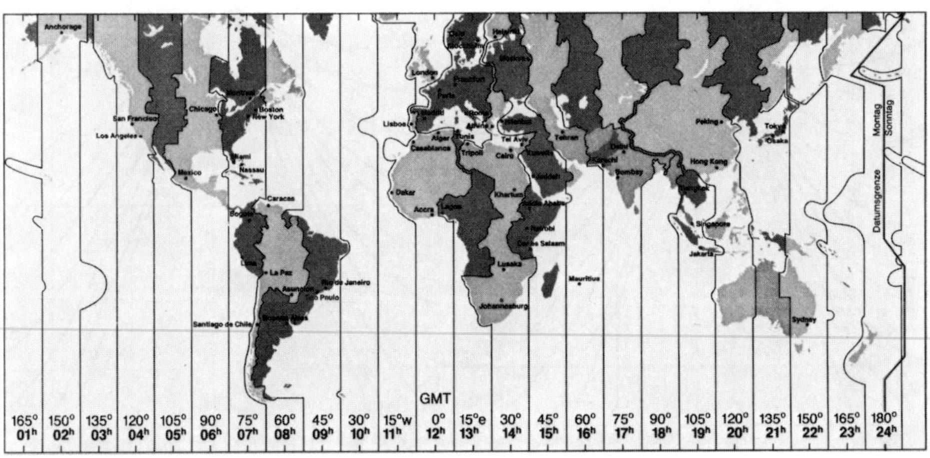

Fig. 48. Universal Time Zones (redrawn after Lufthansa Bordbuch/Log Book 1980) indicates the Universal Time (GMT) Zones which for numerous countries are legally adopted. For the time being, the USSR adds one hour to Universal Time. There are local deviations from the Universal Time system in: Iran, Afghanistan, India, Sri Lanka, Nepal, Burma, Western Malaysia, Singapore, North and South Australia, Guayana, Newfoundland, Suriname, Liberia.
NB During the summer months clocks are put forward one hour in some countries; therefore travellers should be aware of its "summer time" periods.

5.4. Health Tips for the Tourist

When travelling to regions with unfamiliar climates, the tourist should, if possible, rest for the first few days, in order to enable the body to acclimatise itself.

Sun-bathing, too, is a strain on the body, and should be done in moderation, particu-

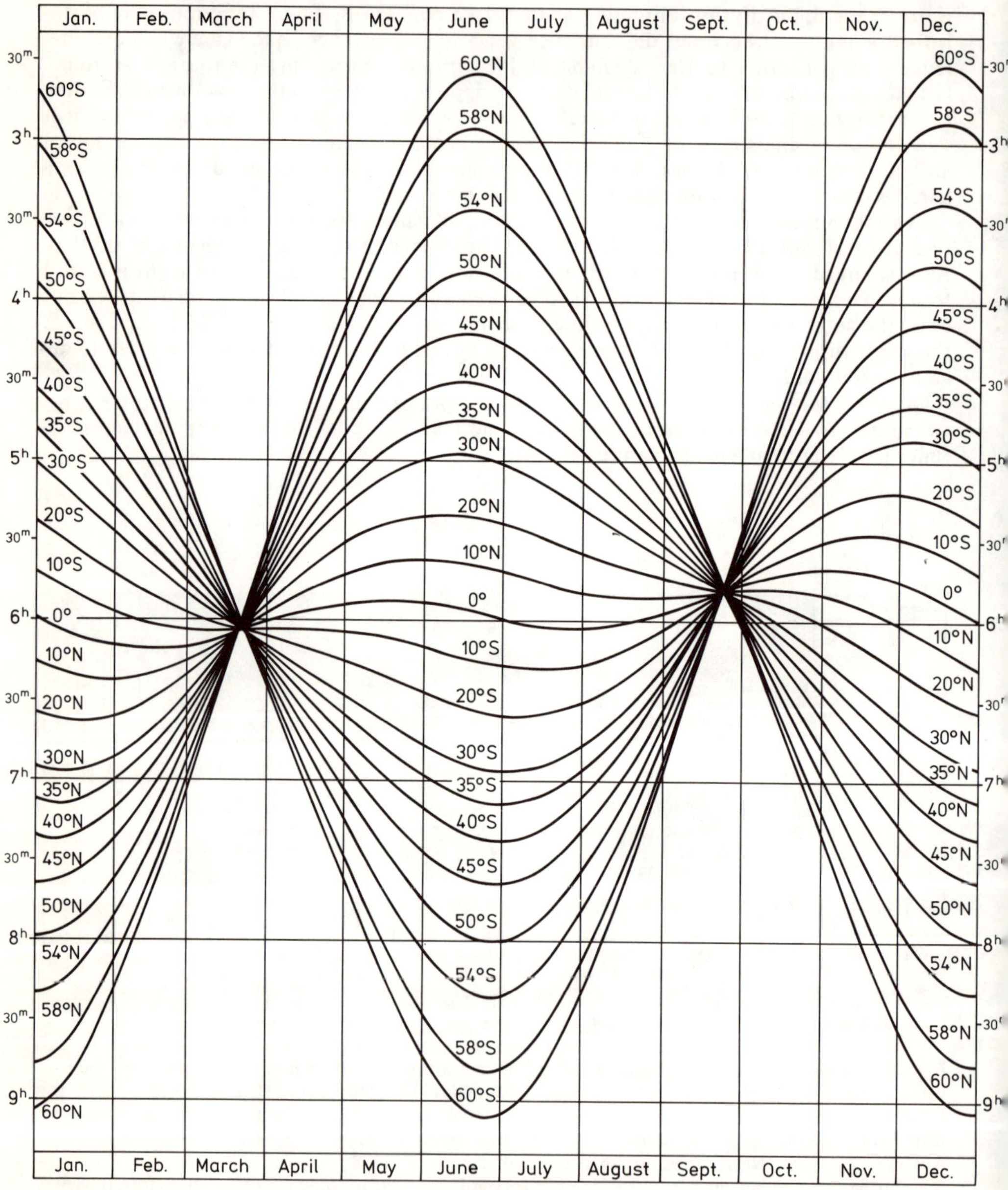

Fig. 49. Time of sunrise for places between 60° North and 60° South

larly to begin with, and increased only slowly from a quarter to half an hour up to over an hour. Protective creams are to be strongly recommended for the tropics and subtropics, as well as for the mountains and the sea. Sun-burn can have quite a deleterious effect on one's well-being.

Sunstroke and heatstroke can have serious consequences; they are often preceded by

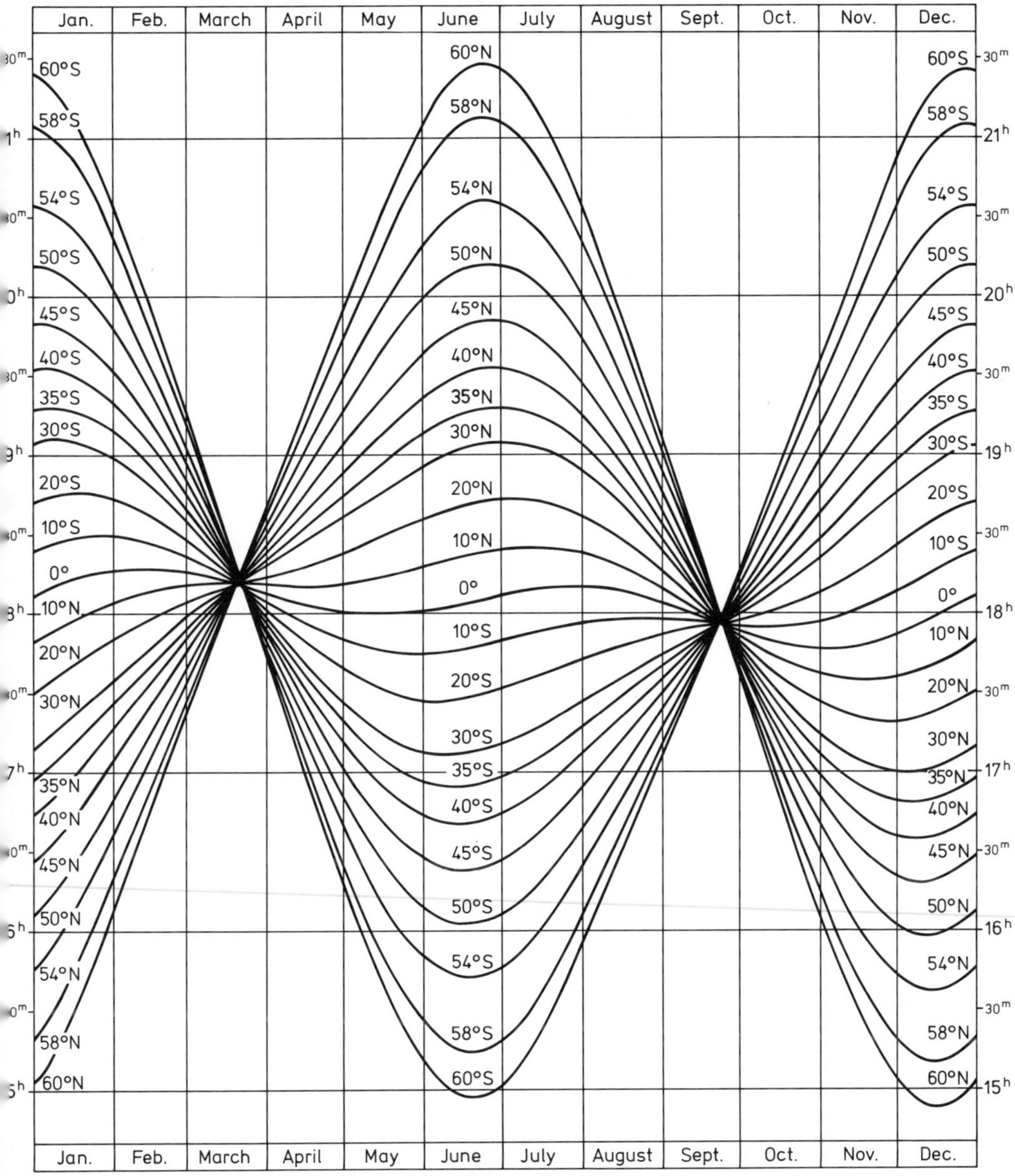

Fig. 50. Time of sunset for places between 60° North and 60° South

headache and nausea. Anything leading to over-heating of the body is thus to be avoided. Airy and shady places should be sought when temperatures are high!

A light head-protection is to be recommended against the direct radiation of the sun. Enough liquid should be drunk, but only if hygienically unexceptionable, and also not ice-cold. Avoid tap-water! and do not slake your thirst with alcohol! The danger of infection is ever-present in many hot countries, particularly in connection with the consumption of uncooked fruit and vegetables or salad. Peel raw fruit yourself; eat only freshly cooked vegetables, and use plenty of salt.

Despite taking great care, it can happen that digestive disorders result from the change in eating habits. Laxatives and remedies for intestinal disorders should always be carried, but at the first sign of any serious trouble a doctor should be consulted.

In the case of long journeys, it is advisable

1. to inquire in good time about the necessary vaccinations (immunisation against tetanus is also to be recommended);
2. to inform one's general practitioner, who will be in a position to prescribe any additional measures necessary, on the grounds of his knowledge of your particular health condition.

List of Countries in alphabetical Order

The number belonging to each state etc. indicates the sub-chapter of chapter 4 and the number of the country or territory.

(States, Groups of Islands, Provinces etc.)

List of Stations in alphabetical Order

The Chapter-No. comprises the sub-chapter of chapter 4 and the table-number of the station, consisting of the number of the country or territory and the place of the station in it.

Name of Station	Country	Chapter-No.	Climate
Aalborg	Denmark	4.1.008-2	DOlo
Abadan	Iran	4.2.039-14	BWil
Abéché	Chad	4.3.095-4	GBSha
Abercorn	Zambia	5.3.110-1	GAwbl
Aberdeen	UK	4.1.004-3	DOlk
Abidjan	Ivory Coast	4.3.103-5	Araa
Abilene	USA	4.4.119-55	GCwhk
Abu Hamed	Sudan	4.3.082-3	BWib
Acajutla	El Salvador	4.4.122-14	Awha
Acapulco	Mexico	4.4.120-13	Awha
Accra	Ghana	4.3.101-3	Awha
Adana	Turkey	4.2.031-23	Cshk
Addis Ababa	Ethiopia	4.3.083-3	GAwbl
Adelaide	Australia	4.6.137-27	Csal
Aden	Arabia	4.2.037-13	BMha
Afyon	Turkey	4.2.031-13	GCsbk
Agadez	Niger	4.3.069-2	GBWhb
Agadir	Morocco	4.3.066-9	BSal
Agalega	Maurititus	4.3.106-2	Arha
Agra	India	4.3.044-7	BSil
Ahmedabad	India	4.3.044-10	BShb
Ajaccio	France	4.1.013-17	Csbk
Akita	Japan	4.2.055-5	DCao
Akola	India	4.2.044-32	Awhb
Akureyri	Iceland	4.1.006-1	EOlo
Akyab	Burma	4.2.056-4	Amhb
Albany	Australia	4.6.137-26	Csbl
Alberville-Kalemia	Zaire	4.3.091-14	GAwab
Albuquerque	USA	4.4.119-74	GWBak
Aldan	USSR	4.1.030-31	ECld
Aldergrove	UK	4.1.004-10	DOlk
Alegrete	Brazil	4.5.127-43	Cral
Aleppo	Syria	4.2.033-2	Cshk
Alert	Canada	4.4.118-1	FTkd
Alexander Bay	South Africa	4.3.113-11	BMbl
Alexandria	Egypt	4.3.081-3	BSal
Alexandrov Gay	USSR	4.1.030-67	DCac
Alexandrovsk	USSR	4.1.030-48	DClc
Algiers	Algeria	4.3.067-2	Csal
Alghero	Italy	4.1.027-16	Csal
Alicante	Spain	4.1.014-5	BSal
Alice Springs	Australia	4.6.137-17	GBWha
Allahabad	India	4.2.044-21	Cwil
Alma Ata	USSR	4.1.030-95	GDCao
Almeria	Spain	4.1.014-6	BSal

Name of Station	Country	Chapter-No.	Climate
Alto, El –	Peru	4.5.129-1	BMab
Alto Tapajos	Brazil	4.5.127-19	Amaa
Ambon	Indonesia	4.2.063-14	Arha
Amini Divi		4.2.048-3	Awha
Amman	Jordan	4.2.036-1	GCsak
Amoy	China	4.2.053-24	Crhl
Amritsar	India	4.2.044-4	BShl
Am-Timan	Chad	4.3.095-7	Awha
Amundsen-Scott		4.5.135-15	HFIde
Anadyr	USSR	4.1.030-41	EClc
Analalava	Madagascar	4.3.108-6	Awha
Anchorage	USA	4.4.119-9	EClc
Ancona	Italy	4.1.027-11	Crak
Andagoya	Colombia	4.5.124-5	Arha
Angkor	Khmer	4.2.060-2	Awha
Angmagsalik	Greenland	4.1.007-7	FTko
Ankara	Turkey	4.2.031-8	GCsak
Annette Island	USA	4.4.119-18	DOlk
Antalaha	Madagascar	4.3.108-1	Arab
Antalya	Turkey	4.2.031-24	Cshl
Antofagasta	Chile	4.5.131-2	BMbl
Aoulef	Algeria	4.3.067-11	BWil
Aparri	Philippines	4.2.064-2	Arha
Apia – Upola	Samoa	4.6.140-8	Araa
Aracaju	Brazil	4.5.127-24	Asaa
Arad	Romania	4.1.025-4	DCbo
Aranyaprathet	Thailand	4.2.057-7	Awha
Araxa	Brazil	4.5.127-32	GAwbl
Archangel'sk	USSR	4.1.030-10	EClc
Arequipa	Peru	4.5.129-4	GBWll
Argentine Island		4.5.135-9	FTkc
Arica	Chile	4.5.131-1	BMbl
Armstrong	Canada	4.4.118-23	EClc
Ašabad	USSR	4.1.030-89	BShk
Asahikawa	Japan	4.2.055-2	DCbo
Asheville	USA	4.4.119-47	GDCak
Asmara	Ethiopia	4.3.083-6	GBSbl
Astrakhan	USSR	4.1.030-88	BSao
Asuncion	Paraguay	4.5.132-3	Arhb
Aswan	Egypt	4.3.081-15	BWhl
Atakpame	Togo	4.3.100-3	Awha
Atbara	Sudan	4.3.083-7	BWib
Athinai	Greece	4.1.029-5	Cshl
Ati	Chad	4.3.095-3	BWha
Atlanta	USA	4.4.119-52	Crak
Atuona – Hiva Oa	Marquesas	4.6.140-7	Araa
Auckland	New Zealand	4.6.136-2	Crbl
Awash	Ethiopia	4.3.083-47	GBSha
Azul	Argentina	4.5.134-35	Crbk
Babolsar	Iran	4.2.039-2	Csak

Name of Station	Country	Chapter-No.	Climate
Badajoz	Spain	4.1.014-9	Csak
Bafata	Guinea-Bissau	4.3.074-1	Awha
Bagampet	India	4.2.044-13	GBShb
Baghdad	Iraq	4.2.038-5	BWil
Baharia	Egypt	4.3.081-8	BWhl
Bahia Blanca	Argentina	4.5.134-37	Crak
Bajo Calima	Colombia	4.5.124-10	Araa
Baker Lake	Canada	4.4.118-15	ECld
Balboa Heights	Canal Zone	4.4.122-21	Awha
Balhaš	USSR	4.1.030-87	BSac
Balikpapan	Indonesia	4.2.063-10	Araa
Balovale	Zambia	4.3.110-5	GAwab
Bamako	Mali	4.3.070-13	Awha
Bambesa	Zaire	4.3.091-1	GAraa
Banana – Moanda	Zaire	4.3.091-15	Awaa
Bandar Abbas	Iran	4.2.039-16	BMhb
Bangalore	India	4.2.044-41	GAwab
Bangassou	Central Africa	4.3.094-11	GAwaa
Bangkok	Thailand	4.2.057-6	Awha
Bangui	Central Africa	4.3.094-12	GAwaa
Barcelona	Spain	4.1.014-3	Crak
Barcelona	Venezuela	4.5.125-3	BSha
Bariloche	Argentina	4.5.134-39	GCslk
Barnaul	USSR	4.1.030-78	DCbc
Barquisimeto Venezuela	Venezuela	4.5.125-4	GBSaa
Barra do Corda	Brazil	4.5.127-16	Awha
Barrow	USA	4.4.119-1	FTkd
Basco	Philippines	4.2.064-1	Arha
Basrah	Iraq	4.2.038-10	BWhl
Bata	Equatorial Guinea	4.3.096-2	Amaa
Barter Island	USA	4.4.119-2	FTkd
Batouri	Cameroun	4.3.097-4	GAwaa
Bathurst	Gambia	4.3.073-1	Awaa
Beaufort West	South Africa	4.3.113-12	GBWal
Béchar	Algeria	4.3.067-7	GBWhk
Beira	Mozambique	4.3.109-4	Awhb
Beirut	Lebanon	4.2.034-4	Csal
Beit Bridge	Rhodesia	4.3.112-6	BWal
Beja	Portugal	4.1.016-5	Csak
Belém	Brazil	4.5.127-3	Araa
Belgaum	India	4.2.044-37	GAwhb
Belize	Belize	4.4.122-1	Arha
Balmullet	Ireland	4.1.055-2	DOlk
Belo Horizonte	Brazil	4.5.127-33	GAwab
Benghazi	Libya	4.3.080-1	Csal
Beni Abbès	Algeria	4.3.067-9	BWil
Benina	Libya	4.3.080-1	Csab
Beograd	Yugoslavia	4.1.023-2	DCak
Berbérati	Central Africa	4.3.094-13	GAwaa
Bergen	Norway	4.1.001-9	DOlk

Name of Station	Country	Chapter-No.	Climate
Bering Island	USSR	4.1.030-44	EOlo
Berlin	West-Berlin	4.1.017-1	DCbk
Bethel	USA	4.4.119-10	EClc
Beyrouth	Lebanon	4.2.034-4	Csal
Biak	Indonesia	4.2.063-9	Araa
Białystok	Poland	4.1.021-5	DCbo
Bikaner	India	4.2.044-1	BWil
Bilma	Niger	4.3.069-1	BWhl
Birao	Central Africa	4.3.094-1	BSha
Birkenhead	UK	4.1.004-5	DOlk
Birni N'Konni	Niger	4.3.069-7	BSha
Bismarck	USA	4.4.119-26	GDCbc
Biskra	Algeria	4.3.067-5	BWhl
Bissau	Guinea-Bissau	4.3.074-2	Awaa
Bistrita	Romania	4.1.025-2	DCbo
Bitam	Gabon	4.3.093-1	GAsaa
Björnöya	Norway	4.1.001-2	FTko
Blagoveščensk	USSR	4.1.030-49	DCbc
Bloemfontein	South Africa	4.3.113-7	GBSbk
Bluefields	Nicaragua	4.4.122-7	Araa
Blue Hill	USA	4.4.119-33	DCbo
Boa Vista	Brazil	4.5.127-1	Arha
Bobo Dioulasso	Upper Volta	4.3.102-5	Awha
Bodö	Norway	4.1.001-6	EOlo
Bogotá	Colombia	4.5.124-3	HAsll
Bogra	Bangladesh	4.2.042-2	Awhb
Bohicon	Benin	4.3.099-6	Awha
Boise	USA	4.4.119-30	GDCao
Bolivar	Venezuela	4.5.125-10	Awha
Bolzano	Italy	4.1.027-1	DObk
Bombari	Central Africa	4.3.094-8	Awha
Bombay	India	4.2.044-34	Awha
Bomnak	USSR	4.1.030-37	ECbd
Bonthe	Sierra Leone	4.3.076-4	Amha
Bora-Bora	Society Is.	4.6.140-12	Araa
Bordeaux	France	4.1.013-11	DObk
Boromo	Upper Volta	4.3.102-4	Awha
Bosaco	Somalia	4.3.085-1	BMia
Bossangoa	Central Africa	4.3.094-6	Awha
Bossembelé	Central Africa	4.3.094-10	GAwaa
Boston	USA	4.4.119-32	DCao
Bouake	Ivory Coast	4.3.103-3	Awha
Bouar	Central Africa	4.3.094-7	GAwaa
Bouârfa	Morocco	4.3.066-7	GBShk
Bouca	Central Africa	4.3.094-4	Awha
Bougouni	Mali	4.3.070-15	Awha
Bourges	France	4.1.013-8	DObk
Bourke	Australia	4.6.137-10	BShl
Bousso	Chad	4.3.095-8	Awha
Bragança	Portugal	4.1.016-1	GCrbk

Name of Station	Country	Chapter-No.	Climate
Brachuap Khirikhan	Thailand	4.2.057-9	Awha
Brazzaville	Congo	4.3.092-10	Awab
Brest	France	4.1.013-5	DOlk
Brest	USSR	4.1.030-64	DCbo
Bria	Central Africa	4.3.094-3	GAwaa
Brindisi	Italy	4.1.027-15	Csak
Brisbane	Australia	4.6.137-28	Cral
Brno	Czechoslavakia	4.1.020-4	DCbo
Brocken	GDR	4.1.018-5	GDOlo
Broken Hill	Australia	4.6.137-12	BSal
Broken Hill	Zambia	4.3.110-6	GAwal
Broome	Australia	4.6.137-7	BShb
Brownsville	USA	4.4.119-62	Crhl
Brusque	Brazil	4.5.127-42	Cral
Brussels	Belgium	4.1.010-2	DObk
Bucuresti	Romania	4.1.025-7	DCao
Budapest	Hungary	4.1.022-3	DCbo
Buenos Aires	Argentina	4.5.134-34	Cral
Buffalo	USA	4.4.119-31	DCbo
Bujumbura	Burundi	4.3.090-1	GAwab
Bukavu	Zaire	4.3.091-10	GAwbb
Bukoba	Tanzania	4.3.087-1	GAmbb
Bulawayo	Rhodesia	4.3.112-5	GBSbl
Bulun	USSR	4.1.030-13	ECle
Burketown	Australia	4.6.137-6	BShb
Burlington	USA	4.4.119-29	DCbo
Bursa	Turkey	4.2.031-6	Csak
Bushehr	Iran	4.2.039-11	BShl
Byrd Station		4.5.135-14	GFIcd
Cabinda	Angola	4.5.105-1	Awab
Cabo Raper	Chile	4.5.131-6	DOlk
Caetité	Brazil	4.5.127-26	GAwab
Cagliari	Italy	4.1.027-18	Csak
Cairns	Australia	4.6.137-3	Amab
Cairo	Egypt	4.3.081-6	BWhl
Cajamarca	Peru	4.5.129-5	HAwll
Calcutta	India	4.2.044-29	Awhb
Calvinia	South Africa	4.3.113-9	GBSak
Camaguey	Cuba	4.4.123-10	Awab
Cameron Highlands	Malaysia	4.2.061-5	GArbb
Camiri	Bolivia	4.5.130-12	GAwal
Campbell Island		4.6.140-27	FTkk
Campo Grande	Brazil	4.5.127-34	GAwab
Canakkale	Turkey	4.2.031-7	Csak
Canberra	Australia	4.6.137-31	GCrbk
Canton	China	4.2.053-34	Cwhl
Canton Island	Phoenix Is.	4.6.140-3	Ashh
Cap Carbon	Algeria	4.3.067-1	Csal
Cape Town	South Africa	4.3.113-13	Csbl
Caracas	Venezuela	4.3.125-6	GAwab

Name of Station	Country	Chapter-No.	Climate
Caravelas	Brazil	4.5.127-9	Arab
Caribou	USA	4.4.119-24	DCbc
Carnarvon	Australia	4.6.137-9	BSal
Car Nicobar		4.2.048-5	Arha
Cartagena	Colombia	4.3.124-1	Awha
Carthage (Tunis)	Tunisia	4.3.068-1	Csal
Casablanca	Morocco	4.3.066-5	Csal
Catacamas	Honduras	4.4.122-16	Awab
Catamarca	Argentina	4.5.134-6	GBShl
Catania	Italy	4.1.027-21	Csal
Cayenne	French Guiana	4.5.126-3	Araa
Cebu	Philippines	4.2.068-8	Arha
Ceres	Argentina	4.5.134-29	Cral
Ceuta	Morocco	4.3.065-1	Csbl
Chahbar	Iran	4.2.039-18	BMhb
Chang Ch'un	China	4.2.053-8	DCac
Chang Sha	China	4.2.053-20	Crhk
Changtu	China	4.2.053-26	HCwlo
Chanthaburi	Thailand	4.2.057-8	Amha
Charana	Bolivia	4.5.130-2	HBSlk
Charleston	USA	4.4.119-52	Cral
Chatham Island		4.6.140-26	Crlk
Cheb	Czechoslovakia	4.1.020-2	DClo
Cheju	Southern Korea	4.2.052-5	Crak
Chengtu	China	4.2.053-28	Cwak
Cherbourg	France	4.1.013-1	DOlk
Cherrapunji	India	4.2.044-25	GAmbl
Chiang Mai	Thailand	4.2.057-1	Awhb
Chicago	USA	4.4.119-34	DCao
Chileka	Malawi	4.3.111-4	GAwab
Chinácota	Colombia	4.5.124-6	GArbb
Chinchina	Colombia	4.5.124-9	GArbb
Chipinga	Rhodesia	4.3.112-3	GAwbl
Chisimaio	Somalia	4.3.085-7	BSha
Chittagong	Bangladesh	4.2.042-6	Amhb
Chiu Chiang	China	4.2.053-17	Crhk
Chos Malal	Argentina	4.5.134-38	GCsb
Christchurch	New Zealand	4.6.136-7	DOlk
Chumphon	Thailand	4.2.057-10	Arha
Chunking	China	4.2.053-29	Cwhk
Churchill	Canada	4.4.118-17	ECld
Cienfuegos	Cuba	4.4.123-9	Awab
Čimbaj	USSR	4.1.030-93	BWao
Čita	USSR	4.1.030-40	ECbd
Claremorris	Ireland	4.1.005-4	DOlk
Cloncurry	Australia	4.6.137-8	BShb
Clones	Ireland	4.1.005-3	DOlk
Cluj	Romania	4.1.025-3	DCbo
Clyde	Canada	4.4.118-7	FTkd
Coari	Brazil	4.5.127-5	Araa

Name of Station	Country	Chapter-No	Climate
Coban	Honduras	4.4.122-3	GArbl
Cobija	Bolivia	4.5.130-5	Awab
Cochabamba	Bolivia	4.5.130-1	HBSbl
Çocos Island	Indonesia	4.2.063-17	Araa
Čokurdah	USSR	4.1.030-14	ECld
Cold Bay	USA	4.4.119-16	EOlo
Colombia	USA	4.4.119-43	DCao
Colombo	Sri Lanka	4.2.045-5	Arha
Columbus	USA	4.4.119-40	DCao
Combolcha	Ethiopia	4.3.083-2	GAwal
Comodore Rivadavia	Argentina	4.5.134-10	BSbk
Conakry	Guinea	4.3.075-3	Amha
Conceiçao do Araguaia	Brazil	4.5.127-21	Awaa
Concepción	Bolivia	4.5.130-7	Awab
Consaca	Colombia	4.5.124-12	GArbb
Coppermine	Canada	4.4.118-9	FTkd
Coquilhatville-Mbandaka	Zaire	4.3.089-4	Araa
Coral Habour	Canada	4.4.118-10	FTkd
Córdoba	Argentina	4.5.134-21	Cwal
Cordova-Mile 13	USA	4.4.119-12	EOlo
Corrientes	Argentina	4.5.134-26	Crhl
Corumba	Brazil	4.5.127-31	Awab
Cotonou	Benin	4.3.099-7	Awha
Cristobal	Canal Zone	4.4.122-9	Araa
Cruzeiro do Sul	Brazil	4.5.127-6	Araa
Cuiaba	Brazil	4.5.127-28	Awaa
Cumberland Bay	South Georgia	4.5.135-5	FTko
Curitiba	Brazil	4.5.127-40	GCrbl
Cuttack	India	4.2.044-31	Awhb
Cuzco	Peru	4.5.129-8	HAwll
Czestochowa	Poland	4.1.021-11	DCbo
Dacca Tejgaon	Bangladesh	4.2.042-3	Awhb
Dagupan	Philippines	4.2.064-3	Awha
Dairen	China	4.2.053-11	DCao
Dakar/Yoff	Senegal	4.3.072-5	BShb
Dakhla	Egypt	4.3.081-13	BWhl
Dalbandin	Pakistan	4.2.041-11	GBWhk
Daltonganj	India	4.2.044-22	Cwhl
Daly Waters	Australia	4.6.137-5	BShb
Damascus	Syria	4.2.033-3	GCshk
Da Nang – Tourane	Viet Nam	4.2.058-3	Amhb
Danmarkshavn	Greenland	4.1.007-3	FTkc
Darbhanga	India	4.2.044-19	Cwhl
Dar El Beida (= Algiers)	Algeria	4.3.067-2	Csal
Dar-Es-Salaam	Tanzania	4.3.087-8	Awha
Darjeeling	India	4.2.044-17	GCwbk
Daru	Sierra Leone	4.3.076-3	Amaa
Darwin	Australia	4.6.137-2	Awha
Dayton	USA	4.4.119-41	DCao
De Bilt	Netherlands	4.1.009-3	DOlk

Name of Station	Country	Chapter-No.	Climate
Debrecen	Hungary	4.1.022-4	DCbo
Deception Island		4.5.135-8	FTko
Deir Ezzor	Syria	4.2.033-4	BShk
Den Helder	Netherlands	4.1.009-2	DOlk
Denver	USA	4.4.119-42	GDCao
Dera Ismail Khan	Pakistan	4.2.040-9	BWil
Des Moines	USA	4.4.119-35	DCao
Dest. Naval Orcadas		4.5.135-7	FTkc
Dezful	Iran	4.2.039-9	BSil
Dhahran	Arabia	4.2.037-2	BMil
Dhubri	India	4.2.044-23	Amhb
Diamond Island	Burma	4.2.056-6	Awha
Diego Garcia	Mauritius	4.3.106-1	Arha
Diego Suarez	Madagascar	4.3.108-5	Awaa
Dijon	France	4.1.013-6	DObk
Dili	Indonesia	4.2.063-21	Awaa
Dimbokro	Ivory Coast	4.3.103-4	Awha
Dire Dawa	Ethiopia	4.3.083-7	GBShb
Diwaniya	Iraq	4.2.038-8	BWhl
Diyarbakir	Turkey	4.2.031-17	GCshk
Djakarta	Indonesia	4.2.063-15	Araa
Djambala	Congo	4.3.092-5	GAwab
Djibouti	Djibouti	4.3.084-1	BMha
Dodoma	Tanzania	4.3.087-12	GBSab
Dolisie	Congo	4.3.092-9	Awab
Dori	Upper Volta	4.3.102-1	BSha
Douala	Cameroun	4.3.097-5	Araa
Dresden	GDR	4.1.018-9	DObo
Dr. Plesman Airport	Curaçao	4.4.123-15	Asha
Dublin Airport	Ireland	4.1.005-6	DOlk
Dueodde	Denmark	4.1.013-4	DOlk
Duluth	USA	4.4.119-25	DCbc
Dunedin	New Zeland	4.6.136-8	DOlk
Durban	South Africa	4.3.113-14	Cral
Dwarka	India	4.2.044-3	BMhb
Dzaoudzi/Pamanzi	Comoro Islands	4.3.107-3	Awaa
Easter Island	Chile	4.5.131-8	Arab
East London	South Africa	4.3.113-15	Crbl
Ed Dueim	Sudan	4.3.082-12	BWha
Edirne	Turkey	4.2.031-3	Csak
Edmonton	Canada	4.4.118-26	GDClc
Eelde	Netherlands	4.1.009-1	DOlk
Egedesminde	Greenland	4.1.007-6	FTko
Eilat	Israel	4.2.035-4	BMhl
El Adem	Libya	4.3.080-5	BWal
El Alto	Peru	4.5.129-1	BMab
El Alto (La Paz)	Bolivia	4.5.130-9	HAwlk
El Fasher	Sudan	4.3.082-13	GBWhb
El Golea	Algeria	4.3.067-8	BWhk
Elisabethville	Zaire	4.3.091-17	GAwbl

Name of Station	Country	Chapter-No.	Climate
El Paso	USA	4.4.119-76	GBWak
El Obeid	Sudan	4.3.082-14	GBWhb
El Recreo	Nicaragua	4.4.122-6	Arha
Ely	USA	4.4.119-71	GBSbo
Emden	FRG	4.1.017-5	DOlk
Encarnación	Paraguay	4.3.132-6	Cral
Enisejsk	USSR	4.1.030-32	ECbc
Eniwetok	Marshal Is.	4.6.139-5	Awha
En Nahud	Sudan	4.3.082-16	GBWhb
Entebbe	Uganda	4.3.088-4	GArbb
Enugu	Nigeria	4.3.098-9	Awha
Equator Station	Kenya	4.3.087-7	HAsll
Erbogačen	USSR	4.1.030-28	ECld
Erfurt	GDR	4.1.018-10	DClo
Erigavo	Somalia	4.3.085-4	GBSbl
Erivan	USSR	4.1.030-91	GBSao
Erzinkan	Turkey	4.2.031-11	GCsao
Erzurum	Turkey	4.3.031-9	GCsbo
Esfahan	Iran	4.2.039-12	GBWhk
Eskdalemuir	UK	4.1.004-4	DOlk
Esmeraldas	Ecuador	4.5.128-2	Asaa
Esperance	Australia	4.6.137-25	Csbl
Essen	FRG	4.1.017-7	DObk
Eureka	Canada	4.4.118-2	FTkd
Faaa (Tahiti)	Society Islands	4.6.140-14	Araa
Fada N Gourma	Upper Volta	4.3.102-3	BSha
Fairbanks	USA	4.4.119-6	EClc
Fanning Island	Line Is.	4.6.139-12	Araa
Faya Largeau	Chad	4.3.095-1	BWhb
Ferkessedougou	Ivory Coast	4.3.103-1	Awha
Fernando Noronha	Brazil	4.5.127-14	Asaa
Fernando Po	Equatorial Guinea	4.3.096-1	Amaa
Fez	Morocco	4.3.066-2	Csak
Fichtelberg	GDR	4.1.018-11	GDOlo
Firenze	Italy	4.1.027-9	Crak
Formosa	Argentina	4.5.134-25	Crhl
Formosa	Brazil	4.5.127-27	GAwab
Forrest	Australia	4.6.137-23	BWal
Fort Archambault-Sarh	Chad	4.3.095-10	Awha
Fort Cochin	India	4.2.044-42	Amha
Fort Dauphin	Madagascar	4.3.108-4	Arab
Fort de France – Desaix	Martinique	4.4.123-4	Araa
Fort Flatters	Algeria	4.3.067-10	BWhl
Fort Gouraud	Mauretania	4.3.071-1	BWhb
Fort Harrison	Canada	4.4.118-12	FTkd
Fort Lamy – Nadjemena	Chad	4.3.095-6	BSha
Fort Nelson	Canada	4.4.118-24	DClc
Fort Sandeman	Pakistan	4.2.041-6	GBShk
Fort Ševčenko	USSR	4.1.030-92	BWao
Fort Smith	Canada	4.4.118-16	ECld

Name of Station	Country	Chapter-No.	Climate
Foz do Iguaçu	Brazil	4.5.127-41	Cral
Franceville	Gabon	4.3.093-8	Awaa
Frankfurt am Main	FRG	4.1.017-9	DObk
Freetown – Lungi	Sierra Leone	4.3.076-2	Amha
Freiburg im Breisgau	FRG	4.1.017-13	DObk
Friedrichshafen	FRG	4.1.017-14	DObk
Frobisher Bay	Canada	4.4.118-11	FTkd
Fukuoka	Japan	4.2.055-16	Crak
Funafuti	Ellice Is.	4.6.140-5	Arha
Funchal (Madeira)	Portugal	4.1.016-9	Csbl
Gabes	Tunisia	4.3.068-2	BSal
Galcayo	Somalia	4.3.085-2	BWha
Gambell	USA	4.4.119-3	FTkc
Gambona	Congo	4.3.092-4	Awaa
Gander	Canada	4.4.118-35	DClc
Gando	Spain	4.1.014-16	BSal
Gao	Mali	4.3.070-4	BWia
Gaoua	Upper Volta	4.3.102-6	Awha
Garissa	Kenya	4.3.086-9	BWha
Garoua	Cameroon	4.3.097-1	Awha
Gauhati	India	4.2.044-20	Cwhl
Gdansk	Poland	4.1.021-1	DCbo
Gebeit	Sudan	4.3.082-4	BWhb
Geneina	Sudan	4.3.082-17	GBShb
Genève	Switzerland	4.1.012-5	DObk
Genova	Italy	4.1.027-8	Crak
George	South Africa	4.3.113-16	Crbl
Georgetown	Guyana	4.5.126-1	Arha
Georgetown	Ascension Island	4.3.077-1	BMha
Ghanzi	Botswana	4.3.115-2	GBSab
Gibraltar (North Front)	Gibraltar	4.1.015-1	Csal
Gisborne	New Zealand	4.6.136-4	Crbl
Gizianter	Turkey	4.2.031-22	GCsak
Goa – Panjim	India	4.2.044-38	Awha
Görlitz	GDR	4.1.018-9	DCbo
Göteborg	Sweden	4.1.002-9	DClo
Goiania	Brazil	4.5.127-30	GAwab
Gondar	Ethiopia	4.3.083-1	GAwab
Goose	Canada	4.4.118-27	DClc
Gorleston	UK	4.1.004-7	DOlk
Gough Island		4.5.135-2	Crll
Grand Junction	USA	4.4.119-72	GBSao
Graz	Austria	4.1.019-6	DCbo
Great Falls	USA	4.4.119-21	GDObo
Greifswald	GDR	4.1.018-2	DCbo
Grytviken	South Georgia	4.5.135-4	FTko
Guam	Mariana Is.	4.6.139-4	Araa
Guanajuato	Mexico	4.4.120-7	GAwbl
Guantanamo	Cuba	4.4.123-16	BSha
Guatemala City	Guatemala	4.4.122-11	GAwbl

Name of Station	Country	Chapter-No.	Climate
Guayamas	Mexico	4.4.120-1	BMhb
Guayaquil	Ecuador	4.5.128-4	Awaa
Guiria	Venezuela	4.5.125-5	Awha
Gulu	Uganda	4.3.088-1	GAwab
Gur'ev	USSR	4.1.030-86	BSac
Gwalior	India	4.2.044-18	Cwil
Györ	Hungary	4.1.022-2	DCbo
H 4 Station	Jordan	4.2.036-2	GBWhk
Habarovsk	USSR	4.1.030-50	DCbc
Habbaniya	Iraq	4.2.038-4	BWil
Hachijojima	Japan	4.2.055-19	Cral
Haifa	Israel	4.2.035-1	Cshl
Halifax	Canada	4.4.118-37	DCbo
Hall Beach	Canada	4.4.118-8	FTkd
Hamadan	Iran	4.2.039-3	GCsao
Hamburg	FRG	4.1.017-4	DOlk
Hankow	China	4.2.053-16	Crhk
Hannover	FRG	4.1.017-6	DOlk
Hanoi	Viet Nam	4.2.058-1	Cwhl
Hanty-Mansijsk	USSR	4.1.030-30	ECbc
Haparanda	Sweden	4.1.002-2	EClc
Harbin	China	4.2.053-7	DCac
Hargeisa	Somalia	4.3.085-5	GBSab
Harkov	USSR	4.1.030-68	DCbo
Härnösand	Sweden	4.1.002-5	DClo
Hasselbough Bay	Macquerie I.	4.6.140-28	FTkk
Hatanga	USSR	4.1.030-12	ECld
Hatteras	USA	4.4.119-50	Crak
Havana	Cuba	4.4.123-8	Awab
Hay	Australia	4.6.137-14	BSak
Heard Island		4.3.107-11	FTko
Helgoland	FRG	4.1.017-3	DOlk
Hellburg	Réunion Is.	4.3.107-6	GArbl
Helsinki	Finland	4.1.003-7	DClo
Hengchow	China	4.2.053-21	Crhk
Hengchu	Taiwan	4.2.054-6	Amhb
Henrique de Carvalho	Angola	4.3.105-3	GAwab
Heraklion	Greece	4.1.029-7	Csal
Herat	Afghanistan	4.2.040-3	GBShk
Hilo	USA	4.4.119-81	Arab
Hiroshima	Japan	4.2.055-14	Crak
Hoang-Sa – Pattle	Viet Nam	4.2.058-2	Awha
Hobart	Australia	4.6.137-36	Crlk
Hokitika	New Zealand	4.6.136-6	DOlk
Holar i Hornafirdi	Iceland	4.1.006-2	EOlk
Hombantota	Sri Lanka	4.2.045-6	Asha
Hombori	Mali	4.3.070-6	BWia
Hon	Libya	4.3.080-6	BWhl
Hong Kong	Hong Kong	4.2.049-1	CWhl
Honiara	Solomon Is.	4.6.140-6	Araa

Name of Station	Country	Chapter-No.	Climate
Honolulu	USA	4.4.119-79	Asab
Horta	Portugal	4.1.016-7	Cral
Houston	USA	4.4.119-59	Crhl
Hsining	China	4.2.053-5	GBSbo
Huancayo	Peru	4.5.129-7	HAwll
Hurghada	Egypt	4.3.081.10	BMhl
Hutts Gate	St. Helena	4.3.078-2	GAsbl
Hwalien	Taiwan	4.2.054-2	Crhl
Hyderabad	Pakistan	4.2.041-14	BWhl
Iasi	Romania	4.1.025-1	DCbc
Ibiza	Spain	4.1.014-13	Csal
Ichang	China	4.2.053-27	Cwhk
Idris	Libya	4.3.080-3	BShl
Iguape	Brazil	4.5.127-39	Cral
Ikeja (Lagos)	Nigeria	4.3.098-8	Amha
Ile Europa		4.3.107-7	BShb
Ilirnej	USSR	4.1.030-17	ECld
Iloilo	Philippines	4.2.064-7	Amha
Ilorin	Nigeria	4.3.098-6	Awha
Imata	Peru	4.5.129-9	HAwkk
Impfondo	Congo	4.3.092-3	Araa
Imhambane	Mozambique	4.3.109-5	Awab
Indore	India	4.2.044-28	GAwhb
Innsbruck	Austria	4.1.019-4	GDCbo
Inogo	Zaire	4.3.091-9	Awaa
Inuvik	Canada	4.4.118-13	ECld
Invercargill	New Zealand	4.6.136-9	DOlk
Iquitos	Peru	4.5.129-12	Araa
Irkutsk	USSR	4.1.030-39	ECbc
Isachsen	Canada	4.4.118-3	FTkd
Isfahan	Iran	4.2.039-12	GBWhk
Isfjord Radio	Norway	4.1.001-1	FTkc
Ishigakijima	Japan	4.2.055-24	Arhb
Isparta	Turkey	4.2.031-19	GCsak
Istanbul	Turkey	4.2.031-5	Csbk
Izmir	Turkey	4.2.031-15	Cshk
Jacksonville	USA	4.4.119-57	Crhl
Jacobabad	Pakistan	4.2.041-12	BWil
Jaffa	Israel	4.2.035-2	Csal
Jagdalpur	India	4.2.044-33	GAwhb
Jaipur	India	4.2.044-8	BShl
Jakarta	Indonesia	4.2.063-15	Araa
Jakobshavn	Greenland	4.1.007-6	FTkc
Jakutsk	USSR	4.1.030-26	ECbe
Jamestown	St. Helena	4.3.078-1	BMab
Jan Mayen	Norway	4.1.001-3	FTko
Jask	Iran	4.2.039-17	BMhb
Jauja	Peru	4.5.129-6	HAwll
Jeddah	Arabia	4.2.037-5	BMha
Jefman	Indonesia	4.2.063-8	Arha

Name of Station	Country	Chapter-No.	Climate
Jerusalem	Israel	4.2.035-3	GCsak
Jessore	Bangladesh	4.2.042-5	Awhb
Jimma	Ethiopia	4.3.083-4	GAwbl
Jodhpur	India	4.2.044-2	BWhl
Jönköping	Sweden	4.1.002-8	DClo
Johannesburg	South Africa	4.3.113-2	GAwbl
Johnston Island		4.6.139-3	Asha
Jokioinen	Finland	4.1.003-5	DClo
Jos	Nigeria	4.3.098-3	GAwab
Juan Fernandez	Chile	4.5.131-9	Crbl
Juba	Sudan	4.3.082-20	Awha
Juneau	USA	4.4.119-14	EOlo
Jungfraujoch	Switzerland	4.1.012-3	HDOoc
Kabala	Sierra Leone	4.3.076-1	Amha
Kabale	Uganda	4.3.088-5	GAwll
Kabul	Afghanistan	4.2.040-1	GCsao
Kagoshima	Japan	4.3.055-18	Crak
Kaimana/Utarom	Indonesia	4.2.063-13	Araa
Kaitaia	New Zealand	4.6.136-1	Crbl
Kajaani	Finland	4.1.003-2	EClc
Kalamai	Greece	4.1.029-6	Csal
Kalat	Pakistan	4.2.041-8	GBSak
Kalemie (Albertville)	Zaire	4.3.091-14	GAwab
Kalgoorlie	Australia	4.6.137-11	BSal
Kaliningrad	USSR	4.1.030-62	DCbo
Kamaran Island	Arabia	4.2.037-11	BMha
Kamina-Baka	Zaire	4.3.091-16	GAwab
Kamishli	Syria	4.3.033-1	Cshk
Kampot	Khmer	4.2.060-5	Awha
Kananga-Luluabourg	Zaire	4.3.091-13	GAwaa
Kanazawa	Japan	4.2.055-9	Craa
Kandahar	Afghanistan	4.2.040-4	GBShk
Kandi	Benin	4.3.099-1	Awha
Kandy	Sri Lanka	4.2.045-3	Araa
Kangnung	South Korea	4.2.052-1	DCao
Kano	Nigeria	4.3.098 1	BShh
Kaolack	Senegal	4.3.072-6	BSha
Kap Tobin	Greenland	4.1.007-5	FTkc
Kapuskasing	Canada	4.4.118-31	DClc
Karachi	Pakistan	4.2.041-16	BMhb
Karaganda	USSR	4.1.030-81	DCbc
Karesuando	Sweden	4.1.002-1	EClc
Karima	Sudan	4.3.082-5	BWhb
Karlstad	Sweden	4.1.002-6	DClo
Karnobat	Bulgaria	4.1.026-5	DCbk
Karonga	Malawi	4.3.111-1	GAwab
Kasama	Zambia	4.3.110-2	GAwbl
Kashgar	China	4.2.053-1	GBWao
Kassala	Sudan	4.3.082-10	GBWha
Kassel	FRG	4.1.017-8	DObk

Name of Station	Country	Chapter-No.	Climate
Kastamonu	Turkey	4.2.031-4	GCsbo
Katmandu	Nepal	4.2.047-1	GCwal
Kaunas	USSR	4.1.030-61	DCbo
Kayes	Mali	4.3.070-9	BSia
Kayseri	Turkey	4.2.031-12	GCsao
Kazan	USSR	4.1.030-59	DCbo
Keetmanshoop	Namibia	4.3.114-4	GWBal
Kéniéba	Mali	4.3.070-12	Awha
Kerkyra	Greece	4.1.029-3	Csal
Kerman	Iran	4.2.039-10	GBSak
Kermanshan	Iran	4.2.039-4	GCsak
Kew (London)	UK	4.1.004-8	DObk
Khanaquin	Iraq	4.2.038-3	BSil
Kharga	Egypt	4.3.081-14	BWhl
Kharkov	USSR	4.1.030-67	DCbo
Khartoum	Sudan	4.3.082-9	BWhb
Khormaksar	Arabia	4.2.037-13	BMha
Kidal	Mali	4.3.070-2	BWhb
Kiev	USSR	4.1.030-66	DCbo
Kigali	Rwanda	4.3.089-1	GAwbb
Kigoma	Tanzania	4.3.087-4	GAwaa
Kisozi-Collina	Burundi	4.3.090-2	GAwll
Kilkenny	Ireland	4.1.005-8	DOlk
Kikwit	Zaire	4.3.091-12	GAwaa
Kimberley	South Africa	4.3.113-6	GBSal
Kindley Air Force Base	Bermuda	4.4.121-1	Cral
Kindu	Zaire	4.3.091-6	Araa
King Salmon	USA	4.4.119-11	EClc
Kingston	Jamaica	4.4.123-13	Awha
Kinshasa (Léopoldville)	Zaire	4.3.091-11	Awab
Kirensk	USSR	4.1.030-34	ECbd
Kirkuk	Iraq	4.2.038-2	Csik
Kirov	USSR	4.1.030-56	DCbc
Kirovabad	USSR	4.1.030-90	BSak
Kisangani	Zaire	4.3.091-3	Araa
Kisaumi	Tanzania	4.3.087-6	Awha
Kisozi	Burundi	4.3.090-2	GAwll
Kitale	Kenya	4.3.086-5	GAwbl
Klagenfurt	Austria	4.1.019-7	DCbo
Klodzko	Poland	4.1.021-15	DClo
København	Denmark	4.1.008-3	DObk
Kochi	Japan	4.2.055-15	Crak
Kodiak	USA	4.4.119-15	EOlo
Kolarovgrad	Bulgaria	4.1.026-2	DCbo
Kolpašev	USSR	4.1.030-33	ECbc
Kompong Cham	Khmer	4.2.060-3	Awha
Konya	Turkey	4.2.031-18	GCsak
Korf	USSR	4.1.030-42	EClc
Koror	Palau Is.	4.6.139-9	Araa
Kosti	Sudan	4.3.082-15	BWha
Koszalin	Poland	4.1.021-2	DClo

Name of Station	Country	Chapter-No.	Climate
Kota	India	4.2.044-9	BSib
Kota Bharu	Malaysia	4.2.061-1	Arha
Kotzebue	USA	4.4.119-5	EClc
Koumac	New Caledonia	4.6.140-18	Awab
Koundja	Cameroun	4.3.097-3	GAwab
Koutiala	Mali	4.3.070-14	Awha
Krakow	Poland	4.1.021-16	DCbo
Krasnoyarsk	USSR	4.1.030-36	ECbc
Krasnovodsk	USSR	4.1.030-94	BWhk
Krems an der Donau	Austria	4.1.019-2	DCbo
Ksara	Lebanon	4.2.034-3	GCsak
Kuala Lumpur	Malaysia	4.2.061-6	Arha
Kuala Trengganu	Malaysia	4.2.061-4	Arha
Kuching (Sarawak)	Malaysia	4.2.061-8	Arha
Kufra	Libya	4.3.080-8	BWhl
Kumasi	Ghana	4.3.101-2	Araa
Kunming	China	4.1.030-31	GCwbl
Kupang/Penfui	Indonesia	4.2.063-22	Awha
Kustanaj	USSR	4.1.030-79	DCbc
Kut-El-Hai	Iraq	4.2.038-7	BWil
Kuwait	Arabia	4.2.037-1	BMil
Kwajalein	Marshall Is.	4.6.139-7	Arha
Kweilin	China	4.2.053-22	Crhk
Laboulaye	Argentina	4.5.134-23	Cwak
La Cola	Costa Rica	4.4.122-8	Araa
La Coruña	Spain	4.1.014-2	Crbl
Lae	Papua-Nuigini	4.6.138-2	Arha
Laghouat	Algeria	4.3.067-6	GBWhk
Lagos/Ikeja	Nigeria	4.3.098-8	Amha
La Guardia Airport (New York)	USA	4.4.119-39	DOak
Lahore City	Pakistan	4.2.041-5	BShl
Lambarene	Gabon	4.3.093-6	Awaa
Lambayeque	Peru	4.5.129-2	BMab
Lanai City	USA	4.4.119-80	Asbb
La Orchila	Venezuela	4.4.123-17	BMha
La Paz	Mexico	4.4.120-2	BMhb
La Paz – El Alto	Bolivia	4.5.130-8	HAwlk
La Quiaca	Argentina	4.5.134-1	GBSlk
La Rioja	Argentina	4.5.134-14	BWal
Larissa	Greece	4.1.029-2	Cshk
Lashio	Burma	4.2.056-2	GCwal
Las Palmas	Spain	4.1.014-16	BSal
Lastourville	Gabon	4.3.093-7	Awab
Las Vegas	USA	4.4.119-73	GBWhk
Launceston	Australia	4.6.137-35	Crbk
Lauthala Bay	Fiji	4.6.140-16	Araa
Le Bourget	France	4.1.013-2	DObk
Legaspi	Philippines	4.2.064-5	Arha
Leh	Kashmir	4.2.043-2	HBWlo
Leipzig	GDR	4.1.018-7	DCbo

Name of Station	Country	Chapter-No.	Climate
Lenkoran	USSR	4.1.030-98	DCak
Leningrad	USSR	4.1.030-53	DCbo
Léopoldville-Kinshasa	Zaire	4.3.091-11	Awab
Lerwick	UK	4.1.004-1	DOlk
Lethbridge	Canada	4.4.118-34	GDCbo
Lhasa (Tibet)	China	4.2.053-6	HBSlo
Libenge	Zaire	4.2.091-7	Awaa
Libreville	Gabon	4.3.093-4	Amaa
Lihue – Kauai	USA	4.4.119-78	Arab
Lilongwe	Malawi	4.3.111-2	GAwal
Lima	Peru	4.5.129-3	BMbl
Limoges	France	4.1.013-9	DObk
Lindi	Tanzania	4.3.087-10	Awha
Linguère	Senegal	4.3.072-4	BSha
Lisala	Zaire	4.3.091-2	GAraa
Lisboa	Portugal	4.1.016-4	Csbl
Little Rock	USA	4.4.119-51	Crhk
Livingstone	Zambia	4.3.110-9	GBSal
Lobito	Angola	4.3.105-9	BMab
Lod Airport	Israel	4.3.035-2	Csal
Lodwar	Kenya	4.3.086-2	BWhh
Lom	Bulgaria	4.1.026-1	DCao
Lomé	Togo	4.3.100-5	Awha
London	UK	4.1.004-8	DObk
Lord Howe Island		4.6.140-25	Cral
Los Angeles	USA	4.4.119-67	Csal
Lourence Marques	Mozambique	4.3.109-6	Awab
Luanda	Angola	4.3.105-8	BSab
Luang Prabang	Laos	4.3.059-1	Awhb
Ludhiana	India	4.2.044-5	BShl
Luga	Malta	4.1.028-1	Csal
Lugano	Switzerland	4.1.012-6	DObk
Lug Ferrandi	Somalia	4.3.085-3	BWhh
Luluabourg (Kananga)	Zaire	4.3.091-13	GAwaa
Lulumbashi (Elisabethville)	Zaire	4.3.091-17	GAwbl
Lunganville	New Hebrides	4.6.140-10	Araa
Lungchow	China	4.2.053-35	Cwhl
Lungi (Freetown)	Sierra Leone	4.3.076-2	Amha
Luonetjärvi	Finland	4.1.003-3	EOlo
Lusaka	Zambia	4.3.110-8	GAwal
Luso	Angola	4.3.105-4	GAwbl
Luxembourg	Luxembourg	4.1.011-1	DOlk
Luxor	Egypt	4.3.081-12	BWhl
L'vov	USSR	4.1.030-69	DCbo
Lyon	France	4.1.013-10	DObk
Ma'an Airport	Jordan	4.2.036-3	GBWak
Macao	Macao	4.3.050-1	Cwhl
Macenta	Guinea	4.3.075-4	GAmha
Macquerie Island		4.6.140-28	FTkk
Madras	India	4.2.044-39	Awha

Name of Station	Country	Chapter-No.	Climate
Madrid	Spain	4.1.014-8	GCsak
Mafeking	South Africa	4.3.113-5	GBSal
Magdeburg	GDR	4.1.018-4	DCbo
Magyaróvár	Hungary	4.1.022-2	DCbo
Mahalapye	Botswana	4.5.115-3	GBSal
Mahe	Seychelles Islands	4.3.107-1	Arha
Mahon	Spain	4.1.014-11	Csal
Maiduguri	Nigeria	4.3.098-2	BShb
Maine Soroa	Niger	4.3.069-10	BShb
Maiquetia	Venezuela	4.5.125-2	BSha
Maintirano	Madagascar	4.3.108-9	Awab
Majunga	Madagascar	4.3.108-7	Awha
Majuro	Marshall Is.	4.6.139-10	Araa
Makassar/Hassunuddin	Indonesia	4.2.063-18	Awha
Makindu	Kenya	4.3.086-9	GAwab
Makokou	Gabon	4.3.093-4	GAsab
Makurdi	Nigeria	4.3.098-7	Awha
Malacca	Malaysia	4.2.061-7	Araa
Malakal	Sudan	4.3.082-18	BSha
Malange	Angola	4.3.105-2	GAwbb
Malatya	Turkey	4.2.031-16	GCsao
Malin Head	Ireland	4.1.055-1	DOlk
Mamou	Guinea	4.3.075-2	GAwaa
Managua	Nicaragua	4.4.122-19	Awha
Mananjary	Madagascar	4.3.108-3	Arab
Manaus	Brazil	4.5.127-4	Arha
Manchester Airport	UK	4.1.004-5	DOlk
Mandalay	Burma	4.2.056-3	Awhb
Mandang	Papua-Nuigini	4.6.138-1	Araa
Mandera	Kenya	4.3.086-1	BWhh
Mangalore	India	4.2.044-40	Amha
Mango	Togo	4.3.100-1	Awha
Manila	Philippines	4.2.064-4	Awha
Mankwari/Rondani	Indonesia	4.2.063-7	Araa
Mannar	Sri Lanka	4.2.045-1	Asha
Manqabad	Egypt	4.3.081-11	BWhl
Manta	Ecuador	4.5.128-5	DMaa
Manzanillo	Mexico	4.4.120-12	Awha
Manzanares	Colombia	4.5.124-8	GArbb
Mao	Chad	4.3.095-2	BWha
Maracaibo	Venezuela	4.5.125-1	BSha
Maracay	Venezuela	4.5.125-7	Awaa
Maradi	Niger	4.3.069-8	BShb
Mar del Plata	Argentina	4.5.134-36	Crbk
Marion Island		4.3.107-9	FTkk
Mariscal Estigarribia	Paraguay	4.5.132-1	Awhb
Marrakech	Morocco	4.3.066-8	BShl
Marseille	France	4.1.013-15	Csak
Maseru	Lesotho	4.3.117-1	GCrbk
Mashad	Iran	4.2.039-7	GBSak

Name of Station	Country	Chapter-No.	Climate
Masindi	Uganda	4.3.088-3	GAwab
Masirah Island	Arabia	4.2.037-9	BMhb
Masulipatnam	India	4.2.044-36	Awha
Matan	Senegal	4.3.072-3	BSha
Maturin	Venezuela	4.5.125-8	Awaa
Maun	Botswana	4.3.115-1	GBSal
Mawashi/Naha	Japan	4.2.055-22	Crhl
Mawson	Antarctic	4.5.135-10	FIkc
Mayumba	Gabon	4.3.093-10	Awab
Mazari-Sherif	Afghanistan	4.2.040-2	BShk
Mazatlan	Mexico	4.4.120-5	Awhb
Mbabane	Ngwane	4.3.116-1	GAwbl
Mbandaka-Coquilhatville	Zaire	4.3.089-4	Araa
Mbeya	Tanzania	4.3.087-9	GAwbl
McGrath	USA	4.4.119-8	EClc
Medan	Indonesia	4.2.063-2	Araa
Medford	USA	4.4.119-23	DObk
Medina	Arabia	4.2.037-4	BWil
Meknès	Morocco	4.3.066-4	GCsal
Melbourne	Australia	4.6.137-32	Crbl
Menado	Indonesia	4.2.063-4	Arha
Menaka	Mali	4.3.070-5	BWia
Mendoza	Argentina	4.5.134-16	GBWak
Meng-Tzu	China	4.2.053-33	GCwal
Merauke	Indonesia	4.2.063-20	Awaa
Mergui	Burma	4.2.056-7	Amha
Merida	Mexico	4.4.120-8	Awha
Merida	Venezuela	4.5.125-9	GAwbb
Mersa Mertruh	Egypt	4.3.081-2	BSal
Messina	Italy	4.1.027-19	Csal
Mexico City (Tacubaya)	Mexico	4.4.120-10	GAwll
Miami	USA	4.4.119-77	Awhb
Milano	Italy	4.1.027-4	DOak
Mildura	Australia	4.6.137-13	BSal
Minamidaitojima	Japan	4.2.055-23	Crhl
Mindelo/Cabo Verde	Portugal	4.1.016-10	BMab
Minicoy	Laccadive Islands	4.2.048-4	Awha
Minna	Nigeria	4.3.098-4	Awha
Minsk	USSR	4.1.030-63	DCbo
Minusinsk	USSR	4.1.030-77	DCbc
Minya	Egypt	4.3.082-9	BWhl
Miskolc	Hungary	4.1.022-1	DCbo
Misurata	Libya	4.3.080-4	BSal
Mitzic	Gabon	4.3.093-2	GAsab
Miyako	Japan	4.2.055-6	DObk
Moanda (Banana)	Zaire	4.3.091-15	Awab
Mobile	USA	4.4.119-56	Crhl
Mocamedes	Angola	4.3.105-10	BMal
Mogadiscio	Somalia	4.3.085-6	BSha
Mohanbari	India	4.2.044-16	Cwhl

Name of Station	Country	Chapter-No.	Climate
Mokpo	South Korea	4.2.052-4	Crab
Mombasa	Kenya	4.3.086-10	Asha
Mongo	Chad	4.3.095-5	BSha
Mongu	Zambia	4.3.110-7	GAwal
Monrovia	Liberia	4.3.104-1	Araa
Monterrey	Mexico	4.4.120-4	GAwhl
Montevideo	Uruguay	4.5.133-3	Crhl
Montreal	Canada	4.4.118-33	DCbc
Mopelia	Society Is.	4.6.140-13	Arha
Moonsonee	Canada	4.4.118-22	EClc
Mopti	Mali	4.3.070-8	BSha
Morelia	Mexico	4.4.120-9	GAwbl
Morogoro	Tanzania	4.3.087-7	GAwab
Morondava	Madagascar	4.3.108-10	BShb
Moroni	Comoro Islands	4.3.107-2	Arab
Moroto	Uganda	4.3.088-2	GAwab
Moshi	Tanzania	4.3.087-3	GAwab
Moskva	USSR	4.1.030-60	DCbc
Mosul	Iraq	4.2.038-1	Cshk
Mouila	Gabon	4.3.093-9	Awaa
Mould Bay	Canada	4.4.118-4	FTkd
Moundou	Chad	4.3.095-11	Awha
Mouyondzi	Congo	4.3.092-8	GAwab
Mpika	Zambia	4.3.110-3	GAwal
Mpouya	Congo	4.3.092-6	Awaa
Mt. Gambier	Australia	4.6.137-33	Crlk
Mugla	Turkey	4.2.031-20	GCsak
Muharraq-Bahrain	Arabia	4.2.037-6	BMhl
Mukden	China	4.2.052-9	DCac
Mukteswar Kumaon	India	4.2.044-15	GCwbk
Mukumari	Zaire	4.3.091-5	GAraa
Mullingar	Ireland	4.1.005-5	DOlk
Multan	Pakistan	4.2.041-10	BWil
München	FRG	4.1.017-12	GDObo
Mundiwindi	Australia	4.6.137-18	GBWhl
Murmansk	USSR	4.1.030-8	EClo
Murree	Pakistan	4.2.041-1	GCrbk
Muscat	Arabia	4.2.037-8	BMhb
Mwanza	Tanzania	4.3.087-2	GAwab
Myitkyina	Burma	4.2.056-1	Cwhl
Mys Čeljuskin	USSR	4.1.030-1	FTkd
Mys Šmidta	USSR	4.1.030-6	FTkd
Mys Uzlen	USSR	4.1.030-7	FTkc
Nagasaki	Japan	4.2.055-17	Crak
Nagoya	Japan	4.2.055-12	Crak
Nagpur	India	4.2.044-30	Awib
Nairobi	Kenya	4.3.086-8	GAwbl
Nakhon Ratchasima	Thailand	4.2.057-5	Awhb
Nakhon Sawan	Thailand	4.2.057-4	Awha
Nanchang	China	4.2.053-19	Crhk

Name of Station	Country	Chapter-No.	Climate
Nancy	France	4.1.013-3	DObk
Nandi – Vitu Levu	Fiji	4.6.140-15	Araa
Nanking	China	4.2.030-14	Crhk
Nantes	France	4.1.013-7	DObk
Nantucket	USA	4.4.119-36	DObk
Nanyuki	Kenya	4.3.086-6	GAwll
Napoli	Italy	4.1.027-14	Csak
Narayanganj	Bangladesh	4.2.042-4	Awhb
Nar'jan-Mar	USSR	4.1.030-9	EClc
Nashville	USA	4.4.119-49	Crak
Nasiriya	Iraq	4.2.038-9	BWhl
Nassau	Bahamas	4.4.123-7	Awhb
Natal	Brazil	4.5.127-18	Asaa
Natitingou	Benin	4.3.099-2	Awha
Nauru	Gilbert Is.	4.6.140-1	Arha
Naze	Japan	4.2.055-21	Crhl
Ndele	Central Africa	4.3.094-2	GAwha
Ndjemena (Fort Lamy)	Chad	4.3.095-6	BSha
Ndola	Zambia	4.3.110-4	GAwal
Nema	Mauretania	4.3.071-5	BWia
Nemuro	Japan	4.2.055-3	DCbo
Neuquén	Argentina	4.5.134-17	BWbk
Neustrelitz	GDR	4.1.018-3	DClo
Nevel'sk	USSR	4.1.030-51	DCbc
New Amsterdam		4.3.107-8	Crll
Newcastle	South Africa	4.3.113-3	GAwbl
New Delhi	India	4.2.044-6	BShl
New Orleans	USA	4.4.119-58	Cral
New Plymouth	New Zealand	4.6.136-3	Crlk
New York	USA	4.4.119-39	DCak
Ngaundere	Cameroon	4.3.097-2	GAwab
N'Guigmi	Niger	4.3.069-4	BWhb
Nha-Trang	Viet Nam	4.2.058-4	Awhb
Niamey	Niger	4.3.069-6	BSha
Nice	France	4.1.013-14	Csak
Nicosia	Gyprus	4.2.032-1	Cshk
Nigro Du Sahel	Mali	4.3.070-7	BSha
Niigate	Japan	4.2.055-8	DOak
Nikolaevsk-na-Amure	USSR	4.1.030-47	DClc
Nimes	France	4.1.013-13	Csak
Nitchequon	Canada	4.4.118-21	GEClc
Nome	USA	4.4.119-7	EClc
Nord	Greenland	4.1.007-2	FTkd
Norfolk Island	Mauretania	4.6.140-23	Crbl
Norman Wells	Canada	4.4.118-14	ECld
North Bay	Canada	4.4.118-32	DCbc
North Front	Gibraltar	4.1.015-1	Csal
North Platte	USA	4.4.119-70	GBSao
Nou Adhibou-Port Etienne	Mauretania	4.3.071-2	BMab
Nouakschott	Mauretania	4.3.071-4	BMhb

Name of Station	Country	Chapter-No.	Climate
Nouméa	New Hebrides	4.6.140-20	Arab
Nova Lisboa	Angola	4.3.105-5	GAwbb
Nürnberg	FRG	4.1.017-10	DObo
Nurwara Elya	Sri Lanka	4.2.045-4	GArll
Novo Pjatigorsk	USSR	4.1.030-73	GDCbo
Oakland	USA	4.4.119-64	Csbk
Obo	Central Africa	4.3.094-9	GAwaa
Ocean Island	Gilbert Is.	4.6.140-2	Arha
Odessa	USSR	4.1.030-71	DCbo
Odienné	Ivory Coast	4.3.102-2	Awha
Ojmjakon	USSR	4.1.030-24	GECle
Okhotsk	USSR	4.1.030-43	EClc
Okiep	South Africa	4.3.113-8	GBSal
Oklahoma	USA	4.4.119-48	DCak
Olenek	USSR	4.1.030-15	ECle
Olinda	Brazil	4.5.127-20	Asaa
Omsk	USSR	4.1.030-75	DCbc
Ondina	Brazil	4.5.127-8	Araa
Onslow	Australia	4.6.137-16	BMhb
Oodnadatta	Australia	4.6.137-20	BWhl
Oostende	Belgium	4.1.010-1	DOlk
O Porto	Portugal	4.1.016-2	Crbk
Oran	Algeria	4.3.067-3	Csal
Orenburg	USSR	4.1.030-65	DCbc
Orlandet	Norway	4.1.001-7	EOlo
Oruro	Bolivia	4.5.130-3	HBSlk
Osaka	Japan	4.2.055-13	Crhk
Oslo	Norway	4.1.001-11	DClo
Östersund	Sweden	4.1.002-4	EObo
Ostrava	Czechoslovakia	4.1.020-3	DCbo
Ostrov Beringa	USSR	4.1.030-44	EOlo
Ostrov Četyreh-stolbavoj	USSR	4.1.030-5	FTkd
Ostrov Dikson	USSR	4.1.030-3	FTkd
Ostrov Kotel'nyi	USSR	4.1.030-2	FTkd
Ostrov Vrangelya	USSR	4.1.030-4	FTkd
Ouagadougou	Upper Volta	4.3.102-2	BSha
Ouallen	Algeria	4.3.067-12	BWil
Oudja	Morocco	4.3.066-6	BSak
Ouesso	Congo	4.3.092-2	Araa
Padang	Indonesia	4.2.063-6	Araa
Pahlavi	Iran	4.2.039-1	Crak
Pakhoi	China	4.2.052-25	Crhl
Paksé	Laos	4.2.059-4	Awha
Pala	Chad	4.3.095-9	Awha
Palma de Mallorca	Spain	4.1.014-12	Csal
Palmira	Colombia	4.5.124-4	GAsaa
Palmyra	Syria	4.2.033-5	BWhk
Pamban	India	4.2.044-43	Awha
Panjgur	Pakistan	4.2.041-13	GBWhl
Paotou	China	4.2.053-3	GBSac

Name of Station	Country	Chapter-No.	Climate
Papeete – Faaa Tahiti	Society Islands	4.6.140-14	Araa
Parachinar	Pakistan	4.2.041-2	GCrak
Parakou	Benin	4.3.099-3	Awha
Paramaribo	Surinam	4.5.126-2	Arha
Paraná	Argentina	4.5.134-31	Cral
Paris (Le Bourget)	France	4.1.013-2	DObk
Pasni	Pakistan	4.2.041-15	BMhb
Paso de los Libres	Argentina	4.5.134-30	Cral
Patna	India	4.2.044-24	Awhb
Patrai	Greece	4.1.029-4	Csal
Pattle (Hoang-Sa)	Viet-Nam	4.2.058-2	Awha
Paucarany	Peru	4.5.129-11	HBSko
Peary Land	Greenland	4.1.007-1	FTkd
Pécs	Hungary	4.1.022-6	DCao
Pei-Yü-Shan	China	4.2.053-18	Crak
Peking	China	4.2.053-10	DCao
Penang	Malaysia	4.2.061-3	Arha
Penghu	Taiwan	4.2.054-4	Cwhl
Penhas Douradas	Portugal	4.1.016-3	GCrlk
Perpignan	France	4.1.013-16	Csak
Perth	Australia	4.6.137-24	Csal
Pescara	Italy	4.1.027-12	Crak
Peshawar	Pakistan	4.2.041-4	BShl
Petropavlovsk	USSR	4.1.030-45	EOlo
Phitsanulok	Thailand	4.2.057-3	Awha
Phnom Penh-Pochentong	Khmer	4.2.060-4	Awha
Phoenix	USA	4.4.119-75	BWhl
Phuket	Thailand	4.2.057-11	Amha
Piarco Airport	Trinidad	4.4.123-6	Araa
Pietersburg	South Africa	4.3.113-4	GBSbl
Pilar	Argentina	4.5.134-22	Cwal
Pisa	Italy	4.1.027-10	Crak
Plaisance	Mauritius	4.3.106-5	Arab
Plesman, Dr.- Airport	Curaçao	4.4.123-25	Asha
Plymouth	UK	4.1.004-9	DOlk
Pochentong (Phnom Penh)	Khmer	4.2.060-4	Awha
Podkamenaya Tunguska	USSR	4.1.030-27	EClc
Podor	Senegal	4.3.072-1	BWha
Pointe-à-Pitre	Guadeloupe	4.4.123-3	Araa
Pointe Noire	Congo	4.3.092-11	Awab
Ponape	Caroline Is.	4.6.129-11	Araa
Ponta Delgada	Portugal	4.1.016-8	Crbl
Ponta Pora	Brazil	4.5.127-10	GAral
Pontianak	Indonesia	4.2.063-5	Arha
Poona	India	4.2.044-12	GBShb
Popayan	Colombia	4.5.124-11	GArbl
Poprad	Czechoslovakia	4.1.020-5	GDClo
Port-au-Prince	Haiti	4.4.123-11	Awha
Port aux Français	Kerguelen	4.3.107-10	FTkk
Port Blair	Andaman Islands	4.2.048-2	Awha

Name of Station	Country	Chapter-No.	Climate
Port Elizabeth	South Africa	4.3.113-17	Crbl
Port Etienne-Nou Adhibou	Mauretania	4.3.071-2	BMab
Port Gentil	Gabon	4.3.093-5	Awaa
Port Harcourt	Nigeria	4.3.098-11	Araa
Port Hardy	Canada	4.4.118-41	DOlk
Port Hedland	Australia	4.6.137-15	BMhb
Portland	USA	4.4.119-22	DObk
Port Moresby	Papua-Nuigini	4.6.138-3	Awha
Porto	Portugal	4.1.016-2	Crbk
Porto Alegre	Brazil	4.5.127-44	Cral
Porto Amélia	Mozambique	4.3.109-1	Awha
Porto Nacional	Brazil	4.5.127-23	Awaa
Porto Velho	Brazil	4.5.127-22	Amaa
Port Said	Egypt	4.3.081-5	BMhl
Port Sudan	Sudan	4.3.082-2	BMhb
Posadas	Argentina	4.5.134-28	Cral
Poznan	Poland	4.1.021-7	DCbo
Praha	Czechoslovakia	4.1.020-1	DCbo
Praia – Sao Tiago/Cabo Verde	Portugal	4.1.016-12	BMab
Presidencia Roque	Argentina	4.5.134-20	Cwal
Pretoria	South Africa	4.3.113-1	GAwbl
Prince George	Canada	4.4.118-25	GDClc
Prins Christian Sund	Greenland	4.1.007-8	FTko
Przemysl	Poland	4.1.021-17	DCbo
Pueblo Bello	Colombia	4.5.124-2	GAwbb
Puerto Baquerizo	Ecuador	4.5.128-6	BSab
Puerto Cabezas	Honduras	4.4.122-4	Araa
Puerto Casado	Paraguay	4.5.132-2	Awhb
Puerto Deseado	Argentina	4.5.134-11	BSlk
Puerto Madryn	Argentina	4.5.134-18	BMbk
Puerto Plata	Dominican Prep.	4.4.123-1	Arab
Puerto Puente Franco	Paraguay	4.5.132-5	Cral
Puerto Suárez	Bolivia	4.5.130-11	Awhb
Puno	Peru	4.5.129-10	HAwlk
Punta Arenas	Chile	4.5.131-7	EOlk
Punta del Este	Uruguay	4.5.133-4	Crbl
Pusan	South Korea	4.2.052-3	Crak
Qualequaychu	Argentina	4.5.134-33	Cral
Quarzazate	Morocco	4.3.066-11	GBWhk
Quedlinburg	GDR	4.1.018-6	DOlk
Quelimane	Mozambique	4.3.109-3	Awhb
Quetta	Pakistan	4.2.041-7	GBSak
Quiaca, La –	Argentina	4.5.134-1	HBSlk
Quito	Ecuador	4.5.128-3	HAwll
Quixerambin	Brazil	4.5.127-15	Asha
Rabat	Morocco	4.3.065-3	Csal
Rabaul	Papua	4.6.140-4	Arha
Rangoon	Burma	4.2.056-5	Amha
Raoul Island	Kermadec Is.	4.6.140-24	Crbl
Rapa	Austral. Isl.	4.6.140-22	Cral

Name of Station	Country	Chapter-No.	Climate
Rapid City	USA	4.4.119-69	GBSao
Rawalpindi	Pakistan	4.2.041-3	Gcrhl
Rayack	Lebanon	4.2.034-2	GCsak
Recife – Olinda	Brazil	4.5.127-20	Asaa
Regina	Canada	4.4.118-28	GDCbc
Remanso	Brazil	4.5.127-38	BSha
Resistencia	Argentina	4.5.134-27	Cral
Resolute	Canada	4.4.118-5	FTkd
Reykjavik	Iceland	4.1.006-3	EOlk
Riberalta	Bolivia	4.5.130-3	Awha
Riga	USSR	4.1.030-57	DClo
Rikitea	Mangareva Island	4.6.140-21	Arab
Rio de Janeiro	Brazil	4.5.127-36	Awab
Rio de Oro	West Sahara	4.2.065-2	BMab
Rio Gallegos	Argentina	4.5.134-13	BSlk
Rivadavia	Argentina	4.5.134-2	BShl
Riyadh	Arabia	4.2.037-3	GBWil
Riyan	Arabia	4.2.037-12	BWha
Rize	Turkey	4.2.031-1	Crak
Roboré	Bolivia	4.5.130-17	Arhb
Rocha	Uruguay	4.5.133-2	Crbl
Roche's Point	Ireland	4.1.005-10	DOlk
Rodriguez	Mauritius	4.3.106-4	Awab
Roma	Italy	4.1.027-13	Csak
Roratonga Island	Cook Is.	4.6.140-19	Arab
Rosario	Argentina	4.5.134-32	Cral
Rosso	Mauretania	4.3.071-6	BWhb
Rostov-Na-Donu	USSR	4.1.030-70	DCao
Rubona-Collina	Rwanda	4.3.089-2	GAwbb
Rurrenabaque	Bolivia	4.5.130-14	Arab
Rutbah	Iraq	4.2.038-6	GBWhk
Sable Island	Canada	4.4.118-38	DCbo
Sachs Habour	Canada	4.4.118-6	FTkd
Sacramento	USA	4.4.119-63	Csak
Sa Da Bandeira	Angola	4.3.105-7	GAwbl
Sagar	India	4.2.044-27	GAwhb
Sagres	Portugal	4.1.016-6	Csal
Saigon	Viet-Nam	4.2.058-5	Awha
Saint Denis	Réunion Is.	4.3.107-5	Awab
Saint Louis	Senegal	4.3.072-2	BWhb
Sal/Cabo Verde	Portugal	4.1.016-11	BMab
Salalah	Arabia	4.2.037-10	BMha
Salehard	USSR	4.1.030-18	EClc
Salina Cruz	Mexico	4.4.120-14	Awha
Salisbury	Zimbabwe	4.3.112-1	GAwbl
Salloum	Egypt	4.3.081-4	BMal
Salvador – Ondina	Brazil	4.5.127-8	Araa
Salt Lake City	USA	4.4.119-38	GDCao
Salta	Argentina	4.5.134-3	GBSbl

Name of Station	Country	Chapter-No.	Climate
Salto	Uruguay	4.5.133-1	Cral
Salzburg	Austria	4.1.019-3	DCbo
Samsun	Turkey	4.2.031-2	Crak
San	Mali	4.3.070-11	BSha
Sana	Arabia	4.2.037-15	GBWhb
Sanae	Antarctic	4.5.135-11	FIod
San Antonio/Texas	USA	4.4.119-60	Crhl
San Antonio	Venezuela	4.5.125-12	Awaa
San Antonio Oeste	Argentina	4.5.134-8	BSbk
San Christobal	Ecuador	4.5.128-6	BSab
Sandakan – Sarawak	Malaysia	4.2.061-2	Araa
San Diego	USA	4.4.119-68	Csbl
Sandomierz	Poland	4.1.021-14	DCbc
Sandspit	Canada	4.4.118-40	DOlk
Sandvig	Denmark	4.1.008-4	DOlk
San Fernando	Venezuela	4.5.125-11	Awha
San Francisco	USA	4.4.119-65	Csbk
San Ignacio de Valasco	Bolivia	4.5.130-8	Awab
San Joaquín	Bolivia	4.5.130-6	Awha
San José	Costa Rica	4.4.122-20	GAwbb
San José	Bolivia	4.5.130-10	Awab
San Juan	Argentina	4.5.134-15	GBWak
San Juan	Puerto Rico	4.4.123-2	Araa
San Lorenzo	Ecuador	4.5.128-1	Araa
San Luis	Argentina	4.5.134-7	GBSak
San Pedro Sula	Honduras	4.4.122-4	Arha
San Salvador	El Salvador	4.4.122-12	GAwab
Santa Cruz (Mercuri)	Brazil	4.5.127-29	Awab
Santa Cruz	Bolivia	4.5.130-16	Arab
Santa Cruz	Argentina	4.5.134-12	BSlk
Santa Cruz de Tenerife	Spain	4.1.014-15	BSal
Santa Cruz Porillo	El Salvador	4.4.122-15	Awha
Santa Elena	Venzuela	4.5.125-14	GArbb
Santa Isabel	Equatorial Guinea	4.3.096-1	Amaa
Santa Maria	USA	4.4.119-66	Csll
Santander	Spain	4.1.014-1	Crbk
Santarém	Brazil	4.5.127-13	Amaa
Santa Rosa de Copan	Honduras	4.4.122-16	GAwbl
Santa Rosa	Argentina	4.5.134-24	Cwak
Santa Tecla	El Salvador	4.4.122-13	GAwbb
Santa Victoria do Palmar	Brazil	4.5.127-45	Crbl
Santiago	Chile	4.5.131-4	GCsbk
Santiago del Estero	Argentina	4.5.134-5	BSal
Santo Domingo	Dominican Rep.	4.4.123-12	Amaa
Säntis	Switzerland	4.1.012-2	GDOko
Santos	Brazil	4.5.127-11	Arab
Saõ Paulo	Brazil	4.5.127-37	GAwbl
Saõ Thomé	Sao Thomé	4.3.079-1	Asha
Sapporo	Japan	4.2.055-4	DCbo
Sarajevo	Yugoslavia	4.1.023-3	GDCbo

Name of Station	Country	Chapter-No.	Climate
Sarh-Fort Archambault	Chad	4.3.095-10	Awha
Sarmiento	Argentina	4.5.134-9	BSlk
Sault Ste. Marie	USA	4.4.119-27	DCbo
Save	Benin	4.3.099-5	Awha
Schleswig	FRG	4.1.017-2	DOlk
Scoresbysund	Greenland	4.1.007-5	FTkc
Scott Base	Antarctic	4.5.135-12	FIod
Seattle	USA	4.4.119-20	DObk
Seawell Airport	Barbados	4.4.123-5	Araa
Sebha	Libya	4.3.080-7	BWhl
Segou	Mali	4.3.070-10	BSha
Seistan	Iran	4.2.039-13	GBWhk
Şejmčan	USSR	4.1.030-25	ECld
Šelagoncy	USSR	4.1.030-19	ECle
Semiplatinsk	USSR	4.1.030-80	DCbc
Sendai	Japan	4.2.055-7	DOak
Seno	Laos	4.2.059-3	Awhb
Seno Madureira	Brazil	4.5.127-7	Araa
Sentani	Indonesia	4.2.063-11	Araa
Seoul	South Korea	4.2.052-2	DCao
Sept Iles	Canada	4.4.118-29	DClc
Serge Frolow	Tromelin Islands	4.3.107-4	Awaa
Serpa Pinto	Angola	4.3.105-6	GAwbl
Sevilla	Spain	4.1.014-10	Csal
Shanghai	China	4.2.053-15	Crak
Shannon Airport	Ireland	4.1.005-7	DOlk
Sharjah	Arabia	4.2.037-7	BMhb
Shemya Island	USA	4.4.119-17	EOlo
Shendi	Egypt	4.3.082-8	BWia
Sheridan	USA	4.4.119-29	GDCbo
Shiraz	Iran	4.2.039-5	GCshk
Shkoder	Albania	4.1.024-1	Crak
Shreveport	USA	4.4.119-54	Crhk
Shuwaikh/Kuwait	Arabia	4.2.037-1	BMil
Sian	China	4.2.053-13	DChk
Sibiti	Congo	4.3.092-7	GAwab
Sibiu	Romania	4.1.025-5	DClo
Sidi Barani	Egypt	4.3.081-1	BSal
Sidi Ifni	Morocco	4.3.066-10	BSbl
Siemreap – Angkor	Cambodia	4.2.060-2	Awha
Signy Island	Southern Orkney Islands	4.5.135-6	FTkc
Siguiri	Guinea	4.3.075-1	Awha
Sikasso	Mali	4.3.070-16	Arha
Silchar	India	4.2.044-26	Amhb
Simferopol'	USSR	4.1.030-72	DCbk
Simla	India	4.2.044-14	GCwbk
Simusir	USSR	4.1.030-46	EOlo
Singapore	Singapore	4.2.062-1	Arha
Sinuiju	North Korea	4.2.051-2	DCao
Sivas	Turkey	4.2.031-10	GCsbo

Name of Station	Country	Chapter-No.	Climate
Siwa	Egypt	4.3.081-7	BWhl
Skopje	Yugoslavia	4.1.023-4	DCak
Sliac	Czechoslovakia	4.1.20-6	DCbo
Smara	West Sahara	4.3.065-1	BWhl
Šniežka	Poland	4.1.021-12	GDCkc
Sochi	USSR	4.1.030-82	Crak
Sodankylä	Finland	4.1.003-1	EClc
Sofia	Bulgaria	4.1.026-4	GDCbo
Sokodé	Togo	4.3.100-2	Awha
Sokotra	Arabia	4.2.037-14	BMha
Songea	Tanzania	4.3.087-11	GAwal
Songkhla	Thailand	4.2.057-12	Amha
Sonnblick	Austria	4.1.019-5	HDCkc
Souanke	Congo	4.3.092-1	GAraa
Split	Yugoslavia	4.1.023-5	Crak
Spokane	USA	4.4.119-19	GDObo
Srinagar	Kashmir	4.2.043-1	GCrak
Stanley	Australia	4.6.137-34	Crlk
Stanley	Falkland Islands	4.5.135-3	FTkk
Stavanger	Norway	4.1.001-10	DOlk
St. Brandon	Mauritius	4.7.106-3	Awha
Stensele	Sweden	1.4.002-3	EClc
St. Louis	USA	4.4.119-45	DCak
St. Moritz	Switzerland	4.1.012-4	GDOlo
Stockholm	Sweden	4.1.002-7	DCbo
Stornoway	UK	4.1.004-2	DOlk
St. Paul Island	USA	4.4.119-4	FTko
St. Pölten	Austria	4.1.019-2	DCbo
Strasbourg	France	4.1.013-4	DObk
Stung Treng	Khmer	4.2.060-1	Awha
Stuttgart	FRG	4.1.017-11	DObk
Sukhumi	USSR	4.1.030-84	Crbk
Sulina	Romania	4.1.025-6	DCao
Surabaja	Indonesia	4.2.063-19	Awha
Surgut	USSR	4.1.030-29	EClc
Surigao	Philippines	4.2.064-9	Arha
Suwalki	Poland	4.1.021-3	DCbc
Sverdlovsk	USSR	4.1.030-75	DCbc
Swan Island		4.4.123-14	Awha
Sydney	Australia	4.6.137-30	Crbl
Sydney	Canada	4.4.118-36	DCbo
Syktyvkar	USSR	4.1.030-11	EClc
Sylhet	Bangladesh	4.2.042-1	Amhb
Szczecin	Poland	4.1.021-4	DCbo
Szeged	Hungary	4.1.022-5	DCao
Table Island	Andamon Islands	4.2.048-1	Awha
Tabligbo	Togo	4.3.100-4	Awha
Tabora	Tanzania	4.3.087-5	GAwab
Tabou	Ivory Coast	4.3.103-6	Araa
Tabriz	Iran	4.2.039-9	GBSak

Name of Station	Country	Chapter-No.	Climate
Tacloban	Philippines	4.2.064-6	Arha
Tacubaya – Mexico City	Mexico	4.4.120-10	GAwll
Tahiti – Faaa	Society Is.	4.6.140-14	Araa
Tahoua	Niger	4.3.069-3	BWha
Taichung	Taiwan	4.2.054-3	Cwhl
Tainan	Taiwan	4.2.054-5	Cwhl
Taipei	Taiwan	4.2.054-1	Crhl
Takaroa	French Polynesia	4.6.140-9	Arha
Takengön	Indonesia	4.2.063-1	GArbb
Takoradi	Ghana	4.3.101-4	Awab
Tallinn	USSR	4.1.030-54	DClo
Tamale	Ghana	4.3.101-1	Awha
Tamanrasset	Algeria	4.3.067-13	GBWhl
Tamatave	Madagascar	4.3.108-2	Arab
Tambacounda	Senegal	4.3.072-7	Awha
Tambora-Soembawa	Indonesia	4.2.063-16	GArab
Tampa	USA	4.4.119-61	Crhl
Tampico	Mexico	4.4.120-6	Awhb
Tananarive	Madagascar	4.3.108-8	GAwbl
Tandjoengpandan	Indonesia	4.2.063-12	Araa
Tarakan	Indonesia	4.2.063-3	Araa
Taranto	Italy	4.1.027-17	Csak
Tarawa	Gilbert Is.	4.6.139-13	Arhh
Taškent	USSR	4.1.030-96	DCak
Tbilisi	USSR	4.1.030-97	DCak
Tchaourou	Benin	4.3.099-4	Awha
Tébessa	Algeria	4.3.067-4	GBSak
Tegucigalpa	Honduras	4.4.122-18	GAwab
Tehran	Iran	4.2.039-8	GBShk
Tela	Honduras	4.4.122-2	Araa
Tel Aviv	Israel	4.2.035-2	Csal
Tenerife	Spain	4.1.014-14	GCsbl
Teng-Chung	China	4.2.053-32	GCwbk
Terezina	Brazil	4.5.127-17	Awha
Tessalit	Mali	4.3.070-1	BWhb
Tete	Mozambique	4.3.109-7	BShb
The Pas	Canada	4.4.118-19	ECbc
Thessaloniki	Greece	4.1.029-1	Csak
Thorshavn	Denmark	4.1.008-1	EOlk
Thule	Greenland	4.1.007-4	FTkd
Thursday Island	Australia	4.6.137-1	Awha
Tidjikja	Mauretania	4.3.071-3	BWhb
Tientsin	China	4.2.053-4	BSao
Tillabery	Niger	4.3.069-5	BSha
Timbuktu	Mali	4.3.070-3	BWha
Timeremos	Venezuela	4.3.123-13	Araa
Tingo Maria	Peru	4.3.129-12	GAraa
Tirana	Albania	4.1.024-2	Crak
Titizima	Bonin Islands	4.6.139-1	Arab
Titograd	Yugoslavia	4.1.023-6	Crak

Name of Station	Country	Chapter-No.	Climate
Tobolsk	USSR	4.1.030-74	DCbc
Tokar	Sudan	4.2.082-6	BWia
Tokyo	Japan	4.2.055-10	Crak
Tombouctou	Mali	4.3.070-3	BWha
Tomsk	USSR	4.1.030-35	EClc
Torgilsbu	Greenland	4.1.007-8	FTko
Toowoomba	Australia	4.6.137-29	GCrbl
Torino	Italy	4.1.027-6	DOak
Torishima	Japan	4.2.055-20	Cral
Toronto	Canada	4.4.118-39	DCbo
Torun	Poland	4.1.021-6	DCbo
Toulouse	France	4.1.013-12	DObk
Tourane : Da Nang	Viet-Nam	4.2.058-3	Amhb
Townsville	Australia	4.6.137-4	Awab
Trapani	Italy	4.1.027-20	Csal
Trelew	Argentina	4.5.137-19	BMbk
Tres Lagos	Brazil	4.5.127-35	Awab
Trieste	Italy	4.1.027-7	Crak
Trincomalee	Sri Lanka	4.2.045-2	Arha
Trinidad	Bolivia	4.5.130-15	Arha
Tripoli	Lebanon	4.2.034-1	Cshl
Tripoli (Idris)	Libya	4.3.080-3	BShl
Tristan da Cunha		4.5.135-1	Crbl
Trivandrum	India	4.2.044-44	Awha
Troickij Priisk	USSR	4.1.030-38	ECld
Tromsö	Norway	4.1.001-5	EOlo
Trondheim	Norway	4.1.001-8	EOlo
Trout Lake	Canada	4.4.118-20	EClc
Truk	Caroline Is.	4.6.139-8	Araa
Tsabong	Botswana	4.3.115-4	GBWal
Tsingtao	China	4.2.053-12	DCao
Tsumeb	Namibia	4.3.114-1	GBSal
Tucumán	Argentina	4.5.134-4	BSal
Tulear	Madagascar	4.3.108-11	BWab
Tunis – Carthago	Tunesia	4.3.068-1	Csal
Tura	USSR	4.1.030-22	ECld
Turgaj	USSR	4.1.030-85	BSac
Turiaçu	Brazil	4.5.127-12	Asha
Turialba	Costa Rica	4.4.122-9	GArab
Turku	Finland	4.1.003-6	DClo
Turukhansk	USSR	4.1.030-20	ECld
Tylstrup	Denmark	4.1.008-2	DOlo
Uaupés	Brazil	4.5.127-2	Araa
Uccle (Brussels)	Belgium	4.1.010-2	DObk
Udine	Italy	4.1.027-2	DOak
Udon Thani	Burma	4.2.057-2	Awhb
Uil	USSR	4.1.030-82	DCac
Ulan Bator	Mongolia	4.2.046-1	GBSld
Umtali	Zimbabwe	4.3.112-2	GAwal
Unggi	North Korea	4.2.051-1	DCbo

Name of Station	Country	Chapter-No.	Climate
Upington	South Africa	4.3.113-10	GBWhl
Urfa	Turkey	4.2.031-21	GCshk
Urumchi	China	4.2.053-2	GDCac
Ushuaia	Argentina	4.5.134-40	FTkk
Usumbura (Bujumbura)	Burundi	4.3.090-1	GAwab
Utiariti	Brazil	4.5.127-25	Awab
Van	Turkey	4.2.031-14	GCsbo
Vaasa	Finland	4.1.003-4	DClo
Valdivia	Chile	4.5.131-5	DOlk
Valentia (Observatory)	Ireland	4.1.005-9	DOlk
Valladolid	Spain	4.1.014-7	GCsbk
Valparaiso	Chile	4.5.131-3	Csbl
Vancouver	Canada	4.4.118-42	DObk
Vardö	Norway	4.1.001-4	EOlo.
Vavau	Tonga Is.	4.6.140-17	Araa
Veliki Luki	USSR	4.1.030-58	DCbo
Venezia	Italy	4.1.027-3	DOak
Veracruz	Mexico	4.4.120-11	Awhb
Veraval	India	4.2.044-11	BShb
Verhoyansk	USSR	4.1.030-16	ECle
Verona	Italy	4.1.027-5	DOak
Victoria Point	Burma	4.2.056-8	Amha
Vidim	Bulgaria	4.1.026-1	DCao
Vientiane	Laos	4.2.059-2	Awhb
Vila Cabral	Mozambique	4.3.109-2	GAwbl
Viljusk	USSR	4.1.030-23	ECbd
Villarica	Paraguay	4.5.132-4	Crhl
Visby	Sweden	4.1.002-10	DClo
Vishakhapatnam	India	4.2.044-35	Awha
Vladivostók	USSR	4.1.030-52	DCbc
Vlone	Albania	4.1.024-3	Crak
Vologda	USSR	4.1.030-55	DClc
Vostok	Antarctic	4.5.135-13	HFIde
Vratza	Bulgaria	4.1.026-3	DCbo
Walvish Bay	Namibia	4.5.114-3	BMbl
Waddington	UK	4.1.004-6	DOlk
Wadi Halfa	Sudan	4.3.082-1	BWhl
Wad Medani	Sudan	4.3.082-11	BWhb
Wajir	Kenya	4.3.086-3	BWha
Wake Island		4.6.139-2	Awha
Wakkanai	Japan	4.2.055-1	DCbo
Wankie	Zimbabwe	4.3.112-4	GBSab
Warnemünde	GDR	4.1.018-1	DObk
Warri	Nigeria	4.3.098-10	Araa
Warszawa	Poland	4.1.021-8	DCbo
Washington D.C.	USA	4.4.119-44	DOak
Watsa	Zaire	4.3.091-8	GAmab
Wau	Sudan	4.3.082-19	Awha
Wellington	New Zealand	4.6.138-5	Crlk
Wernigerode	GDR	4.1.018-6	DOlk

Name of Station	Country	Chapter-No.	Climate
White Horse	Canada	4.4.118-18	GECle
Wichita	USA	4.4.119-46	DCak
Wien	Austria	4.1.019-1	DCbo
Willemstad	Curaçao	4.4.123-15	Asha
William Creek	Austraila	4.6.137-21	BWhk
Willis Island	Willis Islets	4.6.140-11	Awha
Windhuk	Namibia	4.3.114-2	GBSal
Windorah	Australia	4.6.137-19	BWhl
Winnimucca	USA	4.4.119-37	GDCbo
Winnipeg	Canada	4.4.118-30	DCbc
Wonsan	North Korea	4.2.051-3	DCao
Woomera	Australia	4.6.137-22	BWhk
Wrocław	Poland	4.1.021-10	DCbo
Wu–Ch'iu-Hsü	China	4.2.053-23	Cral
Yacuiba	Bolivia	4.5.130-13	GAwal
Yakutat	USA	4.4.119-13	EOlo
Yakutsk	USSR	4.1.030-26	ECbe
Yalinda	Central Africa	4.3.094-5	GAwaa
Yaoundé	Cameroun	4.3.097-6	GArab
Yap	Caroline Is.	4.6.139-6	Arha
Yatung (Tibet)	China	4.2.053-30	HCwlk
Yoff	Senegal	4.3.072-5	BShb
Yola	Nigeria	4.3.098-5	Awha
Yolombo	Colombia	4.5.124-7	GArbb
Yonago	Japan	4.2.055-11	Crak
Yulin	China	4.2.053-36	Awhb
Zacatecas	Mexico	4.4.120-3	HBSll
Zagreb	Yugoslavia	4.1.023-1	DCbk
Zahedan	Iran	4.2.039-15	GBWhk
Zakopane	Poland	4.1.021-18	GDCbo
Zamboanga	Philippines	4.2.064-10	Arha
Zamość	Poland	4.1.021-13	DCbo
Zaragoza	Spain	4.1.014-4	BSak
Zielona Góra	Poland	4.1.021-9	DCbo
Ziguinchor	Senegal	4.3.072-8	Awha
Zinder	Niger	4.3.069-9	BSha
Zomba	Malawi	4.3.111-3	GAwal
Zuara	Libya	4.3.080-2	BSal
Zugspitze	GDR	4.1.017-15	HDOkc
Zuid Limburg	Netzerlands	4.1.009-4	DOlk
Zürich	Switzerland	4.1.012-1	GDObo
Zyrjanka	USSR	4.1.030-21	ECld

Index

World-Climates

To facilitate the usage of this book there is enclosed a book-mark containing the following informations:
(.... World Climates: Vorder- u. Rückseite)

How to use the tables

the data (mean monthly and yearly values) are given in the following order:

tx:	temperature maxima
t:	temperature
tn:	temperature minima
clothing:	kind of clothing to be recommended
stress:	great heat stress as result of sultriness or great heat
p:	amount of precipitation in mm (= liter per sq.m)
dp:	number of days with at least 1 mm precipitation (if there is another limit, it is put just before dp)
sunshine:	total duration in hours
%:	percental ratio of the duration to the duration on cloudless days

Code letters for

clothing		**stress**	
p:	polar	•:	no indication (but see NB)
w:	winter		
n:	normal	o:	occasionally
s:	summer	f:	frequently
t:	tropical	ff:	very frequently

NB: Given the variability of general weather conditions it is strongly advised not only to note the data for the month in question, but always to consider those for the months before and after.
In addition to this advice regarding seasonal changes it is recommended to regard also spatial changes and to consider always the area of a place or station of interest and the neighbourly areas as well. It may possibly be that it is situated near a boundary between two climatic areas. Such boundary normally is not a fixed one of narrow width but a more or less extended transition zone, what has to be appreciated.
Therefore users interested simply and soly in the climatic conditions of a place should nevertheless refer as well to those of the neighbourly area.
The list of countries indicates the subchapter of chapter 4 "climatic descriptions of the countries of the world" and the number of the country the user is interested in. The list of stations indicates in addition the place of the station in the country. As each table is labelled by the country number and place number of the station it will be easy to find both: the country and the station.

World-Climates

Köppen's reformed classification

Class of climates	Description: mean monthly air temperature
A tropical climates:	over 17 °C all months
C subtropical climates:	over 9 °C 8–12 months
D temperate climates:	over 9 °C 4–7 months
E subarctic climates:	over 9 °C 1–3 months
F polar climates:	over 9 °C no month
B dry climates:	evaporation exceeds precipitation

dryness limit: RB = 20. (t – 10 + 300. RS/R) in mm
desert limit: RW = 10. (t – 10 + 300. RS/R) in mm

t:	mean annual air temperature	in °C
R:	mean annual precipitation	in mm
RS:	mean summer precipitation	in mm

If R is less than RB the climate is a steppe climate — BS
if R is less than RW the climate is a desert climate — BW
or a marine desert climate — BM

Main Climates

Ar	tropical rain climate
Am	tropical monsoonal rain climate
Aw	tropical summer rain climate
As	tropical winter rain climate
Cr	subtropical rain climate
Cw	subtropical summer rain climate
Cs	subtropical winter rain climate
DO	temperate oceanic climate
DC	temperate continental climate
EO	subarctic oceanic climate
EC	subarctic continental climate
FT	tundra climate
FI	ice climate
BS	steppe climate
BW	desert climate
BM	marine desert climate

Universal Thermal Scale

Thermal conditions of a mean monthly temperature of °C are termed-coded

35	to	..	severely hot	i
28	to	34	very hot	h
23	to	27	hot	a
18	to	22	warm	b
10	to	17	mild	l
0	to	9	cool	k
–9	to	–1	cold	o
–24	to	–10	very cold	c
–39	to	–25	severely cold	d
–..	to	–40	excessively cold	e

The climate formula

According to Köppen the climate of a place can shortly be described by using the corresponding letters of its main climate. This formula is prefixed by a G (mountain climate) or H (high mountain climate) if the height is at least 500 m or 2500 m. Indicating the thermal conditions of the hottest and coldest month by coding its mean temperature according to the universal thermal scale the climatic formula can be completed in adding the two letters.